Chemical Modelling
Applications and Theory

Volume 4

A Specialist Periodical Report

Chemical Modelling
Applications and Theory

Volume 4

A Review of Recent Literature Published between
June 2003 and May 2005

Editor
A. Hinchliffe, *School of Chemistry, The University of Manchester, Manchester, UK*

Authors
B. Coupez, *Novartis Institutes for Biomedical Research, Basel, Switzerland*
R.A. Lewis, *Novartis Institutes for Biomedical Research, Basel, Switzerland*
H. Möbitz, *Novartis Institutes for Biomedical Research, Basel, Switzerland*
A.J. Mulholland, *University of Bristol, Bristol, UK*
A. Miličević, *The Institute of Medical Research and Occupational Health, Zagreb, Croatia*
D. Pugh, *University of Strathclyde, Glasgow*
D.J. Searles, *Griffith University, Brisbane, Australia*
D.S. Sholl, *Carnegie Mellon University, Pittsburgh, PA, USA*
T.E. Simos, *University of Peloponnese, Athens, Greece*
M. Springborg, *University of Saarland, Saarbrücken, Germany*
B.D. Todd, *Swinburne University of Technology, Victoria, Australia*
N. Trinajstić, *Rudjer Bošković Institute, Zagreb, Croatia*
S. Wilson, *Rutherford Appleton Laboratory, Chilton, Oxfordshire*

RSC Publishing

If you buy this title on standing order, you will be given FREE access to the chapters online. Please contact sales@rsc.org with proof of purchase to arrange access to be set up.

Thank you.

ISBN-10: 085404-243-1
ISBN-13: 978-0-85404-243-2
ISSN 0584-8555

A catalogue record for this book is available from the British Library

Published by The Royal Society of Chemistry,
Thomas Graham House, Science Park, Milton Road,
Cambridge CB4 0WF, UK

Registered Charity Number 207890

For further information see our web site at www.rsc.org

Typeset by Macmillan India Ltd, Bangalore, India
Printed and bound by Henry Ling Ltd, Dorchester, Dorset, UK

Preface

Welcome to Volume 4 of the 'Chemical Modelling' SPR. Naturally, I want to start by thanking my team of authors for the hard work they have put into making this the best and most comprehensive volume so far.

It seems a long time since I wrote the following in my Preface to Volume 1 (1999) . . .

'*Starting a new SPR is never easy, and there was the problem of where the contributors should start their accounts; since time began? five years ago? An SPR should be the first port of call for an up-to-the-minute account of trends in a specialist subject rather than a dull collection of references. My solution was to ask contributors to include enough historical perspective to bring a non-specialist up to speed, but to include all pertinent references through May 1999. Volume 2 will cover the literature from June 1999 to May 2001 and so on. In subsequent Volumes, I shall ask those Contributors dealing with the topics from Volume 1 to start from there. New topics will be given the same generous historical perspective opportunity as Volume 1 but will have to cover the literature to 2001 + n where n = 0, 2, 4, ... This process will continue until equilibrium is reached.*'

I think we have now reached equilibrium; some topics have reached maturity and so don't need cover every Volume, whilst a casual monthly glance at the content pages of JACS, JCP, JPC, CPL, THEOCHEM, Faraday Transactions (to name my favorites, not given in order of merit) reveals growth areas.

As an example of a 'mature' topic, consider Density Functional Theory (DFT). DFT is far from new and can be traced back to the work of John Slater and other solid state physicists in the 1950's, but it was ignored by chemists despite the famous papers by Hohenberg/ Kohn (1964) and Kohn/ Sham (KS) (1965). The HF-LCAO model dominated molecular structure theory from the 1960's until the early 1990s and I guess the turning point was the release of the rather primitive KS-LCAO version of GAUSSIAN. DFT never looked back after that point, and it quickly became the standard for molecular structure calculations. So this Volume of the SPR doesn't have a self contained Chapter on DFT because the field is mature.

As an example of a 'perennial' topic, consider the theory of liquids. Almost every undergraduate physical chemistry text tells us that gases and solids are easy to understand because in the first case we have random motion, whilst in the second rigid structures. The gist of this argument is that liquids are really tricky, as indeed they are. The first computer simulation of a liquid was carried out in 1953 at the Los Alamos National Laboratories. The MANIAC mainframe was much less powerful than the PC I am using to write this Preface but the early work by Metropolis et. al. laid the foundations for modern liquid modeling. David Heyes (Volume 2) and Karl Travis (Volume 3) told you how things were in a few years ago, and the story is continued by Billy Todd and Debra Bernhardt in Volume 4.

My final sentence for Volume 1 was

'I am always willing to listen to convincing ideas for new topics'

as indeed I am. My colleague J Jerry Spivey is Editor for the Catalysis SPR; he took me at my word and as a result it is a pleasure to welcome our first contribution from David S Sholl on Heterogeneous Catalysis.

I haven't space to give glowing descriptions of the remaining contributions from each colleague. We hope you will derive benefit and perhaps even pleasure from our efforts.

On a rare personal note, I should tell you that UMIST and the Victoria University of Manchester recently decided to merge to become the UK's largest University; I'm still sitting at the same desk in the same office but my employer is now 'The University of Manchester' and my e-mail has changed to alan.hinchliffe@manchester.ac.uk

Alan Hinchliffe
Manchester 2006

Contents

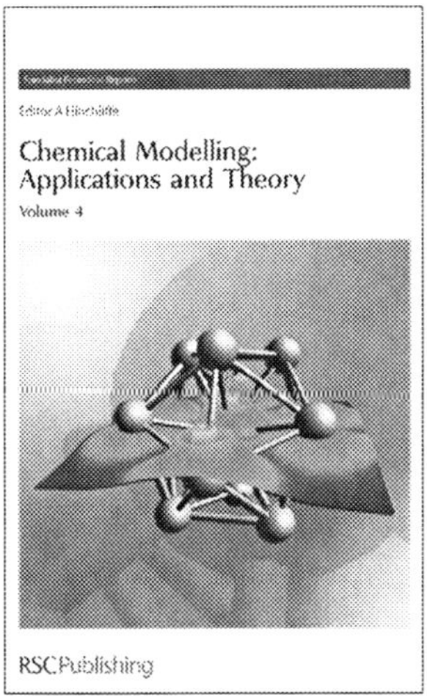

Cover
The icosahedral 'golden fullerene' WAu_{12} reproduced by permission of Pekka Pyykkö, Chemistry Department, University of Helsinki, Finland.

1

Computer-Aided Drug Design 2003–2005

BY BERNARD COUPEZ, HENRIK MÖBITZ AND RICHARD A. LEWIS
Novartis Institutes for Biomedical Research, Basel CH-4002, Switzerland

1 Introduction

The themes for this review again have been driven strongly by the need of the Pharmaceutical industry to make the discovery process quicker and more reliable. Virtual screening in all its forms is at the heart of most research, from bioavailability filters through to rigorous estimations of the free energy of binding. Two areas of relative heat have been docking/scoring, and ADME/Tox. On the other hand, 3D-QSAR and pharmacophores have become quiet. Part of the reason for this may arise from the successes in high-throughput crystallography, delivering more targets and complexes, the relative failure of HTS, and the increase in the amount of high quality data coming from late-phase research/early-phase development concerning the fate of clinical candidates. These trends look set to continue in the future, and the next two years should yield many new breakthroughs.

2 ADME/Tox and Druggability

There has been a fresh impetus to the modelling of ADME, Toxicity and druggability phenomena, partly driven by a desire to understand why such complex phenomena can, apparently, be described so simply, and partly to see if better models can be built, to improve the attrition rate in medicinal chemistry still further.

2.1 Druggability and Bioavailability. – In the continuing debate over what physicochemical properties are required for bioavailability, Vieth *et al.*[1] have surveyed 1729 marketed drugs with respect to their route of administration, h-bonding capability, lipophilicity and flexibility. One conclusion they draw is that these properties have not varied substantially over time, implying that oral bioavailability is independent of target or molecular complexity. Compounds with lower molecular weight, balanced lipophilicity and less flexibility tend to be favoured. Leeson and Davis[2] claim that molecular weight, flexibility, the number of O and N atoms and hydrogen-bond acceptors have risen, by up to

Chemical Modelling: Applications and Theory, Volume 4

29%. This may be partly due to the choice of 1983 as the reference year, or the advent of more complex targets with greater selectivity needs (*e.g.* kinases). In the same vein, a study[3] re-examined the correlation of flexibility and polar surface area (PSA) with bioavailability proposed by Veber *et al.*[4] One conclusion is that there are significant differences in the ways of defining flexibility and PSA, and the correlations depend markedly on the method used (this is not surprising, as neither quantity is precisely definable). A second conclusion was that the limits defined (Number of rotatable bond < 10, PSA < 140 Å^2) excluded a significant number of compounds with acceptable rat bioavailability. In the authors' words, "This observation underscores the potential danger of attempting to generalise a very complicated endpoint and of using that generalisation in a prospective selection application". Despite this, another bioavailability score[5] has been devised, to predict the probability that a compound has $>10\%$ bioavailability in the rat. Compounds are grouped by ionisation class (anions, cations, neutral). It was found that the standard rule-of-5 does well for cations and neutrals (88% of the compounds predicted to have low bioavailability are observed as such). Anionic compounds were better described by PSA limits. Some simple rules are given to compute the bioavailability score. In Abbott laboratories, this score is now routinely computed for all compounds and is used for hit-list triaging. It will be interesting to see if the results can be repeated on other data sets; the paper has certainly sparked much interest in the modelling community. Wegner[6] provides support for the idea that human intestinal absorption correlates with PSA, by generating a classification model. The justification is that the error in the experimental data is 25%, and 80% of the observations occur in the top and bottom quartiles, that is, the data is more binary than evenly spread. In addition to PSA, other descriptors that reflect the electronic character of atoms and their environment also came to the fore.

2.2 Metabolism, Inhibitors and Substrates. – The field of cytochrome modelling is becoming more mature as we begin to understand the limitations of the experimental data and the subtleties of the mechanisms (the whole field of cytochrome P450 modelling, including homology, pharmacophore and 3D-QSAR models has been reviewed in detail recently[7]). Empirical models are still preferred, especially for rapid evaluation of large libraries. In one case, use of a jury system improved prediction accuracy to over 90%.[8] Chohan *et al.*[9] have developed 4 models for Cytochrome P450 (Cyp) 1A2 inhibition, and identified the expected descriptors as being important to the QSAR (lipophilicity, aromaticity, HOMO/LUMO energies). Perhaps a more interesting result in this paper was the use of the *k* index to assess predictive powers of the models using test data.

$$k = \frac{\text{observed agreement-chance agreement}}{\text{total observed-chance agreement}}$$

This index should prove useful for data sets that are diverse and noisy. The validity of QSAR model predictions has also been studied by Guha and Jurs.[10]

The protocol is quite straightforward. The initial QSAR models were built, and the residuals of the compounds in the training set were used to classify the trains set predictions into good and bad. The threshold for the classification is arbitrary. Test compounds were predicted, and the predictions were grouped by substructural similarity to the nearest neighbour in the training set. It was seen that test compounds that had neighbours with low/good residuals were themselves well-predicted, with the reverse being the case for neighbours with high residuals. The success rate for classifying the strength of the prediction was 73% to 94%. The Merck group[11] performed a retrospective study of in-house data sets, and concluded that the distance to the nearest neighbour, and the number of nearest neighbours (local density) were the two most useful measures for predicting prediction quality. They also concluded that distance does not have to be measured in the same descriptor space as was used to build the QSAR model. Topological descriptors combined with a Dice coefficient worked equally well.

A number of groups have been active in the prediction of the most likely sites of metabolism of molecules that are substrates for cytochromes. Singh *et al.*[12] developed a semi-quantitative method based on the energy barrier to the creation of hydrogen radicals as calculated by AM1. Using a set of 50 substrates for Cyp 3A4, they were able to show that only hydrogens with a solvent-accessible surface area over 8 Å^2 are susceptible to attack. The expensive quantum mechanic calculations could be approximated by local neighbourhood descriptors which could be well correlated to the energies (R^2 = 0.98), offering a fast and practical method for screening large libraries. An extension of this concept is embodied in the MetaSite program,[13] which uses propensity to react, accessibility and GRID molecular interaction fields as descriptors. The methodology is more general, and can be applied to any cytochrome structure: in validation experiments, an accuracy of 80% is claimed. It is also important to be able to predict which compounds will be inhibitors as well as substrates, to avoid drug-drug interactions. A classifier based on a support vector machine (SVM)[14] has been created that correctly predicts compounds into high, medium and low affinity at 70% accuracy, even with simple 2D descriptors. The improved accuracy was obtained through a systematic variation and optimisation of the SVM parameters.

Considering the success of surprisingly simple, semiempirical methods in ADME modelling, it is interesting to see whether more advanced methods could bring further improvements. A recent paper of Beck[15] provides a link to the rich literature of DFT studies of hemes and cytochromes. The author uses Fukui functions to gauge the site of highest nucleophilicity of a number of known drugs. The predictions give mixed results and demonstrate that the implicit assumption of Fukui functions, *i.e.* an isotropic electrophilic attack, is flawed, not to mention that their MO-like shape does not allow a ranking of single atoms. In conclusion, the study suggests that it is more important to have an accurate description of the cytochrome-ligand complex than to invest in a high-level description of the chemical reactivity. De Visser *et al.*[16] have used DFT on 10 C–H barriers with reference to bacterial cytochromes, and claim an excellent correlation between bond energy and observed activation energy barriers, so there is still some mileage in this approach.

2.3 Toxicity. – Unacceptable toxicity is still a key source of compound failure in clinical trials. Several groups have developed tools and programs for predicting toxicity for use in early phase, but the question arises about the accuracy of these models, and the levels of false positives and negatives that are acceptable. In research, an overly strict model with no false negatives may cause the discarding of a perfectly reasonable lead series. In development, missing a toxic alert which shows up in a later phase is unacceptable. Similarly, any program that is used by regulatory authorities to screen compounds must be very unforgiving of any flaw. In a recent study by the FDA[17] on maximum human therapeutic dose, rules-based programs managed 64% accuracy, not much over random, giving an indication of the pitfalls in this field; Helma has given an overview of this area.[18] Clearly, the domain of the models is critical and this has been addressed explicitly for QSARs that make toxicity predictions.[19] Another route to predicting ADME properties is to use screening results, as exemplified by the Bioprint approach.[20] 1198 drugs have been assayed against 130 screens, to give an activity fingerprint. QSAR models are then derived using pharmacophore descriptors. New compounds can be run through the models to predict binding affinity in all the screens, compared to the nearest neighbours in the database and finally fingerprinted themselves for confirmation. Using the affinity fingerprint alone, one can again identify similar molecules (sometimes with surprising results) and extrapolate to the potential side-effect profiles. This is very useful when selecting one from several lead series for optimisation.

3 Docking and Scoring

3.1 Ligand Database Preparation. – The ligand database is the basis for virtual screening (VS). Special care must be taken at this stage; accurate and physically relevant tautomeric and protonation states need to be assigned. Often compounds are registered in a database as a tautomer that is not necessarily the most probable state of the molecule and it is difficult to assign the correct state, so all relevant states should be generated. Similarly, as the stereochemistry of chiral centres is often not known, one must generate all stereoisomers. A recent article reveals the impact of pre-processing a database containing both known actives and inactives, where multiple protonated, tautomeric, stereochemical, and conformational states have been enumerated.[21] The authors show that the interplay between 2D representations, stereochemical information, protonation states, and ligand conformation ensembles has a profound effect on the success rates of VS and conclude that the enrichment is highly dependent on the initial treatments used in database construction. In a paper that is bound to become a citation classic for the service that it has provided to the academic modelling community, Irwin and Shoichet describe the creation of the ZINC database of commercially available compounds, available via the web.[22] The resource can be used in virtual screening studies, as the authors have taken care to provide compounds in multiple protonation and tautomer states, even multiple conformations. The paper provides a useful recipe for creating such a database for general use.

3.2 Target Preparation. – Thanks to high throughput crystallography and structural genomics, we have the X-ray structures of many targets of therapeutic interest, with the obvious exception of membrane proteins. When no experimental structure is available, it is possible to generate a 3D structure based on a template protein of similar sequence and a known structure, for example the model of CDK10,[23] based on the CDK2 crystal structure, that was successfully used for a docking study.

If several structures of the target are available, which structures should be used: the apo form, a holo complex, or a homology model? This issue was examined by McGovern *et al.*[24] They docked a large number of small molecules against 10 targets using the apo-, holo-, and modelled forms of the binding site. Using enrichment rates, they found that the holo form gave the best results (70% enrichment) followed by the apo (20%) and then the modelled form (10%). However, the holo form can be over influenced by the ligand in holo complex, if the active site has "collapsed" around the ligand. Then one would get a lower retrieval rate of similar but larger ligands, due to the increased steric constraints; the apo form of an active site can be markedly different from holo form.[25] The conclusion is that VS using any form of the target will do better than random, but the holo form will give a best enrichment. This was also confirmed by Erickson *et al.*[26] which show that the docking accuracy decreases dramatically if one uses an average or apo structure. Another approach is to use softened repulsive terms in the Lennard-Jones potential, to allow a closer approach of ligand and protein atoms that could be later resolved by minimisation.[27] The T4 lysozyme system was used, with the ACD database as the source of ligands. The soft function was worse than the hard function, if multiple protein conformations were used, and vice versa for a single model. It was concluded that soft potential favour the decoys as much as the true ligands, so needs to be used with care.

Like the ligand preparation, the preparation of the target also requires great care. Incorrect protonation states or tautomers of histidines can lead to serious docking errors. For example Polgar *et al.*[28] demonstrate the importance of protonation states in virtual screening for β-secretase (BACE1) inhibitors. They observed improvement of enrichment rates when they assigned different protonation states to catalytic Asp32 and Asp228 residues. Some docking methods require the addition of hydrogens. It is recommended that after the addition of hydrogen atoms to the protein, the positions of the hydrogens are relaxed by energy minimization to avoid any steric clashes. The positioning of hydrogen atoms on hydroxyl groups in the active site should also be checked and changed if necessary. In some instances, hydrogen bonds to crystallographic waters might need to be maintained for the docking.

However increasing the degrees of flexibility also increases the computational complexity and cost. Different methods have been described in the literature to tackle this critical issue (for a review see ref. 29). Often these methods model the flexibility in the binding site exclusively, by sampling the protein conformational space using molecular dynamics or Monte Carlo calculations or rotamer libraries. Another way of treating protein flexibility is to use an ensemble of protein conformations, rather than a single one. In a recent paper, Barril and Morley[30] use all the X-ray structures of cyclin dependant kinase 2 (CDK2) and heatshock protein 90 (HSP90) to assess the performance of flexible receptor docking. They observe that flexible receptor docking performs much better in binding-mode prediction than rigid receptor docking. However, they also noticed that for library screening, ensembles of cavities often result in worse hit rates than rigid docking. This trend can be reversed by selecting those ligands that bind consistently well to many cavities in the ensemble.

3.3 Water Molecules. – Another challenge in protein-ligand docking is the modelling of the water molecules in protein ligand recognition. Water can form hydrogen bonds between the protein and the ligand or can be displaced by the ligand.[31] Recently a new approach that allows this was implemented by Verdonk *et al.* in GOLD.[32] The method allows water molecules to switch on and off and to spin. The explicit inclusion of water molecules in a docking program improves the binding mode when a ligand interacts with a water molecule. A distinction can also be made between the compounds that can displace a water molecules and the compounds that cannot. They claim that their algorithm correctly predicts water mediation/displacement in 93% of their tests and they observe some slight improvements in binding mode quality for water-mediated complexes. Similar results were reported by De Graf *et al.*[33] for cytochrome P450s. The waters were either removed, or the crystallographic waters retained, or waters in GRID minima were used. Surprisingly, the last scenario gave the best results by up to 20% in the number of correct poses that scored highest.

3.4 Comparison of Docking Methods. – The flow of papers performing comparative evaluations of docking/scoring programs continues,[26,34] although there is an increasing feeling that these studies offer only limited insight.[35] Another potential pitfall in studies evaluating docking and scoring functions has been highlighted.[36] Enrichment rates can be artificially boosted by not matching the 1D properties of the decoy set to the true ligands (for example, if the ligand is much larger than the decoys, it will be favoured). It was also observed that incorporating even small amounts of chemical knowledge, in the form of pharmacophoric constraints, could improve the quality of binding modes, and hence the enrichment. The work of Warren *et al.*[37] deserves mention as their protocol did not rely on evaluations performed with default parameter settings, but rather let expert users set up the runs, which is a more realistic scenario. Their conclusion was that no one approach was clearly better than any other for all targets; all failed to predict binding affinity with any confidence. Against that, the study was performed with quite old versions of the software, so whether the same conclusions are valid today is moot.

3.5 Scoring. – Success of VS depends more strongly on the quality of the scoring function, than the method for generating dockings. An imperfect scoring function can mislead by predicting incorrect ligand geometries or by selecting nonbinding molecules over true ligands. Graves *et al.*[38] consider these false-positive hits as decoys and have used them to improve their protein-protein docking algorithms. A new version of the knowledge-based scoring function DrugScore has been published,[39] based on better quality small molecule X-ray data. This has the advantage of being higher resolution and better populated that protein x-ray data sources. For common interactions (for example, C.sp3 – C.sp3), the shapes of the potentials are the same, with more definition from the new potentials, reflecting the higher resolution of the underlying data. When this scoring function was used to dock and score 100 complexes with decoys, the crystallographic pose was ranked in the top 3 for 90% of the complexes, a 57% improvement over the previous version. The rank order coefficient for the prediction of binding energy was improved slightly (0.62), but we are still not doing significantly better than the correlation with molecular weight (0.56). This phenomenon is also observed in high-throughput screening, leading to measures for ligand efficiency[40,41] that correct for molecular weight. A simple method for computing thermodynamic energies of binding,[42] allowing flexibility in the protein side chains via Monte Carlo sampling, and a very simple model for van der waals and electrostatic interactions nonetheless proved to be quite effective in predicting the selectivity of 6 kinase inhibitors when tested against a panel of 20 receptors. The authors identified a strong dependence on a good initial binding pose, and saw that minimisation with the function did not improve the results. Some of the success may have come from working in a target family, when one can assume that many of the errors are consistent, so that the relative energies can be trusted. While free energy methods remain the gold standard for ligand affinity prediction, the associated computational cost prohibits their routine use in the pharmaceutical industry. In a series of papers, Oostenbrink and van Gunsteren[43–45] have tackled this problem and extended the one-step perturbation method with the aim of a fast, accurate prediction of structurally diverse compounds. In principle, through the use of a well-chosen reference state, the computational cost is reduced to a single full simulation. In a study of the estrogen binding receptor, a series of biphenyl compounds were predicted with an error of $< 1\ \text{kcal} \cdot \text{mol}^{-1}$, whereas the predictions for a more diverse set of compounds hint that the method needs further improvement before it can be generally applied. One such improvement is the reparameterization of the underlying GROMOS force-field for the prediction of thermodynamic properties of hydration and solvation.[46]

As no scoring function is perfect and each scoring function has its own strengths and weaknesses, we can combine different scoring functions to balance errors of one single scoring function and improve the probability of identifying 'true' ligands by reducing the false positive rates. This approach is called consensus scoring. However, the potential value of consensus scoring might be limited, if terms in different scoring functions are significantly correlated, which could amplify calculation errors, rather than balance them. The success of the consensus scoring approach was analysed by Yang *et al.*[47] Using data from five scoring systems with two evolutionary docking algorithms on four targets, thymidine kinase, human dihydrofolate reductase, and estrogen receptors in antagonist and agonist conformations, the authors demonstrated that combining multiple scoring functions improves the enrichment of true positives only if each of the individual scoring functions has relatively high performance and if the individual scoring functions are very distinct in their philosophy. Recently an alternative way of combining various scores was proposed by Vigers and Rizzi;[48] this approach called "multiple active site correction" can correct library ranking using scores calculated for several active sites. The corrected score is now high only if compounds are found to score well with the target of interest and not with others.

3.6 New Methods. – New docking methods have been developed during these last two years. Glide,[49] developed by Schrödinger, is one of the most popular. Firstly the properties of the active site are mapped on a grid. Then a set of low energy conformations of the ligand is generated using a Monte Carlo approach. These poses are used as input and the ligand is minimized in the binding site and three to six low energy poses are selected and a Monte Carlo simulation is performed on these. The AFMoC protocol has been further developed,[50] to adapt a scoring function with local knowledge provided by known complexes and measured affinities. The key advances have been to use filtering of grid point variables by Shannon entropy, and to use sensible defaults for potentials that became repulsive under the AFMoC protocol. Using a challenging test set of 66 highly flexible HIV-1 protease inhibitors, they were able to identify a correct binding pose with the top binding score in 75% of cases, an improvement of 14% over native scoring functions. Another twist for knowledge-based scoring functions is to optimise the ligand positions before fitting the scores to experiment.[51] This removes the bias of the x-ray refinement protocol. An accuracy of 2 $kcal \cdot mol^{-1}$ in binding energy prediction is claimed, but the results were not compared to the correlation of score to molecular weight.

A current trend in the field is to focus on the inclusion of various solvation and rotational entropy contributions. However the terms currently used to approximate entropy or desolvation energy provide only incomplete descriptions of these effects on protein–ligand binding. For example, Krammer *et al.*[52] present developed two new empirical scoring functions that possess good predictive accuracy in determining the ligand-receptor binding affinities over a wide range of protein classes. A recently introduced new methodology based on ultrashort (50–100 ps) molecular dynamics simulations with a quantum-refined force-field (QRFF-MD)[53] was evaluated by Ferrara *et al.* using CDK2 kinase.[54] The QRFF-MD method achieves a correlation of 0.55, which is significantly better than that obtained by a number of traditional approaches in virtual screening but only slightly better than that obtained by consensus scoring (0.50). The authors also introduced a new scoring function that combines a QRFF-MD based scoring function with consensus scoring, which resulted in substantial improvement on the enrichment profile.

With the increase of the computational power it is now possible to use more rigorous theoretical and more CPU-intensive approaches. Kuhn *et al.*[55] reported the usefulness of the MM-PBSA approach for VS: they showed that applying the MM-PBSA energy function to a single, relaxed complex structure is an adequate and sometimes more accurate approach than the standard free energy averaging. MM-PBSA can also be used as a post-docking filter for enriching virtual screening results, and for distinguishing between good and weak binders for which $\Delta pIC_{50} \geq 2$–3. Huang *et al.*[56] developed a two-stage virtual screening protocol: first a rapid, grid-based scoring function is used to dock large compound databases to a receptor. In the second step the OPLS all-atom force field and a generalized Born implicit solvent model is used to minimize the ligand in the cavity and to rescore the poses for the top 25% of the ligands from the docking phase.

One well-known strategy for improving throughput and accuracy of docking for hit-finding is to apply some extra screens to reduce the size of the database to be screened. Then one can use more expensive but hopefully more accurate protocols for the docking and scoring. Maiorov and Sheridan[57] started by using a fast docking protocol FLOG, then fed the best 1000 scoring results into ICM-Dock for redocking. They showed a 5-fold improvement of the enrichment obtained by FLOG alone. Use of this two step method meant that the entire MDDR database could be screening in under a day.

The role of fluorine in hydrogen-bonding has been difficult to quantify. In some cases, addition of a fluorine can bring great improvements in binding affinity, in many other cases, it seems to be neutral, even in cases where a positive interaction should take place. The GRID program[58] now includes a potential function for fluorine, based on the new survey of protein-ligand x-ray complexes. Aliphatic fluorines make straighter and shorter hydrogen bonds than aromatic fluorines. Bifurcated bonds are not observed at all. When the new term was added to GRID and the GRID field used as a scoring function for docking, an improvement of about 20% in pose generation and ranking was observed.

3.7 Application of Virtual Screening. – Forino *et al.*[59] present an interesting case study of a difficult target, the protein kinase PKB/Akt, notorious for yielding a mere 2 hits in a HTS campaign. In several schemes that rely solely on docking or consensus scores, the authors report near random hit rates. However, when the final selection was based on a visual inspection of consensus hits for the potential to form hydrogen bonds similar to ATP, the hit rate increased significantly, leading to the identification of 3 μM competitive inhibitors. Although anecdotal, this story may offer comfort to modellers that brains can easily be as productive as brute force methods. Similarly, Huang *et al.*[60] were able to find hits for β-secretase after disappointing results from HTS.

Mozziconacci *et al.*[61] used Cox-2 as their test case. The first part of the paper looks at the selection of the optimal parameters for the docking protocol (here using DOCK), followed by a consensus scoring approach. Having optimised the protocol with known ligands, a large (13,711) virtual library was screened. Of the 12 compounds selected and available for assay, 4 had IC50's $< 1\ \mu$M.

4 *De Novo*, Inverse QSAR and Automated Iterative Design

There is continuing activity in the area of *de novo* design, driven partly by the increased interest in fragment-based screening (FBS). FBS is an experimental method for identifying small (<250 Da) molecules that bind to pockets of an active site, rather than to the site as a whole. Then the fragments should be joined into larger composite structures, with hopefully a large gain in affinity. This is the traditional territory of *de novo* design. Schnieder has reviewed the whole area since its inception.[62] SPROUT has been used to design NK2 antagonists[63] based on a GPCR model. The best structure had an affinity of 2 μM as the racemate. In a less ambitious use of *de novo* design, some D3 agonists were designed based on a CoMFA model.[64]

As new compounds in corporate pipelines gravitate towards higher molecular weight and ClogP, it is interesting to see what small building blocks may have been missed in the vastness of chemical space. Fink *et al.*[65] report their findings from virtual database of 14 million compounds weighing less than 160 Da. The exhaustive enumeration of all possible molecules containing C, N, O, H and F was achieved by a mathematical graph representation of the saturated hydrocarbons, followed by permutation of each core. Connectivity criteria were used to obtain a comparable composition and number of basic cores as present in the 36,000 known compounds in public data bases. Not surprisingly, the authors report a denser coverage of the property space of drug-likeness descriptors. In a virtual screening of three representative targets, a mere 10% of the virtual hits are outside the property space covered by existing compounds. Although some of the example structures do not seem desirable from a medicinal chemistry perspective, there are surprising examples of drug-like small molecules not known in any data bases.

The bridge between inverse QSAR and *de novo* design is neatly illustrated by the work of two groups. The CoG program by Brown *et al.*[66] uses a genetic algorithm to evolve similar molecules to starting structures, with fingerprints as the internal definition, and some QSAR models as the external definition of similarity.[67] The molecules are evolved by simple graph operations within a genetic algorithm framework, to change element types, valency and bond orders. The fitness of the new structures is calculated using Tanimoto similarity to a reference set of molecules. The ranking of the molecules is performed by a Pareto score based on the similarities to all the reference structures, to avoid the generation of highly localized islands around the reference set. In the example given, menthol and camphor were the reference molecules, and the method was able to produce a large number of sensible structures that were intermediate between the two. The experiment was repeated with aqueous solubility as the target, and again the program could evolve molecules with the desired characteristics. In a related approach, Lewis[68] developed a full inverse QSAR protocol, using the mutation of structures to drive towards compounds with better fitness. The key findings of the research were that the palette of reactions used had a strong influence on the quality and improvements found, and that one had to take great care that the fitness function could be applied to the molecules that were generated. Restricting the reactions to functional groups found within the set of molecules used to generate the QSAR model helped greatly, as did imposing a core structure as a constraint. To prevent extrapolation, molecules were kept with the QSAR space using distance to the nearest neighbour in the training set as a strategy. The task of computing extrapolation has been taken up by others, as is discussed elsewhere in this review.[10] In two studies with real QSAR sets, Lewis was able to propose molecules with 1–2 fold improvement in predicted activity, and that were similar to the original series.

5 3D-QSAR

The long awaited validation study for the XED method for molecular similarity has been published.[69] Molecules are described using the maxima and minima in the electrostatic and steric fields around the molecule. These points form a pharmacophore, and so can be used to search databases for alternative chemotypes. As the representation is sparse, several conformations per molecule can be considered. A new chemotype with nanomolar potency for CCK2 and improved excretion properties was found. This is a good concrete example of the power of the XED representation. Other than that, this area has been comparatively quiet, awaiting more developments in alignment methods, as discussed in the next section (Table 1)

6 Pharmacophores

Questions around the quality of our methods for generating pharmacophores are being raised, especially as there have been no major advances since GASP and DISCOtech. Three groups are revisiting some of the fundamental issues

Table 1 *Some representative high-quality 3D-QSAR models*

Target	*Method*	*Alignment*	Q^2
Sigma-1[70]	CoMFA	DISCOtech	0.7
COX-2[71]	COMFA/CoMSIA	Docking	0.74
EGFR[72]	CoMFA	Docking	0.7
Choline acetyltransferase[73]	CoMFA	Reference ligand	0.76
Oxytocin[74]	CoMFA	Docking	0.85
Catechol-O-methyltransferase[75]	CoMFA/GOLPE	Docking	0.6
Androgen receptor[76]	CoMSIA	Docking	0.66
δ, μ, κ-Opioids[77]	CoMFA	Reference ligand	0.67
NMDA[78]	CoMFA	Reference ligand	0.5
PPARa/PPARg[79]	CoMFA	Reference ligand	0.7
PEPT1[80]	CoMSIA	Reference ligand	0.82

Table 2 *Some high quality pharmacophore models published during the review period*

Target	*Method*	*Features*	*Tolerances/Ang*
HIV-1 Integrase[83–86]	MD/LigBuilder	4	1
General kinases[87]	MOE	5	n/a
Biogenic amine GPCRs[88]	Catalyst	3–5	n/a
GABA-A[89]	Catalyst	5 + 3 exclusion volumes	n/a

around sampling of conformational space, the generation of ensembles of solutions, and the scoring of those solutions. The Sheffield group[81] have developed a multi-objective genetic algorithm (MOGA), based on their experiences with GASP. The conflicting objectives are conformational energy and the degree of overlap/similarity of the structures when overlaid according to the pharmacophore hypothesis. One inherent difficulty is that the 'correct' answer is often not known. Most methods can produce several plausible solutions but this may reflect the difficulty the programs have in sampling the search space. The MOGA does find a wider range of solutions than other stochastic approaches. Another advantage is that conformational space is sampled on the fly, rather than relying on a precomputed set of conformers, which will bias the search space. The disadvantage is that the MOGA does not allow for partial matches, so the pharmacophore needs to be built from compounds that all have (similar) high affinity.

Kristam *et al.*[82] looked at the approaches used in the literature to generate Catalyst pharmacophore models. In many cases, they found it hard to reproduce the results, leading to them to propose a template for describing Catalyst models which should aid in reproducibility. In addition, they found that the less-expensive rules-based methods for generating conformational ensembles did just as well as much more expensive methods, in line with the earlier findings of Bostrom (Table 2).[83]

7 Library Design

Multi-objective optimisation has again been linked to combinatorial chemistry. In this work,[90] the chemistry space is described by reaction transforms that contain information both about the reaction and the functional groups that the reaction will work for. The fitness function is a combination of basic bioavailability together with some empirical SAR. These reaction schemes are used to operate on a family of starting structures via a Genetic Algorithm. The number of reactions steps is limited to a small number (3). In the case considered, kinase inhibitors, basic kinase scaffolds could be rapidly decorated and their predicted affinity improved according to this protocol. Validation of the approach was performed by comparison to known kinase inhibitors rather than experimentally. In a similar vein, Brauer *et al.*[91] used evolutionary chemistry to find novel inhibitors of glucose-6-phosphate translocase. They start with a selection of low-affinity compounds from screening. After each round of optimisation, compounds are synthesised and assayed, and the authors claim the discovery of a new class of imidazoles from their work.

8 Cheminformatics and Data Mining

It is a pleasure to cite one review[92] on the whole area of similarity and its application to structure retrieval. Prof. Willett has been a leading figure in the area, and the review marks the award of the 2005 ACS award for computers in Chemical and Pharmaceutical research. A nice application of the similarity principal based on some of his work is provided by Krumrine, Maynard and Lerman[93] for the analysis of virtual screening data. In any type of screening, there is a balance between the rigour/throughput of the assay and the number of acceptable false positives and negatives. Most scoring/assay systems can provide some degree of discrimination between hits and inactives, but the overlap between the two classes can be quite significant. In this approach, each hit/inactive is characterised by the behaviour of neighbouring compounds. These compounds should be similar and there exhibit similar behaviour. A false positive is therefore a hit surrounded by inactives, and vice versa for a false negative. The neighbourhood is defined by a threshold of 70% similarity, as measured by Daylight fingerprints and a Tanimoto coefficient. The probability of a compound being truly active/inactive, given its neighbourhood, can then be computed. This enables more effective selection of compounds for further assay. As a sidebar, the authors note that early application of common filters for bioavailability and ADMET would have resulted in the loss of a number of perfectly reasonable lead series. Caveat emptor!

8.1 Scaffold Hopping. – Scaffold hoping is an attractive road to novel chemotypes, yet there are few routine methods that reliably deliver. An interesting case study is presented by Rush *et al.*[94] which uses the overlap of shape Gaussians to search for chemically diverse analogues of a μM HTS hit in the conformationally expanded data base. To weed out steric clashes, a MM-PBSA scoring function was applied. Among the 30 candidates, 3 hits with comparable binding affinities and better overall-properties were identified. As the 2D-similarity of the hits was rather low, the authors claim that shape alone can be a more powerful descriptor for scaffold-hopping. Another explanation why the method works well in this case is that the molecules are flat and conformationally restricted. It would be interesting to see how this method fares across various target classes. The XED approach for scaffold hopping is discussed above.[69] Rarey *et al.*[95] have extended their feature tree algorithm by devising a scheme to combine single trees into a multiple tree. In a sense, this is the same as deducing a pharmacophore model using just 2D topology. They showed that the multiple trees outperformed single trees and Catalyst models in terms of enrichment during virtual screening, and that alternative scaffolds could also be retrieved. The study however only considered two cases, ACE inhibitors and α1 antagonists.

8.2 Descriptors and Atom Typing. – Bender *et al.*[96] have proposed a new descriptor. Molecules are described using atom environments up to an optimal radius (found to be 2). The most information-rich features are then selected by trying to optimise the separation between active and inactive structures, and these features are used to drive a Bayesian classification algorithm. In the tests described, these targeted features seem to do better than analogous descriptors (*e.g.* MACCS keys or Daylight fingerprints) that are unfiltered, and have equivalent performance to other directed feature algorithms like feature trees.[97] The enrichment factors are between 6.5 and 11 fold.

Labute[98] presents a new algorithm for the automated prediction of atom types from structural data, that is based on a maximum weighted matching rather than knowledge-based rules. From the coordinates, a covariance-matrix is used to determine candidate bonding partners. Based on these connectivities and geometric thresholds, a preliminary hybridization is assigned. The final assignment uses the likelihood rates for higher bond orders estimated from a set of 200k commercial compounds. Over a range of test sets from the PDB, the success rates are reported to be in the low 90 percentiles. Poor perception was encountered in the case of strained, unusual conformations, underlining that visual inspection by chemist remains the gold standard. In the advent of high-throughput crystallography, where the interpretation of results might eventually become the scarcer resource, such a method might prove useful.

A fundamental question for most medicinal chemists, is "what makes a compound attractive". This is particularly important when lists of hits from high-throughput screening are being assessed and the series for optimisation is being chosen. Lajiness *et al.*[99] set out to study how subjective the "chemist's eye" really was by asking a set of chemists to independently score the same set of 250 compounds that had already been rejected by a senior and experienced medicinal chemists. The result was that the chemists rejected compounds on a very inconsistent basis, which has serious implications for hit follow-up studies as an entire chemotype can be dropped for no good reason. Even when the same chemist looked at the same compounds for a second time, the pattern of rejections varied greatly. There was also no relation between experience and consistency of opinion; assessment of attractiveness cannot be learned. Overall the consistency of opinion between two chemists was 24%.

The use of biological fingerprints is part of the emerging area of chemogenomics. A seminal paper in the field is on kinomics.[100] Rather than classifying kinases according to sequence alignment or other bioinformatics methods, the authors propose a scheme based entirely on small molecule selectivity data. This enables one to perform structure-based drug design on kinases that are otherwise seeming unrelated. Naturally, at high levels of homology, the results of the two approaches are the same. When faced with a new kinase target or selectivity profile, one could deduce similar targets and hence screening sets.

Conversely, given a new chemotype, one could also predict which kinases it is likely to hit.

8.3 Tools. – The emergence of several tools focussed on the medicinal chemist has been welcome. In most cases, these tools have been built inside Pharma companies to deal with a pressing need. The selection of bioisoteric heterocycles[101] is a good example. There are many possible heterocycles, and the ones favoured by this program are derived from molecules in Phase II clinical trials or later. This should filter out systems associated with poor toxicity, metabolism and the like. The heterocycles are oriented to a common frame of reference, and relevant descriptors, such as charge, shape size *etc* are computed. The most similar systems to a query ring system can be found and presented quickly to the chemist to seed further ideas for lead optimisation. Although it is not really within the scope of this review, the discovery of ferrocene as an effective bioisostere for benzene in GPCR ligands[102] is included as an example of how far we still have to go in understanding similarity.

9 Structure-Based Drug Design

9.1 Analysis of Active Sites and Target Tractability. – Hajduk *et al.*[103] have looked at a large set of NMR-based screening data to assess the important factors that make a protein pocket druggable. Among a diverse group of 23 targets, the authors identify 57 potential binding sites of which only 28 are targeted by small molecules. The highly consistent hit rate of >90% for known binding sites underscores the general wisdom of targeting substrate binding sites. The single most important factor was found to be the apolar surface area, followed by the size and roughness of the pocket. The authors also present an eight-parameter correlation function which may serve to prioritize among several potential targets. The ability to predict the druggability of a potential protein target will certainly be much appreciated when the genomic era of drug design comes to full fruition.

The hardy problem of protein flexibility has been re-examined using B-factors.[104] Despite the known dangers in equating high B-factors with intrinsic mobility, the authors conclude that, in a study of 800 high-resolution x-ray structures, 71% of atoms in the binding sites become less mobile on binding, but that 29% become more so. Explanation of this observation is not easy, and may reflect loss of water, or induced fit, freeing up side chains. The prediction of binding site flexibility on a fairly extensive test set of both NMR and x-ray structures has been studied by Mancera *et al.*[105] Their algorithm DYNASITE uses an iterative procedure of rotamer generation based on a rotamer library and minimization of cluster representatives to identify alternative conformations of binding site side chains. This method obtains up to 20 clusters of alternative conformations and in all but two cases the experimental alternative conformations were among those. Exceptions were ligands which induced a high-energy conformation of the active site. While the authors have demonstrated that a rotamer-approach reliably generates proximal conformations, the prediction of which conformation prevails for a given ligand in an unknown complex places a heavy burden on the scoring function.

9.2 Kinase Modelling. – Selectivity in kinase inhibition is a topic that is bound to gain importance as more and more kinase targets are discovered. Sheinerman *et al.*[106] describe a 3-step approach to distinguish and predict selectivity profiles. After a structure-guided alignment of representative kinases from all families, they use continuum electrostatics to identify the active site residues that contribute most to binding. The spatial arrangement of the important residues then constitutes what the authors term the "binding site signature". Substitutions are classified as non-conservative or conservative based on polarity and volume of the residue. In some cases, this crude method works surprisingly well, *e.g.*, kinases of the ABL-family that are inhibited by Imatinib in the sub-μM range are distinguished from non-targeted families (IC50 > 10 μM) based on only one non-conservative substitution. In some cases, kinases with conserved signature are not inhibited, prompting the authors to concede that conservation of the signature is a necessary but not sufficient criterion. While one non-conservative substitution is not sufficient in all cases to predict whether a drug is active on a given kinase, the correlation for two non-conservative substitutions is very good. By comparison, sequence similarity appears to be a very poor criterion. In this new field of kinomics,[100] approaches like this which combine a simple methodology with high predictive power are very welcome to digest the steadily growing amount of selectivity data.

Comparison of bindings sites, to examine questions of selectivity, has been tried with alignment-free approaches, for example the work of Vulpetti *et al.*[107] on kinases using GRID/CPCA.[108] By analysing the molecular interaction fields, they were able to identify key residues at the back of the ATP-binding pocket. These differences were exploited to design compounds that were both potent and selective. The advantage of this approach is that it is mostly automatic, and only needs unaligned models of the proteins. In case where there structural data is not available, Bayesian models[109] have been built from experimental inhibition data. The models do show some enrichment, and their speed means that they could be used for screening vendor libraries; however, it is still too early to expect the same power as the structure-based methods.

9.3 GPCR Modelling. – Despite the lack of further structures of GPCRs, modelling of this protein family is still an active area. Rognan *et al.*[110] have published an automated protocol for building models of the seven transmembrane domains. They have created a database of 277 human GPCRs from the secretin, rhodopsin and mGlutamate classes. The most interesting part of the approach is the use of docking to identify good and bad models. For a set of known ligands and decoys, good models can pick out the known ligands from the decoys: bad models can also be identified and refined further. The authors also note that the models can be used as a screen to identify possible receptors for ligands of unknown selectivity. In a study of the power of different techniques for identifying leads for GPCRs, Evers *et al.*[111] showed that pharmacophore, 2D-QSARs and Feature Trees did better than classical docking experiments, when this extra information was available. However, they conclude that docking to GPCRs still gave good levels of enrichment, so that this is still a valid method for hit identification for the target class.

A general QSAR model for GPCRs has also been developed.[112] The activity data was taken from 1939 compounds tested over 40 GPCRs, and the QSAR was built using pharmacophore-based descriptors. Experimental validation of 360 compounds against 21 targets showed that the model gave > 5-fold enrichment even at low threshold for the definition of activity. Enrichment was also observed for targets not in the original assay panel, so that ligands could be designed for orphan receptors based on the data from related GPCRs. Bock and Gough also describe a GPCR QSAR model based on very simple descriptors and the protein sequence.[113] Kratochwil *et al.*[114] have devised a method for abstracting the active site residues from GPCRs into a simple pharmacophore-like model. This representation is useful for comparison of active sites, the interpretation of mutagenesis data and the early assessment of selectivity issues. Lamb *et al.*[115] claim a pharmacophore-based model for GPCR screening, with a hit-rate of 2.6% against the μ-opioid receptor (IC50 < 10 μM). The model was derived from activity classes in the MDDR database, and 3- and 4-centre pharmacophores were used. Screening of virtual libraries is then performed by looking for subsets that overall cover the same pharmacophoric signature as the set of GPCR ligands.

Another study into classification of GPCR ligands used support vector machines and pharmacophore fingerprints.[116] The model was very good at identifying inactives (<99%) and could recall actives at 75%. To provide a stringent test (the space of GPCR ligands is relatively crowded, and there will often be a similar ligand left in the training set for leave-N-out validations), an entire chemotype class was removed. By refinding members of this class, for which the nearest neighbour in the training set had a similarity of 0.38 by Daylight fingerprints, the model showed that it could also be used for chemotype hopping.

Bridging the fields of structure-based drug design and chemogenomics is protein function and similarity. Given a query binding site, can one find proteins with similar function, or given a protein, can one identify the active site(s) by analogy with similar proteins. In addition to the standard CavBase,[117] which can be slow to run, Nussinov *et al.*[118] have developed a much faster method, based in essence on pharmacophore triplets. As always, speed is at the cost of resolution, and the pharmacophore features are crude, and not influenced by any local electrostatic perturbations. Also the representation is a snapshot of a single conformation. In tests, adenine and estradiol binding sites were accurately recognised, as could analogues of fatty-acid binding protein. A sequence-based tool has been developed by Lichtarge *et al.*,[119] based on their evolutionary trace algorithm. A set of analogues to the query protein are retrieved using BLAST, then aligned using ClustalW. The residues in the sequence are given a score depending on the degree of conservation: the number of branches the alignment must be split into before the residue becomes invariant within the branch. Clusters of residues of similar rank indicate some evolutionary focussing *i.e.* an active site. In their hands, the best-ranked residues cluster around active sites, and can be shown to occur at highly significant probabilities over chance. Using this measure they were able to identify 70–90% of active sites from sequence alone. This provides two good tools to complement biological fingerprinting.

10 Conclusions

In the last review period, we highlighted the advances in the field of ADME/Tox and docking. This trend has been sustained. One can speculate that collaborations driven by confidential in-house data have provided the impetus for the fresh attempts on some very hardy problems. Virtual screening in all its forms is at the heart of most research, from bioavailability filters through to rigorous estimations of the free energy of binding. It is pleasing that the field of de novo design is undergoing a small renaissance, as we understand how to control the changes made to molecules more carefully. We can also see the development of target-specific approaches becoming more popular over generalised solutions, so that local knowledge can be properly used, rather than washed out in the noise. Finally, we predict that chemogenomics will become a more common-place and accessible tool, allowing all modellers to mine our wealth of experimental data more fully.

References

1. M. Vieth, M.G. Siegel, R.E. Higgs, I.A. Watson, D.H. Robertson, K.A. Savin, G.L. Durst and P.A. Hipskind, *J. Med. Chem.*, 2004, **47**, 224.
2. P.D. Leeson and A.M. Davis, *J. Med. Chem.*, 2004, **47**, 6338.
3. J.J. Lu, K. Crimin, J.T. Goodwin, P. Crivori, C. Orrenius, L. Xing, P.J. Tandler, T.J. Vidmar, B.M. Amore, A.G.E. Wilson, P.F.W. Stouten and P.S. Burton, *J. Med. Chem.*, 2004, **47**, 6104.
4. D.F. Veber, S.R. Johnson, H.Y. Cheng, B.R. Smith, K.W. Ward and K.D. Kopple, *J. Med. Chem.*, 2002, **45**, 2615.
5. Y.C. Martin, *J. Med. Chem.*, 2005, **48**, 3164.
6. J.K. Wegner, H. Frohlich and A. Zell, *J. Chem. Inf. Comput. Sci.*, 2004, **44**, 931.
7. C. de Graaf, N.P.E. Vermeulen and K.A. Feenstra, *J. Med. Chem.*, 2005, **48**, 2725.
8. S.E. O'Brien and M.J. de Groot, *J. Med. Chem.*, 2005, **48**, 1287.
9. K.K. Chohan, S.W. Paine, J. Mistry, P. Barton and A.M. Davis, *J. Med. Chem.*, 2005, **48**, 5154.
10. R. Guha and P.C. Jurs, *J. Chem. Inf. Model.*, 2005, **45**, 65.
11. R.P. Sheridan, B.P. Feuston, V.N. Maiorov and S.K. Kearsley, *J. Chem. Inf. Comput. Sci.*, 2004, **44**, 1912.
12. S.B. Singh, L.Q. Shen, M.J. Walker and R.P. Sheridan, *J. Med. Chem.*, 2003, **46**, 1330.
13. G. Cruciani, E. Carosati, B. DeBoeck, K. Ethirajulu, C. Mackie, T. Howe and R. Vianello, *J. Med. Chem.*, 2005, **48**, 6970.

14. J.M. Kriegl, T. Arnhold, B. Beck and T. Fox, *J. Comput. -Aided Mol. Des.*, 2005, **19**, 189.
15. M.E. Beck, *J. Chem. Inf. Model.*, 2005, **45**, 273.
16. S.P. deVisser, D. Kumar, S. Cohen, R. Shacham and S. Shaik, *J. Am. Chem. Soc.*, 2004, **126**, 8362.
17. J.F. Contrera, E.J. Matthews, N.L. Kruhlak and R.D. Benz, *Reg. Tox. Pharm.*, 2004, **40**, 185.
18. C. Helma, *Curr. Op. Drug Disc. Dev.*, 2005, **8**, 27.
19. S. Dimitrov, G. Dimitrova, T. Pavlov, N. Dimitrova, G. Patlewicz, J. Niemela and O. Mekenyan, *J. Chem. Inf. Model.*, 2005, **45**, 839.
20. C.M. Krejsa, D. Horvath, S.L. Rogalski, J.E. Penzotti, B. Mao, F. Barbosa and J.C. Migeon, *Curr. Op. Drug Disc. Dev.*, 2003, **6**, 470.
21. A.J.S. Knox, M.J. Meegan, G. Carta and D.G. Lloyd, *J. Chem. Inf. Model.*, 2005, **45**, 1908.
22. J.J. Irwin and B.K. Shoichet, *J. Chem. Inf. Model.*, 2005, **45**, 177.
23. M. Sun, Z.S. Li, Y. Zhang, Q.C. Zheng and C.C. Sun, *Bioorg. Med. Chem. Lett.*, 2005, **15**, 2851.
24. S.L. McGovern and B.K. Shoichet, *J. Med. Chem.*, 2003, **46**, 2895.
25. C. Bissantz, P. Bernard, M. Hibert and D. Rognan, *Proteins*, 2003, **50**, 5.
26. J.A. Erickson, M. Jalaie, D.H. Robertson, R.A. Lewis and M. Vieth, *J. Med. Chem.*, 2004, **47**, 45.
27. A.M. Ferrari, B.Q.Q. Wei, L. Costantino and B.K. Shoichet, *J. Med. Chem.*, 2004, **47**, 5076.
28. T. Polgar and G.M. Keseru, *J. Med. Chem.*, 2005, **48**, 3749.
29. H.A. Carlson, *Current Opinion in Chemical Biology*, 2003, **6**, 01.
30. X. Barril and S.D. Morley, *J. Med. Chem.*, 2005, **48**, 4432.
31. J.E. Ladbury, *Chemistry & Biology*, 1996, **3**, 973.
32. M.L. Verdonk, G. Chessari, J.C. Cole, M.J. Hartshorn, C.W. Murray, J.W.M. Nissink, R.D. Taylor and R. Taylor, *J. Med. Chem.*, 2005, **48**, 6504.
33. C. de Graaf, P. Pospisil, W. Pos, G. Folkers and N.P.E. Vermeulen, *J. Med. Chem.*, 2005, **48**, 2308.
34. M.D. Cummings, R.L. DesJarlais, A.C. Gibbs, V. Mohan and E.P. Jaeger, *J. Med. Chem.*, 2005, **48**, 962.
35. J.C. Cole, C.W. Murray, J.W.M. Nissink, R.D. Taylor and R. Taylor, *Proteins*, 2005, **60**, 325.
36. M.L. Verdonk, V. Berdini, M.J. Hartshorn, W.T.M. Mooij, C.W. Murray, R.D. Taylor and P. Watson, *J. Chem. Inf. Comput. Sci.*, 2004, **44**, 793.
37. G.L. Warren, C.W. Andrews, A.M. Capelli, B. Clarke, J. LaLonde, M.H. Lambert, M. Lindvall, N. Nevins, S.F. Semus, S. Senger, G. Tedesco, I.D. Wall, J.M. Woolven, C.E. Peishoff and M.S. Head, *J. Med. Chem.*, 2005, dx.doi.org/10.1021/jm050362n.
38. A.P. Graves, R. Brenk and B.K. Shoichet, *J. Med. Chem.*, 2005, **48**, 3714.
39. H.F.G. Velec, H. Gohlke and G. Klebe, *J. Med. Chem.*, 2005, **48**, 6296.
40. C. bad-Zapatero and J.T. Metz, *Drug Discovery Today*, 2005, **10**, 464.
41. Y. Pan, N. Huang, S. Cho and A.D. MacKerell Jr., *J. Chem. Inf. Comp. Sci.*, 2003, **43**, 267.
42. W.M. Rockey and A.H. Elcock, *J. Med. Chem.*, 2005, **48**, 4138.
43. C. Oostenbrink and W.F. van Gunsteren, *Proc. Natl. Acad. Sci. U. S. A.*, 2005, **102**, 6750.
44. C. Oostenbrink and W.F. van Gunsteren, *Proteins*, 2004, **54**, 237.

45. C. Oostenbrink and W.F. van Gunsteren, *J. Comput. Chem.*, 2003, **24**, 1730.
46. C. Oostenbrink, A. Villa, A.E. Mark and W.F. van Gunsteren, *J. Comput. Chem.*, 2004, **25**, 1656.
47. J.M. Yang, Y.F. Chen, T.W. Shen, B.S. Kristal and D.F. Hsu, *J. Chem. Inf. Model.*, 2005, **45**, 1134.
48. G.P.A. Vigers and J.P. Rizzi, *J. Med. Chem.*, 2004, **47**, 80.
49. R.A. Friesner, J.L. Banks, R.B. Murphy, T.A. Halgren, J.J. Klicic, D.T. Mainz, M.P. Repasky, E.H. Knoll, M. Shelley, J.K. Perry, D.E. Shaw, P. Francis and P.S. Shenkin, *J. Med. Chem.*, 2004, **47**, 1739.
50. S. Radestock, M. Bohm and H. Gohlke, *J. Med. Chem.*, 2005, **48**, 5466.
51. A.E. Muryshev, D.N. Tarasov, A.V. Butygin, O.Y. Butygina, A.B. Aleksandrov and S.M. Nikitin, *J. Comput. -Aided Mol. Des.*, 2003, **17**, 597.
52. A. Krammer, P.D. Kirchhoff, X. Jiang, C.M. Venkatachalam and M. Waldman, *J. Mol. Graph. Mod.*, 2005, **23**, 395.
53. A. Curioni, T. Mordasini and W. Andreoni, *J. Comput. -Aided Mol. Des.*, 2004, **18**, 773.
54. P. Ferrara, A. Curioni, E. Vangrevelinghe, T. Meyer, T. Mordasini, W. Andreoni, P. Acklin and E. Jacoby, *J. Chem. Inf. Model.*, 2006, **46**, 254.
55. B. Kuhn, P. Gerber, T. Schulz-Gasch and M. Stahl, *J. Med. Chem.*, 2005, **48**, 4040.
56. N. Huang, C. Kalyanaraman, J.J. Irwin and M.P. Jacobson, *J. Chem. Inf. Model.*, 2006, **46**, 243.
57. V. Maiorov and R.P. Sheridan, *J. Chem. Inf. Model.*, 2005, **45**, 1017.
58. E. Carosati, S. Sciabola and G. Cruciani, *J. Med. Chem.*, 2004, **47**, 5114.
59. M. Forino, D. Jung, J.B. Easton, P.J. Houghton and M. Pellecchia, *J. Med. Chem.*, 2005, **48**, 2278.
60. D.Z. Huang, U. Luthi, P. Kolb, K. Edler, M. Cecchini, S. Audetat, A. Barberis and A. Caflisch, *J. Med. Chem.*, 2005, **48**, 5108.
61. J.C. Mozziconacci, E. Arnoult, P. Bernard, Q.T. Do, C. Marot and L. Morin-Allory, *J. Med. Chem.*, 2005, **48**, 1055.
62. G. Schneider and U. Fechner, *Nat. Rev. Drug Disc.*, 2005, **4**, 649.
63. M.A. Ali, N. Bhogal, J.B.C. Findlay and C.W.G. Fishwick, *J. Med. Chem.*, 2005, **48**, 5655.
64. J. Elsner, F. Boeckler, F.W. Heinemann, H. Hubner and P. Gmeiner, *J. Med. Chem.*, 2005, **48**, 5771.
65. T. Fink, H. Bruggesser and J.L. Reymond, *Ang. Chem. Int. Ed*, 2005, **44**, 1504.
66. N. Brown, B. McKay and J. Gasteiger, *J. Comput. -Aided Mol. Des.*, 2004, **18**, 761.
67. N. Brown, B. McKay, F. Gilardoni and J. Gasteiger, *J. Chem. Inf. Comput. Sci.*, 2004, **44**, 1079.
68. R.A. Lewis, *J. Med. Chem.*, 2005, **48**, 1638.
69. C.M.R. Low, I.M. Buck, T. Cooke, J.R. Cushnir, S.B. Kalindjian, A. Kotecha, M.J. Pether, N.P. Shankley, J.G. Vinter and L. Wright, *J. Med. Chem.*, 2005, **48**, 6790.
70. D. Jung, J. Floyd and T.M. Gund, *J. Comp. Chem.*, 2004, **25**, 1385.
71. P.A. Datar and E.C. Coutinho, *J. Mol. Graph. Mod.*, 2004, **23**, 239.
72. C.L. Kuo, H. Assefa, S. Kamath, Z. Brzozowski, J. Slawinski, F. Saczewski, J.K. Buolamwini and N. Neamati, *J. Med. Chem.*, 2004, **47**, 385.
73. V. Chandrasekaran, G.B. McGaughey, C.J. Cavallito and J.P. Bowen, *J. Mol. Graph. Model.*, 2004, **23**, 69.
74. B. Jojart, T.A. Martinek and A. Marki, *J. Comput. -Aided Mol. Des.*, 2005, **19**, 341.
75. A.J. Tervo, T.H. Nyronen, T. Ronkko and A. Poso, *J. Comput. -Aided Mol. Des.*, 2003, **17**, 797.

76. A.A. Soderholm, P.T. Lehtovuori and T.H. Nyronen, *J. Med. Chem.*, 2005, **48**, 917.
77. Y.Y. Peng, S.M. Keenan, Q. Zhang, V. Kholodovych and W.J. Welsh, *J. Med. Chem.*, 2005, **48**, 1620.
78. B.M. Baron, R.J. Cregge, R.A. Farr, D. Friedrich, R.S. Gross, B.L. Harrison, D.A. Janowick, D. Matthews, T.C. McCloskey, S. Meikrantz, P.L. Nyce, R. Vaz and W.A. Metz, *J. Med. Chem.*, 2005, **48**, 995.
79. S. Khanna, M.E. Sobhia and P.V. Bharatam, *J. Med. Chem.*, 2005, **48**, 3015.
80. A. Biegel, S. Gebauer, B. Hartrodt, M. Brandsch, K. Neubert and I. Thondorf, *J. Med. Chem.*, 2005, **48**, 4410.
81. S.J. Cottrell, V.J. Gillet, R. Taylor and D.J. Wilton, *J. Comput. -Aided Mol. Des.*, 2004, **18**, 665.
82. R. Kristam, V.J. Gillet, R.A. Lewis and D. Thorner, *J. Chem. Inf. Model.*, 2005, **45**, 461.
83. J. Bostrom, J.R. Greenwood and J. Gottfries, *J. Mol. Graph. Model.*, 2003, **21**, 449.
84. G.L. Mustata, A. Brigo and J.M. Briggs, *Bioorg. Med. Chem. Lett.*, 2004, **14**, 1447.
85. J.X. Deng, K.W. Lee, T. Sanchez, M. Cui, N. Neamati and J.M. Briggs, *J. Med. Chem.*, 2005, **48**, 1496.
86. A. Brigo, K.W. Lee, F. Fogolari, G.L. Mustata and J.M. Briggs, *Proteins*, 2005, **59**, 723.
87. A.M. Aronov and M.A. Murcko, *J. Med. Chem.*, 2004, **47**, 5616.
88. T. Klabunde and A. Evers, *Chembiochem.*, 2005, **6**, 876.
89. P. Kahnberg, M.H. Howard, T. Liljefors, M. Nielsen, E.O. Nielsen, O. Sterner and I. Pettersson, *J. Mol. Graph. Model.*, 2004, **23**, 253.
90. S.C. Schurer, P. Tyagi and S.A. Muskal, *J. Chem. Inf. Model.*, 2005, **45**, 239.
91. S. Brauer, M. Almstetter, W. Antuch, D. Behnke, R. Taube, P. Furer and S. Hess, *J. Comb. Chem.*, 2005, **7**, 218.
92. P. Willett, *J. Med. Chem.*, 2005, **48**, 4183.
93. J.R. Krumrine, A.T. Maynard and C.L. Lerman, *J. Med. Chem.*, 2005, **48**, 7477.
94. T.S. Rush, J.A. Grant, L. Mosyak and A. Nicholls, *J. Med. Chem.*, 2005, **48**, 1489.
95. G. Hessler, M. Zimmermann, H. Matter, A. Evers, T. Naumann, T. Lengauer and M. Rarey, *J. Med. Chem.*, 2005, **48**, 6575.
96. A. Bender, H.Y. Mussa, R.C. Glen and S. Reiling, *J. Chem. Inf. Comput. Sci.*, 2004, **44**, 170.
97. M. Rarey and J.S. Dixon, *J. Comput. -Aided Mol. Des.*, 1998, **12**, 471.
98. P. Labute, *J. Chem. Inf. Model.*, 2005, **45**, 215.
99. M.S. Lajiness, G.M. Maggiora and V. Shanmugasundaram, *J. Med. Chem.*, 2004, **47**, 4891.
100. M. Vieth, R.E. Higgs, D.H. Robertson, M. Shapiro, E.A. Gragg and H. Hemmerle, *Bio. Biophys. Acta.*, 2004, **1697**, 243.
101. H.B. Broughton and I.A. Watson, *J. Mol. Graph. Model.*, 2004, **23**, 51.
102. K. Schlotter, F. Boeckler, H. Hubner and P. Gmeiner, *J. Med. Chem.*, 2005, **48**, 3696.
103. P.J. Hajduk, J.R. Huth and S.W. Fesik, *J. Med. Chem.*, 2005, **48**, 2518.
104. C.Y. Yang, R.X. Wang and S.M. Wang, *J. Med. Chem.*, 2005, **48**, 5648.
105. A.Y.C. Yang, P. Kallblad and R.L. Mancera, *J. Comput. -Aided Mol. Des.*, 2004, **18**, 235.
106. F.B. Sheinerman, E. Giraud and A. Laoui, *J. Mol. Biol.*, 2005, **352**, 1134.
107. A. Vulpetti, P. Crivori, A. Cameron, J. Bertrand, M.G. Brasca, R. D'Alessio and P. Pevarello, *J. Chem. Inf. Model.*, 2005, **45**, 1282.

108. M.A. Kastenholz, M. Pastor, G. Cruciani, E.E. Haaksma and T. Fox, *J. Med. Chem.*, 2000, **43**, 3033.
109. X.Y. Xia, E.G. Maliski, P. Gallant and D. Rogers, *J. Med. Chem.*, 2004, **47**, 4463.
110. C. Bissantz, A. Logean and D. Rognan, *J. Chem. Inf. Comput. Sci.*, 2004, **44**, 1162.
111. A. Evers, G. Hessler, H. Matter and T. Klabunde, *J. Med. Chem.*, 2005, **48**, 5448.
112. C. Rolland, R. Gozalbes, E. Nicolai, M.F. Paugam, L. Coussy, F. Barbosa, D. Horvath and F. Revah, *J. Med. Chem.*, 2005, **48**, 6563.
113. J.R. Bock and D.A. Gough, *J. Chem. Inf. Model.*, 2005, **45**, 1402.
114. N.A. Kratochwil, P. Malherbe, L. Lindemann, M. Ebeling, M.C. Hoener, A. Muhlemann, R.H.P. Porter, M. Stahl and P.R. Gerber, *J. Chem. Inf. Model.*, 2005, **45**, 1324.
115. M.L. Lamb, E.K. Bradley, G. Beaton, S.S. Bondy, A.J. Castellino, P.A. Gibbons, M.J. Suto and P.D. Grootenhuis, *J. Mol. Graph. Model.*, 2004, **23**, 15.
116. J.C. Saeh, P.D. Lyne, B.K. Takasaki and D.A. Cosgrove, *J. Chem. Inf. Model.*, 2005, **45**, 1122.
117. M. Hendlich, A. Bergner, J. Gunther and G. Klebe, *J. Mol. Biol.*, 2003, **326**, 607.
118. A. Shulman-Peleg, R. Nussinov and H.J. Wolfson, *J. Mol. Biol.*, 2004, **339**, 607.
119. H. Yao, D.M. Kristensen, I. Mihalek, M.E. Sowa, C. Shaw, M. Kimmel, L. Kavraki and O. Lichtarge, *J. Mol. Biol.*, 2003, **326**, 255.

2

Modelling Biological Systems

BY ADRIAN J. MULHOLLAND

Centre for Computational Chemistry, School of Chemistry, University of Bristol, Bristol BS8 1TS, UK

1 Introduction

'Can I believe modelling?'. This is a question that is often asked of modellers by biologists and biochemists. To answer it sensibly requires a critical understanding of the capabilities of current modelling and simulation methods, their strengths and limitations, and ranges of application. Knee-jerk scepticism of all biomolecular modelling is sometimes encountered among experimentalists even today; equally misguided is a blind acceptance of modelling results without critical analysis. However, demonstrations of the practical value of biomolecular modelling and simulation for many types of application to biological systems have led to a growing recognition of its worth. This is a fertile and growing area, with exciting opportunities. Crucially for the modeller, it is important to understand the issues of interest to biologists, the complexity of biological systems, and how to tackle them effectively by modelling. The possible range of applications is huge. The challenge posed by the vast amounts of data provided by biological research in genomics, proteomics, glycomics and structural biology is enormous and pressing. Modelling should help in the effort to use this information in the development of new drugs, therapies and biologically inspired or based technologies. Modelling also has a vital and growing role in the interpretation of experimental data from the range of sophisticated physical techniques that are being applied to the study of biological systems.

Increasingly molecular modelling and simulation methods are demonstrat ing their practical worth in investigations of biomolecular systems, making important and indeed often uniquely detailed contributions to the study of the structure and function of biological macromolecules. Applications include studies of protein folding, conformational changes, association of proteins with small molecules or other proteins, structure-based drug design, modelling and analysis of enzyme catalysis, computation of binding free energies for ligands; modelling the dynamics of ion channels and transport across membranes.

Chemical Modelling: Applications and Theory, Volume 4

Improvements in computer hardware continue to deliver ever-increasing computational power, which, when combined with theoretical and algorithmic developments, have led to an increasing range and depth of applications of molecular modelling in biology. Indeed, the whole field of biomolecular modelling is now too large to be reasonably covered in a single review – perhaps even in a textbook. Here, the focus is on atomistic molecular simulation, with a focus on proteins, and a particular emphasis on the area of modelling enzyme-catalysed reaction mechanisms. This reflects the author's own research interests of course, but also the vitality and current interest in this area and methods developed for the study of enzymic reactions. The area of biological catalysis is fascinating from a chemical point of view, and is an important interface between chemistry and biology. This review aims to highlight some exciting recent applications of molecular modelling and simulation methods to biological systems, and some important current theoretical developments in the field.

2 Empirical Forcefields for Biomolecular Simulation: Molecular Mechanics (MM) Methods

The area of biology where modelling and simulation have perhaps had their greatest impact so far, and where the closest integration of modelling and experiment has built up, is in the study of the dynamics of biological macromolecules.[1] Molecular dynamics simulations of proteins have played an important conceptual role, (along with nuclear magnetic resonance (NMR) spectroscopic methods) in demonstrating that proteins flex and undergo complex internal motions, which in some cases are directly related to function.[2] Molecular dynamics simulations are also very important in simulations of protein folding and unfolding,[3] where they have proved their worth in complementing experimental investigations and in the interpretation of experimental data.[4] Molecular dynamics simulations also assist in the refinement of biomolecular structures in structural investigations by X-ray crystallography[5] and NMR[6] (and also in the analysis of NMR data on spin relaxation[7] and dynamics, for example dynamics of protein sidechains[8]). Other important areas of application of biological molecular dynamics simulation include drug design,[9] studies of protein conformational changes,[10,11] simulations of ion channels and other membrane proteins,[12–14] and studies of functional macromolecular assemblies such as F-1-ATPase.[15]

An interesting recent example demonstrating the synergy between experimental studies and molecular dynamics simulations of biomolecules is provided by work of Dedmon *et al.*[16] These authors mapped long-range interactions using a combination of ensemble molecular dynamics simulations and spin-label NMR. This work studied the protein alpha-synuclein, which is involved in the pathogenesis of Parkinson's disease, and is intrinsically disordered. To test this hypothesis, distance restraints derived from paramagnetic relaxation enhancement NMR spectroscopy were applied to ensemble MD simulations containing twenty protein replicas, with the CHARMM19 force field (see below). The results showed that the native state of alpha-synuclein is made up of a broad distribution of conformers. The ensemble-averaged hydrodynamic radius was found to be significantly smaller than that expected for a simple random coil structure. The structural studies showed that this contraction is driven by interactions between the highly charged C-terminus and a large hydrophobic central region of the protein sequence. The results led the authors to suggest that this type of structure may be responsible for inhibiting the formation of alpha-synuclein aggregates, which are thought to be the cytotoxic species responsible for neurodegeneration in Parkinson's Disease. MD simulations have similarly been used to generate conformations in the experimental determination of an ensemble of structures representing the denatured state of the bovine acyl-coenzyme A binding protein.[17]

The first simulation of biomolecular dynamics was carried out almost thirty years ago, on the small protein bovine pancreatic trypsin inhibitor (BPTI), in the gas phase.[18] In the intervening years, several empirical force fields have been developed for the simulation of proteins, nucleic acids, lipids and other biological molecules. It is important to make the distinction between programs used for biomolecular simulation, and the molecular mechanics parameter sets which have been developed for them, in particular as the names are similar and sometimes used interchangeably. A number of good quality parameter sets have been developed, and may be applied with several different programs, as the functional forms used are often the same or very similar. The quality of the particular parameter set is something to consider independently of the quality of the computer program itself. Of course, it is essential that a particular force field parameter set should be implemented in any program exactly as it was designed to be, and this should be carefully checked. Different protocols may apply in different programs, perhaps with different hidden assumptions. Tests on model systems are important to ensure that interactions are treated consistently and correctly by a given force field.

Among the most widely-used computer *programs* used for biological molecular dynamics simulations particularly in academic research are AMBER,[19] CHARMM,[20] GROMOS,[21] NAMD[22] and TINKER.[23] Several other molecular dynamics simulations packages are available, including commercial and academic programs.

Programs for molecular simulation should not be confused with force fields used, as mentioned above. A force field consists of the energy function used and the parameters. A simple energy function has to be used to allow large systems to be studied for long (multinanosecond) timescales. Current protein force fields use similar (and familiar) potential energy functions, in which for example bonds and valence angles are represented by harmonic terms, electrostatic interactions are included through atomic point partial charges, and dispersion and exchange repulsion are included by a simple Lennard-Jones function (usually of the 12-6 variety).

There are of course important limitations to the simple molecular mechanics representation. For example, electrostatic interactions are represented by a simple model including a point change on each atom (and only on atoms). This simple representation obviously cannot capture the full electrostatic properties (*e.g.* multipole moments) of a molecule, a particular problem for less polar species. The atomic charges are also treated as invariant, that is they do not change in response to changes in the molecular environment or conformation: electronic polarization is not included. The attractive (r^{-6}) component of the Lennard-Jones potential has some physical justification for modelling dispersion interactions. The repulsive (r^{-12}) term is chosen simply for computational convenience to represent exchange repulsion at short distances: it is known that an exponential description is more physically realistic, but in the context of the overall MM description, this is typically a small error for 'organic' type molecules. Simple harmonic terms represent the energy of bond stretching and valence angle bending, with simple periodic terms for torsion angles, and terms for other intramolecular interactions where necessary. Clearly potential functions of this type cannot be applied to model the bond breaking and bond making, and electronic reorganization, involved in a chemical reaction: the bond terms do not allow bond dissociation or formation, and electronic redistribution is not be accounted for. Also, the MM force field parameters are developed based on the properties of stable molecules, and so will usually not be applicable to transition states and intermediates. It is possible to develop MM functions and parameters specifically for reactions, and this has been highly successful in application to organic reactions in solution.[24] However, the parameters are generally applicable only to a particular reaction, or small class of reactions, meaning that reparameterization will be necessary for each problem studied. Also, the form of the potential function imposes important limitations, such as the neglect of electronic polarization.

Force fields for biological macromolecules fall into two classes: united-atom, and all-atom force fields. All-atom force fields, as the name suggests, represent all atoms in a protein explicitly. United-atom force fields, on the other hand, treat only heavy (non-hydrogen) atoms and polar hydrogen atoms explicitly, while nonpolar hydrogen atoms are not represented explicitly, but rather represented as part of the carbon atom to which they are bonded (which will have an enlarged van der Waals radius (Lennard-Jones collision diameter)).

Currently, the most widely used all-atom force fields for proteins are OPLS/AA[25,26] CHARMM22[27] and AMBER (PARM99).[19,20] A number of good reviews of the performance of protein MM force fields have been published recently.[29,30] It has been found that modern protein MM force fields behave comparably in molecular dynamics simulations.[31] Parameterization of these force fields is increasingly based on fitting to experimental condensed phase data (such as free energies of solvation for amino acid sidechains), particularly in the optimization of Lennard-Jones parameters, in contrast to the historically dominant place of gas phase data (*e.g.* *ab initio* calculations of heterodimers) in parameterization. Force fields for other types of biological macromolecules (*e.g.* lipids, nucleic acids[32,33] and saccharides, as well as many small molecules and ligands) consistent with these protein force fields have also been developed, which allow simulations of proteins complexed with DNA, embedded in membranes, *etc*. Examples include the CHARMM27 force field for nucleic acids,[34,35] AMBER nucleic acid parameters,[28,36] CHARMM parameters for lipids[37] and a number of different MM parameterizations of common carbohydrates. For example, Kuttel and coworkers have developed carbohydrate parameters for use with the CHARMM forcefield, suitable for nanosecond molecular dynamics simulations in aqueous solution.[38] Free energy profiles for rotation of the hydroxymethyl group for two monosaccharides (beta-D-glucose and beta-D-galactose) with this parameter set showed equilibrium rotamer populations in excellent agreement with NMR data, and both the primary alcohol rotational frequency in solution, and the gas-phase vibrational frequencies, were found to be in excellent agreement with experiment. Heramingsen and coworkers have tested the performance of twenty different MM carbohydrate force fields, by comparison with (gas phase) *ab initio* and hybrid density functional calculations on monosaccharides.[39] Geometry-optimized structures (B3LYP/6-31G(d)) and relative energies using augmented correlation consistent basis sets were calculated in gas phase for monosaccharide carbohydrate benchmark systems. Among their key findings are that most carbohydrate molecular mechanics force fields calculate the interaction energy of the alpha-D-glucopyranose-H_2O complex erroneously, compared to the *ab initio* (up to the coupled cluster CCSD(T) level) result of 4.9 kcal/mol; no single force field was found to perform consistently better than the others for a variety of the test cases (*e.g.* for conformational energies of methyl 5-deoxy-beta-D-xylofuranoside, methyl alpha D-glucopyranoside and methyl alpha-D-galactopyranoside. A statistical assessment of the performance of the force fields suggested that the force fields CHEAT95 (an extended atom model),[40] and some parameterizations developed based on the AMBER,[41] CFF (consistent force field)[42] and MM3[43,44] force fields have the best overall performance, for the gas phase monosaccharide systems studied in this work. It should be noted that several of these force fields have more sophisticated and complicated potential energy functions than those typically used for protein simulations. Developing MM parameters for (poly)saccharides is notoriously difficult,[45] because of their conformational complexity, large range of possible substitution patterns, and the particular difficulty of balancing inter- and intramolecular interactions (because sugars contain very large numbers of hydrogen bonding groups. There are clear limitations of the invariant atomic point charge model for carbohydrates. A QM/MM approach[46] (see section 3 below), treating the sugar by QM, may be preferable in some cases. Standard semiempirical quantum chemical methods have important shortcomings for carbohydrates, but reparameterized variants have been developed which give better descriptions of carbohydrate conformation (*e.g.* PM3CARB-1).[47]

Examples of united-atom protein force-fields for proteins are GROMOS87 and 96,[21,48] CHARMM PARAM19,[49] OPLS/UA (united atom)[50] and the original force fields developed for the AMBER program.[51] United atom force fields were developed to reduce the computer time required for molecular dynamics simulations by reducing the number of atoms. They are still important today, in studies using either explicit or implicit solvation models. They are particularly widely used in studies of protein folding, often employing a continuum solvation description to reduce computational demands in these long timescale simulations, by avoiding the need to include explicit water molecules. Several implicit solvent models have been developed for use with the CHARMM PARAM19 force field, including EEF1 (effective energy function)[52] and ACE (analytic continuum electrostatics)[53] and models based on the Generalized Born approach[54–56] and other fast implicit solvation models for molecular dynamics simulations.[57] Assessment of the performance (both accuracy and efficiency) of implicit solvent models (*e.g.* by comparison with explicit solvent simulations) is a highly active area of research.

Most biomolecular MM forcefields have been developed with simple point charge models of water, in particular the TIP3P water model[58] and variants thereof. Electronic polarization is included only in an approximate, gross way in models such as TIP3P: for example, the dipole moment of such models is higher than that observed in the gas phase, thus including the effects of polarization in the condensed phase. Similarly, as mentioned above, protein MM force fields only include electronic polarization in an average, and invariant way. Polarizable force fields for biological molecules are the subject of much current research and development effort.[59–66] The next generation of protein MM force fields will probably include electronic polarization explicitly. Other improvements to protein MM force fields include the use of *ab initio* data to correct for the two-dimensional potential energy surface for peptide backbone dihedral angle rotation.[67,68]

An example of how far molecular dynamics simulations of biomolecules has advanced is provided by recent simulations of that 'lab rat' of biological simulation, bovine pancreatic trypsin inhibitor (BPTI). Kim *et al.* have examined the effects of solvent and protein polarizability on the solvation structure and dynamics of bovine pancreatic trypsin inhibitor in explicit water.[69] To study specifically the effects of including of polarizability in a molecular mechanics force field model (on the dynamics and structure of the solvating water in proteins), these authors carried out molecular dynamics simulations of bovine pancreatic trypsin inhibitor (BPTI) in explicit water, using MM force fields that include polarization for *both* the water and the protein. Three model potentials for water and two model potentials for the protein were used, of which two of the water models and one of the protein models were polarizable. Six systems were simulated, covering all combinations of these polarizable and nonpolarizable protein and water force fields. These workers found that all six systems behave similarly in less polar parts of the protein (either hydrophobic or weakly hydrophilic). However, close to parts of the protein in which relatively strong electrostatic fields occur (*i.e.*, near positively or negatively charged residues), they found that water structure and dynamics were clearly dependent on both the model of the protein and the model of the water used.

3 Combined Quantum Mechanics/Molecular Mechanics (QM/MM) Methods

Combined quantum mechanics/molecular mechanics (QM/MM) methods are increasingly important in the modelling of biological systems, particularly in the growing field of computational enzymology – *i.e.* computational modelling of enzyme-catalysed reaction mechanisms.[70] The essence of the QM/MM approach is simple: a small part of the system is treated quantum mechanically, *i.e.* by an electronic structure method, for example at the *ab initio* or semiempirical molecular orbital, or density-functional theory QM level. This QM treatment allows the electronic rearrangements involved in the bond-breaking and making in a chemical reaction, as well as electronic polarization, to be modelled. The QM region in a study of an enzymic reaction mechanism would typically be the enzyme active site, and include the reacting groups of the enzyme, substrate and any co-factors. The large non-reactive part is described more simply by empirical molecular mechanics. Different types of coupling between the QM and MM regions can be employed (see below). For applications to biological macromolecules such as proteins, which are polar, including interactions between the QM and MM regions is important. The combination of the versatility and range of applicability of a QM electronic structure method with the efficiency and speed of the MM force field allows reactions in large systems to be studied. As noted above, modern molecular mechanics methods give a good description of protein structure and interactions, so can ensure that these are treated accurately in the QM/MM approach. With lower levels of QM theory (*e.g.* semiempirical molecular orbital, or approximate density functional methods), QM/MM molecular dynamics simulations are feasible.

A QM/MM method was first applied to an enzyme-catalysed reaction by Warshel and Levitt in their ground-breaking 1976 study of the reaction mechanism of hen egg-white lysozyme.[71] Interest in QM/MM methods has grown rapidly in recent years. Following many recent developments and applications, it is now clear that QM/MM calculations can provide useful insight into the mechanisms of enzyme-catalysed reactions.[72–74] For example, they have demonstrated this in identifying catalytic functions for active site residues (such as a conserved proline in two flavin-dependent monooxygenases[75,76]), investigating questions of mechanism (*e.g.* comparing and differentiating between alternative proposed mechanisms), and suggesting and testing catalytic principles (such as the possible contribution of conformational effects and transition state stabilization in chorismate mutase).[77,78] Even for this apparently simple enzyme reaction (a Claisen rearrangement), there are lively current debates on the origin of catalysis, as discussed below. Modelling and simulation have been crucial here in formulating and testing mechanisms and hypotheses. The structure of the transition state found in recent *ab initio* QM/MM modelling of this enzymic reaction[78] is shown in Figure 1.

Many different QM/MM implementations are available, in a number of widely-used program packages. QM/MM calculations can be carried out at *ab initio*[79,80] or semiempirical[81] molecular orbital, density-functional[82] or

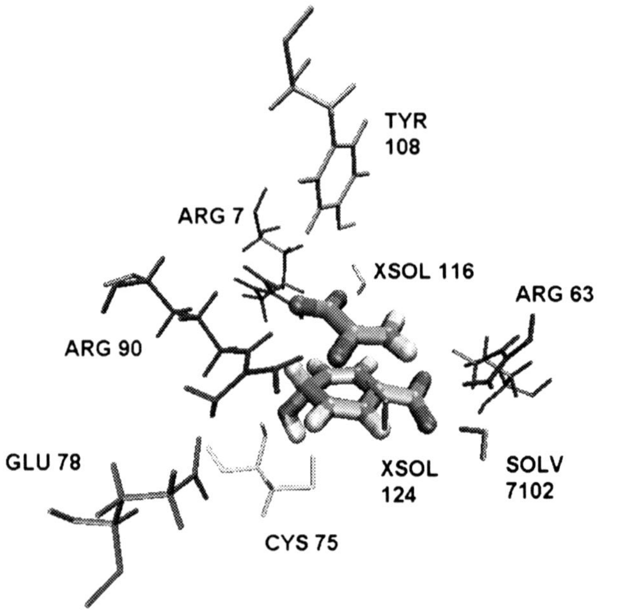

Figure 1 *In a QM/MM calculation, a small region is treated by a quantum mechanical (QM) electronic structure method, and the surroundings treated by simpler, empirical, molecular mechanics. In treating an enzyme-catalysed reaction, the QM region includes the reactive groups, with the bulk of the protein and solvent environment included by molecular mechanics. Here, the approximate transition state for the Claisen rearrangement of chorismate to prephenate (catalysed by the enzyme chorismate mutase) is shown. This was calculated at the RHF/6-31G(d)-CHARMM QM-MM level. The QM region here (the substrate only) is shown by thick tubes, with some important active site residues (treated by MM) also shown. The whole model was based on a 25 Å sphere around the active site, and contained 4211 protein atoms, 24 atoms of the substrate and 947 water molecules (including 144 water molecules observed by X-ray crystallography), a total of 7076 atoms. The results showed specific transition state stabilization by the enzyme. Comparison with the same reaction in solution showed that transition state stabilization is important in catalysis by chorismate mutase*[78].

approximate density functional (*e.g.* the self-consistent charge density functional tight-binding (SCC-DFTB)[83] method combines computational efficiency with reasonable accuracy for many applications) levels of QM electronic structure calculation. Transition state structures can be optimized.[84,85] Molecular dynamics simulations can be carried out with cheaper QM/MM methods.[86] Free energy differences, such as activation free energies can be calculated, as can quantum effects such as tunnelling and zero-point corrections. More approximate, less computer intensive, QM/MM methods (such as semiempirical or SCC-DFTB QM/MM) have an important role as they allow more extensive simulations to be performed (*e.g.* molecular dynamics or Monte Carlo simulations, extensive conformational sampling, and calculation of reaction pathways and Hessians). Specifically parameterized semiempirical methods can give improved accuracy for a particular reaction. High-level QM/MM calculations (*e.g. ab initio* or density functional level QM) are required for some systems but can be extremely computationally demanding. High-level QM/MM calculations also have an important role in testing more approximate (*e.g.* semiempirical QM/MM) methods. Current developments include the use of QM/MM methods in free energy perturbation simulations,[87] *e.g.* to calculate relative binding affinities, and in molecular docking and scoring of binding affinities.[88] QM/MM methods provide several advantages over MM methods in studies of ligands bound to proteins, including potentially a better physical description of a ligand (*e.g.* including electronic polarization), and avoiding the need for time-consuming MM parameterization for the ligand.With increased computer power, and continued theoretical and algorithmic development, QM/MM methods will certainly become increasingly important in practical applications such as drug design, and related areas such as the prediction of drug metabolism and toxicity. Due their growing popularity, and their particular importance in investigations of enzyme-catalysed reaction mechanisms, details of combined quantum mechanics/molecular mechanics approaches are discussed in greater detail below.

3.1 Interactions between the QM and MM Regions. – One of the main differences between various QM/MM models is the type of QM/MM coupling employed *i.e.* in how the interactions (if any) between the QM and MM systems are treated. A useful classification system was put forward by Bakowies and Thiel,[89] in which they differentiate between models of increasing levels of sophistication, from models of types A to D as described below.

A: Type A QM/MM models apply the simplest linking of QM and MM methods. This type of model involves a straightforward 'mechanical' embedding of the QM region in the MM environment. The interactions between the QM and MM regions are treated purely classically by MM, *i.e.* the QM system is represented by (MM) point charges in its interaction with the MM environment, meaning the polarization of the QM region is not included, although extensions allowing variations in atomic charge can easily be envisaged, perhaps most naturally through methods of type B below.

B: QM/MM methods of type B include electrostatic interactions between the QM and MM regions, using atomic partial point charges in the MM region (from the MM force field) that enter the core QM Hamiltonian. Thus the QM region is polarized by its MM environment.

C: QM/MM methods of type C extend beyond type B by also including some polarization of atoms in the MM region by the electric field generated by the QM region.

D: QM/MM methods of type D are the most refined and sophisticated, including the iterative, self-consistent mutual polarization of the MM and QM regions.

Model A is the most straightforward implementation of a combined QM/MM model but in this type of QM/MM approach, the QM and MM regions do not interact in the quantum chemical calculation. In models of type A, the QM/MM energy of the whole system, $E_{\mathrm{TOTAL}}{}^{\mathrm{QM/MM}}$, is calculated in a simple subtractive scheme:

$$E_{\mathrm{TOTAL}}{}^{\mathrm{QM/MM}} = E_{\mathrm{TOTAL}}{}^{\mathrm{MM}} + E_{\mathrm{QM\ region}}{}^{\mathrm{QM}} - E_{\mathrm{QM\ region}}{}^{\mathrm{MM}} \qquad (1)$$

where $E_{\mathrm{TOTAL}}{}^{\mathrm{MM}}$ is the MM energy of the whole system, $E_{\mathrm{QM\ region}}{}^{\mathrm{QM}}$, is the QM energy of the QM region and $E_{\mathrm{QM\ region}}{}^{\mathrm{MM}}$ is the MM energy of the isolated QM region. This simple subtractive approach can be applied to all combinations of theory levels (for example QM/QM) and also forms the basis for the multi-layer ONIOM (Our own N-layered Integrated molecular Orbital and molecular Mechanics) method.[90] An example of the application of this type of method to an enzyme reaction is an ONIOM study (with the MOZYME package) of the mechanism of the enzyme citrate synthase.[91]

The polarization of the QM region by the MM environment is considered to be crucial for most studies of enzyme systems; type A models do not include polarization of the QM system by the atoms in the MM system. Extensions to this simple type of model can certainly be considered, including variation of the point charges of the QM atoms for example to reflect chemical changes or polarization. The approach is probably physically more reasonable for combining different levels of QM treatment. Such a QM/QM calculation involves a high and a low level of QM theory, with a small region treated by a high level and the *entire* system treated at the low level (*e.g.* where the lower level theory is a semiempirical molecular orbital method such as AM1, PM3 or MNDO), polarization is obviously included at the lower level of QM theory.

Type B models have most widely used to date been for QM/MM studies of biological systems, and particularly enzyme reactions. Models of Type B models include polarization of the QM system by the MM system, by directly including the charges of the MM group in the QM calculation. The electronic structure calculation therefore includes the effects of the MM atoms. Electrostatic interactions between the QM and MM regions in such models are accounted for by including the partial atomic charges of the MM atoms in the Hamiltonian for the QM region (through the one-electron integrals). Clearly, no electrons are present on the MM atoms, and so the lack of dispersion and exchange repulsion interactions must be compensated for. This is usually done by including MM (classical) van der Waals terms (for example Lennard-Jones functions, as described above) between QM and MM atoms. It is therefore necessary to assign MM van der Waals radii to the QM atoms. One limitation of current approaches of this type is that the same van der Waals parameters are typically used for the QM atoms throughout a simulation: in modelling a chemical reaction, the chemical nature of the groups involved (treated by QM) may change, altering their interactions, and so the use of unchanging MM parameters may be inappropriate. Riccardi *et al.* have recently systematically tested the importance of van der Waals interactions in QM/MM simulations[92] First, these workers optimized a set of Lennard-Jones/

van der Waals parameters for the self-consistent charge-tight binding density functional (SCC-DFTB) approach (an approximate density functional method), based on interactions in small hydrogen-bonded clusters. The sensitivity of condensed phase observables to the van der Waals parameters used for QM atoms was investigated by SCC-DFTB/MM QM/MM simulations of several model systems. Results using the optimized set were compared with two sets of extreme vdW parameters selected from the CHARMM22 forcefield. The model systems were chosen to represent species in important enzymes and proteins, and included a solvated enediolate model, and a flavin adenine dinucleotide (FAD) molecule in solution. Properties calculated from the QM/MM simulations included the reduction potential of the model FAD, the radial distribution functions of water molecules around the solutes (FAD and enediolate), and the potential of mean force for an intramolecular proton transfer in the enediolate. The different parameter sets gave clear differences in results for gas-phase clusters and solvent structures around the solutes. However, condensed phase thermodynamic quantities (*e.g.* the calculated reduction potential and potential of mean force) were found to be less sensitive to the van der Waals parameters used. The differences between the results with the three different van der Waals parameter sets for QM atoms were found to be due to the effects of the parameter set on solvation. The authors recommend that work to improve the reliability of QM/MM methods for condensed phase energetic properties should focus on factors other than van der Waals interactions between QM and MM atoms, such as the treatment of long-range electrostatic interactions.

The treatment of QM/MM electrostatic interactions is a little less straightforward when semiempirical QM methods are used: semiempirical molecular orbital methods such as AM1 and PM3 treat only valence electrons directly, including the core electrons together with the nucleus as an atomic 'core'. In semiempirical QM/MM methods such as the AM1/CHARMM method of Field *et al.*, the electrostatic interactions between QM and MM atoms are calculated by treating the MM atoms exactly as if they where semiempirical atomic cores.

Type C models go beyond type B by including polarization of the MM region also (for example through a polarizable dipole model[93]). Type D models have not yet been much developed or tested. Models of Type D represent the most complex and sophisticated level of QM/MM coupling, including self-consistent polarization of the MM region through an iterative procedure. At present, extension to models of types C and D are computationally much more intensive and may not always yield better results.[94] Developing QM/MM methods of these types is a significant challenge partly because of the increased computational expense required for the calculation of polarization of the MM system. Also, the MM force fields which have been developed for biological macromolecules (described in section 2 above) do not allow for polarization or indeed any changes in atomic charges. QM/MM methods which include polarization of the MM system have been developed for small molecular systems.[95] QM/MM calculations can assist in the development of polarizable MM force fields, for example in assessing polarization effects for small (QM) regions in large biomolecules.[96]

3.2 Basic Theory of QM/MM Methods. – The underlying theory of the QM/MM approach has been covered by many different authors,[97,98] so only a brief overview is given here. The energy of the whole system, E, can be written in terms of an effective Hamiltonian, $\hat{H}_{\text{eff}}$, and the electronic wavefunction of the QM atoms, ψ:

$$E = \frac{\langle \psi | \hat{H}_{eff} | \psi \rangle}{\langle \psi | \psi \rangle} \tag{2}$$

The effective Hamiltonian can be considered as:

$$\hat{H}_{eff} = \hat{H}_{\text{QM}} + \hat{H}_{\text{MM}} + \hat{H}_{\text{QM/MM}} + \hat{H}_{\text{Boundary}} \tag{3}$$

where $\hat{H}_{\text{QM}}$ is the quantum mechanical Hamiltonian (for the QM system alone), $\hat{H}_{\text{MM}}$ is the MM Hamiltonian, and $\hat{H}_{\text{QM/MM}}$ is the hybrid QM/MM Hamiltonian representing interactions between the QM and MM systems. $\hat{H}_{\text{Boundary}}$ is the Hamiltonian for any boundary terms applied to the simulation system to represent the effects of the bulk surroundings. The total energy of the system is then given by:

$$\hat{H}_{eff}\Psi(r, R_\alpha, R_M) = E(R_\alpha, R_M)\Psi(r, R_\alpha, R_M) \tag{4}$$

where Ψ is the electronic wavefunction of the QM system. The electronic wavefunction is a function of the electronic coordinates, r, and and also depends on the coordinates of the nuclei in the quantum system, R_α, and of the atoms in the MM region, R_M. From the definition of the effective Hamiltonian, $\hat{H}_{eff}$, the total energy of the system can then be written as the sum of four contributions

$$E_{eff} = E_{\text{QM}} + E_{\text{MM}} + E_{\text{QM/MM}} + E_{\text{Boundary}} \tag{5}$$

The energy of the QM atoms, E_{QM}, is calculated in a standard molecular orbital or DFT calculation (for example, the PM3 semiempirical molecular orbital Hamiltonian could be used). The energy of the atoms in the MM region, E_{MM}, is given by a molecular mechanics force field: it is defined by a standard MM potential function, as decribed in Section 2 above, including terms for bond stretching, bond angle bending, dihedral and 'improper' dihedral angles, electrostatic interactions (usually point partial charges from the MM force field represent the MM atoms) and van der Waals interactions (usually by a Lennard-Jones 12-6 potential). The boundary energy, E_{Boundary}, arises (as in MM simulations) because the simulation system can only include a finite number of atoms, so terms to reproduce the effects of the bulk must be included. It may also be necessary to scale/reduce charges at the boundary of the simulation system to include the effects of dielectric screening in a crude sense, and so avoid overestimating the effects of charged groups on the active site.[99] Some methods for more detailed treatment of long-range electrostatic interactions in QM/MM calculations are described in Section 3.4 below.

The QM/MM interaction energy, $E_{QM/MM}$ is found by application of the QM/MM Hamiltonian, which typically consists of terms due to electrostatic interactions and van der Waals interactions, and any bonded interaction terms. MM bonding terms (energies of bond stretching, angle bending, torsion angle rotation, *etc*) are included for all QM/MM interactions which involve at least one MM atom. In an *ab initio* QM/MM calculation, the MM atomic point charges are generally included directly through one-electron integrals as mentioned above. The MM atomic partial charges of course also interact with the nuclei of the atoms in the QM system. On the other hand, QM/MM van der Waals QM/MM interactions are usually calculated by a molecular mechanics procedure (*e.g.* through Lennard-Jones terms), exactly as the corresponding interactions would be calculated between MM atoms not interacting through bonding terms. MM van der Waals parameters must therefore be chosen for each QM atom. The van der Waals terms represent dispersion and exchange-repulsion interactions between QM and MM systems, and are important in differentiating MM atom types in their interactions with the QM system. This is particularly important in differentiating between atoms of the same charge (*e.g.* halide ions) which would otherwise be indistinguishable to the QM system; similarly van der Waals interactions play a crucial role in determining the interaction of the QM system with MM atoms whose charges are close to zero. In general, the van der Waals terms are important at close range, and play an important part determining QM/MM interaction energies and geometries. Often, standard MM van der Waals (Lennard-Jones) parameters optimized for similar MM groups are used for QM atoms in QM/MM calculations. This provides an 'off-the-shelf' convenience, but it is always important to consider whether the van der Waals parameters provide a reliable description of QM/MM interactions. Where necessary, the (MM) van der Waals parameters for the QM atoms can be optimized to reproduce experimental or high level *ab initio* results (*e.g.* structures and interaction energies) for small molecular complexes.

3.3 Treatment of Long-Range Electrostatic Interactions in QM/MM Simulations. – The whole system is typically truncated to reduce the computational effort required: for example, only a part of the whole protein (for example, a rough sphere around the active site) might be included in the simulation. When simulating a truncated protein system it is necessary to include restraints or constraints in the boundary region to force the atoms belonging to it to remain close to their positions in the crystal structure. Typically, harmonic restraints are applied for the atoms towards the edge of the simulation system. Atoms still more distant from the centre of the simulation system under investigation may be held fixed. Molecular dynamics simulations can be carried out for truncated systems by the stochastic boundary molecular dynamics method,[100,101] in which the simulation system is divided into a reaction region, a buffer region and a reservoir region. Typically nowadays, the whole simulation system may include all residues with an atom within a distance of *e.g.* 15–18 Å of an atom in the active site. The buffer region would contain atoms in the outer layer of 5 Å or so. Atoms in the reaction region are treated by standard Newtonion molecular dynamics The atoms in the buffer region are also influenced by mean boundary forces tending to hold them in position (*e.g.* harmonic forces are applied to protein heavy atoms in the buffer, to keep them close to their crystallographic positions, and a solvent deformable boundary potential prevents evaporation of water). Atoms in the buffer follow a Langevin equation of motion: they are subject to frictional and random forces to include approximately the exchange of energy with the surroundings (reservoir region). Those atoms in the reservoir region are usually not treated explicitly as their presence (as fixed atoms) has been found to cause excessive rigidity of the protein. The stochastic boundary molecular dynamics approach makes QM/MM molecular dynamics of large biological systems (such as enzymes) more feasible.

There are clear limitations in not including long-range electrostatic interactions explicitly. Schemes for more reliable treatment of long-range electrostatic interactions in QM/MM simulations have been developed, to allow simulation of periodic systems (periodic boundary conditions are now typical in MM molecular dynamics simulations). For example, Nam *et al.* have developed an efficient linear-scaling Ewald method for long-range electrostatic interactions in QM/MM calculations.[102] This QM/MM-Ewald method is designed to allow efficient evaluation of long-range electrostatic forces in calculations of periodic systems. It is designed to allow application of linear-scaling Ewald methods to QM/MM molecular dynamics simulations of enzyme and ribozyme reactions.It is a linear-scaling electrostatic method that applies the particle mesh Ewald algorithm to calculate interactions of molecular mechanical atomic charges, and a real space multipole expansion for the quantum mechanical electrostatic terms, plus a pairwise periodic correction factor for the QM and QM/MM interactions: importantly, this last term does not need to be reevaluated during the self-consistent field procedure. Nam *et al.* carried out QM/MM molecular dynamics simulations of the association of ammonium chloride and ammonium metaphosphate, and the dissociative phosphoryl transfer of methyl phosphate and acetyl phosphate to test the method. They compared results from periodic boundary molecular dynamics simulations using the QM/MM-Ewald method with periodic simulations using cutoffs for electrostatic interactions, and with results from nonperiodic stochastic boundary molecular dynamics simulations, with cutoffs for electrostatic interactions, and with full electrostatics (no cutoff).

Methods for the reliable treatment of electrostatics for spherical boundary conditions for QM/MM simulations of truncated macromolecules have been examined recently by Schaefer *et al.*[103] These workers have implemented the generalized solvent boundary potential (GSBP) method,[104] for self-consistent-charge density-functional tight-binding QM/MM calculations. Extension to other types of QM treatment should present no difficulties. This method should provide a reliable approach for treating electrostatic interactions for spherical boundary conditions. This retains the practical advantage of treating a truncated system, avoiding having to include the entire macromolecule in a periodic simulation, so avoiding the associated large computational demands. Compared to the popular stochastic boundary approach, the generalized solvent boundary potential method provides a balance between quantum mechanics/molecular mechanics (QM/MM) and MM/MM interactions. The effects of the bulk solvent and macromolecule atoms outside the simulation system are included at the Poisson-Boltzmann level. Schaefer *et al.* applied this method in QM/MM simulations of the enzyme human carbonic anhydrase II, and compared the results with those from stochastic boundary molecular dynamics simulations with a several different treatments of electrostatic interactions. The QM/MM simulations using the generalized solvent boundary potential method were found to be more consistent with available experimental data. Conventional stochastic boundary molecular dynamics simulations produced artefacts, of a number of different sorts depending on the treatment of electrostatic interactions. The results produced by these studies show how important it can be to treat electrostatics carefully and correctly in QM/MM simulations of biomolecules. These authors suggest that the commonly used truncation schemes should not be applied if possible in QM/MM simulations, in particular for simulations that may involve extensive conformational sampling.

3.4 QM/MM Partitioning Methods and Schemes. – In most QM/MM studies of enzymes, it is necessary to separate covalently bonded parts of the system into QM and MM regions. Some amino acid side chains may participate directly in the reaction, undergoing chemical change as part of the mechanism, and must therefore be included in the QM region. Similarly, other side chains will play binding roles, and a MM representation could be inadequate in some cases, for example for particularly strong binding interactions. Similarly, it may be more practical to treat only the reactive parts of large cofactors or substrates by quantum chemical methods. In most enzymes, therefore, there is a need to be able to partition covalently bonded molecules into QM and MM regions. There has been considerable research into methods for QM/MM partitioning of covalently bonded systems. There are two general techniques that can be employed: firstly a 'frozen' bond orbital to satisfy the valence shell of the QM atom at the QM/MM junction, for example the local self-consistent field (LSCF) method[105,106] or the generalized hybrid orbital (GHO) method.[107] Alternatively a QM atom (or QM pseudoatom) can be added to allow a proper bond at the QM/MM frontier, for example the link atom method or the connection atom method. These various methods for QM/MM partitioning are described in more detail below.

The local self-consistent field (LSCF) method[108] provides a clear and consistent framework for treating the boundary between covalently bonded QM and MM atoms. In the LSCF method, a strictly localized bond orbital, also often described as a frozen orbital, describes the electrons of the frontier bond. This frozen orbital is used at the QM/MM boundary, *i.e.* for the QM atom at the frontier between QM and MM regions. The electron density of the orbital is calculated in advance, using small molecular models. The frozen orbitals are then do not change during the QM/MM calculation. The orbitals must be parameterized for each system, basis set and energy calculation method. This approach has been used at the semiempirical[109] and *ab initio*[108] levels. The LSCF approach avoids the need for dummy atoms and provides a reasonable description of the chemical properties of the frontier bond.

The generalized hybrid orbital (GHO) method for QM/MM calculations uses hybrid orbitals as basis functions on the frontier atom of the MM fragment. This method removes the need for extensive specific parameterization, which is necessary with the LSCF method. The classical frontier atom is described by a set of orbitals divided into two sets of auxiliary and active orbitals. The active orbitals are included in the SCF calculation and the auxiliary orbitals generate an effective core potential for the frontier atom. The generalized hybrid orbital method was first developed at the semiempirical molecular orbital (neglect of diatomic differential overlap) level, and parameters for the orbital sets have been developed at this level of theory.[110] The GHO method uses four hybrid orbitals, one of which is included in the in self-consistent field optimization of the QM region, while three are treated as auxiliary orbitals that do not participate in the QM optimization, but provide an effective electric field for interactions. In contrast to the LSCF method, the semiempirical parameters for the frontier atom are optimized to reproduce bonding properties of full QM systems. As a result, the GHO method enables the localized orbitals to be transferred, and does not require specific parameterization of the active orbital for each new system. An analogous solution in DFT based QM/MM calculations is to freeze the electron density at the QM/MM junction.[111]

Garcia-Viloca and Gao have developed a QM/MM method combining the semiempirical PM3 method with the CHARMM MM forcefield using the GHO approach.[112] As they state, a key aspect of the GHO method is that the semiempirical parameters for the boundary atom are transferable. These workers have developed parameters (consistent with the PM3 method) for a carbon boundary atom. They found the combined GHO-PM3/CHARMM model to perform well for on molecular structures and proton affinities for a number of organic molecules.

More recently, the GHO approach has been extended to QM/MM calculations at the *ab initio*,[113] self-consistent-charge density-functional tight-binding (SCC-DFTB)[114] and density functional[115] QM levels, by Pu, Gao and Truhlar. At the *ab initio* level, these workers tested four different approaches to overcome the nonorthogonality between active molecular orbitals (MOs) and auxiliary MOs (projecting the auxiliary hybrid basis functions out of the active QM basis; neglecting the diatomic differential overlap between the auxiliary basis and the active QM basis; constructing hybrid orbitals from Lowdin-type symmetric orthogonalized atomic orbitals; and local Lowdin orthogonalization). These workers derived analytical gradient expressions. They tested the unparametrized GHO-*ab initio* Hartree-Fock method on hydrocarbons with various basis sets, comparing the geometries and charges found with pure QM calculations for ethane, ethyl radical, and n-octane. The method was also tested for the torsion potential in n-butane. Pu *et al.* also developed a parameterization of the GHO-*ab initio* Hartree-Fock method for the MIDI! basis set.

Pu, Gao and Truhlar have also presented a GHO method (together with analytical gradients for geometry optimization or molecular dynamics simulations) for QM/MM calculations at the self-consistent-charge density-functional tight-binding (SCC-DFTB) level of QM theory. A specifically parameterized empirical correction term is added for the GHO boundary atom and its neighbouring QM frontier atom, to give a good bond length between the two. The method was tested by comparing geometries and Mulliken charges with full QM results for a a series of small molecules and ions, including testing the potential energy for C–C stretching in ethane and in propanoate. The torsional barrier in QM/MM partitioned n-butane was also tested, as were a series of proton affinities. The results showed the GHO method to perform acceptably well.

The same group has also developed a generalized hybrid-orbital method for combining density functional theory (and hybrid density functional theory) with molecular mechanics, to allow the inclusion of electron-correlation effects accurately and efficiently in QM/MM GHO calculations. This work used density functional theory (DFT) in the generalized-gradient approximation and also hybrid density functional theory (HDFT), with Gaussian-type basis functions. The total electron density is calculated by including charge densities in the auxiliary hybrid orbitals, and orthonormality of the auxiliary Kohn-Sham orbitals is enforced. In the proposed method, scaling parameters are applied for some of the one-electron integrals. These scaling factors were optimized to give correct geometries and charges at the frontier between the QM and MM regions. The method was tested at the generalized gradient approach (GGA) density functional (BLYP and mPWPW91) and hybrid density functional theory methods (B3LYP mPW1PW91, and MPW1K) for several small molecules, by comparing the pure QM results for atomic charges and geometries. These tests included the calculation of proton affinities, C–C stretching energy, the torsional barriers and conformational energies of alanine dipeptide, and a barrier height for transfer of a hydrogen atom. The density functional-GHO method was found to perform satisfactorily.

The 'dummy junction atom' or 'link atom' approach introduces so-called link atoms to satisfy the valence of the frontier atom in the QM system.[116] Usually this link atom is a hydrogen atom[81] but other atom types have also been used, such as a halogen-like fluorine or chlorine.[117] The link atom approach has been criticized, for example because it introduces additional degrees of freedom associated with the link atom, and the fact that, for example a C–H bond is clearly not chemically exactly equivalent to a C–C covalent bond. However, the simplicity of the link atom approach method has led to its being widely used in QM/MM modelling of proteins and other biological molecules. The results of QM/MM calculations using link atoms are highly dependent on the positioning of the link atom, and also on exactly which MM atoms are excluded from the classical electrostatic field that interacts with the

QM region. Reuter and co-workers compared the LSCF and link atom approaches for semiempirical QM/MM calculations, and found the two methods to give similar results for a variety of molecular properties.[118] From the tests they carried out, these workers further recommended that the link atom should interact with all MM atoms except for those closest to the QM atom to which the link atom is bonded. Given a reasonable selection of the boundary between QM and MM regions, (a good choice would be, for example, for the QM/MM boundary to lie across a carbon-carbon single bond, distant from chemical changes and also from highly charged MM atoms), the link atom method can give good results.

Another approach to treating the boundary between covalently bonded QM and MM systems is the connection atom method,[119,120] in which rather than a link atom, a monovalent pseudoatom is used. This 'connection atom' is parameterized to give the correct behavior of the partitioned covalent bond. The connection atoms interact with the other QM atoms as a (specifically parameterized) QM atom, and with the other MM atoms as a standard carbon atom. This avoids the problem of a supplementary atom in the system, as the connection atom and the classical frontier atom are unified. However, the need to reparameterize for each type of covalent bond at a given level of quantum chemical theory is a laborious task.[121] The connection atom method has been implemented for semiempirical molecular orbital (AM1 and PM3)[119] and density functional theory[120] levels of theory. Tests carried out by Antes and Thiel to validate the connection atom method at the semiempirical level suggested that the connection atom approach is more accurate than the standard link atom approach.[119]

Brooks and co-workers have proposed a 'double link atom' method.[122] The double link atom method extends beyond the standard single link atom method to overcome some of the problems of electrostatic interactions that can arise with the single link atom method. For example, the single link atom method of QM/MM partitioning can give an unphysical overall charge or dipole. These same authors have also developed a Gaussian delocalization method for molecular mechanical charges in QM/MM calculations. This delocalized Gaussian MM charge method, aims to provide an empirical means of treating the delocalization of the electron density that should physically really be present for atoms in the MM region. This can have important effects for QM/MM electrostatic interactions. These authors also suggest that this approach may simplify the rules that have to be used to decide which molecular mechanical interactions to include in the calculation of energies and forces. For example, even at short distances between QM and MM atoms , the delocalized Gaussian MM method does not require the MM host atom charge to be excluded from the QM calculation, as would be necessary when treating it simply as a point charge. The delocalized Gaussian MM method can potentially be combined with a wide variety of QM/MM partitioning methods, such as the link atom, frozen orbital, or pseudopotential methods. Das *et al.* tested the delocalized Gaussian MM and double link atom methods on several small model systems chosen to represent key features of reactions in enzymes, such as proton affinities, torsional rotation barriers and deprotonation energies. These methods were found to give better energetic properties for the model compounds studied compared to point atomic MM charge and single link atom QM/MM methods.

Cui *et al.* have tested a number of different QM/MM partitioning methods based on the link atom approach, for the SCC-DFTB QM method.[123] An important feature of this study is that it included studies of enzyme reactions, not just analysis of the properties of gas-phase molecules. Among the partitioning schemes tested were all the options currently available in the CHARMM program for SCC-DFTB/CHARMM QM/MM simulation. These differ in their treatment of electrostatic interactions with the MM atoms close to the QM/MM frontier. These workers also developed a new method, which they describe as the divided frontier charge approach. In this method, the partial charge associated with the MM atom bonded to the QM atom is evenly distributed to the other MM atoms in the same molecular mechanical group. These various QM/MM link atom schemes were tested for calculation of properties such as deprotonation energies, dipole moments, proton affinities, and energetics of proton transfer reactions. As also found in earlier work, Cui *et al.* established that QM/MM calculated proton affinities and deprotonation energies are highly dependent on the particular link atom scheme employed. They also found that the standard single link atom approach often gives errors of the order of 15 to 20 kcal/mol compared to pure QM calculations. Other schemes were found to give better results, and to be generally comparable. It was found that both activation barriers and reaction energies for proton transfer reactions are fairly insensitive to the particular link atom scheme (for example, to within a range of 2 to 4 kcal/mol) because of cancellation of errors, for reactions in the gas phase and in enzymes. This is encouraging: the effect of using different link atom schemes in QM/MM simulations was found to be relatively small for chemical reactions in which the total charge does not change. It is reasonable to expect that the general overall behaviour would be similar at other (higher) levels of QM/MM calculation. These authors emphasize that other technical details, such as the treatment of long-range electrostatics, are likely to play a more significant role in determining energetics generally, and stress that they must be treated carefully for reliable results to be obtained.

4 Some Comments on Experimental Approaches to the Determination of Biomolecular Structure

A wide variety of experimental methods have provided insight into protein, and in particular enzyme structure, and these structures are in turn the 'raw materials' of much biomolecular modelling. The most important experimental technique for studying protein structure to date has been X-ray crystallography. A well-ordered crystal of the protein is required. Finding appropriate conditions to produce suitable crystals can be challenging, for example particularly for membrane proteins. One indication of the precision of a crystallographic protein structure determined by X-ray crystallography structures is the resolution, ranging from very low resolution where perhaps just the overall shape of the protein may be revealed, to higher resolution (1–2 Å) where most atomic positions can be determined, at least for heavy atoms. However, modellers in particular should bear in mind that the quoted resolution is a measure of global model quality (dependent for example on the nature of the crystal and experimental conditions), and even in high-resolution structures there can be considerable uncertainty due to the dynamic nature of proteins, which can give rise to conformational variability. The molecular models of protein structure provided by crystallography are the product of much subjective human intervention (for example in model building and refinement), and more importantly, represent an average over all the molecules in the crystal and over the whole time course of the experiment. One obvious manifestation of this averaging is the presence of alternative conformations for amino acid sidechains in many protein crystal structures – two or more well-ordered conformations are often observed for some groups. Similarly, some parts of the structure may not be resolved by crystallography, in particular surface loops or terminal regions of the protein – such regions may be very mobile and have no well-defined conformation and position in the large numbers of molecules in the crystal. It is important to realise that protein crystal structures are not the equivalent of small molecule crystal structures. Crystallographic structures of biological macromolecules should not be thought of as the structure of a single molecule – they are the best fit to the available experimental data, which as well as sources of experimental errors contain the effects of both static and dynamic disorder. It is often surprising to modellers to find that a molecular mechanics energy minimization (including the effects of solvation) of a protein crystal structure will typically reduce the energy of a protein crystal structure by a large amount (*e.g.* by relaxing large numbers of close interatomic contacts), altering the structure in subtle but important ways. This does not show that the molecular mechanics method cannot be trusted, but nor does it indicate that the crystal structure is 'wrong'. Molecular mechanics methods aim to give a good structure of a single protein molecule, whereas a crystallographic structure is an average, as described above, and the best fit to experimental diffraction data. Proteins are dynamic entitities, undergoing a wide range of complex internal motions (as molecular dynamics simulations have been central in demonstrating. A crystal structure contains the effects of averaging many different protein conformations produced by these motions, and the effects of the motions themselves during the experiment. The information provided by crystallography about protein dynamics is limited, however: typically, only isotropic temperature factors (Debye- or *B*-factors) can be found, for example, although at very high resolutions (*e.g.* < 1 Å) more detailed information (*e.g.* anisotropic temperature factors) can sometimes be extracted. It is revealing that, for high-resolution structures, combinations of two or more different structures may give a better fit to the experimental data than a single molecular model.

Nuclear magnetic resonance (NMR) methods are increasingly important for protein structure determination, particularly for smaller proteins. In NMR the magnetic spin properties of atomic nuclei are used to build up a list of distance constraints between atoms in the enzyme, from which the three-dimensional structure of the protein can be determined. This method does not require the growth of crystals as it can be used on concentrated protein solutions. Direct determination of structure by NMR is generally restricted to smaller proteins. Recently, high resolution X-ray powder diffraction has been used to solve and refine protein structure.[124] This method shares the advantage of not requiring a protein crystal.

There is a huge and ever-growing number of experimental structures available for biological macromolecules. One of the challenges in modelling biological systems is to make use of this wealth of data. One important current source is the Research Collaboratory for Structural Bioinformatics (RCSB) makes three-dimensional biological macromolecular structural data from all experimental techniques available. The RCSB Protein Data Bank (PDB) is the single worldwide repository for the processing and distribution of three-dimensional structure data of large molecules of proteins and nucleic acids.[125] The Protein Data Bank is a vital resource for biomolecular modelling.

5 Computational Enzymology

The field of computational modelling of enzyme-catalysed reaction mechanisms has advanced considerably over recent years to the stage where it is now realistic and sensible to describe a new field of computational enzymology.[70,126,127] Modelling and simulation methods can address fundamental questions on enzyme mechanism and catalysis that cannot easily be tackled experimentally.[72–74] The past few years have seen a large and continuing increase in the computational analysis of enzyme mechanisms. Reviewing the field of computer modelling of enzyme-catalysed reaction mechanisms in the early to mid-1990s,[128–130] the number of mechanistic studies was really rather small. Today, the number of published computational studies of enzyme mechanisms is so large that it is almost impossible practically to cover the majority (even of recent studies) in a single review. As well as the continuing increase in the computational investigations of enzyme action, there has been an accompanying increase in their general level of sophistication and reliability. As well as identifying likely chemical mechanisms, modelling can address issues of specificity, the effects of mutations or genetic variation, and the derivation of structure-activity relationships, and so make an increasingly important practical contribution to enzymology and biochemistry more widely.[131]

5.1 Goals in Modelling Enzyme Reactions. – Experimental data, particularly structural information (as discussed above), is of course typically an essential starting point for modelling an enzyme-catalysed reaction. Using structural data (usually a crystallographic protein structure, although in some cases a homology model could be sufficiently reliable) modelling can investigate mechanistic and other questions that are difficult to answer by experiment alone.[132] A first, vital step in studying an enzyme-catalysed reaction is to establish its chemical mechanism. This means determining the roles (and even the identities, which may not be certain) of catalytic residues, which are often not obvious. Modelling has the advantage that it can be used to analyse transition states directly. Transition states are obviously central to questions of chemical reactivity and catalysis, and cannot be studied directly in enzymes by experiment because of their extremely short lifetimes. To understand a catalytic mechanism, any specific interactions that stabilize transition states or reactive intermediates should also be identified and analysed. Such interactions may well not be clear from experimental structures. As well as providing detailed understanding of the reactive process in the enzyme, and potentially of the root causes of catalysis, identifying interactions of this type may assist ligand design: these interactions potentially offer enhanced affinity if they can be exploited in designed ligands (*e.g.* pharmaceutical lead compounds), becauss many enzymes show exceptionally high apparent binding affinities for transition states and intermediates. Calculations can identify functional groups and interactions vital to catalysis. A number of examples have been published of key catalytic interactions that have been identified through calculations (for example, a conserved proline residue that specifically stabilizes the transition state for aromatic hydroxylation in the flavin dependent monooxygenases para-hydroxybenzoate hydroxylase and phenol hydroxylase).

Proteins have complex dynamics, exhibiting a wide range of internal motions, some of which are vital to function. Many enzymes show large conformational changes during their reaction cycles,[133] and the function and relationship of these changes to the chemical steps in the reaction (to which they may be intimately related) should be explored.[2] There have been many suggestions that protein dynamics may contribute to enzyme catalysis, but simulations indicate that the direct effect of protein dynamics in determining the chemical reaction rates of enzymes is generally relatively small.[72,74,134] It is certainly important in general to consider the effects of protein conformational fluctuations and variations on enzyme reactions, *i.e.* to consider a representative sample of possible conformations, *e.g.* through molecular dynamics or Monte Carlo simulations. Perhaps more esoterically, quantum effects such as nuclear tunnelling are important in many enzyme reactions involving hydrogen transfer.[135,136] It is also worth emphasizing that, for a full understanding of why an enzyme is an effective *catalyst*, *i.e.* to understand why the reaction in the enzyme reaction proceeds more quickly than the uncatalysed reaction, the enzymic and an equivalent ('reference') solution reactions should be compared (though it may not necessarily be obvious for all enzymes what the appropriate reference should be). In practical terms, often the interest is in being able to predict the effects of a mutation (designed or natural) on activity, or the specificity of an enzyme for alternative substrates. Overall, understanding enzyme mechanism, specificity and catalysis involves many different levels of complexity. This presents a variety of different challenges. Different modelling methods or approaches will be more appropriate to investigate different types of question. Several different types of application with different methods are described below.

5.2 Methods for Modelling Enzyme-Catalysed Reaction Mechanisms. – Perhaps the most obvious challenge in investigating enzyme reactions by computational modelling is posed by the very large size of enzymes, exacerbated by the need to include at least a representative part their environment (*i.e.* the surrounding solvent, perhaps membrane or other proteins, cofactors or DNA which may be bound to the enzyme).

Standard 'molecular mechanics' (MM) methods (*e.g.* the popular force fields developed for AMBER, CHARMM and GROMOS decribed in Section 2 above) provide a good description of protein structure and dynamics, but cannot be used to model chemical reactions. This limitation is due their simple functional forms (*e.g.* harmonic terms for bond stretching) and inability to model changes in electronic polarization (because of the invariant point partial atomic charge used by these molecular mechanics methods to represent electrostatic interactions).

While it is possible to develop MM parameters specifically for reactions, but this is highly laborious, and the resulting parameters may not be transferable. Also, the form of the potential function can impose serious limitations, such as the neglect of electronic polarization. Methods that take the fundamental quantum mechanics of electronic structure into account are more generally applicable. Quantum mechanical methods (*i.e.* methods to calculate molecular electronic structure) are often preferable and can be easier to apply than involved molecular mechanics type approaches. The major problem with electronic structure calculations on enzymes is the large computational resources required, which significantly limits the size of the system that can be treated. Quantum chemical approaches to modelling enzyme reactions are described in the next couple of sections.

5.3 Quantum Chemical Approaches to Modelling Enzyme Reactions: Cluster (or Supermolecule) Approaches, and Linear-Scaling QM Methods. – Quantum chemical methods (for example *ab initio* molecular orbital or density-functional theory calculations) can currently be used practically to study reactions in non-periodic, molecular systems containing of the order of tens of atoms. Small 'cluster' models of around this size can represent key features of an enzyme reaction, and can identify likely mechanisms. This is sometimes described as the 'supermolecule' approach. The active site of an enzyme is a relatively small region where the substrate(s) (and co-factors in cases where they are involved) bind. It contains the residues that are directly involved in the chemical reaction and residues involved in binding. The substrates are typically bound at the active site by multiple weak interactions, such as hydrogen bonds, electrostatic and van der Waals interactions. Active sites are typically found in clefts or crevices in the enzyme. Clusters of small molecules are used to represent important functional groups (for example, key amino acid side chains involved in catalysis or binding the substrate or cofactors, *etc.*) with their positions typically taken from a representative X-ray crystal structure of an enzyme complex. For example, acetate may be chosen to represent an aspartate side chain, imidazole to represent histidine, and so on). Such calculations can be used to examine the nature of interactions between groups at the active site, and can provide useful models of transition states and intermediates. They can also assist greatly in testing the accuracy of different levels of calculations for a given application (*e.g.* comparing the results of semiempirical with *ab initio* molecular orbital calculations, or different levels of *ab initio* treatment, for a particular reaction).

The approach of modelling small clusters has proved particularly fruitful in the modelling of the catalytic reaction mechanisms of metalloenzymes. In many metalloenzymes, all the important chemical steps may take place at one metal centre (or a small number of metal ions bound at one site), and the metal may also hold its ligands in place. This gives the technical advantage of limiting the requirement for applying restraints to maintain the correct active site structure. Reliable, semi-quantitatively accurate calculations have increasingly been made feasible by the development of methods based on density functional theory. Popular functionals, *e.g.* the widely used B3LYP hybrid functional, give good results for many reactions without requiring excessive amounts of computer time, memory or disk space, for clusters of quite large size. The large body of thorough and careful research produced by Siegbahn and collaborators[137] on many enzymes provides an excellent example of the biochemical insight that calculations on small clusters can provide in analysing enzyme mechanisms (*e.g.* for discriminating between alternative proposed mechanisms: a mechanism can be excluded if the calculated barriers for it are significantly higher than the experimentally derived activation energy).

In a cluster model containing various small molecules representing important functional groups, it may be possible to optimize the geometries of complexes representing the reactants, transition state, intermediates and products of steps in the reaction. This can often be sufficient to discriminate between alternative possible mechanisms, as the energy difference between alternative mechanisms is often very large, larger than the likely effects of the environment on the relative energies. A small model, though, might lack some important functional groups, and careful consideration should be given as to which groups to include, always balancing computational feasibility against the desire for a larger, more extensive model. Also, perhaps counterintuitively, a larger cluster model is not always a better model: a larger model will involve greater conformational complexity (conformational changes distant from the reaction centre might artificially affect relative energies along the reaction path), and including unshielded charged groups could also have unrealistically large effects on reaction energies.

Environmental effects, such as solvation and long-range electrostatic interactions, can be included approximately in a calculation on a an active site cluster model by the use of continuum solvation models, but these cannot fully represent the heterogeneous electrostatic environment in an enzyme.[74] An important technical and practical aspect of cluster/supermolecule calculations is that it can often be difficult to optimize the geometry of the model (for example to locate a transition state structure), while at the same time maintaining the correct orientations of the groups in the protein.

More approximate quantum chemical methods, (such as the semiempirical molecular orbital techniques AM1 and PM3), can model larger molecular systems (containing of the order of hundreds of atoms). However, semiempirical methods are well known to be inaccurate for many applications (sometimes subject to very large errors in calculated reaction energies, for example). They also cannot straightforwardly be used for some types of system (*e.g.* for many transition metals). Techniques (*e.g.* 'linear-scaling' methods) have been developed that allow semiempirical electronic structure calculations on whole proteins.[138–140] Considerable steps are also being made in improving the scaling properties of higher level quantum chemical methods, which will allow their application to larger systems.

Typical enzyme-substrate complexes, particularly when modelled using an explicit representation of surrounding solvent, will contain at least thousands of atoms. This places them currently beyond even semiempirical quantum chemical methods for modelling reactions. An equally important consideration in modelling a reaction is that calculation of (single point) energies is not enough: important points (such as transition state structures) and preferably entire reaction pathways should be optimized. Extensive conformational sampling may be required to generate a representative ensemble of structures. These are in themselves significant challenges for large molecules. One should also consider the environment of the enzyme (typically aqueous solution, but some enzymes operate in concentrated solutions, in membranes or in protein or nucleic acid complexes). Protein internal motions are highly complex, many conformational substates can exist and a single structure may not be truly representative.[141] For conformational sampling (*e.g.* to calculate free energy profiles, *i.e.* potentials of mean force[142]), a useful simulation method must be capable of calculating trajectories of many picoseconds at least. One useful approach can be to use MM molecular dynamics simulations (which can run to relatively long, nanosecond timescales) to generate multiple models for mechanism calculations, to ensure wide sampling of possible enzyme configurations. Alternatively, or in some cases, where multiple different crystal structures of the same enzyme are available, these can be used to create different starting models to examine the effects of structural variation on the reaction.

5.4 Empirical Valence Bond Methods. – To examine some important questions relating to enzyme action (*e.g.* to analyse the causes of catalysis, *i.e.* why an enzymic reaction proceeds faster than the equivalent, uncatalysed reaction in solution), it is necessary to use a method that not only captures the essential details of the chemical reaction, but also includes the explicit effects of the enzyme and solvent enviroment. One notable method in this area is the empirical valence bond (EVB) model.[143] In the empirical valence bond approach, resonance structures (for example ionic and covalent resonance forms) are chosen to represent the reaction. The energy of each resonance form is given by a simple empirical force field (with realistic treatment of stretching important bonds, for example). The potential energy is given by solving the related secular equation. The EVB Hamiltonian is calibrated to reproduce experimental data for a known and relevant solution reaction, or alternatively *ab initio* results can be used.[144] The surrounding protein and solution are modelled by an empirical force field, with appropriate treatment of long-range electrostatics. The free energy of activation for the reaction in solution, and in the enzyme, can be calculated using free energy perturbation simulations.[145]

One of the the main advantages of the EVB method is that the free energy surfaces can be calibrated by comparison with experimental data for reference reactions in solution. However, as in any valence bond representation, it is essential that the valence bond forms should represent all the resonance forms that are important in the reaction. An appealing feature of the EVB method is that it makes it straightforward to use a non-geometrical reaction coordinates in modelling a reaction, which may be significantly more accurate for some condensed phase reactions. Energy can be used as a reaction coordinate by following a path between valence bond optima. A mapping procedure is followed which moves gradually from the reactant to the product. In this mapping, the change in both the solute structure and charge is taken into account. This EVB umbrella sampling method locates the correct transition state in the combined solute-solvent reaction coordinate. This allows the evaluation of nonequilibrium solvation effects, for example.[143] Other strengths of the method have been discussed elsewhere.[146] The EVB method is a powerful and useful approach, which has now become a widely adopted tool for studying reactions in condensed phases. Recent investigations with EVB methods have included a study of alternative nucleotide insertion mechanisms for T7 DNA polymerase[147] and a study of the reaction mechanism of human aldose reductase.[148]

5.5 Examples of Recent Modelling Studies of Enzymic Reactions. – A key decision in beginning modelling the mechanism of an enzyme-catalysed reaction is the choice of an appropriate method for the particular system. A modelling method should be capable of delivering a reliable result in a reasonable time. Some key strengths and weaknesses of various current methods have been described above. The capabilities of methods are best illustrated by some recent applications to important enzyme reactions: some selected examples are discussed below. It is essential to point out that this field is still evolving, and it is not yet at the stage where quantitative, exact predictions of (for example) reaction rates or the effects of mutation can routinely be made. For this reason, it is important to try to link with experiment to validate predictions from modelling: prediction of pK_as of functional groups in proteins provides a useful and demanding example of this type of test. Similarly, it can be useful to compare activation barriers for a series of alternative substrates with the activation energies derived from experimental rates: demonstration of a correlation can validate mechanistic calculations as being truly predictive.[149]

5.5.1 Methylamine Dehydrogenase: the Role of Quantum Tunnelling. Most enzyme reactions involve the transfer of hydrogen, *i.e.* proton, hydride or hydrogen atom transfer. Experimental results indicate that quantum mechanical tunnelling is significant in many such enzymic reactions.[150,151] In some cases kinetic isotope effects (KIEs) found on substitution of hydrogen by deuterium or tritium are very large, and in some cases show unusual temperature dependence. There has been considerable debate on the contribution of tunnelling to enzyme catalysis, *i.e.* whether tunnelling is enhanced in the enzyme environment compared to an equivalent reference reaction in solution. The possible role of protein dynamics in 'driving' tunnelling has been widely and hotly debated. Simulations allow the investigation of these important questions, by providing atomic-level analysis of enzyme reactions.

Methylamine dehydrogenase (MADH) catalyses the oxidative conversion of primary amines to aldehyde and ammonia. This enzyme is found in several methylotrophic bacteria that use amines as their principal source of carbon and energy. Experiments show unusually large primary kinetic isotope effects for the rate-limiting proton transfer step in the MADH reaction. These results imply that there is a large contribution to the proton transfer reaction from quantum tunnelling. Experiments also show that there is almost no dependence of the primary kinetic isotope effect on temperature for the methylamine substrate.[151]

MADH has been studied by QM/MM modelling using variational transition state theory with the small curvature approximation for tunnelling corrections. These methods have been developed and applied particularly by Truhlar and co-workers.[142] These QM/MM variational transition state theory/small curvature tunnelling methods (VTST/SCT) allow kinetic isotope effects to be calculated for enzyme reactions. Comparison with experimental kinetic isotope effects allows a direct comparison with experiment. However, it is important to remember that the complexity of multi-step enzyme reactions can make direct comparison difficult. Alhambra *et al.* have carried out VTST/SCT QM/MM calculations of kinetic isotope effects for the proton transfer step in the methylamine-to-formaldehyde reaction in methylamine dehydrogenase.[152] The MM region was treated with the CHARMM22 MM force field,[27] and consisted of 7248 protein atoms together with 1243 water molecules. The semiempirical molecular orbital method PM3 was used to describe the QM region, but with parameters specifically optimized for this reaction. The two regions were linked by means of the generalized hybrid orbital (GHO) QM/MM partitioning method. Umbrella sampling molecular dynamics simulations were used to calculate the classical potential of mean force along an approximate reaction coordinate, defined as the difference of two bond lengths: the breaking C–H bond, and the forming O–H bond. The difference between the quantal and classical vibrational free energies along the path was calculated for the reactants and the variationally optimized transition state.[153] This procedure gives a quantum-corrected potential of mean force.[154] The transmission coefficient, including the effects of tunnelling, was then calculated. The classical activation free energy was calculated as 20.3 kcal mol^{-1}. When quantum mechanical vibrational energy was included, the barrier height was reduced to 17.1 kcal mol^{-1}. Including quantum mechanical tunnelling contributions gave an effective (phenomological) activation energy of 14.6 kcal mol^{-1}, a finding which can be compared directly with experiment, and agrees well with the experimental value of 14.2 kcal mol^{-1}. A hydrogen/deuterium primary KIE of 18.3 was calculated for the perdeuterated substrate, agreeing well with the experimental value of 17.2. Without tunnelling, the calculated KIE was only 5.9. The VTST/SCT approach allows the various different contributions (for example, of tunnelling) to be calculated separately. These results demonstrate the central role of proton tunnelling in the reaction catalysed by MADH.

MADH was the subject of another recent study, which examined the proton transfer for two substrates (methylamine and ethanolamine).[155] This study also applied QM/MM methods and variational transition state theory combined with multi-dimensional tunnelling in the small curvature approximation. These calculations used PM3 with specific reaction parameters (with the Gaussian94 program[156]), with the MM system described by the AMBER force field. The two regions were connected using the link atom approach. The calculated kinetic isotope effects were close to the experimental values, though covering a wide range. Two different modes were found to be possible for the ethanolamine substrate, giving rise to quite different kinetic behaviors. One configuration was found to give rise to considerably more tunnelling than the other. Tresadern *et al.* suggested this conformational behaviour to be the cause of the different temperature behavior found experimentally for the KIEs of the two substrates.

5.5.2 Chorismate Mutase: Analysing Fundamental Principles of Catalysis. Chorismate mutase is at the centre of current debates about enzyme catalysis, and illustrates well how modelling can contribute to such enzymological debates. Chorismate mutase catalyses the Claisen rearrangement of chorismate to prephenate: this makes it an ideal system for analysing the root causes of catalysis, because the reaction does not involve any covalent interaction the enzyme and the substrate, and also because the same reaction occurs in solution with the same reaction mechanism. The activation free energy $\Delta^{\ddagger}G = 15.4$ kcal mol^{-1} ($\Delta^{\ddagger}H = 12.7$ kcal mol^{-1}) in the *Bacillus subtilis* enzyme is found experimentally to be significantly lower than that for the uncatalysed reaction in aqueous solution ($\Delta^{\ddagger}G = 24.5$ kcal mol^{-1}, $\Delta^{\ddagger}H = 20.7$ kcal mol^{-1}).[157] This translates to a rate acceleration of 10^6 by the enzyme ($\Delta\Delta^{\ddagger}G = 9.1$ kcal mol^{-1}). QM/MM calculations (*e.g.* at the semiempirical AM1/CHARMM or *ab initio* QM level) have previously shown TS stabilization by the enzyme.[158–163] The enzyme-bound conformation of chorismate is significantly different from that in solution, and more closely resembles the TS.[164–168] Bruice *et al.* have recently, and controversially, argued that TS stabilization is not involved in chorismate mutase catalysis. Instead, these workers have proposed that catalysis is almost entirely due to the selection of a reactive conformation, described as a near-attack conformation (NAC).[169,170] This proposal, although it has mutated over time, has been vigorously promoted by these workers as vital not only in chorismate mutase, but as a potentially generally important effect in enzyme catalysis. It has attracted a lot of attention and debate. It is similar in many respects to the old 'strain' hypothesis that enzymes function by distorting their substrates into reactive conformations. One central problem with the NAC proposal, though, is that there is no unique definition of a NAC. Bruice has argued trenchantly, though often with a lack of clarity, against any contradictory evidence.

Bruice *et al.* based their estimates of NAC populations largely on unrestrained molecular dynamics simulations (*e.g.* in solution and in the enzyme). These estimates are unreliable and may be significantly in error, as high energy conformation will be sampled too infrequently (even in multi-nanosecond dynamics simulations), thus overestimating the free energy cost for their formation. This leads to a significant overestimation of the catalytic benefit of NAC formation. These same workers carried out thermodynamic integration molecular dynamics simulations to try to estimate the free energy cost of NAC formation in different environments.[171] Important technical details of these simulations are lacking in the published work (making it hard to reproduce or analyse), and as they used yet another different definition of a NAC, it is difficult to assess the relevance of these thermodynamic integration calculations. Also, the accuracy of the methods (*e.g.* for conformational energies and interactions) used by Bruice *et al.* has not been fully tested (most of these simulations applied standard molecular mechanics, though it is not always clear from the published work what parameters were used).

The exact free energy cost of forming a 'NAC' (and the catalytic benefit associated with forming such a conformation, will depend on the particular definition used. As noted above, there is no unique (or general) definition of a 'NAC', and a number of different proposals have been made. Initially Bruice *et al.* suggested that a contact between reacting atoms of the order of a van der Waals contact between reacting atoms would be sufficient to define a NAC, but subsequently they have found it necessary to add restrictions on angles (with a variation in angle of *e.g.* +/− 20 degrees currently allowed, chosen apparently to fit the desired result for catalytic benefit; some definitions of a NAC predict almost infinite catalytic rate accelerations by the enzyme!). While such general statements might have some immediate intuitive appeal, and many definitions can be made, the lack of a general or rigorous definition makes this hypothesis weak and unsatisfactory. There is a danger that the definition of what a NAC may be is fitted to the catalytic effect it is designed to explain, making it a circular definition.

Rather than relying on a subjective definition of a 'NAC', or a definition derived by fitting to the observed catalytic effect, it is more useful to ask what is the catalytic benefit of forming the substrate conformation bound to the enzyme. As noted by many workers, dating back to the first QM/MM study of the enzyme,[158] the conformation of chorismate bound to the enzyme is significantly altered from the conformation in solution (or indeed in the gas phase). Extensive free energy perturbation molecular dynamics methods give a free energy cost of 3.8–4.6 kcal mol^{-1}, or 5 kcal mol^{-1} (by semiempirical

QM/MM (AM1/CHARMM) or empirical valence bond methods, respectively[164,168]), for forcing the conformation of chorismate in solution into the more restricted conformation found in the enzyme. This equates to a catalytic benefit of only around 40–55% of the total $\Delta\Delta^{\ddagger}G$ between enzyme and solvent. The good agreement between these findings, which applied completely different theoretical methods, is striking and suggests that this is a reliable result.

These results imply that catalysis cannot be due solely to binding of a reactive conformation, and therefore that stabilization of the TS (relative to the bound substrate) by the enzyme must be involved, in agreement with earlier QM/MM results for the enzyme-catalysed reaction. The reliability of these lower-level methods has been questioned, however. This key issue of whether the TS is stabilized relative to the bound substrate has recently been examined by Claeyssens *et al.*, using high-level QM/MM methods (B3LYP/6-31G(d)/CHARMM).[172] To study the effect of the enzyme on the reaction, 16 different adiabatic reaction pathways (see Figure 2) were obtained using a combination of the *Jaguar*[173] and Tinker[174] programs for QM/MM calculations[175]. The substrate does not form any covalent bonds and so it was chosen as the QM region, treated at the hybrid density functional B3LYP/6-31G(d) level of theory, which is known to give a good description of this reaction.[160,161] The effects of including some amino acid sidechains (*e.g.* Glu78 and Arg90) in the QM region are relatively small.[176,177] The QM/MM treatment of the active site interactions has been found to be accurate for chorismate mutase, which is because electrostatic interactions dominate.[163] The MM region was a 25 Å radius sphere of protein and solvent, treated with the CHARMM force field.[27] The outer 5 Å was fixed, with all other atoms free to move. Starting structures were taken from semiempirical QM/MM (AM1/CHARMM and PM3/CHARMM) molecular dynamics simulations of the TS. The difference in length between the forming C–C and breaking C–O bonds was used as a reaction coordinate: this has been shown to be a good choice for modelling the reaction.[160,161] Reaction pathways were calculated by restrained optimizations in both directions along the reaction coordinate. The average calculated barrier (Figure 2) was 12.0 kcal mol^{-1} (standard deviation, σ = 1.7 kcal mol^{-1}), in excellent agreement with the experimental activation enthalpy (12.7 kcal mol^{-1}). At the TS, the average length of the breaking C–O bond was 2.02 Å, with a standard deviation, σ, of only 0.03 Å, while the average length of the forming C–C bond was 2.63 Å (σ = 0.03 Å) Å. A relatively large spread in the calculated energy barriers (9 to 15 kcal mol^{-1}) was observed, due almost entirely to differences in the protein environment.

Analysis of catalysis ideally requires comparison of energy profiles in the enzyme and in solution. The barrier in solution, relative to the enzyme-bound conformation, is similar to that in the gas phase,[161–164] so for this enzyme, the gas-phase profiles can be taken as a convenient and meaningful reference. The difference between the gas-phase (QM-only) and QM/MM energy gives the stabilization of the reacting system by the protein environment. This term is large and negative along the whole reaction coordinate, because of favourable Coulombic interactions between the dianionic substrate and the positively

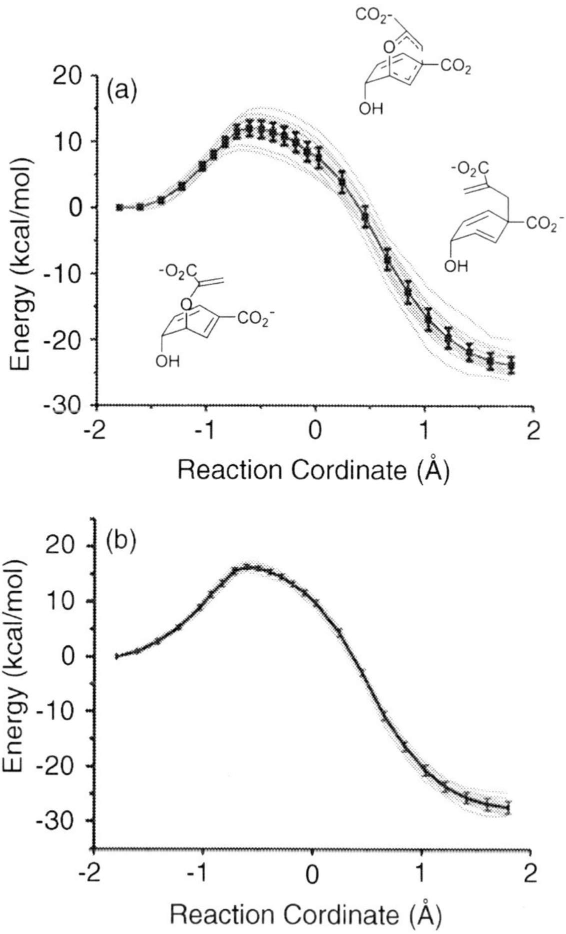

Figure 2 *(a) QM/MM energy profiles for the Claisen rearrangement reaction of chorismate to form prephenate, in the enzyme chorismate mutase. 16 reaction pathways were calculated, from structures derived from QM/MM molecular dynamics simulations. The reaction pathways were optimized at the B3LYP/6-31G(d)CHARMM27 QM/MM level. Black points show the average energies for the 16 paths (with errors bars showing one average deviation), while individual pathways are denoted by grey lines (all energies are shown relative to substrate). Structures of chorismate, prephenate and the transition state (TS) are shown. (b) Relative energies along the same path for the QM region only, i.e. with the effect of the MM (protein and solvent) removed. Comparison of the QM/MM and QM energies allows the effects of the enzyme on the reaction to be analysed, showing significant transition state stabilization*[172].

charged side-chains in the active site. Most importantly, a systematic variation in the stabilization energy relative to the reactant complex was found along the reaction coordinate. In all cases, the TS was calculated to be stabilized significantly more than the reactant, while the product is generally destabilized (relative to the reactant). The stabilization of the TS is quite variable, and correlates very well with the computed barrier height. On average, the enzyme was found to stabilize the TS by 4.2 kcal mol^{-1} more than it stabilizes the reactant. The TS stabilization is overwhelmingly electrostatic, in agreement with previous findings from calculations.[158,163,164] This study provides the most accurate estimate of TS stabilization by the enzyme to date, by applying high level calculations, and including multiple reaction pathways. The correlation of barrier height with TS stabilization shows that the reactivity in the enzyme is determined by the degree of TS stabilization. It appears that conformational effects (*i.e.* binding of a reactive conformation),[166,168] and TS stabilization (relative to the bound substrate) contribute roughly equally to catalysis in this important model enzyme. The calculated average TS stabilization in this work (4.2 kcal mol^{-1})[172] and the previously calculated cost of forming a reactive conformation in the enzyme, compared to solution[164,168] (3.8–5 kcal mol^{-1}) sum to give a value very close to the experimentally observed catalytic rate acceleration by chorismate (experimentally, the difference in activation barriers for the reaction in solution compared to the enzyme is $\Delta\Delta^{\ddagger}G = 9.1$ kcal mol^{-1}), and therefore together account for catalysis by the enzyme.

5.5.3 Cytochrome P450. Cytochrome P450s make up a ubiquitous class of haem enzymes, which function as mono-oxygenases in steroid biosynthesis and a wide range of biological reactions, including the metabolism of potentially toxic hydrocarbons. These enzymes are of great pharmacological interest, because of their roles in drug metabolism.[149] Better understanding of the mechanisms of P450 enzymes will help in predicting biotransformations of pharmaceuticals and other xenobiotics, and should therefore assist drug development.[70] Modelling is playing a central part in analysing the mechanism and determinants of specificity in reactions catalysed by cytochrome P450. A large number of cytochrome P450 proteins are known, with widely varying specificities and activities. P450s carry out vital functions in bioregulation (for example in the detoxification and biosynthesis of sex hormones, and in the metabolism of compounds having anti-inflammatory and anti-hypertensive roles. The active haem is an iron protoporphyrin IX complex, linked covalently to a cysteine residue in the protein. The catalytically active form of the enzyme for oxidation is thought to be a a haem oxoiron (IV) porphyrin radical cation, the so-called Compound I. Among the many reactions catalysed by P450 enzymes are the hydroxylation of alkanes and aromatic compounds, and the epoxidation of alkenes. Many of these reactions are potentially useful in synthetic and other practical applications.

Shaik and co-workers have carried out a number studies using density functional theory based quantum chemical and QM/MM techniques to examine various aspects of the mechanism of alkane hydroxylation by cytochrome P450.[178–181] These studies included, for example, calculation of the potential energy surface for the so-called 'rebound' mechanism with methane as a substrate for two spin states, the high spin (HS) quartet state and low spin (LS) doublet state. In the rebound mechanism, Compound I initially abstracts a hydrogen atom from the alkane. This step is followed by recombination of the hydroxo-radical on the iron with the alkyl radical, generating the ferric-alcohol complex. Calculations were carried out on a model of Compound I, with SH^- used to represent the $SCys^-$ ligand. Methane was used as a model alkane substrate, although it is not itself known to be a substrate of P450. Calculations were performed at the B3LYP hybrid density functional theory level using the *Gaussian*[156] (for both high and low spin states) and *Jaguar*[173] (for the high spin state only) programs, using the Los Alamos effective core potential (ECP) coupled with the double-ξ LACVP basis set for iron and a 6-31G basis for all other atoms (denoted as ECP+LACVP-6-31G). The B3LYP hybrid density functional theory method has been found to predict structures and energetics accurately for many transition metal complexes, and in particular for bio-inorganic systems such as P450 Compound I. Geometries for all the species in the high and low-spin rebound mechanisms were fully optimized. For the first step of the reaction, the two spin states showed nearly identical reactivity (a reaction barrier in both cases of approximately 27 kcal mol^{-1}). However, for the second step of the rebound mechanism, the high spin state was found to have a barrier of approximately 5 kcal mol^{-1} while there was no barrier to reaction for the low-spin state. With ethane as a substrate, the two spin states were found to be nearly degenerate with a barrier to the second step of approximately 5 kcal mol^{-1}.[182]

Shaik and co-workers have also studied the mechanism of ethane epoxidation by Compound I[183,184] with similar techniques. They found that the barrier for the C–O bond formation leading to the radical intermediate was 14–15 kcal mol^{-1} from either the high or low spin state of Compound I. The second step, to form the epoxide, was found to proceed from the low spin state with no barrier, and from the high spin state has a barrier of approximately 3 kcal mol^{-1}. These results indicate that the reaction of the low spin state is effectively concerted while that of the high spin state reaction is stepwise. The concerted reaction in the high spin state was also found to have a barrier only around 4 kcal mol^{-1} higher than with the stepwise reaction. The possibility of intermediates with differing electronic configurations, and significantly different lifetimes, may help to explaining experimental observations such as occasional *cis*/*trans* isomerization and the production of aldehydes

Research on aromatic hydroxylation by cytochrome P450 provides an example of how quantum chemical calculations on small models can help in developing structure-reactivity relationships. Hydroxylation of C–H bonds is a particularly important class of reaction in drug metabolism,[185] which can activate pro-drugs, or affect the bioavailability of pharmaceuticals. For the reliable prediction of pharmaceutical metabolism and toxicology (ADME/TOX) properties, a key aim is the development of structure-activity relationships to predict conversions of drugs. Earlier work has shown that structure-activity relationships based on the structures and properties of substrates alone are of limited utility. There is a need for more detailed models, which can include effects of the reaction mechanism and specificity of different cytochrome P450 isozymes.

Bathelt *et al.*[186,187] (Figure 3) have investigated hydroxylation of simple aromatic compounds by Compound I, in a model consisting of the porphyrin (without side chains), and the cysteinate iron ligand represented by a methyl mercaptide group (CH_3S^-). For the addition of Compound I to benzene, two different possible orientations of the substrate approach were found ('side on, and 'face on', the second with a lower barrier). Both orientations (Figure 3) may be important in the reactions of different drugs in different P450s. The transition state for aromatic hydroxylation was found to have mixed radical and cationic character. This insight from calculations led to the development of new structure-reactivity relationships for substituted aromatics, using a dual-parameter approach combining radical and cationic electronic descriptors.

The reactive properties of the haem group and Compound I may be affected by the particular protein environment of a specific cytochrome P450. Different P450 isoenzymes show very different substrate specificity, and hydroxylation patterns. These could be the result of orientation or binding effects,[188] or the

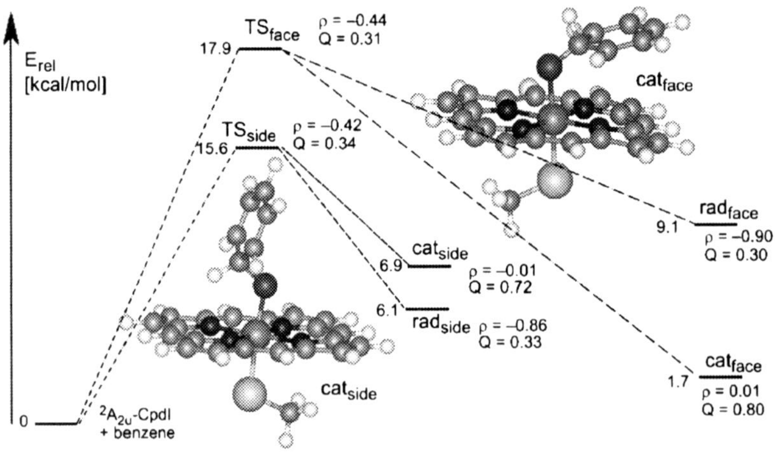

Figure 3 *Quantum chemical calculations on small models of key active site groups can provide insight into the mechanisms of enzyme-catalysed reactions. The cytochrome P450 enzymes, for example, play a crucial role in drug metabolism. The mechanism of hydroxylation of aromatics by cytochrome P450 enzymes has been investigated by B3LYP density functional theory calculations on a realistic prophyrin model system*[186,187]*. These calculations modelled the addition of the reactive Compound I (an oxoiron(IV) FeO porphyrin radical cation) species to benzene, and found two orientations of substrate approach to be possible ('face-on' and 'side-on'), both of which may be important in the metabolism of different drugs. The addition of Compound I model species. The resulting adducts can have either cation-like (cat) or radical-like (rad) character, as shown by calculated atomic charges (Q) and spin (ρ) densities on the benzene moiety. Optimized structures of the cation-like adducts for both modes of approach are shown. Analysis of the electronic properties of the transition state showed it to have mixed radical and cationic character: this insight was crucial in the development of predictive structure-reactivity relationships*[186,187]*.*

intrinsic chemical reactivity of different positions in the substrates. Genetic polymorphisms can also have significant effects, *e.g.* in determining drug metabolism.[189] It is possible also that the electronic properties of Compound I may be modulated by the protein environment, and that this could be a key factor in determining the reactivity of cytochrome P450s. To investigate questions of this sort, methods that include the protein explicitly are needed. This can be achieved by QM/MM modelling. Recent QM/MM studies of bacterial P450$_{cam}$, for example, have raised controversial issues about the factors important in the activity of the enzyme.[190,191] The first QM/MM study of human cytochrome P450 enzymes (including complexes with the drugs diclofenac and ibuprofen) has recently been published, demonstrating the potential for QM/MM methods to contribute directly to practical questions of drug metabolism.[192]

5.5.4 Other Recent Modelling Studies of Enzymes and Enzyme-Catalysed Reactions. The number of modelling studies of enzymes and their reactions continues to increase. Only a few studies can be highlighted here. Interesting and representative studies include density functional modelling of the mechanisms of naphthalene dioxygenase,[193] class III ribonucleotide reductase[194] and 4-hydroxyphenylpyruvate dioxygenase.[195] PM3/CHARMM QM/MM methods have been used to model the formation of the Meisenheimer intermediate in 4-chlorobenzoyl-CoA dehalogenase.[196] Proline isomerization in cylcophilin (and mutant proteins) has been investigated with SCC-DFTB/CHARMM QM/MM methods.[197] QM/MM methods have been used to calculate kinetic isotope effects in chorismate mutase[198] and catechol O-methyltransferase.[199] A combination of QM/MM molecular dynamics simulations and density-functional calculations have been applied to study a metallo beta-lactamase.[200] In 4-oxalocrotonate tautomerase, QM/MM methods have been used to investigate the contribution of the protein backbone in the mechanism.[201] Multiple steered molecular dynamics simulations with a density-functional QM/MM technique have been used to calculate the free energy profile in chorismate mutase.[202] QM/MM Monte Carlo free energy perturbation simulations have been applied to study the mechanism of macrophomate synthase, comparing the Diels-Alder with the Michael-Aldol reaction mechanism.[203] An interesting QM/MM study modelled inhibition mechanisms of neutrophil elastase by peptidyl alpha-ketoheterocyclic inhibitors of human neutrophil elastase, and highlighted the potential of QM/MM calculations in structure-based drug design.[204] Dinner *et al.* demonstrated substrate autocatalysis in uracil DNA-glysosylase by QM/MM modelling.[205] Recent simulations have examined the nature of the proton bottleneck in redox-coupled proton transfer in cytochrome c oxidase.[206]

β-Lactamases enzymes are responsible for most bacterial resistance against β-lactam antibiotics. As such, they are a serious and growing threat to the effectiveness of antibacterial chemotherapy, and are a major threat to human health. The reaction mechanism of a Class A β-lactamase (with benzylpenicillin) has been investigated by illustrative recent QM/MM calculations (see Figure 4 and 5). Glu166 was identified as the base in both acylation and deacylation

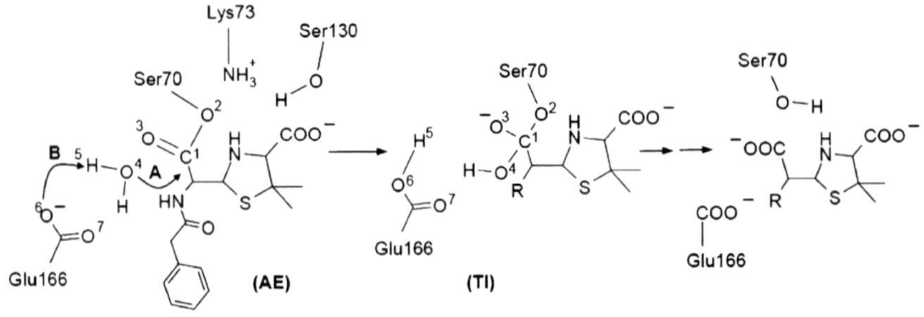

Figure 4 *QM/MM calculations have been used to investigate the mechanism of the first stage of the deacylation reaction of a Class A β-lactamase (the TEM1 enzyme) with benzylpenicillin.*[209] *In the modelled mechanism, Glu166 deprotonates a water molecule, which is the nucleophile for attack on the acylenzyme (AE), forming a tetrahedral intermediate (TI). This was modelled using two geometrical reaction coordinates, R_a (the distance between the attacking water oxygen and the ester carbonyl carbon (R_a=d(O4–C1), and R_b (defined as the difference of the distance between each of the donating and the accepting oxygen atoms and the transferring proton (R_b=d(O4–H5)–d[O6–H5], to model abstraction of a proton form the catalytic water by Glu166). (See also Figure 5).*

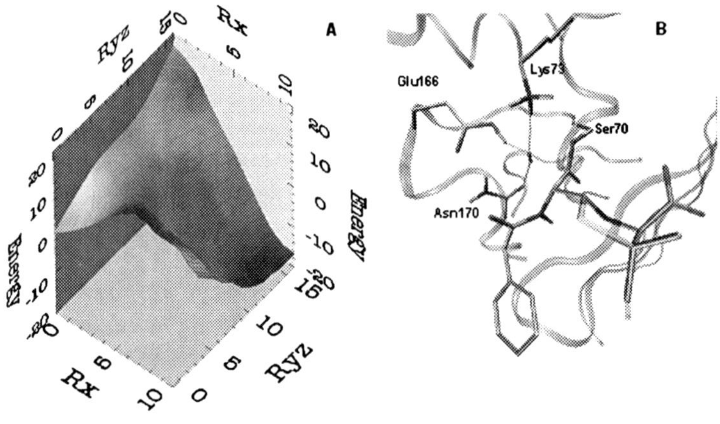

Figure 5 *QM/MM modelling of benzylpencillin deacylation in the TEM1 Class A β-lactamase enzyme. (a) Potential energy surface for formation of the tetrahedral intermediate in deacylation (see Figure 4 for the definition of the reaction coordinates R_a and R_b). QM/MM energies, calculated at the B3LYP/6-31G+(d)//AM1-CHARMM22 level, are shown relative to the acylenzyme (AE) in kcal/mol. The use of high-level (density functional) QM calculations was necessary to give reliable reaction energetics. (b) Structure of the transition state for deacylation (the structure at the point (a1,b8) on the potential energy surface) showing some important residues and hydrogen bonds at the enzyme active site.*[209] *The entire reaction pathway for this enzyme has been modelled by this QM/MM approach, identifying a mechanism in which Glu166 acts as the base in both acylation and deacylation.*[208,209] *The results are consistent with experimental data.*

reactions in the mechanism of breakdown of beta-lactam antibiotics (such as penicillin) in the TEM1 β-lactamase enzyme, by QM/MM modelling with high level (hybrid density functional) QM corrections.[207–209] A similar approach has been applied to study the mechanism of fatty acid amide hydrolase,[210] an enzyme involved in endocannabinoid metabolism, and a promising target for treatment of central and peripheral nervous system disorders.

Bjelic and Åqvist have used a well validated homology model to examine the substrate binding mode and reaction mechanism of a malaria protease with a novel active site.[211] This enzyme (histo-aspartic protease (HAP) from the malaria parasite *P. falciparum*) is a target for anti-malarial drug design, but its three-dimensional structure is not yet known. These workers used a combination of homology modelling, automated docking, and molecular dynamics/reaction free energy profile simulations to predict the structure of the enzyme and the conformation of bound substrate. Finally, these calculations were used to predict the mechanism of the enzymic reaction.[211] The only amino acid residue found to be involved directly in the reaction was a catalytic aspartate, with stabilization by a histidine residue. The calculated reaction rate agreed well with experimental kinetic data for a hexapeptide substrate derived human haemoglobin.

6 *Ab initio* (Car-Parrinello) Molecular Dynamics Simulations

An increasingly important technique in biomolecular simulations[212] is the *ab initio* molecular dynamics technique first proposed by Car and Parrinello around twenty years ago.[213,214] The scheme combines molecular dynamics simulation and density-functional theory: it integrates fictitious wave function coefficient dynamics with classical molecular dynamics in a single extended Lagrangian. Crucially, the electronic wavefunctions are included as dynamical variables. Initially, a converged wavefunction is determined, and the orbitals subsequently evolve simultaneously with the changes in nuclear position. The orbital parameters are included as variables with fictitious masses in the dynamics, analogous to the nuclear positions and masses. The nuclear forces are not exactly correct in dynamics, as the electronic wavefunction is not converged in the orbital parameter space, but this error is controlled by an appropriate choice of dynamic parameters (*e.g.* the fictitious masses). Constraints are applied to the system to ensure that the orbitals remain orthonormal. One interesting and relevant application of these techniques examined the catalytic site of galactose oxidase and a biomimetic catalyst.[215]

Despite the development of highly efficient codes and algorithms, *ab initio* molecular dynamics simulations are extremely computationally expensive, requiring very large amounts of supercomputing time. They provide an advantage over molecular dynamics employing empirical force-fields in that the electronic structure methods are able to describe bond breaking and forming reactions and therefore Car-Parrinello methods can in principle allow the direct simulation of chemical reactions. Similarly, they overcome other limitations of molecular mechanics forcefields: for example, electronic polarization effects are included naturally. The major practical limitation is the size of the systems that can be simulated, and the timescale of feasible dynamics simulations, are relatively limited. For this reason, combined QM/MM approaches are attractive for *ab initio* molecular dynamics simulations also. For example, Parrinello and coworkers have developed a scheme for Car-Parrinello molecular dynamics simulations with a QM/MM method, with the CPMD and EGO programs.[216] Using these interfaced programs, efficient and consistent QM/MM Car-Parrinello simulations of large systems can be performed, including the steric and electrostatic effects are of the protein and its solvent environment explicitly.

7 Conclusions

The modelling of biological systems is one of the most exciting and rapidly developing areas of molecular modelling and simulation. This review has concentrated on recent developments in molecular mechanics methods for atomistic molecular dynamics simulations of biological macromolecules, and on the growing field of computational enzymology, namely the investigation of the mechanisms of enzyme-catalysed reactions by modelling. The whole field of biomolecular modelling is vast, and it has not been possible to describe many areas here. One notably important field is structure-aided drug design. Recent developments in structure-based lead optimization have been nicely reviewed recently by Joseph-McCarthy;[217] ligand docking methods have been covered among others by Taylor *et al.*[218] Carlson has reviewed methods for docking including protein flexibility,[219] Teague has also discussed the effects of protein flexibility in drug design,[220] and Gerstein *et al.* their effects in protein-protein inteactions.[221] Lamb has reviewed drug design applications to the important target class of protein kinases.[222] Raha and Merz have reviewed the field of calculations of protein-ligand binding free energies, including covering scoring functions for ranking binding affinities in such complexes.[223] One exciting new area in which molecular modelling plays a vital part is protein design:[224,225] developments here, relying on practical and reliable modelling methods, promise a route to new catalysts and components for biologically inspired nanotechnology and molecular medicine. It is also worth pointing out some theoretical developments, such as in the calculation of free energies by non-equilibrium approaches, such as through the application of the Jarzynski relation to calculate free energies from steered molecular dynamics simulations.[202,226] Altogether, the field of molecular modelling of biological systems is thriving and growing, and its importance looks certain to increase in future.

Acknowledgements

The author thanks his coworkers on the work described here in which he has been involved. He also thanks BBSRC, EPSRC, Vernalis plc and the IBM High Performance Computing Life Sciences Outreach Programme for support.

References

1. M. Karplus and J. Kuriyan, *Proc. Nat. Acad. Sci. USA*, 2005, **102**, 6679.
2. M. Karplus, Y.Q. Gao, J.P. Ma, A. van der Vaart and W. Yang, *Phil. Trans. Roy. Soc. London Ser A*, 2005, **363**, 331.
3. V. Daggett and A. Fersht, *Nature Reviews Mol. Cell Biol.*, 2003, **4**, 497.
4. U. Mayor, N.R. Guydosh, C.M. Johnson, J.G. Grossmann, S. Sato, G.S. Jas, S.M.V. Freund, D.O.V. Alonso, V. Daggett and A.R. Fersht, *Nature*, 2003, **421**, 863.
5. A.T. Brünger and P.D. Adams, *Acc. Chem. Res.*, 2002, **35**, 404.
6. M. Fossi, H. Oschkinat, M. Nilges and L.J. Ball, *J. Magnetic Resonance*, 2005, **175**, 92.
7. D.A. Case, *Acc. Chem. Res.*, 2002, **35**, 325.
8. R.B. Best, J. Clarke and M. Karplus, *J. Mol. Biol.*, 2005, **349**, 185.
9. C.F. Wong and J.A. McCammon, *Protein simulations: Advances Prot. Chem.*, 2003, **66**, 87.
10. C.J. Woods, M.H. Ng, S. Johnston, S.E. Murdock, B. Wu, K. Tai, H. Fangohr, P. Jeffreys, S. Cox, J.G. Frey, M.S.P. Sansom and J.W. Essex, *Phil. Trans. Roy. Soc. A*, 2005, **363**, 2017.
11. R. Elber, *Curr. Opin. Struct. Biol.*, 2005, **15**, 151.
12. M.S.P. Sansom, P.J. Bond, S.S. Deol, A. Grottesi, S. Haider and Z.A. Sands, *Biochem. Soc. Trans.*, 2005, **33**, 916.
13. B. Roux, *Acc. Chem. Res.*, 2002, **35**, 366.
14. J. Gumbart, Y. Wang, A. Aksimentiev, E. Tajkhorshid and K. Schulten, *Curr. Opin. Struct. Biol.*, 2005, **15**, 423.
15. M. Dittrich, S. Hayashi and K. Schulten, *Biophys. J.*, 2004, **87**, 2954.
16. M.M. Dedmon, K. Lindorff-Larsen, J. Christodoulou, M. Vendruscolo and C.M. Dobson, *J. Am. Chem. Soc.*, 2005, **127**, 476.
17. K. Lindorff-Larsen, S. Kristjansdottir, K. Teilum, W. Fieber, C.M. Dobson, F.M. Poulsen and M.J. Vendruscolo, *Am. Chem. Soc.*, 2004, **126**, 3291–3299.
18. J.A. McCammon, B.R. Gelin and M. Karplus, *Nature (London)*, 1977, **267**, 585.
19. D.A. Case, T.E. Cheatham III, T. Darden, H. Gohlke, R. Luo, K.M. Merz Jr., A. Onufriev, C. Simmerling, B. Wang and R. Woods, *J. Computational. Chem.*, 2005, **26**, 1668–1688(see http://amber.scripps.edu/).
20. B.R. Brooks, R.E. Bruccoleri, B.D. Olafson, D.J. States, S. Swaminathan and M. Karplus, *J. Computational Chem.*, 1983, **4**, 187(see also http://www.charmm.org).
21. W.R.P. Scott, P.H. Hunenberger, I.G. Tironi, A.E. Mark, S.R. Billeter, J. Fennen, A.E. Torda, T. Huber, P. Kruger and W.F. van Gunsteren, *J. Phys. Chem. A*, 1999, **103**, 3596(see http://www.igc.ethz.ch/grooms/grooms.html).
22. J.C. Phillips, R. Braun, W. Wang, J. Gumbart, E. Tajkhorshid, E. Villa, C. Chipot, R.D. Skeel, L. Kale and K. Schulten, *J. Computational Chem.*, 2005, **26**, 1781(see http://www.ks.uiuc.edu/Research/namd/).
23. J.W. Ponder and F.M. Richards, *J. Computational Chem.*, 1987, **8**, 1016(see http://dasher.wustl.edu/tinker/).
24. D. Lim, J. Jenson, M.P. Repasky and W.L. Jorgensen, in *Transition State Modeling for Catalysis*, D.G. Truhlar and K. Morokuma (eds), American Chemical Society, Washington, DC, 1999, pp. 74–85. (ACS Symposium Series 721).
25. W.L. Jorgensen, D.S. Maxwell and J. Tirado-Rives, *J. Am. Chem. Soc.*, 1996, **118**, 11225.
26. G. Kaminski, R.A. Friesner, J. Tirado-Rives and W.L. Jorgensen, *J. Phys. Chem. B*, 2001, **105**, 6474.

27. A.D. MacKerell Jr., D. Bashford, M. Bellott, R.L. Dunbrack Jr., J.D. Evanseck, M.J. Field, S. Fischer, J. Gao, H. Guo, S. Ha, D. Joseph-McCarthy, L. Kuchnir, K. Kuczera, F.T.K. Lau, C. Mattos, S. Michnick, T. Ngo, D.T. Nguyen, B. Prodhom, W.E. Reiher III, B. Roux, M. Schlenkrich, J.C. Smith, R. Stote, J. Straub, M. Watanabe, J. Wiórkiewicz-Kuczera, D. Yin and M. Karplus, *J. Phys. Chem. B*, 1998, **102**, 3586.
28. W.D. Cornell, P. Cieplak, C.I. Bayly, I.R. Gould, K.M. Merz, D.M. Ferguson, D.C. Spellmeyer, T. Fox, J.W. Caldwell and P.A. Kollman, *J. Am. Chem. Soc.*, 1995, **117**, 5179.
29. A.D. Mackerell Jr., *Ann. Rep. Comp. Chem.*, 2005, **1**, 91.
30. J.W. Ponder and D.A. Case, *Adv. Protein Chem.*, 2003, **66**, 77.
31. D.J. Price and C.L. Brooks III, *J. Comput. Chem.*, 2002, **23**, 1045.
32. T.E. Cheatham III, *Ann. Reports Comp. Chem.*, 2005, **1**, 75.
33. T.E. Cheatham III, *Curr. Opin. Struct. Biol.*, 2004, **14**, 360.
34. N. Foloppe and A.D. MacKerell Jr., *J. Computational Chem.*, 2000, **21**, 86.
35. A.D. MacKerell Jr. and N. Banavali, *J. Computational Chem.*, 2000, **21**, 105.
36. T.E. Cheatham, P. Cieplak and P.A. Kollman, *J. Biomol. Struct. Dyn.*, 1999, **16**, 845.
37. S.E. Feller, D. Yin, R.W. Pastor and A.D. MacKerell Jr., *Biophysical J.*, 1997, **73**, 2269.
38. M. Kuttel, J.W. Brady and K.J. Naidoo, *J. Comput. Chem.*, 2002, **23**, 1236.
39. L. Heramingsen, D.E. Madsen, A.L. Esbensen, L. Olsen and S.B. Engelsen, *Carbohydrate Res.*, 2004, **339**, 937.
40. M.L.C.E. Kouwijzer and P.D.J. Grootenhuis, *J. Phys. Chem.*, 1995, **99**, 13426.
41. H. Senderowitz, *J. Am. Chem. Soc.*, 1996, **118**, 2078.
42. H.C. Siebert, S. Andre, J.L. Asensio, F.J. Canada, X. Dong, J.F. Espinosa, M. Frank, M. Gilleron, H. Kaltner, T. Kozar, N.V. Bovin, C.W. von der Lieth, J.F.G. Vliegenthart, J. Jimenez-Barbero and H.J. Gabius, *Chem. Bio. Chem.*, 2000, **1**, 181.
43. N.L. Allinger, M. Rahman and J.H. Lii, *J. Am. Chem. Soc.*, 1990, **112**, 8293.
44. C.A. Stortz and A.S. Cerezo, *Carbohydrate Res.*, 2003, **338**, 1679.
45. A. Imberty and S. Perez, *Chem. Rev.*, 2000, **100**, 4567.
46. A.D. French, G.P. Johnson, A.M. Kelterer, M.K. Dowd and C.J. Cramer, *Intl. J. Quantum Chem.*, 2001, **84**, 416.
47. J.P. McNamara, A.-M. Muslim, H. Abdel-Aal, H. Wang, M. Mohra, I.H. Hillier and R.A. Bryce, *Chem. Phys. Lett.*, 2004, **394**, 429.
48. L.D. Schuler, X. Daura and W.F. van Gunsteren, *J. Comput. Chem.*, 2001, **22**, 1205.
49. E. Neria, S. Fischer and M. Karplus, *J. Chem. Phys.*, 1996, **105**, 1902.
50. W.L. Jorgensen and J. Tirado-Rives, *J. Am. Chem. Soc.*, 1988, **110**, 1657.
51. S.J. Weiner, P.A. Kollman, D.A. Case, U.C. Singh, C. Ghio, G. Alagona, S. Profeta and P. Weiner, *J. Am. Chem. Soc.*, 1984, **106**, 765.
52. T. Lazaridis and M. Karplus, *Proteins*, 1999, **35**, 133.
53. M. Schaefer, C. Bartels, F. LeClerc and M. Karplus, *J. Comput. Chem.*, 2001, **22**, 1857.
54. M.S. Lee, F.R. Salsbury Jr. and C.L. Brooks III, *J. Chem. Phys.*, 2002, **116**, 10606.
55. M.S. Lee, M. Feig, F.R. Salsbury Jr. and C.L. Brooks III, *J. Comput. Chem.*, 2003, **24**, 1348.
56. W. Im, M.S. Lee and C.L. Brooks III, *J. Comput. Chem.*, 2003, **24**, 1691.
57. P. Ferrara, J. Apostolakis and A. Caflisch, *Proteins*, 2002, **46**, 24.
58. W.L. Jorgensen and J. Tirado-Rives, *Proc. Natl. Acad. Sci. USA*, 2005, **102**, 6665.

59. N. Gresh, J.P. Piquemal and M. Krauss, *J. Comput. Chem.*, 2005, **26**, 1113.
60. I.V. Vorobyov, V.M. Anisimov and A.D. MacKerell, *J. Phys. Chem. B*, 2005, **109**, 18988.
61. V.M. Anisimov, G. Lamoureux, I.V. Vorobyov, N. Huang, B. Roux B and A.D. MacKerell, *J. Chem. Theory Comput.*, 2005, **1**, 153.
62. S. Patel, A.D. Mackerell and C.L. Brooks III, *J. Comput. Chem.*, 2004, **25**, 1504.
63. E. Harder, B.C. Kim, R.A. Friesner and B.J. Berne, *J. Chem. Theory Comput.*, 2005, **1**, 169.
64. G.A. Kaminski, H.A. Stern, B.J. Berne and R.A. Friesner, *J. Phys. Chem. A*, 2004, **108**, 621.
65. G.A. Kaminski, H.A. Stern, B.J. Berne, R.A. Friesner, Y.X.X. Cao, R.B. Murphy, R.H. Zhou and T.A. Halgren, *J. Comput. Chem.*, 2002, **23**, 1515.
66. P.Y. Ren and J.W. Ponder, *J. Phys. Chem. B*, 2003, **107**, 5933.
67. A.D. Mackerell Jr., M. Feig and C.L. Brooks III, *J. Am. Chem. Soc.*, 2004, **126**, 698.
68. A.D. Mackerell Jr., M. Feig and C.L. Brooks III, *J. Comput. Chem.*, 2004, **25**, 1400.
69. B.C. Kim, T. Young, E. Harder, R.A. Friesner and B.J. Berne, *J. Phys. Chem. B*, 2005, **109**, 16529.
70. A.J. Mulholland, *Drug Discovery Today*, 2005, **10**, 1393.
71. A. Warshel and M. Levitt, *J. Mol. Biol.*, 1976, **103**, 227.
72. M. Garcia-Viloca, J. Gao, M. Karplus and D.G. Truhlar, *Science*, 2004, **303**, 186.
73. M.J. Field, *J. Computational Chem.*, 2002, **23**, 48.
74. A. Shurki and A. Warshel, *Protein Simulations: Advances in Protein Chem.*, 2003, **66**, 249.
75. L. Ridder, L.J.N. Harvey, I.M.C.M. Rietjens, J. Vervoort and A.J. Mulholland, *J. Phys. Chem. B*, 2003, **107**, 2118.
76. L. Ridder, A.J. Mulholland, I.M.C.M. Rietjens and J. Vervoort, *J. Am. Chem. Soc.*, 2000, **122**, 8728.
77. K.E. Ranaghan and A.J. Mulholland, *Chem. Commun.*, 2004, 1238.
78. K.E. Ranaghan, L. Ridder, B. Szefczyk, W.A. Sokalski, J.C. Hermann and A.J. Mulholland, *Organic and Biomolecular Chemistry*, 2004, **2**, 968.
79. A.J. Mulholland, P.D. Lyne and M. Karplus, *J. Am. Chem. Soc.*, 2000, **122**, 534.
80. H.L. Woodcock, M. Hodoscek, P. Sherwood, Y.S. Lee, H.F. Schaefer and B.R. Brooks, *Theor. Chem. Acc.*, 2003, **109**, 140.
81. M.J. Field, P.A. Bash and M. Karplus, *J. Comp. Chem.*, 1990, **11**, 700.
82. P.D. Lyne, M. Hodoscek and M. Karplus, *J. Phys. Chem. A*, 1999, **103**, 3462.
83. Q. Cui, M. Elstner, E. Kaxiras, T. Frauenheim and M. Karplus, *J. Phys. Chem. B*, 2001, **105**, 569.
84. S. Marti and V. Moliner, *J. Chem. Theory and Computation*, 2005, **1**, 1008.
85. X. Prat-Resina, J.M. Bofill, A. Gonzalez-Lafont and J.M. Lluch, *Intl. J. Quantum Chem.*, 2004, **98**, 367.
86. L. Ridder, I.M.C.M. Rietjens, J. Vervoort and A.J. Mulholland, *J. Am. Chem. Soc.*, 2002, **124**, 9926.
87. D. Riccardi, P. Schaefer and Q. Cui, *J. Phys. Chem. B.*, 2005, **109**, 17715.
88. K. Raha and K.M. Merz Jr., *J. Am. Chem. Soc.*, 2004, **126**, 1020.
89. D. Bakowies and W. Thiel, *J. Phys. Chem.*, 1996, **100**, 10580.
90. M. Svensson, S. Humbel, R.D.J. Froese, T. Matsubara, S. Sieber and K. Morokuma, *J. Phys. Chem.*, 1996, **100**, 19357.
91. W. Yang and D.G. Drueckhammer, *J. Phys. Chem. B*, 2003, **107**, 5986.

92. D. Riccardi, G.H. Li and Q. Cui, *J. Phys. Chem. B*, 2004, **108**, 6467.
93. S.R. Gooding, P.J. Winn, R.I. Maurer, G.G. Ferenczy, J.R. Miller, J.E. Harris, D.V. Griffiths and C.A. Reynolds, *J. Comp. Chem.*, 2000, **21**, 478.
94. I. Antes and W. Thiel, *J. Phys. Chem. A*, 1999, **103**, 9290.
95. L. Jensen and P.T. van Duijnen, *J. Chem. Phys.*, 2005, **123**, Art No. 074307.
96. S.P. Greatbanks, J.E. Gready, A.C. Limaye and A.P. Rendell, *Proteins*, 1999, **37**, 157.
97. A.J. Mulholland, in *'Theoretical Biochemistry'*, ed. L.A. Erikkson, Elsevier, Amsterdam, 2001 Ch. 14, p. 597–653.
98. R.A. Friesner and V. Guallar, *Annu. Rev. Phys. Chem.*, 2005, **56**, 389.
99. Q. Cui and M. Karplus, *J. Am. Chem. Soc.*, 2002, **124**, 3093.
100. T.D. Poulsen, M. Garcia-Viloca, J.L. Gao and D.G. Truhlar, *J. Phys. Chem. B*, 2003, **107**, 9567.
101. C.L. Brooks III, M. Karplus and B.M. Pettitt, *'Proteins, A Theoretical Perspective of Dynamics, Structure and Thermodynamics'*, Wiley, New York, 1988.
102. K. Nam, J.L. Gao and D.M. York, *J. Chem. Theory Comput.*, 2005, **1**, 2.
103. P. Schaefer, D. Riccardi and Q. Cui, *J. Chem. Phys.*, 2005, **123**, Art No. 014905.
104. W. Im, *J. Chem. Phys.*, 2001, **114**, 2924.
105. G. Monard, M. Loos, V. Théry, K. Baka and J.-L. Rivail, *Int. J. Quantum Chem.*, 1996, **58**, 153.
106. X. Assfeld and J.-L. Rivail, *Chem. Phys. Lett.*, 1996, **263**, 100.
107. J. Gao, P. Amara, C. Alhambra and M. J. Field, *J. Phys. Chem. A*, 1998, **102**, 4714.
108. N. Ferre, X. Assfeld and J.L. Rivail, *J. Comput. Chem.*, 2002, **23**, 610.
109. S. Antonczak, G. Monard, M.F. Ruiz-Lopez and J.L. Rivail, *J. Am. Chem. Soc.*, 1998, **120**, 8825.
110. J. Gao, P. Amara, C. Alhambra and M.J. Field, *J. Phys. Chem. A*, 1998, **102**, 4714.
111. T.A. Wesolowski and A. Warshel, *J. Phys. Chem.*, 1993, **97**, 8050.
112. M. Garcia-Viloca and J.L. Gao, *Theor. Chem. Acc.*, 2004, **111**, 280.
113. J.Z. Pu, J.L. Gao and D.G. Truhlar, *J. Phys. Chem. A*, 2004, **108**, 632.
114. J.Z. Pu, J.L. Gao and D.G. Truhlar, *J. Phys. Chem. A*, 2004, **108**, 5454.
115. J.Z. Pu, J.L. Gao and D.G. Truhlar, *Chem. Phys. Chem.*, 2005, **6**, 1853.
116. P. Amara and M.J. Field, *Theor. Chem. Acc.*, 2003, **109**, 43.
117. HyperChem Users Manual, 2002, HyperCube, Inc., Waterloo, Ontario, Canada.
118. N. Reuter, A. Dejaegere, B. Maigret and M. Karplus, *J. Phys. Chem. A*, 2000, **104**, 1720.
119. I. Antes and W. Thiel, *J. Phys. Chem. A*, 1999, **103**, 9290.
120. Y. Zhang, T.-S. Lee and W. Yang, *J. Chem. Phys.*, 1999, **110**, 46.
121. G. Monard, X. Prat-Resina, A. Gonzalez-Lafont and J.M. Lluch, *Int. J. Quantum Chem.*, **2003 93**, 229.
122. D. Das, K.P. Eurenius, E.M. Billings, P. Sherwood, D.C. Chatfield, M. Hodoscek and B.R. Brooks, *J. Chem. Phys.*, 2002, **117**, 10534.
123. P.H. Konig, M. Hoffmann, T. Frauenheim and Q. Cui, *J. Phys. Chem. B*, 2005, **109**, 9082.
124. R.B. Von Dreele, P.W. Stephens, G.D. Smith and R.H. Blessing, *Acta Crystallogr. D*, 2000, **56**, 1549.
125. H.M. Berman, J. Westbrook, Z. Feng, G. Gilliland, T.N. Bhat, H. Weissig, I.N. Shindyalov and P.E. Bourne, *Nucleic Acids Res.*, 2000, **28**, 235(See http://www.rcsb.org/pdb/.
126. M.A. Cunningham and P.A. Bash, *Biochimie*, 1997, **79**, 687.
127. T.C. Bruice and K. Kahn, *Curr. Opin. Chem Biol.*, 2000, **4**, 540.

128. A.J. Mulholland, G.H. Grant and W.G. Richards, *Protein Engng.*, 1993, **6**, 133.
129. A.J. Mulholland and M. Karplus, *Biochem. Soc. Trans.*, 1996, **24**, 247.
130. J. Åqvist and A. Warshel, *Chem. Rev.*, 1993, **93**, 2523.
131. F. Perruccio L. Ridder and A.J. Mulholland in 'Quantum Medicinal Chemistry', Editors P. Carloni and F. Alber, Chapter 6, pp. 177–198, (2003) Wiley-VCH.
132. S. Martí, M. Roca, J. Andrés, V. Moliner, E. Silla, I. Tuñón and J. Bertrán, *Chem. Soc. Rev.*, 2004, **33**, 98.
133. A. Fersht, *'Structure and Mechanism in Protein Science. A Guide to Enzyme Catalysis and Protein Folding'*, Freeman, New York, 1999.
134. M.H.M. Olsson and A. Warshel, *J. Am. Chem. Soc.*, 2004, **126**, 15167.
135. A. Kohen, R. Cannio, S. Bartolucci and J.P. Klinman, *Nature*, 1999, **399**, 496.
136. L. Masgrau, J. Basran, P. Hothi, M.J. Sutcliffe and N.S. Scrutton, *Arch. Biochem. Biophys.*, 2004, **428**, 41.
137. F. Himo and P.E.M. Siegbahn, *Chem. Rev.*, 2003, **103**, 2421.
138. A. Van der Vaart, V. Gogonea, S.L. Dixon and K.M. Merz Jr., *J. Comput. Chem.*, 2000, **21**, 1494.
139. J. Khandogin and D.M. York, *Proteins-Struct. Funct. Bioinf.*, 2004, **56**, 724.
140. J. Khandogin, K. Musier-Forsyth and D.M. York, *J. Mol. Biol.*, 2003, **330**, 993.
141. Y. Zhang, J. Kua and J.A. McCammon, *J. Phys. Chem. B*, 2003, **107**, 4459.
142. J.L. Gao and D.G. Truhlar, *Ann. Rev. Phys. Chem.*, 2002, **53**, 467.
143. A. Warshel, *Ann. Rev. Biophys. Biomol. Struct.*, 2003, **32**, 425.
144. J. Bentzien, R.P. Muller, J. Florian and A. Warshel, *J. Phys. Chem. B*, 1998, **102**, 2293.
145. A. Warshel, *'Computer modeling of Chemical Reactions in Enzymes and Solutions'*, John Wiley & Sons, New York, 1997.
146. J. Villà and A. Warshel, *J. Phys. Chem. B*, 2001, **105**, 7887.
147. J. Florian, M.F. Goodman and A. Warshel, *J. Am. Chem. Soc.*, 2003, **125**, 8163.
148. P. Varnai and A. Warshel, *J. Am. Chem. Soc.*, 2000, **122**, 3849.
149. L. Ridder and A.J. Mulholland, *Curr. Topics Medicinal Chem.*, 2003, **3**, 1241.
150. S.C. Tsai and J.P. Klinman, *Biochemistry*, 2001, **40**, 2303.
151. J. Basran, S. Patel, M.J. Sutcliffe and N.S. Scrutton, *J. Biol. Chem.*, 2001, **276**, 6234.
152. C. Alhambra, M.L. Sanchez, J.C. Corchado, J. Gao and D.G. Truhlar, *Chem. Phys. Lett.*, 2002, **355**, 388.
153. M. Garcia-Viloca, C. Alhambra, D.G. Truhlar and J. Gao, *J. Chem. Phys.*, 2001, **114**, 9953.
154. C. Alhambra, J. Corchado, M.L. Sanchez, M. Garcia-Viloca, J. Gao and D.G. Truhlar, *J. Phys. Chem. B*, 2001, **105**, 11326.
155. G. Tresadern, H. Wang, P.F. Faulder, N.A. Burton and I.H. Hillier, *Mol. Phys.*, 2003, **101**, 2775.
156. M.J. Frisch, G.W. Trucks, H.B. Schlegel, G.E. Scuseria, M.A. Robb, J.R. Cheeseman, V.G. Zakrzewski, J.A. Montgomery, R.E. Stratmann, J.C. Burant, S. Dapprich, J.M. Millam, A.D. Daniels, K.N. Kudin, M.C. Strain, O. Farkas, J. Tomasi, V. Barone, M. Cossi, R. Cammi, B. Mennucci, C. Pomelli, C. Adamo, S. Clifford, J. Ochterski, G.A. Petersson, P.Y. Ayala, Q. Cui, K. Morokuma, D.K. Malick, A.D. Rabuck, K. Raghavachari, J.B. Foresman, J. Cioslowski, J.V. Ortiz, A.G. Baboul, B.B. Stefanov, G. Liu, A. Liashenko, P. Piskorz, I. Komaromi, R. Gomperts, R.L. Martin, D.J. Fox, T. Keith, M.A. Al-Laham, C.Y. Peng, A. Nanayakkara, C. Gonzalez, M. Challacombe, P.M.W. Gill, B. Johnson, W. Chen, M.W. Wong, J.L. Andres, C. Gonzalez, M. Head-Gordon, E.S. Replogle and J.A.

Pople Gaussian 98, Revision A.7 (1998) Gaussian, Inc, Pittsburgh PA. See www.gaussian.com.
157. P. Kast, M. Asif-Ullah and D. Hilvert, *Tetrahedron Lett.*, 1996, **37**, 2691.
158. P.D. Lyne, A.J. Mulholland and W.G. Richards, *J. Am. Chem. Soc.*, 1995, **117**, 345.
159. S. Martí, J. Andrés, V. Moliner, E. Silla, I. Tuñón and J. Bertrán, *J. Phys. Chem. B*, 2000, **104**, 11308.
160. K.E. Ranaghan, L. Ridder, B. Szefczyk, W.A. Sokalski, J.C. Hermann and A.J. Mulholland, *Mol. Phys.*, 2003, **101**, 2695.
161. K.E. Ranaghan, L. Ridder, B. Szefczyk, W.A. Sokalski, J.C. Hermann and A.J. Mulholland, *Org. Biomol. Chem.*, 2004, **2**, 968.
162. S. Martí, J. Andrés, V. Moliner, E. Silla, I. Tuñón and J. Bertrán, *Theor. Chem. Acc.*, 2001, **105**, 207.
163. B. Szefczyk, A.J. Mulholland, K.E. Ranaghan and W.A. Sokalski, *J. Am. Chem. Soc.*, 2004, **126**, 16148.
164. M. Strajbl, A. Shurki, M. Kato and A. Warshel, *J. Am. Chem. Soc.*, 2003, **125**, 10228.
165. H. Guo, Q. Cui, W.N. Lipscomb and M. Karplus, *Proc. Natl. Acad. Sci. USA*, 2001, **98**, 9032.
166. S. Martí, J. Andrés, V. Moliner, E. Silla, I. Tuñón and J. Bertrán, *Chem. Eur. J.*, 2003, **9**, 984.
167. C.R.W. Guimarães, M.P. Repasky, J. Chandrasekhar, J. Tirado–Rives and W.L. Jorgensen, *J. Am. Chem. Soc.*, 2003, **125**, 6892.
168. K.E. Ranaghan and A.J. Mulholland, *Chem. Commun.*, 2004, 1238.
169. S. Hur and T.C. Bruice, *J. Am. Chem. Soc.*, 2003, **125**, 1472.
170. S. Hur and T.C. Bruice, *J. Am. Chem. Soc.*, 2003, **125**, 5964.
171. S. Hur and T.C. Bruice, *J. Am. Chem. Soc.*, 2003, **125**, 10540.
172. F. Claeyssens, K.E. Ranaghan, F.R. Manby, J.N. Harvey and A.J. Mulholland, *Chem. Commun.*, 2005, **40**, 5068.
173. Jaguar, 4.0, Schrödinger, Inc., Portland, Oregon, 2001.
174. J.W. Ponder, TINKER: Software Tools for Molecular Design, v4.0, Saint Louis, MO, 2003.
175. J.N. Harvey, *Faraday Discuss.*, 2004, **127**, 165.
176. Y.S. Lee, S.E. Worthington, M. Krauss and B.R. Brooks, *J. Phys. Chem. B*, 2002, **106**, 12059.
177. A. Crespo, D.A. Scherlis, M.A. Marti, P. Ordejon, A.E. Roitberg and D.A. Estrin, *J. Phys. Chem. B*, 2003, **107**, 13728.
178. B. Meunier, S.P. de Visser and S. Shaik, *Chem. Rev.*, 2004, **104**, 3947.
179. S. Shaik, D. Kumar, S.P. de Visser, A. Altun and W. Thiel, *Chem. Rev.*, 2005, **105**, 2279.
180. F. Ogliaro, N. Harris, S. Cohen, M. Filatov, S.P. de Visser and S. Shaik, *J. Am. Chem. Soc.*, 2000, **122**, 8977.
181. N. Harris, S. Cohen, M. Filatov, F. Ogliaro and S. Shaik, *Angew. Chem. Int. Ed.*, 2000, **39**, 2003.
182. K. Yoshizawa, T. Kamachi and Y. Shiota, *J. Am. Chem. Soc.*, 2001, **123**, 980.
183. S.P. de Visser, F. Ogliaro, N. Harris and S. Shaik, *J. Am. Chem. Soc.*, 2001, **123**, 3037.
184. S.P. de Visser, F. Ogliaro and S. Shaik, *Chem. Commun.*, 2001, 2322.
185. F.P. Guengerich, *Chem. Res. Toxicol.*, 2001, **14**, 611.
186. C.M. Bathelt, L. Ridder, A.J. Mulholland and J.N. Harvey, *J. Am. Chem. Soc.*, 2003, **125**, 15004.

187. C.M. Bathelt, L. Ridder, A.J. Mulholland and J.N. Harvey, *Organic and Biomolecular Chemistry*, 2004, **2**, 2998.
188. M.J. de Groot, S.B. Kirton and M.J. Sutcliffe, *Curr. Topics Med. Chem.*, 2004, **4**, 1803.
189. M. Pirmohamed and B.K. Park, *Toxicology*, 2003, **192**, 23.
190. J.C. Schoneboom, S. Cohen, H. Lin, S. Shaik and W. Thiel, *J. Am. Chem. Soc.*, 2004, **126**, 4017.
191. V. Guallar, M.-H. Baik, S.J. Lippard and R.A. Friesner, *Proc. Natl. Acad. Sci. USA*, 2003, **100**, 6998.
192. C.M. Bathelt, J. Zurek, A.J. Mulholland and J.N. Harvey, *J. Am. Chem. Soc.*, 2005, **127**, 12900.
193. A. Bassan, M.R.A. Blomberg and P.E.M. Siegbahn, *J. Biol. Inorg. Chem.*, 2004, **9**, 439.
194. K.-B. Cho, V. Pelmenschikov, A. Gräslund and P.E.M. Siegbahn, *J. Phys. Chem. B*, 2004, **108**, 2056.
195. T. Borowski, A. Bassan and P.E.M. Siegbahn, *Biochemistry*, 2004, **43**, 12331.
196. D. Xu, Y. Wei, J. Wu, D. Dunaway-Mariano, H. Guo, Q. Cui and J. Gao, *J. Am. Chem. Soc.*, 2004, **126**, 13649.
197. G.H. Li and Q. Cui, *J. Am. Chem. Soc.*, 2003, **125**, 15028.
198. S. Martí, V. Moliner, I. Tuñón and I.H. Williams, *J. Phys. Chem. B*, 2005, **109**, 3707.
199. G.D. Ruggiero, I.H. Williams, M. Roca, V. Moliner and I. Tuñón, *J. Am. Chem. Soc.*, 2004, **126**, 8634.
200. H. Park, E.N. Brothers and K.M. Merz Jr., *J. Am. Chem. Soc.*, 2005, **127**, 4232.
201. G.A. Cisneros, M. Wang, P. Silinski, M.C. Fitzgerald and W. Yang, *Biochemistry*, 2004, **43**, 6885.
202. A. Crespo, M.A. Martí, D.A. Estrin and A.E. Roitberg, *J. Am. Chem. Soc.*, 2005, **127**, 6940.
203. C.R.W. Guimarães, M. Udier-Blagovi and W.L. Jorgensen, *J. Am. Chem. Soc.*, 2005, **127**, 3577.
204. M.P. Gleeson, I.H. Hillier and N.A. Burton, *Org. Biomol. Chem.*, 2004, **2**, 2275.
205. A.R. Dinner, G.M. Blackburn and M. Karplus, *Nature*, 2001, **413**, 752.
206. M.H.M. Olsson, P.K. Sharma and A. Warshel, *FEBS Lett.*, 2005, **579**, 2026–2034.
207. J.C. Hermann, L. Ridder, A.J. Mulholland and H.-D. Höltje, *J. Am. Chem. Soc.*, 2003, **125**, 9590.
208. J.C. Hermann, C. Hensen, L. Ridder, A.J. Mulholland and H.-D. Höltje, *J. Am. Chem. Soc.*, 2005, **127**, 4454.
209. J.C. Hermann, L. Ridder, H.-D. Höltje and A.J. Mulholland, *Org. Bioml. Chem.*, 2006, **2**, 206–210.
210. A. Lodola, M. Mor, J.C. Hermann, G. Tarzia, D. Piomelli and A.J. Mulholland, *Chem. Commun.*, 2005, 4399.
211. S. Bjelic and J. Åqvist, *Biochemistry*, 2004, **43**, 14521.
212. P. Carloni, U. Röthlisberger and Michele Parrinello, *Acc. Chem. Res.*, 2002, **35**, 455.
213. R. Car and M. Parrinello, *Phys. Rev. Lett.*, 1985, **55**, 2471.
214. D.K. Remler and P.A. Madden, *Mol. Phys.*, 1990, **70**, 921.
215. U. Röthlisberger, P. Carloni, K. Doclo and M. Parrinello, *J. Biol. Inorg. Chem.*, 2000, **5**, 236.
216. M. Eichinger, P. Tavan, J. Hutter and M. Parrinello, *J. Chem. Phys.*, 1999, **110**, 10452.

217. D. Joseph-McCarthy, *Ann. Rep. Comp. Chem.*, 2005, **1**, 169.
218. R.D. Taylor, P.J. Jewsbury and J.W. Essex, *J. Comput. Aided Mol. Design*, 2002, **16**, 151.
219. H.A. Carlson, *Curr. Opin. Chemical Biol.*, 2002, **6**, 447.
220. S.J. Teague, *Nature Reviews Drug Discovery*, 2003, **2**, 527.
221. C.-S. Goh, D. Milburn and M. Gerstein, *Curr. Opin. Structural Biol.*, 2004, **14**, 104.
222. M.L. Lamb, *Ann. Rep. Comp. Chem.*, 2005, **1**, 185.
223. K. Raha and K.M. Merz Jr., *Ann. Rep. Comp. Chem.*, 2005, **1**, 113.
224. S. Park and J.G. Saven, *Ann. Rep. Comp. Chem.*, 2005, **1**, 245.
225. B. Kuhlman, G. Dantas, G.C. Ireton, G. Varani, B.L. Stoddard and D. Baker, *Science*, 2003, **21**, 1364.
226. A.E. Roitberg, *Ann. Rep. Comp. Chem.*, 2005, **1**, 103.

3

Polarizabilities, Hyperpolarizabilities and Analogous Magnetic Properties

BY DAVID PUGH
Department of Pure and Applied Chemistry, 295 Cathedral Street, Glasgow, G1 1XL

1 Introduction

Molecular polarizabilities and hyperpolarizabilities are now routinely calculated in many computational packages and reported in publications that are not primarily concerned with these properties. Very often the calculated values are not likely to be of quantitative accuracy when compared with experimental data. One difficulty is that, except in the case of very small molecules, gas phase data is unobtainable and some allowance has to be made for the effect of the molecular environment in a condensed phase. Another is that the accurate determination of the nonlinear response functions requires that electron correlation should be treated accurately and this is not easy to achieve for the molecules that are of greatest interest. Very often the higher-level calculation is confined to zero frequency and the results scaled by using a less complete theory for the frequency dependence. Typically, *ab initio* studies use coupled-cluster methods for the static values scaled to frequencies where the effects are observable with time-dependent Hartree-Fock theory. Density functional methods require the introduction of specialized functions before they can cope with the hyperpolarizabilities and higher order magnetic effects.

Less fundamental approaches continue to be useful in making comparative studies of larger molecules and complex systems where there is a potential application of the nonlinear response.

In view of these considerations the review has necessarily to be rather selective. While attempting, at least, to mention a wide range of work that might turn out to be significant, the current state of the art can perhaps only be appreciated by looking at a few cases in more detail. It might be considered that a disproportionate amount of space has been devoted to a review of work on water in the gas and liquid phases, but there have been a number of high level studies of the subject in the last few years which seem to encapsulate some of the special difficulties in the field, including the rather unsatisfactory state of some of the experimental data and its interpretation and the introduction of rigorous

Chemical Modelling: Applications and Theory, Volume 4

methods of dealing with the liquid medium. Some new work on inert gas compounds leads to interesting analogies with the standard push-pull interpretation of large first hyperpolarizabilities in donor-acceptor organic molecules.

A review volume edited by Maroulis[1] appeared in 2004 and provides extensive coverage of much of the field.

2 Electric Field Related Effects

2.1 Atoms. – In principle polarizabilities can be calculated through standard time-dependent perturbation theory by summing over the contributions of all virtual excited states. While this method is actually used in some molecular cases where it is thought that a few special low-lying excited states provide the major contribution, it is not generally expected to provide a means of calculating precise values, even for atoms and small molecules. For example, there is usually a substantial contribution from continuum states. Chernov *et al.*[2] have put forward a method for atomic dipole polarizabilities in which the Green function quantum defect procedure is used to deal with higher excited and continuum states while the results are modified by representing the ground and lower excited states in terms of high level *ab initio* wavefunctions. Good overall agreement with experiment has been achieved for the static and dynamic polarizabilities of a large number of atoms. The method depends on the availability of good spectroscopic data to set up the quantum defect procedure, but, with this proviso, it is claimed that the results are of an accuracy comparable with the best *ab initio* calculations.

It is essential to have reliable values for the polarizabilities of atoms in order to be able to interpret molecular beam experiments and understand the behaviour of clusters. The usual method of determining atomic α values is through the deflection of atomic beams in an electric field gradient. Usually such experiments will be calibrated against the known polarizability of a reference atom and the absolute determination of such reference values is therefore important. In particular Lupinetti and Thakkar[3] and Fuentealba[4] have independently carried out *ab initio* calculations up to CCSD(T) level on the Al atom. Their results are in close agreement with each other and both groups suggest that the experimentally determined value must be in error. Lupinetti and Thakkar have applied the same methods to a number of second row elements, and the comparison of their results on argon with those of numerous other experimental and theoretical studies allows them to make a realistic assessment of the accuracy of their methods. They have also calculated the mean second hyperpolarizability (γ).

Thakkar and Lupinetti[5] have used the coupled-cluster method in conjunction with the Douglas-Kroll relativistic Hamiltonian to obtain a very accurate value for the static dipole polarizability of the sodium atom. Their revised value for $\alpha(\mathrm{Na}) = 162.88 \pm 0.6$ au resolves a previous discrepancy between theory and experiment and when combined with an essentially exact value for lithium, establishes the ratio $\alpha(\mathrm{Li})/\alpha(\mathrm{Na}) = 1.0071 \pm 0.0037$, so that, because of the relativistic effect, lithium is more polarizable than sodium. The lithium measurement, however, was made in 1974 by the E-H gradient balance technique and has an estimated error of ±3.4 au, while the value obtained for sodium in the same experiment was 159.3 ± 3.4 au. The theoretical figure for Li is probably much more accurate and, coupled with the more recent molecular beam determination of the sodium value, allows the accurate determination of the ratio.

In another study of the polarizability and hyperpolarizability of the Si atom Maroulis and Pouchan[6] used the finite field method with correlation effects estimated through Moeller-Plesset perturbation theory. Correlation effects are found to be small.

Comparisons between some of the static α data from the above papers are made in Tables 1 and 2.

Working to similar levels of accuracy, Pawlowski *et al.*[7] have calculated the static and frequency-dependent linear polarizability and second hyperpolarizability of the Ne atom using coupled-cluster methods with first order relativistic corrections. Good agreement with recent experimental results is achieved. Klopper *et al.*[8] have applied an implementation of the Dalton code that enables

Table 1 *Recent calculations of atomic polarizabilities*

$\bar{\alpha}$/au	Al	Si	P	S	Cl	Ar
Chernov(QDGF)[a]	52.8	37.3	25.0	18.5	-	10.7
Lupinetti & Thakkar (UHF)[b]	61.0	38.5	25.5	19.1	14.2	10.8
(CCSD(T)[b]	57.8	37.2	24.9	19.4	14.6	11.1
Fuentealba (RHF)[c]	62.1	-	-	-	-	-
(CCSD(T)[c]	58.4	-	-	-	-	-
Maroulis & Pouchan UHF[d]		38.5				
(CCSD(T)[d]		37.4				
Experiment	45.9[e]	-	-	-	-	11.1[f]

[a] ref 2; [b] ref 3; [c] ref 4; [d] ref 6; [e] P. Milani, I. Moullet and W.A. de Heer, *Phys. Rev. A*, 1990, **42**, 5150. [f] D.R. Johnston, G.J. Oudemans and R.H. Cole, *J. Chem. Phys.*, 1960, **33**, 1310, U. Hohm and K. Kerl, *Mol. Phys.*, 1990, **69**, 819.

Table 2 *Comparison of non-relativistic and relativistic values for the polarizabilities of Li and Na*

α/au	Li	Na
Non-relativistic	164.11[a]	163.90[b]
relativistic	164.04 ± 0.01[c]	162.88 ± 0.6[b]
experimental	164.0 ± 3.4[d]	162.7 ± 0.8[e]

[a] Z.-C. Yan, J.F. Babb, A. Dalgarno and G.W.F. Drake, *Phys. Rev. A.*,1996, **54**, 2824.; [b] ref 5; [c] I.S. Lim, M. Pernpointer, M. Seth, J.K. Laerdahl, P. Schwerdtfeger, P. Neogrady and M. Urban, *Phys. Rev. A.*, 1999, **60**, 2822; [d] R.W. Molof, H.L. Schwartz, T.M. Miller and B. Bederson, *Phys. Rev. A.*, 1974, **10**, 1131; [e] C.R. Ekstrom, J. Schmiedmayer, M.S. Chapman, T.D. Hammond and D.E. Pritchard, *Phys. Rev. A.*, 1995, **51**, 3883.

electric response properties to be calculated with first order relativistic corrections at coupled-cluster level. They calculate the α and γ static values for Ne to illustrate possible applications of the code. Chong *et al.*[9] have extended the concept of field-induced polarization functions in GTO form to include all atoms up to Kr. These functions may be of use in polarizability and hyperpolarizability calculations. Ren and Hsue[10] have addressed the problem of the dipole and quadrupole polarizabilities of rare gas and alkaline-earth atoms by a DFT method with an optimized effective potential and an explicit self-interaction term. With these modifications their results compare well with those obtained by correlated *ab initio* methods. Adamov *et al.*[11] have calculated the dynamic dipole polarizability and a number of spectroscopic properties of Ar, Ca, Sr and Xe using the Hartree-Fock-Roothaan method.

There is increasing interest in relativistic calculations for heavier atoms and ions. The relativistic procedures are often used in conjunction with density functional (DFT) techniques. Lim and Schwerdtfeger[12] employ the Douglas-Kroll four component and scalar relativistic Hamiltonians as the basis for their study of the static dipole polarizabilities of the neutral atoms and the singly and doubly charged positive ions of all the group II elements from Ca to Ra. The finite field procedure and specifically optimized Gaussian basis sets are used. The relationship between polarizability and ionization potential is explored. The ions of heavier elements have also been the subject of some relativistic studies. Clavaguera and Dognon[13] have investigated the lanthanide ions (La^{3+} to Lu^{3+}) and have shown that, in order to obtain reliable results, it is necessary to include scalar relativistic effects and to use flexible basis sets. The authors provide a useful database of reference values for these ions. A pseudopotential method adapted for open shell systems was used.

The effect of increasing the number of basis functions in correlation–consistent augmented sets approaching the Hartree-Fock limit for calculations of the second hyperpolarizability, γ, has been studied by Ding and Liu.[14] Using the Coupled-Perturbed-Hartree-Fock approach, the authors identify smooth convergence up to a point, followed by divergence in the results if too many diffuse functions are added. If the γ-values are plotted against the number of diffuse functions, the curve exhibits a point of inflexion and comparison of the results with other calculations indicates that the values obtained near the point of inflexion are good approximations. Results have been obtained for He, Ne, H_2, FH, CO and OH^-. The authors have also studied open-shell systems,[15] approached from the restricted open shell Hartree-Fock (ROHF) approximation where the behaviour follows the same pattern as found in closed–shell systems. Results have been obtained for small atoms and OH and OH^+. Having examined their basis sets in this way the authors apply them to CCSD(T) correlated calculations of γ for Ne using a finite-field approach and obtain a result in good agreement with experiment.

Ghosh and Biswas[16] have further exploited their method for the calculation of the absolute radii of atoms and ions and attempt to correlate the computed radii with a number of properties including the polarizability.

2.2 Diatomic Molecules: Non-Relativistic. – Kobus *et al.*[17] have continued their work on a comparison of finite-difference solutions of the HF equations for diatomics with those obtained by the conventional method of expansion in terms of basis functions. New calculations are reported for LiH, FH and BH. In the last case the hyperpolarizability is shown to be very sensitive both to the type of grid used in the finite difference method and to the choice of basis set in the expansion method. More information on the HF molecule[18] has been provided by Maroulis who calculates values of the electric multipole moments and the polarizability and hyperpolarizability derivatives using his finite field method with large basis sets and high level correlation methods. Maroulis[19] has also provided a study of the bonding, polarizability and hyperpolarizability of the Na_2 dimer, using coupled-cluster correlated *ab initio* and DFT methods. He finds that low level *ab initio* methods underestimate the anisotropy of the polarizability tensor and that the inclusion of electron correlation also has a drastic effect on the components of the hyperpolarizability tensor; and in a study of Cu_2[20] he demonstrates the difficulty of getting reliable results for the γ–hyperpolarizability.

Merawa *et al.*[21,22] have carried out *ab initio* calculations on the ground and lowest excited states of LiH and NaH and on NO^+ using the time-dependent gauge invariant method (TDGI). For some of the states a CCSD(T) method has also been used for comparison. Results have been obtained as a function of internuclear separation and are found to be in good agreement with experiment. Work using multi-reference CI methods have also been reported: Minaev[23] has made a study of the ground state properties of the O_2 molecule using Multi-Configurational -SCF (MCSCF). Electric and magnetic response functions are calculated. (see Section 3); and Giese and York[24] have made an extensive study of the energy, dipole and polarizability surfaces of the alkali halides (LiF, LiCl, NaF, NaCl) with and without applied fields along the molecular axis.

Vibrational contributions to the α and β response functions of NaF and NaCl have been calculated by Andrade *et al.*[25] at HF, MP and CC levels. The results obtained from perturbation theory are in agreement with those from the finite field method and demonstrate that the inclusion of vibrational effects is essential to get reliable electric response functions in these molecules.

2.3 Diatomic Molecules: Relativistic. – Application of relativistic methods are becoming much more widespread. In addition to being essential for work on heavier atoms it is becoming more apparent that the high level of accuracy attainable on small molecules with modern powerful correlated procedures is such that comparison with experiment at the precision attainable will require careful relativistic corrections to be applied.

Ilias *et al.*[26] have made a detailed study of the electric dipole moment and static dipole polarizability of HI. Relativistic correlated CCSD(T) calculations have been made. The authors believe that the remaining discrepancies between theory and experiment, particularly in the polarizability anisotropy, cannot be reconciled and that the experimental data must be inaccurate.

Norman and Jensen[27] have implemented a method for obtaining second order response functions within the four–component (relativistic) time-dependent Hartree-Fock scheme. Results are presented for the first–order hyperpolarizabilities for second harmonic generation, $\beta(-2\omega;\omega,\omega)$ for CsAg and CsAu. A comparison of the results with those of non-relativistic calculations implies that the nonrelativistic results are over-estimated by 18% and 66% respectively. In this method transitions that are weakly-allowed relativistically can lead to divergences in the frequency-dependent response, which would be removed if the finite lifetimes of the excited states could be taken into account.

Wang and Liu[28] have shown, in a study of the molecular properties of Cu_2, Ag_2 and Au_2 using four component relativistic DFT, that accurate results for the longitudinal component of the dipole polarizability can be obtained with basis sets containing up to g-type angular functions. Higher angular functions have a negligible effect. The authors make an estimate of the basis set limit for calculations using their DFT functional.

New DFT techniques have been introduced by Salek, Helgaker and Saue[29] and by Bauer and Neuhauser.[30] The former report an implementation of a relativistic DFT theory based on the 4-component Dirac-Coulomb Hamiltonian. The method allows the calculation of frequency-dependent response functions through the quasi-energy (Floquet theory) formulation. The method has been applied to Hg, AuH and PtH_2. Baer and Neuhauser have employed an exact representation of the exchange-correlation functional to develop approximations that have correct long-range asymptotic behaviour. A simple theory is produced which leads to encouraging results when applied, *inter alia*, to several first row atoms and diatomic molecules.

2.4 Atom-Atom Interactions. – The methods applied, usually to interactions in the inert gases, are a natural extension of diatomic molecule calculations. From the interaction potentials observable quantities, especially the virial coefficients can be calculated. Maroulis *et al.*[31] have applied the *ab initio* finite field method to calculate the interaction polarizability of two xenon atoms. A sequence of new basis sets for Xe, especially designed for interaction studies have been employed. It has been verified that values obtained from a standard DFT method are qualitatively correct in describing the interaction polarizability curves. Haskopoulos *et al.*[32] have applied similar methods to calculate the interaction polarizability of the Kr-Xe pair. The second virial coefficients of neon gas have been computed by Hattig *et al.*,[33] using an accurate CCSD(T) potential for the Ne–Ne van der Waals potential and interaction-induced electric dipole polarizabilities and hyperpolarizabilities also obtained by CCSD calculations. The refractivity, electric-field induced SHG coefficients and the virial coefficients were evaluated. The authors claim that the results are expected to be more reliable than current experimental data.

2.5 Inert Gas Compounds. – A recent development in the investigation of noble gas compounds has been the synthesis by Kriachtchev *et al.*[34] of the compound HArF by the photolysis of hydrogen fluoride in a solid Argon matrix. The same group carried out *ab initio* (MP2 and CCSD(T)) studies that indicated that the molecule should have a linear structure and Runeberg *et al.*[35] suggested that it should be stable in the gas phase. The calculated electron distribution is such that the molecule appears to be strongly ionic. Avramopoulos *et al.*[36] have made an extensive theoretical study of the electric properties of the molecule, including calculations of the electronic and vibrational contributions to the dipole moment, polarizability and first hyperpolarizability. The values obtained for the electronic contributions to the longitudinal components, α_{zz} and β_{zzz} are very large in comparison with hydrogen fluoride and indeed when compared with much larger conjugated compounds (see Table 3).

A study of the effects of basis size and level of correlation has been made and in the table results at the SCF level and at the highest recommended correlated level in the paper concerned are included. Results for hydrogen fluoride from recent work by Maroulis are also included. The frequency dependent quantities (at ω = 0.072 au) are obtained by scaling TDHF calculations using the correlated zero-frequency result. The authors attempt a semi-quantitative explanation of the large β values by invoking the two-state model which predicts that the hyperpolarizability should be proportional to $3\Delta\mu.\mu_{ge}^2/E_{ge}^2$, where the difference between the dipole moments in the excited and ground states is $\Delta\mu = \mu_e - \mu_g$, μ_{ge} is the transition dipole moment and E_{ge} the electronic excitation energy. First electronic excited state energies and the matrix elements were obtained from the CIS/Pol model, both for HF and HArF so that a comparison with the SCF β results is appropriate. Table 4 shows the quantities used in the two state model calculation and it can be seen that the lower excitation energy and larger change in dipole moment both act to increase β. The ratio, $[\beta_{zzz}(0;0,0)]_{HArF}/[\beta_{zzz}(0;0,0)]_{HF}$ is found from the 2-state model to be approximately 60 while the SCF calculations give a ratio of 75. More accurate calculations from the CASVB algorithm of Li and McWeeny[37] show that the excited state is largely covalent, as opposed to the ground state, which is dominated by the resonance structure H-Ar$^+$F$^-$. The picture that emerges is similar to that provided by the two-state model for push-pull organic molecules. In the ground state the HAr group has a substantial positive charge and the fluorine is negative, so that the molecule contains one largely covalent and

Table 3 *Comparison of the properties of HF and the inert gas compound HArF. The polarizabiliies and first (SHG) hyperpolarizabilities are shown at ω = 0 and ω =0.072 au*

molecule	*method*	μ	$\alpha_{zz}(0,0)$	$\alpha_{zz}(-\omega;\omega)$	$\beta_{zz}(-0;0,0)$	$\beta_{zz}(-2\omega;\omega,\omega)$
HF	SCF/Pol[a]	0.744	5.59		−7.39	
HF	SCF[b]	0.756	5.75		−8.4	
HF	Corr[b]	0.7043	6.36		−9.4	
HArF	SCF/Pol[a]	3.47	34.25	35.69	−561.5	−835.4
HArF	MP2/Pol[a]	2.78	59.24	65.57	−1443.4	−3723.9

[a] ref 36; [b] ref 18. 'Pol' refers to the Pol basis sets of Sadlej and 'Corr' to the highest level of correlation in ref 18.

Table 4 *Two state model parameters for HF and HArF. From ref 36 using the HF/Pol method for the ground state properties and CIS/Pol for the excited state. All quantities in atomic units*

	HF	*HArF*
μ_g	0.745	3.473
μ_e	−0.907	−0.814
μ_{ge}	−0.611	1.419
$\Delta\mu$	1.652	4.287
E_{ge}	0.570	0.276

one largely ionic bond. The increase in the ground state dipole and the reduction in the excitation energy are in part simply a consequence of the greater extension of the molecule; and the combination of factors occurring in the two-state formula then account for an enlargement factor for β_{zzz} that is close to that obtained from the high level calculations. McDowell[38] has extended the theoretical work to include the three compounds HHeF, HArF and HKrF. The krypton compound has also been produced experimentally. McDowell calculates the polarizability tensor, α, defined for linear molecules by the mean polarizability, $\bar{\alpha}$ and the polarizability anisotropy, $\Delta\alpha$. He compares the results of SCF, MP2, QCISD(Quadratic CISD, a method derived from CISD by including higher order terms to improve size-extensivity and which usually gives results very similar to those of CCSD) and DFT methods. If it is accepted that the most reliable of these procedures is QCISD then the MP2 results are nearest to this standard while the DFT and SCF results are respectively small and rather larger underestimates. Possible semi-quantitative interpretations suggest themselves, in terms of two state models and using the polarizabilities of hydrogen fluoride and the inert gases as reference points. The quantitative differences between McDowell's results for the HArF compound and those of Avropoulos *et al.* are attributable to his choice of basis set.

2.6 Water. – There has been continuing interest in calculations on water. The main objective of the more recent work has been to extend the application of high level correlated *ab initio* and density functional methods into the condensed phase, although refined data on the isolated molecule has also been obtained in the course of these studies. In order to clarify the objectives of the calculations on the liquid and the criteria by which they must be assessed a brief résumé of the most relevant experimental data is included. The theoretical work to be described relating to liquid water is directed toward understanding the properties in the optical range of frequencies, where the very large low frequency effects that increase the relative permittivity to a value of about 80 are not effective. In the optical regime it is still possible to regard the problem as that of the behaviour of one water molecule surrounded by an environment that may be represented as a continuum or by the average of some ensemble of molecules perturbing the central one and it is under these conditions that the use of high level correlated procedures is feasible. An alternative approach is to carry out calculations on clusters of various sizes where a particular central molecule is not given a special status.

2.6.1 Refractive Index Experimental Data for Gas and Liquid. From a measured refractive index it is always possible to extract formally an average linear dipole polarizability, $\bar{\alpha}$, from the Lorentz-Lorenz equation,

$$\left(\frac{n^2-1}{n^2+2}\right)\frac{M}{\rho}=\frac{N_{av}\bar{\alpha}}{3\varepsilon_0} \tag{1}$$

Table 5 shows the experimental specific refractivities, $K(\lambda) = [n(\lambda)-1]/\rho$, and the average polarizability as calculated from equation (1) at a number of frequencies for liquid and vapour phases. The values of the specific refractivity of the vapour have been obtained from the Cauchy dispersion formula of Zeiss and Meath.[39] In this paper the authors assess the results of a number of experimental determinations of the refractive index of water vapour and its variation with frequency. Even after some normalization of the data to harmonize the absolute values from different determinations there is a one or two percent spread of results at any one wavelength. Extrapolation of the renormalized data for five independent sets of data leads to zero frequency values of $K(\lambda)$ within the range (2.985–3.013) $\times$ 10^{-4} m^3 kg^{-1}, giving, *via* equation (1), $\bar{\alpha}_{LL} = 9.63 \pm 0.10$ au. Extrapolation of the earlier refractive index data of Cuthbertson and Cuthbertson[40] by Russell and Spackman[41] from 8 values of frequency between 0.068 and 0.095 au, leads to a zero frequency value, of $\bar{\alpha}_{LL}(0) = 9.83$ au. While the considerable variation between the raw experimental data reported in different determinations is cause for some uncertainty, it appears that the most convincing analysis to date is that of

Table 5 *Experimental data for the linear response of water. The gas phase data is from the Cauchy formula of Zeiss and Meith.[39] The Cuthbertson and Cuthbertson[40] value of $\alpha(0)$ is shown in brackets. The figures in heavy type lie in the range containing the data used in deriving the Zeiss-Meath formula. The liquid phase data is from P. Schiebener, J. Straub, J.M.H. Levelt Sengers and J.S. Gallagher, J. Chem. Phys. Ref. Data, 1990,* **19**, *677,1617. Polarizability values in all cases have been derived from the refractive index data using equation (1)*

λ/nm	226.50	361.05	404.1	514.5	589.00	632.0	1013.98	∞
ω/au	0.201	0.126	0.113	0.089	0.077	0.072	0.045	0
K(liq)/10^{-4} m^3 kg^{-1}	3.941	3.486	3.438	-	3.340	3.327	3.258	-
K(gas)/10^{-4} m^3 kg^{-1}	3.650	**3.196**	**3.152**	**3.090**	**3.067**	**3.057**	3.021	2.998
$\bar{\alpha}_{LL}$(liq)/au	11.52	10.33	10.20	-	9.93	9.90	9.71	-
$\bar{\alpha}_{LL}$(gas)/au	11.71	**10.26**	**10.12**	**9.92**	**9.84**	**9.81**	9.70	9.62 (9.83)

Zeiss and Meath. However a number of recent theoretical papers take the Zeiss-Meath value at 514.5 nm as the basis for the experimental value of the polarizability anisotropy and the Cuthbertson value for the limiting zero frequency polarizability. Examination of Table 5 and Figure 1 shows that this would imply that the dispersion at the longer wavelength end of the optical range is very small.

2.6.2 The Mean Polarizability, $\bar{\alpha}$: Gas Phase. There have been numerous calculations of the average polarizability for water vapour. The SCF/FF calculation of Maroulis[42] (1998), giving $\bar{\alpha}(0) = 8.53$ is claimed to be close to the Hartree-Fock limit, while the CCSD work of Sienko and Bartlett[43] gave a value about 1 au higher. Recent high level results[44–49] for $\bar{\alpha}(0)$ are summarized in Table 6.

The calculations give the electronic part of the polarizability. It is thought that the ZPVA correction is the major part of the vibrational contribution and has a value of about 0.29 au at zero frequency. In Table 6 this correction has been added to the electronic part to give the values in the final column, which can be compared with the two possible experimental values in the first two rows. The most accurate of the CCSD type work should be that of Christiansen

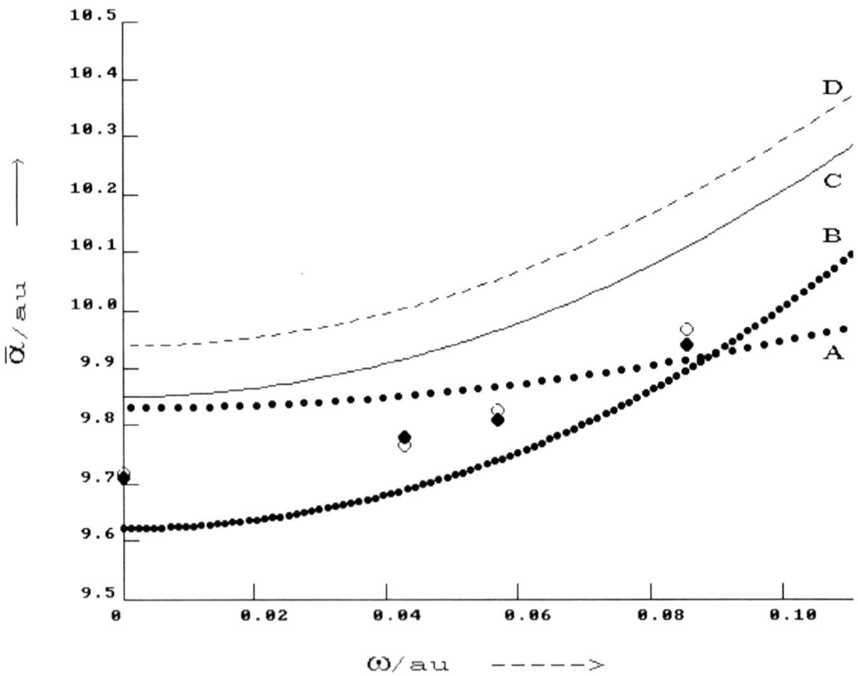

Figure 1 *Dynamic Polarizability, $\bar{\alpha}(\omega)$ for H_2O in the gas phase. Curves A and B are alternative interpretations of the experimental data (see text); C and D computed curves from Christiansen et al.*[44] *and Kongsted et al.*[45]*; filled points are from Poulsen et al.*[46]*; open points from Jensen et al.*[48] *All the theoretical values have been increased at all frequencies by 0.29 au, the estimated ZPVA correction*

Table 6 *Calculated and experimental values of $\alpha^{gas}_{H_2O}(0)$. The basis sets referred to are, X_T = aug-cc-pVTZ, X_Q = aug-cc-pVQZ, dX_T = d-aug-cc-pVTZ, dX_Q = d-aug-cc-pVQZ. A vibrational correction of 0.29 au has been added in the final column*

	method	$\bar{\alpha}^{el}$	$\bar{\alpha}$
Cuthbertson[40]	Experiment		9.83
Zeiss & Meath[39]	Experiment		9.63
Christiansen *et al.*[44]	CC3/dX_T + CCSD/dX_Q-CCSD/dX_T	9.56	9.85
Poulsen *et al.*[46]	MCSCF/X_Q	9.42	9.71
Kongsted *et al.*[45]	CCSD/X_T	9.47	9.76
Osted *et al.*[47]	CCSD/dX_T	9.65	9.94
Jensen *et al.*[48]	DFT/SAOP/STO	9.44	9.73
Jensen *et al.*[49]	DFT/BP-GRAC/STO	9.90	10.19

et al.[44] who have included a CC3 type correction to the usual CCSD(T) approach and used the largest basis sets. It would be expected that this result should be more reliable than that of Kongsted *et al.*,[45] who use CCSD with a smaller basis set. Christiansen *et al.'s* value is similar to that of Osted *et al.*,[47] whose basis set is extended as compared to that of Kongsted *et al.* When the vibrational correction is added Christiansen *et al.* obtain a value close to the experimental result derived from the Cuthbertson data. The MCSCF work of Poulsen *et al.*[46] leads to a corrected value of 9.71 which is rather closer to the Zeiss-Meath experimental value. One of the DFT functionals[48] produces a similar result but the other is clearly an over-estimate. It appears that neither the theoretical nor the experimental work has yet produced a definitive value for $\bar{\alpha}(0)$, although it is true that almost all the more sophisticated *ab initio* calculations get within a few percent of any reasonable experimental estimate. The HF value, on the other hand, is in error by more than 10%.

The dynamic polarizability has also been extensively studied and an attempt is made to represent the relationship between theory and experiment graphically in Figure 1.

Curves A and B are alternative interpretations of the experimental situation. Curve B is a plot of the 6 term Cauchy dispersion formula derived by Zeiss and Meath, while curve A is a simple quadratic interpolation (2-term Cauchy formula) between the static value of Cuthbertson[40] and the Zeiss-Meath[39] value at 514.5 nm (the only point where the polarizability anisotropy has been measured). Theoreticians appear to have taken these two values to heart. Curves C and D are plots of similar formulae [$\bar{\alpha}(\omega) = A(1 + B\omega^2)$] derived theoretically by Christiansen *et al.*[44] and Kongsted *et al.*[45] respectively, using the methods shown in Table 6 with suitable time-dependent procedures. The points obtained from the MCSCF[46] work and the DFT/SAOP method[48] are also plotted. The ZPVA correction of 0.29 au has been added at all theoretical points at all frequencies.

There appear to be two possible interpretations of the developing relationship between theory and experiment. In the first it would be assumed that the Zeiss-Meath formula is correct, in which case it looks as though the coupled-cluster approaches are destined to lead to values that are too high, while the MCSCF method may emerge with a more satisfactory answer. The second interpretation is that the very flat Cuthbertson-Zeiss-Meath curve represents the experimental data so that the elaborate CC3/CCSD approach of Christiansen *et al.* has produced the correct answer at zero frequency. The lack of agreement with regard to the dispersion could be accounted for if the magnitude of the vibrational contribution were to diminish rapidly with frequency- which is likely to be the case. These questions are still to be resolved.

2.6.3 The Gas Phase Polarizability Anisotropy. Murphy[50] has measured the depolarization ratio for Rayleigh scattering, ρ_R, and analysed the intensity distribution in the rotational Raman spectrum of the vapour at 514.5 nm. The ratio R_{20} of the invariants of the $\alpha_{ij}\alpha_{kl}$ tensor can be determined by fitting the rotational Raman distribution. and $\bar{\alpha}$ is known (from the Zeiss-Meath formula). Knowledge of the three quantities, $\bar{\alpha}$, ρ_R and R_{20}, allows the polarizability anisotropy, $\Delta\alpha$, and the three principal values of the tensor to be calculated. The polarizability anisotropy invariant is numerically equal to the quantity,

$$\Delta\alpha = \left\{ \frac{(\alpha_{xx} - \alpha_{yy})^2 + (\alpha_{yy} - \alpha_{zz})^2 + (\alpha_{zz} - \alpha_{xx})^2}{2} \right\}^{1/2} \tag{2}$$

provided (x,y,z) are the principal axes (*i.e.* $\alpha_{xx} = \alpha_x$ *etc.*). Only the square of the polarizability anisotropy parameter is uniquely determined from the depolarization experiments. The choice of sign in taking the square root is not experimentally determinable and the choice has been made on the basis of theoretical molecular calculations. These earlier calculations are reviewed by Thomsen and Swanstrom.[51] The positive root, which best fits these results, gives $\Delta\alpha = 0.667$ au and this value has been used in calculating the principal values. The data is summarized in Table 7.

Many of the recent calculations report values for all the tensor components and the anisotropy as a function of frequency. In Table 8 the values obtained in some of the work referred to above are shown at zero frequency and, when they are available, at or near the frequency of the measurement.

Table 7 *The linear polarizability tensor and polarizability anisotropy of the H_2O molecule as determined by Murphy*[50]

	$\alpha(H_2O)$ at 514.5 nm	
	$K(514.5\ nm) = 3.0896 \times 10^{-4}\ m^3\ kg^{-1}$	$\Delta\alpha = 0.667$ au
	$10^{-24}\ cm^3$	au
$\bar{\alpha}$	1.4703 ± 0.0025	9.92 ± 0.02
α_{xx}	1.528 ± 0.013	10.31 ± 0.09
α_{yy}	1.415 ± 0.013	9.54 ± 0.09
α_{zz}	1.468 ± 0.003	9.90 ± 0.02

Table 8 *Recent calculations of the polarizability tensor and anisotropy of H_2O in the gas phase*

		ω/au	$\bar{\alpha}$	α_{xx}	α_{yy}	α_{zz}	$\Delta\alpha$
Sekino and Bartlett[43]	CCSD	0	9.55	9.89	-	9.49	0.56
Maroulis[42]	SCF/FF	0	8.53	9.18	7.90	8.52	1.11
Maroulis[42]	CCSD(T)/FF	0	9.61	9.93	9.33	9.59	0.52
Christiansen *et al.*[44]	CC3/CCSD	0	9.56	9.89	9.26	9.54	0.54
Gubskaya[52]	MP2	0	9.85	10.02	9.70	9.84	0.47
Gubskaya[52]	MP4	0	9.94	10.11	9.78	9.93	0.29
Poulsen *et al.*[46]	MCSCF	0	9.42	9.74	9.15	9.36	0.43
Jensen *et al.*[48]	DFT/SAOP	0	9.44	9.88	8.92	9.52	0.84
Jensen *et al.*[48]	DFT/BP-GRAC	0	9.90	10.12	9.68	9.91	0.38
Kongsted *et al.*[45]	CCSD	0	9.47	9.94	9.05	9.40	0.78
Christiansen *et al.*[44]	CC3/CCSD	0.088	9.83	-	-	-	0.41
Poulsen *et al.*[46]	MCSCF	0.0856	9.65	9.92	9.45	9.58	0.34
Jensen *et al.*[48]	DFT/SAOP	0.0856	9.69	10.08	9.23	9.76	0.75
Jensen *et al.*[48]	DFT/ BP-GRAC	0.0856	10.20	10.33	10.07	10.19	0.22
Kongsted *et al.*[45]	CCSD	0.0856	9.71	10.13	9.37	9.63	0.67

The values in Table 8 refer to the electronic contribution. If, as is sometimes assumed, a correction of approximately 0.29 au is still to be added, then the lower electronic anisotropies calculated by Poulsen *et al.*[46] and Jensen[48] would again be in close agreement with the experimental result.

2.6.4 Theoretical Methods for the Liquid. The approaches to be described all attempt to calculate the response functions for a central water molecule influenced by the surrounding medium. The older methods represent the surrounding medium through a continuum parametrized by a dielectric function but more recently the medium has been described by molecular mechanics (MM). In both cases the central molecule is the subject of a full correlated quantum mechanical treatment.

Gubskaya and Kusalik[52] introduced a variant of the finite field method in which a local field was used to mimic the electrostatic interaction with the liquid environment while the reference molecule was treated at MP2 and MP4 levels; Sylvester-Hvid *et al.*[53] worked at Hartree-Fock level and included the effects of the medium through three models, continuum, semi-continuum and by a supermolecule calculation; Poulsen *et al.*[46] used a MCSCF/MM scheme, Kongsted *et al.*[45,54,55] CC2, CCSD/DC, MM systems and Jensen *et al.*[48,49] two DFT/MM methods (where DC means 'dielectric continuum' and MM 'Molecular Mechanics'). The methods have been applied to the linear response and to the hyperpolarizabilities. The scheme used by Kongsted *et al.* in three recent papers covering linear and nonlinear properties is probably the most sophisticated. Their most recent method has emerged from a number of previous studies on the water monomer, clusters and liquid water. References can be found in the cited articles. The method uses coupled-cluster calculations (at various levels) applied to a water molecule interacting with surrounding molecules described through molecular mechanics. The technique of introducing a

Lagrangian incorporating the equations of motion and varying the quasi-energy is used to allow variational calculations to be carried out on the interacting system within the coupled-cluster framework (which, when applied straightforwardly, is non-variational).[56–58]

2.6.5 Liquid Phase Calculations of the Linear Response. The data in Table 5 for the isotropic polarizability, derived formally *via* the Lorentz-Lorenz equation (1) from the measured refractive index, shows that the assumption that individual molecular properties are largely retained at high frequency in the liquid is very reasonable. While the specific susceptibilities for the gas and liquid phases differ, once the correction for the polarization of the surface of a spherical cavity, which is the essential feature of the Lorentz-Lorenz equation, has been applied, it is clear that the average molecular polarizabilities in the gas and liquid have values which always agree within 5 or 10%.

Given the uncertainty in the vapour phase data and the incomplete data available for the vibrational effects it seems unlikely that any definitive theoretical account of the differences between the polarizability of the molecule in the vapour and liquid can be achieved at present. In fact the absolute values obtained for $\bar{\alpha}$ for the liquid vary from method to method by far more than the differences between liquid and vapour. Some representative values are shown in Table 9.

The application of the Lorentz-Lorenz equation gives a convincing demonstration of the general similarity of the linear response in gas and liquid but its application in the liquid introduces an approximation which has not yet been quantified. A more precise objective for the theory would be to calculate the frequency dependent susceptibility or refractive index directly. For a continuum model this may lead to a polarizability rigorously defined through the Lorentz-Lorenz equation as shown in treatments of the Ewald-Oseen theorem (see, for example Born and Wolf, p100),[59] but the polarizability defined in this way need not refer to one molecule and would not be precisely related to the gas parameters.

2.6.6 Experimental Hyperpolarizabilities for Gas and Liquid. The usual sources of hyperpolarizability data are EFISH (Electric Field Induced Second Harmonic), the static and optical Kerr effect (KE and OKE) and hyper-Rayleigh scattering (HRS). The extraction of molecular hyperpolarizabilities from the EFISH signal requires careful analysis of the second harmonic output signal

Table 9 *Average polarizability, $\bar{\alpha}(\omega)$ for H_2O in the liquid state*

ω/au	*CC/MM*[a]	*CC/MM*[b]	*DFT/SAOP*[c]	*MCSCF/MM*[d]	*Exp*[e]
0	9.629			9.70	
0.0428	9.701	10.037	9.79	9.74	9.71
0.0570	9.761	10.088	9.84	9.79	9.79
0.0856	9.808	10.241	9.99	9.93	9.99

[a] *Kongsted et al. aug-cc-pVTZ*[45]; [b] *Kongsted et al. d-aug-ccpVTZ*[45]; [c] *Jensen et al.*[48]; [d] *Poulsen et al.*[46]; [e] *calculated from refractivity data in Schiebener et al.*(*see* Table 5)

and its comparison with a standard signal produced by the same optical laser, usually in quartz. Over the past two decades the absolute value of the quartz standard, which must be known if *a priori* theoretical calculations are to be assessed, and the precise definition of the response functions measured in the EFISH signal have both been sources of uncertainty in the interpretation of the measurements. Essentially the measured quantity (before any response functions have been defined) is the induced polarization per unit volume at a frequency of 2ω, $\mathbf{P}(2\omega)$ and this scales as the second order susceptibility component, d_{11}(quartz), for which the currently accepted value is (0.30 ± 0.02) pmV^{-1} as opposed to the earlier 0.335 pmV^{-1}. The different possible definitions of the molecular response functions involved have been exhaustively discussed by Willets *et al.*,[60] but it can still sometimes be difficult to be certain which has been applied when interpreting a quoted experimental determination. It is advisable to adhere to the convention used in all *a priori* theoretical work and define the molecular response functions in terms of a Taylor series:

$$\delta\mu_i = \alpha_{ij}E_j^L + \frac{1}{2}\beta_{ijk}E_j^L E_k^L + \frac{1}{6}\gamma_{ijkl}E_j^L E_k^L E_l^L + \ldots . \tag{3}$$

where the local fields acting on the molecule include the appropriate field factors.

This expansion must be regarded as a shorthand notation if frequency dependent effects are to be treated. For a composite field such as the one acting in an EFISH experiment

$$E_z = E_z(0) + E_z(\omega)\cos\omega t, \tag{4}$$

each coefficient has to be assigned a different frequency dependence when the brackets are expanded. This procedure can be justified by Fourier transforming real time non-local response functions. Then,

$$P_z(2\omega) = N\,\langle\delta\mu_z(2\omega)\rangle \tag{5}$$

where $\langle\delta\mu_z(2\omega)\rangle$ is the average induced dipole in the direction of the applied fields (z).

Inserting equation (3) for the induced dipole, using (4) for the total field and extracting the terms in 2ω gives,

$$\langle\delta\mu_z(2\omega)\rangle = \frac{1}{4}\gamma f_0 f_\omega^2 f_{2\omega} E(0)E(\omega)^2 E(2\omega) \tag{6}$$

where, after averaging over the partly orientated dipole distribution,

$$\gamma = \left\{\gamma_e + \frac{\mu\beta_{//}}{3kT}\right\} = \left\{\gamma_e + \frac{\mu\beta_z}{5kT}\right\} \tag{7}$$

The f functions are the Lorentz local field factors, N is the number of molecules in unit volume and,

$$\beta_z = \frac{5}{3}\beta_{//} = \frac{1}{3}\left(\beta_{zjj} + 2\beta_{jzj}\right), \quad \gamma_e = \frac{1}{15}\left(\gamma_{iijj} + \gamma_{ijji}\right) \tag{8}$$

The EFISH experiment allows a value of γ to be found and, if the temperature can be varied over a large enough range, $\beta_{//}$ and γ_e can be determined independently.

There are two reported determinations of the hyperpolarizabilities of water vapour in the literature, by Ward and Miller[61] and Kaatz *et al.*[62] (see Table 10).

The only EFISH data for the liquid phase is contained in a paper by Levine and Bethea (1976)[63] on associating liquids. In this paper the macroscopic nonlinearity, Γ, is defined by the equation,

$$P(2\omega) = \Gamma E(0)\ E(\omega)^2 \tag{9}$$

and the following values are given:-

$$\Gamma = 17.6 \times 10^{-14}\text{ esu} \quad \gamma = 1.44 \times 10^{-36}\text{ esu} \quad \beta = 0.46 \times 10^{-31}\text{ esu}$$
$$(\gamma = 1.38 \times 10^{-36}\text{ esu})$$

However, recalculating the value of γ using the method described in the paper for the field factors, gives the value in brackets. The unbracketed value, for the overall microscopic nonlinearity, converts to 2859 au. In the case of associating liquids the authors argue that equation (7) can be used in modified form with the inclusion of a factor, g, which they deduce from the Kirkwood-Frohlich modification of the Onsager theory,

$$\gamma = \left\{\gamma_e + \frac{g\mu\beta_z}{5kT}\right\} \tag{10}$$

For water they deduce that $g = 2.94$. If one uses this equation to calculate the contribution from the β term to the overall third order nonlinearity, one finds,

$$\frac{g\mu\beta}{5kT} = 1.18 \times 10^{-36}\text{ esu. leaving } \gamma_e = 0.20 \times 10^{-36}\text{ esu.}$$

The convention used by Levine and Bethea to define the response functions omits the Taylor series factors in the series for the induced dipoles but includes a factor of (3/2) implicitly in the definition of the macroscopic quantity. Their β is equivalent to β_z. Hence to relate their results to the more usual conventions, the β-value must be multiplied by $4 \times (3/2) \times (3/5) = 18/5$ and the γ value by $4 \times (3/2) = 6$. Finally a factor (0.30/0.335) must be applied to allow for the change in the quartz standard. Carrying out these operations and converting to atomic units gives the values in Table 10.

Table 10 *Experimental hyperpolarizabilities of H_2O in the gas and liquid phases. The earlier result of Ward and Miller has been rescaled by Shelton and Rice to allow for the change in the quartz standard*

phase	λ/*nm*	ω/*au*	$\beta_{//}$/*au*	γ_e/*au*	*reference*
gas	1064	0.0428	-19.2 ± 0.9	1800 ± 150	Kaatz *et al.*[62]
gas	694.3	0.0656	-22.0 ± 0.9	2310 ± 120	Ward and Miller[61]
liquid	1064	0.0428	19.2	2134	Levine & Bethea[63]

These are the values that several recent theoretical studies use to compare their calculations with experiment. They should, however, be treated with caution. The error estimate, based only on the accuracy of the optical and field measurements is about 15 or 30%, but there are considerable additional uncertainties in the application of the theory of the field factors in the strongly associated liquid; the value of g, in particular, is critical for the accurate separation of the two terms in equation (10).

Triple harmonic generation (THG) has also been measured in liquid water (Kajzar and Messier (1985)[64]) where $\chi(-3\omega; \omega, \omega, \omega)$ was found to have a value of 1.29×10^{-14} esu at a frequency of 0.0428 au. This quantity should be of the same order as γ_e, defined above, and the fact that it is very much smaller than the overall γ value from the EFISH experiment is an indication that the major contribution to the latter comes from the β term. The most striking difference between the gas and liquid values is that β and, consequently, the overall effective γ have opposite signs in the two phases.

2.6.7 Gas Phase Hyperpolarizability Calculations. In the absence of directly measured static values of the tensor components and the internal consistency amongst the higher level *ab initio* results, current work is directed to obtaining accurate numbers for the frequency-dependant parameters. It is in this area that most progress has been made in recent years. In the context of the *ab initio* methods, the refinement of the coupled-cluster method to the point where the poles of the cluster expansion are in reasonable agreement with the molecular electronic excitation energies, has perhaps been the most significant development. Simultaneously. There has been a rapid development of the time - dependent density functional approach.

Only the EFISH hyperpolarizabilities, $\beta_{//}(\omega)$ and $\gamma(\omega)$ will be considered here. In principle the most accurate calculation of $\beta_{//}(\omega)$ of the coupled-cluster genre is that of Christiansen *et al.*[44] using a CC3 correction. Their results for the electronic contribution, which show surprisingly small dispersion, are represented by the formula,

$$\beta_{//}^{SHG}(\omega) = -21.28(1 + 7.00\omega + 65\omega^4) \tag{11}$$

An estimate of the vibrational correction based on earlier MP2[65] and MCSCF[66] calculations has been made which leads to a final value of −22.4 au for $\beta_{//}^{SHG}$ at 694.3 nm in comparison with the experimental value of -22.0 ± 0.9 au quoted above. The agreement with the more recent experimental determination at 1064 nm (−18.86 au compared to experimental −19.81 au) is not as close but an extrapolation of the electronic contribution based on equation (11) has been made and there is no vibrational data directly referring to the 1064 nm wavelength.

Other CCSD calculations show considerable variations especially with the level of correlation (see Kongsted *et al.*[54], Jensen[49]) but are generally within about 20% or better of the experimental value. The Maroulis[42] finite field static values with very large basis sets and CCSD(T) correlation are still an important reference point. Calculations of γ^{SHG} appear to attain an almost similar degree of accuracy. Kongsted *et al.*[55] obtain a value of 2355 au for the electronic contribution at 0.07 au. After making vibrational corrections (which involves plausible but unsubstantiated assumptions about their dispersion) they arrive at a final estimate of 2413 au to compare with the experimental 2310 au at 694 nm.

2.6.8 Liquid Phase Hyperpolarizability Calculations. In contrast to the case of the linear polarizability, the differences in hyperpolarizabilities between gas and liquid phases are very marked. The most characteristic feature is the reversal of the sign of the second order response functions on going from the gas to the liquid. Some results from recent calculations are shown in Table 11.

While all the studies reproduce the sign change in the first hyperpolarizability the numerical values are rather inconsistent. The QM/MM method appears to be superior to the QM/ Continuum approaches. Silvester-Hvild *et al.*[53] describe an extension of their approach using three different models, representing the system respectively as a continuum, semi-continuum or supermolecular. They show that the sign change is not reproduced in the continuum model and only the semicontinuum model, where some specific intermolecular interaction at smaller distances is combined with a continuum model, gave results approaching quantitative accuracy.

2.6.9 Other Work on Water–Related Systems. Sonoda *et al.*[67] have simulated a time-resolved optical Kerr effect experiment. In this model, which uses molecular dynamics to represent the behaviour of the extended medium, the principle intermolecular effects are generated by the dipole-induced-dipole (DID) mechanism, but the effect of the second order molecular response is also include through terms involving the static molecular β tensor, calculated by an MP2 method. Weber *et al.*[68] have applied *ab initio* linear scaling response theory to water clusters. Skaf and Vechi[69] have used MP2/6-311 ++ G(d,p) calculation of the α and γ tensors of water and dimethylsulfoxide (DMSO) to carry out a molecular dynamics simulation of DMSO/Water mixtures. Frediani *et al.*[70] have used a new development of the polarizable continuum model to study the polarizability of halides at the water/air interface.

Table 11 *Liquid phase hyperpolarizability calculations*

	Method	$\beta_{//}^{SHG}(0.0428)$	γ^{SHG} *(0.0428)*
Levine and Bethea[63]	EFISH experiment	19.2	2134
Gubskaya & Kusalik[52]	Pt. Charge field/MP4	28.9	2677
Kongsted *et al.*[54,55]	CC2/Cont	3.31	7632
	CCSD/Cont	4.51	4784
	CC2/MM	12.85	2914
	CCSD/MM	12.22	2169
Jensen[49]	DFT/BP-GRAC	6.82	2180

2.7 Small Polyatomic Molecules. – The results of valence bond (VB) and localized MO approaches have suggested that the linear OCS molecule might have two different polarizability components normal to the molecular axis and also two distinct bending modes. Essentially, this arises from the substantially different C=O and C=S π-bonds. Sabzyan and Bamdad[71] have examined this question using RHF, CIS, MP4 *ab initio* methods and DFT(B3LYP) and find no isolated molecular orbitals and degeneracy consistent with the cylindrical symmetry of the overall Hamiltonian. Modifications to the usual VB theory are suggested.

Haskopoulos and Maroulis[72] have published a further study of ozone in which they investigate the effect of a symmetric stretch on the dipole polarizability (α) using *ab initio* and DFT methods with large basis sets specially designed for polarizability calculations. Karamanis and Maroulis[73] have also published a study of the $(CN)_2$ dimer in which they calculate the static α and γ tensors for the three linear dimers (CN–CN, CN–NC and NC–CN). Medved *et al.*[74] have investigated electron correlation and vibrational effects on the longitudinal nonlinear optical properties of C_2H_2, HCCF and C_2F_2 using MP2, coupled cluster and DFT (B3LYP,B98) methods. Saal and Ouamerali[75] have calculated the electronic and vibrational contributions to the polarizability and first hyperpolarizability of a number of substituted acetylenes (H–C≡C–X). In some cases (X = NH_2, SiH_3 and SH) it is predicted that the first hyperpolarizability is almost entirely due to the vibrational effect.

Karamanis and Maroulis[76] report the static α and γ tensors for diacetylene (H–C≡C–C≡C–H). They have previously studied the α and β tensors for fluorodiacetylene (F–C≡C–C≡C–H),[77] which is noncentrosymmetric. The values obtained for the two molecules using the CCSD(T) method are shown in Table 12. The basis sets are discussed in the previous SPR and in the papers.

There is a striking similarity between the α tensors in the two molecules. The variation of the properties with the bond lengths has also been investigated and it is found that the dependencies on the single and triple bonds are markedly different.

The response functions of short conjugated chains (substituted or unsubstituted) have been the subject of a number of studies: Cao and Lin[78] (alkenes), Park and Cho[79] (triazine), Choytun *et al.*[80] (donor/acceptor substituted azines), Rosseto *et al.*[81] (substituted alkynes).

A knowledge of the frequency-dependent polarizability functions and their gradients allows the Raman scattering cross sections and depolarization ratios

Table 12 *Polarizability of diacetylene and fluorodiacetylene.*[76,77] $\bar{\alpha} = (\alpha_{zz} + 2\alpha_{xx})$, $\bar{\beta} = \frac{3}{5}(\beta_{zzz} + 2\beta_{zxx})$, $\bar{\gamma} = \frac{1}{15}(3\gamma_{zzzz} + 8\gamma_{xxxx} + 12\gamma_{xxzz})$. *All values in au*

	α_{zz}	α_{XX}	$\bar{\alpha}$	$\bar{\beta}$	$\bar{\gamma}$
HC≡CH	85.40	30.95	49.10	0	15 151
FC≡CH	86.04	30.35	48.91	−235.05	Not calculated

to be calculated – at least within the semi-classical Placzek theory. Vidal and Vasquez[82] examine the methodology of such calculations and explain how the polarizability gradients can be derived from the dynamic polarizabilities calculated at the excitation frequencies using *ab initio* response theory. They demonstrate their method in HF, MP2 and CCSD level applications to H_2O and NH_3. They confirm that high level correlated calculations are necessary to obtain satisfactory agreement between theory and experiment. Rissi *et al.*[83] have carried out *ab initio* calculations on an acetonitrile molecule hydrogen bonded to a water molecule ($CH_3CN–H_2O$) They find that $\bar{\alpha}$ has the value 39.67 au and the polarizability anisotropy is 21.78 au. The changes in intramolecular vibrational frequencies are analysed as are the Rayleigh and Raman light scattering activities. Substantial changes in the Raman intensities and depolarization ratios are predicted.

Quinet *et al.*,[84] in have interpreted their measurements of the hyper-Raman spectrum of carbon tetrachloride through *ab initio* TDHF simulations and find satisfactory agreement between theory and experiment.

Scattering measurements of various kinds in gases also provide information on collision induced effects that depend on multipolar polarizabilities. Molecular modelling calculations are increasingly used to interpret these experiments. For example, Bancewicz *et al.*[85] have determined values of the dipole-quadrupole and dipole octupole polarizabilities for methane from experimental studies of the Rayleigh scattering. Recent *ab initio* calculations are in agreement with their values. Collision induced light scattering in osmium tetroxide has provided values for the same two response functions and the high level DFT calculations of Holm and Maroulis[86] are in agreement with the results. El-Kader *et al.*[87] have measured the dipole-octupole polarizability of SF_6 from collision-induced rotational Raman effect and obtain a result in good agreement with recent calculations.

When attempting to treat ensembles of molecules it is essential to be able to represent the intermolecular interactions, at least beyond some cut-off distance, by pre-determined functions of distance so that high level quantum mechanical calculations do not have to be extended to larger and larger clusters. Masia *et al.*[88] have addressed the question of how the interaction of a water or carbon tetrachloride molecule interacting with a point charge can be represented. They monitor the molecular dipole as a function of the charge-molecule distance for various orientations of the molecule as calculated by *ab initio* methods. They find that the most satisfactory method is to represent the molecule by a small number of induced point dipoles with different orientations. In the case of water the *ab initio* induced dipoles are reproduced at all distances.

Lamanna *et al.*[89,90] attempt to find suitable STO basis sets for the computation of the response functions, taking the first and second hyperpolarizabilities of H_2O, CH_4 and NH_3 as calculated by TDHF theory as examples.

2.8 Medium Sized Organic Molecules. – 4-nitroaniline (pNA), the prototype for the enormous number of organic π-conjugated intramolecular donor/acceptor molecules that have been studied in the expectation of finding large

Table 13 *Comparison of the electronic and ZPVA contributions to the SHG hyperpolarizability of 4-nitroaniline, from ref. 91*

$\hbar\omega/au$	$\beta^{el}_{//}(-2\omega;\omega,\omega)/au$	$\beta^{ZPVA}_{//}(-2\omega;\omega,\omega)/au$
0	625.88	−27.92
0.04	809.67	−23.33
0.08	2658.76	465.47

β-values, has been the subject of further scrutiny. The ZPVA (Zero Point Vibrational Average) contribution to $\beta(-2\omega;\ \omega,\ \omega)$ has been calculated by Quinet *et al.*[91] using analytical methods and the (2n + 1)-rule at the TDHF level with a 6-31G basis set. A comparison of the electronic and ZPVA contributions at three frequencies is shown in Table 13.

At a frequency of 0.04 au (approximately equal to the YAG laser frequency (1.064 nm)) the ZPVA contribution is about 3% of the electronic value, but the ratio rises to 17% at twice this frequency. Rashid[92] has questioned the assumptions made in previous work about the geometry of pNA and has carried out optimizations of the structure of the free molecule in the ground and charge-transfer excited state. He finds that the amino group is much more pyramidal as compared with its nearly planar form in the crystal. The dipole moment in the charge transfer state does not increase by as much as it does in the planar structure and it would therefore be expected that the hyperpolarizability would be reduced. It is unclear which form is prevalent in solutions but future studies should take the possible variations into account. The paper contains useful references to previous semi-empirical and *ab initio* calculations on pNA. Wang *et al.*[93] have also investigated the response of pNA in solution, optimizing the molecular structure in each solvent environment (represented by a continuum model) and calculating the first hyperpolarizability by the TDHF method. As in earlier work they have attempted to represent the calculated dispersion by a two state formula. They find substantial differences in the hyperpolarizabilities in different solvents. Cammi *et al.*[94] have investigated the linear response of pNA in solution at the MCSCF level using a polarisable continuum model (PCM) to represent the solvent effects. The model allows the dynamics of the continuum to be taken into account(non-equilibrium PCM) so that time–dependent effects and, in particular, excitation energies can be properly treated. The behaviour of the molecule can be rationalized by including the effect of the zwitterion structure.

A systematic study of the effects of amine donors and conjugation length in a series of dipolar molecules using the CPHF/6-31G method has been conducted by Park *et al.*.[95] There is a variation in the order of merit of the donors in increasing β when the conjugation length is increased as compared with pNA.

The relative importance of electronic and vibrational effects has been a prominent topic in recent years, but reliable calculations of the vibrational contributions to the hyperpolarizabilities can only be achieved by using correlated methods with fairly large basis sets and time consuming computational procedures. The extension of such computations to medium sized organic molecules is only beginning. Torrent-Sucarrat *et al.*[96] have attempted to carry out such a study on three conjugated linear molecules (structures **I**). They have calculated the electronic and vibrational parts of the longitudinal α, β and γ with several frequency dependences. Following the procedures of Bishop and Kirtman the vibrational contribution can be written as,

$$P = P^{nr} + P^{c-ZPVA} + P^{ZPVA} \tag{12}$$

where the nuclear relaxation term, P^{nr}, contains the lowest order anharmonicity terms, P^{ZPVA} is the zero point vibrational average and P^{c-ZPVA} contains all other contributions. When low frequency fields are involves the P^{nr}term makes the largest contribution and is often comparable or greater than the electronic contribution. Some of the results are shown in Table 14.

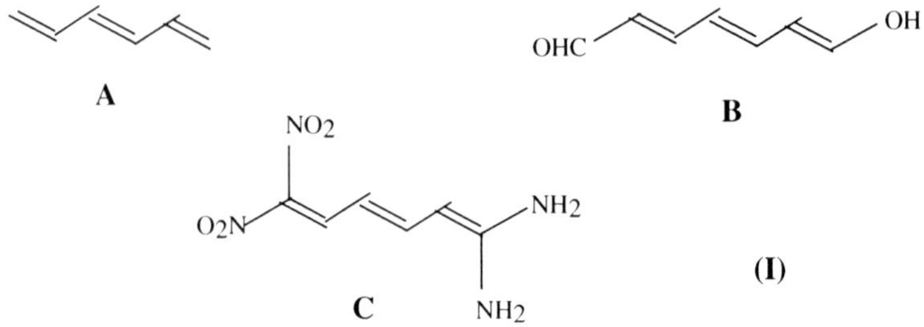

(I)

Only the static values of the electronic contributions are available at the same level of theory. These are compared with the static, electro-optic effect β and field induced SHG γ, which are experimentally relevant quantities.

In compound C the static nuclear relaxation term is nearly seven times greater than the static electronic term, although in the frequency-dependent effects its value is greatly reduced while the electronic effect is expected to get substantially greater. Generally the correlated and HF vibrational values show similar trends; the differences for the electronic contributions are much greater.

Suitieri[97] has also evaluated the anharmonic contributions to the nuclear relaxation γ for some push-pull polyenes using analytical methods in a valence bond charge transfer model. Saal and Ouamerali[98] have investigated the vibrational β of N-fluorophemyl-2,5-dimethypyrrole in the double harmonic

Table 14 *Electronic and nuclear relaxation contributions to the polarizabilities and hyperpolarizabilities of structures* **I**, *ref 96. Static and infinite frequency limit values are given for* $\beta^{nr}(-\omega;\omega,0)$

	A		B		C	
	HF	*MP2*	*HF*	*MP2*	*HF*	*MP2*
$\alpha^{e}(0,0)/10^2$	1.41	1.45	2.15	2.15	2.94	3.69
$\alpha^{nr}(0,0)/10^2$	0.0379	0.0255	0.420	0.296	3.44	2.34
$\beta^{e}(0;0,0)/10^3$			1.79	4.24	2.56	10.8
$\beta^{nr}(-0;0,0)/10^3$			4.50	3.29	112	68.8
$\beta^{nr}(-\omega;\omega,0)/10^3$			1.22	1.06	8.26	10.3

oscillator approximation at HF/6-31G level and find that its contribution is less than 10% of the total value.

A discrepancy between experimental and theoretical values for the nonlinear polarizabilities of the heavier members of the series of furan homologues (C_4H_4X, X = O, S, Se,Te) has attracted some interest. Jansik *et al.*[99] have calculated dipole moments, α, β and γ response functions including relativistic effects at three levels of theory: time-dependent Dirac-Hartree-Fock (TDDHF), TDHF with a Douglas-Kroll transformed one-component Hamiltonian and TDHF using effective core potentials. It is concluded that relativistic effects are describable with comparable accuracy at all three levels and that non-scalar effects (which are included only at time-dependent DHF level) are of minor importance, while frequency dispersion and relativistic contributions are found to be additive at the single-determinant level. However, the authors conclude that neither the relativistic corrections nor the inclusion of vibrational effects can account for the discrepancies between theory and experiment. Qin *et al.*[100] have calculated frequency-dependent α and γ response functions for furan homologues (C_4H_4X, X = O, S, Se,Te) using TDDFT and interpreted the results in terms of a two state model. The second hyperpolarizability of pyrrole homologues (C_4H_4XH,X = N, P, As, Sb) has been studied by the same group[101] and using *ab initio* HF + MP2 and TDDFT. In both cases the second hyperpolarizability increases markedly with the atomic number of the heteroatom. Fu *et al.*[102,103] have found a similar heavy atom effect in squaric acid homologues. Wang *et al.*[104] have optimized the structures of 1,3-substituted squairaines at 6-31G* level and calculated the electronic spectra and the first hyperpolarizability using CPHF, again with the 6-31G* basis set and by a variety of semi-empirical methods. The results indicated that the hyperpolarizability is greatly affected when five membered hetero-rings are introduced. Yue *et al.*[105] have investigated squaraine derivatives linked with borazine.

The effects of heavy atoms and a consideration of relativistic effects are also the subject of an investigation of the cyclopentadiene homologous series, $C_4H_4XH_2$(X = C, Si, Ge, Sn) by Alparone *et al.*[106,107] The α and γ responses are found to increase monotonically with the atomic number of the heavy atom. Vibrational contributions are also calculated and compared with calculations on furan.

Park and Cho[108] have found that the first hyperpolarizabilities of 1,3,5-tricyano-2,4,6-tris(styryl)benzene derivatives attain a maximum value of 262 $\times$ 10^{-30} esu when an oxyanion is used as donor. A correlation with the Hammett constants is explored.

Lanata *et al.*[109] have argued that molecules with quinoid ground states are particularly useful for the production of large linearities on account of the conjugation through the whole of the molecule. In certain cases, such as the merocyanine dyes in solution (see below in the section on open shell and ionic structures) the zwitterionic forms are more favourable.

Acebal *et al.*[110] have used *ab initio* correlated methods to demonstrate that the oxidation of Donor-π-Donor might be an effective mechanism for producing high values of β.

Mendis and de Silva[111] have investigated a class of novel charge transfer molecules (**II**) using the semi-empirical MOPAC6.0/AM1 method. Large values of the static β are found which depend critically on the twist angle −1234.

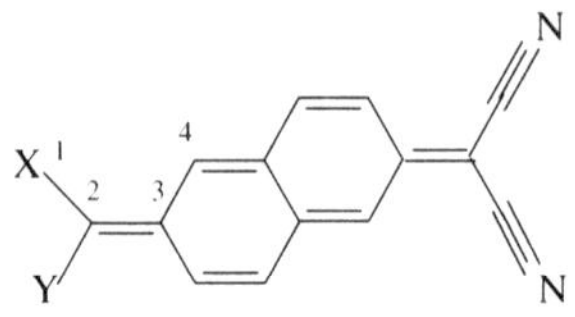

(II)

De Silva and co-workers[112,113] have made theoretical investigations of self-organising donor-acceptor aromatic systems.

A large number of reports have been published of calculations on medium sized organic molecules, often using *ab initio* methods (with basis sets roughly of a size inversely proportional to the size of the molecule) and sometimes supplemented by semi-empirical calculations, especially for frequency dependent properties:-

N-(3-fluorophenyl)naphthaldimide,[114] donor/acceptor substituted Schiff bases[115,116] and barbituric acid derivatives substituted with a Schiff base[117] or aniline,[118] 2-hydroxy-3-methoxy-N-(2-chloro-benzyl)-benzaldehyde-imine,[119] derivatives of salicylaldimine based ligands,[120] centrosymmetric pyrazine derivatives,[121] substituted stilbene, azoarene and related push-pull molecules,[122] effects of fluorine atoms on nonlinear response of stilbene derivatives,[123] s-triazine derivatives,[124] pyrazolo-quinoline derivatives,[125] 2-(4′-amino-2′-hydroxyphenyl)-6-nitrobenzmidazole(LEN),[126] tetrathia-[7]-helicenes,[127] monocyclic azines,[128] 26 derivatives of 1,5-diphenylpenta-2,4-dien-1-one (DDO) and 18 chalcones,[129] 2,3-diketobenzopiperazine,[130] hemicyanine derivatives,[131] tautomeric forms of uracil[132] and its resonance Raman spectrum,[133] methylene blue, thionine and coumarin dyes,[134] benzothiaziole derived push-pull dyes[135] donor/acceptor systems containing 1,3-heteroaromatic π-bridging units(oxazole, imidazole and thiazole),[136–138] TPA materials(diphenylacetylene, stilbene and azobenzene),[139,140] azulene derivatives[141] N-helicenes and N-phenylenes,[142] push-pull porphyrins,[143] azo-enaminone compounds,[144] pyrromethene dyes,[145] furamic and thiophenic ethane-1,2-diols.[146]

Mandal *et al.*[147] have made a TDHF study of α and β for amino and nitro substituted chromophores containing two hetero aromatic rings. The dependence on the twist angle in the bridging rings has been explored. Qin *et al.*[148] have employed AM1/FF and ZINDO/S-CI methods to design new chromophores with low ground state dipole and large β. The low dipole is thought to favour crystallization in non-centrosymmetric form.

The β-hyperpolarizability of octupolar molecules has continued to be a subject of study by Zyss and co-workers. The sulphthalocyanines[149] are an example with potential applications.

2.9 Organo-Metallic Complexes. – A series of ferrocenyl complexes with the ferrocenyl group as electron donor and 2-dicyanomethylene-3-cyano-4-methyl-2,5-dihydrofuran (TCF) derivatives as acceptors has been investigated by Liao *et al.*[150] The hyper–Rayleigh β has been experimentally determined (relative to pNA) and compared with DFT calculations. Agreement between experiment and theory, in systems with a constant donor and varying bridges and acceptors, has confirmed that a linear relation exists between the hyperpolarizability and the bond length alternation in the π-bonded bridges. One of the complexes incorporated into a polymer matrix gives rise to an electro-optic coefficient, r_{33} = 25 pm V^{-1}, which is larger than previously reported values.

Elmali *et al.*[151] have synthesized a Schiff base-Fe(III) complex which gives strong solvatochromism, implying that β might be large. Finite field MP2 calculations have been used to compute its value. Other work includes studies of Schiff base-Zn(II) complexes,[152] work on effective core potentials(ECP)[153] and on Schiff base Ni(II) complexes.[154] Insuasty *et al.*[155] have studied electronic and structural effects on the nonlinear optical behaviour of push-pull TTF/tricarbonyl chromium arene complexes; copper(II)-gadolinium complexes are the basis of work by Marget *et al.*[156]; Xue *et al.*[157] have studied metal complexes containing phenothiazine ligand. Johansson[158] has made *ab initio* calculations of the Raman spectra of pyridine and pyridine –metal complexes with silver. Ohnishi *et al.*[159] have used the elongation FF method to calculate α and β for donor/acceptor substituted polymer chains and for some block copolymers. Complexes of ruthenium(II) with ammine donor and N-methylpyridinium acceptor[160] and tetra-aza coordinated nickel complexes[161] have also been investigated.

2.10 Open Shells and Ionic Structures. – Nakano *et al.*[162,163] have previously investigated the effect of increasing the bond length in H_2 on its γ-hyperpolarizability. As the bond length is increased the electronic configuration changes from a predominantly covalent single determinant to an open shell ionic function and it is found that at intermediate lengths, where both structures make comparable contributions and their proportions are easily changed by the application of an electric field, γ attains a maximum value. This intermediate region is associated with bond breaking. More recently Nakano et al[164] have investigated the C_5H_7 radical in the doublet, quartet and sextet spin states. They find that γ increases with the spin multiplicity and that the occurrence of such states can again be related to bond breaking. The mixture of ionic and covalent states in these intermediate regions can only be properly represented by methods that give an adequate account of the electron correlation. With these considerations in mind, Nakano *et al.*[165] have investigated the p-quinodimethane molecule (structure **III**) when the R_1 bond length is varied subject to arbitrary constraints on the ring bond lengths, R_2 and R_3. While uncorrelated and MP2 methods show $|\gamma|$ values increasing monotonically with the diradical character the unrestricted coupled cluster method (UCCSD(T)) indicates, as would be expected by comparison with the H_2 results, that it attains a maximum value in the intermediate region. It is pointed out that twisted ethylene is also expected to show this effect and that there are possibilities for simulating it in other organic structures and thereby enhancing γ-values for potential applications.

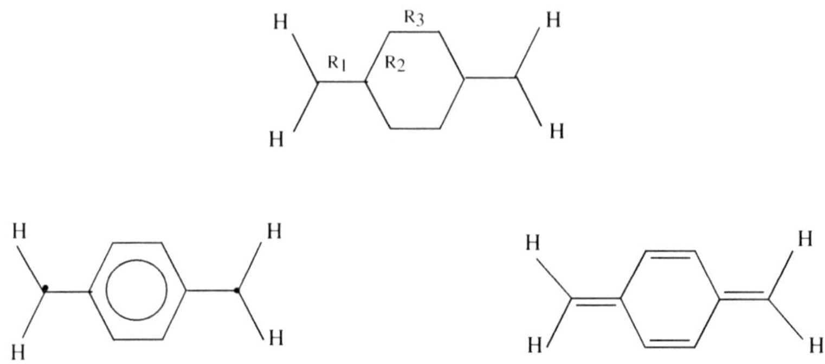

Diradical Quinoid

(III)

While diradical systems may be primarily of interest in connection with γ, zwitterionic molecules have long been investigated as a source of high β materials. A study by Ray[166] has addressed the structures of merocyanine dyes in solvents of increasing polarity (represented through the dielectric constant in a continuum model) and found that for the more polar solvents the ground states are zwitterionic, in contrast to the gas phase where calculations have predicted quinoidal forms. He shows, using TDHF/SOS procedures that the prevalence of the zwitterionic forms leads to a reversal of the sign of the principal component of β and a large increase in its magnitude. Orimoto and Aoki[167] predict very large values of β some zwitterionic molecules with a σ-bonded donor-acceptor system. Geskin *et al.*[168] have performed MP2/6-31G structural optimizations and INDO/SOS and finite-field calculations of the NLO response functions on ammonio/borato diphenyl polyene zwitterions and find that the phenylene groups are responsible for the high response. Calculations on salts includes work on lithium bis[1,2-benzodiolato-O,O'] borate.[169] Charged push-pull polyenes in solution have been investigated by Laage *et al.*[170]

Botek and Champagne[171] have devised and implemented a semi-empirical TDUHF scheme based on the AM1 hamiltonian in the MOPAC 2000 package. The procedure has been tested on some standard small and medium sized open shell molecular systems. There are discrepancies with good *ab initio* calculations on the small systems, but it is hoped that the procedure may be of use for larger open shell molecules.

Ray and Leszcynski[172] have investigated the hyperpolarizabilities of ionic octupolar compounds including solvent effects and Ray[173] has also analysed the molecular structure and β hyperpolarizabilities of a series of ionic organic and organometallic compounds.

2.11 Clusters, Intermolecular and Solvent Eeffects, Fullerenes, Nanotubes. – π-stacked dimers and trimers of 2-methyl-4-nitroaniline(mNA) have been investigated using TDDFT/B3LYP/3-21G+ and SOS methods[174]

A combination of HF and MM methods has been used by Tongraar and Rode[175] to investigate anion-water hydrogen bonds in aqueous solution. Chen *et al.*[176] have used *ab initio* methods to study the β–hyperpolarizability of $(HCN)_nLi$ clusters, Datta and Pati[177] on $Al_{14}M_4$ (M = Li,Na, K), Zhang *et al.*[178] and Pouchan *et al.*[179] on Si-9 and Si-12. Chandrakumar *et al.*[180] have used *ab initio* results as the basis for establishing relationships between the ionization potential, polarizability and softness of Li and Na metal clusters. Wang *et al.*[181] have calculated the static α for ((Cu-n, Ag-n and Au-n clusters, n ≤ 9).

Investigations of fullerenes have included calculations of the static α for $B_{36}N_{36}$,[182] $C_{50}Cl_{110}$,[183] $C_{48}N_{12}$,[184] C-36 structures($C_{34}X_2$,=B,N);[185] size scaling of the tubular fullerenes using TDDFT,[186] static polarizability of carbon nanotubes,[187,188] TPA in C-60,[189] dielectric response of carbon fullerenes up to 3840 atoms using TDDFT.[190]

Yamada *et al.* have continued their development of the theory of dendritic systems[191] (see also previous SPR).

2.12 One and Two Photon Absorption, Luminescence etc. – Papers primarily concerned with the calculation of spectroscopic properties closely related to the polarizability and hyperpolarizability are listed here:-

Zojer *et al.*[192] have discussed the limitations of the essential-state model in calculating the TPA of bis(dioxaborine-)-substituted chromophores. Lukes and Breza[193] have made an AM1-ZINDO/SOS study of spirobifluorene type molecules. Jha *et al.*[194] have employed their method of obtaining model exact solutions in the PPP approximation to calculate TPA cross-sections for *trans*-stilbene and 7,8-disubstituted stilbenes. Zhou *et al.*[195] and Liu *et al.*[196] have studied one and two photon absorption in octupolar compounds using DFT and ZINDO/SOS methods and the group has applied similar methods to trigonal dehydrobenzo[18]annulenes.[197]

Other work includes a study of 3D chromophores base on [2,2]-paracyclophane,[198] a new carbazole and a new amine, each containing phenyl and pyridine rings.[199]

2.13 Theoretical Developments. – In this section are listed a small number of papers which are more concerned with developing new theoretical methods than with obtaining results for particular substances.

Kobko *et al.*[200] have used a third order response function formalism with TDHF and TDDFT to assess different levels of theory for calculations of excited state structure and nonlinear optical responses in donor-donor and donor-acceptor π-conjugated molecules. They make suggestions for numerically efficient approximations.

The multi-reference coupled electron pair approximation (MRCEPA) has a long pedigree as one of the more rigorous quantum chemical procedures. A recent development, the state specific version (SS-MRCEPA) has been exploited by Chattopadhyay[201] who has, *inter alia*, used the method to calculate accurate an dipole moment and polarizability for the ground state and an excited state of trapezoidal H_4.

Moad and Simpson[202] have developed a method for simplifying the molecular interpretation of nonlinear optical and multiphoton phenomena in molecules. Essentially the method consists of re-writing general SOS perturbation formulae in terms of simple products of lower order effects. This can be done exactly without sacrificing mathematical rigour.

A correct form of the treatment of the long-range asymptotic behaviour of DFT is essential for the investigation of extended systems and this problem has been addressed by Baer and Neuhauser.[30] Jansik *et al.*[203] have extended the treatment time-dependent DFT to deal with cubic response functions and have applied their method to nitrogen, benzene and C-60 fullerene.

2.14 Oligomers and Polymers. – The distinction between oligomers and small or medium sized molecules is arbitrary, but the oligomers of conjugated polymer chains are often the subject of investigation as a preliminary stage in treating the polymer. Some of these studies are grouped in this section. They include work on excitation energies of π-conjugated oligomers within TDDFT.[204]

HF and DFT methods have been used by Smith *et al.*[205] to calculate the dipole polarizabilities, α, of the *all trans* polyenes up to $C_{18}H_{20}$, and polyacenes to $C_{18}H_{12}$ and their molecular ions. The static values and the dynamic response at 800 nm have been calculated. For the smaller molecules the results have been compared with those of correlated methods. It is found that the uncorrelated results are about 20% higher than those of the correlated methods for the neutral molecules but are very similar for the molecular ions. General conclusions about the scaling and frequency dependence of α values obtained by simpler methods are drawn.

Esteves *et al.*[206] have calculated the longitudinal polarizabilities of fluorinated polyacetylene chains and Shukla[207] has used semi-empirical methods to investigate the NLO properties of phenyl-substituted polyacetylenes. Torres *et al.*[208] have calculated the longitudinal α and γ for polydifluoroacetylene at MP2/6-31G level. Jacquemin and co-workers have investigated the geometry and β of polymethineimine including solvent effects and find that convergence of β with chain length is much more rapid than in the gas phase and that the final value is reduced by a factor of two.[209] The effects of electron correlation in polymethinemine have been assessed.[210] They have also studied polysila-acetylene.[211] Jacquemin[212] has also investigated the β-hyperpolarizability of polyaminoborane and polyiminoborane oligomers.

Ohnishi *et al.*[213] have used the elongation FF method to calculate α and β for donor/acceptor substituted polymer chains and for some block copolymers.

2.15 Molecules in Crystals. – Theoretical work on crystalline structures is not generally covered in this review, but papers predominantly concerned with crystal properties sometimes include new computations on the properties of the constituent molecules.

The particular problems involved in treating the polarizability (and *a fortiori* the hyperpolarizabilities) in molecular crystals are exhibited in work on acetanilide and 1,3,5-trinitro-1,3,5-triazacyclohexane (RDX) by Tsiaousis *et al.*[214] where DFT/B3LYP/6-311 ++ G** calculations show that the dipoles of the in-crystal molecular structures are respectively 21% and 15% greater than in the gas phase while the polarizabilities are about 3% smaller. However the in-crystal polarizabilities deduced from the crystal optical properties are larger than those of the gas phase. It therefore appears that effects other than those attributable to a molecular geometry change are present in the crystal.

Skwara *et al.*[215] have also discussed the effect of intermolecular interactions, represented through multipole moments, on the second order response of molecular crystals.

The well-known NLO molecular crystal POM (3-methyl-4-nitropyridine-1-oxide) is simulated through cluster calculations by Guillaume *et al.*[216] Semi-empirical and MP2 *ab initio* results are considered and comparisons of the NLO response with those obtained from the usual oriented gas model are made. POM is also selected by Hamzaoui *et al.*[217] as an example of an NLO molecular crystal on which to test their procedure for relating the polarizabilities to the multipolar components of the ground state charge distribution determined by X-ray diffraction methods.

The TeO_2, TeO_3 and TeO_4 structural units in TeO_2 based glasses are the subject of *ab initio* static finite field work by Suehara *et al.*[218] The results provide the basis for a study of the Raman spectra, and response functions of the glass.[219,220]

Work has also been reported on methyl 3(4-methoxy-phenyl)prop-2-enoate)[221]

3 Magnetic Effects

There have been several relevant general reviews[222–225]

3.1 Inert Gases, Atoms, Diatomics. – An analytical solution for the relativistic energy of a hydrogenic atom has allowed Rutkowski and Poszwa[226] to determine its magnetizability to very high accuracy.

The magnetizabilities and Cotton-Mouton effect constants in the inert gases continue to attract a great deal of attention, theoretically and experimentally. Precise agreement between theory and experiment for the Cotton-Mouton parameter is currently achievable only in these simple systems. New measurements for He, Ne and Ar[227,228] and for Kr and Xe[229] have been reported and a high level coupled-cluster study has produced new theoretical data for Ne and Ar.[230] Pecul *et al.*[231] have investigated the variation of magnetic shielding constants with electric field for all the inert gas atoms up to Xe. Relativistic 4-component HF values have been compared with non-relativistic results. Due to a partial cancellation of effects the mean shielding constant is far less affected by the relativistic corrections than the individual components. Pagola *et al.*[232]

have treated the effect of strong magnetic fields on the magnetizability of He, Ne, Ar and Kr. They find small positive values for the fourth rank hypermagnetizability so that the absolute magnitude of the diamagnetic magnetizability is slightly reduced.

Rizzo and Coriani[233] have used coupled cluster *ab initio* methods to calculate the Jones birefringence of He, Ne, Ar and Kr (and also for H_2, N_2, C_2H_2 and CO) and compare the values with the birefringence induced by the Cotton-Mouton effect. They find that the latter is between 100 and 3500 times larger than the former.

Hanni *et al.*[234] have investigated the magnetic properties of the Xe_2 dimer, which provides data for corrections to ideal inert gas measurements at finite pressure.

3.2 Molecular Magnetisabilities, Nuclear Shielding and Aromaticity, Gauge Invariance. – Prediction of NMR shielding constants to an accuracy that might be useful in interpreting experimental results is an objective that motivates a great deal of computational work. For larger molecules DFT methods are usually employed and a number of papers attempt to interpret the results of diamagnetic screening in terms of currents represented by current density maps. Discussions of aromaticity and anti-aromaticity are sometimes included where appropriate. There are several popular methods for attempting to deal with gauge invariance. The work of Lazzeretti and collaborators has been prominent in the exploration of alternative methods. In connection with magnetic properties it is of crucial importance to establish that gauge invariance has been achieved to a reasonable approximation and the Lazzeretti group have developed sum rules for third rank tensor magnetic properties and applied them to make numerical checks on their wave-functions.[235,236] In recent work[237–239] the relationships between computed current densities, magnetizability, nuclear shielding and aromaticity have been investigated for molecules consisting of two fused five membered rings (bis-heteropentalenes). Results obtained using several methods of obtaining approximate gauge invariance have been compared. The outcome of the study shows that the ring current and the magnetizability and shielding constants are well correlated, but that there is no connection with aromaticity defined in terms of the relative stability of the molecules. Faglioni *et al.*[240] discuss this conclusion in more general terms. Pelloni *et al.*[241] have investigated dithiines, which are formally anti-aromatic, but find no evidence of this in the magnetic properties or the induced current density topology. The Lazzeretti group has recently utilized large gaugeless basis sets with a common origin for the vector potential. If the basis set is large enough for the wave function to approach the exact solution closely enough then gauge invariance is automatically satisfied. This approach has been adopted in work by Pagola *et al.*[242,243] on calculations of the fourth rank hypermagnetisability of some small molecules and benzene.

Fowler, Steiner *et al.*[244,245] have used a specific method of selecting the origin of the vector potential (the ipsocentric choice, where the induced current density at each point is calculated with that point as origin) to develop a method of analysing induced current densities in π-electron systems. The method has been applied to a study of the current densities and aromaticity of a set of annelated pentalenes.

The search for better DFT functionals for the calculation of magnetic properties (and particularly NMR shielding factors) continues. Arbuznikov and Kaupp[246–249] are concerned with the self-consistent implementation of exchange-correlation functionals depending on the local kinetic energy density. Mothana *et al.*[250] have compared ^{15}N and ^{13}C shielding constants calculated with a number of functionals and basis sets. Migda and Rys[251] have used a GIAO/DFT method to evaluate the ^{13}C chemical shifts for some acetals.

Nonlinear magnetic effects in molecules have been investigated by Rizzo *et al.*[252] who have compared the results from a number of DFT functionals with those from HF and coupled-cluster methods for calculations of a range of field induced birefringences. (Kerr, Cotton-Mouton, Buckingham, Jones and magnetoelectric) in gaseous benzene and hexafluorobenzene.[253] Cappelli *et al.*[254] have used DFT to calculate the tensor components of the polarizability and the magnetizability at 632 nm and the static hypermagnetizability and, using a polarisable continuum model, have obtained the Cotton-Mouton constant in liquid and gas phases for furan, thiophene and selenophene. They compare the computed values with experimental values obtained at the Grenoble High magnetic Field laboratory.

The work of Pagola *et al.*,[242,243] on the 4th rank hyperpolarizability of benzene, has been mentioned above in connection with methods of treating gauge invariance. The calculation was carried out the at the coupled-HF level of approximation adopting the common origin approach to gauge invariance using gaugeless basis sets of increasing size and flexibility.

Hoggan[255,256] advocates the use of hydrogenic atomic orbitals for the calculation of sensitive molecular properties such as NMR chemical shifts. The correct shielding of the nucleus, as given by the radial hydrogenic factors, is essential. He lists the packages available for STOs and other combinations of exponentially decaying functions.

A number of magnetic properties have been calculated *inter alia* in a MCSCF study of the ground state properties of molecular oxygen by Minaev.[23]

Solvent effects on NMR shielding have been discussed by Cossi and Crescenzi.[257,258]

The magnetizabilities and shielding tensor for [8]-cyclacene and derivatives and the shielding tensor of a ^{3}He atom passing through the ring have been calculated using a DFT method by Tonmunphean *et al.*[259] and the ring currents and magnetic properties of some fullerenes by Viglione and Zanasi.[260]

References

1. G. Maroulis (Editor), *J. Comp. Meth. Sci. Eng.*, 2004, **4**, issues 3 and 4.
2. V.E. Chernov, D.L. Dorofeev, I.Y. Kretinin and B.A. Zon, *Phys. Rev. A*, 2005, **71**, 022505.
3. C. Lupinetti and A.J. Thakkar, *J. Chem. Phys.*, 2005, **122**, 044301.

4. P. Fuentealba, *Chem. Phys. Lett.*, 2004, **397**, 459.
5. A.J. Thakkar and C. Lupinetti, *Chem. Phys. Lett.*, 2005, **402**, 270.
6. G. Maroulis and C. Pouchan, *J. Phys. B Mol. Opt. Phys.*, 2003, **36**, 2011.
7. F. Pawlowski, P. Jorgensen and C. Hattig, *Chem. Phys. Lett.*, 2004, **391**, 27.
8. W. Klopper, S. Coriani, T. Helgaker and P. Jorgensen, *J. Phys. B-At. Mol. Phys.*, 2004, **37**, 3753.
9. D.L.P. Chong, M. Gruning and E.J. Baerends, *J. Comp. Chem.*, 2003, **24**, 1582.
10. C.Y. Ren and C.S. Hsue, *Chin J. Phys.*, 2004, **42**, 162.
11. M.N. Adamov, Y.B. Malykhanov, V.V. Meshkov and R.M. Chadin, *Opt. Spectrosc.*, 2004, **96**, 192.
12. I.S. Lim and P. Schwerdtfeger, *Phy. Rev. A*, 2004, **70**, 062501.
13. C. Clavaguera and J.P. Dognon, *Chem. Phys.*, 2005, **311**, 169.
14. F.J. Ding and L.X. Liu, *Acta Chim. Sin.*, 2004, **62**, 471.
15. L.X. Liu and F.J. Ding, *Acta Chim. Sin.*, 2004, **62**, 1607.
16. D.C. Ghosh and R. Biswas, *Int. J. Mol. Sci.*, 2003, **4**, 379.
17. J. Kobus, D. Moncrieff and S. Wilson, *J. Phys. B-At. Mol. Opt. Phys.*, 2004, **37**, 571.
18. G. Maroulis, *Theochem-J. Mol. Struct.*, 2003, **633**, 177.
19. G. Maroulis, *J. Chem. Phys.*, 2004, **121**, 10519.
20. G. Maroulis, *J. Phys. Chem.*, 2003, **107**, 6495.
21. M. Merawa, D. Begue and A. Dargelos, *J. Phys. Chem. A*, 2003, **107**, 9628.
22. M. Merawa, D. Begue and C. Pouchan, *Theochem-J. Mol. Struct.*, 2003, **633**, 157.
23. B.F. Minaev, *Spectroc.Acta A*, 2004, **60**, 1027.
24. T.J. Geise and D.M. York, *J. Chem. Phys.*, 2004, **120**, 7939.
25. O.P. Andrade, A. Aragao, O.A.V. Amaral, T.L. Fonseca and M.A. Castro, *Chem. Phys. Lett.*, 2004, **392**, 270.
26. M. Ilias, V. Kello, T. Fleig and M. Urban, *Theor. Chem. Acc.*, 2003, **110**, 176.
27. P. Norman and H.J.A. Jensen, *J. Chem. Phys.*, 2004, **121**, 6145.
28. F. Wang and W.J. Liu, *Chem. Phys.*, 2005, **311**, 63.
29. P. Salek, T. Helgaker and T. Saue, *Chemical Physics*, 2005, **311**, 187.
30. R. Bauer and D. Neuhauser, *Phys. Rev. Lett.*, 2005, **94**, 043002.
31. G. Maroulis, A. Haskopoulos and D. Xenides, *Chem. Phys. Lett.*, 2004, **396**, 59.
32. A. Haskopoulos, D. Xenides and G. Maroulis, *Chem. Phys.*, 2005, **309**, 271.
33. C. Hattig, J.L. Cacheiro, B. Fernandez and A. Rizzo, *Mol. Phys.*, 2003, **101**, 1983.
34. L. Kriachtchev, M. Pettersson, N. Runeberg, J. Lundel and M. Rasanen, *Nature*, 2000, **406**, 874.
35. N. Runeberg, M. Pettersson, L. Kriachtchev, J. Lundel and M. Rasanen, *J. Chem. Phys.*, 2001, **114**, 836.
36. A. Avramopoulos, H. Reis, J. Li and M.G. Papadopoulos, *J. Am. Chem. Soc.*, 2004, **126**, 6179.
37. J. Li and R. McWeeny, *Int. J. Quant. Chem.*, 2002, **89**, 208.
38. A.C. McDowell, *Chem. Phys. Lett.*, **396**, 346.
39. G.D. Zeiss and W.J. Meath, *Mol. Phys.*, 1975, **30**, 161.
40. C. Cuthbertson and M. Cuthbertson, *Phil. Trans. Roy. Soc. A.*, 1913, **113**, 1.
41. Russell and Spackman, *Mol. Phys.*, 1995, **84**, 1239.
42. G. Maroulis, *Chem. Phys. Lett*, 1998, **289**, 403.
43. H. Sienko and R.J. Bartlett, *J. Chem. Phys.*, 1993, **98**, 3022.
44. O. Christiansen, J. Gauss and J.F. Stanton, *Chem. Phys. Lett.*, 1999, **305**, 147.
45. J. Kongsted, A. Osted, K.V. Mikkelsen and O. Christiansen, *J. Chem. Phys.*, 2003, **118**, 1620.

46. T.D. Poulsen, P.R. Ogilby and K.V. Mikkelsen, *J. Chem. Phys.*, 2002, **116**, 3730.
47. A. Osted, J. Kongsted, K.V. Mikkelsen and O. Christiansen, *J. Phys. Chem.A*, 2004, **108**, 8646.
48. L. Jensen, P.Th. van Duijnen and J.G. Snijders, *J. Chem. Phys.*, 2003, **119**, 3800.
49. L. Jensen, M. Swart and P.Th. van Duijnen, *J. Chem. Phys.*, 2005, **122**, 034103.
50. Murphy, *J. Chem. Phys.*, 1977, **67**, 5877.
51. K. Thomsen and P. Swanstróm, *Mol. Phys.*, 1973, **26**, 751.
52. A.V. Gubskaya and P.G. Kusalik, *Mol. Phys.*, 2001, **99**, 1107.
53. K.O. Sylvester-Hvild, K.V. Mikkelsen, P. Norman, D. Jonsson and H. Ågren, *J. Phys. Chem. A.*, 2004, **108**, 8961.
54. J. Kongsted, A. Osted, K.V. Mikkelsen and O. Christiansen, *J. Chem. Phys.*, 2003, **119**, 10519.
55. J. Kongsted, A. Osted and K.V. Mikkelsen, *J. Chem. Phys.*, 2004, **120**, 3787.
56. O. Christiansen, P. Jorgensen and C. Hattig, *Int. J. Quantum Chem.*, 1998, **68**, 1.
57. T. Helgaker and P. Jorgensen, *Theoret. Chim. Acta*, 1989, **75**, 111.
58. H. Koch, J.J.A. Jensen, P. Jorgensen, T. Helgaker, G.E. Scuseria and H.F. Schaefer, *J. Chem. Phys.*, 1990, **92**, 4924.
59. M. Born and E. Wolf, *Principles of Optics, 6th Edition, CUP*,1997.
60. A. Willets, J.E. Rice, D.M. Burland and D.P. Shelton, *J. Chem. Phys.*, 1992, **97**, 7590.
61. J.F. Ward and K. Miller, *Phys. Rev. A.*, 1979, **19**, 826.
62. P. Kaatz, E.A. Donley and D.P. Shelton, *J. Chem. Phys.*, 1998, **108**, 849.
63. B.F. Levine and C.G. Bethea, *J. Chem. Phys.*, 1976, **65**, 2429.
64. F. Kajzar and J. Messier, *Phys. Rev. A.*, 1985, **32**, 2352.
65. D.M. Bishop, B. Kirtman, H.A. Kurtz and J.E. Rice, *J. Chem. Phys.*, 1993, **98**, 8024.
66. Y. Luo, H. Agren, O. Vahtras, P. Jorgensen, V. Spirko and H. Hettema, *J. Chem. Phys.*, 1993, **98**, 7159.
67. M.T. Sonoda, S.M. Vechi and M.S. Skaf, *PCCP*, 2005, **7**, 1176.
68. V. Weber, A.M.N. Niklasson and M. Challacombe, *Phys. Rev. Lett.*, 2004, **92**, 193002.
69. M.S. Skaf and S.M. Vechi, *J. Chem. Phys.*, 2003, **119**, 2181.
70. L. Frediani, B. Mennucci and R. Cammi, *J. Phys. Chem. B.*, 2004, **108**, 13796.
71. H. Sabzyan and M. Bamdad, *J. Mol. Structure (Theochem)*, 2004, **712**, 109.
72. A. Haskopoulos and G. Maroulis, *Chem. Phys. Lett.*, 2004, **397**, 253.
73. P. Karamanis and G. Maroulis, *Mol. Phys.*, 2004, **102**, 13.
74. M. Medved, J. Noga, D. Jacquemin and E.A. Perpete, *Int. J. Quantum Chem.*, 2005, **102**, 209.
75. A. Saal and O. Ouamerali, *Int. J. Quantum Chem.*, 2004, **96**, 333.
76. P. Karamanis and G. Maroulis, *Chem. Phys. Letters*, 2003, **376**, 403.
77. P. Karamanis and G. Maroulis, *J. Mol. Struct.(Theochem)*, 2003, **621**, 157.
78. C.Z. Cao and Y.B. Lin, *Chin. J. Chem. Phys.*, 2004, **17**, 703.
79. G. Park and B.R. Cho, *J. Phys. Org. Chem.*, 2004, **17**, 169.
80. D.D. Choytun, L.D. Langlois, T.P. Johansson, G.W. Leach, N. Weinberg and J.A.C. Clyburne, *Chem. Comm.*, 2004, **16**, 1842.
81. R. Rosseto, M.D. Vargas and J. Del Nero, *Int. J. Quantum Chem.*, 2003, **95**, 137.
82. L.N. Vidal and P.A.M. Vasquez, *Quim. Nova.*, 2003, **26**, 507.
83. E. Rissi, E.E. Fileti and S. Canuto, *Theor. Chem. Acc.*, 2003, **110**, 360.
84. O. Quinet, B. Champagne and V. Rodriguez, *J. Chem. Phys.*, 2004, **121**, 4705.
85. T. Bancewicz, K. Nowicka, J.L. Godet and Y. Le Duff, *Phys. Rev. A.*, 2004, **69(6)**.

86. U. Holm and G. Maroulis, *J. Chem. Phys.*, 2004, **121**, 10411.
87. M.S.A. El-Kader, S.M. El-Sheikh and M. Omran, *PCCP*, 2004, **218**, 1197.
88. M. Masia, M. Probst and R. Rey, *J. Chem. Phys.*, 2004, **121**, 7362.
89. U.T. Lamanna, C. Guidotti, N. Durante and G.P. Arrighini, *Theor. Chem. Acc.*, 2004, **111**, 414.
90. U.T. Lamanna, C. Guidotti, N. Durante, G.P. Arrighini and S. Bruzonne, *Theor. Chem. Acc.*, 2004, **112**, 135.
91. O. Quinet, B. Champagne and B. Kirtman, *J. Mol. Struct.(Theochem)*, 2003, **633**, 199.
92. A.N. Rashid, *J. Mol. Struct.*, 2004, **681**, 57.
93. Y.H. Wang, Y.H. Sun and C.K. Wang, *Chin J. Chem. Phys.*, 2004, **17**, 518.
94. R. Cammi, L. Frediani, B. Mennucci and K. Ruud, *J. Chem. Phys.*, 2003, **119**, 5818.
95. G. Park, C.S. Ra and B.R. Cho, *Bull. Korean Chem. Soc.*, 2003, **24**, 1671.
96. M. Torrent-Sucarrat, M. Sola, M. Duran, J.M. Luis and B. Kirtman, *J. Chem. Phys.*, 2004, **120**, 6346.
97. E. Suitieri, *J. Phys. Org. Chem.*, 2004, **17**, 131.
98. A. Saal and O. Ouamerali, *Struct. Chem.*, 2003, **14**, 479.
99. B. Jansik, B. Schimmelpfennig, P. Norman, P. Macak, H. Agren and K. Ohta, *J. Mol. Struct. (Theochem)*, 2003, **633**, 237.
100. C.S. Qin, G.C. Yang, Z.M. Su, Y.L. Zhu and Z.Y. Zhou, *Chem. J. Chime. Univ.-Chin.*, 2005, **26**, 290.
101. G.C. Yang, D. Shi, Z.M. Su and C.S. Qin, *Acta Chim. Sin.*, 2005, **63**, 184.
102. W.W. Fu, L.X. Zhou and H.P. Wan, *Chin. J. Struct. Chem.*, 2004, **23**, 480.
103. W.W. Fu and L.X. Zhou, *Chin. J. Chem.*, 2004, **22**, 333.
104. C.M. Wang, F.L. Bei, X.J. Yang, L.D. Lu and X. Wang, *Chin. J. Chem. Phys.*, 2005, **18**, 33.
105. S.M. Yue, K. Tan, M. Zhang, Y.Q. Lan and Z.M. Su, *Chem. J. Chin. Univ.-Chin.*, 2003, **24**, 2231.
106. A. Alparone, A. Millefiori and S. Millefiori, *Chem. Phys.*, 2004, **298**, 75.
107. A. Alparone, A. Millefiori and S. Millefiori, *J. Mol. Struct.*, 2003, **640**, 123.
108. G. Park and B.R. Cho, *J. Phys. Org. Chem.*, 2005, **18**, 264.
109. M. Lanata, C. Bertarelli, M.C. Gallazzi, A. Bianco, M. Del Zoppo and G. Zerbi, *Synth. Met.*, 2003, **138**, 357.
110. P. Acebal, S. Blaya and L. Carretero, *Chem. Phys. Lett.*, 2004, **394**, 76.
111. B.A.S. Mendis and K.M.N. de Silva, *J. Mol. Struct. (Theochem)*, 2004, **678**, 31.
112. A.S. Coorey and K.M.N. de Silva, *J. Mol. Struct(Theochem)*, 2004, **678**, 223.
113. A.K. Jeewandarw and K.M.N. de Silva, *J. Mol. Struct. (Theochem)*, 2004, **686**, 131.
114. H. Unver, A. Karakas, A. Elmani and T.N. Durfu, *J. Mol. Struc.*, 2005, **737**, 131.
115. A. Karakas, H. Unver and A. Elmali, *J. Mol. Struct(Theochem)*, 2004, **712**, 117.
116. I. Sheikshoaie, M. Hossein, Mashhadizadeh and S. Saeid-Nia, *J. Coord. Chem.*, 2004, **57**, 417.
117. G. Sun, Y.Q. Qiu, H.Z. Sun, Z.M. Su, J.D. Feng and Y.L. Zhu, *Chin. J. Chem.*, 2004, **22**, 425.
118. L. Zhao, Z.M. Su, Y.H. Kan, Y.L. Zhu, A.M. Ren and J.K. Feng, *Acta Chim. Sin.*, 2003, **61**, 1197.
119. H. Unver, A. Karakas and A. Elmali, *J. Mol. Struct.*, 2004, **702**, 49.
120. A. Karakas, A. Elmali, H. Unver and I. Svoboda, *J. Mol. Struct.*, 2004, **702**, 103.
121. Y.Q. Qiu, H.Y. Yu, Z.M. Su, L. Zhao, R.S. Wang and Y.L. Zhu, *Chem. J. Chin. Univ.-Chinese*, 2004, **25**, 2335.

122. A. Hinchliffe, B. Nikolaidi and H.J.S. Machado, *Int. J. Mol. Sci.*, 2004, **5**, 224.
123. P. Romaniello and F. Lelj, *J. Fluor. Chem.*, 2004, **125**, 145.
124. L. Han, M.Z. Xue, L. Xiang and H.B. Qian, *Opt. Mater.*, 2004, **27**, 235.
125. M. Makowska-Janusik, E. Gondek, I.V. Kityk, J. Wisla, J. Sanetra and A. Danel, *Chem. Phys.*, 2004, **306**, 265.
126. F.S. Rodembusch, T. Buckup, M. Segala, L. Tavares, R.R.B. Correia and V. Stefani, *Chem. Phys.*, 2004, **305**, 115.
127. B. Champagne, J.M. Andre, E. Botek, E. Licandro, S. Maiorana, A. Bossi, K. Clays and A. Persoons, *Chem. Phys. Chem.*, 2004, **5**, 1438.
128. H. Soscun, Y. Bermudez, O. Castellano and J. Hernandez, *Chem. Phys. Lett.*, 2004, **396**, 117.
129. D. Wu, B. Zhao and Z.H. Zhou, *J. Mol. Struct(Theochem)*, 2004, **682**, 83.
130. F.F. Jian and P.S. Zhao, *J. Mol. Struct.*, 2004, **705**, 133.
131. H.P. Li, K. Han and Q. Wang, *Acta Physico-Chimica Sinica*, 2004, **20**, 806.
132. S. Millefiori and A. Alparone, *Chem. Phys.*, 2004, **303**, 27.
133. J. Neugebauer and B.A. Hess, *J. Chem. Phys.*, 2004, **120**, 11564.
134. P. Acebal, S. Blaya and L. Carretero, *Bol. Soc. Esp. Ceram. Vidr.*, 2004, **43**, 467.
135. P. Hrobarik, P. Zahradnik and W.M.F. Fabian, *PCCP*, 2004, **6**, 495.
136. C.S. Ra, S.C. Kim and G. Park, *J. Mol. Struct.(Theochem)*, 2004, **677**, 173.
137. M. Spassova and V. Enchev, *Chem. Phys.*, 2004, **298**, 29.
138. G. Park and C.S. Ra, *Bull.Kor. Chem. Soc.*, 2003, **24**, 1051.
139. L. Yang, A.M. Ren, J.K. Fen, X.J. Liu and X. Zhou, *Chem. J. Chin. Univ-Chin.*, 2004, **25**, 526.
140. A.M. Ren, J.K. Feng and X.J. Liu, *Chin. J. Chem.*, 2004, **22**, 243.
141. Z. Xin, A. M. Ren, J.K. Fen, X.J. Liu and C.C. Shu, *Chin. J. Chem.*, 2004, **22**, 38.
142. E. Botek, B. Champagne, M. Turki and J.M. Andre, *J. Chem. Phys.*, 2004, **120**, 2042.
143. A.D. Tillekaratne, R.M. de Silva and K.M.N. de Silva, *J. Mol. Struct.(Theochem)*, 2003, **638**, 169.
144. H.C.B. de Oliveira, T.L. Fonseca, O.A.V. Amaral and S. Cunha, *J. Chem. Phys.*, 2003, **119**, 8417.
145. P. Acebal, S. Blaya and L. Carretero, *J. Phys. B.*, 2003, **36**, 2445.
146. V. Lukes, M. Breza, D. Vegh, P. Hrdlovic and V. Laurinc, *Synthet. Metal*, 2003, **138**, 399.
147. K. Mandal, T. Kar, F. Nandi and S.P. Battacharyya, *Chem. Phys. Lett.*, 2003, **376**, 116.
148. A.J. Qin, F.L. Bai and C. Ye, *J. Mol. Struct.(Theochem)*, 2003, **631**, 79.
149. C.G. Claesens, D. Gonzalex-Rodriguez, T. Torres, G. Martin, F. Agullo-Lopez, I. Ledoux, J. Zyss, V.R. Ferro and J.M. G de la Vega, *J. Phys. Chem. B.*, 2005, **109**, 3800.
150. Y. Liao, B.E. Eichinger, K.A. Firestone, M. Haller, J.D. Luo, W. Kaminsky, J.B. Benedict, P.J. Reid, A.K.Y. Jen, L.R. Dalton and B.H. Robinson, *JACS*, 2005, **127**, 2758.
151. A. Elmali, A. Karakas and H. Unver, *Chem. Phys.*, 2005, **309**, 251.
152. L.K. Yan, Z.M. Su, Y.H. Kan, Y.Q. Qiu, D.X. Zhu and Y. Wang, *Chem. J. Chin. Univ-Chin.*, 2003, **24**, 1876.
153. C.P. Liu, K.C. Wu, T. Hong, C.Y. Mang, M.X. Zhang, Z.F. Zhou and B.T. Zhuang, *Acta. Chim. Sin.*, 2005, **63**, 465.
154. J.P. Costes, J.F. Lamere, C. Lepetit, P.G. Lacroix and F. Dahan, *Inorg. Chem.*, 2005, **44**, 1973.

155. B. Insuasty, C. Atienza, C. Seoane, N. Martin, J. Garin, J. Orduna, R. Alcala and B. Villacampa, *J. Org. Chem.*, 2004, **69**, 6986.
156. O. Marget, P.G. Lacroix, J.P. Costes, B. Donnadieu and C. Lepetit, *Inorg. Chem.*, 2004, **43**, 4743.
157. Z.M. Xue, X.J. Zhang and Y.P. Tian, *Spectrosc. Spectr. Ann.*, 2004, **24**, 704.
158. P. Johansson, *PCCP*, 2005, **7**, 475.
159. S. Ohnishi, F.L. Gu, K. Naka, A. Imamura, B. Kirtman and Y. Aoki, *J. Phys. Chem. A.*, 2004, **108**, 8478.
160. B.J. Coe, L.A. Jones, J.A. Harris, B.S. Brunschwig, I. Asselberghs, K. Clays, A. Persoons, J. Garin and J. Orduna, *JACS*, 2004, **126**, 3880.
161. S. di Bella and I. Fragala, *Eur. J. Inorg. Chem.*, 2003, **14**, 2606.
162. M. Nakano, H. Nagao and K. Yamaguchi, *Phys. Rev. A.*, 1997, **55**, 1503.
163. M. Nakano, S. Yamada, R. Kishi, M. Takahata, T. Nitta and K. Yamaguchi, *J. Nonlinear. Opt. Phys. Mater.*, 2004, **13**, 411.
164. M. Nakano, N. Tomoshige, K. Yamaguchi, B. Champagne and E. Botek, *J. Chem. Phys. A*, 2004, **108**, 4105.
165. M. Nakano, R. Kishi, T. Nitta, T. Kubo, K. Nakasuji, K. Kamada, K. Ohta, B. Champagne, E. Botek and K. Yamaguchi, *J. Chem. Phys. A.*, 2005, **109**, 885.
166. P.C. Ray, *Chem. Phys. Lett.*, 2004, **395**, 269.
167. Y. Orimoto and Y. Aoki, *Phys. Rev. A.*, 2003, **68**, 063802.
168. V.M. Geskin, C. Lambert and J.L. Bredas, *JACS*, 2003, **125**, 15651.
169. Z.M. Xue and C.H. Chen, *Electrochim. Acta*, 2004, **49**, 5167.
170. D. Laage, W.H. Thompson, M. Blanchard-Desce and J.T. Hynes, *J. Phys. Chem. A.*, 2003, **107**, 6032.
171. E. Botek and B. Champagne, *Chemical Physics Lett.*, 2004, **387**, 130.
172. P.C. Ray and J. Leszcynski, *Chem. Phys. Letters*, 2004, **399**, 162.
173. P.C. Ray, *Chem. Phys. Lett.*, 2004, **394**, 354.
174. W.D. Cheng, D.S. Wu, H. Zhang, X.D. Li, D.G. Chen, Y.Z. Lang, Y.C. Zhang and Y.J. Gong, *J. Phys. Chem. B.*, 2004, **108**, 12658.
175. A. Tongraar and B.M. Rode, *Chem. Phys. Lett.*, 2005, **403**, 314.
176. W. Chen, Z.R. Li, D. Wu, R.Y. Li and C.C. Sun, *J. Phy. Chem. B.*, 2005, **109**, 601.
177. A. Datta and S.K. Pati, *J. Phys. Chem. A.*, 2004, **108**, 9527.
178. D.Y. Zhang, D. Begue and C. Pouchan, *Chem. Phys. Lett.*, 2004, **398**, 283.
179. C. Pouchan, D. Begue and D.Y. Zhang, *J. Chem. Phys.*, 2004, **121**, 4628.
180. K.R.S. Chandrakumar, T.K. Ganty and S.K. Ghosh, *J. Phys. Chem.*, 2004, **108**, 6661.
181. H.Y. Wang, X.B. Li, Y.J. Tang, H.P. Mao and Z.H. Zhu, *Chin J. Chem. Phys.*, 2005, **18**, 50.
182. R.R. Zope, T. Baruah, M.R. Pederson and B.I. Dunlap, *Phys. Rev. A*, 2005, **71**, 025201.
183. Y. Yang, F.H. Wang, Y.S. Zhou, L.F. Yuan and J.L. Yang, *Phys. Rev. A.*, 2005, **71**, 013202.
184. F.L. Gu, Z.F. Chen, H.J. Jiao, W.Q. Tian, Y. Aoki, W. Thiel and P.V. Schleyer, *PCCP*, 2004, **6**, 4566.
185. X.D. Li, W.D. Cheng, D.S. Wu, Y.Z. Lan, H. Zhang, Y.J. Gong, F.F. Li and J. Shen, *J. Chem. Phys.*, 2004, **121**, 5885.
186. M. van Faassen, L. Jensen, J.A. Berger and P.L. de Boeij, *Chem. Phys. Lett.*, 2004, **395**, 274.
187. G.Y. Guo, K.C. Chu, D.S. Wang and C.G. Duan, *Comput. Mater. Sci.*, 2004, **30**, 269.

188. G.Y. Guo, K.C. Chu, D.S. Wang and C.G. Duan, *Phys. Rev. B.*, 2004, **69**, 205416.
189. X. Zhou, A.M. Ren and J.K. Feng, *J. Mol. Struct.(Theochem)*, 2004, **680**, 237.
190. G.K. Gueorguiev, J.M. Pacheco and D. Tomanek, *Phys. Rev. Lett.*, 2004, **92**, 215501.
191. S. Yamada, M. Nakano, M. Takahata, T. Nitta and K. Yamaguchi, *J. Nonlinear Opt. Phys. Mat.*, 2004, **13**, 417.
192. E. Zojer, W. Wenseleers, P. Pacher, S. Barlow, M. Halik, C. Grasso, J.W. perry, S.R. Marder and J.L. Bredas, *J. Phys. Chem. B.*, 2004, **108**, 8641.
193. V. Lukes and M. Breza, *J. Mol. Struct.*, 2004, **699**, 93.
194. P.C. Jha, M. Das and S. Ramasesha, *J. Phys. Chem. A*, 2004, **108**, 6279.
195. X. Zhou, J.K. Feng and A.M. Ren, *Chem. Phys. Lett.*, 2005, **403**, 7.
196. X.J. Liu, J.K. Feng, A.M. Ren, X. Zhou and H. Cheng, *Chem. Phys.*, 2004, **307**, 61.
197. X. Zhou, A.M. Ren, J.K. Feng and X.J. Liu, *Can J. Chem.-Rev. Can. Chim.*, 2004, **82**, 1172.
198. G.P. Bartholomew, M. Rumi, S.J.K. Pond, J.W. perry, S. Tretiak and G.C. Bazan, *JACS*, 2004, **126**, 11529.
199. Y.H. Sun, K. Zhao, C.K. Wang, Y. Luo, Y.X. Yan, X.T. Tao and M.H. Jiang, *Chem. Phys. Lett.*, 2004, **394**, 176.
200. N. Kobko, A. Masunov and S. Tretiak, *Chem. Phys. Lett.*, 2004, **392**, 444.
201. S. Chattopadhyay, *J. Phys. B.*, 2004, **37**, 1783.
202. A.J. Moad and G.J. Simpson, *J. Phys. Chem. A.*, 2005, **109**, 1316.
203. B. Jansik, P. Salek, D. Jonsson, O. Vahtras and H. Agren, *J. Chem. Phys.*, 2005, **122**, 054107.
204. M. van Faassen and P.L. de Boeij, *J. Chem. Phys.*, 2004, **121**, 10707.
205. S.M. Smith, A.N. Markevitch, D.A. Romanov, X.S. Li, R.J. Levis and H.B. Schlegel, *J. Phys. Chem. A.*, 2004, **108**, 11063.
206. C.S. Esteves, T.L. Fonseca, M.A. Castro and O.A.V. Amaral, *Chem. Phys.*, 2004, **303**, 291.
207. A. Shukla, *Phys. Rev. B.*, 2004, **69**, 165218.
208. E.M. Torres, T.L. Fonseca, C.S. Esteves, O.A.V. Amaral and M.A. Castro, *Chem. Phys. Lett.*, 2005, **403**, 268.
209. D. Jacquemin, X. Assfeld and E.A. Perpete, *J. Mol. Struct.(Theochem)*, 2004, **710**, 13.
210. D. Jacquemin, E.A. Perpete and J.M. Andre, *J. Chem. Phys.*, 2004, **120**, 10317.
211. D. Jacquemin, J.M. Andre and E.A. Perpete, *J. Chem. Phys.*, 2004, **121**, 4389.
212. D. Jacquemin, *J. Phys. Chem. A.*, 2004, **108**, 9260.
213. S. Ohnishi, F.L. Gu, K. Naka, A. Imamura, B. Kirtman and Y. Aoki, *J. Phys. Chem. A.*, 2004, **108**, 8478.
214. D. Tsiaousis, R.W. Munn, P.J. Smith and P.L.A. Popelier, *Chem. Phys.*, 2004, **305**, 317.
215. B. Skwara, W. Bartkowiak and J. Leszcynski, *Struct. Chem.*, 2004, **15**, 363.
216. M. Guillaume, E. Botek, B. Champagne, F. Castet and L. Ducasse, *J. Chem. Phys.*, 2004, **121**, 7390.
217. F. Hamzaoui, A. Zanoun and G. Vergoten, *J. Mol. Struct.*, 2004, **697**, 17.
218. S. Suehara, P. Thomas, A. Mirgorodsky, T. Merle-Mejean, J.C. Champarnaud-Mesjard, T. Aizawa, S. Hishita, S. Todoroki, T. Konishi and S. Inoue, *J. Non-Crst. Solids*, 2004, **345–46**, 730.
219. O. Noguera, M. Smirnov, A.P. Mirgorodsky, T. Merle-Mejean, P. Thomas and J.C. Champarnaud-Mesjard, *J. Non-Cryst. Solids*, 2004, **345–46**, 734.

220. S. Suehara, P. Thomas, A. Mirgorodsky, T. Merle-Mejean, J.C. Champarnaud-Mesjard, T. Aizawa, S. Hishita, S. Todoroki, T. Konishi and S. Inoue, *Phys. Rev. B.*, 2004, **70**, 205121.
221. D. Sajan, J. Binoy, I.H. Joe, V.S. Jayakumar and J. Zaleski, *J. Raman Spectrosc.*, 2005, **36**, 221.
222. S.P.A. Sauer, H.J.A. Jensen and J.F. Ogilvie, *Advances in Quantum Chemistry*, 2005, **48**, 319.
223. L. Visscher, *Advances in Quantum Chemistry*, 2005, **48**, 369.
224. P. Lazzeretti, *Advances in Quantum Chemistry*, 2004, **47**, 177.
225. J. Autschbach, *Struct. Bond.*, 2004, **112**, 1.
226. A. Rutkowski and A. Poszwa, *Phys. Scr.*, 2005, **71**, 484.
227. K. Muroo, N. Ninomiya, M. Yoshino and Y. Takubo, *J. Opt. Soc. Am B.*, 2003, **20**, 2249.
228. M. Bregante, G. Cantatore, S. Carusotto, R. Cimino, F. Della Valle, G. Di Domenico, U. Gastaldi, M. Karuza, E. Milotti, E. Polacco, G. Ruoso, E. Zavattini and G. Zavattini, *Chem. Phys. Lett.*, 2005, **410**, 288.
229. M. Bregante, G. Cantatore, S. Carusotto, R. Cimino, F. Della Valle, G. Di Domenico, U. Gastaldi, M. Karuza, E. Milotti, E. Polacco, G. Ruoso, E. Zavattini and G. Zavattini, *Chem. Phys. Lett.*, 2004, **392**, 276.
230. A. Rizzo, M. Kallay, J. Gauss, F. Pawlowski, P. Jorgensen and C. Hattig, *J. Chem. Phys.*, 2004, **121**, 9461.
231. M. Pecul, T. Saue, K. Ruud and A. Rizzo, *J. Chem. Phys.*, 2004, **121**, 3051.
232. G.I. Pagola, M.C. Caputo, M.B. Ferraro and P. Lazzeretti, *Chem. Phys. Lett.*, 2004, **400**, 133.
233. A. Rizzo and S. Coriani, *J. Chem. Phys.*, 2003, **119**, 11064.
234. M. Hanni, P. Lanto, N. Runeberg, J. Jokisaari and J. Vaara, *J. Chem. Phys.*, 2004, **121**, 5908.
235. M.C. Caputo and P. Lazzeretti, *Chem. Phys.*, 2003, **288**, 281.
236. G.I. Pagola, M.C. Caputo, M.B. Ferraro and P. Lazzeretti, *Chem. Phys. Lett.*, 2005, **408**, 403.
237. I.G. Cuest, R.S. Jartin, A.S. de Meras and P. Lazzeretti, *J. Chem. Phys.*, 2003, **119**, 5518.
238. I.G. Cuesta, R.S. Jartin, A.S. de Meras and P. Lazzeretti, *J. Chem. Phys.*, 2004, **120**, 6542.
239. I.G. Cuesta, R.S. Jartin, A. De Meras and P. Lazzeretti, *Mol. Phys.*, 2005, **103**, 789.
240. F. Faglioni, A. Ligabue, S. Pelloni, A. Soncini, R.G. Viglione, M.B. Ferraro, R. Zanasi and P. Lazzerretti, *Org. Lett.*, 2005, **7**, 3457.
241. S. Pelloni, F. Faglioni, A. Soncini, A. Ligabue and P. Lazerretti, *Chem. Phys. Lett.*, 2003, **375**, 583.
242. G.I. Pagola, M.C. Caputo, M.B. Ferraro and P. Lazzeretti, *J.Chem. Phys.*, 2004, **120**, 9556.
243. G.I. Pagola, M.C. Caputo, M.B. Ferraro and P. Lazzeretti, *J. Chem. Phys.*, 2005, **122**, 074318.
244. E. Steiner and P.W. Fowler, *PCCP*, 2004, **6**, 261.
245. P.W. Fowler, E. Steiner, R.W.A. Havenith and L.W. Jenneskens, *Magn. Reson. Chem.*, 2004, **42**, S68.
246. A.V. Arbuznikov and M. Kaupp, *Chem. Phys. Lett.*, 2003, **381**, 495.
247. A.V. Arbuznikov and M. Kaupp, *Chem. Phys. Lett.*, 2004, **386**, 8.
248. A.V. Arbuznikov and M. Kaupp, *Chem. Phys. Lett.*, 2004, **391**, 16.

249. A.V. Arbuznikov and M. Kaupp, *Int. J. Quantum Chem.*, 2005, **104**, 261.
250. B. Mothana, F.Q. Ban and R.J. Boyd, *Chem. Phys. Lett.*, 2005, **401**, 7.
251. W. Migda and B. Rys, *Magn. Reson. Chem.*, 2004, **42**, 459.
252. A. Rizzo, C. Cappelli, B. Jansik, D. Jonsson, P. Salek and H. Agren, *J. Chem. Phys.*, 2004, **121**, 8814.
253. A. Rizzo, C. Cappelli, B. Jansik, D. Jonsson, P. Salek, S. Coriani, D.J.D. Wilson, T. Helgaker and H. Agren, *J. Chem. Phys.*, 2005, **122**, 234314.
254. C. Cappelli, A. Rizzo, B. Mennucci, J. Tomasi, R. Cammi, G.L.J.A. Rikken, R. Mathevet and C. Rizzo, *J. Chem. Phys.*, 2003, **118**, 10712.
255. P.E. Hoggan, *J. Theor. Comput. Chem.*, 2004, **3**, 163.
256. P.E. Hoggan, *Int. J. Quantum Chem.*, 2004, **100**, 214.
257. M. Cossi and O. Crescenzi, *J. Chem. Phys.*, 2003, **118**, 8863.
258. M. Cossi and O. Crescenzi, *Theor. Chem. Acc.*, 2004, **111**, 162.
259. S. Tonmunphean, A. Wijitkosoom, Y. Tantirungrotechai, N. Nuttavut and J. Limtrakul, *Bull. Che. Soc. Jpn.*, 2003, **76**, 1537.
260. R.G. Viglione and R. Zanasi, *PCCP*, 2004, **6**, 295.

4

Applications of Density Functional Theory to Heterogeneous Catalysis

BY DAVID S. SHOLL

Department of Chemical Engineering, Carnegie Mellon University, Pittsburgh, PA 15213, USA

1 Introduction

Density Functional Theory (DFT) has had an immense influence on our ability to perform quantitative quantum chemistry calculations for complex materials, a fact celebrated by the award of the Nobel Prize in Chemistry in 1998 to Walter Kohn.[1] The aim of this review is to examine recent progress in the application of DFT to heterogeneous catalysis. The widespread availability of efficient software packages for DFT calculations has made these calculations the source of a great deal of scientific activity. Setting out to review all of this activity exhaustively, even for a limited time frame, would be at a minimum a foolhardy activity. I have therefore written this review with the more restricted aim of compiling some of the recent directions in using DFT for problems related to heterogeneous catalysis that seem most likely to shape the community's thinking in coming years.

A number of other reviews exist that cover the use of DFT in heterogeneous catalysis.[2–9] In most cases, I have concentrated in this article on work that has been published since 2003. This recent work necessarily expands on earlier methodological and conceptual developments, so these previous reviews are invaluable for providing a full historical picture of the rapid development of this field. In the same vein, I have concentrated on theoretical studies, which means that typically only the key experimental studies on the same topics have been described. This approach is in no way meant to suggest that theoreticians are independent of or ahead of experimenters in this area; nothing could be further from the truth! I hope it will be evident to all readers by the conclusion of this review, if they need further convincing of this point, that the most significant contributions from DFT and other theoretical methods to heterogeneous catalysis arise when these theoretical efforts are pursued in intimate collaboration with high precision experiments.

I will not describe here the fundamental theory underlying Density Functional Theory. Several books have been devoted to this topic.[10–14] The review

Chemical Modelling: Applications and Theory, Volume 4

paper by Greeley, Mavrikakis, and Nørskov gives a useful overview of the application of DFT to materials of interest to heterogeneous catalysis.[4] The review by Payne *et al.* provides an excellent introduction to the calculation methods that form the core of plane wave DFT calculations.[15] The reviews by Bell and Friesner both give useful comparison between DFT calculations and other *ab initio* methods.[5,16]

Plane wave calculations using periodically replicated supercell geometries now make up the great majority of DFT calculations in the area of surface chemistry and heterogeneous catalysis. In the remainder of this review, the reader should assume unless a note is made to the contrary that all calculations that are discussed are from calculations using periodic geometries. Several popular software packages exist for these calculations including VASP,[17] DACAPO,[18] CASTEP,[19] and SIESTA.[20] The first three of these packages perform plane wave calculations, while SIESTA is based on a linear combination of numerical atomic orbitals. The individuals who contributed to the development of these packages are owed a great debt of gratitude by the large community of people who use them. DFT calculations based on non-periodic (*i.e.* cluster) geometries remain important, particularly for materials that remain awkward to deal with in periodic calculations because of the large numbers of atoms required to construct a consistent supercell or where calculations with higher level quantum chemistry approaches are desirable.[21]

The most important distinguishing feature of a plane wave DFT calculation is the functional used to represent electron exchange and correlation effects. Calculations based on the local density approximation (LDA) overestimate molecular bond energies, cohesive energies of solids, and adsorption energies on surfaces.[4,22] These effects are significantly reduced when calculations use the generalized gradient approximation (GGA),[4,22] and the great majority of recent DFT calculations associated with heterogeneous catalysis use this approach. The two most common functionals that are used within the GGA are the Perdew-Wang 91 (PW91) functional[23] and the Revised Perdew-Burke-Ernzerhof (RPBE) functional.[22] The RPBE functional was introduced by Hammer *et al.*[22] following the functional introduced by Perdew, Burke, and Erzenhof[24] and subsequent revisions of that functional by Zhang and Yang.[25] PW91 and RPBE give molecular bond energies and adsorption energies on metal surfaces that are more quantitatively accurate than LDA calculations, with the RPBE functional generally yielding the most accurate predictions.[4,22]

The GGA functionals just discussed are distinct from the hybrid DFT functionals that are widely preferred for DFT calculations of molecular systems.[16,26] Hybrid functionals such as the B3LYP functional[27] use a combination of exact Hartree-Fock exchange with GGA-based DFT functionals. This approach gives results that are in general more reliable than GGA functionals alone. Unfortunately, the exact exchange contributions are difficult to evaluate in plane wave basis sets,[16] so the option of using hybrid DFT within plane wave calculations is not routinely available. Friesner recently reviewed the comparative performance of GGA and hybrid functionals, as well as other *ab initio* methods.[16]

Most plane wave calculations use ultrasoft pseudopotentials (USPP),[15,28] which describe the core electrons of atoms in a mathematically efficient form that greatly reduces the computational cost associated with heavy atoms. An increasing number of calculations used the projector augmented wave (PAW) approach instead.[28] In most circumstances where both approaches can be used, the differences between USPP and PAW calculations are minor. Some exceptions to this observation include transition metals with large magnetic moments (*e.g.*, Fe) and alkali metals.[28]

In comparing or connecting DFT calculations with experiments, it is helpful to consider three different sources of uncertainty in calculated results. First, the use of an approximate exchange-correlation functional represents the primary physical approximation separating DFT calculations from the true solution of the Schrödinger equation. Second, the mathematical problem posed by a DFT calculation must of course be solved numerically, so there are many factors that control the level to which a numerical solution approaches the "true" solution of this mathematical problem. For plane wave calculations, these factors include the choice of an energy cutoff for the plane wave expansion, the number of *k*-points used to sample reciprocal space, the convergence criteria chosen for iterative construction of the ground state energy and for terminating geometry optimizations, and so on. It seems reasonable to say that these aspects of mathematical convergence are well understood and are easily controlled. Thorough reports of DFT calculations typically include some numerical evidence that these factors have been addressed in an appropriate way.

The third source of uncertainty in comparing DFT calculations with experiments arises from the intrinsic complexity of heterogeneous catalysts. The range of physical environments and effects that coexist in practical catalysts, even in carefully controlled experiments with model materials, is too large to allow every possible situation to be examined with a DFT calculation. Even nominally single-crystal surfaces are populated by steps and other defects, which in at least some circumstances can play a crucial role in catalytic processes. Industrial materials typically use small catalyst particles dispersed on supports, often in the presence of low concentrations of promoters, operating in chemically complex environments. This situation means that any effort to "explain" an effect via DFT calculations should be linked to experimental results as closely as possible. Of course, the complexity associated with heterogeneous catalysis presents an equally great challenge to experimental efforts to characterize and control practical materials. In many cases, DFT calculations can provide atomic-scale insight that would be difficult or impossible to obtain with experiments alone.

The review is broken into three broad sections, each containing three separate topics. The first section highlights three dramatic successes in the application of DFT to heterogeneous catalysis, followed by a discussion of the accuracy of DFT. The second section reviews three areas within heterogeneous catalysis that have attracted large volumes of theoretical effort in recent years, namely *ab initio* thermodynamics, the catalytic activity of nanoclusters of gold, and the development of bimetallic catalysts. The third section provides recent highlights from three areas that seem poised for intensive theoretical activity and progress in the near future, catalysts for use in reversible storage of hydrogen, electrocatalysis, and catalysis in zeolites. Specialists in the field can of course read or skip these sections in any order. Readers who are new to the field or who are aiming to gain a rapid overview of the key trends in the field are recommended to at least read all of the section on success stories and also the section on *ab initio* thermodynamics.

2 Success Stories

This section describes in detail three topics in heterogeneous catalysis to which DFT calculations have recently been applied with great effect, the prediction of CO oxidation rates over RuO_2(110), the prediction of ammonia synthesis rates by supported nanoparticles of Ru, and the DFT-based design of new selective catalysts for ethylene epoxidation. All three examples involve the careful application of DFT calculations and other appropriate theoretical methods to make quantitative predictions about the performance of heterogeneous catalysts under realistic operating conditions.

2.1 Success Story Number One: CO Oxidation over RuO_2(110). – The oxidation of CO over Ru is one of the best known examples of the so-called pressure gap in which qualitatively different behavior is seen in ultra high vacuum (UHV) experiments and catalysts under practical conditions. In UHV, the activity of Ru for CO oxidation is the lowest of the late transition metals, but under oxygen pressure this situation is reversed and Ru is superior to Pt, Rh, and Pd.[29] Over and Muhler recently reviewed experimental efforts to understand and bridge this gap.[29] A key outcome of both experimental and theoretical efforts has been that under high pressure conditions, the surface of Ru is better described as RuO_2. From this perspective, it is not so surprising that UHV experiments performed with Ru surfaces have quite different outcomes to higher pressure experiments.[30] Another important experimental finding is that the activity of RuO_2(110) towards CO oxidation is dominated by adsorption sites on the ideal, defect-free surface, not by surface defects such as steps.[29]

Reuter, Frenkel, and Scheffler have recently used DFT-based calculations to predict the CO turnover frequency on RuO_2(110) as a function of O_2 pressure, CO pressure, and temperature.[31] This was an ambitious undertaking, and as we will see below, remarkably successful. Much of this work was motivated by the earlier success of *ab initio* thermodynamics, a topic that is reviewed more fully below in section 3.1. The goal of Reuter *et al.*'s work was to derive a lattice model for adsorption, dissociation, surface diffusion, surface reaction, and desorption on defect-free RuO_2(110) in which the rates of each elementary step were calculated from DFT via transition state theory (TST). As mentioned above, experimental evidence strongly indicates that surface defects do not play a dominant role in this system, so neglecting them entirely is a reasonable approach. The DFT calculations were performed using a GGA full-potential linear augmented plane wave method. Calculations for the adsorption energy of CO and O as the surface coverage was varied indicated that adsorbate-adsorbate interactions were smaller than 0.15 eV. Because this value was quite small relative to the desorption and reaction energies, adsorbate-adsorbate interactions were neglected. This reasonable approximation provides an enormous simplification to the task of performing DFT calculations for "all" the relevant microscopic processes, since isolated adsorbates or pairs of adsorbates in a relatively small surface unit cell could be used.

CO and O both have two possible adsorption sites on $RuO_2(110)$, so DFT calculations were used to determine the adsorption energies for these sites and activation energies for diffusion between them. All diffusion processes were assumed to have a prefactor of 10^{12} s^{-1}. This assumption could in principle be avoided by computing the vibrational frequencies needed to derive these prefactors within harmonic TST, but this is unlikely to be necessary unless surface diffusion is found in cases of special interest to be a rate-limiting factor.

The key step in the derivation by Reuter *et al.* of their lattice model is the use of detailed balance to determine the sticking coefficients for each species on each type of site.[31] The total adsorption rate at a particular site can be expressed as $\Gamma_{ad} = \tilde{S}I(p,T)$, where $\tilde{S}$ is the local sticking coefficient and $I(p,T)$ is the impingement rate of the species of interest from a gas phase with partial pressure p and temperature T. At steady state, the total adsorption and desorption rates must satisfy the detailed balance condition $\Gamma_{des}/\Gamma_{ad} = \exp[(F_b - \mu(T,p))/kT]$, where F_b is the free energy of the adsorbed species and $\mu(T,p)$ is the chemical potential of the gas phase species. The adsorption free energy is well approximated by the adsorption enthalpy, which is simply the adsorption energy calculated by a DFT calculation. This approach provides a direct link between the adsorption and desorption rates and the pressure and temperature of the bulk gas phase.

Having determined the microscopic rates needed to define a lattice model from DFT calculations, Reuter *et al.* performed Kinetic Monte Carlo (KMC) simulations of the dynamics of the model to determine the steady-state CO oxidation rate of a $RuO_2(110)$ surface over a broad range of temperatures and partial pressures of O_2 and CO.[31] KMC provides a numerically efficient way to simulate the evolution of lattice models that can be rigorously interpreted as evolving in physical time provided that the lattice model itself includes rates that are defined using meaningful time dynamics such as TST. A variety of numerical algorithms are available for KMC simulations. The relative numerical efficiency of these algorithms has been carefully compared by Reese, Raimondeau, and Vlachos.[32]

The steady state reaction rates predicted by Reuter *et al.* at 600 K in terms of the turnover frequency (TOF) and a summary of the surface structures at this temperature are shown in Figure 1. For most combinations of the CO and O_2 partial pressures, the surface is dominated by one adsorbed species and as a result the CO oxidation rate is low. However, if the partial pressures are chosen appropriately, a dynamic equilibrium between adsorbed O, CO, and empty sites exists and the oxidation rate can be large. These regions are shown in white

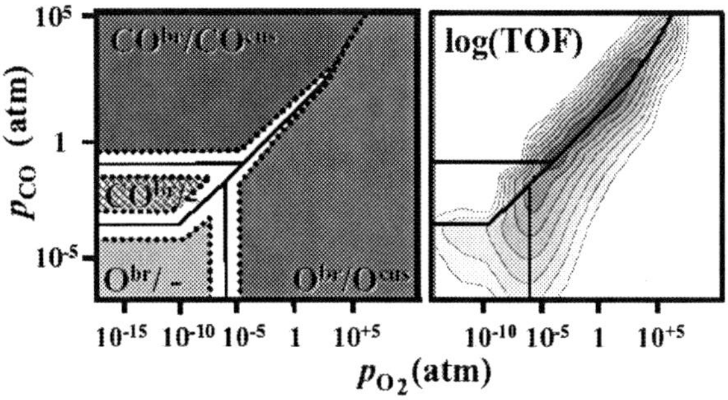

Figure 1 *The steady state surface coverage (left panel) and CO turnover frequency (right panel) for CO oxidation over $RuO_2(110)$ at 600 K as functions of the O_2 and CO partial pressure as predicted by Reuter et al.*[84] *In the left panel, the shaded regions indicate surfaces with >90% of available sites covered by the indicated species. In the right panel, each contour line indicates an increase in the TOF of an order of magnitude. The darkest region in the right panel has a TOF $> 10^{17}$ $cm^{-2}s^{-1}$. [Reproduced with permission from APS]*

in the left panel of Figure 1. Reuter *et al.* highlighted the interesting observation that the reactions with the fastest microscopic rates do not necessarily dominate the overall reaction because the sites required for these reactions to proceed are not necessarily the most likely adsorption sites on the surface under reaction conditions. The magnitude of the predicted TOF is in good agreement with the experimental results of Peden and Goodman;[33] the experiments give TOFs as high as 8×10^{16} $cm^{-2}s^{-1}$ at 600 K for CO pressures of 16 torr while the DFT-based method predicts that the highest TOF for these conditions is slightly greater than 10^{16} $cm^{-2}s^{-1}$. Considering that the TOF varies by more than 7 orders of magnitude over the conditions shown in Figure 1, this agreement is extraordinary. The O_2 pressure predicted by the DFT-based approach to give the largest TOF at these conditions is ~0.01 torr; this is somewhat smaller than the O_2 pressure observed by Peden and Goodman to maximize the TOF.

Reuter, Frenkel, and Scheffler also compared their DFT-based predictions with experimental measurements by Wang *et al.* at 350 K.[30] This comparison is shown in Figure 2. Similar to the TOF map shown in Figure 1, the overall reaction rate is a sensitive function of the gas phase pressures. The accuracy of the DFT-based approach in predicting the conditions that give the largest oxidation rates is impressive.

There are of course a number of directions in which the lattice model derived by Reuter *et al.* could be made more rigorous. The influence of adsorbate-adsorbate interactions, while small, could be quantified, as could the prefactors used in the TST description of diffusion. Reuter *et al.* noted that a small contribution to the total reaction rate may arise from an Eley-Rideal mechanism that is not included in their model. A more challenging task would be to account for the potential role of surface defects. These directions, however, do not diminish the importance of this work, which can reasonably be thought of

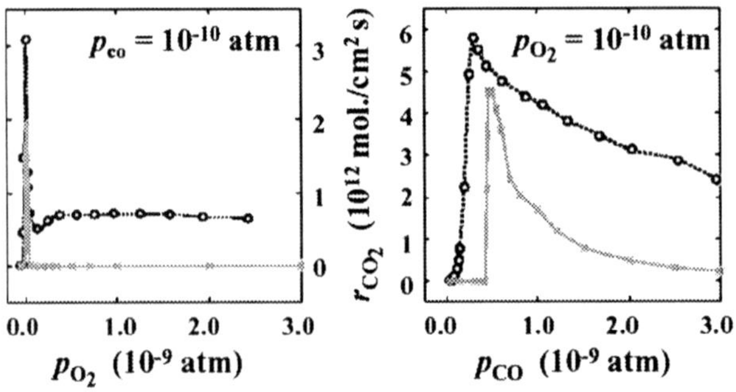

Figure 2 *CO_2 formation rate over $RuO_2(110)$ as predicted by Reuter et al.*[84] *(solid curves) and measured experimentally by Wang et al.*[30] *(open circles and dotted curves) at 350 K. [Reproduced with permission from APS]*

as one of the first examples to bridge the formidable pressure gap for a complete catalytic process theoretically.

2.2 Success Story Number Two: Ammonia Synthesis on Ru Catalysts. – *2.2.1 Rate Determining Step for Ammonia Synthesis on Ru.* The recent work by Nørskov and co-workers on Ru-catalyzed ammonia synthesis[34,35] is an excellent example of how the complexities of real catalysts can be understood with the aid of DFT calculations. In this section, I review this work in some detail and explore some of the related issues that are beginning to emerge in the recent literature.

Commercially, the catalytic synthesis of ammonia from hydrogen and nitrogen remains of great importance almost ninety years after its initial development. Although Fe is the most widely used catalyst for this process, Ru is known to be more active than Fe.[34] Ammonia synthesis using Fe catalysts is relatively well understood. The rate limiting process in the overall Fe-catalyzed reaction is the initial dissociation of N_2. The overall reaction can be described as follows:

$$N_2 + 2^* \leftrightarrow 2\,N^*$$

$$H_2 + 2^* \leftrightarrow 2\,H^*$$

$$NH_x^* + H^* \leftrightarrow NH_{x+1}^*\ (x = 0, 1, 2)$$

$$NH_3^* \leftrightarrow NH_3,$$

where * denotes a surface species or an empty surface site. The first effort to describe this collection of reaction steps theoretically for a Ru catalysts was a series of plane wave DFT calculations by Rod *et al.* for Ru(0001).[36] This work noted that dissociation of H_2 on Ru(0001) is facile. The transition state for N_2 dissociation on Ru(0001) was calculated to lie ~1.75 eV higher in energy than gas phase N_2. The transition state for N_2 dissociation on a stepped Ru(0001) surface was also computed. This transition state lay ~0.3 eV above the energy of gas phase N_2, strongly suggesting that surface steps play a crucial role in breaking the N_2 bond. By comparing the energies of the intermediates in the reactions above, Rod *et al.* concluded that N_2 dissociation is the rate limiting step for the overall reaction.[36]

The identification of N_2 dissociation as the rate limiting step for ammonia synthesis on Ru was reexamined by Zhang, Liu, and Hu.[37] This work complemented the earlier work by Rod *et al.* by determining transition states for the stepwise addition reactions in the reaction scheme above. A surprising result from these calculations was that the activation energies of these stepwise addition reactions were almost as large as the barrier for N_2 dissociation on the flat surface. Zhang *et al.* also examined the potential role of surface steps in accelerating the rate of stepwise addition reactions. For $NH^* + H^* \leftrightarrow NH_2^*$, they found an activation energy of 0.81 eV at a step edge, compared to 1.28 eV on the flat surface.[37] That is, the step edge reduced the activation energy for this addition reaction by 0.47 eV. While this is certainly not an inconsequential amount of energy, it is considerably less than the ~1.4 eV reduction in the activation energy for N_2 dissociation at a step edge computed by Rod *et al.*[36] On this basis, Zhang *et al.* made the qualitative conclusions that step sites will play a less important role for addition reactions in the reaction scheme above than they will in the initial dissociation of N_2 and that this dissociative reaction cannot necessarily be identified as the rate limiting step in the overall reaction.

Logadóttir and Nørskov extended the calculations of Zhang *et al.* to provide a complete description of the reaction pathway listed above on both flat and stepped Ru(0001).[34] The resulting potential energy diagram is shown in Figure 3. These plane-wave calculations used the RPBE functional. The surface step was modeled by using a (4 × 2) unit cell of Ru(0001) with two adjacent rows of Ru added on top of this surface. Several geometrically distinct types of surface steps can exist on metal surfaces,[38–40] so this approach makes the implicit assumption that the straight step chosen for the calculations is representative of the steps that appear on real materials. One role of the surface step is to stabilize all of the reaction intermediates relative to their adsorption configurations on flat Ru(0001). For example, the energy required for molecular desorption of ammonia was computed to be 1.32 eV on Ru(0001) and 1.76 eV when NH_3 adsorbed on a step site. The other impact of the surface step that was already noted by the previous studies cited above is the reduction of the energy barriers to each step in the overall reaction. Logadóttir and Nørskov calculated the barrier height for N_2 dissociation to be 1.9 eV on Ru(0001) and 0.4 eV on the stepped surface. The three stepwise additions of H were found to have barriers of 1.2, 1.3, and 1.4 eV on the flat surface and 1.1, 1.3, and 1.2 eV on the stepped surface. One of these addition reactions on the stepped surface was also examined by Zhang *et al.*, who reported an activation energy of 0.81 eV.[37] Logadóttir and Nørskov explain this apparent discrepancy by noting that their calculated transition state was defined relative to an initial state that was more stable than the initial state used by Zhang *et al.* The barrier computed

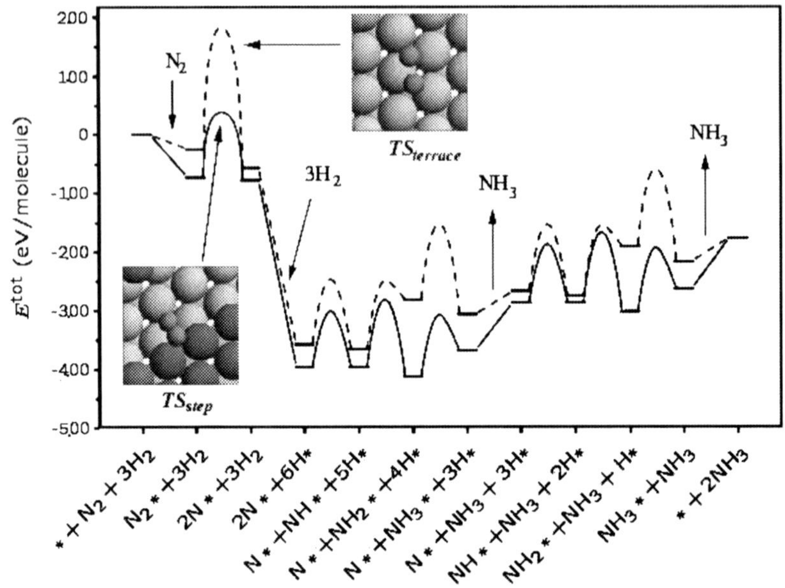

Figure 3 *The calculated potential energy diagram for NH_3 synthesis from N_2 and H_2 over Ru(0001) (dashed curve) and stepped Ru(0001) (solid curve).*[34,35] *[Reproduced with permission from AAAS]*

by Logadóttir and Nørskov therefore provides a more useful estimate of the relevant barrier in the overall reaction process.

The calculations of Logadóttir and Nørskov give activation energies of 0.4 eV for N_2 dissociation (relative to gaseous N_2) and 1.3 eV for $NH^* + H^* \leftrightarrow NH_2^*$ on stepped Ru(0001). The relative size of these two activation energies appears to provide a strong challenge to the idea that N_2 dissociation is the rate limiting step in ammonia synthesis on Ru. But it is of course the relative rate of the reaction that is important in making this judgement, not just the activation energy. To compare the rates of these two reactions, Logadóttir and Nørskov estimated the rates of each reaction.[34] They expressed the overall rate of the N_2 dissociation reaction by

$$r_{diss} = k_{diss}\theta_*^2 P_{N2}/P_0,$$

where θ_* is the density of free surface sites and P_0 is the standard pressure. Similarly, the rate of H addition to NH was expressed as

$$r_{add} = k_{add}\,\theta_{NH}\,\theta_H.$$

The rate constants k_{diss} and k_{add} were estimated using transition state theory. The crucial difference between the two processes is that while the partition functions of the reactant and transition states are quite similar for the addition reaction, the same is not true for N_2 dissociation. Writing each rate as $k = A\exp(-E_a/kT)$, where E_a is the activation energy, Logadóttir and Nørskov found that at 700 K, $A_{add} = 10^{13}$ s^{-1}, while $A_{diss} = 0.24$ s^{-1}. That is, the prefactors for these two rate constants differ by twelve orders of magnitude. At 700 K, this implies that $k_{diss}/k_{add} \approx 7.2 \times 10^{-8}$, meaning that the rate constant for N_2 dissociation is enormously slower than the rate constant for the addition reaction, despite its smaller activation energy. Logadóttir and Nørskov found that $k_{diss}/k_{add} < 1$ for all temperatures higher than $\sim$350 K.

2.2.2 Predicting the Total Reaction Rate for an Ru Catalyst. Based on the information above, Honkala, Nørskov and coworkers took on the extremely challenging task of seeking to predict the overall reaction rate for ammonia synthesis on a real catalyst with essentially no experimental input.[35] To do so, they had to tackle two principal hurdles to connect their previous DFT results with a real material: the fact that a real material is typically covered by a complex arrangement of adsorbates and the need to describe the number of step sites on a realistic catalyst quantitatively. An important note for an in depth reading of the paper by Honkala *et al.* is that much of the technical detail is included in the supplementary information associated with the online version of the paper.

To account for the effect of co-adsorbates on N_2 dissociation on step edges on Ru(0001), Honkala *et al.* first used DFT to determine the activation energy for this process while varying the adsorbates in nearby sites on the surface.[35] By defining the neighborhood of the dissociating molecule as including four distinct surface sites, it was possible to describe all possible local environments. The resulting activation energies varied from 0.49 eV on the bare surface to 1.06 eV when NH_2 is adsorbed on the upper step adjacent to the dissociating molecule.

In addition to knowing the activation energy for N_2 dissociation in the presence of co-adsorbates, it is necessary to predict the probability of observing each possible local configuration of adsorbates. To do this, DFT calculations were used to compute pairwise interaction energies between adsorbed H, N, NH, NH_2, and NH_3 species.[35] These interaction energies vary from weakly attractive for H–NH_2 and N–NH_3 pairs to relatively repulsive, for example for NH_2–NH_2 pairs. Interactions between adsorbates on different terraces of the stepped Ru surface were neglected. These interactions define a lattice model for the co-adsorption of H, N, NH, NH_2, and NH_3 on a stepped Ru surface. The thermodynamic equilibrium states of this model were established by performing Grand Canonical Monte Carlo (GCMC) simulations.[35] In these GCMC simulations the bulk gas phase was assumed to be ideal, a good approximation at the temperatures and pressures of interest. Once the probabilities of observing each type of local environment, P_i, were obtained from these GCMC simulations, the net reaction rate for ammonia production was expressed as[35]

$$r(T, p_{N2}, p_{H2}, p_{NH3}) = \left(1 - \frac{p_{NH3}^2}{p_{H2}^3 p_{N2} K_g}\right) \sum_i P_i k_i p_{N2},$$

where the p's are the partial pressures of the gas phase species, K_g is the gas phase equilibrium constant, and k_i is the N_2 dissociation rate in environment i. This expression assumes that all the reactions in the reaction scheme outlined above are in equilibrium except for the N_2 dissociation reaction and reflects the fact that the reaction must stop once gas phase equilibrium is established.

The expression above is only applicable to N_2 dissociation at step edges, so to compare it to real catalysts, the density of step edges on a real material must be estimated. Honkala used a very creative approach to this problem. DFT calculations were used to predict the surface energy of all Ru(*hkl*) faces with $h + k + l < 4$, and these energies were used to determine the favored nanoparticle shape for Ru via the Wulff construction.[35] The resulting particle shape, while qualitatively comparable to TEM images of Ru catalyst particles,[35] exposes no step sites of the type implicated as being the active sites in the DFT calculations discussed above. Honkala *et al.* showed, once again using DFT calculations, that a plausible reconstruction of Ru(103) reveals alternating terraces of Ru(100) and Ru(101) edges.[35] On this basis, they modified their model of Ru nanoparticles by removing atoms at (001)/(101) edges to reveal monoatomic steps. This description predicts a precise relationship between the radius of a Ru nanoparticle and the number of step sites available for N_2 dissociation. For particles smaller than 1.94 nm in diameter, the number of active sites is predicted to be zero. This approach clearly involves some important approximations: the Wulff shape derived from DFT is assumed to be temperature independent, interactions of the nanoparticles with the catalyst support are ignored, the existence of step sites from sources other than the intersection of (001)/(101) edges is not included, and the tendency of adsorbates to alter the relative energies of surfaces and step edges is not included.[35] An example of the nanoparticle shape predicted by Honkala *et al.* with the step edge atoms highlighted is shown in Figure 4.

With their DFT-based model for the number of active sites as a function of nanoparticle radius, the only experimental input Honkala *et al.* needed to compare their predictions with experiments was the particle size distribution of the experimental catalyst. The catalyst used in the experimental portion of this work was 0.2 g of an 11.1 wt% $Ru/MgAl_2O_4$ material. The particle size distribution was established by examining ~1000 nanoparticles using TEM.[35] With this information, Honkala *et al.* compared their DFT-based rate expression with experimental data over a range of operating conditions. It is fair to describe this comparison of theory and experiment as a first principles comparison, since no information from the catalyst under operating conditions was used to fit the theoretical data. Remarkably, the theory does an excellent job of predicting the ammonia reaction rate. The experimentally observed rate was underpredicted by a factor of 3–20.[35]

Honkala *et al.* supplemented their comparison between their DFT-based results and experiments with two revealing pieces of sensitivity analysis. First, they arbitrarily decreased the stability of adsorbed H atoms relative to NH_x species by 0.06 eV and repeated their comparison. This change in energy produced essentially perfect agreement between the model predictions and

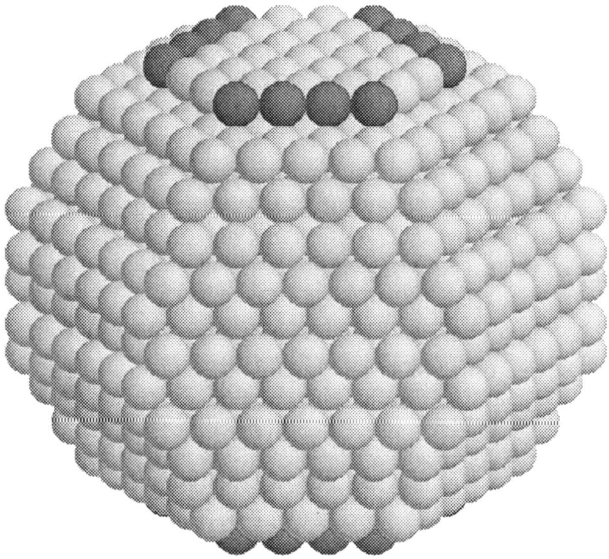

Figure 4 *A Ru nanoparticle with diameter 2.9 nm predicted by the DFT-based Wulff construction of Honkala et al.*[35] *The active B_5 sites are highlighted at the top and bottom of the nanoparticle. [Reproduced with permission from AAAS]*

experiment. This calculation does not demonstrate that the original DFT calculations gave H–NH_x energies that are 0.06 eV in error; there are many other potential contributions to the observed differences between theory and experiment.

A second very useful (and undoubtedly very time consuming) piece of sensitivity analysis was to recompute all the DFT parameters in the DFT-based lattice model using the PW91 GGA functional rather than the RPBE functional. Using the PW91 function had some dramatic effects on the individual rates in the reaction scheme. Using PW91, both the energy of molecularly adsorbed N_2 and the barrier to N_2 dissociation on the bare stepped surface were computed to be 0.6 eV lower than with RPBE. This implies (*cf.* Figure 3) that dissociation of N_2 is not activated with respect to gaseous N_2 according to the PW91 functional. PW91 also predicts the chemisorption energy of H_2 on the surface to be larger than the RPBE result, by 0.16 eV/molecule. These results means that the N_2 dissociation rate predicted by PW91 on the bare surface is considerably faster than the RPBE prediction. At the same time, however, the surface coverage of free sites predicted by PW91 is considerably less than the RPBE prediction; an effect which reduces the net N_2 dissociation rate and hence the ammonia production. As a result of this compensatory effect, the difference in overall reaction rate between the RPBE and PW91 calculations is much smaller than might be thought at first. In the example shown by Honkala *et al.*, the PW91-derived reaction rate is approximately 10 times smaller than the RPBE-derived reaction rate.

2.2.3 Tunneling Contributions in Ammonia Synthesis on Ru. The success of the DFT-based predictions above relies on rate constants calculated using classical transition state theory (TST). Several recent papers have examined the validity of TST for key steps in ammonia synthesis on Ru with particular focus on possible contributions to reaction rates due to quantum mechanical tunneling. There have been reports in the literature based on empirical potentials that suggested that N_2 dissociation on metal surfaces may be dominated by tunneling.[41] van Harrevelt *et al.* have recently reexamined this issue for N_2 dissociation on Ru(0001).[42] In this work, plane wave DFT was used to compute a six-dimensional potential energy surface for N_2 on a rigid Ru(0001) substrate. Time-dependent wavefunctions for N_2 were then propagated on this potential energy surface using an efficient numerical scheme. The principal result of this work is that the contribution to the N_2 dissociation rate due to tunneling is very small. At 300 K, TST underestimates the total rate by $<20\%$. At 700 K, TST is accurate to within 2%. Because of the great computational cost associated with calculations of this sort, it is not currently practical to apply them to situations with more degrees of freedom such as calculations in which metal atoms are allowed to relax. The calculations also have not been applied to N_2 dissociation at stepped sites on Ru(0001). Nevertheless, the results from rigid Ru(0001) argue persuasively that neglecting tunneling contributions to N_2 dissociation during ammonia synthesis on Ru is not an approximation that should be of great concern.

It is reasonable to suspect that the reactions involving H atom transfer during the overall synthesis of ammonia may potentially be more strongly influenced by tunneling than N_2 dissociation. Clary and co-workers have performed detailed studies of $N^* + H^* \leftrightarrow NH^*$ on Ru(0001) using potential energy surfaces computed from plane wave DFT[43] and from hybrid DFT applied to a Ru cluster.[44] As in the tunneling calculations of van Harrevelt *et al.*, the metal atoms were held rigid in these two studies to reduce the number of degrees of freedom needed to define the relevant potential energy surface. In both studies, several alternative methods for computing the total reaction rate from the potential energy surface were explored. At room temperature, classical TST was found to underestimate the full reaction rate by a factor of 20–70, depending on the potential energy surface used.[43,44] This effect becomes considerably weaker at higher temperatures. At 700 K, classical TST underestimated the net rate by a factor of 2–4.

A difficulty with the detailed calculations of tunneling contributions described above is that they are extremely computationally intensive. Other methods have been used to examine H tunneling on potential energy surfaces derived from DFT calculations in interstitial sites in metals and on metal surfaces, but these methods are also computationally demanding.[45,46] It may be useful in assessing the importance of tunneling within complex catalytic reactions to use methods that can estimate tunneling effects with relatively modest computational effort. One promising approach of this type is the semiclassical formalism of Fermann and Auerbach, which calculates a tunneling correction to harmonic TST using only information on the harmonic vibrational frequencies at the potential energy minimum and saddle point.[47] If one is applying harmonic TST to determine a net rate constant, these vibrational frequencies already must be computed,[35] so no additional effort is required to apply Ferrman and Auerbach's approach. Bhatia and Sholl have recently applied this method to probe H diffusion in ordered AB_2 intermetallics and in bcc CuPd using plane wave DFT, with encouraging results.[48] An interesting future direction for work of this type will be to compare this approximate method with detailed calculations of the type described above by van Harrevelt *et al.*[42] and Clary *et al.*[43,44]

2.2.4 Ammonia Decomposition Reactions. The opposite of ammonia synthesis, ammonia decomposition, has attracted interest as a potential route for delivering hydrogen as a fuel source.[49,50] Vlachos and co-workers have presented microkinetic models for ammonia decomposition using Ru catalysts, based in part on DFT calculations.[51,52] In this work, plane wave GGA calculations with the PW91 functional were used to examine the influence of N and H coverage on the chemisorption energy of these species on Ru(0001).[51] Information on these interactions was then used within a microkinetic model for the overall reaction based largely on experimental data. This model suggests that because lateral interactions between adsorbed N atoms reduce the barrier to N_2 desorption, the rate limiting step in ammonia decomposition on Ru(0001) is the decomposition of NH_2* to NH* and H*.[51]

An interesting point that has not yet been examined in the context of ammonia decomposition on Ru is the role of step sites. We saw above that step sites have been shown to be crucial to understanding ammonia synthesis. It is clear from the calculations of Logadóttir and Nørskov that step sites can dramatically change the stabilities and energy barriers to reaction of the surface species relevant to ammonia decomposition. Extending the DFT-based approach used by Honkala *et al.* to examine ammonia synthesis on Ru nanoparticles[35] to ammonia decomposition would provide an interesting complement to the existing models of this process.

2.2.5 Promoters. Experimentally, electropositive alkali metals are known to act as effective promoters for ammonia synthesis with Ru catalysts.[53] Logadóttir and Nørskov estimated the effect of electropositive promoters on the reaction pathway shown for flat and stepped Ru(0001) in Figure 3.[34] To do this, they assumed that the interaction energy between an adsorbed species and a promoter arises from the dipole moment of the adsorbate interacting with the electric field induced by the promoter atom. On both the flat and stepped surface, this interaction lowers the energy of the transition state for N_2 dissociation while increasing the energy of adsorbed NH_x species.[34] These results indicate that electropositive promoters increase the rate of ammonia synthesis both by enhancing the rate of N_2 dissociation and by increasing the number of empty sites on the catalyst surface available for reaction. These ideas support similar conclusions that were drawn earlier based on experimental studies.[54]

2.3 Success Story Number Three: Ethylene Epoxidation. – DFT calculations have played a crucial role in the recent dramatic advances in the catalytic epoxidation of ethylene by Barteau and co-workers.[55] Epoxidation of ethylene to form ethylene oxide (EO) has been performed commercially using Ag as a catalyst for decades. Many other olefin epoxides are also of commercial value, but until recently EO was the only product made using direct oxidation. In the 1990s, Eastman Chemical commercialized production of 3,4-epoxy-1-butene via oxidation of butadiene over a silver catalyst. An obvious question of interest in relation to these reactions is why Ag is effective as an epoxidation catalyst while other metals are much less useful. Providing a convincing answer to this question must be preceded by knowing the reaction mechanism of the Ag-catalyzed reaction. The DFT calculations reviewed below, in tandem with a range of surface science experiments, have achieved this goal. Once epoxidation by Ag has been tackled in an explanatory way, it is natural to consider the predictive power of calculations to seek improved catalysts. This challenge has recently been tackled by Linic, Jankowiak, and Barteau, who used DFT calculations to seek a bimetallic catalyst with Ag as the majority component.[56] Subsequent experiments by the same authors using supported bimetallic catalysts confirmed that a catalyst whose selectivity for epoxidation of ethylene was higher than pure Ag had been identified. Because this work provides a superb example of connecting DFT calculations with an important heterogeneously catalyzed reaction under realistic conditions, I have reviewed it in some detail below.

We first consider the mechanism of epoxidation of ethylene to EO over Ag catalysts, drawing primarily from the results of Linic and Barteau.[57] Although the large experimental literature on this reaction had not lead to a firm conclusion concerning the mechanism of the reaction, several points had been well established experimentally. For example, it is widely accepted that it is adsorbed atomic oxygen that participates in the oxidation reaction, not molecular oxygen.[58,59] As a result, dissociation of molecular oxygen is a necessary step in the overall reaction mechanism. Second, stereospecific studies of reactions with deuterated ethylene argued that the reaction cannot proceed by a concerted addition of surface oxygen to C=C.[60,61]

Surface oxametallacycles have been shown to exist as stable surface intermediates during ethylene epoxidation. These oxametallacycles consist of an O–C–C (or similar) backbone that is attached at both ends to a metal surface. Although oxametallacycles may act as intermediates in a variety of reactions, only a limited number of direct observations of these species have been made.[62] Linic and Barteau established that if EO was adsorbed on Ag(111) within narrow range of temperatures, a stable surface intermediate was formed.[62] To establish the identity of this intermediate, cluster-based DFT calculations were performed for two potential oxametallacycle structures and the predicted vibrational frequencies were compared with HREELS experiments. This comparison strongly indicated that the intermediate was an oxametallacycle in which the O and C atoms bonded to the surface could be thought of as being bonded to the same metal atom on the surface.[62] This is obviously not the only possible adsorption configuration; on Ag(110) it is known that the same species exists while binding to two distinct surface atoms.[63] DFT calculations were also critical in establishing the identity of this surface species.[63]

The information outlined above provided the starting point for the calculations by Linic and Barteau of a complete reaction coordinate for the formation of EO via epoxidation of ethylene on Ag(111).[57] This reaction coordinate is illustrated in Figure 5. The proposed mechanism involves two distinct reactions, first dissociation of molecular oxygen on the surface and then reaction of surface oxygen with ethylene. The barrier and reaction energy shown in Figure 5 come from plane-wave RPBE-DFT calculations by Nørskov *et al.*[64] These calculations established that O_2 dissociation proceeded with a much lower barrier on step sites on Ag(111) than on Ag(111) terraces, and it is the result from the step sites that is shown in Figure 5. The reaction coordinate in Figure 5 does not show the barriers that must be overcome for diffusion of adsorbed O from a step site to the terrace sites before the second portion of the reaction can take place; these barriers were implicitly assumed by Linic and Barteau to be small relative to the other barriers relevant to the reaction coordinate.

The second subreaction begins with ethylene interacting with an adsorbed O atom. Ethylene desorbs molecularly from Ag surfaces at low temperatures,[65] suggesting that ethylene is extremely mobile on the surface under reaction conditions. Cluster and plane wave DFT calculations were used to identify a

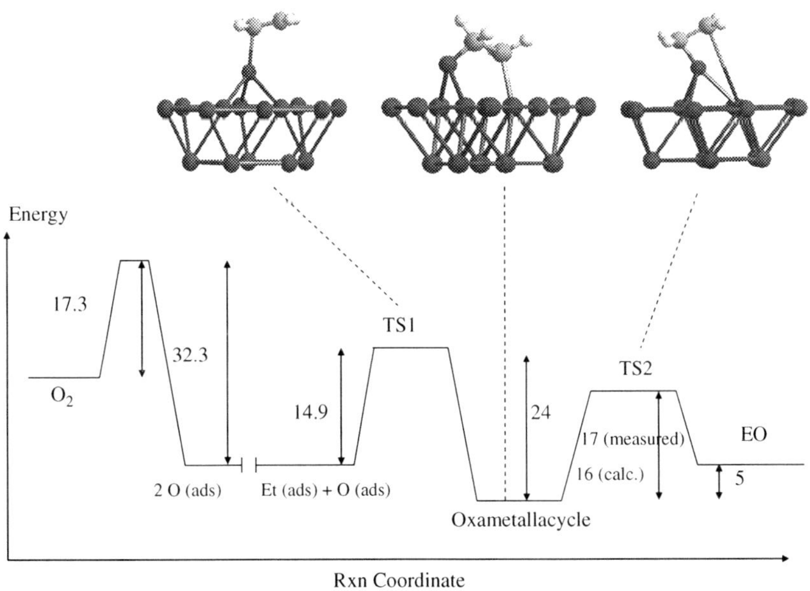

Figure 5 *The reaction coordinate determined by Linic and Barteau for the epoxidation of ethylene over Ag(111), with energies in kcal/mol.*[57] *The discontinuity in the reaction coordinate arises from a change in the oxygen molarity. [Reproduced with permission from Elsevier]*

transition state separating the adsorbed ethylene and oxygen from a stable oxametallacycle. The geometry of this oxametallacycle, which arose from calculations following the local reaction path from the preceding transition state, is slightly different from the one identified previously on Ag(111).[62] Linic and Barteau suggest that the barrier to interconversion between these two distinct oxametallacycle geometries is small.

Cluster DFT calculations were also used to identify a transition state for the formation of ethylene oxide. In this transition state, the Ag–O bonds are elongated relative to the oxametallacycle. The product of this reaction is gaseous EO. The activation energy for this step had previously been determined by Linic and Barteau experimentally.[62] The predicted activation barrier from DFT calculations, 16 kcal/mol, is in very good agreement with the experimental result of $\sim$17 kcal/mol.

Although Figure 5 defines a complete reaction pathway for ethylene epoxidation in Ag(111), it is not necessarily clear from this pathway what the rate-limiting step in the process is. To address this issue, Linic and Barteau constructed a microkinetic model for this pathway that accounts for the reaction rates of each surface species.[57] Models of this type, while they do not account for potential complication arising from lateral interactions amongst adsorbed species or the diffusion of species between relevant surface sites, are useful because they allow a range of relevant reaction conditions to be examined relatively rapidly. Linic and Barteau analyzed their microkinetic model with two complementary methods for establishing the relative importance of kinetic limitations in reactions networks, a method based on De-Donder relationships put forward by Dumesic[66,67] and the degree of rate control method of Campbell.[68] Both methods gave similar conclusions, namely that both the dissociation of oxygen on the surface and the addition of ethylene to adsorbed O were steps that provided significant kinetic limitations to the overall reaction.

The reaction pathway summarized in Figure 5 allows a number of conclusions about why Ag is such an effective ethylene epoxidation catalyst.[62] If this pathway is also representative of other potential catalysts, then a desirable catalyst in terms of activity is one that does not have large barriers for dissociation of molecular oxygen or the formation of oxametallacycles. By invoking the "universal" relations that exist between the reaction energies and reaction activation energies on metal surfaces,[64] then the conditions for a desirable catalyst can be restated as requiring weak bonding of both atomic oxygen and oxametallacycles. Both of these conditions have been evaluated using DFT calculations. Nørskov *et al.* used DFT to examine the binding of atomic O on 8 densely packed metal surfaces and a similar number of surface steps.[64] These calculations identified Ag as the metal that bound surface O atoms most weakly, apart from Au, for which formation of surface O from gaseous O_2 is energetically unfavorable. Mavrikakis, Doren, and Barteau used cluster-based DFT calculations to study the relative stability of oxametallacycles for 14 transition metals.[69] Although most of these calculations were performed for clusters containing only two metal atoms, a series of calculations for larger clusters of Ag and Cu suggested that the smaller clusters provided a useful way to rank the relative stabilities of the different oxametallacycles. Of the 14 transition metals considered in these calculations, only the oxametallacycles of Au and Cd had smaller binding energies than Ag.[69] Linic and Barteau pointed out that the relatively weak interaction of ethylene oxide with Ag may also play a role in the effectiveness of Ag as a catalyst, since once ethylene oxide forms it will rapidly desorb from the surface under reaction conditions.[57] On many other metal surfaces that interact with ethylene oxide more strongly, decomposition of adsorbed ethylene oxide will compete strongly with molecular desorption.[62] One topic that still remains to be fully explored in the context of this reaction is the role of surface oxygen coverage. *Ab initio* thermodynamics calculations of the type reviewed below in section 3.1 indicate that under practical operating conditions, Ag catalysts are likely to be covered with a thin oxide layer.[70]

The discussion above of the effectiveness of Ag as a catalyst centered on the activity of the catalyst. The value of Ag as an epoxidation catalyst, however, comes as much from its selectivity as from its activity. In discussing the selectivity of Ag, it is crucial to notice that Figure 5 only shows the pathway leading to the formation of ethylene oxide. Once the oxametallacycle has been formed in this pathway, a competing reaction is the formation of adsorbed acetaldehyde.[56] It is known experimentally that once acetaldehyde forms on Ag surfaces it readily oxidizes to form acetates that further oxidize to form CO_2.[71] Linic and Barteau used cluster and plane wave DFT to determine the activation energies for this competing reaction on Ag(111), finding that both the desirable pathway leading to ethylene oxide and the undesirable pathway forming acetaldehyde had activation energies of 16–17 kcal/mol.[72] The difference in Gibbs' free energies between the two transition states was calculated to be 0.3 kcal/mol, with the EO pathway having the lower of the two barriers. This prediction suggests that the selectivity for the EO reaction will be $\sim$40% at 400–500 K, which is consistent with experimental observations.[72] Additional support for the mechanism suggested in these calculations comes from the effect of using deuterated reactants. Linic and Barteau calculated the Gibbs' free energies for the transition states for deuterated reactants, finding that deuteration should increase the reaction selectivity. The predicted increase in selectivity due to deuteration was in good agreement with previous experimental observations of this effect.[72] This work strongly suggests that the selectivity of Ag as an ethylene epoxidation catalyst is solely controlled by the transition states associated with the oxametallacycle that acts as an intermediate in the formation of EO. An interesting result of this observation is that the selectivity and activity of the overall catalytic reaction are controlled by different reactions within the overall reaction pathway.

The work reviewed above is a wonderful example of the explanatory use of DFT calculations. Prior to these calculations, the precise mechanism of EO formation on Ag had remained unclear, despite many years of applied and fundamental experimental studies. Having said this, I hasten to add that it was only through a close interaction between experimental studies and theoretical calculations that success with this topic was finally possible.

Once the reaction pathway for ethylene epoxidation on Ag(111) was determined, a logical direction was to use this knowledge to identify catalysts with better performance than Ag. This task was tackled recently by Linic, Jankowiak, and Barteau,[56] who decided to focus on the selectivity of the reaction rather than the activity of the catalyst. Linic *et al.* considered bimetallic catalysts in which the majority component was Ag. A first consideration in any effort of this type is whether stable alloys of the compositions of interest exist and whether the bulk alloys of interest also form stable surface alloys. The phase diagrams of binary metal alloys are available from a number of reference sources.[73] The surface segregation energies for a wide range of binary alloys where one metal is present as a dilute component have been computed using DFT and tabulated by Ruban, Skriver, and Nørskov.[74] Linic *et al.* augmented the calculations of Ruban *et al.*, which examined bare surfaces, with surfaces in the presence of adsorbed O to gain insight into the thermodynamically stable state of bimetallic surfaces under reaction conditions. Based on these considerations, Linic *et al.* examined three potential bimetallics, Ag/Cu, Ag/Pd, and Ag/Au. In each case, plane wave DFT calculations were performed using four layers of Ag(111), each containing a 2×2 surface unit cell, with one metal atom in the top layer substituted for the second component of the bimetallic. On each surface, the structure of the oxametallacycle identified as the reaction intermediate in Figure 5 was optimized and the transition states leading to the formation of EO and acetaldehyde were located. The selectivity of a given surface was characterized by the difference in these two reactions' activation energies, ΔE_a. The change in selectivity of the bimetallics relative to pure Ag was characterized by $\Delta\Delta E_a = \Delta E_a$(bimetallic)-$\Delta E_a$(Ag(111)). With this definition, a positive value of $\Delta\Delta E_a$ indicates that the bimetallic will be more selective for EO production than pure Ag. The calculations for Ag/Cu, Ag/Pd, and Ag/Au gave $\Delta\Delta E_a = +1.8$ kcal/mol, +0.7 kcal/mol, and −0.3 kcal/mol, respectively. The individual reaction activation energies were also different on the bimetallic surfaces than on pure Ag. For example, ΔE_a for EO production on Ag/Cu was determined to be ~2 kcal/mol higher than on Ag(111).[75] While at first glance this may seem problematic, it is important to remember that analysis of the reaction pathway for EO production showed that the conversion of the surface oxametallacycle into EO is not a rate determining step in the overall catalytic process.[57] That is, moderate changes in ΔE_a will not change the overall activity of a catalyst, but changes in $\Delta\Delta E_a$ can change the product selectivity of the process. The final outcome from these calculations is the prediction that small concentrations of Cu in the surface of Ag should enhance the selectivity of EO formation, while small concentrations of Pd or Au would have weaker or undesirable effects.

Having identified Cu as a potential additive to Ag for ethylene epoxidation catalysis, it is useful to examine the properties of Cu in Ag–Cu alloys more closely. The phase diagram of bulk Ag–Cu alloys shows that at almost all compositions, mixtures of Ag and Cu will phase separate into an alloy that is very rich in Ag and an alloy that is very rich in Cu.[73] At 200°C, the Ag-rich phase is ~99 at.% Ag and the Cu-rich phase contains < 1 at.% Ag. This situation is very different from the bimetallic alloys of Ag–Pd and Ag–Au, both of which are miscible at all compositions.[73] The DFT calculations of Linic *et al.* indicate that in the presence of surface O, isolated Cu atoms prefer to reside in the (111) surface rather than in the bulk of Ag(111).[56] No calculations have been performed to examine the propensity for Cu to cluster on the surface under these conditions, but the very low concentration of Cu that can exist in the bulk Ag-rich alloy suggests that surface clustering effects are unlikely to be crucial.

To test the prediction that Cu would improve the selectivity of Ag as an ethylene epoxidation catalyst, Linic, Jankowiak, and Barteau tested several Cu/Ag catalysts on porous α-Al_2O_3 monoliths.[56] The Cu concentrations in the catalysts varied from 0 to 0.8 at.%, so only a single bimetallic phase was expected. The catalysts were tested under conditions where the temperature, the feed composition of ethylene and oxygen, and the conversion of ethylene were held constant. Under these conditions, the selectivity of the catalyst increased significantly as the Cu content was increased from 0 to 0.2 at.%. This is a dramatic success; the DFT calculations of transition states for competing reactions on bimetallic catalysts lead directly to the experimental identification of an improved catalyst.

Linic *et al.* also used XPS to probe the surface composition of their experimental catalysts.[56] These experiments were performed on catalyst samples after exposing them to reaction conditions at 500 K.[75] For a sample with 0.1 at.% Cu in the bulk, the surface composition was found to be ~9 at.% Cu. A sample with a higher bulk concentration of Cu, 0.8 at.%, had ~43 at.% Cu in the surface. In both cases, the enrichment of Cu in the surface is very large relative to the bulk concentration. This is qualitatively consistent with the DFT prediction that in the presence of O, isolated Cu atoms have an energetic preference to segregate to the surface of Ag(111).[56] The calculations to date by Linic *et al.* have examined oxametallacycles and reactive transition states in which just one of the relevant surface atoms is a Cu atom. Their XPS results suggest that many reactive events on the catalysts they examined experimentally may in fact involve configurations that interact with multiple surface Cu atoms. Using DFT to understand the spatial correlations between Cu atoms on Cu/Ag catalyst surfaces under reaction conditions and to then probe the implications of these correlations for reaction selectivity seems to be a fruitful direction for further study.

One important component of practical epoxidation catalysis for which many questions remain open is the role of promoters. Industrially, both Cs and Cl are used as promoters.[76] Saravanan *et al.* used cluster DFT calculations to examine the interactions between adsorbed Cs and oxametallacycles on Ag(111).[77] These calculations suggested that both neutral Cs and Cs^+ gave similar outcomes and that the promoter atom made the formation of a surface oxametallacycle less energetically favorable than on the bare Ag surface. More recently, Linic and Barteau used plane wave DFT calculations to probe the effect of adsorbed Cs on the transition states controlling the formation of EO and acetaldehyde from oxametallacycles on Ag(111).[78] The role of Cs was found to be quite similar to that of surface Cu in the Cu/Ag bimetallics discussed above; when Cs was close to an oxametallacycle it stabilized the transition state for EO formation relative to the transition state for the competing pathway. This result is consistent with the experimental observation that Cs enhances the selectivity of ethylene epoxidation in favor of EO.[76] Analysis of the calculated results showed that the Cs acts via dipole interactions with the transition states. This is an encouraging result, but much work remains before the role of promoters is completely understood. A very significant complication in any effort to understand the role of promoters is that at low promoter concentrations it seems likely that the great majority of promoter atoms on an active catalyst will decorate step edges and other defects rather than surface terraces. The adsorption of Cl on Ag(111) has been observed to have dramatic consequences for the structure of steps on these surfaces.[79] Moreover, as noted by Serafin *et al.*, "concentrations of . . . promoters in industrial catalysts are finely adjusted at the ppm level, with differences of as little as 100 ppm or less between commercially acceptable and unacceptable formulations".[76]

2.3.1 The Unreasonable Effectiveness of DFT. Wigner famously wrote about the "unreasonable effectiveness of mathematics" for describing the physical world.[80] A reasonable response to the examples of DFT calculations described above is to be pleasantly surprised at the unreasonable effectiveness of plane wave DFT calculations for describing heterogeneous catalysis. The systematic inaccuracies in plane wave DFT calculations are well documented.[4,16,22] PW91-GGA calculations typically overbind small molecules such as CO and NO on metal surfaces by ~0.5 eV.[22] RPBE-DFT calculations reduce this overbinding considerably, but the root mean square deviations between calculated and experimental binding energies for O, CO, and NO on metal surfaces are still ~0.2–0.3 eV. GGA-DFT calculations are less accurate than hybrid DFT results (or higher level *ab initio* approaches) for small molecules.[16] In some well known examples, DFT incorrectly predicts the adsorption sites of small molecules on metal surfaces when there are small energy differences between the sites.[21,81,82] Moreover, DFT does not correctly describe van der Waals interactions.[83]

Despite this litany of problems, it is quite evident from the examples presented above (and others mentioned later) that plane wave DFT calculations are able to make useful *quantitative* predictions about real materials of interest for heterogeneous catalysis. The question of why this is possible was considered by both Honkala *et al.*[35] and Reuter *et al.*[84] Several factors appear to contribute to the success of DFT calculations. It is reasonable to expect that DFT calculations describe the relative energies of surface processes more accurately than the absolute adsorption energies. The overall rate of a catalytic process is rarely determined by the energetics of a single process, instead, a combination of processes usually contribute to the overall rate, so these relative energies are often more important to overall accuracy than absolute energies.

At the same time, activation energies for reactions and bond energies for the reacting species in catalytic processes are often observed to vary in a concerted manner.[4,35,64,85] This gives rise to compensating influences that greatly reduces the overall sensitivity of calculated catalytic rates to absolute errors in energy.

It is useful to highlight one class of materials for which standard DFT approaches fail quite severely, namely materials in which electron correlation is strong. One example of materials with this characteristic are the transition metal oxides and sulfides.[86–88] In the last several years, efforts have begun to treat phenomena relevant to catalysis on such materials using DFT+*U* methods. These methods combine LDA or GGA DFT with a Hubbard Hamiltonian for the Coulomb and exchange interaction, introducing a single numerical parameter to account for the strength of the Hubbard Hamiltonian.[86–88] Recent applications of DFT+*U* to the adsorption of small molecules on NiO surfaces[86,87] and the structure of Fe_2O_3 and $Cr_2O_3(0001)$ surfaces[89] have suggested that these calculations can perform very well in correcting the deficiencies of traditional DFT for these materials. DFT+*U* methods play an important role in contemporary treatments of multifunctional materials.[90–93] Because DFT+*U* is now widely available via its implementation in VASP,[86–88] it is likely that applications of this approach will rapidly grow in popularity.

The development and implementation of quantum chemistry tools that are applicable to the materials of interest in heterogeneous catalysis that improve upon the accuracy of current plane wave DFT methods will remain an important area for the foreseeable future. Nevertheless, as illustrated by the success stories above, it is not necessary to achieve "chemical accuracy"[16] in order for DFT calculations to make substantial quantitative contributions to our understanding of existing catalysts and the search for improved materials. Theoretical treatments of materials with real world applications are always performed with a tension between the desire to use the most accurate computational methods, with the seemingly inevitable consequence of increased computational effort, and the need to consider materials that are complex enough to be representative of real-world phenomena. Given the very great complexity of realistic heterogeneous catalysts, it can be reasonably argued that DFT calculations currently form the best tradeoff between these two competing desires.

3 Areas of Recent Activity

The success stories reviewed above are, of course, just several of a larger number of areas of heterogeneous catalysis in which DFT-based calculations have made valuable contributions in recent years. In this section, three topics that have attracted considerable attention are described, *ab initio* thermodynamics, the catalytic activity of nanoclusters of gold, and computational efforts to screen bimetallic catalysts. Again, the choice of these topics does not indicate that they are the only topics of interest in the community, but they give a good representation of current approaches.

3.1 *Ab Initio* Thermodynamics. – Almost anyone who uses DFT to describe real materials has heard after a seminar or from a referee the common objection that DFT is "only a zero temperature description". This view stems from the correct observation that DFT calculations typically only give information on the ground state of the system of interest in a local energy minimum of the global potential energy surface. One response to the "zero temperature" complaint is that in many situations, transition state theory accurately describes the rate at which a system will make transitions between local minima over an intervening transition state, so knowledge of the minima and transition states gives accurate information about finite temperature rates. A less trivial aspect of this objection is that the physical structure of many materials changes in a non-trivial way with temperature and other environmental variables (*e.g.* humidity, pressure *etc.*), so predictions made with the zero temperature structure simply might not be relevant to the real material. To give just one illustrative example, the (001) surface of the widely studied perovskite $SrTiO_3$ exhibits at least three distinct surface reconstructions under oxidizing conditions and others under reducing conditions.[94] To accurately predict the properties of any material with similar behavior using DFT calculations, the thermodynamic stability of the material's structure at the experimental conditions of interest must be established.

Predicting the thermodynamic stability of materials, or equivalently, the material's phase diagram, from DFT can be achieved by what has become known as *ab initio* thermodynamics. Broadly, these methods use ground state energies obtained from DFT calculations in combination with statistical mechanical information to calculate the relative free energies of a set of potential phases. This concept goes back at least as far as the work by Kaxiras *et al.*[95] and Qian *et al.*[96] in the late 1980s on GaAs surfaces. In the past few years, *ab initio* thermodynamics has been widely used to examine the structure of a large number of technologically relevant oxides under practical reaction conditions. Much of the credit for making these methods an indispensable tool for studying heterogeneous catalysis should go to Scheffler and his co-workers.[97–99] An extensive review of *ab initio* thermodynamics as applied to surface oxides was recently given by Stampfl.[100] For this reason, my overview of recent work in this area is less detailed than some of the other topics covered in this article. While applications of *ab initio* thermodynamics relevant to heterogeneous catalysis have focused almost exclusively on surfaces, the same general ideas are being developed in other contexts such as predicting the phase diagrams of bulk materials.[101,102]

Detailed descriptions of the theory underlying *ab initio* thermodynamics are available in the review by Stampfl,[100] as well as in the papers by Zhang *et al.*[103] and Reuter and Scheffler.[99] Rather than reiterating these discussions, it is more useful to present a simple pedagogical example, which I have modeled on the calculations by Greeley and Mavrikakis of H adlayers on Ni(111).[104] Similar calculations for H on Pt(111) have been performed by Légaré.[105] Consider the situation where DFT calculations have been used to minimize the energy, E, associated with n different adlayer structures of H on a Ni surface. We would like to know which one of these structures is the most stable at a specified temperature, T and hydrogen pressure, p_{H2}. The grand potential of structure i is defined by

$$\Omega_i(T, \mu_H, \mu_{Ni}) = E_i - TS_i - \mu_H N_{H,i} - \mu_{Ni} N_{Ni,i} \tag{1}$$

where S_i is the entropy of structure i, μ_a is the chemical potential of species a, and $N_{a,i}$ is the number of atoms of species a in structure i. The most stable structure is the one that minimizes the grand potential at the conditions of interest. If we consider only structures that all have the same number of metal atoms, the last term in Eq. (1) is a constant for all structures. In many circumstances for surface structures, differences in the entropic term, TS_i, between potential structures are small relative to the other contributions, so we neglect this term. This approximation and methods for treating the entropic contributions accurately are discussed by Reuter and Scheffler.[99] Examples are known for bulk materials where vibrational entropic contributions are not negligible.[106] For the situation we are considering, the grand potential can be written as

$$\Omega_i(T, \mu_H) = E_i - \mu_H N_{H,i} + \Omega^0, \tag{2}$$

where Ω^0 is a constant for all the structures being considered. The chemical potential of H is related to the H_2 pressure by first noting that the adsorbed H must be in thermodynamic equilibrium with the gas phase H_2, so $\mu_H = (1/2)\mu_{H2}$. The chemical potential of gas phase H_2 is, for conditions where the ideal gas approximation is reasonable,

$$\mu_{H2} = \mu^0_{H2} + kT \ln(p_{H2}/p^0_{H2}).$$

Here the superscript 0 denotes the standard state, T=298 K and p=1 atm. The standard state chemical potential, μ^0_{H2}, can be evaluated using thermodynamic tables. At higher pressures, more accurate equations of state than the ideal gas law can be used.

Figure 6 shows a schematic example of this approach for a situation where structures with N_H = 1, 1/2, 1/4, and 0 are considered, the latter being a clean surface. Here, N_H is the number of H atoms per surface unit cell. This diagram

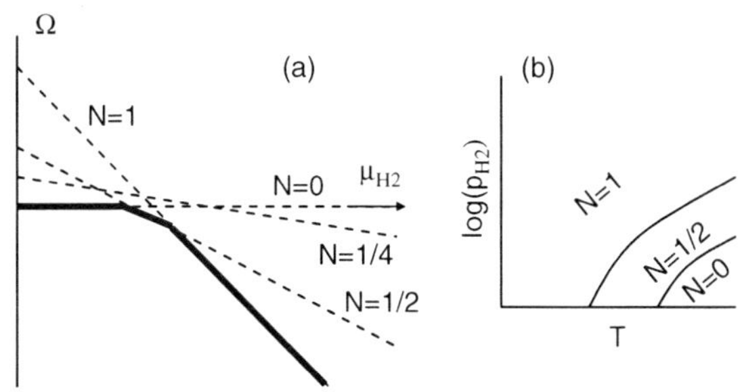

Figure 6 *(a) A schematic illustration of the determination of the stable adsorption configuration of H on a surface and (b) the resulting phase diagram. See text for details*

is very similar to the one computed by Greeley and Mavrikakis for H adlayers on Ni(111).[104] The dashed lines in Figure 6(a) show the grand potential for each of the structures listed above as a function of μ_{H2}. At each value of the chemical potential, the structure that minimizes the grand potential is the equilibrium structure. These structures are denoted in Figure 6(a) by a thick solid line. It can be seen that for low enough values of μ_{H2}, the clean surface ($N_H = 0$) is favored. The structure with $N_H = 1/4$ is not favored under any conditions. Replotting the stable structure as a function of temperature and p_{H2} gives the phase diagram shown in Figure 6(b).

Ab initio thermodynamics has now been successfully applied to a large collection of materials relevant to heterogeneous catalysis. The existence of oxide layers on transition metal surfaces has received much attention from *ab initio* thermodynamics calculations.[31,100,107,108] A number of papers have examined surface oxide formation on Ag(111).[70,109–111] Surface oxide formation on RuO_2(110) has similarly been studied[99] including the effect of humidity[112] and the presence of O_2 and CO.[113] The work of Sun, Reuter, and Scheffler[112] includes a useful discussion of the importance of vibrational entropy in determining phase diagrams for systems of this type. Todorova *et al.* and Rogal *et al.* examined oxide formation on Pd,[114,115] and Légaré has performed similar calculations for Pt(111).[116] Tang, Van der Ven, and Trout have also reported a detailed phase diagram for O on Pt(111) with coverages up to half a monolayer.[117]

The surface structure of α-Al_2O_3(0001) (sapphire) as function of oxygen chemical potential was first examined with *ab initio* thermodynamics by Wang, Chaka, and Scheffler.[98] Recent calculations have extended this initial work to establish the surface structure of six α-Al_2O_3 surfaces as functions of oxygen and hydrogen pressure[118] and to probe the structures of H and C overlays on α-Al_2O_3(0001) as functions of the H and C chemical potentials.[119] Calculations have also recently been reported for the thermodynamically stable structure of ultrathin aluminum oxide films on NiAl(110).[120] Zhang *et al.* have examined ultrathin Al_xO_y films on Al-doped Cu(111) as function of O_2 pressure and Al activity at 1200 K.[103]

The surface phase diagram of Cr_2O_3(0001) as a function of O_2 pressure and temperature has been calculated by Wang and Smith.[121] Kresse and co-workers have examined the surface terminations of V_2O_3(0001) as a function of O_2 pressure[122] and V_xO_y films grown on Pd(111).[123] The structure of V_2O_5 surfaces has been examined by Ganduglia-Pirovano and Sauer.[124,125] The surface structure of α-Fe_2O_3(0001) as a function of oxygen chemical potential[97] and the structure of hydrated α-Fe_2O_3(0001) as a function of oxygen chemical potential with μ_H corresponding to 1 bar of H_2O[126] have both been treated with *ab initio* thermodynamics methods. Loffreda and Sautet have used DFT methods to examine the reconstructions of Au(110) as a function of temperature and CO pressure.[127] Eichler and Kresse used chemical potential-based arguments to aid in describing the surface terminations of pure and yttria-doped zirconia surfaces.[128] Bottin *et al.* studied the surface phase diagram of $SrTiO_3$(110) as function of Sr and O chemical potential.[129]

As mentioned briefly above, *ab initio* thermodynamics have not only been applied to surfaces, but to bulk materials. Stampfl and Freeman have investigated the structure of tantalum nitride in this way.[102] These methods are well suited for predicting the solubility of atoms in bulk materials. Calculations of this type have been reported for the solubility of O in Pd[114] and H solubility in CuPd alloys.[130]

Ab initio thermodynamics methods can also be applied to nanoparticles, a topic that has potential to make a significant impact on heterogeneous catalysis. The main example of this approach to date has been the study of MoS_2 nanoparticles. MoS_2 is of great importance as a hydrodesulfurization catalyst. A large number of DFT calculations have been applied to examine various aspects of HDS catalysis on MoS_2.[131–142] The edge sites of MoS_2 are the active sites for catalysis, while the MoS_2 basal plane is essentially inactive. As a result, understanding the detailed structure of the edge sites of MoS_2 nanoparticles is crucial if an atomic-scale understanding of this catalyst is to be achieved. Bollinger, Jacobsen, and Nørskov first used *ab initio* thermodynamics to describe the structure of MoS_2 nanoparticles as functions of temperature, H_2 and H_2S pressure.[131] Comparisons made between calculated structures and experimental STM images allowed the structure of the experimentally observed nanoparticles to be established. Further comparisons between the predicted structures and STM images and refinements of the *ab initio* phase diagram have since been made by the same group.[143] The influence of the catalyst's support on the MoS_2 nanocluster morphology has recently been studied with *ab initio* thermodynamics methods by Arrouvel *et al.*[144]

Two caveats must be kept in mind when interpreting results from *ab initio* thermodynamics. First, these methods predict the thermodynamically stable state among a catalog of known structures. If the structure (or process underlying a structure) that appears in reality is not included in the catalog then its presence will of course not be predicted on a phase diagram. One interesting example of this type is seen in the kinetic modeling of Reuter *et al.* of CO oxidation over $RuO_2(110)$ that was reviewed above in section 2.1. Because the *ab initio* thermodynamics description of this system[113] did not allow for the removal of adsorbed O via CO_2 formation, the thermodynamic phase diagram overestimates the range of conditions where the surface is occupied by both O and CO at steady state.[84] An approach that is likely to be important in the future is the use of catalogs of DFT calculations to derive interaction parameters for general lattice models via cluster expansions.[100,117] This technique can reduce the uncertainty that arises from basing *ab initio* thermodynamics calculations on a limited catalog of structures. The second caveat is also reasonably obvious; if processes in real systems are strongly influenced by kinetic limitations then the outcome in these systems may differ significantly from the thermodynamic equilibrium state. An interesting example of this situation was reported recently by Lundgren *et al.*, who showed that the initial formation of oxide layers on Pd(100) at ambient pressures is strongly kinetically limited.[145]

3.2 Catalytic Activity of Supported Gold Nanoclusters. – Gold is a noble metal, so the realization that nanometer-sized clusters of gold can act as very effective catalysts for low temperature oxidation of CO and a number of other reactions has created immense interest in the last decade.[146–151] Until quite recently, the origin of the catalytic activity in supported gold catalysts has been strongly debated. In the past several years, DFT calculations by a number of groups have provided great insight into the issues surrounding gold catalysis. Here, we focus on the oxidation of CO, since essentially all theoretical work on heterogeneous gold catalysis has been applied to this reaction. CO oxidation is of course convenient for theoretical work because it involves only a small number of reacting atoms. But this reaction is not only of academic interest; as pointed out by Campbell, the development of supported catalysts that efficiently oxidize CO at low temperature (below $\sim 200°C$) could have a dramatic impact on reducing automobile pollution.[150]

Before considering catalysis by supported Au nanoclusters, I first review efforts to apply theory and experiment to gas phase Au clusters. An important feature of work on small gas phase Au clusters is that it allows high level *ab initio* methods to be compared to the DFT methods that must be used when considering heterogeneous catalysis. I then consider two specific examples of supported Au clusters in some detail, namely Au supported on MgO(100) and $TiO_2(110)$. Following these specific examples, I turn to the general question of why Au nanoclusters can act as active catalysts and highlight some future challenges for theoretical work in this area.

Pyykkö has reviewed theoretical efforts to describe the chemistry of gold, with a focus on molecular systems,[152] highlighting the importance of relativistic effects in the chemistry of gold. A recent example of the importance of using relativistic approaches when performing calculations with Au are the DFT calculations and experiments for the gas phase anion Ag_{55}^- by Häkkinen *et al.*[153] The ultrasoft pseudopotentials used in most plane wave DFT calculations[28,154] and the more recent projector augmented wave (PAW) potentials[28,155] are based on scalar relativistic all electron calculations of atoms, so these approaches include scalar relativistic effects.

3.2.1 Gas Phase Au Clusters. Before considering the catalytic activity of supported Au nanoparticles, we first review recent activities associated with the structure and adsorption properties of Au clusters in the gas phase.

Experiments and DFT calculations have been used to examine the binding of O_2 with gas phase Au_n^- anions.[156–159] For the gas phase anions, O_2 readily binds to clusters with even n but clusters with odd n are relatively inert for $n<20$.[157,158] In the gas phase, Au_n^- anions are found to be planar for $n<13$.[159]

The interaction of O_2 and CO with neutral Au_n clusters with $n = 5–10$ has been recently explored in DFT calculations by Fernández *et al.*[160] The O_2 adsorption energy exhibits odd-even effects. CO prefers to bind on the least coordinated Au atom except for $n = 5$ and 7.[160]

Franceschetti *et al.* used DFT to examine interactions of O_2 with Au_n and Au_n^- for $n = 3–6$.[161] They concluded that chemisorbed states where the two O atoms where spatially separated (*i.e.* O_2 was dissociated) were in general more stable than configurations with molecularly adsorbed O_2, and that O_2 binds more strongly to anionic clusters than to neutral clusters.

Phala *et al.* used DFT to probe H and CO adsorption on neutral Au_n clusters with $n = 1–13$.[162] 3D clusters were used for $n > 6$. One interesting result from this work was the observation that the adsorption energy of H on Au_6 on which CO was already adsorbed was 29 kJ/mol smaller than the H adsorption energy on the bare cluster.

Olson *et al.* have compared plane wave DFT with several higher level *ab initio* methods for predicting the structure of neutral Au_6 and Au_8.[163] The highest level *ab initio* calculations used MP2 geometry optimization with large basis sets, followed by single point CCSD(T) energy calculations. This approach was compared to plane wave DFT calculations performed with the PW91 functional and two different pseudopotentials. For Au_6, CCSD(T) and DFT both predict a planar structure. The energy difference between the lowest and second lowest energy isomer was calculated to be 12.4 kcal/mol with CCSD(T) and 10.5 kcal/mol with plane wave DFT. The agreement is not so satisfactory for Au_8, where CCSD(T) predicts a nonplanar structure as being most stable with a planar structure less favored by 4.7 kcal/mol. Plane wave DFT predicts the reverse, with the planar structure favored by 7.0 kcal/mol. Olson *et al.* point out that they are currently unable to geometry optimize the Au_8 clusters using CCSD(T), so the effect of this optimization on the relative cluster energies is not known. It is clear, however, that the triples contributions in the CCSD(T) single point calculations stabilize the nonplanar isomers of Au_8 relative to the planar structure; an electron correlation effect that cannot be captured correctly by DFT. Varganov *et al.* compared CCSD(T) and plane wave DFT calculations for the binding and dissociation of H_2 on neutral Au_2 and Au_3.[164] The calculated energy barrier relative to gas phase H_2 for dissociation of H_2 on Au_2 was 0.55 eV with CCSD(T), 0.34 eV with DFT using the PW91 functional, and 0.47 with DFT using the RPBE functional.[164] Varaganov *et al.* have also compared plane wave DFT and CCSD(T) results for the interaction of O_2 with neutral and anionic Au_2 and Au_3.[165] For these systems, there are several notable examples where CCSD(T) and plane wave DFT with the PW91 functional give qualitatively different results. For example, the binding energy of one O_2 molecule to Au_2^- is found to be 1.07 eV with CCSD(T) and 1.40 with PW91. The binding of second O_2 molecule to Au_2^- is found to be unfavorable by 0.56 eV with CCSD(T) but favored by 0.71 eV with PW91. This example is a severe failure of plane wave DFT. For neutral Au_3, the binding energy of one O_2 molecule was found to be 0.08 eV with CCSD(T) and 0.90 with PW91. It was concluded that treating O_2 with DFT is problematic, and that the deviations between DFT and CCSD(T) cannot be quantified by any simple rule.

There are considerable differences between the properties of supported Au clusters and gas phase clusters, particularly when the supported clusters bind near surface defects that can tightly bind Au. For example, no evidence for odd-even effects are seen in studies of CO oxidation on size resolved Au clusters on Mg(001),[166] and the 2D to 3D transition may occur for small clusters on supports than in the gas phase.[167] While the properties of gas phase clusters may not be directly connected with the details of supported catalyst particles, the careful comparisons described above between DFT calculations and high level *ab initio* calculations for small Au clusters should be carefully considered by anyone who applies DFT to study supported catalysts.

3.2.2 Au Clusters Supported on MgO(100). Yan *et al.* recently reported experiments demonstrating a strong correlation between the density of F-centers on an MgO support and the catalytic activity for CO oxidation of Au clusters on this support.[168] Del Vitto *et al.* used plane wave DFT calculations to examine the diffusion and trapping of Au atoms and dimers on MgO(100) terraces, steps, F centers, F^+ centers, and divacancies, concluding the F-centers are the dominant structure associated with trapping of Au atoms and formation of nanoclusters.[169]

Häkkinen *et al.* used experiments to explore the catalytic activity for CO oxidation of very small Au clusters bound on F-centers on MgO(100).[167] These experiments give superb size-resolved information. Gold clusters with less than 8 atoms were essentially inactive, but Au_8 clusters showed strong activity. Including a single Sr atom in the clusters dramatically changed this result; Au_nSr clusters with $n \geq 3$ were catalytically active, although CO oxidation occurs at higher temperatures on these clusters than on active pure Au clusters.[167] Experiments have also shown that Au_8 clusters deposited on F-center-free MgO(100) are essentially inactive.[170]

Häkkinen *et al.* also used DFT calculations to investigate the structure and reactivity of Au_8, Au_4, and Au_3Sr bound on F-centers on MgO(100).[167] The calculations showed that the clusters were strongly bound at the defect sites relative to non-defective MgO(100). Interactions of the clusters with the surface included charge transfer to the clusters of 0.3–0.5 e. The binding of O_2 to Au_4 was found to be weak (0.18 eV) and to only slightly perturb the O–O bond length compared to the gas phase molecule. These observations were used to account for the inactivity of this cluster for CO oxidation. The binding of O_2 to Au_3Sr was much stronger (1.94 eV) and takes place primarily through interactions of the adsorbate with the Sr atom.

Häkkinen *et al.* examined two isomers of Au_8 bound on an F-center of MgO(100),[167] a quasi-planar structure and a two-layered structure that is similar to the structure of neutral gas phase Au_8. After O_2 adsorption, the quasi-planar structure is the more stable of the two by 0.10 eV. Based on the concept that clusters of both types must exist in real materials, the bonding of O_2 was primarily examined with respect to the slightly less stable two-layered structure. O_2 adsorbed on this cluster is accompanied by an extension of the O–O bond length by 0.18 Å compared to gaseous O_2 and the adsorbed molecule has zero net spin, unlike the gas phase triplet state. The adsorption of O_2 on this cluster is also accompanied by considerable structural rearrangement of the cluster itself.

The coadsorption of CO and O_2 on the metastable Au_8 cluster described above has been further examined in a series of experiments and calculations by Yoon *et al.*[170] These calculations compared the Au_8 cluster bound on an F-center on Mg(001) and on a defect-free surface. Experimentally, the former is active for CO oxidation while the latter is not.[170] The calculations confirm that the cluster is much more strongly bound on the F-center than on the defect-free surface. The net charge transfer to the cluster-adsorbate complex was ~1.5e (1e) on the F-center (defect-free surface). A key point of this combined experimental and DFT study was that shifts in CO stretching frequencies on these clusters could be used as a means to probe the charging of the clusters.

A somewhat surprising result from the recent calculations of Yoon *et al.* is that the binding energies of CO and O_2 do not vary dramatically between Au_8 clusters on F-centers and defect-free surfaces. The energy required to detach CO and O_2 from these clusters varied only from 1.09–1.16 eV, regardless of the spin state of the adsorbed O_2. All the clusters examined also showed considerable extension of the O_2 bond length upon adsorption. It is therefore unclear from these calculations to date why such a marked difference in catalytic activity is observed experimentally between Au_8 clusters on F-centers and on defect-free surfaces. The reaction mechanism for CO oxidation on Au_8 clusters on Mg(001) was studied several years ago by Sanchez *et al.*[166] DFT calculations indicated that both Eley-Rideal and Langmuir-Hinshelwood mechanisms can proceed with small or no energy barrier on clusters on both F-centers and on defect-free surfaces. These mechanisms involve O_2 bound on top of the cluster. By comparison with temperature programmed reaction experiments, it appears that this low temperature reaction channel only accounts for a small amount of the overall CO oxidation for clusters bound at F-centers.[166] When O_2 is bound near the base of the cluster, a Langmuir-Hinshelwood mechanism can allow CO oxidation with a barrier of ~0.5 eV (~0.8 eV) on the F-center (defect-free surface). It was this state that was considered in the recent report by Yoon *et al.*[170] The calculated reaction barrier for the cluster on the defect-free surface[166] is comparable to the barrier for CO desorption.[170] When the cluster is bound on an F-center, however, the computed reaction barrier is somewhat smaller than the CO desorption barrier. These observations suggest, at least qualitatively, that the F-center bound cluster will be more active for CO oxidation than the cluster bound on the defect-free surface, as observed experimentally.

The work of Landman and co-workers has provided great insight into the details of CO oxidation on Au_8 clusters on MgO(100). Several open questions remain with respect to this system, however. What is the role of the thermodynamically stable quasi-planar Au_8 cluster[167] in CO oxidation? The focus of work until now has been on the two-dimensional Au_8 cluster, which is predicted by DFT calculations to be slightly less stable than the quasi-planar cluster. Can the information that is available on reaction pathways[166] be extended to quantitatively examine the differences in reactivities between clusters bound at F-centers and on defect-free surfaces? Dramatic differences in reactivity between clusters of this type are observed experimentally.[166,168]

Molina and Hammer have performed extensive DFT calculations of CO oxidation by Au on defect-free MgO(100).[171,172] The objective of this work was to provide insight into the behavior of a range of Au clusters sizes, rather than the highly specific cluster size studied by Landman and co-workers. They note that formation of (100)-oriented Au nanoparticles is favored on MgO(100) by the small mismatch between these materials and that the relatively low adhesion energies between Au and MgO promote 3D growth of Au clusters.[172] This is a relatively common situation for metals deposited on metal oxide surfaces with small lattice mismatches.[173,174] DFT calculations using the PW91 and RPBE functionals were used to examine several structures characteristic of edge sites on Au nanoclusters on Mg(100).[171,172] Molina and Hammer stress that their approach does not describe corner sites, which are expected to make a sizeable contribution to the overall reactivity of small Au clusters. An especially useful feature of Molina and Hammer's work is that they considered the intermediates and transition states involved in a complete catalytic cycle of CO oxidation. They find that the favorable reaction pathway first involves adsorption of CO, followed by a reaction with gas phase O_2 to form a $CO \cdot O_2$ adsorbed intermediate:

$$CO + * \rightarrow CO^*,$$

$$O_2 + CO^* \leftrightarrow CO \cdot O_2^*.$$

Here, * represents a vacant surface site or an adsorbed species. The adsorbed intermediate then reacts to form CO_2 that readily desorbs:

$$CO \cdot O_2^* \rightarrow CO_2 + O^*.$$

This process leaves adsorbed atomic O on the surface, a species that is not spontaneously generated by the dissociation of adsorbed O_2 because of the large energy barriers associated with this process.[172] The catalytic cycle is completed by the reaction of adsorbed CO with the adsorbed atomic O:

$$CO^* + O^* \rightarrow CO_2.$$

This reaction mechanism is distinct from the reaction mechanisms studied on Au_8 clusters by Sanchez *et al.*[166] Although the barriers to each step in this reaction cycle were found to be relatively small on each of the Au environments examined, Molina and Hammer argue that only sites where CO adsorption is relatively strong and there is a sizeable energetic driving force for the formation of the $CO \cdot O_2^*$ intermediate will make the overall pathway feasible. These conditions were met only for sites adjacent to the MgO support, for in these sites the $CO \cdot O_2^*$ intermediate is stabilized by favorable interactions with the support.

To examine the role of corner sites in small Au clusters, Molina and Hammer performed RPBE calculations of CO, O_2, and $CO \cdot O_2$ adsorption on an Au_{34} cluster on defect-free MgO(100).[175] CO and $CO \cdot O_2$ adsorption was found to be favored at corner sites relative to other possible adsorption sites. The binding of CO was ~0.2 eV stronger on corner sites than on other types of edge sites. O_2 adsorption was less sensitive to site coordination. These results support the concept that a significant fraction of the activity of very small Au clusters can arise from reactions associated with adsorption on corner sites.

Molina and Hammer have also studied the activity of a Au_{20} cluster, both in the gas phase and supported on MgO(100).[176] The activity of the supported cluster, as characterized by the binding energy for adsorbed O_2, is comparable for the cluster on a defect-free support, on a surface F-center, and on a surface F^+-center. This result is initially surprising when compared to the gas phase behavior of the same cluster; the gas phase neutral cluster binds O_2 very weakly while the gas phase anion binds O_2 strongly.[176] This difference between the neutral and anionic cluster would seem to imply that similar differences should appear between the cluster supported on defect-free and defect sites on MgO(100). Molina and Hammer showed that in fact when Au_{20} is located on an F^+-center, the extra electron that would be expected to make the cluster similar to the anionic gas phase cluster is pinned at the surface defect, so charge transfer to the active sites on the cluster is minimal. In the same paper, Molina and Hammer used DFT to test several monovalent atoms as potential additives to Au_{20} clusters as a means to enhance the activity of these clusters.[176] These calculations suggested that the presence of a single Na atom in a Au_{20} cluster may significantly enhance the cluster's catalytic activity.

3.2.3 Au Clusters Supported on $TiO_2(110)$. The rutile surface of TiO_2, $TiO_2(110)$, is the most stable surface of this common oxide support. Several groups have used DFT to examine the binding of Au atoms, clusters, and layers on defect-free $TiO_2(110)$, most recently Lopez and Norskov,[177] Wang and Hwang,[178] Okazaki *et al.*,[179] and Vijay *et al.*[180] Earlier calculations had made conflicting predictions about the preferred binding site for Au on this surface.[181–183] It is now clear that multiple potential binding sites for isolated Au adatoms exist with relatively similar binding energies, and that the preferred binding site changes as the surface coverage is increased.[177–180] As Au clusters grow on $TiO_2(110)$, the Au(111) plane is parallel to the surface plane.[179] Some expansion/compression of the Au lattice spacings along orthogonal directions in the plane of the surface are necessary to make Au(111) commensurate with $TiO_2(110)$.[177] The relatively weak interactions between Au and the defect-free $TiO_2(110)$ surface mean that 3D cluster growth is favored for thick Au films.[177]

Similar to the case of MgO discussed above, defects are common on $TiO_2(110)$, especially oxygen vacancies. Wang and Hwang,[178] Okazaki *et al.*,[179] and Vijay *et al.*[180] all used DFT to probe the binding of Au to O vacancies on $TiO_2(110)$ and concluded that Au binds much more strongly to these defects than on the stoichiometric surface. The most detailed examination of the role of surface defects has been a combined experimental and theoretical effort by Wahlström *et al.*[184] In this work, scanning tunneling microscopy was used to directly show that Au adatoms strongly bind to oxygen vacancies on $TiO_2(110)$ and these defects are shown to play a key role in the nucleation and growth of Au clusters on the surface. An interesting feature of the DFT calculations in this work was the observation that very little charge transfer occurs during binding of Au adatoms on O vacancies, a situation that is rather different from Au adsorption on defects on MgO(100). A similar conclusion was reached by Vijay *et al.*[180]

Liu *et al.* used GGA calculations with the PBE functional to probe adsorption and reaction of O_2 and CO on Au supported on defect-free $TiO_2(110)$.[185] These calculations used a bilayer of Au forming a continuous strip across the support surface. O_2 adsorption was found to be favorable only for adsorption on Au adjacent to the support surface. An analysis of the bonding character of these states indicated that the support enhances charge transfer from Au to O_2 for Au atoms in close proximity to the support. CO oxidation via a reaction between adsorbed CO and adsorbed O_2 at the Au/support interface was found to have a small activation barrier (0.1 eV). CO oxidation pathways involving adsorbed atomic oxygen were not examined, although adsorbed O is created by the pathway mentioned above. Also, dissociation of adsorbed O_2 was found to occur with a barrier of 0.5 eV.

Molina, Rasmussen, and Hammer have also used DFT to probe CO oxidation on strips of Au on $TiO_2(110)$.[186] These calculations considerably extend the earlier report by Liu *et al.*,[185] especially because they compare Au clusters on defect-free surfaces as well as on and next to oxygen vacancies. As with their similar work on MgO(100),[171,172] the calculations of Molina *et al.* provide insight into the edges of Au nanoclusters that are likely to appear in experiments but cannot give similar insight into adsorption or reactions involving corner atoms on nanoclusters. Two distinct types of O_2 adsorption configurations were found, one in which O_2 binds to a Ti atom directly adjacent to a Au cluster, and another where O_2 "leans" on the cluster, forming bonds to both the cluster and to the adjacent support. The former situation binds O_2 more strongly than the latter, and having the cluster bound to a surface defect strongly enhances the binding of O_2 in both cases. Examples of these two structures are shown in Figure 7. In the absence of the adsorbed Au, O_2 does not bind to defect-free $TiO_2(110)$, although it will bind strongly to O vacancies.[186] Although placing O_2 on the support adjacent to a Au cluster gives a stronger binding energy than the leaning configuration, Molina *et al.* argue persuasively that the leaning configurations are likely to exist under practical experimental conditions and that it is these configurations that are likely to be involved in oxidation reactions since the O_2 is only moderately bound. CO was determined to only bind with appreciable strength to the least coordinated Au atoms in the nanoclusters examined. From this location, the activation energy for the adsorbed CO to react with O_2 in the leaning configuration was found to be only 0.15 eV for Au clusters adsorbed either on or adjacent to an oxygen vacancy in the support. This barrier is quite similar to the barrier identified by Liu *et al.* for the similar process on a defect-free support.[185] In the reaction mechanisms calculated by Liu *et al.*[185] and Molina *et al.*,[186] $CO \cdot O_2$ is not formed as a stable reaction intermediate as it is in the reaction pathways examined by Molina *et al.* for Au/MgO(001).[171,172]

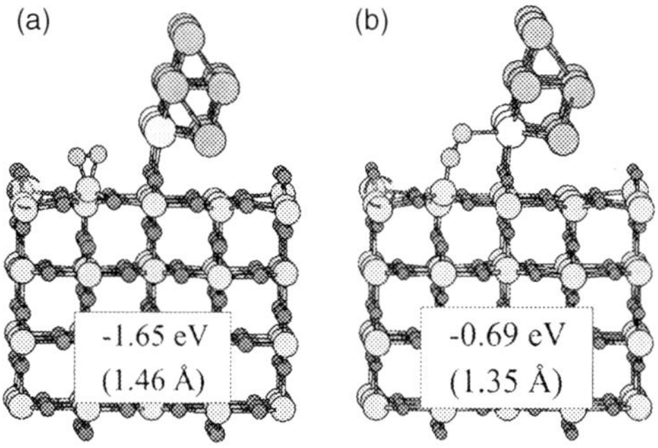

Figure 7 *Examples of O_2 adsorption (a) on a TiO_2 surface adjacent to a Au nanoparticle and (b) bridging between the TiO_2 surface and a Au nanoparticle from the calculations of Molina, Rasmussen and Hammer.*[186] *The insets show that adsorption energy and O_2 bond length. [Reproduced with permission from AIP]*

The calculations by Liu *et al.* and Molina *et al.* reviewed above all dealt with two dimensional strips of Au on $TiO_2(110)$. The only DFT calculations for a finite-sized Au cluster on $TiO_2(110)$ to date are from a recent paper by Remediakis, Lopez, and Nørskov.[187] These calculations examined an Au_{10} cluster bound on three adjacent oxygen vacancies on $TiO_2(110)$. Many of the results are similar to those found previously for two dimensional strips of Au, for example, the most favorable binding site for O_2 involves a site in which the O_2 forms a bridge between a Ti atom in the support and a Au atom at the base of the cluster. A low energy reaction pathway involving this adsorbed O_2 and CO adsorbed on the cluster was found, which is similar in character to those known previously. The more significant outcome from this work was the identification of a second reaction pathway that involves only adsorbates on undercoordinated sites on the cluster, with no involvement of species partially adsorbed on the support. The activation energy for both pathways was found to be $\sim$0.40 eV relative to coadsorbed CO and O_2.

3.2.4 A General View of the Activity of Au Nanoclusters. The calculations reviewed above represent considerable progress in understanding the origin of catalytic activity in supported Au clusters. There are now multiple examples of reactive pathways for CO oxidation on Au clusters that have been identified using DFT calculations, and it should no longer be considered surprising that very small Au clusters are catalytically active on a wide range of supports.

An important direction for continuing work is to classify the effects that contribute most strongly to this activity. This concept was addressed by Lopez *et al.* in 2004, who compiled experimental data from a range of supported gold clusters and discussed the expected activity of Au clusters in terms of under-coordination of Au atoms, charge transfer and strain, among other effects.[188] Lopez *et al.* suggested that the dominant effect is the presence of low-coordinated sites on small clusters. The observation that CO and O_2 bind more tightly to less coordinated Au atoms has been made in multiple DFT studies, including studies of gas phase clusters,[189] Au surfaces,[127] and supported Au clusters.[175,176,186,188] This argument does not mean that charge transfer and other effects are unimportant, just that these effects are less dominant than the effects of undercoordinated Au atoms. Charge transfer clearly enhances the binding of O_2 to gas phase Au clusters, but detailed studies to date seem to indicate that charge transfer from supports to Au clusters is not dramatic.[167,176] The role of charge transfer has also been challenged by recent experiments by Guzman and Gates that show spectroscopically that both neutral and cationic gold is present in working supported catalysts on MgO supports.[190]

The main effect not considered in the discussion by Lopez *et al.*[188] is the relative role of support-mediated reactions in the activity of Au clusters. This point of view is quite reasonable, for the aim of the paper by Lopez *et al.* was to focus on effects that occur across a wide range of supports. Nevertheless, it is interesting to note that most of the reaction pathways that have been specifically identified to date by DFT calculations for CO oxidation on supported gold invoke species that are adsorbed on the support or on the support/cluster interface.[167,171,172,185,186] Only the work of Sanchez *et al.* for Au_8 on MgO(100)[166] and calculations of Remediakis *et al.* for Au_{10} on reduced $TiO_2(110)$[187] have demonstrated a CO oxidation pathway that involves only adsorbates in contact with the Au cluster. In both of these cases, competing pathways that involve the support/cluster interface were also observed.

To resolve the relative importance of support/cluster and cluster-mediated reactions, it would be very interesting to take up the challenge of predicting the overall reaction rate for CO oxidation for a specific Au cluster using DFT methods. This would certainly not be a trivial task, but the recent successes in predicting the overall reactivity of other catalysts that were discussed above in section 2[35,84] provide hope that it is feasible. To make this effort worthwhile, it would ideally be performed for a material that can be accurately characterized experimentally. Two materials appear attractive from this point of view. First, size-selected experiments for cluster activity on MgO(100) supports exist, as do extensive DFT studies of Au clusters on this support (see section 3.2.2). Another very interesting material of more recent origin comes from the experimental work of Chen and Goodman, who have examined highly ordered Au thin films on ultrathin TiO_2 films that were in turn grown on Mo(112).[151] In view of the comments above, it is interesting to note that for this material the highest activity for CO oxidation was observed for a film that completely covers the support, apparently ruling out Au-support-mediated reactions. This material exposes a regular array of undercoordinated Au atoms and appears well suited for examination with theoretical methods.

3.3 Bimetallic Catalysts. – It is hardly surprising that metallic catalysts containing more than one species of metal are more chemically versatile than monometallic catalysts. The fact that an enormous variety of bimetallic (and more complex) catalysts can be made with relatively simple synthetic techniques poses both a great opportunity and a challenge for the development of improved catalysts. A key challenge in this area is to effectively screen large numbers of potential materials in the search for highly effective catalysts. The last several years have seen a surge of activity in the use of DFT calculations to complement longstanding experimental approaches to the search for effective bimetallic catalysts. One example has already been mentioned as one of the three "success stories" earlier in this review; the DFT-driven development by Linic, Barteau, and co-workers of a bimetallic catalyst with improved selectivity for ethylene epoxidation compared to elemental Ag.[56]

Theoretical and experimental studies of model bimetallic catalysts in recent years have distinguished between thermodynamically stable bulk alloys and so-called near surface alloys. Near surface alloys are materials where the top few surface layers are created in a chemically heterogeneous way, for example, by depositing a monolayer of one metal on top of another metal. These structures are often not the thermodynamic equilibrium states of the material. To give one example, Ni and Pt form an fcc bulk solid solution under most (but not all) conditions,[73] so if a monolayer of Ni is deposited on Pt and the system comes to equilibrium, all of the deposited Ni will dissolve into the bulk. There is, however, a considerable kinetic barrier to this process, so the near surface alloy of a monolayer on Ni on Pt(111) is quite stable provided a moderate temperature is used.[191] If the deposited monolayer in systems of this type has a tendency to segregate away from the surface, a common near surface alloy structure is the formation of a subsurface layer of the deposited metal.[85] The deposition of V on Pd(111) is one example of this behavior.[192]

Near surface alloys offer a fascinating means to tune the chemical properties of catalytic surfaces. It is useful to consider this statement via several specific examples. Schlapka *et al.* used a combination of experiments and DFT to characterize heteroepitaxial Pt layers on Ru(0001).[193] The binding energy of CO on this surface varies from ~1.1 eV when there is 1 layer of Pt to ~1.4 eV after 8 layers of Pt had been deposited. This was the first study that was able to decouple the electronic effects of the surface layers, that is, changes in the electronic state of the surface layer due to the chemical difference between the layer and the substrate, and strain effects due to the expansion or contraction of the surface layer due to the substrate. The electronic effect is also frequently referred to as the ligand effect. Klötzer *et al.* showed that incorporating around 1/3 of a layer of V into the subsurface layer of Pt(111) or Rh(111) significantly reduced the binding energy of CO on the surface.[192] In an example that is more directly relevant to catalysis, Hwu and co-workers performed experiments for the hydrogenation of cyclohexene on Pt(111), Ni(111), and a near surface alloy with one layer of Ni on Pt(111).[194] Complete decomposition of cyclohexene to atomic C and H is the preferred reaction pathway on Ni(111), but this pathway is negligible on the surface alloy. Zellner *et al.* recently used DFT calculations to predict that a layer of Pd on Mo(110) would bind ethylene more weakly than clean Mo(110) or a thick Pd film, a prediction that was confirmed experimentally.[195]

A rich variety of phenomena can arise in bimetallics that do not need to be considered for monometallic catalysts. Surface segregation can be strong in metal alloys at thermodynamic equilibrium.[196] DFT-based methods have been used to generate databases suitable for characterizing surface segregation tendencies of dilute substituents to close-packed metal surfaces,[74] but methods to accurately predict the surface composition of alloys with less simple compositions are not well developed. Even in the absence of surface segregation, local disorder in bimetallic alloys can play a role in the material properties. Examples have been noted in DFT calculations where the dominant adsorption sites for the surfaces of partially disordered alloys are not present in the most common ordered approximations of the same alloys.[197] Many examples exist where the crystal structure of alloys is fundamentally different than the crystal structure of the parent metals. For example, Cu–Pd alloys with composition close to 50 at.% Cu form a bcc structure for temperatures below 600 K, although both Cu and Pd are fcc crystals. This structural change has large consequences for use of Cu–Pd alloys as H_2 purification membranes.[130] A large range of complex intermetallic crystal structures exist for alloys with compositions such as AB_2 or AB_5. DFT methods have been applied recently to understand H diffusion through alloys of this type.[48,198,199] Accurately accounting for these phenomena will create many opportunities for fruitful research in the future. The most common approach in DFT-based studies of bimetallics to date has been to sidestep the issues raised by these phenomena by studying the properties of prototypical ordered structures or near surface alloys or, in a smaller number of example, surfaces of bulk alloys. This approach is quite reasonable, since it can be coupled with available information regarding surface segregation to at least provide an indication of which of the prototypical structures can be expected to be experimentally stable. It is worth noting that almost nothing is currently known about the catalytic activity of defect sites on bimetallics. As we have seen repeatedly in the material reviewed above, defects play a crucial role in many catalytic processes. Understanding the structure and activity of these sites in bimetallic systems may become an important topic in the future.

One approach in using DFT to study bimetallic surfaces is to look at a set of adsorbed intermediates that are relevant to an overall catalytic reaction on a small number of surfaces. Sheth, Neurock, and Smith performed a thorough study of 14 intermediates relevant to acetylene hydrogenation on Pd(111) and Pd(111) with 1/4 or 1/2 a layer of Ag incorporated into the top layer.[200] The addition of Ag was found to reduce the binding energy of all the C_1 and C_2 intermediates that were examined. Neurock and co-workers have also studied ethylene hydrogenation on Pd/Au bimetallic surfaces[201] and nitric oxide decomposition on Pt-Au(100) surfaces.[202] Both of these studies involved using DFT-based Kinetic Monte Carlo simulations to simulate the overall reaction kinetics on the materials of interest. Gokhale *et al.* performed a similar thorough study of H, N, O, S, C, NO, CO, and OH on Cu(111) and CuSn(0001).[203] This approach is likely to be of great value in understanding overall catalytic reactions on bimetallics, but it has the short term disadvantage that large amounts of effort are required to examine even a handful of surfaces.

A complementary approach to the task of using DFT calculations to probe bimetallic surfaces is to use one or two relatively simple adsorbates that are indicative of the outcome of a reaction of general interest to characterize a large number of surfaces. Kitchin and co-workers and Mavrikakis and co-workers have used the dissociative binding energy of H_2 and O_2 to compare the properties of a large number of near surface alloys.[85,204–206] The dissociation of O_2 is of great interest in the development of catalysts for fuel cells.[207] There is still considerable debate over the mechanism controlling the oxygen reduction reaction under practical conditions in fuel cells,[207] so an important topic for efforts to use DFT to screen materials for this application is to make a connection between properties that are readily accessible with DFT and real electrocatalysis. This topic is discussed below in section 4.2. Very promising results on this idea have been recently reported by Zhang *et al.*, who compared electrocatalysis experiments in 0.1 M $HClO_4$ for electrodes with one layer of Pt deposited on Ru(0001), Ir(111), Rh(111), Au(111), and Pd(111) with DFT calculations of O and OH binding on the same surfaces in vacuum.[208] The DFT calculations indicated that a trade-off exists in vacuum between lowering the dissociation barrier for O_2 and the formation barrier for an OH bond, that is, lowering one barrier is correlated with the other barrier increasing. The calculations suggest that the Pt/Pd surface alloy should have the best combination of these barriers and exhibit only a moderate binding energy for O. This prediction correlated well with the experimental observation that this surface alloy has the highest O_2 reduction current of the materials tested.

There is considerable interest in the development of simple physical principles or correlations that can connect accessible properties of bimetallic surfaces with their catalytic function. The *d*-band model of Hammer and Nørskov[4,209,210] has been widely and successfully used to understand trends in chemisorption on metallic surfaces, and variations of this model have also been developed for bimetallic surfaces.[204] In materials where a heterogeneous population of binding sites is available, a model that characterizes the *d* band of these sites is needed. Pallassana *et al.* introduced an expression for weighting the *d*-bands of the various atoms that contribute to a binding site,[211] and this expression has been used in a variety of studies.[85,197,204] Tang and Trout have recently introduced a different expression for this weighting that captures the variations they observed in SO_2 adsorption on a large collection of bimetallics more accurately than the expression of Pallassana *et al.*[212]

Brønsted-Evans-Polanyi energy correlations have been applied with success to connect the binding energy of surface species with the activation energies associated with reactions of these species for monometallic catalysts.[64] Yu, Ruban, and Mavrikakis have shown that similar correlations accurately describe the dissociation of O_2 on a variety of Pt–Co and Pt–Fe surfaces. When applicable, these correlations are of great benefit to DFT calculations, because they mean that reaction activation energies can be characterized simply by studying the binding energy of the reactants. The exceptions from these correlations, however, may prove to be more interesting than the rule. Greeley and Mavrikakis examined H binding and H_2 dissociation barriers on a large number of near surface alloys with DFT, and identified several materials with moderate H binding energies and low H_2 dissociation barriers.[85] These desirable materials are at odds with the typical Brønsted-Evans-Polanyi behavior, where weaker binding implies higher dissociation barriers. Further advances in understanding of the factors underlying these deviations from "typical" behavior will undoubtedly lead to the discovery of other interesting materials for catalytic applications in the future.

An area that does not appear to have been explored in current calculations is the potential use of catalysts with more than two metallic components. These might be dubbed multimetallic catalysts. There are obvious complications in considering these materials, but the same general argument that suggests that bimetallic catalysts should be more versatile than monometallic catalysts also suggests that multimetallic catalysts could be even more versatile. One recent attempt has been made to begin using DFT calculations to screen bulk ternary alloys for H_2 purification using metal membranes.[213]

4 Areas Poised for Future Progress

To conclude this review, I highlight three topics within heterogeneous catalysis for which DFT-based calculations appear poised to allow rapid progress in the near future, the development of catalysts for use in reversible hydrogen storage, electrocatalysis for fuel cell applications, and catalysis in zeolites.

4.1 Catalysis in Reversible Hydrogen Storage. – One of the major impediments to the widespread use of H_2 as a mobile fuel source is the lack of lightweight materials that reversibly store large amounts of H_2. The technological issues associated with reversible hydrogen storage have been discussed extensively[214–218] and I will not reiterate them here. Perhaps the most promising class of materials for achieving effective reversible hydrogen storage are complex metal hydrides. The most widely studied example to date is sodium alanate, $NaAlH_4$, which decomposes in two steps to release a total of 5.5 wt% H_2:[216]

$$3NaAlH_4 \leftrightarrow Na_3AlH_6 + 2Al + H_2,$$

$$2Na_3AlH_6 \leftrightarrow 6NaH + Al + 3H_2.$$

Many other similar complex hydrides made from light metals also exist that in principle store large amounts of hydrogen. The sticking point for all of these materials is the absolute necessity for a practical storage material to *reversibly* take up and release hydrogen at mild conditions.[216] The key discovery that has created so much interest in sodium alanate was the work by Bogdanovic and Schwickardi showing that adding small amounts of Ti to $NaAlH_4$ greatly accelerated the hydrogenation and dehydrogenation reactions above.[219] This success lead Grochala and Edwards to conclude in their recent review of metal hydrides for hydrogen storage that "catalysis will play a pivotal role in the development of attractive hydrogen-storage media in the future".[216]

This pivotal role for catalysis will of course become more readily achievable if the mechanisms underlying the effectiveness of Ti-catalyzed sodium alanate are understood. A large number of groups have used DFT calculations to assess the physical and electronic structure of sodium alanate and other similar complex metal hydrides.[220–230] Several recent papers have begun to explore the catalytic role of Ti. Work to date has focused on understanding the location of Ti in $NaAlH_4$. Iñiguez *et al.* initially reported DFT calculations suggesting that substitution of Ti at Na sites in bulk $NaAlH_4$ was energetically feasible,[231] but subsequent calculations by the same group[232] and by Løvvik and Opalka[233] contradicted this suggestion, indicating instead that Ti is much more likely to reside at or near the surface of $NaAlH_4$. Two significant complications that exist in studies of this kind are that Ti is typically introduced into $NaAlH_4$ via ball-milling, a highly energetic process that could create deeply metastable atomic configurations, and the fact that very large scale mass transfer (from an atomic point of view) is required when passing through an entire dehydrogenation-hydrogenation cycle. Løvvik has recently used DFT calculations to examine Ti near the surfaces of $LiBH_4$, another promising complex hydride, concluding that it prefers subsurface interstitial sites to surface sites.[228] The studies have focused on the first dehydrogenation step in the reaction scheme shown above. Chaudhuri and Muckerman have used DFT calculations to examine the first step in the hydrogenation of NaH and Al.[234] Their calculations support the view that Ti atoms are responsible for chemisorption of hydrogen, as both NaH and Al surfaces are inactive for H_2 dissociation. They also suggest that diffusion of hydride species on the surface of the metallic Al phase is important to the next step in the overall hydrogenation reaction. The application of DFT-based methods to understand and screen potential catalysts for complex metal hydrides is likely to grow rapidly in the next several years.

4.2 Electrocatalysis. Catalysis plays a central role in any application of fuel cells, so there is great worldwide interest in the optimization and discovery of catalysts in these applications. Electrocatalysis poses a great challenge for detailed theoretical approaches, because in addition to all the usual complications associated with heterogeneous catalysis, the processes take place in solution in the presence of non-trivial electric fields. The possibilities for theory to contribute to this area, however, are large. This area was recently reviewed by Koper, who highlighted the opportunities existing for theory by noting that "even a cursory look into most (recent) electrochemistry textbooks reveals a conspicuous lack of molecular detail".[235,236] Rather than reiterate the ground that is covered by Koper's review, I just highlight here several recent trends within the application of DFT to electrocatalysis.

Perhaps the most ambitious efforts to include solvent effects in DFT calculations of metal surface catalysis has been the work by Desai and Neurock on Pd(111)[237] and a Pt-Ru alloy.[238] In these papers, plane wave DFT calculations were performed with the space between the surfaces filled with H_2O at the density of bulk water. The results hint at the complexity of these problems;

water dissociation is found to occur via a concerted action involving both the metal surface and the solution phase.[238] An important drawback of applying DFT to systems as complex as these is that the number of physical states that can be probed within a reasonable amount of computational time is strongly limited. In this context, it is important to note that the apparently simpler problem of understanding the structure of molecularly-thin layers of water on metal surfaces under vacuum conditions remains an area of vigorous activity.[239–251]

In addition to studies of the structure of water on catalytic metals, a smaller number of papers have examined the impact of electric fields. Wang and Balbuena used Car-Parrinello molecular dynamics of electroreduction of O_2 on Pt(111) to compare a variety of possible mechanisms for this reaction.[252,253] Hyman and Medlin characterized the adsorption and dissociation of O_2 on cluster and slab models of Pt(111) in the presence of homogeneous electric fields.[254] The energy change related to the presence of an electric field was calculated to be much smaller in the slab models than with cluster calculations. Roudgar and Gross used periodic GGA calculations to assess possible structures of water bilayers on a Au(111) surface covered with a single layer of Pd(111), finding that the relative stability of the bilayer structures considered could be reversed by applying a negative electric field.[255]

4.3 Zeolite Catalysis. – All of the work reviewed above dealt with heterogeneous catalysts in which the catalyst is, at an atomic level, a solid surface or nanoparticle. There are many practical applications where heterogeneous catalysis is performed using microporous materials such as zeolites.[256,257] It is fair to say that a detailed atomic-scale understanding of catalysis in zeolites lags behind catalysis on solid surfaces. On the experimental side, the range of scanning probe microscopies and surface sensitive spectroscopies that can be brought to bear on solid surfaces are typically not of use for probing inside the pores of microporous materials. On the theoretical side, the structural complexity of zeolites creates challenges that are not present in dealing with at least atomically flat surfaces. One useful way to understand this situation is that zeolites are somewhat analogous to solid nanoparticles in that they intrinsically present a multiplicity of adsorption and reaction sites and cannot be easily represented in calculations using a small number of atoms. Despite these inherent challenges, the practical value of zeolite catalysis is large and a thriving community of theoreticians has pursued ideas that complement experimental studies of this topic.

I restrict my attention here to work that has applied periodic DFT calculations to structure or reactivity of zeolites. The great majority of work on this topic has focused on a small number of materials that have small unit cells, including chabazite (structure code CHA), which has 36 framework atoms per unit cell, mordenite (structure code MOR), which has 146 atoms per unit cell, ZSM-22 (structure code TON), with 72 atoms per unit cell, and ferrierite (structure code FER), with 108 atoms per unit cell. Notably, this list does not include ZSM-5, a material with 288 framework atoms per unit cell, which is the zeolite with the most widespread use in practical applications. This list represents only a small fraction of the zeolite structures that are known.[258]

The large unit cells required to represent zeolites in periodic DFT calculations have motivated many quantum chemistry cluster studies of zeolites. A complication in these studies is that long-range electrostatic contributions from the zeolite framework and steric constraints from the zeolite pore shape are often significant.[259] This complication can be overcome by using embedding methods in which quantum chemistry calculations are coupled with classical force fields,[260] but the fact that periodic DFT calculations have now become computationally accessible for these systems suggests that these calculations will widely supplant cluster approaches in the future.

The large size of the unit cells needed to describe zeolites makes them a promising area for the application of linear-scaling DFT calculations. Poulet, Sautet, and Artacho recently presented a careful comparison between plane wave DFT calculations performed with VASP and linear-scaling calculations based on localized basis sets performed with SIESTA for two aluminophosphate structures.[261] Although the computer time associated with the initial total energy calculation with the linear-scaling method is considerably larger than in the plane wave approach, subsequent energy calculations as part of geometry optimization were considerably faster than the plane wave approach. The linear-scaling method also used significantly less memory than the plane wave calculations. A complication with the latter approach is that optimization of the localized basis set had to be performed prior to the structural calculations. It seems reasonable to hope that as basis sets of this type are introduced for a variety of relevant materials the use of linear-scaling methods for large unit cell materials might become more widespread in the near future.

Quantum chemistry approaches to zeolites are complemented by an active research community that uses classical force-field methods to study molecular adsorption and diffusion in zeolites and similar materials. This topic was comprehensively reviewed by Keil, Krishna, and Coppens in 2000.[262] For more recent examples of activity in this area, see References 263–270. Examples of impressive agreement between adsorption isotherms and molecular diffusivities predicted with calculations of this type and experimental data are available.[271,272] There appear to be many future opportunities for linking the detailed understanding of multi-component adsorption and diffusion that is now emerging from this area with detailed quantum chemistry approaches to reactivity at active sites inside zeolites.

4.3.1 Zeolite Structure. While understanding the crystal structure of zeolites is well developed experimentally, many challenges remain in achieving a detailed understanding of the preferred locations of framework substituents and the accompanying cationic species that create acid sites in zeolites. A number of DFT studies have examined this issue in variations of the chabazite structure. Poulet, Sautet, and Tuel used GGA calculations to study the structure of hydrated $AlPO_4$-34, the aluminophosphate analog of chabazite.[261] The structure of this material distorts considerably upon hydration. Simulated annealing using *ab initio* molecular dynamics were used to probe the mobility of non-framework water molecules. Corà *et al.* used calculations based on an unrestricted Hartree-Fock Hamiltonian to study the structure of framework substitutions for Si and Al atoms in chabazite and $AlPO_4$-34 by low-valence dopant ions.[273,274] In this work, acid strength was characterized by calculating the OH stretching frequency associated with protonation of a framework O atom adjacent to the framework substituent. Lo and Trout used GGA calculations to examine the acid strength of sites in Al-substituted chabazite by comparing deprotonation energies, adsorption energies of several molecular bases, OH vibrational frequencies, and visualization of the electron localization function.[275] The comparison of these different characterizations of acid strength gives a clear example of why a simple universal scale for measuring zeolite acidity is elusive. The site preferences and relative energies for cation exchange in alkai-exchanged chabazite have been studied using B3LYP-DFT calculations in a periodic geometry by Civalleri *et al.*[276]

Astala, Auerbach, and Monson[277] used plane wave LDA calculations to examine structural characteristics of five all-silica zeolites: SOD, LTA, CHA, MOR, and MFI. The first four of these materials can be examined in calculations with 36–72 atoms in a periodic unit cell, but the MFI structure is considerably larger and requires 288 atoms in a single unit cell. Astala *et al.* argue that LDA calculations give more accurate results for the structure and energies of silica polymorphs than GGA calculations. Demuth *et al.* used LDA and GGA calculations to examine the properties of Al substitutions in mordenite both in a protonated form and in the presence of Na cations.[278]

DFT calculations are one useful way to consider novel microporous structures with the aim of broadening the range of zeolite-like materials that are known. An interesting recent example of this direction is the work of Astala and Auerbach, who examined how bridging O atoms in the framework of sodalite and LTA zeolites can potentially be replaced by methylene or amine groups.[279] This work was motivated by experimental studies by Yamamoto *et al.* that created zeolites that contained a combination of Si–CH_2–Si bridges and Si–CH_3 groups.[280] Astala and Auerbach's calculations suggest that an interesting class of materials with only Si–CH_2–Si bridges exist that are mechanically stable and have relatively small amounts of strain.

4.3.2 Catalysis in Zeolites. Rozanska *et al.* used plane wave GGA calculations to study isomerization of toluene and xylenes by protonated sites in mordenite (MOR).[281] The MOR unit cell contains 146 atoms. They noted that the adsorption energy of molecules in zeolites are often dominated by van der Waals interactions that are not treated corrected by GGA-DFT, so adjustments were made to the adsorption energies using an empirical Lennard-Jones potential. Comparison of the periodic DFT results with cluster calculations suggested that cluster approaches tended to greatly overestimate the activation energies of isomerization reactions. The transition states of these reactions tend to be charged or carbocationic, so stabilization of these species by the electric field of the full zeolite structure can be considerable. Cluster DFT results were found to correctly predict the relative order of activation energies for sets of reactions, indicating they are useful to qualitatively comparing reactivities. Rozanska, Barbosa, and van Santen recently reported similar calculations for step-by-step alkylation of benzene with propene in mordenite.[259] In this case, cluster DFT calculations for the same reaction pathway led to qualitatively different reaction energy diagrams because of the difficulty of accounting for framework stabilization of transition states in cluster calculations.

Rozanska *et al.* compared plane wave LDA and GGA calculations for the chemisorption of propylene in chabazite.[282] They concluded that allowing zeolite atoms to relax upon chemisorption made a significant impact on the computed results, but relaxation of the zeolite's unit cell shape and volume had considerably less effect. The same group performed plane wave GGA calculations for the chemisorption of isobutene in three different zeolites, chabazite, ZSM-22 (TON), and mordenite.[283]

Benco *et al.* compared periodic LDA and GGA calculations for the adsorption of linear hydrocarbons (C_1–C_6) in gmelinite, both as a purely siliceous material and with one or two protonated sites.[83] GGA calculations predict only a weak dependence of the adsorption energy on the hydrocarbon chain length, while LDA gives the more realistic result that increasing the chain length also increases the adsorption energy. Although no experimental data is available for gmelinite, the trend of the LDA adsorption energies matches experimental results from a variety of other zeolites.[284,285] These calculations confirm the well known observation that GGA-DFT performs poorly in describing van der Waals interactions.

Lo *et al.* used plane wave calculations to examine methanol coupling near a protonated site in chabazite.[286,287] Constrained Car-Parrinello molecular dynamics were used to probe potential reaction coordinates for the reaction. These calculations suggest a reaction path that proceeds via formation of stable intermediates of methane and formaldehyde. A novel feature of this work was the use of transition path sampling, a technique to efficiently search for transition states that requires little *a priori* information regarding the configuration of the final products of a reaction. Methanol coupling in ferrierite (FER) has been examined with periodic DFT calculations by Govind *et al.*[288]

Bucko, Hafner, and Benco performed a detailed study of the Beckmann rearrangement of cyclohexanone oxime to ε-caprolactam at acid sites in mordenite, both with and without the presence of water or methanol.[289]

4.3.3 External Surfaces of Zeolites. All of the work reviewed above dealt with the internal pores of zeolites. In situations where zeolite particles are large and molecules of interest are easily accommodated in the zeolite pores, it is entirely appropriate to focus exclusively on the internal pores. There is growing interest, however, in situations where the external surfaces of zeolites can play a role in catalytic and separation processes. Molecules entering or exiting zeolite pores at an external surface can in some cases experience large resistances to transport that can contribute to the net resistance to diffusion experienced by mobile molecules.[290–293] The total surface area of nanocrystalline forms of zeolites includes large contributions from external surfaces.[294,295] Deliberate modification of the external surfaces of zeolites offers one route to tuning their performance in separation and catalysis applications.[296,297]

Knowledge of the external surfaces of zeolites is poorly developed. Insight into the properties of zeolite surfaces from periodic DFT calculations has to date been limited to a series of fascinating papers by Bucko, Hafner, and Benco. This group first used GGA calculations to examine the surface structure of the (001) surface of mordenite, both in a pure silica form and in the presence of Al framework substitutions.[298] The relaxed surface structure was found to differ only moderately to the bulk terminated structure, although the acidity of some terminal OH groups, as characterized by the OH stretching frequency, was found to be comparable to acid sites inside the zeolite pores. Several pathways to the formation of surface defects have been identified.[299] The same group compared the Beckmann rearrangement of cyclohexanone oxime to ε-caprolactam at surface defect sites on mordenite(001) with acid sites in mordenite pores.[289] At least for this reaction, the surface sites were found to be inactive. Calculations by the same group have compared the vibrational spectra of adsorbed CO and ammonia on mordenite(001) with adsorption of these molecules inside mordenite pores.[300,301]

5 Conclusion and Outlook

This article has reviewed recent applications of Density Functional Theory to heterogeneous catalysis, with an emphasis on identifying topics that represent successful examples of this approach. It is worth reiterating my disclaimer from the Introduction that this review was not intended to be exhaustive. This disclaimer when taken in combination with the number of references listed below gives a useful measure of the level of worldwide activity in this field. The great complexity of real heterogeneous catalysts and their enormous technological significance guarantees that quantum chemistry based methods such as DFT will continue to play a vital role within the overall field of heterogeneous catalysis for the foreseeable future.

Acknowledgements

Conversations with many colleagues around the globe contributed greatly to this review. Comments on a draft manuscript by S. Alapati, B. Bhatia, J. James, R. Rankin, and D. Sorescu were extremely helpful. The hospitality of the Division of Chemical Engineering at the University of Queensland, where this review was written, was much appreciated. My own work in this area has been supported by the NSF (CTS-0216170), the DOE Catalysis Futures Program, the National Energy Technology Laboratory, and the donors of the American Chemical Society Petroleum Research Fund.

References

1. W. Kohn, *Rev. Mod. Phys.*, 1999, **71**, 001253.
2. R.A. van Santen and M. Neurock, *Catal. Rev. Sci. Eng.*, 1995, **37**, 557.
3. B. Hammer and J.K. Nørskov, *Adv. Catal.*, 2000, **45**, 71.
4. J. Greeley, J.K. Nørskov and M. Mavrikakis, *Annu. Rev. Phys. Chem.*, 2002, **53**, 319.
5. A.T. Bell, *Mol. Phys.*, 2004, **102**, 319.
6. A.A. Gokhale, S. Kandoi, J.P. Greeley, M. Mavrikakis and J.A. Dumesic, *Chem. Eng. Sci.*, 2004, **59**, 4679.
7. M. Neurock, S.A. Wasileski and D. Mei, *Chem. Eng. Sci.*, 2004, **59**, 4703.
8. B. Hammer, *Top. Catal.*, 2006, *in press*.
9. K.M. Neyman and F. Illas, *Catal. Today*, 2005, **105**, 2.
10. R.G. Parr and Y. Weitao, *Density-Functional Theory of Atoms and Molecules*, Oxford University Press, Oxford, 1994.
11. *Modern Density Functional Theory: A Tool For Chemistry*, eds. J.M. Seminario, P. Politzer, Elsevier, Amsterdam, 1995.
12. W. Koch and M.C. Holthausen, *A Chemist's Guide to Density Functional Theory*, Wiley-VCH, Berlin, 2001.
13. *A Primer in Density Functional Theory*, eds. C. Fiolhais, F. Noguera, M.A.L. Marques, Springer-Verlag, Berlin, 2003.
14. R.M. Martin, *Electronic Structure: Basic Theory and Practical Methods*, Cambridge University Press, Cambridge, 2004.
15. M.C. Payne, M.P. Teter, D.C. Allan, T.A. Arias and J.D. Joannopoulos, *Rev. Mod. Phys.*, 1992, **64**, 1045.
16. R.A. Friesner, *Proc. Nat. Acad. Sci.*, 2005, **102**, 6648.
17. http://cms.mpi.univie.ac.at/vasp/.
18. http://dcwww.camp.dtu.dk/campos/Dacapo/.
19. http://www.tcm.phy.cam.ac.uk/castep/.
20. http://www.uam.es/departamentos/ciencias/fismateriac/siesta/.
21. G. Pacchioni, C.D. Valentin, D. Dominguez-Ariza, F. Illas, T. Bredow, T. Klüner and V. Staemmler, *J. Phys. Cond. Mat.*, 2004, **16**, S2497.
22. B. Hammer, L.B. Hansen and J.K. Nørskov, *Phys. Rev. B*, 1999, **59**, 7413.
23. J.P. Perdew, J.A. Chevary, S.H. Vosko, K.A. Jackson, M.R. Pederson, D.J. Singh and C. Fiolhais, *Phys. Rev. B*, 1992, **46**, 6671.
24. J.P. Perdew, K. Burke and M. Erzenhof, *Phys. Rev. Lett.*, 1996, **77**, 3865.
25. Y. Zhang and W. Yang, *Phys. Rev. Lett.*, 1998, **80**, 890.
26. R. Janoschek, *Pure Appl. Chem.*, 2001, **73**, 1521.
27. B.G. Johnson, P.M.W. Gill and J.A. Pople, *J. Chem. Phys.*, 1993, **98**, 5612.
28. G. Kresse and D. Joubert, *Phys. Rev. B*, 1999, **59**, 1758.
29. H. Over and M. Muhler, *Prog. Surf. Sci.*, 2003, **72**, 3.
30. J. Wang, C.Y. Fan, K. Jacobi and G. Ertl, *J. Phys. Chem. B*, 2002, **106**, 3422.
31. K. Reuter and M. Scheffler, *Phys. Rev. Lett.*, 2003, **90**, 046103.
32. J.S. Reese, S. Raimondeau and D.G. Vlachos, *J. Comp. Phys.*, 2001, **173**, 302.
33. C.H.F. Peden and D.W. Goodman, *J. Phys. Chem.*, 1986, **90**, 1360.
34. Á. Logadóttir and J.K. Nørskov, *J. Catal.*, 2003, **220**, 273.
35. K. Honkala, A. Hellman, I.N. Remedeiakis, Á. Logadóttir, A. Carlsson, S. Dahl, C.H. Christensen and J.K. Nørskov, *Science*, 2005, **307**, 555.
36. T.H. Rod, A. Logadottir and J.K. Nørskov, *J. Chem. Phys.*, 2000, **112**, 5343.
37. C. Zhang, Z.-P. Liu and P. Hu, *J. Chem. Phys.*, 2001, **115**, 609.

38. D.S. Sholl, A. Asthagiri and T.D. Power, *J. Phys. Chem. B*, 2001, **105**, 4771.
39. R.M. Hazen and D.S. Sholl, *Nature Materials*, 2003, **2**, 367.
40. S.J. Pratt, S.J. Jenkins and D.A. King, *Surf. Sci. Lett.*, 2005, **585**, L159.
41. G. Haase, M. Asscher and R. Kosloff, *J. Chem. Phys.*, 1989, **90**, 3346.
42. R. van Harrevelt, K. Honkala, J.K. Norskov and U. Manthe, *J. Chem. Phys.*, 2005, **122**, 234702.
43. A. Volpi and D.C. Clary, *J. Phys. Chem. B*, 2004, **108**, 336.
44. C.S. Tautermann and D.C. Clary, *J. Chem. Phys.*, 2005, **122**, 134702.
45. P.G. Sundell and G. Wahnström, *Phys. Rev. Lett.*, 2004, **92**, 155901.
46. P.G. Sundell and G. Wahnström, *Phys. Rev. B*, 2004, **70**, 081403.
47. J.T. Fermann and S. Auerbach, *J. Chem. Phys.*, 2000, **112**, 6787.
48. B. Bhatia and D.S. Sholl, *Phys. Rev. B*, 2005, **72**, 224302.
49. T.V. Choudhary, C. Sivadinarayana and D.W. Goodman, *Catal. Lett.*, 2001, **72**, 197.
50. J.C. Ganley, E.G. Seebauer and R.I. Masel, *AIChE J.*, 2004, **50**, 829.
51. A.B. Mhadeshwar, J.R. Kitchin, M.A. Barteau and D.G. Vlachos, *Catal. Lett.*, 2004, **96**, 13.
52. S.R. Deshmukh, A.B. Mhadeshwar and D.G. Vlachos, *Ind. Eng. Chem. Res.*, 2004, **43**, 2986.
53. K.-I. Aika, T. Takano and S. Murata, *J. Catal.*, 1992, **136**, 126.
54. O. Hinrichsen, F. Rosowski, A. Hornung, M. Muhler and G. Ertl, *J. Catal.*, 1997, **165**, 33.
55. M.A. Barteau, *Top. Catal.*, 2003, **22**, 3.
56. S. Linic, J. Jankowiak and M.A. Barteau, *J. Catal.*, 2004, **224**, 489.
57. S. Linic and M.A. Barteau, *J. Catal.*, 2003, **214**, 200.
58. R.B. Grant and R. Lambert, *J. Catal.*, 1985, **92**, 364.
59. R.A. van Santen and H.P.C.E. Kuipers, *Adv. Catal.*, 1987, **35**, 265.
60. N.W. Cant and W.K. Hall, *J. Catal.*, 1978, **52**, 81.
61. W.F. Richey, *J. Phys. Chem.*, 1972, **76**, 213.
62. S. Linic and M.A. Barteau, *J. Am. Chem. Soc.*, 2002, **124**, 310.
63. G.S. Jones, M. Mavrikakis, M.A. Barteau and J.M. Vohs, *J. Am. Chem. Soc.*, 1998, **120**, 3196.
64. J.K. Nørskov, T. Bligaard, Á. Logadóttir, S. Bahn, L.B. Hansen, M. Bollinger, H. Bengaard, B. Hammer, Z. Sljivancanin, M. Mavrikakis, Y. Xu, S. Dahl and C.J.H. Jacobsen, *J. Catal.*, 2002, **209**, 275.
65. D. Stacchiola, G. Wu, M. Kaltchev and W.T. Tysoe, *Surf. Sci.*, 2001, **486**, 9.
66. J.A. Dumesic, *J. Catal.*, 1999, **185**, 496.
67. J.A. Dumesic, *J. Catal.*, 2001, **204**, 525.
68. C.T. Campbell, *J. Catal.*, 2001, **204**, 520.
69. M. Mavrikakis, D.J. Doren and M.A. Barteau, *J. Phys. Chem. B*, 1998, **102**, 394.
70. A. Michaelides, M.-L. Bocquet, P. Sautet, A. Alavi and D.A. King, *Chem. Phys. Lett.*, 2003, **367**, 344.
71. E.M. Cordi and J.L. Falconer, *Appl. Catal. A*, 1997, **151**, 179.
72. S. Linic and M.A. Barteau, *J. Am. Chem. Soc.*, 2003, **125**, 4034.
73. H. Okamoto, *Desk Handbook Phase Diagrams for Binary Alloys*, ASM International, Materials Park, OH, 2000.
74. A.V. Ruban, H.L. Skriver and J.K. Norskov, *Phys. Rev. B*, 1999, **59**, 15990.
75. S. Linic, *private communication.*
76. J.G. Serafin, A.C. Liu and S.R. Seyedmonir, *J. Mol. Catal. A*, 1998, **131**, 157.
77. C. Saravanan, M.R. Salazar, J.D. Kress and A. Redondo, *J. Phys. Chem. B*, 2000, **104**, 8685.

78. S. Linic and M.A. Barteau, *J. Am. Chem. Soc.*, 2004, **126**, 8086.
79. E.R. Frank and R.J. Hamers, *J. Catal.*, 1997, **172**, 406.
80. E.P. Wigner, *Comm. Pure Appl. Math.*, 1960, **13**, 1.
81. P.J. Feibelman, B. Hammer, J.K. Nørskov, F. Wagner, M. Scheffler and R. Stumpf, *J. Phys. Chem. B*, 2001, **108**, 4018.
82. M. Gajdos and J. Hafner, *Surf. Sci.*, 2005, **590**, 117.
83. L. Benco, T. Demuth, J. Hafner, F. Hutschka and H. Toulhoat, *J. Chem. Phys.*, 2001, **114**, 6327.
84. K. Reuter, D. Frenkel and M. Scheffler, *Phys. Rev. Lett.*, 2004, **93**, 116105.
85. J. Greeley and M. Mavrikakis, *Nature Materials*, 2004, **3**, 810.
86. A. Rohrbach, J. Hafner and G. Kresse, *Phys. Rev. B*, 2004, **69**, 075413.
87. A. Rohrbach and J. Hafner, *Phys. Rev. B*, 2005, **71**, 045405.
88. A. Rohrbach, J. Hafner and G. Kresse, *J. Phys. Cond. Mat.*, 2003, **15**, 979.
89. A. Rohrbach, J. Hafner and G. Kresse, *Phys. Rev. B*, 2004, **70**, 125426.
90. J.B. Neaton, C. Ederer, U.V. Waghmare, N.A. Spaldin and K.M. Rabe, *Phys. Rev. B*, 2005, **71**, 014113.
91. P. Baettig, C.F. Schelle, R. LeSar, U.V. Waghmare and N.A. Spaldin, *Chem. Mater.*, 2005, **17**, 1376.
92. N.A. Spaldin and M. Fiebig, *Science*, 2005, **309**, 391.
93. P. Baettig and N.A. Spaldin, *Appl. Phys. Lett.*, 2005, **86**, 012505.
94. N. Erdman, K.R. Poeppelmeier, M. Asta, O. Warschkow, D.E. Ellis and L.D. Marks, *Nature*, 2002, **419**, 55.
95. E. Kaxiras, Y. Bar-Yam, J.D. Joannopoulos and K.C. Pandey, *Phys. Rev. B*, 1987, **35**, 9636.
96. G.-X. Qian, R.M. Martin and D.J. Chadi, *Phys. Rev. B*, 1988, **38**, 7649.
97. X.-G. Wang, W. Weiss, S.K. Shaikhutdinov, M. Ritter, M. Petersen, F. Wagner, R. Schlögl and M. Scheffler, *Phys. Rev. Lett.*, 1998, **81**, 1038.
98. X.-G. Wang, A. Chaka and M. Scheffler, *Phys. Rev. Lett.*, 2000, **84**, 3650.
99. K. Reuter and M. Scheffler, *Phys. Rev. B*, 2002, **65**, 035406.
100. C. Stampfl, *Catal. Today*, 2005, **105**, 17.
101. C. Wolverton, X.Y. Yan, R. Vijayaraghavan and V. Ozolins, *Acta Materialia*, 2002, **50**, 2187.
102. C. Stampfl and A.J. Freeman, *Phys. Rev. B*, 2005, **71**, 024111.
103. W. Zhang, J.R. Smith and X.-G. Wang, *Phys. Rev. B*, 2004, **70**, 024103.
104. J. Greeley and M. Mavrikakis, *Surf. Sci.*, 2003, **540**, 215.
105. P. Légaré, *Surf. Sci.*, 2004, **559**, 169.
106. V. Ozolins, B. Sadigh and M. Asta, *J. Phys. Cond. Matt.*, 2005, **17**, 2197.
107. M. Todorova, W.X. Li, M.V. Ganduglia-Pirovano, C. Stampfl, K. Reuter and M. Scheffler, *Phys. Rev. Lett.*, 2002, **89**, 096103.
108. M.V. Ganduglia-Pirovano, K. Reuter and M. Scheffler, *Phys. Rev. B*, 2002, **65**, 245426.
109. W.-X. Li, C. Stampfl and M. Scheffler, *Phys. Rev. B*, 2003, **68**, 165412.
110. W.-X. Li, C. Stampfl and M. Scheffler, *Phys. Rev. lett.*, 2003, **90**, 256102.
111. W.-X. Li, C. Stampfl and M. Scheffler, *Phys. Rev. B*, 2003, **67**, 045408.
112. Q. Sun, K. Reuter and M. Scheffler, *Phys. Rev. B*, 2003, **67**, 205424.
113. K. Reuter and M. Scheffler, *Phys. Rev. B*, 2003, **68**, 045407.
114. M. Todorova, K. Reuter and M. Scheffler, *Phys. Rev. B*, 2005, **71**, 195403.
115. J. Rogal, K. Reuter and M. Scheffler, *Phys. Rev. B*, 2004, **69**, 075421.
116. P. Légaré, *Surf. Sci.*, 2005, **580**, 137.
117. H. Tang, A. Van der Ven and B.L. Trout, *Phys. Rev. B*, 2004, **70**, 045420.

118. A. Marmier and S.C. Parker, *Phys. Rev. B*, 2004, **69**, 115409.
119. X.-G. Wang and J.R. Smith, *Phys. Rev. B*, 2004, **70**, 081401(R).
120. G. Kresse, M. Schmid, E. Napetschnig, M. Shishkin, L. Köhler and P. Varga, *Science*, 2005, **308**, 1440.
121. X.-G. Wang and J.R. Smith, *Phys. Rev. B*, 2003, **68**, 201402(R).
122. G. Kresse, S. Surnev, J. Schoiswohl and F.P. Netzer, *Surf. Sci.*, 2004, **555**, 118.
123. G. Kresse, S. Surnev, M.G. Ramsey and F.P. Netzer, *Surf. Sci.*, 2001, **492**, 329.
124. M.V. Ganduglia-Pirovano and J. Sauer, *Phys. Rev. B*, 2004, **70**, 045422.
125. M.V. Ganduglia-Pirovano and J. Sauer, *J. Phys. Chem. B*, 2005, **109**, 374.
126. T.P. Trainor, A.M. Chaka, P.J. Eng, M. Newville, G.A. Waychunas, J.G. Catalano and G. E. Brown Jr., *Surf. Sci.*, 2004, **573**, 204.
127. D. Loffreda and P. Sautet, *J. Phys. Chem. B*, 2005, **109**, 9596.
128. A. Eichler and G. Kresse, *Phys. Rev. B*, 2004, **69**, 045402.
129. F. Bottin, F. Finocchi and C. Noguera, *Phys. Rev. B*, 2003, **68**, 035418.
130. P. Kamakoti, B.D. Morreale, M.V. Ciocco, B.H. Howard, R.P. Killmeyer, A.V. Cugini and D.S. Sholl, *Science*, 2005, **307**, 569.
131. M.V. Bollinger, K.W. Jacobsen and J.K. Nørskov, *Phys. Rev. B*, 2003, **67**, 085410.
132. D.C. Sorescu, D.S. Sholl and A.V. Cugini, *J. Phys. Chem. B*, 2003, **107**, 1988.
133. D.C. Sorescu, D.S. Sholl and A.V. Cugini, *J. Phys. Chem. B*, 2004, **108**, 239.
134. J.A. Spirko, M.L. Neiman, A.M. Oelker and K. Klier, *Surf. Sci.*, 2004, **2004**, 191.
135. T. Zeng, X.-D. Wen, Y.-W. Li and H. Jiao, *J. Phys. Chem. B*, 2005, **109**, 13704.
136. B. Hinnemann, J.K. Nørskov and H. Topsøe, *J. Phys. Chem. B*, 2005, **109**, 2245.
137. M.Y. Sun, A.E. Nelson and J. Adjaye, *J. Catal.*, 2005, **233**, 411.
138. M.Y. Sun, A.E. Nelson and J. Adjaye, *Catal. Today*, 2005, **105**, 36.
139. H. Schweiger, P. Raybaud and H. Toulhoat, *J. Catal.*, 2002, **212**, 33.
140. H. Schweiger, P. Raybaud, G. Kresse and H. Toulhoat, *J. Catal.*, 2002, **207**, 76.
141. P. Raybaud, J. Hafner, G. Kresse, S. Kasztelan and H. Toulhoat, *J. Catal.*, 2000, **190**, 128.
142. P. Raybaud, J. Hafner, G. Kresse, S. Kasztelan and H. Toulhoat, *J. Catal.*, 2000, **189**, 129.
143. J.V. Lauritsen, M.V. Bollinger, E. Lægsgaard, K.W. Jacobsen, J.K. Nørskov, B.S. Clausen, H. Topsøe and F. Besenbacher, *J. Catal.*, 2004, **221**, 510.
144. C. Arrouvel, M. Breysse, H. Toulhoat and P. Raybaud, *J. Catal.*, 2005, **232**, 161.
145. E. Lundgren, J. Gustafson, A. Mikkelson, J.N. Andersen, A. Stierle, H. Dosch, M. Todorova, J. Rogal, K. Reuter and M. Scheffler, *Phys. Rev. Lett.*, 2004, **92**, 046101.
146. M. Haruta, *Catal. Today*, 1997, **36**, 153.
147. T. Hayashi, K. Tanaka and M. Haruta, *J. Catal.*, 1998, **178**, 566.
148. M. Valden, X. Lai and D.W. Goodman, *Science*, 1998, **281**, 1647.
149. Q. Fu, H. Saltsburg and M. Flytzani-Stephanopoulos, *Science*, 2003, **301**, 945.
150. C.T. Campbell, *Science*, 2004, **306**, 234.
151. M.S. Chen and D.W. Goodman, *Science*, 2004, **306**, 252.
152. P. Pyykkö, *Angew. Chemie Int. Ed.*, 2004, **43**, 4412.
153. H. Häkkinen, M. Moseler, O. Kostko, N. Morgner, M.A. Hoffman and B.V. Issendorff, *Phys. Rev. Lett.*, 2004, **93**, 093401.
154. D. Vanderbilt, *Phys. Rev. B*, 1990, **41**, R7892.
155. P.E. Blöchl, *Phys. Rev. B*, 1994, **50**, 17953.
156. B. Yoon, H. Häkkinen and U. Landman, *J. Phys. Chem. A*, 2003, **107**, 4066.
157. D. Stolcic, M. Fischer, G. Ganteför, Y.D. Kim, Q. Sun and P. Jena, *J. Am. Chem. Soc.*, 2003, **125**, 2848.
158. Y.D. Kim, M. Fischer and G. Ganteför, *Chem. Phys. Lett.*, 2003, **377**, 170.

159. F. Furche, R. Ahlrichs, P. Weis, C. Jacobs, S. Gilb, T. Bierweiler and M.M. Kappes, *J. Chem. Phys.*, 2002, **117**, 6982.
160. E.M. Fernández, P. Ordejón and L.C. Balbás, *Chem. Phys. Lett.*, 2005, **408**, 252.
161. A. Franceschetti, S.J. Pennycook and S.T. Pantelides, *Chem. Phys. Lett.*, 2003, **374**, 471.
162. N.S. Phala, G. Klatt and E. van Steen, *Chem. Phys. Lett.*, 2004, **395**, 33.
163. R.M. Olson, S. Varganov, M.S. Gordon, H. Metiu, S. Chretien, P. Piecuch, K. Kowalski, S.A. Kucharski and M. Musial, *J. Am. Chem. Soc.*, 2005, **127**, 1049.
164. S.A. Varganov, R.M. Olson, M.S. Gordon, G. Mills and H. Metiu, *J. Chem. Phys.*, 2004, **120**, 5169.
165. S.A. Varganov, R.M. Olson, M.S. Gordon and H. Metiu, *J. Chem. Phys.*, 2003, **119**, 2531.
166. A. Sanchez, S. Abbet, U. Heiz, W.-D. Schneider, H. Häkkinen, R.N. Barnett and U. Landman, *J. Phys. Chem. A*, 1999, **103**, 9573.
167. H. Häkkinen, S. Abbet, A. Sanchez, U. Heiz and U. Landman, *Angew. Chemie Int. Ed.*, 2003, **42**, 1297.
168. Z. Yan, S. Chinta, A.A. Mohamed, J.P. Fackler Jr. and D.W. Goodman, *J. Am. Chem. Soc.*, 2005, **127**, 1604.
169. A. Del Vitto, G. Pacchioni, F. Delbecq and P. Sautet, *J. Phys. Chem. B*, 2005, **109**, 8040.
170. B. Yoon, H. Häkkinen, U. Landman, A.W. Wörz, J.-M. Antonietti, S. Abbet, K. Judai and U. Heiz, *Science*, 2005, **307**, 403.
171. L.M. Molina and B. Hammer, *Phys. Rev. Lett.*, 2003, **90**, 206102.
172. L.M. Molina and B. Hammer, *Phys. Rev. B*, 2004, **69**, 155424.
173. A. Asthagiri, C. Niederberger, A.J. Francis, L.M. Porter, P.A. Salvador and D. S. Sholl, *Surf. Sci.*, 2003, **537**, 134.
174. A. Asthagiri and D.S. Sholl, *Surf. Sci.*, 2005, **581**, 66.
175. L.M. Molina and B. Hammer, *Appl. Catal. A*, 2005, in press.
176. L.M. Molina and B. Hammer, *J. Catal.*, 2005, **233**, 399.
177. N. Lopez and J.K. Norskøv, *Surf. Sci.*, 2002, **515**, 175.
178. Y. Wang and G.S. Hwang, *Surf. Sci.*, 2003, **542**, 72.
179. K. Okazaki, Y. Morikawa, S. Tanaka, K. Tanaka and M. Kohyama, *Phys. Rev. B*, 2004, **69**, 235404.
180. A. Vijay, G. Mills and H. Metiu, *J. Chem. Phys.*, 2003, **118**, 6536.
181. L. Thien-Nga and A.T. Paxon, *Phys. Rev. B*, 1998, **58**, 13233.
182. Z. Yang, R. Wu and D.W. Goodman, *Phys. Rev. B*, 2000, **61**, 14066.
183. L. Giordano, G. Pacchioni, T. Bredow and J.F. Sanz, *Surf. Sci.*, 2001, **471**, 21.
184. E. Wahlström, N. Lopez, R. Schaub, P. Thostrup, A. Ronnau, C. Africh, E. Lægsgaard, J.K. Nørskov and F. Besenbacher, *Phys. Rev. Lett.*, 2003, **90**, 026101.
185. Z.-P. Liu, X.-Q. Gong, J. Kohanoff, C. Sanchez and P. Hu, *Phys. Rev. Lett.*, 2003, **91**, 266102.
186. L.M. Molina, M.D. Rasmussen and B. Hammer, *J. Chem. Phys.*, 2004, **120**, 7673.
187. I.N. Remediakis, N. Lopez and J.K. Nørskov, *Angew. Chemie Int. Ed.*, 2005, **44**, 1824.
188. N. Lopez, T.V.W. Janssens, B.S. Clausen, Y. Xu, M. Mavrikakis, T. Bligaard and J.K. Norskøv, *J. Catal.*, 2004, **223**, 232.
189. G. Mills, M.S. Gordon and H. Metiu, *J. Chem. Phys.*, 2003, **118**, 4198.
190. J. Guzman and B.C. Gates, *J. Am. Chem. Soc.*, 2004, **126**, 2672.
191. J.R. Kitchin, N.A. Khan, M.A. Barteau, J.G. Chen, B. Yakshinskiy and T.E. Madey, *Surf. Sci.*, 2003, **544**, 295.

192. B. Klötzer, W. Unterberger and K. Hayek, *Surf. Sci.*, 2003, **532–535**, 142.
193. A. Schlapka, M. Lischka, A. Groß, U. Käsberger and P. Jakob, *Phys. Rev. Lett.*, 2003, **91**, 016101.
194. H.H. Hwu, J.E. Jr. and J.G. Chen, *J. Am. Chem. Soc.*, 2002, **124**, 702.
195. M.B. Zellner, A.M. Goda, O. Skoplyak, M.A. Barteau and J.G. Chen, *Surf. Sci.*, 2005, **583**, 281.
196. S. Muller, *J. Phys. Condens. Mat*, 2002, **15**, R1429.
197. D.R. Alfonso, A.V. Cugini and D.S. Sholl, *Surf. Sci.*, 2003, **546**, 12.
198. B. Bhatia, X. Luo, C.A. Sholl and D.S. Sholl, *J. Phys. Cond. Mat.*, 2004, **16**, 8891.
199. S. Hong and C.L. Fu, *Phys. Rev. B*, 2002, **66**, 094109.
200. P.A. Sheth, M. Neurock and C.M. Smith, *J. Phys. Chem. B*, 2005, **109**, 12449.
201. D.H. Mei, E.W. Hansen and M. Neurock, *J. Phys. Chem. B*, 2003, **107**, 798.
202. L.D. Kieken, M. Neurock and D.H. Mei, *J. Phys. Chem. B*, 2005, **109**, 2234.
203. A.A. Gokhale, G.W. Huber, J.A. Dumesic and M. Mavrikakis, *J. Phys. Chem. B*, 2004, **108**, 14062.
204. J.R. Kitchin, J.M. Nørskov, M.A. Barteau and J.G. Chen, *Phys. Rev. Lett.*, 2004, **93**, 156801.
205. J.R. Kitchin, J.K. Nørskov, M.A. Barteau and J.G. Chen, *J. Chem. Phys.*, 2004, **120**, 10240.
206. Y. Xu, A.V. Ruban and M. Mavrikakis, *J. Am. Chem. Soc.*, 2004, **126**, 4717.
207. J.L. Fernández, D.A. Walsh and A.J. Bard, *J. Am. Chem. Soc.*, 2005, **127**, 357.
208. J. Zhang, M.B. Vukmirovic, Y. Xu, M. Mavrikakis and R.R. Adzic, *Angew. Chemie Int. Ed.*, 2005, **44**, 2132.
209. B. Hammer, Y. Morikawa and J.K. Nørskov, *Phys. Rev. Lett.*, 1996, **76**, 2141.
210. B. Hammer and J.K. Nørskov, *Surf. Sci.*, 1995, **343**, 211.
211. V. Pallassana, M. Neurock, L.B. Hansen and J.K. Nørskov, *J. Chem. Phys.*, 2000, **112**, 5435.
212. H. Tang and B.L. Trout, *J. Phys. Chem. B*, 2005, **109**, 6948.
213. P. Kamakoti and D.S. Sholl, *J. Membrane Sci.*, 2005, submitted.
214. L. Schlapbach and A. Züttel, *Nature*, 2001, **414**, 353.
215. A. Züttel, *Mater. Today*, 2003, **6**, 24.
216. W. Grochala and P.P. Edwards, *Chem. Rev.*, 2004, **104**, 1283.
217. J.J. Vajo, F. Mertens, C.C. Ahn, C. Robert, J. Bowman and B. Fultz, *J. Phys. Chem. B*, 2004, **108**, 13977.
218. J.J. Vajo, S.L. Skeith and F. Meters, *J. Phys. Chem. B*, 2005, **109**, 3719.
219. B. Bogdanovic and M. Schwickardi, *J. Alloys Compounds*, 1997, **253**, 1.
220. S.-C. Chung and H. Morioka, *J. Alloys Compounds*, 2004, **372**, 92.
221. K. Miwa, N. Ohba, S.-i. Towata, Y. Nakamori and S.-I. Orimo, *Phys. Rev. B*, 2004, **69**, 245120.
222. K. Miwa, N. Ohba, S.-i. Towata, Y. Makamori and S.-I. Orimo, *Phys. Rev. B*, 2005, **71**, 195109.
223. O.M. Løvvik, *Phys. Rev. B*, 2005, **71**, 144111.
224. X. Ke and I. Tanaka, *Phys. Rev. B*, 2005, **71**, 024117.
225. E.H. Majzoub, K.F. McCarty and V. Ozolins, *Phys. Rev. B*, 2005, **71**, 024118.
226. A. Peles, J.A. Alford, Z. Ma, L. Yang and M.Y. Chou, *Phys. Rev. B*, 2004, **70**, 165105.
227. P. Vajeeston, P. Ravindran, A. Kjekshus and H. Fjellvåg, *Phys. Rev. B*, 2004, **69**, 020104.
228. O.M. Løvvik, S.M. Opalka, H.W. Brinks and B.C. Hauback, *Phys. Rev. B*, 2004, **69**, 134117.

229. A. Aguayo and D.J. Singh, *Phys. Rev. B*, 2004, **69**, 155103.
230. Q. Ge, *J. Phys. Chem. A*, 2004, **108**, 8682.
231. J. Iñiguez, T. Yildirim, T.J. Udovic, M. Sulic and C.M. Jensen, *Phys. Rev. B*, 2004, **70**, 060101.
232. J. Iñiguez and T. Yildirim, *Appl. Phys. Lett.*, 2005, **86**, 103109.
233. O.M. Løvvik and S.M. Opalka, *Phys. Rev. B*, 2005, **71**, 054103.
234. S. Chaudhuri and J.T. Muckerman, *J. Phys. Chem. B*, 2005, **109**, 6952.
235. M.T.M. Koper, *J. Electroanalytical Chem.*, 2005, **574**, 375.
236. C. Hartnig, P. Vassilev and M.T.M. Koper, *Electrochimica Acta*, 2003, **48**, 3751.
237. S.K. Desai, V. Pallassana and M. Neurock, *J. Phys. Chem. B*, 2001, **105**, 9171.
238. S.K. Desai and M. Neurock, *Phys. Rev. B*, 2003, **68**, 075420.
239. P.J. Feibelman, *Science*, 2002, **295**, 99.
240. P.J. Feibelman, *Phys. Rev. Lett.*, 2003, **90**, 186103.
241. P.J. Feibelman and A. Alavi, *J. Phys. Chem. B*, 2004, **108**, 14362.
242. V.A. Ranea, A. Michaelides, R. Ramirez, P.L. de Andres, J.A. Verges and D.A. King, *Phys. Rev. Lett.*, 2004, **92**, 136104.
243. A. Michaelides, V.A. Ranea, P.L. de Andres and D.A. King, *Phys. Rev. B*, 2004, **69**, 075409.
244. A. Michaelides, A. Alavi and D.A. King, *Phys. Rev. B*, 2004, **69**, 113404.
245. A. Tilocca and A. Selloni, *J. Phys. Chem. B*, 2004, **108**, 4743.
246. J. Cerda, A. Michaelides, M.L. Bocquet, P.J. Feibelman, T. Mitsui, M. Rose, E. Fomin and M. Salmeron, *Phys. Rev. Lett.*, 2004, **93**, 116101.
247. K. Andersson, A. Nikitin, L.G.M. Pettersson, A. Nilsson and H. Ogasawara, *Phys. Rev. Lett.*, 2004, **93**, 196101.
248. P.J. Feibelman, *Chem. Phys. Lett.*, 2005, **410**, 120.
249. P. Vassilev, R.A. van Santen and M.T.M. Koper, *J. Chem. Phys.*, 2005, **122**, 054701.
250. G. Materzanini, G.F. Tantardini, P.J.D. Lindan and P. Saalfrank, *Phys. Rev. B*, 2005, **71**, 155414.
251. G. Cicero, J.C. Grossman, A. Catallani and G. Galli, *J. Am. Chem. Soc.*, 2005, **127**, 6830.
252. Y.X. Wang and P.B. Balbuena, *J. Phys. Chem. B*, 2004, **108**, 4376.
253. Y.X. Wang and P.B. Balbuena, *J. Phys. Chem. B*, 2005, **109**, 14896.
254. M.P. Hyman and J.W. Medlin, *J. Phys. Chem. B*, 2005, **109**, 6304.
255. A. Roudgar and A. Gross, *Chem. Phys. Lett.*, 2005, **409**, 157.
256. J. Kärger and D. Ruthven, *Diffusion in Zeolites and Other Microporous Materials*, John Wiley & Sons, New York, 1992.
257. N.Y. Chen, T.F. Degnan and C.M. Smith, *Molecular transport and reaction in zeolites: design and application of shape selective catalysts*, VCH, New York, 1994.
258. http://www.iza-structure.org/databases/.
259. X. Rozanska, L.A.M.M. Barbosa and R.A. van Santen, *J. Phys. Chem. B*, 2005, **109**, 2203.
260. M. Sierka and J. Sauer, *J. Phys. Chem. B*, 2001, **105**, 1603.
261. G. Poulet, P. Sautet and A. Tuel, *J. Phys. Chem. B*, 2002, **106**, 8599.
262. F.J. Keil, R. Krishna and M.O. Coppens, *Rev. Chem. Eng.*, 2000, **16**, 71.
263. B. Smit and R. Krishna, *Chem. Eng. Sci.*, 2003, **58**, 557.
264. A.I. Skoulidas and D.S. Sholl, *J. Phys. Chem. B*, 2001, **105**, 3151.
265. A.I. Skoulidas, D.M. Ackerman, J.K. Johnson and D.S. Sholl, *Phys. Rev. Lett.*, 2002, **89**, 185901.
266. A.I. Skoulidas and D.S. Sholl, *J. Phys. Chem. B*, 2002, **106**, 5058.

267. A.I. Skoulidas, D.S. Sholl and R. Krishna, *Langmuir*, 2003, **19**, 7977.
268. A.I. Skoulidas and D.S. Sholl, *J. Phys. Chem. B*, 2003, **107**, 10132.
269. T. Duren, L. Sarkisov, O.M. Yaghi and R.Q. Snurr, *Langmuir*, 2004, **20**, 2683.
270. L. Sarkisov, T. Duren and R.Q. Snurr, *Mol. Phys.*, 2004, **102**, 211.
271. T.J.H. Vlugt, W. Zhu, F. Kapteijn, J.A. Moulijn, B. Smit and R. Krishna, *J. Am. Chem. Soc.*, 1998, **120**, 5599.
272. S.-S. Chong, H. Jobic, M. Plazanet and D.S. Sholl, *Chem. Phys. Lett.*, 2005, **408**, 157.
273. F. Corà, C.R.A. Catlow, B. Civalleri and R. Orlando, *J. Phys. Chem. B*, 2003, **107**, 11866.
274. F. Corà and C.R.A. Catlow, *J. Phys. Chem. B*, 2003, **107**, 11861.
275. C. Lo and B.L. Trout, *J. Catal.*, 2004, **227**, 77.
276. B. Civalleri, A.M. Ferrari, M. Llunell, R. Orlando, M. Mérawa and P. Ugliengo, *Chem. Mater.*, 2003, **15**, 3996.
277. R. Astala, S.M. Auerbach and P.A. Monson, *J. Phys. Chem. B*, 2004, **108**, 9208.
278. T. Demuth, J. Hafner, L. Benco and H. Toulhoat, *J. Phys. Chem. B*, 2000, **104**, 4593.
279. R. Astala and S.M. Auerbach, *J. Am. Chem. Soc.*, 2004, **126**, 1843.
280. K. Yamamoto, Y. Sakata, Y. Nohara, Y. Takahashi and T. Tatsumi, *Science*, 2003, **300**, 470.
281. X. Rozanska, R.A. van Santen, F. Hutschka and J. Hafner, *J. Am. Chem. Soc.*, 2001, **123**, 7655.
282. X. Rozanska, T. Demuth, F. Hutschka, J. Hafner and R.A. van Santen, *J. Phys. Chem. B*, 2002, **106**, 3248.
283. X. Rozanska, R.A. van Santen, T. Demuth, F. Hutschka and J. Hafner, *J. Phys. Chem. B*, 2003, **107**, 1309.
284. M.S. Sun, O. Talu and D.B. Shah, *J. Phys. Chem. B*, 1996, **100**, 17276.
285. J.F. Denayer, G.V. Baron, J.A. Martens and P.A. Jacobs, *J. Phys. Chem. B*, 1998, **102**, 3077.
286. C. Lo, C.A. Giurumescu, R. Radhakrishnan and B.L. Trout, *Mol. Phys.*, 2004, **102**, 281.
287. C.S. Lo, R. Radhakrishnan and B.L. Trout, *Catal. Today*, 2005, **105**, 93.
288. N. Govind, J. Andzelm, K. Reindel and G. Fitzgerald, *Int. J. Mol. Sci.*, 2002, **3**, 423.
289. T. Bucko, J. Hafner and L. Benco, *J. Phys. Chem. A*, 2004, **108**, 11388.
290. D.A. Newsome and D.S. Sholl, *J. Phys. Chem. B*, 2005, **109**, 7237.
291. M. Chandross, E.B. Webb, G.S. Grest, M.G. Martin, A.P. Thompson and M.W. Roth, *J. Phys. Chem. B*, 2001, **105**, 5700.
292. O. Geier, R.Q. Snurr, F. Stallmach and J. Karger, *J. Chem. Phys.*, 2004, **120**, 367.
293. E.B. Webb and G.S. Grest, *J. Chem. Phys.*, 2002, **116**, 6311.
294. W. Song, R.E. Justice, C.A. Jones, V.H. Grassian and S.C. Larsen, *Langmuir*, 2004, **20**, 8301.
295. W. Song, J.F. Woodworth, V.H. Grassian and S.C. Larsen, *Langmuir*, 2005, **21**, 7009.
296. M. Hong, J.L. Falconer and R.D. Noble, *Ind. Eng. Chem. Res.*, 2005, **44**, 4035.
297. W. Chen, *J. Mol. Catal. A*, 2002, 41.
298. T. Bucko, L. Benco, T. Demuth and J. Hafner, *J. Chem. Phys.*, 2002, **117**, 7295.
299. T. Bucko, L. Benco and J. Hafner, *J. Chem. Phys.*, 2003, **118**, 8437.
300. T. Bucko, J. Hafner and L. Benco, *J. Chem. Phys.*, 2004, **120**, 10263.
301. T. Bucko, J. Hafner and L. Benco, *J. Phys. Chem. B*, 2005, **109**, 7345.

5

Numerical Methods in Chemistry

BY T.E. SIMOS

University of Peloponnese, 26 Menelaou St, Amfithea-Palea Faliron, Gr-175 64, Athens, Greece

1 Introduction

In this paper we will present the recent developments on the numerical integration of the Schrödinger equation and related problems.

More specifically we will present the recent advances for the numerical integration of the radial time-independent Schrödinger equation and related problems and of the coupled differential equations of the Schrödinger type.

The radial Schrödinger equation has the form:

$$y''(r) = [l(l + 1)/r^2 + V(r) - k^2]\, y(r). \tag{1}$$

Models of this type, which represent a boundary value problem, occur frequently in theoretical physics and chemistry, (see for example 1–4).

Here are some notes for (1):

- The function $W(r) = l(l+1)/r^2 + V(r)$ denotes *the effective potential*. This satisfies $W(r) \to 0$ as $r \to \infty$
- k^2 is a real number denoting *the energy*
- l is a given integer representing *angular momentum*
- V is a given function which denotes the potential.
- The boundary conditions are:

$$y(0) = 0$$

- and a second boundary condition, for large values of r, determined by physical considerations.

It is known from the literature that the last decades many numerical methods have been developed for the numerical solution of the Schrödinger equation (see 5,6). The aim and the scope of the above activity was the development of fast and reliable methods.

Chemical Modelling: Applications and Theory, Volume 4

The developed methods can be divided into two main categories:

- Methods with constant coefficients
- Methods with coefficients dependent on the frequency of the problem.[†]

In the present review we shall present recent advances in the construction of numerical methods for the numerical integration of the Schrödinger equation and related problems. We will also present remarks for improvement of some methods. Finally for the most of the case we will give software for the construction of the methods.

In Section 2 we analyse a new category of methods named partitioned multistep methods. More specifically we derived exponentially-fitted partitioned multistep methods. For these methods numerical results on the problem of Stiefel and Bettis is presented.

In Section 3 we present the Dispersion and Dissipation properties for Runge-Kutta methods. Based on these properties we have constructed:

- A dispersive-fitted and dissipative-fitted explicit Runge-Kutta
- A Runge-Kutta method with minimal dispersion error
- A Runge-Kutta method with minimal dissipation error

For these methods numerical results on the resonance problem of the radial Schrödinger equation are given and analysed.

In Section 4 we present Four-step P-stable Methods with minimal Phase-Lag. We give a new procedure for the construction of such methods. This procedure is based on the requirement that the roots of the characteristic equation associated with the methods must have specific forms. For these methods numerical results on the resonance problem of the radial Schrödinger equation are given and analysed.

In Section 5 trigonometrically fitted Fifth algebraic order Runge-Kutta methods are presented. For these methods we present the construction and the error analysis from which one can see that the classical method is dependent on the third power of energy, the first exponentially-fitted methods is depended on a second power of the energy and finally the second exponentially-fitted methods is depended on first power of the energy.

In Section 6 we present four-step P-stable trigonometrically-Fitted methods. The procedure for the satisfaction of P-stability is similar with this mentioned in section 4 and is based on the requirement that the roots of the characteristic equation associated with the methods must have specific forms. For these methods numerical results on the resonance problem of the radial Schrödinger equation are given and analysed.

Finally, in Section 7 we present extended comments on the recent bibliography with emphasis to the last 2 years.

[†] In the case of the Schrödinger equation the frequency of the problem is equal to: $\sqrt{|l(l+1)/r^2 + V(r) - k^2|}$

In Appendix A the Maple programme for the construction of Partitioned Multistep Methods mentioned in Section 2 is presented. In Appendix B the Maple Programme for the development of Dispersive-fitted and dissipative-fitted explicit Runge-Kutta method is presented. In Appendix C we present the Maple Programme for the development of explicit Runge-Kutta method with minimal Dispersion. In Appendix D we present the Maple Programme for the development of explicit Runge-Kutta method with minimal Dissipation. In Appendix E the Maple Programme for the development of the New Four-Step P-stable method with minimal Phase-Lag is presented. In Appendix F we present the Maple Programme for the development of the Trigonometrically Fitted Fifth-Order Runge-Kutta Methods. Finally, in the Appendix G the Maple Programme for the development of the New Four-Step P-stable Trigonometrically-Fitted method is presented.

2 Partitioned Trigonometrically-Fitted Multistep Methods

The following partitioned multistep method is considered

$$y_{n+3} - y_{n-2} = h^2\,[b_0 f_{n+2} + b_1 f_{n+1} + b_2 f_n + b_3 f_{n-1} \tag{2}$$

$$z_{n+2} - z_{n-2} = h^2\,[c_0 g_{n+1} + c_1 g_n + c_2 g_{n-1}], \tag{3}$$

where $y_{n\pm i} = y(x \pm ih)$, $i = -2{,}3$, $f_{n\pm j} = y'(x \pm jh)$, $j = -1(1)2$, $z_{n\pm k} = z(x \pm kh)$, $k = -2.2$, $g_{n\pm 1} = z'(x \pm lh)$, $l = -1(1)1$, h is the step size b_i, $i = 0(1)3$ and $c_i, j = 0(1)2$ are the parameters of the method, in order to be trigonometrically-fitted.

2.1 First Method of the Partitioned Multistep Method. – *2.1.1 Exponential Fitting of First Order.*

We want the method (2) to integrate exactly any linear combination of the functions:

$$\{1,\ x,\ x^2,\ \cos(\pm vx),\ \sin(\pm vx)\}. \tag{4}$$

For every linear combination of the functions given above, the appropriate parameters of the new method are the solution of a system of equations, which is produced in the following way:

We calculate $y_{n\pm m} = y(x \pm mh)$, $m = -2{,}3$ and $f_{n\pm j} = y'(x \pm jh)$, $j = -1(1)2$ for $y(x) = x^n$, $n = 0, \ldots 2$ and for $y(x) = \exp\,(ivx)$ where $i = \sqrt{-1}$. The following system of equations is obtained:

$$5 = b_0 + b_1 + b_2 + b_3, \tag{5}$$

$$5 = 4b_0 + 2b_1 - 2b_3, \tag{6}$$

$$-\cos(2w) + \cos(3w) = -w[b_0 \sin(2w) + b_1 \sin(w) - b_3 \sin(w)], \tag{7}$$

$$\sin(3w) + \sin(2w) = w[b_0 \cos(2w) + b_1 \cos(w) + b_2 + b_3 \cos(w)], \tag{8}$$

where $w = vh$.

Solving the system of Eqs. (5)–(8), we obtain the parameters of the method, which are:

$$b_0 = \frac{-5w + 2\sin(3w) - 5\cos(w)w + 2\sin(2w)}{-2w[1 - \cos(2w)]},$$

$$b_1 = \frac{5w\cos(2w) - 2\sin(3w) + 5w\cos(w)w - 2\sin(2w)}{-2w[1 - \cos(2w)]},$$

$$b_2 = b_1, \quad b_3 = b_0 \tag{9}$$

The above parameters converted into their Taylor series expansions are given below:

$$\begin{aligned} b_0 &= \frac{55}{24} - \frac{95}{288}w^2 + \frac{65}{4032}w^4 - \frac{47}{96768}w^6 - \frac{89}{19160064}w^8 - \frac{38923}{34871316480}w^{10} \\ &\quad - \frac{3253}{29889699840}w^{12} + \ldots, \\ b_1 &= \frac{5}{24} + \frac{95}{288}w^2 - \frac{65}{4032}w^4 + \frac{47}{96768}w^6 + \frac{89}{19160064}w^8 + \frac{38923}{34871316480}w^{10} \\ &\quad + \frac{3253}{29889699840}w^{12} + \ldots. \end{aligned} \tag{10}$$

It can be seen (for more details see 8) that for $3.13 < w < 3.15$ and for $6.27 < w < 6.29$ is better to use the Taylor series expansion.

The local truncation error of this method is:

$$LTE_1(h) = \frac{95h^5}{144}\left(y_n^{(5)} + v^2 y_n^{(3)}\right), \tag{11}$$

where $y_n^{(q)} = \frac{d^q y_n}{dx^q}, \quad q = 1, 2, \ldots.$

2.1.2 Exponential Fitting of Second Order. We want the method (2) to integrate exactly any linear combination of the functions:

$$\{1, \cos(\pm vx), \sin(\pm vx), x\cos(\pm vx), x\sin(\pm vx)\}. \tag{12}$$

For every linear combination of the functions given above, the appropriate parameters of the new method are the solution of a system of equations, which is produced in the following way:

We calculate $y_{n\pm m} = y(x \pm mh)$, $m = -2{,}3$ and $f_{n\pm j} = y'(x \pm jh)$, $j = -1(1)2$ for $y(x) = \exp(ivx)$ and for $y(x) = x\exp(ivx)$ where $i = \sqrt{-1}$. The following system of equations is obtained:

$$-\cos(2w) + \cos(3w) = -w[b_0\sin(2w) + b_1\sin(w) - b_3\sin(w)], \tag{13}$$

$$\sin(3w) + \sin(2w) = w[b_0\cos(2w) + b_1\cos(w) + b_2 + b_3\cos(w)], \tag{14}$$

$$- \cos(2w)(x - 2h) + \cos(3w)(x + 3h) = -h \left\{ \begin{array}{l} b_0[\sin(2w)vx - \cos(2w) + 2\sin(2w)w] + \\ +b_1[\sin(w)vx - \cos(w) + \sin(w)w] - b_2 - \\ -b_3[\sin(w)vx + \cos(w) - \sin(w)w] \end{array} \right\}, \tag{15}$$

$$\sin(3w)(x + 3h) + \sin(2w)(x - 2h) = h \left\{ \begin{array}{l} b_0[\cos(2w)vx + \sin(2w) + 2\cos(2w)w] + \\ +b_1[\cos(w)vx + \sin(w) + \cos(w)w] + b_2vx + \\ +b_3[\cos(w)vx - \sin(w) - \cos(w)w] \end{array} \right\}, \tag{16}$$

where $w = vh$.

Solving the system of Eqs. (5)–(8), we obtain the parameters of the method, which are:

$$b_0 = \frac{wT(w) + \sin(4w) - \sin(2w) - \sin(3w) + \sin(w)}{w^2[\sin(3w) - 3\sin(w)]},$$

$$b_1 = \frac{2 + \cos(6w) - 2\cos(5w)5 + 3\cos(3w)w - 3\cos(2w) - \cos(w) + wC(w)}{w^2[4\cos(2w) - \cos(4w) - 3]},$$

$$b_2 = b_1, \quad b_3 = b_0 \tag{17}$$

where

$$\begin{aligned} T(w) &-4\cos(2w) - 2\cos(4w) - 3\cos(w) + \cos(3w) \\ C(w) &= 4\sin(w) + 6\sin(3w) + \sin(6w) - 9\sin(2w) \end{aligned}$$

The above parameters converted into their Taylor series expansions are given below:

$$\begin{aligned} b_0 =& \frac{55}{24} - \frac{95}{144}w^2 + \frac{505}{24192}w^4 - \frac{97}{48384}w^6 - \frac{2953}{19160064}w^8 \\ & \frac{1118269}{52306974720}w^{10} - \frac{174541}{64377815040}w^{12} + \ldots, \\ b_1 =& \frac{5}{24} + \frac{95}{144}w^2 - \frac{8485}{24192}w^4 + \frac{1657}{48384}w^6 + \frac{51299}{19160064}w^8 \\ & + \frac{1467709}{52306974720}w^{10} + \frac{1163597}{119558799360}w^{12} + \ldots. \end{aligned} \tag{18}$$

From Figures 1–8 it can be seen that for $3.13 < w < 3.15$ and for $6.27 < w < 6.29$ is better to use the Taylor series expansion.

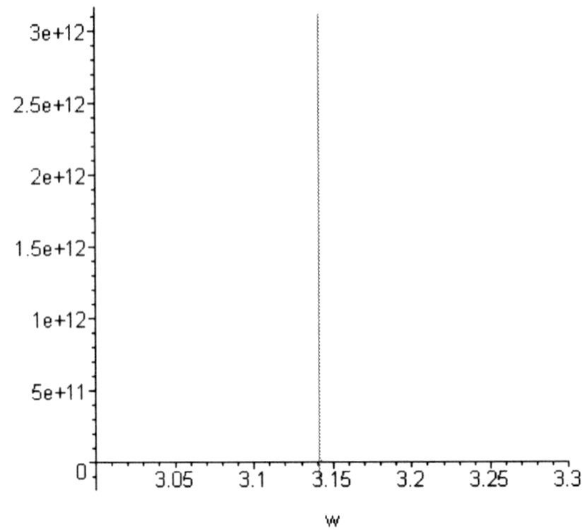

Figure 1 *Behavior of the coefficient b_0 of the exponentially-fitted method of second exponential order for the interval [3.0,3.3]*

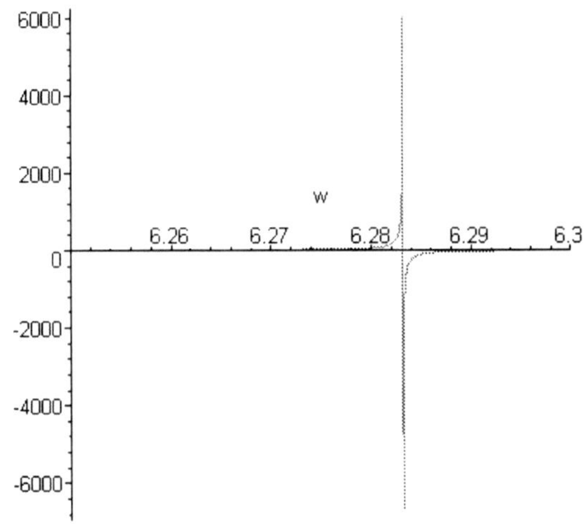

Figure 2 *Behavior of the coefficient b_0 of the exponentially-fitted method of second exponential order for the interval [6.25,6.30]*

The local truncation error of this method is:

$$LTE_3(h) = \frac{95h^5}{144}\left(y_n^{(5)} + 2v^2 y_n^{(3)} + v^4 y_n'\right), \tag{19}$$

where $y_n^{(q)} = \frac{d^q y_n}{dx^q}$, $q = 1, 2, \ldots$.

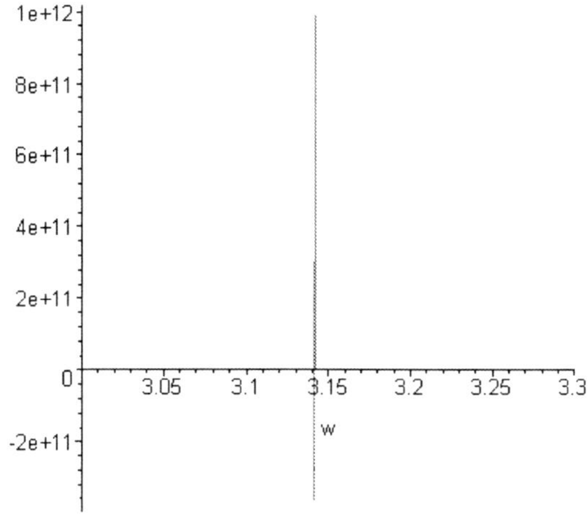

Figure 3 *Behavior of the coefficient b_1 of the exponentially-fitted method of second exponential order for the interval [3.0,3.3]*

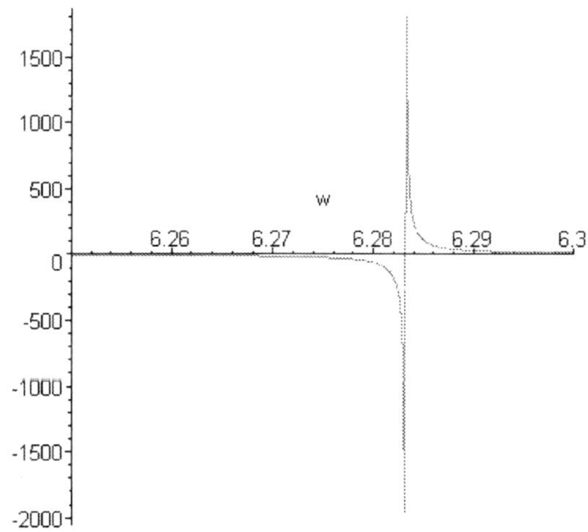

Figure 4 *Behavior of the coefficient b_1 of the exponentially-fitted method of second exponential order for the interval [6.25,6.30]*

2.2 Second Method of the Partitioned Mutistep Method. – *2.2.1 Exponential Fitting of First Order*. We want the method (3) to integrate exactly any linear combination of the functions (4).

The above requirement leads to the following system of equations, which is produced in the following way:

We calculate $z_{n\pm m} = y(x \pm mh)$, $m = -2,2$ and $g_{n\pm j} = z'(x \pm jh)$, $j = -1(1)1$ for $z(x) = x^n$, $n = 0,\ldots, 2$ and for $z(x) = \exp(ivx)$ where $i = \sqrt{-1}$. The following

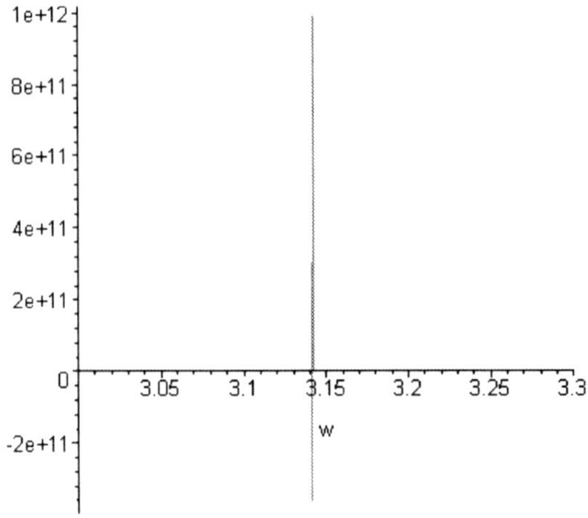

Figure 5 *Behavior of the coefficient b_2 of the exponentially-fitted method of second exponential order for the interval [3.0,3.3]*

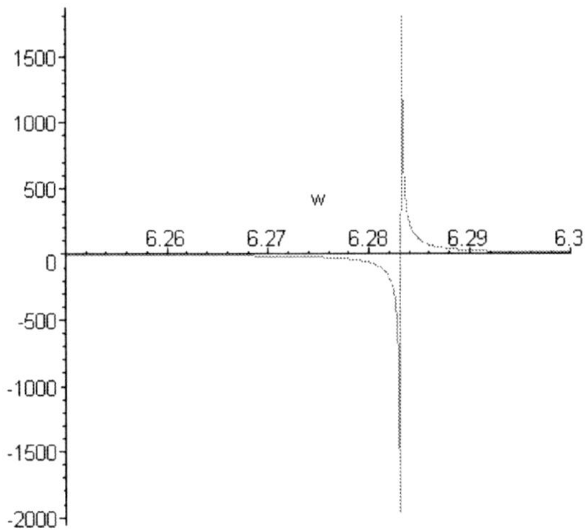

Figure 6 *Behavior of the coefficient b_2 of the exponentially-fitted method of second exponential order for the interval [6.25,6.30]*

system of equations is obtained:

$$4 = c_0 + c_1 + c_2, \tag{20}$$

$$0 = c_0 - c_2, \tag{21}$$

$$0 = -w[c_0 \sin(w) + c_2 \sin(w)], \tag{22}$$

$$2 \sin(2w) = w[c_0 \cos(w) + c_1 + c_2 \cos(w)], \tag{23}$$

where $w = vh$.

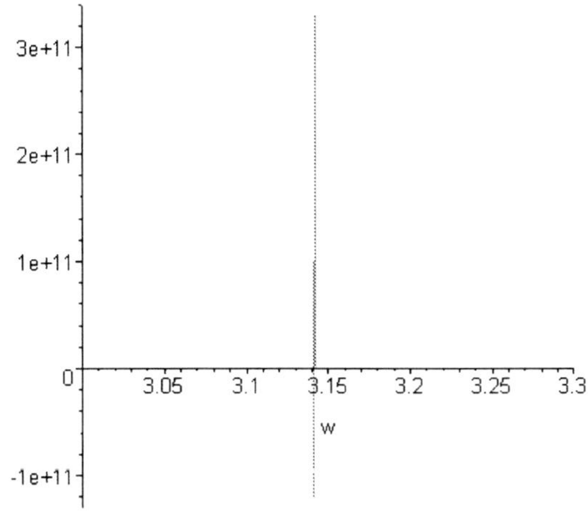

Figure 7 *Behavior of the coefficient b_3 of the exponentially-fitted method of second exponential order for the interval [3.0,3.3]*

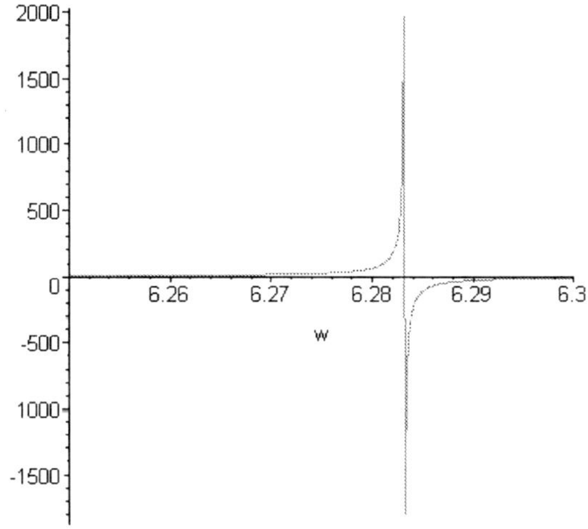

Figure 8 *Behavior of the coefficient b_3 of the exponentially-fitted method of second exponential order for the interval [6.25,6.30]*

Solving the system of Eqs. (20)–(23), we obtain the parameters of the method, which are:

$$c_0 = \frac{-2w + \sin(2w)}{w[\cos(w) - 1]},$$

$$c_1 = \frac{4w\cos(w) - 2\sin(2w)}{w[\cos(w) - 1]}, \; c_2 = c_0. \tag{24}$$

The above parameters converted into their Taylor series expansions are given below.

$$c_0 = \frac{8}{3} - \frac{14}{45}w^2 + \frac{11}{630}w^4 - \frac{1}{2700}w^6 + \frac{19}{2993760}w^8 - \frac{337}{13621608000}w^{10} + \frac{1}{934053120}w^{12} + \dots,$$

$$c_1 = -\frac{4}{3} + \frac{28}{45}w^2 - \frac{11}{315}w^4 + \frac{1}{1350}w^6 - \frac{19}{1496880}w^8 + \frac{337}{6810804000}w^{10} - \frac{1}{467026560}w^{12} + \dots. \tag{25}$$

It is easy to see (see 8) that for $3.13 < w < 3.15$ and $6.27 < w < 6.29$ is better to use the Taylor series expansion.

The local truncation error of this method is:

$$LTE_2(h) = \frac{14h^5}{45}\left(y_n^{(5)} + v^2 y_n^{(3)}\right). \tag{26}$$

2.2.2 Exponential Fitting of Second Order. We want the method (3) to integrate exactly any linear combination of the functions (12).

The above requirement leads to the following system of equations, which is produced in the following way:

We calculate $z_{n\pm m} = y(x \pm mh)$, $m = -2,2$ and $g_{n\pm j} = z'(x \pm jh)$, $j = -1(1)1$ for $z(x) = \exp(ivx)$ and for $z(x) = x\exp(ivx)$ where $i = \sqrt{-1}$. The following system of equations is obtained:

$$0 = -w[c_0 \sin(w) + c_2 \sin(w)], \tag{27}$$

$$2\sin(2w) = w[c_0\cos(w) + c_1 + c_2\cos(w)], \tag{28}$$

$$4\cos(2w) = -\left\{\begin{array}{l} c_0[-\cos(w) + vx\sin(w) + w\sin(w)]- \\ -c_1 - c_2\cos(w) - c_2 vx\sin(w) + c_2 w\sin(w) \end{array}\right\} \tag{29}$$

$$2\sin(2w).x = h\left\{\begin{array}{l} c_0[\sin(w) + vx\cos(w) + w\cos(w)]+ \\ +c_1 vx - c_2[\sin(w) - vx\cos(w) + w\cos(w)] \end{array}\right\} \tag{30}$$

where $w = vh$.

Solving the system of Eqs. (27–30), we obtain the parameters of the method, which are:

$$c_0 = \frac{-2w\cos(2w) + \sin(2w)}{w^2\sin(w)},$$

$$c_1 = \frac{3w\cos(w) + w\cos(3w) - \sin(3w) - \sin(w)}{w^2\sin(w)}, \tag{31}$$

$$c_2 = c_0.$$

The above parameters converted into their Taylor series expansions are given below.

$$c_0 = \frac{8}{3} - \frac{28}{45}w^2 + \frac{5}{189}w^4 - \frac{11}{9450}w^6 - \frac{241}{7484400}w^8 - \frac{90367}{20432412000}w^{10} - \frac{7157}{16345929600}w^{12} + \dots$$

$$c_1 = -\frac{4}{3} + \frac{56}{45}w^2 - \frac{344}{945}w^4 + \frac{176}{4725}w^6 - \frac{1096}{467775}w^8 + \frac{49072}{638512875}w^{10} - \frac{16}{5108103}w^{12} + \dots. \quad (32)$$

From Figures 9–14, it is easy to see that for $3.13 < w < 3.15$ and $6.27 < w < 6.29$ is better to use the Taylor series expansion.

The local truncation error of this method is:

$$LTE_2(h) = \frac{14h^5}{45}\left(y_n^{(5)} + 2v^2 y_n^{(3)} + v^4 y_n'\right). \quad (33)$$

In Figure 15 we present a flow chart for the new methods for the numerical solution of the system of equations:

$$q' = k(q, z),\ z' = g(q, z). \quad (34)$$

In this flow chart x_e is the end-point of integration, x_s is the start-point of integration and NSTEP is the number of steps. The computation of q_i, z_i, $i = 1(1)3$ is based on the Runge-Kutta-Nyström method of Dormand and Prince 8(7) (see 9–10).

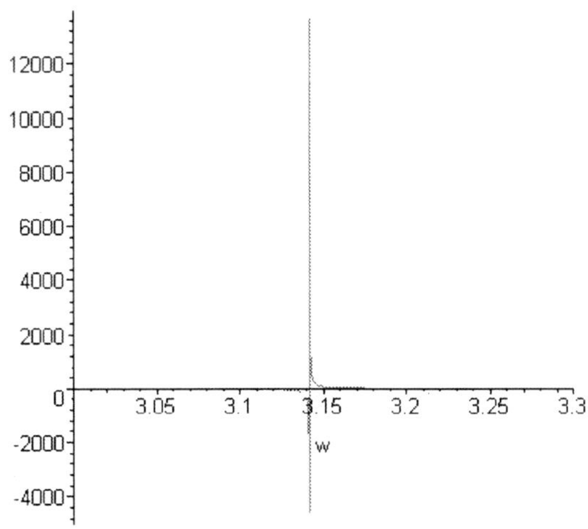

Figure 9 *Behavior of the coefficient c_0 of the exponentially-fitted method of second exponential order for the interval [3.0,3.3]*

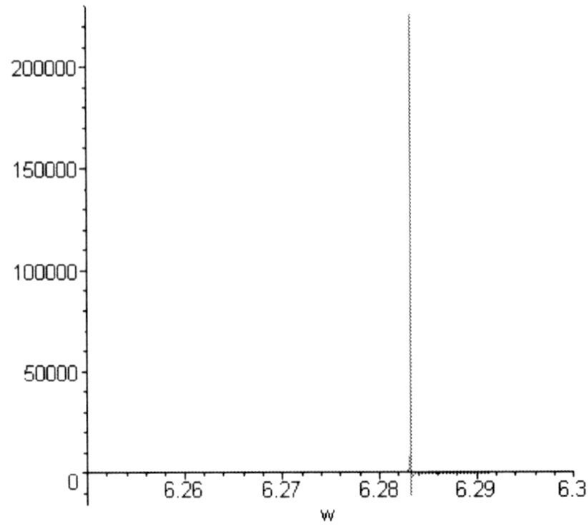

Figure 10 *Behavior of the coefficient c_0 of the exponentially-fitted method of second exponential order for the interval [6.25,6.30]*

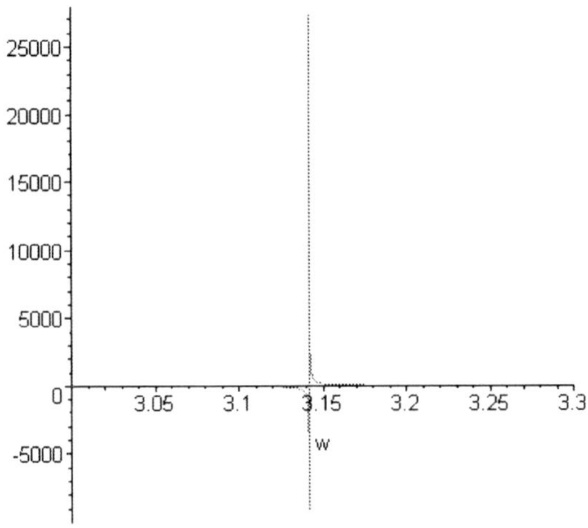

Figure 11 *Behavior of the coefficient c_1 of the exponentially-fitted method of second exponential order for the interval [3.0,3.3]*

2.3 Numerical Results. – We apply the new methods to the well known "almost" periodic orbit problem studied by Stiefel and Bettis [11].

2.3.1 A Problem by Stiefel and Bettis. The "almost" periodic orbit problem studied by Stiefel and Bettis [11] is the problem, which is considered.

$$y'' + y = 0.001e^{ix},\ y(0) = 1,$$
$$y'(0) = 0.9995i,\ y \in C, \quad (35)$$

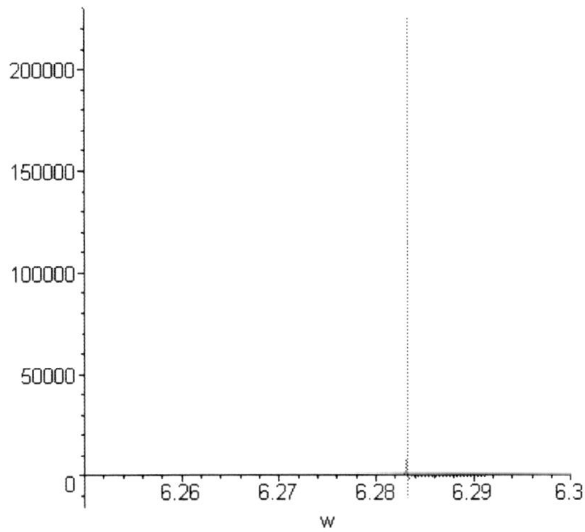

Figure 12 *Behavior of the coefficient c_1 of the exponentially-fitted method of second exponential order for the interval [6.25,6.30]*

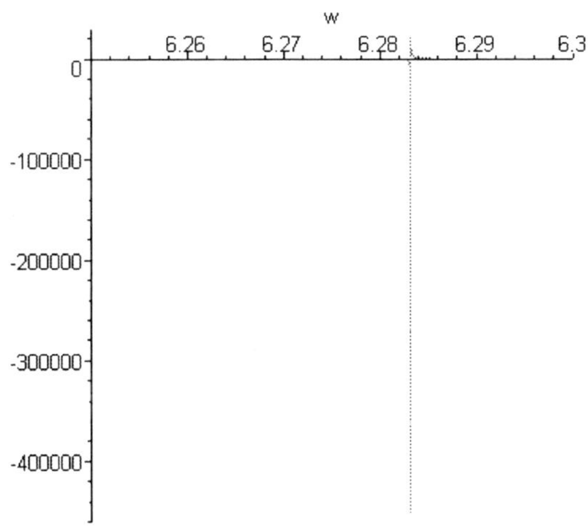

Figure 13 *Behavior of the coefficient c_2 of the exponentially-fitted method of second exponential order for the interval [3.0,3.3]*

whose equivalent form is:

$$u'' + u = 0.001\cos(x),\ u(0) = 1,\ u'(0) = 0, \tag{36}$$

$$\upsilon'' + \upsilon = 0.001\sin(x),\ \upsilon(0) = 0,\ \upsilon'(0) = 0.9995. \tag{37}$$

The analytical solution of the problem (35) is following

$$y(x) = u(x) + i\upsilon(x), \quad u, \upsilon \in R, \tag{38}$$

$$u(x) = \cos(x) + 0.0005x\sin(x) \tag{39}$$

$$\upsilon(x) = \sin(x) - 0.0005x\cos(x) \tag{40}$$

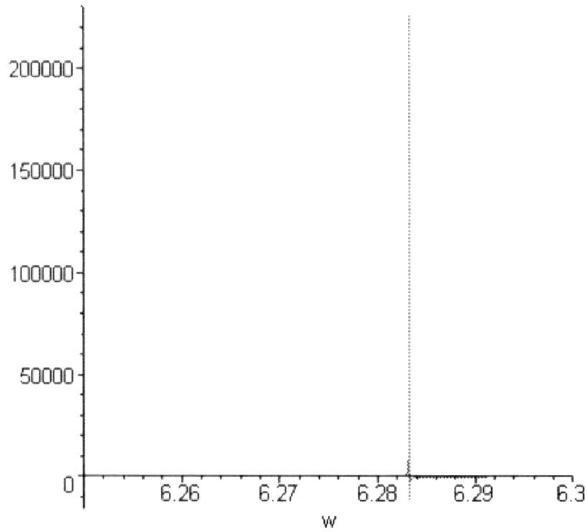

Figure 14 *Behavior of the coefficient c_2 of the exponentially-fitted method of second exponential order for the interval [6.25,6.30]*

The solution of Eqs. (38)–(40) represents motion of a perturbation of a circular orbit in the complex plane.

The system of Eqs. (36) and (37) has been solved for $0 \leq x \leq 56340\pi$ using the following methods:

(i) The Runge-Kutta–Nyström Symplectic method of algebraic order four developed by Calvo Sanz-Serna and Calvo [12] (which is denoted as Method [a]),
(ii) The embedded Runge-Kutta method of Dormand and Prince 5(4)[‡] (see Hairer *et al.*, 13) (which is denoted as Method as Method [b]),
(iii) The classical partinioned multistep method[§] (which is denoted as Method [c]),
(iv) The method developed in paragraph 2 with the coefficients given by (9)–(10) and (24)–(25) (Which is denoted as Method [d]),
(v) The method developed in paragraph 2 with the coefficients given by (17)–(18) and (31)–(32) (Which is denoted as Method [d])

For the problem studied by Stiefel and Bettis[11] we have that $v = 1$. The numerical results obtained for the three methods, with the same number of function evaluations, were compared with the analytical solution. Figure 16 shows the absolute errors Errmax for the same number of function evaluations.

2.3.2 Conclusions. From the results presented in Figure 16 we have the following conclusions:

- The new developed exponentially-fitted methods are the most efficient ones

[‡] Which is of fifth algebraic order and uses the fourth algebraic order formula in order to control the error.

[§] With the term "classical" we defined the method (2)–(3) withconstant coefficients.

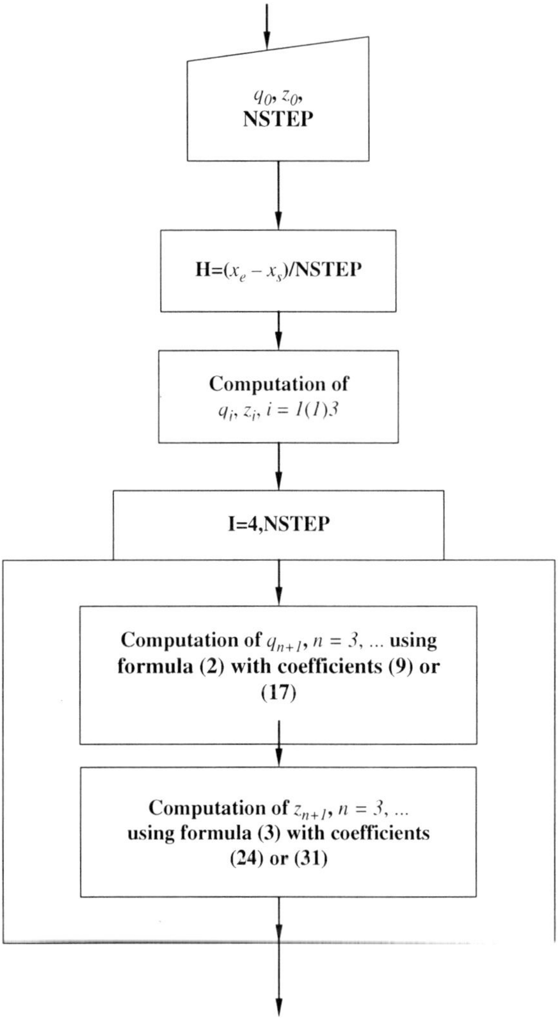

Figure 15 *Flow chart for the new developed methods. In this flow chart x_e is the end-point of integration, x_5 is the start-point of integration and NSTEP is the number of steps. The computation of q_i z_i, $i = 1(1)3$ is based on the Runge-Kutta-Nyström method of Dormand and Prince 8(7) (see Dormand et al., 1987)*

- The exponentially-fitted method developed in Section 2.3 is much more efficient than the exponentially-fitted method developed in Section 2.2.
- The behavior of the Runge-Kutta-Nyström Symplectic method of algebraic order four developed by Sanz-Serna and Calvo[12] and the behavior of the classical partitioned multistep method is similar. These methods are much more efficient that the embedded Runge-Kutta method of Dormand and Prince 5(4) (see 13).

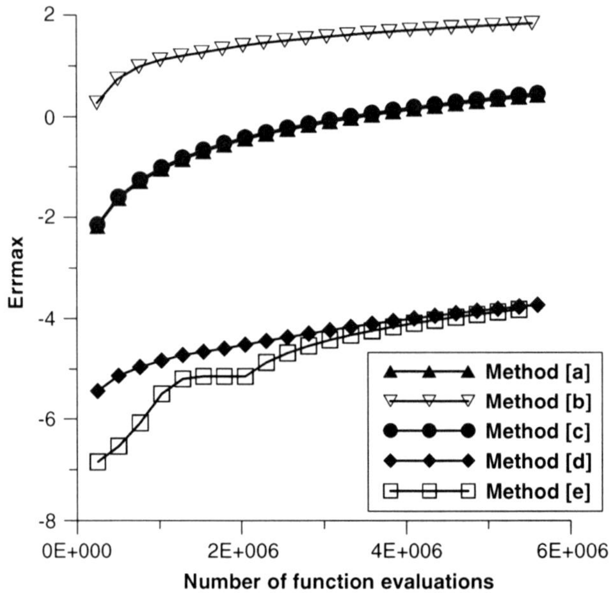

Figure 16 *Values of Errmax as a function of values of the number of function evaluations required for the integration of the orbit problem studied by Stiefel and Bettis*

3 Dispersion and Dissipation Properties for Explicit Runge-Kutta Methods

3.1 Basic Theory. – *3.1.1 Explicit Runge-Kutta Form-Order Conditions.* An s–stage explicit Runge-Kutta method used for the computation of the approximation of $y_{n+1}(x)$ in problem (1), when $y_n(x)$ is known, can be expressed by the following relations:

$$y_{n+1} = y_n + \sum_{i=1}^{s} b_i k_i$$

$$k_i = hf\left(x_n + c_i h, y_n + h\sum_{j=1}^{i-1} a_{ij} k_j\right), \; i = 1, \ldots, s. \tag{41}$$

The method mentioned previously can also be presented using the Butcher table below

$$\begin{array}{c|ccccc}
0 & & & & & \\
c_2 & a_{21} & & & & \\
c_3 & a_{31} & a_{32} & & & \\
\vdots & \vdots & \vdots & & & \\
c_s & a_{s1} & a_{s2} & \ldots & a_{s,s-1} & \\
\hline
 & b_1 & b_2 & \ldots & b_{s-1} & b_s
\end{array} \tag{42}$$

Coefficients $c_2, \ldots, c_s$ must satisfy the equations $c_i = \sum_{j=1}^{s} a_{ij}, \quad i = 2, \ldots, s.$

$$(43)$$

Definition 1 (see 13). A Runge-Kutta method has algebraic order p when the method's series expansion agrees with the Taylor series expansion in the p first terms

$$y^{(n)}(x) = y_n{}^{(n)}(x), \quad n = 1, 2, \ldots, p.$$

A Runge-Kutta method must satisfy a number of equations, in order to have a certain algebraic order. These equations will be shown later in this paper.

3.1.2 Phase-Lag Analysis of Runge-Kutta Methods. We consider the scalar test equation:

$$y' = iwy, \ w \text{ real.} \tag{44}$$

Application of the Runge-Kutta method described in (41)–(43) to the scalar test Eq. (44) obtains the numerical solution

$$y_{n+1} = a_*^n y_n, \ a_* = A_s(\upsilon^2) + i\upsilon B_s(\upsilon^2) \tag{45}$$

where $\upsilon = wh$ and A_s, B_s are polynomials in υ^2 completely defined by Runge-Kutta parameters a_{ij} b_i and ci, as shown in (3).

Definition 2 (see 5). In the explicit s–stage Runge-Kutta method, presented in (41)–(43), the quantities

$$t(\upsilon) = \upsilon - \arg[a_*(\upsilon)], \quad a(\upsilon) = 1 - |a_*(\upsilon)| \tag{46}$$

are, respectively, called the phase-lag or dispersion error and the dissipative error. If $t(\upsilon) = O(\upsilon^{q+1})$ and $a(\upsilon) = O(\upsilon^{r+1})$ then the method is said to be of dispersive order q and dissipative order r.

3.2 Construction of Runge-Kutta Methods which is Based on Dispersion and Dissipation Properties.

– *3.2.1 A dispersive-fitted and dissipative-fitted explicit Runge-Kutta.* We consider a 6-Stage explicit Runge-Kutta method:

$$\begin{array}{c|cccccc}
0 & & & & & & \\
c_2 & a_{21} & & & & & \\
c_3 & a_{31} & a_{32} & & & & \\
c_4 & a_{41} & a_{42} & a_{43} & & & \\
c_5 & a_{51} & a_{52} & a_{53} & a_{54} & & \\
c_6 & a_{61} & a_{62} & a_{63} & a_{64} & a_{65} & \\
\hline
1 & b_1 & b_2 & b_3 & b_4 & b_5 & b_6
\end{array}$$

$$(47)$$

Table 1 *Order Equations for 6-Stage explicit Runge-Kutta method*

1st Alg. Order (1 equation)

$$\sum_{i=1}^{6} b_i = 1$$

2nd Alg. Order (2 equations)

$$\sum_{i=1}^{6} b_i c_i = \frac{1}{2}$$

3rd Alg. Order (4 equations)

$$\sum_{i=1}^{6} b_i c_i^2 = \frac{1}{3}$$

$$\sum_{i,j=1}^{6} b_i a_{ij} c_j = \frac{1}{6}$$

4th Alg. Order (8 equations)

$$\sum_{i=1}^{6} b_i c_i^3 = \frac{1}{4}$$

$$\sum_{i,j=1}^{6} b_i c_i a_{ij} c_j = \frac{1}{8}$$

$$\sum_{i,j=1}^{6} b_i a_{ij} c_j^2 = \frac{1}{12}$$

$$\sum_{i,j,k=1}^{6} b_i a_{ij} a_{jk} c_k = \frac{1}{24}$$

5th Alg. Order (17)

$$\sum_{i=1}^{6} b_i c_i^4 = \frac{1}{5}$$

$$\sum_{i,j=1}^{6} b_i c_i^2 a_{ij} c_j = \frac{1}{10}$$

$$\sum_{i,j=1}^{6} b_i c_i a_{ij} c_j^2 = \frac{1}{15}$$

$$\sum_{i,j,k=1}^{6} b_i c_i a_{ij} a_{jk} c_k = \frac{1}{30}$$

$$\sum_{i,j=1}^{6} b_i a_{ij} c_j^3 = \frac{1}{20}$$

$$\sum_{i,j=1}^{6} b_i a_{ij} c_j = \frac{1}{6}$$

$$\sum_{i,j,k=1}^{6} b_i c_i a_{ij} a_{jk} c_k^2 = \frac{1}{60}^{(i)}$$

$$\sum_{i,j,k,l=1}^{6} b_i a_{ij} a_{jk} a_{kl} c_l = \frac{1}{120}^{(ii)}$$

$$\sum_{i,j,k=1}^{6} b_i a_{ij} c_j a_{ik} c_k = \frac{1}{20}$$

In Table 1 we present the algebraic order equations up the algebraic order five. There are 21 unknowns totally[¶]. In order the method to be of 4th algebraic order eight equations must be satisfied while in order the method to be of 5th algebraic order seventeen equations must be satisfied.

[¶] We do not consider c_i as unknowns, because they always depend on $a_{i,j}$ through (43).

Two more equations must be satisfied in order to achieve zero dispersion and zero dissipation. These are shown in Definition 2. By requiring

$$A_s(\upsilon^2) = \cos(\upsilon) \text{ and } \upsilon B_s\,(\upsilon^2) = \sin(\upsilon), \tag{48}$$

Where A_s and B_s are defined in (46), we achieve both zero dispersion and zero dissipation.

We define:

$$b_2 = 0,\ c_2 = 321/1000000, \quad c_3 = 13/20, \quad c_5 = 1/6, \quad c_6 = 1 \tag{49a}$$

From the system of equations presented in Table 1 only 15 out of 17 can be satisfied (all except (i) and (ii) as marked in Table 1).

After satisfying these 15 equations plus the two equations from (48), equation (ii) of Table 1 becomes:

$$(ii) = -\frac{1}{120}\frac{120\upsilon + \upsilon^5 - 20\upsilon^3 - 120\sin(\upsilon)}{\upsilon^5}, \tag{49b}$$

and $\lim_{\upsilon\to 0}^{(ii)} = 0$. The equation (i) of Table 1 takes a much more complicated form, but its Taylor series expansion is given by

$$(i) = -\frac{5194757}{13104000000}\upsilon^2 + \frac{109148747}{19813248000000}\upsilon^4 + \ldots, \tag{50}$$

also has $\lim_{\upsilon\to 0}^{(i)} = 0$

So, decreasing the step-length, therefore $\upsilon = wh$ also decreases when w remains constant, the absolute error of the 5th algebraic order decreases and tends to zero. So we have an "almost" fifth-order method.

The new method is shown below.

$$\begin{array}{c|cccccc}
0 & & & & & & \\
\frac{321}{1000000} & \frac{321}{1000000} & & & & & \\
\frac{13}{20} & \frac{4220827}{6420} & \frac{211250}{321} & & & & \\
\frac{1}{6} & a_{41} & a_{42} & a_{43} & & & \\
\frac{1}{6} & \frac{168842389}{450684} & \frac{3250000}{8667} & \frac{35}{351} & \frac{5}{12} & & \\
1 & \frac{168837574}{95979} & \frac{13000000}{7383} & \frac{11060}{8671} & a_{64} & a_{65} & \\
\hline
\mathbf{1} & \frac{1}{78} & 0 & \frac{4000}{7917} & 0 & \frac{54}{145} & \frac{23}{210}
\end{array} \tag{51}$$

where:

$$a_{41} = -\frac{249037}{5778} + \frac{649679}{321} a_{43},$$

$$a_{42} = \frac{125000}{2889} - \frac{650000}{321} a_{43},$$

$$a_{64} = \frac{945}{677} - a_{65},$$

$$a_{65} = \frac{1890}{677} \frac{4563 a_{43} \upsilon^5 - 11600\upsilon^3 + 69600\upsilon - 69600\sin(\upsilon)}{\upsilon^5(-580 + 4563 a_{43})},$$

$$\begin{aligned} A =& -3\upsilon^{10} + 100\upsilon^8 + 72\cos(\upsilon)\upsilon^6 - 1128\upsilon^6 + 480\sin(\upsilon)\upsilon^5 + 5508\upsilon^4 - 1152\cos(\upsilon)\upsilon^4 \\ & - 3960\sin(\upsilon)\upsilon^3 - 8604\upsilon^2 - 900(\cos(\upsilon))^2\upsilon^2 + 9504\cos(\upsilon)\upsilon^2 \\ & - 4320\sin(\upsilon)\cos(\upsilon)\upsilon + 4320\sin(\upsilon)\upsilon + 5184(\cos(\upsilon))^2 - 10368\cos(\upsilon) + 5184, \end{aligned}$$

$$a_{43} = \frac{1160}{4563}\left(8\upsilon^4 - 66\upsilon^2 + 30\sin(\upsilon)\upsilon + 72 - 72\cos(\upsilon) - \sqrt{A}\right)/\upsilon^6.$$

More details for the construction of the above methods can be found in 63.

3.2.2 An Explicit Runge-Kutta with Minimal Dispersive or Dissipative Error. Here we give a brief explanation on the procedure followed in order to derive methods with maximum finite order and constant coefficients. We describe the procedure:

- After satisfying all 17 algebraic conditions but two plus the two equations from dispersion and dissipation, we all coefficients set except for a_{65} and a_{43}.
- Since we care for methods with constant coefficients, we expand the two algebraic conditions into their Taylor series over υ, where $\upsilon = wh$, w is the frequency, h is the step-length.
- After this we want to nullify as many as possible terms beginning with the smallest powers of υ, since they produce the largest error (h tends to zero). We see that the principal term of the dispersion is equal to the first algebraic conditions out of two. So we can not omit it. For the other coefficient we take two cases:

(i) to maximize dispersion order,
(ii) to maximize dissipation order.

In these cases the methods produced have the following properties:

(i) 9th dispersion order and 6th dissipation order,
(ii) 7th dispersion order and 8th dissipation order.

The produced methods are give by the following formulae

Runge-Kutta with Maximun Dispersion Order

$$\begin{array}{c|cccccc}
\frac{321}{100000} & \frac{321}{100000} & & & & & \\
\frac{13}{20} & \frac{4220827}{6420} & \frac{211250}{321} & & & & \\
\frac{1}{6} & \frac{308291341}{6835374} & \frac{11825000}{262899} & \frac{464}{10647} & & & \\
\frac{1}{6} & \frac{168842389}{450684} & \frac{3250000}{8667} & \frac{35}{351} & \frac{5}{12} & & \\
1 & \frac{168837574}{95979} & \frac{13000000}{7383} & \frac{11060}{8671} & \frac{945}{667} & \frac{1890}{667} & \\
\hline
 & \frac{1}{78} & 0 & \frac{4000}{7917} & 0 & \frac{54}{145} & \frac{23}{210}
\end{array} \qquad (52)$$

Runge-Kutta with Maximun Dissipation Order

$$\begin{array}{c|cccccc}
\frac{321}{100000} & \frac{321}{100000} & & & & & \\
\frac{13}{20} & \frac{4220827}{6420} & \frac{211250}{321} & & & & \\
\frac{1}{6} & \frac{175189297}{2929446} & \frac{6725000}{112671} & \frac{232}{4563} & & & \\
\frac{1}{6} & \frac{168842389}{450684} & \frac{3250000}{8667} & \frac{35}{351} & \frac{5}{12} & & \\
1 & \frac{168837574}{95979} & \frac{13000000}{7383} & \frac{11060}{8671} & \frac{945}{667} & \frac{1890}{667} & \\
\hline
 & \frac{1}{78} & 0 & \frac{4000}{7917} & 0 & \frac{54}{145} & \frac{23}{210}
\end{array} \qquad (53)$$

We compare these two methods to the corresponding method with infinite order of dispersion and dissipation and to some classical Runge-Kutta methods.

3.3 Numerical Results. – *3.3.1 The Methods.* In order to measure the efficiency of the new constructed method we compare it to a wide range of already known methods (some of which are higher order *e.g.* sixth).

- Runge-Kutta Fehlberg method (5th order) from 14. Fehlberg I, Fehlberg 4th, Kutaa-Nyström, England II and England I have also been tested, but had slightly to moderately worse results. Fehlberg II had similar results with Fehlberg 6(5)8.

- Runge-Kutta Fehlberg I method (5th order) from 14.
- Runge-Kutta Kutaa-Nyström method (5th order) from 14.
- Runge-Kutta Butcher method (6th order) from 14.
- Runge-Kutta Fehlberg method (4th order) from 14.
- Runge-Kutta England I method (4th order) from 14.
- Runge-Kutta Gill method (4th order) from 14.
- New developed Runge-Kutta method with infinite order of dispersion and dissipation
- New developed Runge-Kutta method with minimal dispersion
- New developed Runge-Kutta method with minimal dissipation

In this section we present some numerical results to illustrate the performance of our new methods. Consider the numerical integration of the Schrödinger equation (1) using the well-known Woods-Saxon potential (see 1, 4–6, 7,8) which is given by

$$V(r) = V_w(r) = \frac{u_0}{(1+z)} - \frac{u_0 z}{[a(1+z)^2]}$$

with $z = \exp[(r - R_0)/a]$, $u_0 = -50$, $a = 0.6$ and $R_0 = 7.0$.

In the case of negative eigenenergies (*i.e.* when $E \in [-50, 0]$) we have the well-known **bound-states problem** while in the case of positive eigenenergies (*i.e.* when $E \in (0,1000]$) we have the well-known **resonance problem** (see 5, 6 and 15).

3.3.2 Resonance Problem. In the asymptotic region the equation (1) effectively reduces to

$$y''(x) + (k^2 - \frac{l(l+1)}{x^2})y(x) = 0,$$

for x greater than some value X.

The above equation has linearly independent solutions $kxj_l\ (kx)$ and $kxn_l(kx)$, where $j_l(kx)$,$n_l(kx)$ are the **spherical Bessel and Neumann functions** respectively. Thus the solution of equation (1) has the asymptotic form (when $x \to \infty$)

$$y(x) \approx Akxj_l(kx) - Bn_l(kx) \approx D[\sin(kx - \pi l/2) + \tan\delta_l \cos(kx - \pi l/2)$$

where δ_l is the **phase shift** which may be calculated from the formula

$$tan\delta_l = \frac{y(x_2)S(x_1) - y(x_1)S(x_2)}{y(x_1)C(x_2) - y(x_2)C(x_1)}$$

for x_1 and x_2 distinct points on the asymptotic region (for which we have that x_1 is the right hand end point of the interval of integration and $x_2 = x_1 - h$, h is the stepsize) with $S(x) = kxj_l\ (kx)$ and $C(x)\ kxn_l(kx)$.

As a test for the accuracy of our methods we consider the numerical integration of the Schrödinger equation (1) with $l = 0$ in the well-known case where the potential $V(r)$ is the Woods-Saxon one.

One can investigate the problem considered here, following two procedures. The first procedure consists of finding the **phase shift** $\delta(E) = \delta_l$ for $E \in [1,1000]$. The second procedure consists of finding those E, for $E \in [1,1000]$, at which δ equals $\pi/2$. In our case we follow the first procedure *i.e.* we try to find the phase shifts for given energies. The obtained phase shift is then compared to the analytic value of $\pi/2$.

The above problem is the so-called resonance problem when *the positive eigenenergies lie under the potential barrier*. We solve this problem, using the technique fully described in 5,6.

The boundary conditions for this problem are:

$$y(0) = 0,$$
$$y(x) \approx cos[\sqrt{E}x] \text{ for large } x.$$

The domain of numerical integration is [0, 15].

For comparison purposes in our numerical illustration we use the methods mentioned above.

The numerical results obtained for the above methods, were compared with the analytic solution of the Woods-Saxon potential resonance problem, rounded to six decimal places. In Figure 17 we show the errors $Err = log_{10}|E_{calculated} - E_{analytical}|$ of the highest eigenenergy $E_3 = 989.701916$ for several values of the number of function evaluations.

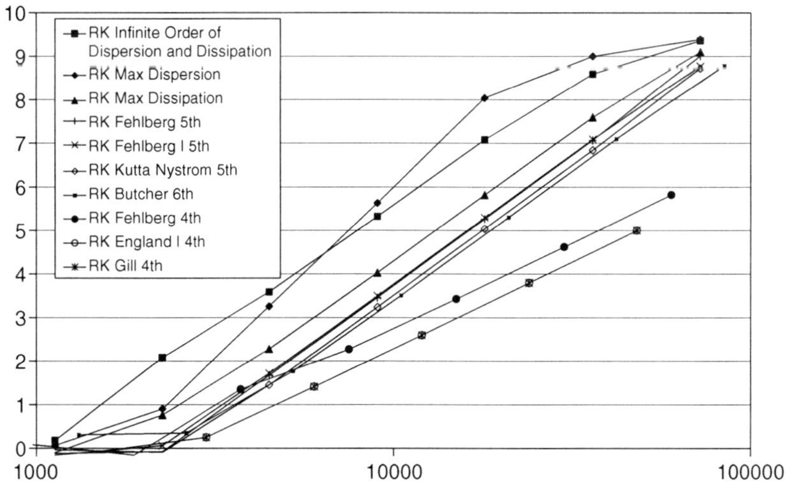

Figure 17 *Comparison of the maximum errors Err in the computation of the resonance E_3 = 989.701916 using the Methods I-X mentioned in the legend. The values of Err have been obtained based on the Number of Function Evaluations*

In Appendix B we present a Maple programme for the development of Dispersive-fitted and dissipative-fitted explicit Runge-Kutta method. In Appendix C we present a Maple programme for the development of explicit Runge-Kutta method with minimal dispersion. In Appendix D we present a Maple programme for the development of explicit Runge-Kutta method with minimal dissipation.

3.3.3 The Bound-States Problem. For negative energies we solve the so-called bound-states problem, *i.e.* the equation (1) with $l = 0$ and boundary conditions given by

$$y(0) = 0,$$
$$y(x) \approx exp(-\sqrt{-E}x) \text{ for large } x.$$

In order to solve this problem numerically we use a strategy which has been proposed by Cooley[21] and has been improved by Blatt[22]. This strategy involves integrating forward from the point $x = 0$, backward from the point $x_b = 15$ and matching up the solution at some internal point in the range of integration. As initial conditions for the backward integration we take:

$$y(x_b) = exp(-\sqrt{-E}x_b) \text{ and } y(x_b - h) = exp[-\sqrt{-E}(x_b - h)] ,$$

where h is the steplength of integration of the numerical method.

The true solutions to the Woods-Saxon bound-states problem were obtained correct to nine decimal places using the analytic solution and the numerical results obtained for the six methods mentioned above were compared to this true solution. The results are similar with those of resonance problem.

3.3.4 Remarks and Conclusions. Based on the results presented in the Figure 17 we can concluded the following:

- The New developed Runge-Kutta method with infinite order of dispersion and dissipation is the most efficient for large stepsizes
- The New developed Runge-Kutta method with minimal dispersion is the most efficient for small stepsizes
- The New developed Runge-Kutta method with minimal dissipation is worse than the other new developed methods but is better than all the other compared methods
- The Runge-Kutta Fehlberg and Runge-Kutta Fehlberg I have the same behavior and are worse than the new developed methods but better than all the other compared methods.
- The Runge-Kutta Kutaa-Nyström method has better behavior than the Runge-Kutta Butcher method
- The Runge-Kutta Fehlberg 4th order method has better behavior than the Runge-Kutta England I 4th order method and the Runge-Kutta Gill method but is worse than the Runge-Kutta Butcher method

- Finally, Runge-Kutta England I 4th order method has the same behavior than the Runge-Kutta Gill method

4 Four-Step P-Stable Methods with Minimal Phase-Lag

4.1 Phase-Lag Analysis of General Symmetric 2k – Step, $k \in N$ Methods..

– When a symmetric $2k$ method is applied to the scalar test equation:

$$y'' = -w^2 y \tag{54}$$

a difference equation of the form

$$A_k(H)y_{n+k} + \ldots + A_1(H)y_{n+1} + A_0(H)y_n + A_1(H)y_{n-1} + \ldots + A_k(H)y_{n-k} = 0 \tag{55}$$

is obtained, where $H = wh$, h is the step length and $A_0(H)$, $A_1(H)$, . . . ,+ $A_k(H)$ are polynomials of H and y_n is the computed approximation to $y(nh)$, $n = 0,1,2,\ldots$

The characteristic equation associated with (55) is given by

$$A_k(H)s^k + \ldots + A_1(H)s + A_0(H) + A_1(H)s^{-1} + \ldots + A_k(H)s^{-k} = 0 \tag{56}$$

From Lambert and Watson[16] we have the following definitions

Definition 3 (see 6) A symmetric $2k$ - step method with the characteristic equation given by (55) is said to have an **interval of periodicity** (H_0^2, H_1^2) if, for all $H \in (H_0^2, H_1^2)$, the roots s_i, $i = 1, \ldots, 2k$ of (56) satisfy:

$$s_1 = e^{i\theta(H)},\ S_2 = e^{i\theta(H)},\ \text{and}\ |S_i| \leq 1, \quad i = 3, \ldots 2k \tag{57}$$

where $\theta(H)$ is a real function of H.

Theorem 1 (see 5) For all H in the interval of periodicity, the symmetric $2k$ - step method with characteristic equation given by (56) has phase-lag order q and phase-lag constant c given by

$$-cH^{q+2} + O(H^{q+3}) = \frac{2A_k(H)\cos(kH) + \ldots + 2A_j(H)\cos(jH) + \ldots + A_0(H)}{2k^2 A_k(H) + \ldots + 2j^2 A_j(H) + \ldots + 2A_1(H)} \tag{58}$$

Theorem 2 (see 17) All symmetric four-step methods have a non-empty interval of periodicity $(0, H^2)$, if, for all $H \in [0, H^2]$

$$P_j(H) \geq 0, \quad i = 1(1)3, \quad N(H) = P_2(H)^2 - 4P_1(H)\,P_3(H) \geq 0 \tag{59}$$

where

$$P_1(H) = 2A_2(H) - 2A_1(H) + A_0(H), \qquad P_2(H) = 12A_2(H) - 2A_0(H)$$

$$P_3\,(H) = 2A_2(H) + 2A_1\,(H) + A_0\,(0)$$

4.2 Development of the New Method. – Consider the following four-step method:

$$\begin{aligned} &\bar{y}_n = y_n - a_0 h^2 \left(y''_{n+1} - 2y''_n + y''_{n-1}\right) \\ &\bar{\bar{y}}_n = y_n - a_1 h^2 \left(y''_{n+2} - 4y''_{n+1} + 6\bar{y}''_n + y''_{n-1}\right) \\ &\tilde{y}_n = y_n - a_2 h^2 \left(y''_{n+2} - 4y''_{n+1} + 6\bar{\bar{y}}''_n + y''_{n-1}\right) \\ &y_{n+2} - 2y_{n+1} + 2y_n - 2y_{n-1} + y_{n-2} = \\ &\frac{h^2}{120}\left[9\left(y''_{n+2} + y''_{n-2}\right) + 104\left(y''_{n+1} + y''_{n-1}\right) + 14\tilde{y}''_n\right] \end{aligned} \tag{60}$$

The local truncation error of this method is give by:

$$\mathrm{LTE} = -\frac{h^8}{30240}\left(95y_n^{(8)} + 3528a_2 y_n^{(6)}\right) + \ldots$$

where $y_n^{(t)}$ is the t-th derivative of the function y at x_n.

Applying the above method to the scalar test equation (54) we obtain the difference equation (55) with $k = 2$ and the associated characteristic equation (56) with also $k = 2$

Applying the Theorem 1 and formula (58) to the method (60) we have that the phase-lag is given by:

$$\begin{aligned} \text{phase-lag} = &\left(\frac{7}{240}a_2 - \frac{19}{24192}\right)H^8 + \left(\frac{587}{1036800} + \frac{7}{40}a_2 a_1 - \frac{7}{320}a_2\right. \\ &\left.-\frac{21}{20}a_2 a_1 a_0\right)H^{10} + \cdots \end{aligned} \tag{61}$$

In order the new method to have minimal phase-lag the following equation must hold:

$$\frac{7}{240}a_2 - \frac{19}{24192} = 0 \tag{62}$$

i.e. the value $a_2 = \frac{95}{3528}$ must hold.

In order the above method to be P-stable we apply the following approach.

Based on the Definition 3, in order a method to be P-stable, the roots of the characteristic equation (56) must satisfy the relation (57) for $k = 2$.

We choose as roots of the characteristic equation the following

$$S_1 = e^{iH}, \quad S_2 = e^{-iH}, \quad S_3 = -e^{iH}, \quad S_4 = -e^{-iH} \tag{63}$$

In order the characteristic equation given by (56) for $k = 2$ to have the roots given by (63), the following relations must hold:

$$\begin{aligned}
&4\left(\cos(H) - \cos(H)^2 - 2\cos(H)^3 + 2\cos(H)^4\right) \\
&+ H^2\left(\frac{1}{30} - \frac{26}{15}\cos(H) - \frac{11}{30}\cos(H)^2 + \frac{52}{15}\cos(H)^3 + \frac{3}{5}\cos(H)^4\right) \\
&+ H^4\left(-\frac{19}{1512} + \frac{19}{756}\cos(H) + \frac{19}{1512}\cos(H)^2 - \frac{19}{378}\cos(H)^3 + \frac{19}{756}\cos(H)^4\right) \\
&+ H^6\left(-\frac{19}{252}a_1 + \frac{19}{126}\cos(H)a_1 + \frac{19}{252}\cos(H)^2 a_1 - \frac{19}{63}\cos(H)^3 a_1 + \frac{19}{126}\cos(H)^4 a_1\right) \\
&+ H^8\left(\frac{19}{84}a_1 a_0 - \frac{19}{84}\cos(H)a_1 a_0 - \frac{19}{42}\cos(H)^2 a_1 a_0 + \frac{19}{42}\cos(H)^3 a_1 a_0\right) = 0
\end{aligned}$$

$$\begin{aligned}
&-30240\left(\cos(H) - \cos(H)^2\right) + H^2\left(-252 + 13104\cos(H) + 2268\cos(H)^2\right) \\
&+ H^4\left(95 - 190\cos(H) + 95\cos(H)^2\right) + H^6\left(570a_1 - 1140\cos(H)a_1 + 570\cos(H)^2 a_1\right) \\
&- 1710H^8 a_1 a_0(1 - \cos(H)) = 0
\end{aligned}$$

(64)

$$\begin{aligned}
&4\left(-\cos(H) - \cos(H)^2 + 2\cos(H)^3 + 2\cos(H)^4\right) \\
&+ H^2\left(\frac{1}{30} + \frac{26}{15}\cos(H) - \frac{11}{30}\cos(H)^2 - \frac{52}{15}\cos(H)^3 + \frac{3}{5}\cos(H)^4\right) \\
&+ H^4\left(-\frac{19}{1512} - \frac{19}{756}\cos(H) + \frac{19}{1512}\cos(H)^2 + \frac{19}{378}\cos(H)^3 + \frac{19}{756}\cos(H)^4\right) \\
&+ H^6\left(-\frac{19}{252}a_1 - \frac{19}{126}\cos(H)a_1 + \frac{19}{252}\cos(H)^2 a_1 + \frac{19}{63}\cos(H)^3 a_1 + \frac{19}{126}\cos(H)^4 a_1\right) \\
&+ H^8\left(\frac{19}{84}a_1 a_0 + \frac{19}{84}\cos(H)a_1 a_0 - \frac{19}{42}\cos(H)^2 a_1 a_0 - \frac{19}{42}\cos(H)^3 a_1 a_0\right) = 0
\end{aligned}$$

$$\begin{aligned}
&-30240\left(\cos(H) + \cos(H)^2\right) + H^2\left(252 + 13104\cos(H) - 2268\cos(H)^2\right) \\
&+ H^4\left(-95 - 190\cos(H) - 95\cos(H)^2\right) + H^6\left(-570a_1 - 1140\cos(H)a_1 - 570\cos(H)^2 a_1\right) \\
&+ 1710H^8 a_1 a_0(1 + \cos(H)) = 0
\end{aligned}$$

Solving the system of equations given by (64) we have:

$$a_1 = \frac{30240(1 - \cos(2H)) - 756H^2(37 + 3\cos(2H)) + 95H^4(1 - \cos(2H))}{570H^6(\cos(2H) - 1)}$$

$$a_0 = \frac{-10080(1 - \cos(2H)) + 840H^2(7\cos(2H) + 17)}{-30240H^2(1 - \cos(2H)) + 756H^4(37 + 3\cos(2H)) - 95H^6(1 - \cos(2H))} \qquad (65)$$

For small values of w the above formulae are subject to heavy cancellations. In this case we use the following Taylor series expansions:

$$a_1 = \frac{16}{57} + \frac{56}{1425}H^2 + \frac{16}{3135}H^4 + \frac{11056}{17506125}H^6 + \frac{32}{423225}H^8 + \frac{28936}{3273645375}H^{10} + \cdots \tag{66}$$

$$a_0 = \frac{1}{6} - \frac{1}{72}H^2 + \frac{1}{2160}H^4 + \frac{121}{5443200}H^6 + \frac{493}{133056000}H^8 + \frac{18675581}{7355668320000}H^{10} + \cdots \tag{63}$$

The stability polynomials $P_j(H) \geq 0 \quad i = 1(1)3$, and $N(H)$ are given by:

$$\begin{aligned} P_1(H) &= P_3(H) = 2H^2 \\ P_2(H) &= \frac{-4H^2(\cos(2H)+3)}{\cos(2H)-1} \\ N(H) &= \frac{128H^4(\cos(2H)+1)}{(\cos(2H)-1)^2} \end{aligned} \tag{67}$$

In Figure 18 we present the behaviour of the above polynomials. It is easy for one to see that the above polynomials are greater or equal to 0 for all $H^2 \in (0, \infty)$, *i.e.* the method is P-stable. In the Appendix E we present the Maple programme for the construction of this new method.

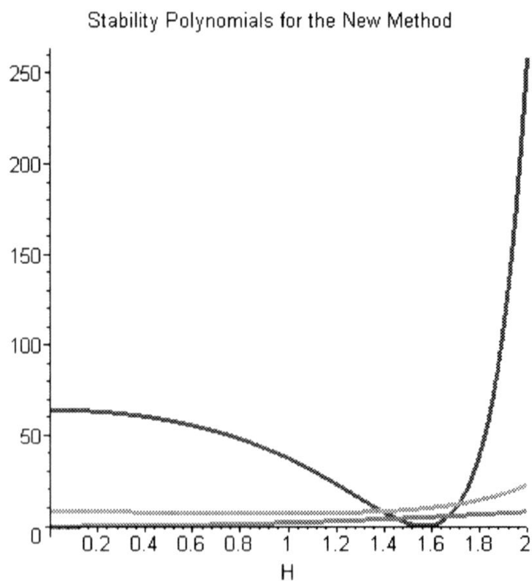

Figure 18 *The polynomials $P_j(H) \geq 0$, $i = 1(1)3$, and $N(H)$ of (59)*

4.3 Numerical Results. – We consider the numerical integration of the Schrödinger equation (1) using the well-known Woods-Saxon potential (see 1, 4–6, 8) which is given by

$$V(r) = V_w(r) = \frac{u_0}{(1+z)} - \frac{u_0 z}{[a(1+z)^2]}$$

with $z = \exp[(r - R_0)/a]$, $u_0 = -50$, $a = 0.6$ and $R_0 = 7.0$.

We solve the resonance problem described above in the paragraph 3.3.

4.3.1 The Methods. In order to measure the efficiency of the new developed method we compare it to a wide range of already known methods.

- Explicit Four-Step Method (indicated as Method [a]).
- Explicit Numerov-Type Method developed by Chawla from 18 (indicated as Method [b]).
- Four-step method developed in 19 (indicated as Method [c]).
- Exponentially-Fitted Method developed by Raptis [20] (indicated as Method [d]).
- Exponentially-Fitted Method developed by Raptis and Allison [21] (indicated as Method [e]).
- Exponentially-Fitted Method developed by Ixaru and Rizea [22] (indicated as Method [f]).
- New developed four-step P-stable method with minimal phase-lag (indicated as Method [g]).

4.3.2 The Methods. In this section we present some numerical results to illustrate the performance of our new method. Consider the numerical integration of the Schrödinger equation (1) using the well-known Woods-Saxon potential (see 1, 4–6, 8) which is given by

$$V(r) = V_w(r) = \frac{u_0}{(1+z)} - \frac{u_0 z}{[a(1+z)^2]}$$

with $z = \exp[(r - R_0)/a]$, $u_0 = -50$, $a = 0.6$ and $R_0 = 7.0$.

We have solved the resonance problem described in paragraph 3.3. The domain of numerical integration is [0, 15].

For comparison purposes in our numerical illustration we use the methods mentioned above.

The numerical results obtained for the above methods, were compared with the analytic solution of the Woods-Saxon potential resonance problem, rounded to six decimal places. In Figure 19 we show the errors $Err = log_{10}|E_{calculated} - E_{analytical}|$ of the highest eigenenergy $E_3 = 989.701916$ for several values of the number of function evaluations.

In Appendix E we present a Maple programme for the development of the new P-stable four-step method.

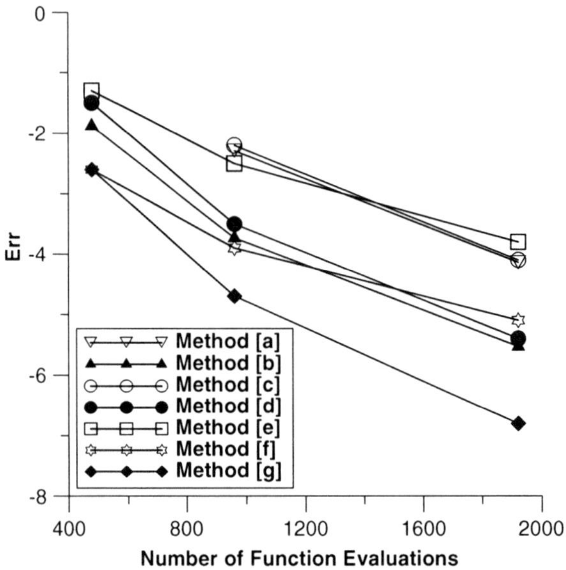

Figure 19 *Comparison of the maximum errors Err in the computation of the resonance E_3 = 989.701916 using the Methods [a]-[g] mentioned in the legend. The values of Err have been obtained based on the Number of Function Evaluations*

4.3.3 Remarks and Conclusions. Based on the results presented in the Figure 19 we can concluded the following:

- The New developed four-step method with minimal phase-lag is the most efficient one.
- The method of Ixaru and Rizea [22] is the second most efficient method for the two step sizes, while for the smallest step size is less efficient than the method of Chawla [18] and the method of Raptis [20].
- The method of Chawla [18] is the third most efficient method with the remark mentioned in the previous conclusion.
- The method of Raptis [20] is the fourth most efficient method with the remark mentioned in the previous conclusion.
- Finally, the Explicit Four-Step Method developed, the Four-step method developed in [19] and the Exponentially-Fitted Method developed by Raptis and Allison [21] are less efficient than the others.

5 Trigonometrically Fitted Fifth-Order Runge-Kutta Methods for the Numerical Solution of the Schrödinger Equation

5.1 Explicit Runge-Kutta Methods for the Schrödinger Equation. – An s– stage explicit Runge-Kutta method used for the computation of the approximation of $y_{n+1}(x)$, when $y_n(x)$,is known, can be expressed by the relations (41) where $f(x, y(x)) = (W(x) - E)y(x)$.

Actually to solve the second-order ODE (1) using first-order numerical method (41), (1) becomes

$$z'(x) = (W(x) - E)y(x),\ y'(x) = z(x) \tag{68}$$

while we use two pairs of equations (41): one for y_{n+1} and one for z_{n+1}. The method shown above can also be presented using the Butcher table (42). Coefficients $c_2, \ldots, c_s$ must satisfy the equations (43)

Tree theory is a convenient way to obtain a certain algebraic order equations.

5.2 Exponentially Fitted Runge-Kutta Methods. – The method (41) is associated with the operator,

$$\begin{aligned} L(x) &= u(x+h) - u(x) - h\sum_{i=1}^{s} b_i u'(x + c_i h, U_i), \\ U_i &= u(x) + h\sum_{j=1}^{i-1} a_{ij} u'\left(x + c_j h, U_j\right), \quad i = 1, \ldots, s, \end{aligned} \tag{69}$$

Where u is a continuously differentiable function.

Definition 4. (See 23.) The method (69) is called exponential of order p if the associated linear operator L vanishes for any linear combination of the linearly independent functions $\exp(\upsilon_0 x)$, $\exp(\upsilon_1 x)$, . . . , $\exp(\upsilon_p x)$, where $\upsilon_i | i = 0(1)p$ are real or complex numbers.

Remark 1. (See 24.) If $\upsilon_l = \upsilon$ for $\iota = 0,1, \ldots, n$, $n \leq p$, then the operator L vanishes for any linear combination of $\exp(\upsilon x)$, $x \exp(\upsilon x)$, $x^2 \exp(\upsilon x)$, . . . ,x^n $\exp(\upsilon x)$, $\exp(\upsilon_{n+1} x)$,. . .,$\exp(\upsilon_p x)$.

Remark 2. (See 24.) Every exponentially fitted method corresponds in a unique way to an algebraic method (by setting $\upsilon_\iota = 0$ for all i).

Definition 5. (See 23.) The corresponding algebraic method is called the classical method.

5.3 Construction of Trigonometrically-Fitted Runge-Kutta Methods. – Consider the explicit Runge-Kutta method Kutta-Nyström 14, which has fifth algebraic order and six stages. The coefficients are shown in (70).

$$
\begin{array}{c|ccccc|c}
0 & & & & & & \\
\frac{1}{3} & \frac{1}{3} & & & & & \\
\frac{2}{5} & \frac{4}{25} & \frac{6}{25} & & & & \\
1 & \frac{1}{4} & \frac{12}{4} & \frac{15}{4} & & & \\
\frac{2}{3} & \frac{6}{81} & \frac{90}{81} & \frac{50}{81} & \frac{8}{81} & & \\
\frac{4}{5} & \frac{6}{75} & \frac{36}{75} & \frac{10}{75} & \frac{8}{75} & 0 & \\
\hline
1 & \frac{23}{192} & 0 & \frac{125}{192} & 0 & -\frac{81}{192} & \frac{125}{192}
\end{array}
\tag{70}
$$

We will present two trigonometrically fitted methods. The construction is based on 25.

5.3.1 First Trigonometrically-Fitted Method. We demand the method (70) to integrate exactly the functions,

$$\{1,\, x,\, x^2,\, x^3,\, x^4,\, \exp(Iwx)\} \text{ or equivalently, } \{1,\, x,\, x^2,\, x^3,\, x^4,\, \cos(wx),\, \sin(wx)\} \tag{71}$$

where w is real number and it is called frequency and $I = \sqrt{-1}$.

In order to satisfy (71), we put as free parameters the coefficients b_5 and b_6 and leave the other coefficients the same as the classical method. Then we demand the approximate solution $y_{app.}$ to integrate exactly $\exp(Iwx)$ for the real and the imaginary part. This demand produces the following equations:

$$
\begin{aligned}
\cos[w(x-h)] &= \frac{\cos(wx)}{6750}\big(6750(1-b_4 v^2) - 2250 b_2 v^2 - 2700 b_3 v^2 + 2025 b_4 v^4 \\
&\quad -4500 b_5 v^2 - 5400 b_6 v^2 + 432 b_6 v^4\big) \\
&\quad + \frac{\sin(wx)}{6750}\left(6750 v \sum_{i=1}^{6} b_i + 216 b_6 v^5 - 1500 b_5 v^3 - 3375 b_4 v^3 + 200 b_5 v^5 - 2160 b_6 v^3 - 540 b_3 v^3\right)
\end{aligned}
\tag{72}
$$

$$-\sin[w(x-h)] = \frac{\cos(wx)}{6750}\left(6750\upsilon\sum_{i=1}^{6} b_i + 216b_6\upsilon^5 - 1500b_5\upsilon^3 - 3375b_4\upsilon^3 + 200b_5\upsilon^5 - 2160b_6\upsilon^3 - 540b_3\upsilon^3\right)$$
$$-\frac{\sin(wx)}{6750}\left(6750\left(1-b_4\upsilon^2\right) - 2250b_2\upsilon^2 - 2700b_3\upsilon^2 + 2025b_4\upsilon^4 - 4500b_5\upsilon^2 - 5400b_6\upsilon^2 + 432b_6\upsilon^4\right)$$

where again $\upsilon = wh$, w is the frequency and h is the step length used.

From these two equations we derive b_5 and b_6. These coefficients are given in 25.

5.3.2 Second Trigonometrically-Fitted Method. For the second method we demand to integrate exactly the functions,

$$\begin{gathered}\{1, x, x^2, \exp(Iwx), x\exp(Iwx)\} \text{ or equivalently,} \\ \{1, x, x^2, \cos(wx), \sin(wx), x\cos(xw), x\sin(wx)\}\end{gathered} \tag{73}$$

In order to satisfy (73), we put as free parameters the coefficients b_3, b_4, b_5 and b_6 and leave the other coefficients the same as the classical method. Then we demand the approximate solution y_{app} to integrate exactly exp(Iwx) and x exp(Iwx) for the real and the imaginary part. This demand produces the following equations:

$$\begin{aligned}\cos(\upsilon) &= 1 - \frac{1}{15}\upsilon^2(5b_2 + 6b_3 + 15b_4 + 10b_5 - 12b_6) + \frac{1}{250}\upsilon^4(75b_4 + 16b_6) \\ \sin(\upsilon) &= \upsilon\sum_{j=1}^{6} b_j - \frac{1}{225}\upsilon^3(50b_5 + 72b_6 + 125b_4 + 18b_3) + \frac{1}{250}\upsilon^4(75b_4 + 16b_6)\end{aligned} \tag{74}$$

$$(x+h)\cos(\upsilon) = x + h\sum_{j=1}^{6} b_j - \frac{1}{15}\upsilon^2 x(5b_2 + 6b_3 + 15b_4 + 10b_5 + 12b_6)$$
$$-\frac{1}{150}h\upsilon^2(36b_3 + 225b_4 + 100b_5) + \frac{1}{250}\upsilon^4 x(75b_4 + 16b_6) + \frac{1}{675}\upsilon^4 h(100b_5 + 108b_6)$$

$$\begin{aligned}(x+h)\sin(\upsilon) = {}& x\upsilon\sum_{j=1}^{6} b_j + \frac{1}{15}\upsilon h(10b_2 + 8b_3 + 30b_4 + 20b_5 + 24b_6) \\ & - \frac{1}{450}x\upsilon^3(36b_3 + 225b_4 + 100b_5 + 144b_6) - \frac{1}{125}\upsilon^3 h(150b_4 + 32b_6) \\ & + \frac{1}{3375}\upsilon^5 x(100b_5 + 108b_6)\end{aligned}$$

where again $\upsilon = wh$, w is the frequency and h is the step length used.

From these two equations we derive b_3, b_4, b_5, and b_6. These coefficients are given in 25.

For small values of υ the coefficients are subject to heavy cancellations, thus, we expand the coefficients over the Taylor series around zero.

The algebraic order equations are given in 25.

5.3.3 Error Analysis. In this paragraph we will study the behavior of the error and for this reason we have used the local truncation error (LTE), that is the difference between the theoretical and the approximate solution. We have studied the analytic form of the local truncation error for the three cases of

(i) the classical Kutta-Nyström method,
(ii) the first trigonometrically fitted method, and
(iii) the second trigonometrically fitted method.

The errors correspond to the ODE (4) and has two parts: one for $y(x)$ and one for $z(x)$. To calculate the errors of methods (b) and (c) we need to determine the frequency w. The formula for w as it used during calculations for the resonance problem is

$$w = \sqrt{E - \bar{W}}$$

and this is also used during the error analysis.

Based on the results presented in 25 we have the following conclusions:

The classical method (a) includes at both $y(x)$ and $z(x)$ the third power of energy, *i.e.* the LTE is a function of (E^3, E^3). In the error of method (b), the maximum power of energy is decreased from three to two at both $y(x)$ and $z(x)$ *i.e.* the LTE is a function of (E^3, E^2). In the error of method (c), the maximum power is one for $y(x)$ and two for $z(x)$ *i.e.* the LTE is a function of (E, E^2).

These conclusions are very critical for large values of energy, because the error becomes significantly smaller. We note that it is not the maximum power of the two functions $y(x)$, $z(x)$ that plays critical role for the error propagation rather than each of the maximums separately. That happens because the new value of the derivative y'_{n+1} needs the value of z_{n+1} and the derivative z'_{n+1} needs y_{n+1} as seen in (4). This explains the higher efficiency of method (c) opposite to method (b).

In the Appendix F we present the Maple programme for the construction of the presented methods.

6 Four-Step P-Stable Trigonometrically-Fitted Methods

6.1 Development of the New Method. – Consider the following four-step Method:

$$\begin{aligned}
&\bar{y}_n = y_n - a_0 h^2 \left(y''_{n+1} - 2y''_n + y''_{n-1}\right) \\
&\bar{\bar{y}}_n = y_n - a_1 h^2 \left(y''_{n+2} - 4y''_{n+1} + 6\bar{y}''_n + y''_{n-1}\right) \\
&y_{n+2} - 2y_{n+1} + 2y_n - 2y_{n-1} + y_{n-2} = \\
&h^2\left[b_0\left(y''_{n+2} + y''_{n-2}\right) + b_1\left(y''_{n+1} + y''_{n-1}\right) + b_2 \bar{\bar{y}}''_n\right]
\end{aligned} \tag{75}$$

We require the part of the method:

$$y_{n+2} - 2y_{n+1} + 2y_n - 2y_{n-1} + y_{n-2} = \\ h^2\left[b_0\left(y''_{n+2} + y''_{n-2}\right) + b_1\left(y''_{n+1} + y''_{n-1}\right) + b_2\bar{\bar{y}}''_n\right] \tag{76}$$

to integrate exactly the functions:

$$\{1,\ x,\ x^2,\ \cos(wx),\ \sin(wx),\ x\cos(xw),\ x\sin(wx)\} \tag{77}$$

This requirement leads to the system of equations:

$$\begin{aligned} &4\cos(w)[\cos(w) - 1] = -w^2\left[4b_0\cos(w)^2 + 2b_1\cos(w) + b_2 - 2b_0\right] \\ &4\sin(w)[2\cos(w) - 1] = \\ &-2w[4b_0\cos(w)(w\sin(w) - \cos(w)) + b_1(w\sin(w) - 2\cos(w)) - b_2 + 2b_0] \\ &2 = 2b_0 + 2b_1 + b_2 \end{aligned} \tag{78}$$

The solution of the above system is given by

$$\begin{aligned} b_0 &= \frac{2\mathrm{Tw} - w\mathrm{Cw}}{w^3(5\sin(w) + \sin(3w) - 4\sin(2w))} \\ b_1 &= \frac{2\mathrm{Sw} + 2w\mathrm{Qw}}{w^3(5\sin(w) + \sin(3w) - 4\sin(2w))} \\ b_2 &= \frac{4\mathrm{Dw} + 2w\mathrm{Fw} + 2w^3\mathrm{Gw}}{w^3(5\sin(w) + \sin(3w) - 4\sin(2w))} \end{aligned} \tag{79}$$

where

$$\begin{aligned} \mathrm{Tw} &= 4 - 7\cos(w) + 4\cos(2w) - \cos(3w), \\ \mathrm{Cw} &= 5\sin(w) + \sin(3w) - 4\sin(2w) - 2w^2\sin(w) \\ \mathrm{Sw} &= \cos(4w) - 1 - 2\cos(3w) + 2\cos(w) \\ \mathrm{Qw} &= 5\sin(w) - 4\sin(2w) - 2w^2\sin(2w) \\ \mathrm{Dw} &= -3 - 4\cos(2w) - \cos(4w) + 3\cos(3w) + 5\cos(w) \\ \mathrm{Fw} &= 4\sin(2w) - \sin(3w) - 5\sin(w) \\ \mathrm{Gw} &= \sin(3w) + 3\sin(w) \end{aligned}$$

For small values of w the above coefficients are subject to heavy cancellations, thus, we expand the coefficients over the Taylor series around zero:

$$b_0 = \frac{3}{40} - \frac{19}{3024}w^2 + \frac{139}{259200}w^4 + \frac{5771}{119750400}w^6 + \frac{271391}{594397440000}w^8 + \cdots$$

$$b_1 = \frac{13}{15} - \frac{19}{756}w^2 + \frac{113}{113400}w^4 - \frac{293}{29937600}w^6 - \frac{8213}{81729648000}w^8 + \cdots$$

$$b_2 = \frac{7}{60} + \frac{19}{504}w^2 - \frac{103}{33600}w^4 - \frac{73}{950400}w^6 - \frac{973199}{108972864000}w^8 + \cdots \tag{80}$$

The local truncation error of the method (75) with coefficients given by (79)–(80) is give by:

$$\mathrm{LTE} = -\frac{h^8}{30240}\left(95y_n^{(8)} + \left(3528a_1 + 190w^2\right)y_n^{(6)} + \left(21168a_1a_0 + 95w^4\right)y_n^{(4)}\right) + \cdots$$

where $y_n^{(t)}$ is the t-th derivative of the function y at x_n.

Applying the above method to the scalar test equation

$$y'' = -v^2 y \tag{81}$$

we obtain the difference equation (55) with $k = 2$ and the associated characteristic equation (56) with also $k = 2$ with $H = vh$.

In order the above method (75) to be P-stable we apply the following approach.

Based on the Definition 3, in order a method to be P-stable, the roots of the characteristic equation (56) must satisfy the relation (57) for $k = 2$.

We choose as roots of the characteristic equation the given by (63).

In order the characteristic equation given by (56) for $k = 2$ to have the roots given by (63), the following relations must hold:

$$\begin{aligned}
&4\left(\cos(H) - \cos(H)^2 - 2\cos(H)^3 + 2\cos(H)^4\right)\\
&+ H^2\left(2b_0 - b_2 - 2b_1\cos(H) + 2b_2\cos(H)^2 - 8b_0\cos(H)^2 + 4b_1\cos(H)^3 + 8b_0\cos(H)^4\right)\\
&+ H^4\left(-4b_2a_1 + 8b_2a_1\cos(H) + 4b_2a_1\cos(H)^2 - 16b_2a_1\cos(H)^3 + 8b_2a_1\cos(H)^4\right)\\
&+ H^6\left(12b_2a_1a_0 - 12b_2a_1a_0\cos(H) - 24b_2a_1a_0\cos(H)^2 + 24b_2a_1a_0\cos(H)^3\right) = 0
\end{aligned}$$

$$\begin{aligned}
&-4\left(\cos(H) - \cos(H)^2\right) + H^2\left(-2b_0 + b_2 + 2b_1\cos(H) + 4b_0\cos(H)^2\right)\\
&+ H^4\left(4b_2a_1 - 8b_2a_1\cos(H) + 4b_2a_1\cos(H)^2\right) + 12H^6b_2a_1a_0(-1 + \cos(H)) = 0
\end{aligned} \tag{82}$$

$$4\left(-\cos(H)-\cos(H)^2+2\cos(H)^3+2\cos(H)^4\right)$$
$$+H^2\left(2b_0-b_2+2b_1\cos(H)+2b_2\cos(H)^2-8b_0\cos(H)^2-4b_1\cos(H)^3+8b_0\cos(H)^4\right)$$
$$+H^4\left(-4b_2a_1-8b_2a_1\cos(H)+4b_2a_1\cos(H)^2+16b_2a_1\cos(H)^3+8b_2a_1\cos(H)^4\right)$$
$$+H^6\left(12b_2a_1a_0+12b_2a_1a_0\cos(H)-24b_2a_1a_0\cos(H)^2-24b_2a_1a_0\cos(H)^3\right)=0$$

$$-4\left(\cos(H)+\cos(H)^2\right)+H^2\left(2b_0-b_2+2b_1\cos(H)-4b_0\cos(H)^2\right)$$
$$-H^4\left(4b_2a_1+8b_2a_1\cos(H)+4b_2a_1\cos(H)^2\right)+12H^6b_2a_1a_0(1+\cos(H))=0$$

Solving the system of equations given by (82) we have:

$$\begin{aligned}a_1&=\frac{2(1-\cos(2H))-H^2[2b_1+2b_0\cos(2H)+b_2]}{2H^4b_2(\cos(2H)-1)}\\a_0&=\frac{-2(1-\cos(2H))+H^2[b_1(\cos(2H)+3)+2(2b_0\cos(2H)+b_2)]}{-6H^2(1-\cos(2H))+3H^4[2b_1+2b_0\cos(2H)+b_2]}\end{aligned}\tag{83}$$

For small values of w the above formulae are subject to heavy cancellations. In this case we use the following Taylor series expansions:

$$\begin{aligned}a_1=&-\frac{704425}{1210104}+\frac{1107770309}{3913476336}H^2-\frac{9944130643501}{99715377041280}H^4\\&+\frac{3535406414283643}{87948962550408960}H^6-\frac{742538888913446318801}{48352580630963838028800}H^8+\cdots\end{aligned}\tag{84}$$

$$a_0=\frac{1}{6}-\frac{1}{72}H^2+\frac{1}{2160}H^4-\frac{1}{120960}H^6+\frac{1}{10886400}H^8+\cdots$$

The stability polynomials $P_j(H)\geq 0,\quad i=1(1)3,\quad$ and $N(H)$ are given by:

$$\begin{aligned}P_1(H)&=P_2(H)=2H^2\\N(H)&=\frac{128H^4(\cos(2H)+1)}{(\cos(2H)-1)^2}\end{aligned}\tag{85}$$

$$\begin{aligned}P_2(H)=&\,(16016\,H^2+18616\,H^3\,\sin(H)+10968\,H^3\,\sin(3H)-21792\,H^3\,\sin(2H)\\&+10248\,H^2\,\cos(7H)+1024\,H^3\,\sin(8H)+16448\,H^2\,\cos(2H)+140\,H^5\,\sin(9H)\\&-128\,H^5\,\sin(4H)-448\,H^5\,\sin(8H)+88\,H^2\,\cos(11H)-1760\,H^5\,\sin(6H)+4\,H^5\,\sin(11H)\\&-32\,H^5\,\sin(10H)-3704\,H^5\,\sin(3H)+7488\,H^5\,\sin(2H)-6632\,H^5\,\sin(H)\\&+1060\,H^5\,\sin(7H)+1768\,H^2\,\cos(9H)-8\,H^2\,\cos(12H)-276\,H^3\,\sin(9H)\\&-11128\,H^2\,\cos(4H)+5136\,H^3\,\sin(6H)-27760\,H^2\,\cos(H)+17464\,H^2\,\cos(5H)\\&-15968\,H^2\,\cos(6H)+1804\,H^5\,\sin(5H)-480\,H^2\,\cos(10H)-4\,H^3\,\sin(11H)\\&+48\,H^3\,\sin(10H)-1808\,H^2\,\cos(3H)-4880\,H^2\,\cos(8H)-2716\,H^3\,\sin(7H)\end{aligned}$$

$-5996\ H^3\ \sin(5H) + 1024\ H^3\ \sin(4H))/(572 - 930\ \cos(4H) - 964\ \cos(H)$

$+ 2210\ H\ \sin(H) + 170\ H\ \sin(3H) - 2040\ H\ \sin(2H) - 806\ H^3\ \sin(H) - 18\ H^3\ \sin(3H)$

$+ 688\ H^3\ \sin(2H) - 82\ \cos(7H) + 48\ H^3\ \sin(8H) + 61\ H\ \sin(9H) + 464\ \cos(2H)$

$+ 188\ H\ \sin(6H) + 260\ \cos(3H) + 104\ \cos(10H) - 27\ H^3\ \sin(9H) + 175\ H\ \sin(7H)$

$- 160\ H\ \sin(8H) - 22\ \cos(11H) - 72\ H^3\ \sin(6H) - 266\ \cos(9H) + 356\ \cos(8H) - H^3\ \sin(11H)$

$- 12\ H\ \sin(10H) + H\ \sin(11H) + 8\ H^3\ \sin(10H) - 33\ H^3\ \sin(7H) + 1074\ \cos(5H)$

$- 901\ H\ \sin(5H) + 1088\ H\ \sin(4H) - 568\ \cos(6H) + 2\ \cos(12H)$

$+ 269\ H^3\ \sin(5H) - 352\ H^3\ \sin(4H))$

In Figure 20 we present the behaviour of the above polynomials. It is easy for one to see that the above polynomials are greater or equal to 0 for all $H^2 \in (0, \infty)$, *i.e.* the method is P-stable. In the Appendix G we present the Maple programme for the construction of this new method.

6.2 Numerical Results. – We consider the numerical integration of the Schrödinger equation (1) using the well-known Woods-Saxon potential (see 1, 4–6, 8) which is given by

$$V(r) = V_w(r) = \frac{u_0}{(1+z)} - \frac{u_0 z}{[a(1+z)^2]}$$

with $z = \exp[(r - R_0)/a]$, $u_0 = -50$, $a = 0.6$ and $R_0 = 7.0$.

We solve the resonance problem described above in the paragraph 3.3.

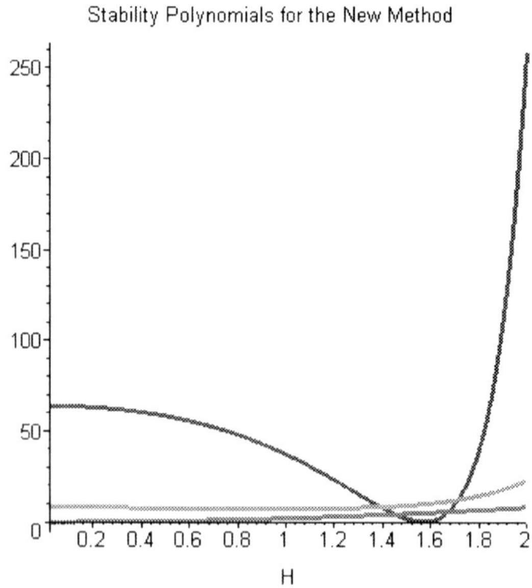

Figure 20 *The polynomials $P_j(H) \geq 0$, $i = 1(1)3$, and $N(H)$ of (85)*

6.2.1 The Methods. In order to measure the efficiency of the new developed method we compare it to a wide range of already known methods.

- Explicit Four-Step Method (indicated as Method [a]).
- Four-step method developed in 19 (indicated as Method [b]).
- Exponentially-Fitted Four-Step Method developed by Raptis [20] (indicated as Method [c]).
- Exponentially-Fitted Four-Step Method (indicated as Method [d]).
- New developed four-step P-stable trigonometrically-fitted method (indicated as Method [e]).

In this section we present some numerical results to illustrate the performance of our new method. We consider the same problem as in paragraph 4.

For comparison purposes in our numerical illustration we use the methods mentioned above.

The numerical results obtained for the above methods, were compared with the analytic solution of the Woods-Saxon potential resonance problem, rounded to six decimal places. In Figure 21 we show the errors $Err = log_{10}|$

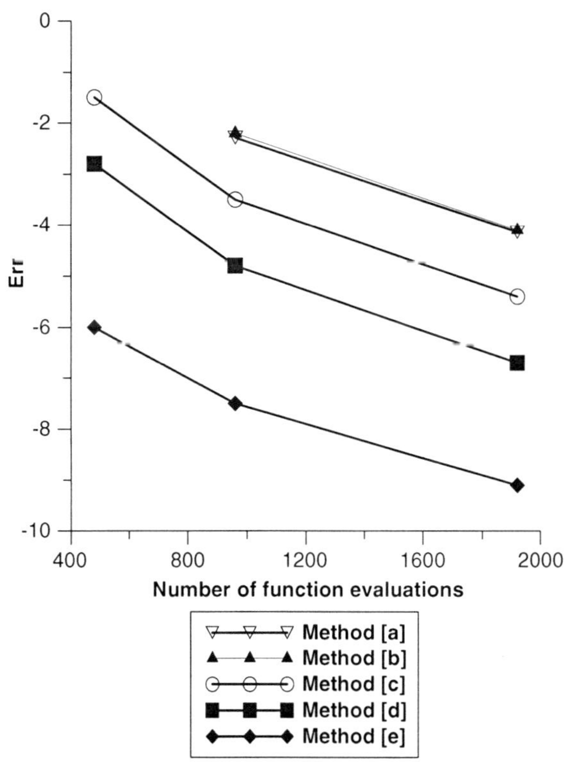

Figure 21 *Comparison of the maximum errors Err in the computation of the resonance E_3 = 989.701916 using the Methods [a]-[e] mentioned in the legend. The values of Err have been obtained based on the Number of Function Evaluations*

$E_{calculated} - E_{analytical}|$ of the highest eigenenergy $E_3 = 989.701916$ for several values of the number of function evaluations.

6.2.2 Remarks and Conclusions. Based on the results presented in the Figure 19 we can concluded the following:

- The new developed four-step P-stable trigonometrically-fitted method is the most efficient one.
- The classical exponentially-fitted four-step method is the second most efficient method
- The Exponentially-Fitted Four-Step Method developed by Raptis [20] is the third most efficient method
- Finally, the Explicit Four-Step Method developed and the Four-step method developed in 19 are less efficient than the others.

7 Comments on the Recent Bibliography

In 26 the authors have developed a new trigonometrically-fitted predictor-corrector (P-C) scheme based on the Adams-Bashforth-Moulton P-C methods. In particular, the predictor is based on the fifth algebraic order Adams-Bashforth scheme and the corrector on the sixth algebraic order Adams-Moulton scheme. More specifically the new developed scheme integrates exactly any linear combination of the functions:

$$\{1, \cos(wx), \sin(wx), x\cos(wx), x\sin(wx), x^2\cos(wx), x^2\sin(wx)\}$$

They have investigated the stability of the new scheme. They have compared the efficiency of this new scheme against well known methods and the numerical illustrations showed that their method is more efficient, even when compared to methods that have been specially designed for the solution of the radial Schrödinger equation.

In 27 the authors have developed trigonometrically-fitted Runge-Kutta method for the numerical integration of orbital problems. The developed method is based on the Runge-Kutta Zonneveld method. More specifically is based on a modification of the Butcher table for the Runge-Kutta methods showed below:

$$\begin{array}{c|cccccc}
0 & & & & & & \\
c_2 & d_2 & a_{21} & & & & \\
c_3 & d_3 & a_{31} & a_{32} & & & \\
\vdots & \vdots & \vdots & \vdots & & & \\
c_m & a_m & a_{m,1} & a_{m,2} & \cdots & a_{m,m-1} & \\
\hline
 & & b_1 & b_2 & \cdots & b_{m-1} & b_m
\end{array} \tag{86}$$

The new method is developed in order to integrate exactly any linear combination of the functions:

$$\{\cos(\pm w), \sin(\pm w)\}$$

Detailed errors analysis is presented. Theoretical and numerical results obtained for some well-known orbital problems show the efficiency of the method.

In 28 the author have developed a family of trigonometrically fitted Runge-Kutta methods for the numerical integration of the radial Schrödinger equation. The developed method is based on the Runge-Kutta Zonneveld method. More specifically the new methods are developed in order to integrate exactly any linear combination of the functions:

$$\{1, x, x^2, x^3, x^4 \cos(\pm w), \sin(\pm w)\} \text{ and}$$

$$\{1, x, x^2, \cos(\pm w), \sin(\pm w), x\cos(\pm w), x\sin(\pm w)\}$$

Theoretical and numerical results obtained for the radial Schrödinger equation and for the well known Woods-Saxon potential and for the coupled differential equations of the Schrödinger type show the efficiency of the new methods.

In 29 the authors have developed a trigonometrically-fitted multiderivative method for the numerical solution of the radial Schrödinger equation. The methods are called multiderivative since uses derivatives of order two and four. The method has the general form:

$$\begin{aligned} \bar{y}_{n+1} &= 2y_n - y_{n-1} + a_0 h^2 y''_n + a_1 h^4 y_n^{(4)} \\ y_{n+1} - 2y_n + y_{n-1} &= h^2\left(c_0 y''_n + c_1\left(y''_{n+1} + y''_{n-1}\right)\right) + h^4\left(c_2 y_n^{(4)} + c_3\left(y_{n+1}^{(4)} + y_{n-1}^{(4)}\right)\right) \end{aligned}$$

The coefficients of the above method are determined in order the method integrates exactly any linear combination of the functions:

$$\{1, x, x^2, x^3, x^4, x^5, \cos(\pm w), \sin(\pm w), x\cos(\pm w), x\sin(\pm w)\}$$

A numerical illustration of the new developed method to the resonance problem of the radial Schrödinger equation shows its efficiency compared with other similar well known methods of the literature.

In 30 the authors have developed trigonometrically fitted Adams-Bashforth-Moulton predictor-corrector (P-C) methods. It is the first time in the literature that these methods are applied for the efficient solution of the resonance problem of the Schrödinger equation. The new trigonometrically fitted P-C schemes are based on the well known Adams-Bashforth-Moulton methods. In particular, they are based on the fourth order Adams-Bashforth scheme (as predictor) and on the fifth order Adams-Moulton scheme (as corrector). More specifically the new method has the form:

$$\bar{y}_{n+1} = y_n + h\sum_{i=0}^{3} a_i f_{n-i}$$

$$y_{n+1} = y_n + h\left(c_0\bar{f}_{n+1}\sum_{i=1}^{4} c_i f_{n-i+1}\right)$$

The coefficients of the above method are determined in order the method integrates exactly any linear combination of the functions:

$$\{1, x, x^2 \cos(\pm w), \sin(\pm w)\} \text{ and}$$

$$\{1, x, \cos(\pm w), \sin(\pm w), x\cos(\pm w), x\sin(\pm w)\}$$

They have studied in detail the stability of the new developed schemes. They have tested the efficiency of their newly developed schemes against well known methods, with efficient results. The numerical illustrations indicate that at least one of their methods is significantly more efficient compared to other methods from which some of them are specially designed for the numerical solution of the Schrödinger equation.

In 31 the authors present a review on multistep methods for the numerical solution of the Schrödinger equation. More specifically, they have presented a simple way that permits production of Bettis-Cowell methods for any algebraic and trigonometric order. Their study is based on the general method:

$$y_{n+1} - 2y_n + y_{n-1} = h^2\sum a_k\nabla^k f_{m+1} \tag{87}$$

where $\nabla^k f_m$ are the backward differences which are defined by the relation:

$$\nabla^{k+1} f_m = \nabla^k f_m - \nabla^k f_{m-1}, \quad \nabla^0 f_m = f_m$$

and the coefficients a_k are the coefficients of the McLaurin expansion of the function:

$$\xi/(1 - \xi)\log(1 - \xi)$$

The methods of Bettis-Cowell associated with the methods (87) can be obtained by simply substituting the coefficients a_k with the new ones β_k with a procedure described in 31.

Numerical comparisons on resonance problems and bound-states problems are also described and show the efficiency of the new developed methods.

In 32 the authors consider the solution of the two-dimensional time-independent Schrödinger equation by partial discretization. The discretized problem is treated as a problem of the numerical solution of system of ordinary differential equations and has been solved numerically by symplectic methods. The problem is then transformed into an algebraic eigenvalue problem involving real, symmetric, large sparse matrices. As numerical illustrations the authors have found the eigenvalues of the two-dimensional harmonic oscillator the two-dimensional Henon-Heils potential and the helium atom using the methods developed and other well known methods.

In 33 the authors have developed a family of trigonometrically fitted Runge-Kutta methods for the numerical integration of orbital problems. The developed method is based on the Runge-Kutta Fehlberg I method (see 14). More specifically the new method is developed in order to integrate exactly any linear combination of the functions:

$$\{1, x, x^2, x^3, x^4, \cos(\pm w), \sin(\pm w)$$

The method's efficiency is measured while integrating well known orbital problems and in some cases it is extremely more efficient compared to classical methods.

In 34 the eigenvalue problem of the one-dimensional time-independent Schrödinger equation is studied. Exponentially fitted and trigonometrically fitted symplectic integrators are developed, by modification of the first and second order Yoshida symplectic methods. Numerical results are presented for the one-dimensional harmonic oscillator and Morse potential.

In 35 the numerical solution of the two-dimensional time-independent Schrödinger equation is studied using the method of partial discretization. The discretized problem is treated as a problem of the numerical solution of a system of ordinary differential equations and Numerov type methods are used to solve it. More specifically the classical Numerov method, the exponentially and trigonometrically fitting modified Numerov methods of Vanden Berghe *et al.* and the minimum phase-lag method of Rao *et al.* are applied to this problem. The methods are applied for the calculation of the eigenvalues of the two-dimensional harmonic oscillator and the two-dimensional Henon-Heils potential. The results are compared with the results produced by full discretization. Conclusions are presented.

In 36 the authors have developed a family of trigonometrically fitted Runge-Kutta methods for the numerical integration of orbital problems. The developed method is based on the fifth algebraic order Runge-Kutta England II method (see 14). More specifically the new methods are developed in order to integrate exactly any linear combination of the functions:

$$\{1, x, x^2, x^3, x^4, \cos(\pm w), \sin(\pm w)\} \text{ and}$$

$$\{1, x, x^2, \cos(\pm w), \sin(\pm w), x\cos(\pm w), x\sin(\pm w)\}$$

The methods have been used for the integration of the radial Schrödinger equation and have high efficiency as the results show. The efficiency is higher when using higher energy and this can be explained by the error analysis of the methods presented in the paper. More specifically the new methods have lower powers of the energy in the local truncation error expressions and that keeps the error at lower values.

In 37 the authors have developed trigonometrically fitted Adams-Bashforth-Moulton predictor-corrector (P-C) methods. The new trigonometrically fitted P-C schemes are based on the well known Adams-Bashforth-Moulton methods. In particular, they are based on the fifth order Adams-Bashforth scheme (as predictor) and on the sixth order Adams-Moulton scheme (as corrector). More specifically the new method has the form:

$$\bar{y}_{n+1} = y_n + h \sum_{i=0}^{4} a_i f_{n-i}$$

$$y_{n+1} = y_n + h \left(c_0 \bar{f}_{n+1} \sum_{i=1}^{5} c_i f_{n-i+1} \right)$$

The coefficients of the above method are determined in order the method integrates exactly any linear combination of the functions:

$$\{1,\ x,\ x^2, x^3,\ x^4,\ \cos(\pm w),\ \sin(\pm w)\} \text{ and}$$

$$\{1,\ x,\ x^2,\ \cos(\pm w),\ \sin(\pm w),\ x\cos(\pm w),\ x\sin(\pm w)\}$$

They have studied in detail the stability of the new developed schemes. They have tested the efficiency of their newly developed schemes against well known methods, with efficient results. The numerical illustrations indicate that at least one of their methods is significantly more efficient compared to other methods for the numerical solution of the Schrödinger equation.

In 38 the authors have developed a family of multiderivative methods with minimal phase-lag for the numerical solution of the Schrödinger equation. The method has the general form:

$$\begin{aligned}
\bar{y}_{n+1} &= 2y_n - y_{n-1} + a_0 h^2 y_n'' + a_1 h^4 y_n^{(4)} \\
\bar{y}_{n,i} &= y_n - b_i h^2 \left(\bar{y}_{n+1}'' - 2y_n'' + y_{n-1}''\right), \quad i = 1(1)3 \\
\bar{y}_{n,4} &= y_n - b_4 h^2 \left(\bar{y}_{n+1}'' - 2y_n'' + y_{n-1}''\right) \\
\widehat{y}_n &= y_n - b_5 h^2 \left(\widehat{y}_{n+1}^{(4)} - 2y_n^{(4)} + y_{n-1}^{(4)}\right) \\
\tilde{y}_{n+1} &= 2y_n - y_{n-1} + a_0 h^2 \widehat{y}_n + a_1 h^4 \widehat{y}_n^{(4)} + a_2 h^6 \widehat{y}_n^{(6)}
\end{aligned}$$

$$\begin{aligned}
y_{n+1} - 2y_n + y_{n-1} = {} & h^2 \left(c_0 y_n'' + c_1 \left(\tilde{y}_{n+1}'' + y_{n-1}''\right)\right) \\
& + h^4 \left(c_2 y_n^{(4)} + c_3 \left(\tilde{y}_{n+1}^{(4)} + y_{n-1}^{(4)}\right)\right)
\end{aligned}$$

The methods are called multiderivative since uses derivatives of order two, four or six. The parameters of the method are computed in order to have eighth algebraic order and minimal phase-lag. Finally, a family of eighth algebraic order multiderivative methods with phase-lag of order 12(2)18 is developed. Numerical application of the new obtained methods to the Schrödinger equation shows their efficiency compared with other similar well known methods of the literature.

In 39 the author has developed an explicit symmetric eight-step method which is trigonometrically-fitted and is of algebraic order eight. More specifically, the new method has the form:

$$y_{n+4} - \frac{3}{2}y_{n+3} + \frac{3}{2}y_{n+2} - y_{n+1} - y_{n-1} + \frac{3}{2}y_{n-2} - \frac{3}{2}y_{n-3} + y_{n-4} =$$
$$h^2[b_0(f_{n+3} + f_{n-3}) + b_1(f_{n+2} + f_{n-2}) + b_2(f_{n+1} + f_{n-1}) + b_3 f_n]$$

The coefficients of the new method are determined in order the method to integrate exactly any linear combination of the functions:

$$\{1,\ x,\ x^2,\ x^3,\ x^4,\ x^5,\ x^6,\ x^7,\ \cos(\pm w),\ \sin\ (\pm w)\}$$

The Local Truncation Error of the method is given by:

$$LTE = \frac{10009}{161280} h^{10} \left(y_n^{(10)} + w^2 y_n^{(8)} \right)$$

A stability analysis of the new method is also presented. Numerical results from its application to well-known periodic orbital problems show the efficiency of the new methods.

In 40 the authors present a new explicit Runge-Kutta method with algebraic order four, minimum error of the fifth algebraic order (the limit of the error is zero, when the step-size tends to zero), infinite order of dispersion and eighth order of dissipation *i.e.* they present an optimized explicit Runge-Kutta method of fourth order. The efficiency of the newly developed method is shown through the numerical illustrations of a wide range of methods when these are applied to well-known periodic orbital problems.

In 41 the authors have developed trigonometrically fitted Adams-Bashforth-Moulton predictor-corrector (P-C) methods. The new trigonometrically fitted P-C schemes are based on the well known Adams-Bashforth-Moulton methods. In particular, they are based on the third order Adams-Bashforth scheme (as predictor) and on the fourth order Adams-Moulton scheme (as corrector). More specifically the new method has the form:

$$\bar{y}_{n+1} = y_n + h \sum_{i=0}^{3} a_i f_{n-i}$$
$$y_{n+1} = y_n + h \left(c_0 \bar{f}_{n+1} \sum_{i=1}^{4} c_i f_{n-i+1} \right)$$

The coefficients of the above method are determined in order the method integrates exactly any linear combination of the functions:

$$\{1,\ x,\ \cos(\pm w),\ \sin\ (\pm w)\}$$

They have studied in detail the stability of the new developed schemes.

We tested the efficiency of our newly obtained scheme against well known methods, with excellent results. The numerical illustration showed that our method is considerably more efficient compared to well known methods used for the numerical solution of initial value problems with oscillating solutions.

In 42 the authors have developed a family of multiderivative methods with minimal phase-lag for the numerical solution of the Schrödinger equation. The method has the general form:

$$\begin{aligned} \bar{y}_{n+1} &= 2y_n - y_{n-1} + a_0 h^2 y''_n + a_1 h^4 y_n^{(4)} \\ \bar{y}_n &= y_n - b_c h^2 \left(\bar{y}''_{n+1} - 2y''_n + y''_{n-1}\right) \\ \tilde{y}_{n+1} &= 2y_n - y_{n-1} + a_0 h^2 \bar{y}''_n + a_1 h^4 \bar{y}_n^{(4)} + a_2 h^6 \bar{y}_n^{(6)} \end{aligned}$$

$$\begin{aligned} y_{n+1} - 2y_n + y_{n-1} = {} & h^2 \left(c_0 y''_n + c_1 \left(\tilde{y}''_{n+1} + y''_{n-1}\right)\right) \\ & + h^4 \left(c_2 y_n^{(4)} + c_3 \left(\tilde{y}_{n+1}^{(4)} + y_{n-1}^{(4)}\right)\right) \end{aligned}$$

The methods are called multiderivative since uses derivatives of order two, four or six. The parameters of the method are computed in order to have eighth algebraic order and minimal phase-lag. Finally, a family of eighth algebraic order multiderivative methods with phase-lag of order ten is developed. Numerical application of the new obtained methods to the resonance problem of the one-dimensional Schrödinger equation shows their efficiency compared with other similar well known methods of the literature.

In 43 the authors have developed a new Runge-Kutta method with minimal dispersion and dissipation error. More specifically the Runge-Kutta is of fourth algebraic order and the coefficients are determined in order to have fourth order of dispersion and ninth of dissipation or eighth order of dispersion and fifth of dissipation. The Chebyshev pseudospectral method is utilized using spatial discretization and a new fourth-order six-stage Runge-Kutta scheme is used for time advancing. The proposed scheme is more efficient than the existing ones for specific problems.

In 44 the author develops exponentially fitted multiderivative methods for the numerical solution of the one-dimensional Schrödinger equation. The methods are called multiderivative since uses derivatives of order two and four. An application to the resonance problem of the one-dimensional Schrödinger equation indicates that the new method is more efficient than other similar well known methods of the literature.

In 45 the authors develop a family of explicit Runge-Kutta methods of 5th algebraic order, one of which has variable coefficients, for the efficient solution of problems with oscillating solutions. Emphasis has been given on the phase-lag property. Basic theory of Runge-Kutta methods, phase-lag analysis and construction of the new methods are presented. Numerical results produced for known problems show the efficiency of the new methods when they are compared with known methods in the literature. Furthermore we note that the method with variable coefficients has much higher accuracy, which gets close to double precision, when the product of the frequency with the step-length approaches certain values. These values are constant and independent of the problem solved and depend only on the method used and more specifically on the expressions used to achieve higher algebraic order.

In 46 a dissipative trigonometrically-fitted two-step explicit hybrid method is developed. This method is based on a dissipative explicit two-step method developed recently by Tsitouras [64]. More specifically the method has the form:

$$y_a = (1 - c_1)y_n + c_1 y_{n-1} + h^2(d_{11}f_{n-1} + d_{12}f_n)$$
$$y_b = (1 - c_2)y_n + c_2 y_{n-1} + h^2(d_{21}f_{n-1} + d_{22}f_n + g_{21}f_a)$$
$$y_c = (1 - c_3)y_n + c_3 y_{n-1} + h^2(d_{31}f_{n-1} + d_{32}f_n + g_{31}f_a + g_{32}f_b)$$

$$y_{n+1} - 2y_n + y_{n-1} = h^2(w_1 f_{n-1} + w_2 f_n + b_1 f_a + b_2 f_b + b_3 f_c)$$

Numerical illustrations show that the procedure of trigonometric fitting is an efficient way to produce numerical methods for the solution of second-order linear initial value problems (IVPs) with oscillating solutions.

In 47 the author investigates the connection between closed Newton-Cotes, trigonometrically-fitted differential methods and symplectic integrators. It is known from the literature that several one-step symplectic integrators have been produced based on symplectic geometry. However, the development of multistep symplectic integrators is very poor. Zhu *et al.*(1) presented the well known open Newton-Cotes differential methods as multilayer symplectic integrators. Chiou and Wu(2) also studied the construction of multistep symplectic integrators based on the open Newton-Cotes integration methods. The author investigated the closed Newton-Cotes formulae and has written them as symplectic multilayer structures. He has also constructed trigonometrically-fitted symplectic methods which are based on the closed Newton-Cotes formulae. The applications of the symplectic schemes have shown their efficiency. An very important remark is that the Hamiltonian energy of the system remains almost constant as integration proceeds.

In 48 multiderivative methods are developed for the numerical integration of the one dimensional Schrödinger equation. The method is called multiderivative since uses derivatives of order two and four. Application of these methods to the resonance problem of the radial Schrödinger equation indicates that the new developed method is more efficient than the Numerov method and other well known methods of the literature.

In 49 the authors have presented a simple technique that allows to limit the error growth in the Ion-term (long-time) numerical integration of perturbed multi-dimensional oscillators, while using highly efficient and accurate special multistep codes. The author has studied theoretically their behaviour. The new technique has been illustrated with some numerical examples, including a case with non-resonant frequencies.

In 50 the author presents a further investigation of the frequency evaluation techniques which are recently proposed by Ixaru *et al.* for exponentially fitted multistep algorithms for the solution of first-order ordinary differential equations (ODEs). These studies have a scope which is to maximize the benefits of the exponentially-fitted methods via the evaluation of the frequency of the problem. The proposed by Ixaru and co-workers method for frequency evaluation algorithm has been successfully applied in a direct way to a second-order exponentially fitted Runge-Kutta method of collocation type. This seems to be impossible for the higher order exponentially-fitted methods. To overcome this difficulty the author develops an efficient extension of Ixaru's frequency evaluation algorithm for the exponentially fitted RadauIIA method of third order. Numerical illustration confirm the properties of the developed algorithm.

In 51 the author presents a new procedure for constructing efficient embedded modified Runge-Kutta methods for the numerical solution of the Schrödinger equation. The methods of the embedded scheme have algebraic orders five and four. Applications of the new pair to the radial Schrödinger equation and to coupled Schrödinger equations show the efficiency of the approach.

In 52 the author develops a symplectic exponentially fitted modified Runge-Kutta-Nyström method. The method of development was based on the development of symplectic exponentially fitted modified Runge-Kutta-Nyström method by Simos and Vigo-Aguiar. The new method is a two-stage second-order method with FSAL-property (first step as last).

In 53 the authors present a new P-stable Obrechkoff four-step method, which greatly improves the performance of previous Obrechkoff four-step method and extends its application range. Using the requirement that the characteristic equations to be exact for specific roots which have been chosen in order the definition of P-stability to be satisfied, they extend the interval of periodicity to infinity and at the same time, they keep all its advantages in the accuracy and efficiency.

In 54 the authors first have studied a recent method of Psihoyios and Simos. After that they introduce P-stable four-step, six-step and eight-step methods with similar way that have described in 55 (see below).

In 55 the authors present a new procedure for construction of P-stable linear multistep methods for periodic initial-value problems. This procedure is based on the requirement the characteristic equations produced by the methods to have roots of specific form.

In 56–59 and 61 the authors presents efficient Obrechkoff methods.

In 60 the author presents a modified Numerov P-stable method. The construction of this method is based on the procedure mentioned above (see comment on 53).

In 62 the authors present a detailed Mathematica program which is used to integrate a radial Schrödinger equation by a new two-step multiderivative numerical method with fourth- and sixth-order derivatives. In the program the authors use an efficient algorithm to calculate the first-order derivative and avoid unnecessarily repeated calculation resulting from the multi-derivatives.

Finally, in 65 there is for the first time a detailed presentation of the exponential fitting. This is an excellent book in which one can find reference work for the exponential fitting applied to differentiation, to integration and to the solution of differential equations. In chapter 2, some mathematical properties are studied and the mathematical theory of exponential fitting is presented.

In chapter 3, the construction of exponential fitting formulae is presented. In chapter 4, applications of exponential fitting to differentiation, to integration and to interpolation are presented. In chapter 5, application of exponential fitting to multistep methods for the solution of differential equations is presented. Finally, in chapter 6, application of exponential fitting to Runge-Kutta methods for the solution of differential equations is presented.

References

1. L. Gr. Ixaru and M. Micu, *Topics in Theoretical Physics*, Central Institute of Physics, Bucharest, 1978.
2. L.D. Landau and F.M. Lifshitz, *Quantum Mechanics*, Pergamon, New York, 1965.
3. I. Prigogine, Stuart Rice (Eds): Advances in Chemical Physics Vol. 93: New Methods in Computational Quantum Mechanics, John Wiley & Sons, 1997.
4. G. Herzberg, *Spectra of Diatomic Molecules*, Van Nostrand, Toronto, 1950.
5. T.E. Simos, Atomic Structure Computations in Chemical Modelling Applications and Theory (Editor: A Hinchliffe UMIST), *The Royal Society of Chemistry*, 2000, 38–142.
6. T.E. Simos, Numerical methods for 1D, 2D and 3D differential equations arising in chemical problems, *Chemical Modelling: Application and Theory*, The Royal Society of Chemistry, 2002, **2**, 170–270.
7. T.E. Simos, Numerical Methods in Chemistry, *Chemical Modelling: Application and Theory*, The Royal Society of Chemistry, 2004, **3**, 271–378.
8. T.E. Simos, *New Astronomy*, 2004, **9**, 409.
9. J.R. Dormand and P.J. Prince, *J. Comput. Appl. Math.*, 1980, **6**, 19.
10. J.R. Dormand, M.E.A. El-Mikkawy and P.J. Prince, *IMA J. Numer. Anal.*, 1987, **7**, 423.
11. E. Stiefel and D.G. Bettis, *Numer. Math.*, 1969, **13**, 154.
12. J.M. Sanz-Serna and M.P. Calvo, *Numerical Hamiltonian Problem*, Chapman and Hall, London, 1994.
13. E. Hairer, S.P. Nørsett and G. Wanner, *Solving Ordinary Differential equations I, Nonstiff Problems*, Springer, Berlin, Heidelberg, 1993.
14. G. Engeln-Müllges and F. Uhlig, *Numerical Algorithms with Fortran*, Springer Berlin, Heidelberg, 1996.
15. E. Forest and R. D. Ruth, *Physica D*, 1990, **43**, 105.
16. J.D. Lambert and I.A. Watson, Symmetric multistep methods for periodic initial values problems, *J. Inst. Math. Appl.*, 1976, **18**, 89.
17. T.E. Simos and P.S. Williams, *J. Comput. Appl. Math.*, 1997, **79**, 189.
18. M. M. Chawla and P.S. Rao, *J. Comput. Appl. Math.*, 1986, **15**, 329.
19. P. Henrici, *'Discrete Variable Methods in Ordinary Differential Equations'*, John Wiley and Sons, New York, USA, 1962.
20. A.D. Raptis, *Computer Physics Communications*, 1983, **28**, 427.
21. D. Raptis and A.C. Allison, *Computer Physics Communications*, 1978, **14**, 1.
22. L.Gr. Ixaru and M. Rizea, *Computer Physics Communications*, 1980, **19**, 23.
23. T.E. Simos, *Computer Physics Communications*, 1998, **115**, 1.
24. T. Lyche, *Num. Math.*, 1972, **19**, 65.
25. Z.A. Anastassi and T.E. Simos, *Mathematical and Computer Modelling*, 2005, **42**, 877.

26. G. Psihoyios and T.E. Simos, *Mathematical and Computer Modelling*, 2005, **42**, 887.
27. D.P. Sakas and T.E. Simos, *Mathematical and Computer Modelling*, 2005, **42**, 903.
28. T.E. Simos, *Computational Materials Science*, 2005, **34**, 342.
29. D. P. Sakas and T.E. Simos, *MATCH-Communications in Mathematical and in Computer Chemistry*, 2005, **53**, 299.
30. G. Psihoyios and T.E. Simos, *MATCH-Communications in Mathematical and In Computer Chemistry*, 2005, **53**, 321.
31. J. Vigo-Aguiar and T.E. Simos, *International Journal of Quantum Chemistry*, 2005, **103**, 278.
32. T. Monovasilis and T.E. Simos, *Chemical Physics*, 2005, **313**, 293.
33. Z.A. Anastassi and T.E. Simos, *New Astronomy*, 2005, **10**, 301.
34. T. Monovasilis, Z. Kalogiratou and T.E. Simos, *Journal of Mathematical Chemistry*, 2005, **37**, 263.
35. Z. Kalogiratou, T. Monovasilis and T.E. Simos, *Journal of Mathematical Chemistry*, 2005, **37**, 271.
36. Z.A. Anastassi and T.E. Simos, *Journal of Mathematical Chemistry*, 2005, **37**, 281.
37. G. Psihoyios and T.E. Simos, *Journal of Mathematical Chemistry*, 2005, **37**, 295.
38. D.P. Sakas and T.E. Simos, *Journal of Mathematical Chemistry*, 2005, **37**, 317.
39. T.E. Simos, *Mathematical and Computer Modelling*, 2004, **40**, 1263.
40. Z.A. Anastassi and T.E. Simos, *Journal of Computational and Applied Mathematics*, 2005, **175**, 1.
41. G. Psihoyios and T.E. Simos, *Journal of Computational and Applied Mathematics*, 2005, **175**, 137.
42. D.P. Sakas and T.E. Simos, *Journal of Computational and Applied Mathematics*, 2005, **175**, 161–172.
43. K. Tselios and T.E. Simos, *Journal of Computational and Applied Mathematics*, 2005, **175**, 173.
44. T.E. Simos, *Journal of Mathematical Chemistry*, 2004, **36**, 13.
45. Z.A. Anastassi and T.E. Simos, *International Journal of Modern Physics C*, 2004, **15**, 1.
46. T.E. Simos, *Applied Mathematics Letters*, 2004, **17**, 601.
47. T.E. Simos, *International Journal of Modern Physics C*, 2003, **14**, 1061.
48. T.E. Simos, *MATCH-Communications in Mathematical and in Computer Chemistry*, 2004, **50**, 7.
49. J. Vigo-Aguiar, T.E. Simos and J.M. Ferrandiz, Proceedings of the Royal Society of London Series A-Mathematical, *Physical and Engineering Sciences*, 2004, **460**, 561.
50. H. Van De Vyver, *Journal of Computational and Applied Mathematics*, 2005, **184**, 442.
51. H. Van De Vyver, *International Journal of Modern Physics C*, 2005, **16**, 879.
52. H. Van De Vyver, *New Astronomy*, 2005, **10**, 261.
53. Y.M. Dai, Z.C. Wang and D.M. Wu, *Journal of Computational and Applied Mathematics*, 2006, **187**, 192.
54. Z.C. Wang, *New Astronomy*, 2005, **11**, 90.
55. Z.C. Wang, *Computer Physics Communications*, 2005, **171**, 162.
56. Z.C. Wang and Y. Wang, *Computer Physics Communications*, 2005, **171**, 79.
57. D.Y. Zhao, Z.C. Wang, Y.M. Dai and Y. Wang, *International Journal of Modern Physics C*, 2005, **16**, 1027.
58. Z.C. Wang and Q.M. Chen, *Computer Physics Communications*, 2005, **170**, 49.

59. Z.C. Wang, D.Y. Zhao, Y.M. Dai and D.M. Wu, *Proceedings of the Royal Society A-Mathematical Physical, and Engineering Sciences*, 2005, **461**, 1639.
60. Z.C. Wang, *Computer Physics Communications*, 2005, **167**, 1.
61. Y.M. Dai, Z.C. Wang, D.Y. Zhao and X.L. Song, *Computer Physics Communications*, 2005, **165**, 110.
62. Z.C. Wang, Y.H. Ge, Y.M. Dai and D.Y. Zhao, *Computer Physics Communications*, 2004, **160**, 23.
63. Z.A. Anastassi and T.E. Simos, *New Astronomy*, 2004, **10**, 31.
64. Ch. Tsitouras, *Computers Math. Applic.*, 2002, **43**, 943.
65. L. Gr. Ixaru and G. Vanden Barghe, Exponential Fitting, Kluwer Academic Publishers, Series: Mathematics and its Applications – Volume 568, The Netherlands, 2004.

Appendix A

Partitioned Multistep Methods – Maple Program of Construction of the Methods

```
>        Method I;
>

> restart;
> dnp3:=diff(z(x+3*h),x$1);
> dnp2:=diff(z(x+2*h),x$1);
> dnp1:=diff(z(x+h),x$1);
> dn:=diff(z(x),x$1);
> dnm1:=diff(z(x-h),x$1);
> dnm2:=diff(z(x-2*h),x$1);
> dnm3:=diff(z(x-3*h),x$1);
> eqa:=simplify(z(x+2*h)-z(x-2*h)=h*(c[0]*dn+c[1]*dn+c[2]*dnm1));
> z:=x->exp(I*v*x);
> eqa:=simplify(eqa);
> eqa:=simplify(eqa/exp(I*v*x));
> lr:=simplify(evalc(Re(2*I*sin(2*v*h))));
> li:=simplify(evalc(Im(2*I*sin(2*v*h))));
> rr:=simplify(evalc(Re(h*v*(c[0]*exp(h*v*I)+c[1]+c[2]*exp(-I*v*h))*I)));
> ri:=simplify(evalc(Im(h*v*(c[0]*exp(h*v*I)+c[1]+c[2]*exp(-I*v*h))*I)));
> eq1:=lr=rr;
> eq2:=li=ri;
  >
> restart;
> dnp3:=diff(z(x+3*h),x$1);
> dnp2:=diff(z(x+2*h),x$1);
> dnp1:=diff(z(x+h),x$1);
> dn:=diff(z(x),x$1);
> dnm1:=diff(z(x-h),x$1);
> dnm2:=diff(z(x-2*h),x$1);
> dnm3:=diff(z(x-3*h),x$1);
> eqa:=simplify(z(x+2*h)-z(x-2*h)=h*(c[0]*dnp1+c[1]*dn+c[2]*dnm1));
  >
> z:=x->x^n;
> eqa:=simplify(eqa);
```

```
> eq3:=simplify(subs(n=0,eqa));
> eq3:=simplify(subs(n=1,eqa));
> eq4:=simplify(subs(n=2,eqa));
> eq4:=simplify(eq4-2*x*eq3);
> eq 1:=0=-v*h*sin(v*h)*(c[0]-c[2]);
> eq2:=2*sin(2*v*h)=v*h*(c[0]*cos(v*h)+c[1]+c[2]*cos(v*h));
> eq3:=4*h=h*(c[0]+c[1]+c[2]);
> eq4:=0=2*c[0]*h^2-2*c[2]*h^2;
> solut:=solve({eq1,eq2,eq3,eq4},{c[0],c[1],c[2]});
> assign(solut);
> c[0]:=simplify(c[0]);
> c[1]:=simplify(c[1]);
> c[2]:=simplify(c[2]);
> c0t:=convert(taylor(c[0],v=0,19),polynom);
> c1t:=convert(taylor(c[1],v=0,19),polynom);
> c2t:=convert(taylor(c[2],v=0,19),polynom);
>
> restart;
> dnp3:=diff(y(x+3*h),x$1);
> dnp2:=diff(y(x+2*h),x$1);
> dnp1:=diff(y(x+h),x$1);
> dn:=diff(y(x),x$1);
> dnm1:=diff(y(x-h),x$1);
> dnm2:=diff(y(x-2*h),x$1);
> dnm3:=diff(y(x-3*h),x$1);
>eqa:=simplify(y(x+3*h)-y(x-2*h)=h*(b[0]*dnp2+b[1]*dnp1+b[2]*dn+b[3]*dnm1));
> y:=x->exp(I*v*x);
> eqa:=simplify(eqa);
> eqa:=simplify(eqa/exp(I*v*x));
> lr:=simplify(evalc(Re(exp(3*I*h*v)-exp(-2*I*h*v))));
> li:=simplify(evalc(Im(exp(3*I*h*v)-exp(-2*I*h*v))));
> rr:=simplify(evalc(Re(h*v*(b[0]*exp(2*I*h*v)+b[1]*exp(h*v*I)+b[2]+b[3]*exp(-I*h*v))*I)));
> ri:=simplify(evalc(Im(h*v*(b[0]*exp(2*I*h*v)+b[1]*exp(h*v*I)+b[2]+b[3]*exp(-I*h*v))*I)));
> eq1:=lr=rr;
> eq2:=li=ri;
>
> restart;
> dnp3:=diff(y(x+3*h),x$1);
> dnp2:=diff(y(x+2*h),x$1);
> dnp1:=diff(y(x+h),x$1);
> dn:=diff(y(x),x$1);
> dnm1:=diff(y(x-h),x$1);
> dnm2:=diff(y(x-2*h),x$1);
> dnm3:=diff(y(x-3*h),x$1);
>eqa:=simplify(y(x+3*h)-y(x-2*h)=h*(b[0]*dnp2+b[1]*dnp1+b[2]*dn+b[3]*dnm1));
>
> y:=x->x^n;
> eqa:=simplify(eqa);
> eq3:=simplify(subs(n=0,eqa));
> eq3:=simplify(subs(n=1,eqa));
> eq4:=simplify(subs(n=2,eqa));
> eq4:=simplify(eq4-2*x*eq3);
> eq1:=cos(3*v*h)-cos(2*v*h)=-v*h*(b[0]*sin(2*v*h)+b[1]*sin(v*h)-b[3]*sin(v*h));
>eq2:=sin(3*v*h)+sin(2*v*h)=v*h*(b[0]*cos(2*v*h)+b[1]*cos(v*h)+b[2]+b[3]*cos(v*h));
> eq3:=5*h=h*(b[0]+b[1]+b[2]+b[3]);
```

```
> eq4:=5*h^2=4*b[0]*h^2+2*b[1]*h^2-2*b[3]*h^2;
> solut:=solve(eq1,eq2,eq3,eq4,}{b[0],b[1],b[2]b[3]});
> assign(solut);
> b[0]:=combine(b[0]);
> b[1]:=combine(b[1]);
> b[2]:=combine(b[2]);
> b[3]:=combine(b[3]);
> b0t:=convert(taylor(b[0],v=0,19),polynom);
> b1t:=convert(taylor(b[1],v=0,19),polynom);
> b2t:=convert(taylor(b[2],v=0,19),polynom);
> b3t:=convert(taylor(b[3],v=0,19),polynom);
>
>         Method II;
> restart;
> dnp3:=diff(z(x+3*h),x$1);
> dnp2:=diff(z(x+2*h),x$1);
> dnp1:=diff(z(x+h),x$1);
> dn:=diff(z(x),x$1);
> dnm1:=diff(z(x-h),x$1);
> dnm2:=diff(z(x-2*h),x$1);
> dnm3:=diff(z(x-3*h),x$1);
> eqa:=simplify(z(x+2*h)-z(x-2*h)=h*(c[0]*dnp1+c[1]*dn+c[2]*dnm1));
> z:=x->exp(I*v*x);
> eqa:=simplify(eqa);
> eqa:=simplify(eqa/exp(I*v*x));
> lr:=simplify(evalc(Re(2*I*sin(2*v*h))));
> li:=simplify(evalc(Im(2*I*sin(2*v*h))));
> rr:=simplify(evalc(Re(h*v*(c[0]*exp(h*v*I)+c[1]+c[2]*exp(-I*v*h))*I)));
> ri:=simplify(evalc(Im(h*v*(c[0]*exp(h*v*I)+c[1]+c[2]*exp(-I*v*h))*I)));
> eq1:=lr=rr;
> eq2:=li=ri;
>
> restart;
> dnp3:=diff(z(x+3*h),x$1);
> dnp2:=diff(z(x+2*h),x$1);
> dnp1:=diff(z(x+h),x$1);
> dn:=diff(z(x),x$1);
> dnm1:=diff(z(x-h),x$1);
> dnm2:=diff(z(x-2*h),x$1);
> dnm3:=diff(z(x-3*h),x$1);
> eqa:=simplify(z(x+2*h)-z(x-2*h)=h*(c[0]*dnp1+c[1]*dn+c[2]*dnm1));
>
> z:=x->x*exp(I*v*x);
> eqa:=simplify(eqa);
> eqa:=simplify(eqa/exp(I*v*x));
> lr:=simplify(evalc(Re(exp(2*I*v*h)*x+2*exp(2*I*v*h)*h-exp(-2*I*v*h)*x+2*exp(-2*I*v*h)*h)));
> li:=simplify(evalc(Im(exp(2*I*v*h)*x+2*exp(2*I*v*h)*h-exp(-2*I*v*h)*x+2*exp(-2*I*v*h)*h)));
>                                                    rr:=simplify(evalc(Re(h*(c[0]*exp(v*-
h*I)+c[0]*v*exp(v*h*I)*x*I+c[0]*v*exp(v*h*I)*h*I+c[1]+c[1]*x*v*I+c[2]*exp(-I*v*h)+c[2]*v*exp(-
I*v*h)*x*I-I*c[2]*v*exp(-I*v*h)*h))));
>
ri:=simplify(evalc(Im(h*(c[0]*exp(v*h*I)+c[0]*v*exp(v*h*I)*x*I+c[0]*v*exp(v*h*I)*h*I+c[1]+c[1]*-
x*v*I+c[2]*exp(-I*v*h)+c[2]*v*exp(-I*v*h)*x*I-I*c[2]*v*exp(-I*v*h)*h))));
> eq3:=lr=rr;
> eq4:=li=ri;
```

```
> eq1:=0=-v*h*sin(v*h)*(c[0]-c[2]);
> eq2:=2*sin(2*v*h)=v*h*(c[0]*cos(v*h)+c[1]+c[2]*cos(v*h));
> eq3:=4*cos(2*v*h)*h=-h*(-c[0]*cos(v*h)+c[0]*v*sin(v*h)*x+c[0]*v*sin(v*h)*h-c[1]-
c[2]*cos(v*h)-c[2]*v*sin(v*h)*x+c[2]*v*sin(v*h)*h);
> eq4:=2*sin(2*v*h)*x=h*(c[0]*sin(v*h)+c[0]*v*cos(v*h)*x+c[0]*v*cos(v*h)*h+c[1]*x*v-
c[2]*sin(v*h)+c[2]*v*cos(v*h)*x-c[2]*v*cos(v*h)*h);
> solut:=solve(eq1,eq2,eq3,eq4,},{c[0],c[1],c[2]});
> assign(solut);
> c[0]:=combine(c[0]);
> c[1]:=combine(c[1]);
> c[2]:=combine(c[2]);
> c0t:=convert(taylor(c[0],v=0,19),polynom);
>
> c1t:=convert(taylor(c[1],v=0,19),polynom);
> c2t:=convert(taylor(c[2],v=0,19),polynom);
> h:=1;
> v:=w;
> plot(c[0],w=3..3.3);
> plot(c[0],w=6.25..6.3);
> plot(c[1],w=3..3.3);
> plot(c[1],w=6.25..6.3);
> plot(c[2],w=3..3.3);
> plot(c[2],w=6.25..6.3);
>
> restart;
> dnp3:=diff(y(x+3*h),x$1);
> dnp2:=diff(y(x+2*h),x$1);
> dnp1:=diff(y(x+h),x$1);
> dn:=diff(y(x),x$1);
> dnm1:=diff(y(x-h),x$1);
> dnm2:=diff(y(x-2*h),x$1);
> dnm3:=diff(y(x-3*h),x$1);
> eqa:=simplify(y(x+3*h)-y(x-2*h)=h*(b[0]*dnp2+b[1]*dnp1+b[2]*dn+b[3]*dnm1));
> y:=x->exp(I*v*x);
> eqa:=simplify(eqa);
> eqa:=simplify(eqa/exp(I*v*x));
> lr:=simplify(evalc(Re(exp(3*I*h*v)-exp(-2*I*h*v))));
> li:=simplify(evalc(Im(exp(3*I*h*v)-exp(-2*I*h*v))));
> rr:=simplify(evalc(Re(h*v*(b[0]*exp(2*I*h*v)+b[1]*exp(h*v*I)+b[2]+b[3]*exp(-I*h*v))*I)));
> ri:=simplify(evalc(Im(h*v*(b[0]*exp(2*I*h*v)+b[1]*exp(h*v*I)+b[2]+b[3]*exp(-I*h*v))*I)));
> eq1:=lr=rr;
> eq2:=li=ri;
>
> restart;
> dnp3:=diff(y(x+3*h),x$1);
> dnp2:=diff(y(x+2*h),x$1);
> dnp1:=diff(y(x+h),x$1);
> dn:=diff(y(x),x$1);
> dnm1:=diff(y(x-h),x$1);
> dnm2:=diff(y(x-2*h),x$1);
> dnm3:=diff(y(x-3*h),x$1);
> eqa:=simplify(y(x+3*h)-y(x-2*h)=h*(b[0]*dnp2+b[1]*dnp1+b[2]*dn+b[3]*dnm1));
>
> y:=x->x*exp(I*v*x);
> eqa:=simplify(eqa);
> eqa:=simplify(eqa/exp(I*v*x));
```

```
> lr:=simplify(evalc(Re(exp(3*I*v*h)*x+3*exp(3*I*v*h)*h-exp(-2*I*v*h)*x+2*exp(-2*I*v*h)*h)));
> li:=simplify(evalc(Im(exp(3*I*v*h)*x+3*exp(3*I*v*h)*h-exp(-2*I*v*h)*x+2*exp(-2*I*v*h)*h)));
> rr:=simplify(evalc(Re(h*(b[0]*exp(2*I*v*h)+b[0]*v*exp(2*I*v*h)*-
x*I+2*I*b[0]*v*exp(2*I*v*h)*h+b[1]*exp(v*h*I)+b[1]*v*exp(v*h*I)*x*I+b[1]*v*exp(v*h*I)*h*I+b[2]-
b[2]*x*v*I+b[3]*exp(-I*v*h)+b[3]*v*exp(-I*v*h)*x*I-I*b[3]*v*exp(-I*v*h)*h))));
>
ri:=simplify(evalc(Im(h*(b[0]*exp(2*I*v*h)+b[0]*v*exp(2*I*v*h)*x*I+2*I*b[0]*v*exp(2*I*v*h)*h+-
b[1]*exp(v*h*I)+b[1]*v*exp(v*h*I)*x*I+b[1]*v*exp(v*h*I)*h*I+b[2]+b[2]*x*v*I+b[3]*exp(-
I*v*h)+b[3]*v*exp(-I*v*h)*x*I-I*b[3]*v*exp(-I*v*h)*h))));
> eq3:=lr=rr;
> eq4:=li=ri;
>
> eq1:=cos(3*v*h)-cos(2*v*h)=-v*h*(b[0]*sin(2*v*h)+b[1]*sin(v*h)-b[3]*sin(v*h));
> eq2:=sin(3*v*h)+sin(2*v*h)=v*h*(b[0]*cos(2*v*h)+b[1]*cos(v*h)+b[2]+b[3]*cos(v*h));
> eq3:=cos(3*v*h)*x+3*cos(3*v*h)*h-cos(2*v*h)*x+2*cos(2*v*h)*h=-h*(-
b[0]*cos(2*v*h)+b[0]*v*sin(2*v*h)*x+2*b[0]*v*sin(2*v*h)*h-b[1]*cos(v*h)+b[1]*v*-
sin(v*h)*x+b[1]*v*sin(v*h)*h-b[2]-b[3]*cos(v*h)-b[3]*v*sin(v*h)*x+b[3]*v*sin(v*h)*h);
> eq4:=sin(3*v*h)*x+3*sin(3*v*h)*h+sin(2*v*h)*x-2*sin(2*v*h)*h=h*(b[0]*sin(2*v*h)+b[0]*v*-
cos(2*v*h)*x+2*b[0]*v*cos(2*v*h)*h+b[1]*sin(v*h)+b[1]*v*cos(v*h)*x+b[1]*v*-
cos(v*h)*h+b[2]*x*v-b[3]*sin(v*h)+b[3]*v*cos(v*h)*x-b[3]*v*cos(v*h)*h);
> solut:=solve({eq1,eq2,eq3,eq4,},{b[0],b[1],b[2]b[3]});
> assign(solut);
> b[0]:=combine(b[0]);
>
>
> b[1]:=combine(b[1]);
> b[2]:=combine(b[2]);
> b[3]:=combine(b[3]);
> b0t:=convert(taylor(b[0],v=0,19),polynom);
> b1t:=convert(taylor(b[1],v=0,19),polynom);
> b2t:=convert(taylor(b[2],v=0,19),polynom);
> b3t:=convert(taylor(b[3],v=0,19),polynom);
>
> h:=1;
> v:=w;
> plot(b[0],w=3..3.3);
> plot(b[0],w=6.25..6.3);
> plot(b[1],w=3..3.3);
> plot(b[1],w=6.25..6.3);
> plot(b[2],w=3..3.3);
> plot(b[2],w=6.25..6.3);
> plot(b[3],w=3..3.3);
> plot(b[3],w=6.25..6.3);
> restart;
> ynp3:=convert(taylor(y(x+3*h),h=0,11),polynom);
> ynp2:=convert(taylor(y(x+2*h),h=0,11),polynom);
> ynp1:=convert(taylor(y(x+h),h=0,11),polynom);
> yn:=y(x);
> ynm1:=convert(taylor(y(x-h),h=0,11),polynom);
> ynm2:=convert(taylor(y(x-2*h),h=0,11),polynom);
> ynm3:=convert(taylor(y(x-3*h),h=0,11),polynom);
> dnp3:=convert(taylor(diff(y(x+3*h),x$1),h=0,11),polynom);
> dnp2:=convert(taylor(diff(y(x+2*h),x$1),h=0,11),polynom);
> dnp1:=convert(taylor(diff(y(x+h),x$1),h=0,11),polynom);
> dn:=D(y)(x);
> dnm1:=convert(taylor(diff(y(x-h),x$1),h=0,11),polynom);
```

```
> dnm2:=convert(taylor(diff(y(x-2*h),x$1),h=0,11),polynom);
> dnm3:=convert(taylor(diff(y(x-3*h),x$1),h=0,11),polynom);
>
> c[0]:=8/3-14/45*v^2*h^2+11/630*v^4*h^41/2700*v^6*h^6+19/2993760*v^8*h^8337/
13621608000*v^10*h^10+1/934053120*v^12*h^12+1681/83364240960000*v^14*h^14;
> c[1]:=-4/3+28/45*v^2*h^2-11/315*v^4*h^4+1/1350*v^6*h^619/1496880*v^8*h^8+337/
6810804000*v^10*h^10-1/467026560*v^12*h^12-1681/41682120480000*v^14*h^14;
> c[2]:=8/3-14/45*v^2*h^2+11/630*v^4*h^41/2700*v^6*h^6+19/2993760*v^8*h^8-337/
13621608000*v^10*h^10+1/934053120*v^12*h^12+1681/83364240960000*v^14*h^14;
> lte[1]:=simplify(ynp2-ynm2-(h*(c[0]*dnp1+c[1]*dn+c[2]*dnm1)));
> coeff(lte[1],h,5);
> b[0]:=55/24-95/288*v^2*h^2+65/4032*v^4*h^4-47/96768*v^6*h^689/19160064*v^8*h^8-
38923/34871316480*v^10*h^103253/29889699840*v^12*h^12-1894267/
170729965486080*v^14*h^14;
> b[1]:=5/24+95/288*v^2*h^265/4032*v^4*h^4+47/96768*v^6*h^6-89/19160064*v^8*h^8-
38923/34871316480*v^10*h^10-3253/29889699840*v^12*h^12;
> b[2]:=5/24+95/288*v^2*h^2-65/4032*v^4*h^4+47/96768*v^6*h^6+89/
19160064*v^8*h^8+38923/34871316480*v^10*h^10+3253/29889699840*v^12*h^12;
> b[3]:=55/24-95/288*v^2*h^2+65/4032*v^4*h^4-47/96768*v^6*h^6-89/19160064*v^8*h^8-
38923/34871316480*v^10*h^10-3253/29889699840*v^12*h^12;
>
> lte[2]:=simplify(ynp3-ynm2-(h*(b[0]*dnp2+b[1]*dnp1+b[2]*dn+b[3]*dnm1)));
> coeff(lte[2],h,5);
>
>
> c[0]:=8/3-28/45*v^2*h^2+5/189*v^4*h^4-11/9450*v^6*h^6-241/7484400*v^8*h^8-90367/
20432412000*v^10*h^10-7157/16345929600*v^12*h^12-168361/3789283680000*v^14*h^14;
> c[1]:=-4/3+56/45*v^2*h^2-344/945*v^4*h^4+176/4725*v^6*h^6-1096/
467775*v^8*h^8+49072/638512875*v^10*h^10-16/5108103*v^12*h^12-7328/
162820783125*v^14*h^14;
> c[2]:=8/3-28/45*v^2*h^2+5/189*v^4*h^4-11/9450*v^6*h^6-241/7484400*v^8*h^8-90367/
20432412000*v^10*h^10-7157/16345929600*v^12*h^12-168361/3789283680000*v^14*h^14;
> lte[3]:=simplify(ynp2-ynm2-(h*(c[0]*dnp1+c[1]*dn+c[2]*dnm1)));
> coeff(lte[3],h,5);
>
>
> b[0]:=55/24-95/144*v^2*h^2+505/24192*v^4*h^4-97/48384*v^6*h^6-2953/
19160064*v^8*h^8-1118269/52306974720*v^10*h^10-174541/64377815040*v^12*h^12;
> b[1]:=5/24+95/144*v^2*h^2-8485/24192*v^4*h^4+1657/48384*v^6*h^6-51299/
19160064*v^8*h^8+1467709/52306974720*v^10*h^10-1163597/119558799360*v^12*h^12;
> b[2]:=5/24+95/144*v^2*h^2-8485/24192*v^4*h^4+1657/48384*v^6*h^6-51299/
19160064*v^8*h^8+1467709/52306974720*v^10*h^10-1163597/119558799360*v^12*h^12;
> b[3]:=55/24-95/144*v^2*h^2+505/24192*v^4*h^4-97/48384*v^6*h^6-2953/
19160064*v^8*h^8-1118269/52306974720*v^10*h^10-174541/64377815040*v^12*h^12;
> lte[4]:=simplify(ynp3-ynm2-(h*(b[0]*dnp2+b[1]*dnp1+b[2]*dn+b[3]*dnm1)));
> coeff(lte[4],h,5);
>
```

Appendix B

Maple Program for the development of Dispersive-fitted and dissipative-fitted explicit Runge-Kutta method

```
> restart:
```

Tree Theory

Simplifying Assumptions

```
> b[2]:=0:
c[2]:=321/1000000;
c[3]:=13/20;
c[5]:=1/6;
c[6]:=1;
```

Input

```
> Digits:=20:
> stage_max:=6: #Enter number of stages
> alg_max:=17: #Enter number of algebraic order equations (maximum 17 for 5th order)
> alg_act:=8: #Enter number of active algebraic order equations (eg 8 for 4th order, max 17)
> phl_act:=10: #Enter number of active phase lag order equations (usually greater than #number of indeterminates because some coefficients vanish while using the #algebraic equations)
```

Definitions

```
> c[]:=array[1..stage_max];
c[1]:=0;
a[]:=array[1..stage_max, 1..stage_max-1];
alg[]:=array[37];
for i to stage_max do
 for j to stage_max do
  if j>i then a[i,j]:=0; fi;
 od:
od:
for i from 1 to 37 do alg[i]:=0 od:
>
> g(1):=1: #1st Order
g(2):=1/2: #2nd Order
g(3):=1/3: g(4):=1/6: #3rd Order
g(5):=1/4: g(6):=1/8: g(7):=1/12: g(8):=1/24: #4th Order
g(9):=1/5: g(10):=1/10: g(11):=1/15: g(12):=1/30: g(13):=1/20: g(14):=1/40: g(15):=1/60:
g(16):=1/120: g(17):=1/20: #5th Order
g(18):=1/6: g(19):=1/12: g(20):=1/18: g(21):=1/24: g(22):=1/36: g(23):=1/24: g(24):=1/48:
g(25):=1/36: g(26):=1/144: g(27):=1/72: g(28):=1/72: g(29):=1/30: g(30):=1/60: g(31):=1/90:
g(32):=1/180: g(33):=1/120: g(34):=1/120: g(35):=1/240: g(36):=1/360: g(37):=1/720: #6th
```

Equations Computation

```
> for i to stage_max do:
 alg[1]:=alg[1]+b[i]:
 alg[2]:=alg[2]+b[i]*c[i]:
 alg[3]:=alg[3]+b[i]*c[i]^2:
 alg[5]:=alg[5]+b[i]*c[i]^3:
 alg[9]:=alg[9]+b[i]*c[i]^4:
 alg[18]:=alg[18]+b[i]*c[i]^5:
 for j to i-1 do:
  alg[4]:=alg[4]+b[i]*a[i,j]*c[j]:
  alg[6]:=alg[6]+b[i]*c[i]*a[i,j]*c[j]:
  alg[7]:=alg[7]+b[i]*a[i,j]*c[j]^2:
  alg[10]:=alg[10]+b[i]*c[i]^2*a[i,j]*c[j]:
  alg[11]:=alg[11]+b[i]*c[i]*a[i,j]*c[j]^2:
  alg[13]:=alg[13]+b[i]*a[i,j]*c[j]^3:
  alg[19]:=alg[19]+b[i]*c[i]^3*a[i,j]*c[j]:
  alg[20]:=alg[20]+b[i]*c[i]^2*a[i,j]*c[j]^2:
  alg[23]:=alg[23]+b[i]*c[i]*a[i,j]*c[j]^3:
```

```
  alg[29]:=alg[29]+b[i]*a[i,j]*c[j]^4:
  for k to j-1 do:
   alg[8]:=alg[8]+b[i]*a[i,j]*a[j,k]*c[k]:
   alg[12]:=alg[12]+b[i]*c[i]*a[i,j]*a[j,k]*c[k]:
   alg[14]:=alg[14]+b[i]*a[i,j]*c[j]*a[j,k]*c[k]:
   alg[15]:=alg[15]+b[i]*a[i,j]*a[j,k]*c[k]^2:
   alg[22]:=alg[22]+b[i]*c[i]^2*a[i,j]*a[j,k]*c[k]:
   alg[24]:=alg[24]+b[i]*c[i]*a[i,j]*c[j]*a[j,k]*c[k]:
   alg[28]:=alg[28]+b[i]*c[i]*a[i,j]*a[j,k]*c[k]^2:
   alg[30]:=alg[30]+b[i]*a[i,j]*c[j]^2*a[j,k]*c[k]:
   alg[31]:=alg[31]+b[i]*a[i,j]*c[j]*a[j,k]*c[k]^2:
   alg[34]:=alg[34]+b[i]*a[i,j]*a[j,k]*c[k]^3:
   for l to k-1 do:
    alg[16]:=alg[16]+b[i]*a[i,j]*a[j,k]*a[k,l]*c[l]:
    alg[26]:=alg[26]+b[i]*c[i]*a[i,j]*a[j,k]*a[k,l]*c[l]:
    alg[32]:=alg[32]+b[i]*a[i,j]*c[j]*a[j,k]*a[k,l]*c[l]:
    alg[35]:=alg[35]+b[i]*a[i,j]*a[j,k]*c[k]*a[k,l]*c[l]:
    alg[36]:=alg[36]+b[i]*a[i,j]*a[j,k]*a[k,l]*c[l]^2:
    for m to l-1 do:
     alg[37]:=alg[37]+b[i]*a[i,j]*a[j,k]*a[k,l]*a[l,m]*c[m]:
    od:
   od:
  od:


  for jk to j-1 do:
   for jl to j-1 do:
    alg[33]:=alg[33]+b[i]*a[i,j]*a[j,jk]*c[jk]*a[j,jl]*c[jl]:
   od:
  od:
 od:
 for ij to i-1 do:
  for ik to i-1 do:
   alg[17]:=alg[17]+b[i]*a[i,ij]*c[ij]*a[i,ik]*c[ik]:
   alg[21]:=alg[21]+b[i]*c[i]*a[i,ij]*c[ij]*a[i,ik]*c[ik]:
   alg[25]:=alg[25]+b[i]*a[i,ij]*c[ij]*a[i,ik]*c[ik]^2:
   for kl to ik-1 do:
    alg[27]:=alg[27]+b[i]*a[i,ij]*c[ij]*a[i,ik]*a[ik,kl]*c[kl]:
   od:
  od:
 od:
od:
>
> for i to alg_max do
  alg[i]:=simplify(alg[i]-g(i)):
od:
```

Equations Presentation

```
> for i to alg_act do alg[i]:=simplify(alg[i]) od;
```

Ci Substitution

```
> for i to alg_max do alg[i]:=simplify(alg[i]) od:
```

6stage, 4th algebraic (10 indeterminates)

Definitions

```
> stage[]:=array[stage_max]:
> eq[]:=array[1..phl_act]:
> for i from 2 to stage_max do calg[i]:=c[i]-add(a[i,j],j=1..i-1) od;
```

Simplified Algebraic Equations

```
> for i to alg_act do alg[i]:=simplify(alg[i]) od:
```

Stages Computation

```
> for i to stage_max do
  stage[i]:=yn+I*v* add(a[i,j]*stage[j],j=1..i-1):
od:
> stagef:=yn+I*v* add(b[i]*stage[i],i=1..stage_max):
stagef:=collect(simplify(stagef),v):
```

Phase lag Computation

```
> dis_fac:=simplify(stagef/yn):
> dis_A:=simplify(evalc(Re(dis_fac))):
> dis_B:=simplify(evalc(Im(dis_fac)/v)):
> dis_Re:=collect(simplify(dis_A),v):
> dis_Im:=collect(simplify(v*dis_B),v):
> #phl:=tan(v)*dis_Re-dis_Im:
> #dissipation:=1-simplify(sqrt((dis_Im)^2+(dis_Re)^2)):
>
```

Phase lag Equations

```
> for i to phl_act do eq[i]:=simplify(coeff(phl,v,2*i-1)): od:
>
```

Parameters Computation

```
> for i from 1 to alg_max do alg[i]:=simplify(alg[i]): od;
for i from 2 to stage_max do calg[i]:=simplify(calg[i]): od;
> solution:=solve({alg[1]},{b[1]}); assign(solution);
for i from 1 to alg_act do alg[i]:=simplify(alg[i]): od:
for i from 2 to stage_max do calg[i]:=simplify(calg[i]): od:
dis_Re:=simplify(dis_Re): dis_Im:=simplify(dis_Im):
for i to stage_max do; for j to i-1 do;
a[i,j]:=simplify(a[i,j]); od: od:
for i to stage_max do
b[i]:=simplify(b[i]): c[i]:=simplify(c[i]): od:
> solution:=solve({alg[2]},{b[3]}); assign(solution);
for i from 1 to alg_act do alg[i]:=simplify(alg[i]): od:
for i from 2 to stage_max do calg[i]:=simplify(calg[i]): od:
dis_Re:=simplify(dis_Re): dis_Im:=simplify(dis_Im):
for i to stage_max do; for j to i-1 do;
a[i,j]:=simplify(a[i,j]); od: od:
for i to stage_max do
b[i]:=simplify(b[i]): c[i]:=simplify(c[i]): od:
> solution:=solve({alg[3]},{b[4]}); assign(solution);
for i from 1 to alg_act do alg[i]:=simplify(alg[i]): od:
for i from 2 to stage_max do calg[i]:=simplify(calg[i]): od:
dis_Re:=simplify(dis_Re): dis_Im:=simplify(dis_Im):
for i to stage_max do; for j to i-1 do;
a[i,j]:=simplify(a[i,j]); od: od:
for i to stage_max do
b[i]:=simplify(b[i]): c[i]:=simplify(c[i]): od:
> solution:=solve({alg[5]},{b[5]}); assign(solution);
for i from 1 to alg_act do alg[i]:=simplify(alg[i]): od:
for i from 2 to stage_max do calg[i]:=simplify(calg[i]): od:
dis_Re:=simplify(dis_Re): dis_Im:=simplify(dis_Im):
for i to stage_max do; for j to i-1 do;
a[i,j]:=simplify(a[i,j]); od: od:
for i to stage_max do
```

```
b[i]:=simplify(b[i]): c[i]:=simplify(c[i]): od:
> solution:=solve({alg[4]},{a[5,2]}); assign(solution);
for i from 1 to alg_act do alg[i]:=simplify(alg[i]): od:
for i from 2 to stage_max do calg[i]:=simplify(calg[i]): od:
dis_Re:=simplify(dis_Re): dis_Im:=simplify(dis_Im):
for i to stage_max do; for j to i-1 do;
a[i,j]:=simplify(a[i,j]); od: od:
for i to stage_max do
b[i]:=simplify(b[i]): c[i]:=simplify(c[i]): od:
> solution:=solve({alg[6]},{a[6,3]}); assign(solution);
for i from 1 to alg_act do alg[i]:=simplify(alg[i]): od:
for i from 2 to stage_max do calg[i]:=simplify(calg[i]): od:
dis_Re:=simplify(dis_Re): dis_Im:=simplify(dis_Im):
for i to stage_max do; for j to i-1 do;
a[i,j]:=simplify(a[i,j]); od: od:
for i to stage_max do
b[i]:=simplify(b[i]): c[i]:=simplify(c[i]): od:
> solution:=solve({calg[2]},{a[2,1]}); assign(solution);
for i from 1 to alg_act do alg[i]:=simplify(alg[i]): od:
for i from 2 to stage_max do calg[i]:=simplify(calg[i]): od:
dis_Re:=simplify(dis_Re): dis_Im:=simplify(dis_Im):
for i to stage_max do; for j to i-1 do;
a[i,j]:=simplify(a[i,j]); od: od:
for i to stage_max do
b[i]:=simplify(b[i]): c[i]:=simplify(c[i]): od:
> solution:=solve({calg[3]},{a[3,1]}); assign(solution);
for i from 1 to alg_act do alg[i]:=simplify(alg[i]): od:
for i from 2 to stage_max do calg[i]:=simplify(calg[i]): od:
dis_Re:=simplify(dis_Re): dis_Im:=simplify(dis_Im):
for i to stage_max do; for j to i-1 do;
a[i,j]:=simplify(a[i,j]); od: od:
for i to stage_max do
b[i]:=simplify(b[i]): c[i]:=simplify(c[i]): od:
> solution:=solve({calg[4]},{a[4,1]}); assign(solution);
for i from 1 to alg_act do alg[i]:=simplify(alg[i]): od:
for i from 2 to stage_max do calg[i]:=simplify(calg[i]): od:
dis_Re:=simplify(dis_Re): dis_Im:=simplify(dis_Im):
for i to stage_max do; for j to i-1 do;
a[i,j]:=simplify(a[i,j]); od: od:
for i to stage_max do
b[i]:=simplify(b[i]): c[i]:=simplify(c[i]): od:
> solution:=solve({calg[5]},{a[5,1]}); assign(solution);
for i from 1 to alg_act do alg[i]:=simplify(alg[i]): od:
for i from 2 to stage_max do calg[i]:=simplify(calg[i]): od:
dis_Re:=simplify(dis_Re): dis_Im:=simplify(dis_Im):
for i to stage_max do; for j to i-1 do;
a[i,j]:=simplify(a[i,j]); od: od:
for i to stage_max do
b[i]:=simplify(b[i]): c[i]:=simplify(c[i]): od:
> solution:=solve({calg[6]},{a[6,1]}); assign(solution);
for i from 1 to alg_act do alg[i]:=simplify(alg[i]): od:
for i from 2 to stage_max do calg[i]:=simplify(calg[i]): od:
dis_Re:=simplify(dis_Re): dis_Im:=simplify(dis_Im):
for i to stage_max do; for j to i-1 do;
a[i,j]:=simplify(a[i,j]); od: od:
```

```
for i to stage_max do
b[i]:=simplify(b[i]): c[i]:=simplify(c[i]): od:
> solution:=solve({alg[9]},{b[6]}); assign(solution);
for i from 1 to alg_act do alg[i]:=simplify(alg[i]): od:
for i from 2 to stage_max do calg[i]:=simplify(calg[i]): od:
dis_Re:=simplify(dis_Re): dis_Im:=simplify(dis_Im):
for i to stage_max do; for j to i-1 do;
a[i,j]:=simplify(a[i,j]); od: od:
for i to stage_max do
b[i]:=simplify(b[i]): c[i]:=simplify(c[i]): od:
> solution:=solve({alg[11]},{a[5,3]}); assign(solution);
for i from 1 to alg_act do alg[i]:=simplify(alg[i]): od:
for i from 2 to stage_max do calg[i]:=simplify(calg[i]): od:
dis_Re:=simplify(dis_Re): dis_Im:=simplify(dis_Im):
for i to stage_max do; for j to i-1 do;
a[i,j]:=simplify(a[i,j]); od: od:
for i to stage_max do
b[i]:=simplify(b[i]): c[i]:=simplify(c[i]): od:
> solution:=solve({alg[13]},{a[6,4]}); assign(solution);
for i from 1 to alg_act do alg[i]:=simplify(alg[i]): od:
for i from 2 to stage_max do calg[i]:=simplify(calg[i]): od:
dis_Re:=simplify(dis_Re): dis_Im:=simplify(dis_Im):
for i to stage_max do; for j to i-1 do;
a[i,j]:=simplify(a[i,j]); od: od:
for i to stage_max do
b[i]:=simplify(b[i]): c[i]:=simplify(c[i]): od:
> solution:=solve({alg[7]},{a[6,4]}); assign(solution);
for i from 1 to alg_act do alg[i]:=simplify(alg[i]): od:
for i from 2 to stage_max do calg[i]:=simplify(calg[i]): od:
dis_Re:=simplify(dis_Re): dis_Im:=simplify(dis_Im):
for i to stage_max do; for j to i-1 do;
a[i,j]:=simplify(a[i,j]); od: od:
for i to stage_max do
b[i]:=simplify(b[i]): c[i]:=simplify(c[i]): od:
> solution:=solve({alg[8]},{a[4,2]}); assign(solution);
for i from 1 to alg_act do alg[i]:=simplify(alg[i]): od:
for i from 2 to stage_max do calg[i]:=simplify(calg[i]): od:
dis_Re:=simplify(dis_Re): dis_Im:=simplify(dis_Im):
for i to stage_max do; for j to i-1 do;
a[i,j]:=simplify(a[i,j]); od: od:
for i to stage_max do
b[i]:=simplify(b[i]): c[i]:=simplify(c[i]): od:
> solution:=solve({alg[10]},{a[3,2]}); assign(solution);
for i from 1 to alg_act do alg[i]:=simplify(alg[i]): od:
for i from 2 to stage_max do calg[i]:=simplify(calg[i]): od:
dis_Re:=simplify(dis_Re): dis_Im:=simplify(dis_Im):
for i to stage_max do; for j to i-1 do;
a[i,j]:=simplify(a[i,j]); od: od:
for i to stage_max do
b[i]:=simplify(b[i]): c[i]:=simplify(c[i]): od:
> solution:=solve({alg[12]},{a[5,4]}); assign(solution);
for i from 1 to alg_act do alg[i]:=simplify(alg[i]): od:
for i from 2 to stage_max do calg[i]:=simplify(calg[i]): od:
dis_Re:=simplify(dis_Re): dis_Im:=simplify(dis_Im):
for i to stage_max do; for j to i-1 do;
```

```
a[i,j]:=simplify(a[i,j]); od: od:
for i to stage_max do
b[i]:=simplify(b[i]): c[i]:=simplify(c[i]): od:
> solution:=solve({alg[14]},{c[4]}); assign(solution);
for i from 1 to alg_max do alg[i]:=simplify(alg[i]): od:
for i from 2 to stage_max do calg[i]:=simplify(calg[i]): od:
dis_Re:=simplify(dis_Re): dis_Im:=simplify(dis_Im):
for i to stage_max do; for j to i-1 do;
a[i,j]:=simplify(a[i,j]); od: od:
for i to stage_max do
b[i]:=simplify(b[i]): c[i]:=simplify(c[i]): od:
> for i from 1 to alg_max do alg[i]:=simplify(alg[i]): od;
>
> for i to stage_max do;
 for j to i-1 do;
  a[i,j]:=sort(a[i,j],v):
  print(a||i||j, a[i,j]);
  latex(a[i,j]);
 od;
od;
> for i to stage_max do
 b[i]:=simplify(b[i]):
 latex(b[i]);
od;
> for i to stage_max do
 c[i]:=simplify(c[i]):
 latex(c[i]);
od;
>
> dis_Re:=simplify(dis_Re);
dis_Im:=simplify(dis_Im);
> solution:=solve({dis_Im-sin(v)},{a[6,5]}); assign(solution);
for i from 1 to alg_act do alg[i]:=simplify(alg[i]): od:
for i from 2 to stage_max do calg[i]:=simplify(calg[i]): od:
dis_Re:=simplify(dis_Re): dis_Im:=simplify(dis_Im):
for i to stage_max do; for j to i-1 do;
a[i,j]:=simplify(a[i,j]); od: od:
for i to stage_max do
b[i]:=simplify(b[i]): c[i]:=simplify(c[i]): od:
> a[6,5]:=sort(a[6,5],v);
> latex(a[6,5]);
> dis_Re:=simplify(dis_Re);
dis_Im:=simplify(dis_Im);
> solution:=solve({dis_Re-cos(v)},{a[4,3]}); assign(solution);
for i from 1 to alg_act do alg[i]:=simplify(alg[i]): od:
for i from 2 to stage_max do calg[i]:=simplify(calg[i]): od:
dis_Re:=simplify(dis_Re): dis_Im:=simplify(dis_Im):
for i to stage_max do; for j to i-1 do;
a[i,j]:=simplify(a[i,j]); od: od:
for i to stage_max do
b[i]:=simplify(b[i]): c[i]:=simplify(c[i]): od:
> a[4,3]:=sort(simplify(a[4,3]),v);
> latex(a[4,3]);
> dis_Re:=simplify(dis_Re);
dis_Im:=simplify(dis_Im);
```

```
>
```

Solution Confirmation

```
> phl:=simplify(tan(v)*dis_Re-dis_Im);
dissipation:=simplify(1-sqrt((dis_Im)^2+(dis_Re)^2));
> for i from 1 to alg_max do alg[i]:=simplify(alg[i]): od;
for i from 2 to stage_max do calg[i]:=simplify(calg[i]): od;
```

Parameters Presentation

```
> talg[15]:=convert(series(alg[15],v=0,31),polynom):
talg[16]:=convert(series(alg[16],v=0,31),polynom):
talg[15]:=convert(series(alg[15],v=0,11),polynom):
talg[16]:=convert(series(alg[16],v=0,11),polynom):
talg[15]:=convert(series(alg[15],v=0,31),polynom):
talg[16]:=convert(series(alg[16],v=0,31),polynom):
talg[15]:=convert(series(alg[15],v=0,11),polynom);
talg[16]:=convert(series(alg[16],v=0,11),polynom);
> a[4,3]:=simplify(a[4,3]);
> s2taylor(a[4,3],v,17);
> s2taylor(a[6,5],v,17);
>
```

Appendix C

Maple Program for the development of explicit Runge-Kutta method with minimal Dispersion.

```
> restart:
```

Tree Theory

Simplifying Assumptions

```
> b[2]:=0:
c[2]:=321/1000000;
c[3]:=13/20;
c[5]:=1/6;
c[6]:=1;
```

Input

```
> Digits:=20:
> stages:=6: #Enter number of stages
> alg_max:=17: #Enter number of algebraic order equations (maximum 17 for 5th order)
> alg_act:=8: #Enter number of active algebraic order equations (eg 8 for 4th order, max 17)
> phl_act:=10: #Enter number of active phase lag order equations (usually greater than #number of indeterminates because some coefficients vanish while using the #algebraic equations)
```

Definitions

```
> c[]:=array[1..stages];
c[1]:=0;
a[]:=array[1..stages, 1..stages-1];
alg[]:=array[37];
for i to stages do
 for j to stages do
  if j>=i then a[i,j]:=0; fi;
 od:
od:
```

```
for i from 1 to 37 do alg[i]:=0 od:
>
> g(1):=1: #1st Order
g(2):=1/2: #2nd Order
g(3):=1/3: g(4):=1/6: #3rd Order
g(5):=1/4: g(6):=1/8: g(7):=1/12: g(8):=1/24: #4th Order
g(9):=1/5: g(10):=1/10: g(11):=1/15: g(12):=1/30: g(13):=1/20: g(14):=1/40: g(15):=1/60:
g(16):=1/120: g(17):=1/20: #5th Order
g(18):=1/6: g(19):=1/12: g(20):=1/18: g(21):=1/24: g(22):=1/36: g(23):=1/24: g(24):=1/48:
g(25):=1/36: g(26):=1/144: g(27):=1/72: g(28):=1/72: g(29):=1/30: g(30):=1/60: g(31):=1/90:
g(32):=1/180: g(33):=1/120: g(34):=1/120: g(35):=1/240: g(36):=1/360: g(37):=1/720: #6th
```

Equations Computation

```
> for i to stages do:
 alg[1]:=alg[1]+b[i]:
 alg[2]:=alg[2]+b[i]*c[i]:
 alg[3]:=alg[3]+b[i]*c[i]^2:
 alg[5]:=alg[5]+b[i]*c[i]^3:
 alg[9]:=alg[9]+b[i]*c[i]^4:
 alg[18]:=alg[18]+b[i]*c[i]^5:
 for j to i-1 do:
  alg[4]:=alg[4]+b[i]*a[i,j]*c[j]:
  alg[6]:=alg[6]+b[i]*c[i]*a[i,j]*c[j]:
  alg[7]:=alg[7]+b[i]*a[i,j]*c[j]^2:
  alg[10]:=alg[10]+b[i]*c[i]^2*a[i,j]*c[j]:
  alg[11]:=alg[11]+b[i]*c[i]*a[i,j]*c[j]^2:
  alg[13]:=alg[13]+b[i]*a[i,j]*c[j]^3:
  alg[19]:=alg[19]+b[i]*c[i]^3*a[i,j]*c[j]:
  alg[20]:=alg[20]+b[i]*c[i]^2*a[i,j]*c[j]^2:
  alg[23]:=alg[23]+b[i]*c[i]*a[i,j]*c[j]^3:
  alg[29]:=alg[29]+b[i]*a[i,j]*c[j]^4:
  for k to j-1 do:
   alg[8]:=alg[8]+b[i]*a[i,j]*a[j,k]*c[k]:
   alg[12]:=alg[12]+b[i]*c[i]*a[i,j]*a[j,k]*c[k]:
   alg[14]:=alg[14]+b[i]*a[i,j]*c[j]*a[j,k]*c[k]:
   alg[15]:=alg[15]+b[i]*a[i,j]*a[j,k]*c[k]^2:
   alg[22]:=alg[22]+b[i]*c[i]^2*a[i,j]*a[j,k]*c[k]:
   alg[24]:=alg[24]+b[i]*c[i]*a[i,j]*c[j]*a[j,k]*c[k]:
   alg[28]:=alg[28]+b[i]*c[i]*a[i,j]*a[j,k]*c[k]^2:
   alg[30]:=alg[30]+b[i]*a[i,j]*c[j]^2*a[j,k]*c[k]:
   alg[31]:=alg[31]+b[i]*a[i,j]*c[j]*a[j,k]*c[k]^2:
   alg[34]:=alg[34]+b[i]*a[i,j]*a[j,k]*c[k]^3:
   for l to k-1 do:
    alg[16]:=alg[16]+b[i]*a[i,j]*a[j,k]*a[k,l]*c[l]:
    alg[26]:=alg[26]+b[i]*c[i]*a[i,j]*a[j,k]*a[k,l]*c[l]:
    alg[32]:=alg[32]+b[i]*a[i,j]*c[j]*a[j,k]*a[k,l]*c[l]:
    alg[35]:=alg[35]+b[i]*a[i,j]*a[j,k]*c[k]*a[k,l]*c[l]:
    alg[36]:=alg[36]+b[i]*a[i,j]*a[j,k]*a[k,l]*c[l]^2:
    for m to l-1 do:
     alg[37]:=alg[37]+b[i]*a[i,j]*a[j,k]*a[k,l]*a[l,m]*c[m]:
     od:
    od:
  od:
  for jk to j-1 do:
   for jl to j-1 do:
```

```
      alg[33]:=alg[33]+b[i]*a[i,j]*a[j,jk]*c[jk]*a[j,jl]*c[jl]:
     od:
   od:
 od:

 for ij to i-1 do:
   for ik to i-1 do:

     alg[17]:=alg[17]+b[i]*a[i,ij]*c[ij]*a[i,ik]*c[ik]:
     alg[21]:=alg[21]+b[i]*c[i]*a[i,ij]*c[ij]*a[i,ik]*c[ik]:
     alg[25]:=alg[25]+b[i]*a[i,ij]*c[ij]*a[i,ik]*c[ik]^2:

     for kl to ik-1 do:
       alg[27]:=alg[27]+b[i]*a[i,ij]*c[ij]*a[i,ik]*a[ik,kl]*c[kl]:
     od:
   od:
 od:
od:
>
> for i to alg_max do
 alg[i]:=simplify(alg[i]-g(i)):
 od:
```

Equations Presentation

```
> for i to alg_act do alg[i]:=simplify(alg[i]) od;
```

Ci Substitution

```
> for i to alg_max do alg[i]:=simplify(alg[i]) od:
```

6stage, 4th algebraic (10 indeterminates)

Definitions

```
> stage[]:=array[stages]:
> eq[]:=array[1..phl_act]:
> for i from 2 to stages do calg[i]:=c[i]-add(a[i,j],j=1..i-1) od;
```

Simplified Algebraic Equations

```
> for i to alg_act do alg[i]:=simplify(alg[i]) od:
```

Stages Computation

```
> for i to stages do
 stage[i]:=yn+I*v* add(a[i,j]*stage[j],j=1..i-1):
od:
> stagef:=yn+I*v* add(b[i]*stage[i],i=1..stages):
stagef:=collect(simplify(stagef),v):
```

Phase lag Computation

```
> dis_fac:=simplify(stagef/yn):
> dis_A:=simplify(evalc(Re(dis_fac))):
> dis_B:=simplify(evalc(Im(dis_fac)/v)):
> dis_Re:=collect(simplify(dis_A),v):
> dis_Im:=collect(simplify(v*dis_B),v):
> #phl:=tan(v)*dis_Re-dis_Im:
> #dissipation:=1-simplify(sqrt((dis_Im)^2+(dis_Re)^2)):
>
```

Phase lag Equations

```
> for i to phl_act do eq[i]:=simplify(coeff(phl,v,2*i-1)): od:
>
```

Parameters Computation

```
> for i from 1 to alg_max do alg[i]:=simplify(alg[i]): od;
for i from 2 to stages do calg[i]:=simplify(calg[i]): od;
> solution:=solve({alg[1]},{b[1]}); assign(solution);
for i from 1 to alg_act do alg[i]:=simplify(alg[i]): od:
for i from 2 to stages do calg[i]:=simplify(calg[i]): od:
dis_Re:=simplify(dis_Re): dis_Im:=simplify(dis_Im):
for i to stages do; for j to i-1 do;
a[i,j]:=simplify(a[i,j]); od: od:
for i to stages do
b[i]:=simplify(b[i]): c[i]:=simplify(c[i]): od:
> solution:=solve({alg[2]},{b[3]}); assign(solution);
for i from 1 to alg_act do alg[i]:=simplify(alg[i]): od:
for i from 2 to stages do calg[i]:=simplify(calg[i]): od:
dis_Re:=simplify(dis_Re): dis_Im:=simplify(dis_Im):
for i to stages do; for j to i-1 do;
a[i,j]:=simplify(a[i,j]); od: od:
for i to stages do
b[i]:=simplify(b[i]): c[i]:=simplify(c[i]): od:
> solution:=solve({alg[3]},{b[4]}); assign(solution);
for i from 1 to alg_act do alg[i]:=simplify(alg[i]): od:
for i from 2 to stages do calg[i]:=simplify(calg[i]): od:
dis_Re:=simplify(dis_Re): dis_Im:=simplify(dis_Im):
for i to stages do; for j to i-1 do;
a[i,j]:=simplify(a[i,j]); od: od:
for i to stages do
b[i]:=simplify(b[i]): c[i]:=simplify(c[i]): od:
> solution:=solve({alg[5]},{b[5]}); assign(solution);
for i from 1 to alg_act do alg[i]:=simplify(alg[i]): od:
for i from 2 to stages do calg[i]:=simplify(calg[i]): od:
dis_Re:=simplify(dis_Re): dis_Im:=simplify(dis_Im):
for i to stages do; for j to i-1 do;
a[i,j]:=simplify(a[i,j]); od: od:
for i to stages do
b[i]:=simplify(b[i]): c[i]:=simplify(c[i]): od:
> solution:=solve({alg[4]},{a[5,2]}); assign(solution);
for i from 1 to alg_act do alg[i]:=simplify(alg[i]): od:
for i from 2 to stages do calg[i]:=simplify(calg[i]): od:
dis_Re:=simplify(dis_Re): dis_Im:=simplify(dis_Im):
for i to stages do; for j to i-1 do;
a[i,j]:=simplify(a[i,j]); od: od:
for i to stages do
b[i]:=simplify(b[i]): c[i]:=simplify(c[i]): od:
> solution:=solve({alg[6]},{a[6,3]}); assign(solution);
for i from 1 to alg_act do alg[i]:=simplify(alg[i]): od:
for i from 2 to stages do calg[i]:=simplify(calg[i]): od:
dis_Re:=simplify(dis_Re): dis_Im:=simplify(dis_Im):
for i to stages do; for j to i-1 do;
a[i,j]:=simplify(a[i,j]); od: od:
for i to stages do
b[i]:=simplify(b[i]): c[i]:=simplify(c[i]): od:
> solution:=solve({calg[2]},{a[2,1]}); assign(solution);
for i from 1 to alg_act do alg[i]:=simplify(alg[i]): od:
for i from 2 to stages do calg[i]:=simplify(calg[i]): od:
```

```
dis_Re:=simplify(dis_Re): dis_Im:=simplify(dis_Im):
for i to stages do; for j to i-1 do;
a[i,j]:=simplify(a[i,j]); od: od:
for i to stages do
b[i]:=simplify(b[i]): c[i]:=simplify(c[i]): od:
> solution:=solve({calg[3]},{a[3,1]}); assign(solution);
for i from 1 to alg_act do alg[i]:=simplify(alg[i]): od:
for i from 2 to stages do calg[i]:=simplify(calg[i]): od:
dis_Re:=simplify(dis_Re): dis_Im:=simplify(dis_Im):
for i to stages do; for j to i-1 do;
a[i,j]:=simplify(a[i,j]); od: od:
for i to stages do
b[i]:=simplify(b[i]): c[i]:=simplify(c[i]): od:
> solution:=solve({calg[4]},{a[4,1]}); assign(solution);
for i from 1 to alg_act do alg[i]:=simplify(alg[i]): od:
for i from 2 to stages do calg[i]:=simplify(calg[i]): od:
dis_Re:=simplify(dis_Re): dis_Im:=simplify(dis_Im):
for i to stages do; for j to i-1 do;
a[i,j]:=simplify(a[i,j]); od: od:
for i to stages do
b[i]:=simplify(b[i]): c[i]:=simplify(c[i]): od:
> solution:=solve({calg[5]},{a[5,1]}); assign(solution);
for i from 1 to alg_act do alg[i]:=simplify(alg[i]): od:
for i from 2 to stages do calg[i]:=simplify(calg[i]): od:
dis_Re:=simplify(dis_Re): dis_Im:=simplify(dis_Im):
for i to stages do; for j to i-1 do;
a[i,j]:=simplify(a[i,j]); od: od:
for i to stages do
b[i]:=simplify(b[i]): c[i]:=simplify(c[i]): od:
> solution:=solve({calg[6]},{a[6,1]}); assign(solution);
for i from 1 to alg_act do alg[i]:=simplify(alg[i]): od:
for i from 2 to stages do calg[i]:=simplify(calg[i]): od:
dis_Re:=simplify(dis_Re): dis_Im:=simplify(dis_Im):
for i to stages do; for j to i-1 do;
a[i,j]:=simplify(a[i,j]); od: od:
for i to stages do
b[i]:=simplify(b[i]): c[i]:=simplify(c[i]): od:
> solution:=solve({alg[9]},{b[6]}); assign(solution);
for i from 1 to alg_act do alg[i]:=simplify(alg[i]): od:
for i from 2 to stages do calg[i]:=simplify(calg[i]): od:
dis_Re:=simplify(dis_Re): dis_Im:=simplify(dis_Im):
for i to stages do; for j to i-1 do;
a[i,j]:=simplify(a[i,j]); od: od:
for i to stages do
b[i]:=simplify(b[i]): c[i]:=simplify(c[i]): od:
> solution:=solve({alg[11]},{a[5,3]}); assign(solution);
for i from 1 to alg_act do alg[i]:=simplify(alg[i]): od:
for i from 2 to stages do calg[i]:=simplify(calg[i]): od:
dis_Re:=simplify(dis_Re): dis_Im:=simplify(dis_Im):
for i to stages do; for j to i-1 do;
a[i,j]:=simplify(a[i,j]); od: od:
for i to stages do
b[i]:=simplify(b[i]): c[i]:=simplify(c[i]): od:
> solution:=solve({alg[13]},{a[6,4]}); assign(solution);
```

```
for i from 1 to alg_act do alg[i]:=simplify(alg[i]): od:
for i from 2 to stages do calg[i]:=simplify(calg[i]): od:
dis_Re:=simplify(dis_Re): dis_Im:=simplify(dis_Im):
for i to stages do; for j to i-1 do;
a[i,j]:=simplify(a[i,j]); od: od:
for i to stages do
b[i]:=simplify(b[i]): c[i]:=simplify(c[i]): od:
> solution:=solve({alg[7]},{a[6,2]}); assign(solution);
for i from 1 to alg_act do alg[i]:=simplify(alg[i]): od:
for i from 2 to stages do calg[i]:=simplify(calg[i]): od:
dis_Re:=simplify(dis_Re): dis_Im:=simplify(dis_Im):
for i to stages do; for j to i-1 do;
a[i,j]:=simplify(a[i,j]); od: od:
for i to stages do
b[i]:=simplify(b[i]): c[i]:=simplify(c[i]): od:
> solution:=solve({alg[8]},{a[4,2]}); assign(solution);
for i from 1 to alg_act do alg[i]:=simplify(alg[i]): od:
for i from 2 to stages do calg[i]:=simplify(calg[i]): od:
dis_Re:=simplify(dis_Re): dis_Im:=simplify(dis_Im):
for i to stages do; for j to i-1 do;
a[i,j]:=simplify(a[i,j]); od: od:
for i to stages do
b[i]:=simplify(b[i]): c[i]:=simplify(c[i]): od:
> solution:=solve({alg[10]},{a[3,2]}); assign(solution);
for i from 1 to alg_act do alg[i]:=simplify(alg[i]): od:
for i from 2 to stages do calg[i]:=simplify(calg[i]): od:
dis_Re:=simplify(dis_Re): dis_Im:=simplify(dis_Im):
for i to stages do; for j to i-1 do;
a[i,j]:=simplify(a[i,j]); od: od:
for i to stages do
b[i]:=simplify(b[i]): c[i]:=simplify(c[i]): od:
> solution:=solve({alg[12]},{a[5,4]}); assign(solution);
for i from 1 to alg_act do alg[i]:=simplify(alg[i]): od:
for i from 2 to stages do calg[i]:=simplify(calg[i]): od:
dis_Re:=simplify(dis_Re): dis_Im:=simplify(dis_Im):
for i to stages do; for j to i-1 do;
a[i,j]:=simplify(a[i,j]); od: od:
for i to stages do
b[i]:=simplify(b[i]): c[i]:=simplify(c[i]): od:
> solution:=solve({alg[14]},{c[4]}); assign(solution);
for i from 1 to alg_max do alg[i]:=simplify(alg[i]): od:
for i from 2 to stages do calg[i]:=simplify(calg[i]): od:
dis_Re:=simplify(dis_Re): dis_Im:=simplify(dis_Im):
for i to stages do; for j to i-1 do;
a[i,j]:=simplify(a[i,j]); od: od:
for i to stages do
b[i]:=simplify(b[i]): c[i]:=simplify(c[i]): od:
> for i from 1 to alg_max do alg[i]:=simplify(alg[i]): od;
>
> for i to stages do;
 for j to i-1 do;
  a[i,j]:=sort(a[i,j],v):
  print(a||i||j, a[i,j]);
  latex(a[i,j]);
 od;
```

```
od;
> for i to stages do
 b[i]:=simplify(b[i]):
 latex(b[i]);
od;
> for i to stages do
 c[i]:=simplify(c[i]):
 latex(c[i]);
od;
>
> dis_Re:=simplify(dis_Re);
dis_Im:=simplify(dis_Im);
> phl:=tan(v)*dis_Re-dis_Im;
> dissipation:=1-simplify(sqrt((dis_Im)^2+(dis_Re)^2));
> phl:=staylor(phl,v,12);
> dissipation:=staylor(dissipation,v,12);
> solution:=solve({coeff(phl,v,5)},{a[6,5]}); assign(solution);
for i from 1 to alg_act do alg[i]:=simplify(alg[i]): od:
for i from 2 to stages do calg[i]:=simplify(calg[i]): od:
dis_Re:=simplify(dis_Re): dis_Im:=simplify(dis_Im):
for i to stages do; for j to i-1 do;
a[i,j]:=simplify(a[i,j]); od: od:
for i to stages do
b[i]:=simplify(b[i]): c[i]:=simplify(c[i]): od:
> solution:=solve({coeff(phl,v,7)},{a[4,3]}); assign(solution);
for i from 1 to alg_act do alg[i]:=simplify(alg[i]): od:
for i from 2 to stages do calg[i]:=simplify(calg[i]): od:
dis_Re:=simplify(dis_Re): dis_Im:=simplify(dis_Im):
for i to stages do; for j to i-1 do;
a[i,j]:=simplify(a[i,j]); od: od:
for i to stages do
b[i]:=simplify(b[i]): c[i]:=simplify(c[i]): od:
>
```

Solution Confirmation

```
> phl;
  dissipation;
> for i from 1 to alg_max do alg[i]:=simplify(alg[i]): od;
for i from 2 to stages do calg[i]:=simplify(calg[i]): od;
```

Parameters Presentation

```
> for i to stages do;
 for j to i-1 do;
  a[i,j]:=sort(a[i,j],v):
  print(allillj, a[i,j]);
  latex(a[i,j]);
 od;
od;
> for i to stages do
 b[i]:=simplify(b[i]):
 latex(b[i]);
od;
> for i to stages do
 c[i]:=simplify(c[i]):
 latex(c[i]);
od;
```

```
>
> B:={seq(add(b[i]*c[i]^(k-1), i=1..stages)=1/k, k=1..stages)}:
> C:=seq(seq(add(a[i,j]*c[j]^(k-1), j=1..stages)=1/k*c[i]^k, k=1..stages), i=1..stages)}:
> solution:=solve(B union C):
> assign(solution);
> printf("A=[\n");
for i to stages do
 for j to stages do
  printf("%19.17g,\t", evalf(a[i,j]));
 od;
 printf("\n");
od;
printf("];\n \nb=[");
for i to stages do
 printf("%19.17g,\t", evalf(b[i]));
od;
printf("];\n \nc=[");
for i to stages do
  printf("%19.17g,\t",evalf(c[i]));
od;
printf("];");
>
```

Appendix D

Maple Program for the development of explicit Runge-Kutta method with minimal Dissipation

```
> restart;
```

Tree Theory

Simplifying Assumptions

```
> b[2]:=0:
c[2]:=321/1000000;
c[3]:=13/20;
c[5]:=1/6;
c[6]:=1;
```

Input

```
> Digits:=20:
> stages:=6: #Enter number of stages
> alg_max:=17: #Enter number of algebraic order equations (maximum 17 for 5th order)
> alg_act:=8: #Enter number of active algebraic order equations (eg 8 for 4th order, max 17)
> phl_act:=10: #Enter number of active phase lag order equations (usually greater than #number of indeterminates because some coefficients vanish while using the #algebraic equations)
```

Definitions

```
> c[]:=array[1..stages];
c[1]:=0;
a[]:=array[1..stages, 1..stages-1];
alg[]:=array[37];
for i to stages do
 for j to stages do
  if j>=i then a[i,j]:=0; fi;
```

```
 od:
od:
for i from 1 to 37 do alg[i]:=0 od:
>
> g(1):=1: #1st Order
g(2):=1/2: #2nd Order
g(3):=1/3: g(4):=1/6: #3rd Order
g(5):=1/4: g(6):=1/8: g(7):=1/12: g(8):=1/24: #4th Order
g(9):=1/5: g(10):=1/10: g(11):=1/15: g(12):=1/30: g(13):=1/20: g(14):=1/40: g(15):=1/60:
g(16):=1/120: g(17):=1/20: #5th Order
g(18):=1/6: g(19):=1/12: g(20):=1/18: g(21):=1/24: g(22):=1/36: g(23):=1/24: g(24):=1/48:
g(25):=1/36: g(26):=1/144: g(27):=1/72: g(28):=1/72: g(29):=1/30: g(30):=1/60: g(31):=1/90:
g(32):=1/180: g(33):=1/120: g(34):=1/120: g(35):=1/240: g(36):=1/360: g(37):=1/720: #6th
```

Equations Computation

```
> for i to stages do:
 alg[1]:=alg[1]+b[i]:
 alg[2]:=alg[2]+b[i]*c[i]:
 alg[3]:=alg[3]+b[i]*c[i]^2:
 alg[5]:=alg[5]+b[i]*c[i]^3:
 alg[9]:=alg[9]+b[i]*c[i]^4:
 alg[18]:=alg[18]+b[i]*c[i]^5:
for j to i-1 do:
  alg[4]:=alg[4]+b[i]*a[i,j]*c[j]:
  alg[6]:=alg[6]+b[i]*c[i]*a[i,j]*c[j]:
  alg[7]:=alg[7]+b[i]*a[i,j]*c[j]^2:
  alg[10]:=alg[10]+b[i]*c[i]^2*a[i,j]*c[j]:
  alg[11]:=alg[11]+b[i]*c[i]*a[i,j]*c[j]^2:
  alg[13]:=alg[13]+b[i]*a[i,j]*c[j]^3:
  alg[19]:=alg[19]+b[i]*c[i]^3*a[i,j]*c[j]:
  alg[20]:=alg[20]+b[i]*c[i]^2*a[i,j]*c[j]^2:
  alg[23]:=alg[23]+b[i]*c[i]*a[i,j]*c[j]^3:
  alg[29]:=alg[29]+b[i]*a[i,j]*c[j]^4:
  for k to j-1 do:
   alg[8]:=alg[8]+b[i]*a[i,j]*a[j,k]*c[k]:
   alg[12]:=alg[12]+b[i]*c[i]*a[i,j]*a[j,k]*c[k]:
   alg[14]:=alg[14]+b[i]*a[i,j]*c[j]*a[j,k]*c[k]:
   alg[15]:=alg[15]+b[i]*a[i,j]*a[j,k]*c[k]^2:
   alg[22]:=alg[22]+b[i]*c[i]^2*a[i,j]*a[j,k]*c[k]:
   alg[24]:=alg[24]+b[i]*c[i]*a[i,j]*c[j]*a[j,k]*c[k]:
   alg[28]:=alg[28]+b[i]*c[i]*a[i,j]*a[j,k]*c[k]^2:
   alg[30]:=alg[30]+b[i]*a[i,j]*c[j]^2*a[j,k]*c[k]:
   alg[31]:=alg[31]+b[i]*a[i,j]*c[j]*a[j,k]*c[k]^2:
   alg[34]:=alg[34]+b[i]*a[i,j]*a[j,k]*c[k]^3:
   for l to k-1 do:
    alg[16]:=alg[16]+b[i]*a[i,j]*a[j,k]*a[k,l]*c[l]:
    alg[26]:=alg[26]+b[i]*c[i]*a[i,j]*a[j,k]*a[k,l]*c[l]:
    alg[32]:=alg[32]+b[i]*a[i,j]*c[j]*a[j,k]*a[k,l]*c[l]:
    alg[35]:=alg[35]+b[i]*a[i,j]*a[j,k]*c[k]*a[k,l]*c[l]:
    alg[36]:=alg[36]+b[i]*a[i,j]*a[j,k]*a[k,l]*c[l]^2:
    for m to l-1 do:
     alg[37]:=alg[37]+b[i]*a[i,j]*a[j,k]*a[k,l]*a[l,m]*c[m]:
    od:
   od:
  od:
```

```
  for jk to j-1 do:
   for jl to j-1 do:
    alg[33]:=alg[33]+b[i]*a[i,j]*a[j,jk]*c[jk]*a[j,jl]*c[jl]:
   od:
  od:
 od:
 for ij to i-1 do:
  for ik to i-1 do:
   alg[17]:=alg[17]+b[i]*a[i,ij]*c[ij]*a[i,ik]*c[ik]:
   alg[21]:=alg[21]+b[i]*c[i]*a[i,ij]*c[ij]*a[i,ik]*c[ik]:
   alg[25]:=alg[25]+b[i]*a[i,ij]*c[ij]*a[i,ik]*c[ik]^2:
   for kl to ik-1 do:
    alg[27]:=alg[27]+b[i]*a[i,ij]*c[ij]*a[i,ik]*a[ik,kl]*c[kl]:
   od:
  od:
 od:
od:
>
> for i to alg_max do
 alg[i]:=simplify(alg[i]-g(i)):
od:
```

Equations Presentation

```
> for i to alg_act do alg[i]:=simplify(alg[i]) od;
```

Ci Substitution

```
> for i to alg_max do alg[i]:=simplify(alg[i]) od:
6stage, 4th algebraic (10 indeterminates)
```

Definitions

```
> stage[]:=array[stages]:
> eq[]:=array[1..phl_act]:
> for i from 2 to stages do calg[i]:=c[i]-add(a[i,j],j=1..i-1) od;
```

Simplified Algebraic Equations

```
> for i to alg_act do alg[i]:=simplify(alg[i]) od:
```

Stages Computation

```
> for i to stages do
 stage[i]:=yn+I*v* add(a[i,j]*stage[j],j=1..i-1):
od:
> stagef:=yn+I*v* add(b[i]*stage[i],i=1..stages):
stagef:=collect(simplify(stagef),v):
```

Phase lag Computation

```
> dis_fac:=simplify(stagef/yn):
> dis_A:=simplify(evalc(Re(dis_fac))):
> dis_B:=simplify(evalc(Im(dis_fac)/v)):
> dis_Re:=collect(simplify(dis_A),v):
> dis_Im:=collect(simplify(v*dis_B),v):
> #phl:=tan(v)*dis_Re-dis_Im:
> #dissipation:=1-simplify(sqrt((dis_Im)^2+(dis_Re)^2)):
>
```

Phase lag Equations

```
> for i to phl_act do eq[i]:=simplify(coeff(phl,v,2*i-1)): od:
>
```

Parameters Computation

```
> for i from 1 to alg_max do alg[i]:=simplify(alg[i]): od;
for i from 2 to stages do calg[i]:=simplify(calg[i]): od;
```

```
> solution:={solve(alg[1]},{b[1]}); assign(solution);
for i from 1 to alg_act do alg[i]:=simplify(alg[i]): od:
for i from 2 to stages do calg[i]:=simplify(calg[i]): od:
dis_Re:=simplify(dis_Re): dis_Im:=simplify(dis_Im):
for i to stages do; for j to i-1 do;
a[i,j]:=simplify(a[i,j]); od: od:
for i to stages do
b[i]:=simplify(b[i]): c[i]:=simplify(c[i]): od:
> solution:=solve({alg[2]},{b[3]}); assign(solution);
for i from 1 to alg_act do alg[i]:=simplify(alg[i]): od:
for i from 2 to stages do calg[i]:=simplify(calg[i]): od:
dis_Re:=simplify(dis_Re): dis_Im:=simplify(dis_Im):
for i to stages do; for j to i-1 do;
a[i,j]:=simplify(a[i,j]); od: od:
for i to stages do
b[i]:=simplify(b[i]): c[i]:=simplify(c[i]): od:
> solution:=solve({alg[3]},{b[4]}); assign(solution);
for i from 1 to alg_act do alg[i]:=simplify(alg[i]): od:
for i from 2 to stages do calg[i]:=simplify(calg[i]): od:
dis_Re:=simplify(dis_Re): dis_Im:=simplify(dis_Im):
for i to stages do; for j to i-1 do;
a[i,j]:=simplify(a[i,j]); od: od:
for i to stages do
b[i]:=simplify(b[i]): c[i]:=simplify(c[i]): od:
> solution:=solve({alg[5]},{b[5]}); assign(solution);
for i from 1 to alg_act do alg[i]:=simplify(alg[i]): od:
for i from 2 to stages do calg[i]:=simplify(calg[i]): od:
dis_Re:=simplify(dis_Re): dis_Im:=simplify(dis_Im):
for i to stages do; for j to i-1 do;
a[i,j]:=simplify(a[i,j]); od: od:
for i to stages do
b[i]:=simplify(b[i]): c[i]:=simplify(c[i]): od:
> solution:=solve({alg[4]},{a[5,2]}); assign(solution);
for i from 1 to alg_act do alg[i]:=simplify(alg[i]): od:
for i from 2 to stages do calg[i]:=simplify(calg[i]): od:
dis_Re:=simplify(dis_Re): dis_Im:=simplify(dis_Im):
for i to stages do; for j to i-1 do;
a[i,j]:=simplify(a[i,j]); od: od:
for i to stages do
b[i]:=simplify(b[i]): c[i]:=simplify(c[i]): od:
> solution:=solve({alg[6]},{a[6,3]}); assign(solution);
for i from 1 to alg_act do alg[i]:=simplify(alg[i]): od:
for i from 2 to stages do calg[i]:=simplify(calg[i]): od:
dis_Re:=simplify(dis_Re): dis_Im:=simplify(dis_Im):
for i to stages do; for j to i-1 do;
a[i,j]:=simplify(a[i,j]); od: od:
for i to stages do
b[i]:=simplify(b[i]): c[i]:=simplify(c[i]): od:
> solution:=solve({calg[2]},{a[2,1]}); assign(solution);
for i from 1 to alg_act do alg[i]:=simplify(alg[i]): od:
for i from 2 to stages do calg[i]:=simplify(calg[i]): od:
dis_Re:=simplify(dis_Re): dis_Im:=simplify(dis_Im):
for i to stages do; for j to i-1 do;
a[i,j]:=simplify(a[i,j]); od: od:
for i to stages do
```

```
b[i]:=simplify(b[i]): c[i]:=simplify(c[i]): od:
> solution:=solve({calg[3]},{a[3,1]}); assign(solution);
for i from 1 to alg_act do alg[i]:=simplify(alg[i]): od:
for i from 2 to stages do calg[i]:=simplify(calg[i]): od:
dis_Re:=simplify(dis_Re): dis_Im:=simplify(dis_Im):
for i to stages do; for j to i-1 do;
a[i,j]:=simplify(a[i,j]); od: od:
for i to stages do
b[i]:=simplify(b[i]): c[i]:=simplify(c[i]): od:
> solution:=solve({calg[4]},{a[4,1]}); assign(solution);
for i from 1 to alg_act do alg[i]:=simplify(alg[i]): od:
for i from 2 to stages do calg[i]:=simplify(calg[i]): od:
dis_Re:=simplify(dis_Re): dis_Im:=simplify(dis_Im):
for i to stages do; for j to i-1 do;
a[i,j]:=simplify(a[i,j]); od: od:
for i to stages do
b[i]:=simplify(b[i]): c[i]:=simplify(c[i]): od:
> solution:=solve({calg[5]},{a[5,1]}); assign(solution);
for i from 1 to alg_act do alg[i]:=simplify(alg[i]): od:
for i from 2 to stages do calg[i]:=simplify(calg[i]): od:
dis_Re:=simplify(dis_Re): dis_Im:=simplify(dis_Im):
for i to stages do; for j to i-1 do;
a[i,j]:=simplify(a[i,j]); od: od:
for i to stages do
b[i]:=simplify(b[i]): c[i]:=simplify(c[i]): od:
> solution:=solve({calg[6]},{a[6,1]}); assign(solution);
for i from 1 to alg_act do alg[i]:=simplify(alg[i]): od:
for i from 2 to stages do calg[i]:=simplify(calg[i]): od:
dis_Re:=simplify(dis_Re): dis_Im:=simplify(dis_Im):
for i to stages do; for j to i-1 do;
a[i,j]:=simplify(a[i,j]); od: od:
for i to stages do
b[i]:=simplify(b[i]): c[i]:=simplify(c[i]): od:
> solution:=solve({alg[9]},{b[6]}); assign(solution);
for i from 1 to alg_act do alg[i]:=simplify(alg[i]): od:
for i from 2 to stages do calg[i]:=simplify(calg[i]): od:
dis_Re:=simplify(dis_Re): dis_Im:=simplify(dis_Im):
for i to stages do; for j to i-1 do;
a[i,j]:=simplify(a[i,j]); od: od:
for i to stages do
b[i]:=simplify(b[i]): c[i]:=simplify(c[i]): od:
> solution:=solve({alg[11]},{a[5,3]}); assign(solution);
for i from 1 to alg_act do alg[i]:=simplify(alg[i]): od:
for i from 2 to stages do calg[i]:=simplify(calg[i]): od:
dis_Re:=simplify(dis_Re): dis_Im:=simplify(dis_Im):
for i to stages do; for j to i-1 do;
a[i,j]:=simplify(a[i,j]); od: od:
for i to stages do
b[i]:=simplify(b[i]): c[i]:=simplify(c[i]): od:
> solution:=solve({alg[13]},{a[6,4]}); assign(solution);
for i from 1 to alg_act do alg[i]:=simplify(alg[i]): od:
for i from 2 to stages do calg[i]:=simplify(calg[i]): od:
dis_Re:=simplify(dis_Re): dis_Im:=simplify(dis_Im):
for i to stages do; for j to i-1 do;
a[i,j]:=simplify(a[i,j]); od: od:
```

```
for i to stages do
b[i]:=simplify(b[i]): c[i]:=simplify(c[i]): od:
> solution:=solve({alg[7]},{a[6,2]}); assign(solution);
for i from 1 to alg_act do alg[i]:=simplify(alg[i]): od:
for i from 2 to stages do calg[i]:=simplify(calg[i]): od:
dis_Re:=simplify(dis_Re): dis_Im:=simplify(dis_Im):
for i to stages do; for j to i-1 do;
a[i,j]:=simplify(a[i,j]); od: od:
for i to stages do
b[i]:=simplify(b[i]): c[i]:=simplify(c[i]): od:
> solution:=solve({alg[8]},{a[4,2]}); assign(solution);
for i from 1 to alg_act do alg[i]:=simplify(alg[i]): od:
for i from 2 to stages do calg[i]:=simplify(calg[i]): od:
dis_Re:=simplify(dis_Re): dis_Im:=simplify(dis_Im):
for i to stages do; for j to i-1 do;
a[i,j]:=simplify(a[i,j]); od: od:
for i to stages do
b[i]:=simplify(b[i]): c[i]:=simplify(c[i]): od:
> solution:=solve({alg[10]},{a[3,2]}); assign(solution);
for i from 1 to alg_act do alg[i]:=simplify(alg[i]): od:
for i from 2 to stages do calg[i]:=simplify(calg[i]): od:
dis_Re:=simplify(dis_Re): dis_Im:=simplify(dis_Im):
for i to stages do; for j to i-1 do;
a[i,j]:=simplify(a[i,j]); od: od:
for i to stages do
b[i]:=simplify(b[i]): c[i]:=simplify(c[i]): od:
> solution:=solve({alg[12]},{a[5,4]}); assign(solution);
for i from 1 to alg_act do alg[i]:=simplify(alg[i]): od:
for i from 2 to stages do calg[i]:=simplify(calg[i]): od:
dis_Re:=simplify(dis_Re): dis_Im:=simplify(dis_Im):
for i to stages do; for j to i-1 do;
a[i,j]:=simplify(a[i,j]); od: od:
for i to stages do
b[i]:=simplify(b[i]): c[i]:=simplify(c[i]): od:
> solution:=solve({alg[14]},{c[4]}); assign(solution);
for i from 1 to alg_max do alg[i]:-simplify(alg[i]): od:
for i from 2 to stages do calg[i]:=simplify(calg[i]): od:
dis_Re:=simplify(dis_Re): dis_Im:=simplify(dis_Im):
for i to stages do; for j to i-1 do;
a[i,j]:=simplify(a[i,j]); od: od:
for i to stages do
b[i]:=simplify(b[i]): c[i]:=simplify(c[i]): od:
> for i from 1 to alg_max do alg[i]:=simplify(alg[i]): od;
>
> for i to stages do;
 for j to i-1 do;
  a[i,j]:=sort(a[i,j],v):
  print(allillj, a[i,j]);
  latex(a[i,j]);
 od;
od;
> for i to stages do
 b[i]:=simplify(b[i]):
 latex(b[i]);
od;
```

```
> for i to stages do
 c[i]:=simplify(c[i]):
 latex(c[i]);
od;
>
> dis_Re:=simplify(dis_Re);
dis_Im:=simplify(dis_Im);
> phl:=tan(v)*dis_Re-dis_Im;
> dissipation:=1-simplify(sqrt((dis_Im)^2+(dis_Re)^2));
> phl:=staylor(phl,v,12);
> dissipation:=staylor(dissipation,v,12);
> solution:=solve(coeff(dissipation,v,6),); assign(solution);
for i from 1 to alg_act do alg[i]:=simplify(alg[i]): od:
for i from 2 to stages do calg[i]:=simplify(calg[i]): od:
dis_Re:=simplify(dis_Re): dis_Im:=simplify(dis_Im):
for i to stages do; for j to i-1 do;
a[i,j]:=simplify(a[i,j]); od: od:
for i to stages do
b[i]:=simplify(b[i]): c[i]:=simplify(c[i]): od:
> solution:=solve(alg[15],); assign(solution);
for i from 1 to alg_act do alg[i]:=simplify(alg[i]): od:
for i from 2 to stages do calg[i]:=simplify(calg[i]): od:
dis_Re:=simplify(dis_Re): dis_Im:=simplify(dis_Im):
for i to stages do; for j to i-1 do;
a[i,j]:=simplify(a[i,j]); od: od:
for i to stages do
b[i]:=simplify(b[i]): c[i]:=simplify(c[i]): od:
> alg[16];
```

Solution Confirmation

```
> phl:=staylor(phl,v,11);
dissipation:=staylor(dissipation,v,11);
> for i from 1 to alg_max do alg[i]:=simplify(alg[i]): od;
for i from 2 to stages do calg[i]:=simplify(calg[i]): od;
>
```

Parameters Presentation

```
> for i to stages do;
 for j to i-1 do;
  a[i,j]:=sort(a[i,j],v):
  print(allillj, a[i,j]);
  latex(a[i,j]);
 od;
od;
> for i to stages do
 b[i]:=simplify(b[i]):
 latex(b[i]);
od;
> for i to stages do
 c[i]:=simplify(c[i]):
 latex(c[i]);
od;
>
> B:=seq (add(b[i]*c[i]^(k-1), i=1..stages)=1/k, k=1..stages):
> C:=seq (seq (add(a[i,j]*c[j]^(k-1), j=1..stages)=1/k*c[i]^k, k=1..stages), i=1..stages):
> solution:=solve(B union C):
```

```
> assign(solution);
> printf("A=[\n");
for i to stages do
 for j to stages do
  printf("%19.17g,\t", evalf(a[i,j]));
 od;
 printf("\n");
 od;
printf("];\n \nb=[");
for i to stages do
 printf("%19.17g,\t", evalf(b[i]));
od;
printf("];\n \nc=[");
for i to stages do
 printf("%19.17g,\t",evalf(c[i]));
od;
printf("];");
>
```

Appendix E

Maple Program for the development of the New Four-Step P-stable method with minimal Phase-Lag

```
> restart;
> qnp2:=convert(taylor(q(x+2*h),h=0,13),polynom);
> qnp1:=convert(taylor(q(x+h),h=0,13),polynom);
> qnm1:=convert(taylor(q(x-h),h=0,13),polynom);
> qnm2:=convert(taylor(q(x-2*h),h=0,13),polynom);
> snp2:=convert(taylor(diff(q(x+2*h),x$2),h=0,13),polynom);
> snp1:=convert(taylor(diff(q(x+h),x$2),h=0,13),polynom);
> snm1:=convert(taylor(diff(q(x-h),x$2),h=0,13),polynom);
> snm2:=convert(taylor(diff(q(x-2*h),x$2),h=0,13),polynom);
> qn:=q(x);
> sn:=diff(q(x),x$2);
> lte[1]:=simplify(a[0]*h^2*(snp1-2*sn+snm1));
> lte[2]:=simplify(a[1]*h^2*(snp2-4*snp1+6*(sn+lte[1])-4*snm1+snm2));
> lte[3]:=simplify(a[2]*h^2*(snp2-4*snp1+6*(sn+lte[2])-4*snm1+snm2));
>                                         lte:=simplify(qnp2-2*qnp1+2*qn-2*qnm1+qnm2-h^2/
120*(9*(snp2+snm2)+104*(snp1+snm1)+14*(sn+lte[3])));
> coeff(lte,h,8);
> cc:=-7/60*a[2]*'@@'(D,6)(q)(x)-19/6048*'@@'(D,8)(q)(x);
> 5*6048;
> 5*6048*cc;
> restart;
> ypn:=y[n]+a[0]*H^2*(y[n+1]-2*y[n]+y[n-1]);
> yppn:=y[n]+a[1]*H^2*(y[n+2]-4*y[n+1]+6*ypn-4*y[n-1]+y[n-2]);
> ypppn:=y[n]+a[2]*H^2*(y[n+2]-4*y[n+1]+6*yppn-4*y[n-1]+y[n-2]);
>                  stab:=y[n+2]-2*y[n+1]+2*y[n]-2*y[n-1]+y[n-2]+H^2/120*(9*(y[n+2]+y[n-
2])+104*(y[n+1]+y[n-1])+14*ypppn);
```

```
> A[2]:=simplify(coeff(stab,y[n+2]));
> A[1]:=simplify(coeff(stab,y[n+1]));
> A[0]:=simplify(coeff(stab,y[n]));
> pl:=convert(taylor((2*A[2]*cos(2*H)+2*A[1]*cos(H)+A[0])/(8*A[2]+2*A[1]),H=0,20),polynom);
> eq1:=7/240*a[2]-19/24192;
> solut:=solve({eq1});
> assign(solut);
> pl;
> A[2];
> A[1];
> A[0];
> chr:=A[2]*(z^4+1)+A[1]*(z^3+z)+A[0]*z^2;
> chr1:=simplify(subs(z=exp(I*H),chr));
> eqs 1:=simplify(evalc(Re(chr1)));
> eqs 2:=simplify(evalc(Im(chr1)));
> chr2:=simplify(subs(z=exp(-I*H),chr));
> eqs 3:=simplify(evalc(Re(chr2)));
> eqs 4:=simplify(evalc(Im(chr2)));
> chr3:=simplify(subs(z=-exp(I*H),chr));
> eqs 5:=simplify(evalc(Re(chr3)));
> eqs 6:=simplify(evalc(Im(chr3)));
> chr4:=simplify(subs(z=-exp(-I*H),chr));
> eqs 7:=simplify(evalc(Re(chr4)));
> eqs 8:=simplify(evalc(Im(chr4)));
> solut1:=solve({eqs 1,eqs 2,eqs 5,eqs 6},{a[0],a[1]});
> assign(solut1);
> a[2];
> a[1]:=combine(a[1]);
>
> a[0]:=combine(a[0]);
> a1t:=convert(series(a[1],H=0,28),polynom);
> a0t:=convert(taylor(a[0],H=0,18),polynom);
> P[1]:=simplify(2*A[2]-2*A[1]+A[0]);
> P[3]:=simplify(2*A[2]+2*A[1]+A[0]);
> P[2]:=12*A[2]-2*A[0];
> combine(P[2]);
> NH:=simplify(P[2]^2-4*P[1]*P[3]);
> plot([P[1],P[2],P[3],NH], H=0..2, style=[line,line,line,line],thickness=[3,3,3,3],symbol=[box,circle,diamond,cross],symbolsize=[20,20,20,20],title="Stability Polynomials for the New Method");
```

Appendix F

Maple Program for the development of the Trigonometrically Fitted Fifth-Order Runge-Kutta Methods

First Trigonometrically-Fitted Method

```
> restart;
>
> stages:=6; #Number of stages
> Digits:=50; #Number of decimal digits for floating-point operations
```

```
> alg_max:=17; #Number of algebraic order equations
> #Initialization of algebraic order conditions
alg[]:=array[alg_max]:
for i from 1 to alg_max do
 alg[i]:=0;
od:
>
> #Initialization of Runge-Kutta coefficients
for i to stages do;
 for j to stages do;
  a[i,j]:=0;
 od;
od;
c[1]:=0:
>
> #Coefficients of Explicit Runge-Kutta method of England
> c[2]:=1/3;
> c[3]:=2/5;
> c[4]:=1;
> c[5]:=2/3;
> c[6]:=4/5;
> a[2,1]:=1/3;
> a[3,1]:=4/25;
> a[3,2]:=6/25;
> a[4,1]:=1/4;
> a[4,2]:=-12/4;
> a[4,3]:=15/4;
> a[5,1]:=6/81;
> a[5,2]:=90/81;
> a[5,3]:=-50/81;
> a[5,4]:=8/81;
> a[6,1]:=6/75;
> a[6,2]:=36/75;
> a[6,3]:=10/75;
> a[6,4]:=8/75;
> a[6,5]:=0;
>
> #Exponential Fitting-1st Order
> #1st Stage
f1:=I*w*exp(I*w*x);
> #2nd Stage
f2:=I*w*(exp(I*w*x)+a[2,1]*h*f1);
> #3rd Stage
f3:=I*w*(exp(I*w*x)+h*(a[3,1]*f1+a[3,2]*f2));
> #4th Stage
f4:=I*w*(exp(I*w*x)+h*(a[4,1]*f1+a[4,2]*f2+a[4,3]*f3));
> #5th Stage
f5:=I*w*(exp(I*w*x)+h*(a[5,1]*f1+a[5,2]*f2+a[5,3]*f3+a[5,4]*f4));


> #6th Stage
f6:=I*w*(exp(I*w*x)+h*(a[6,1]*f1+a[6,2]*f2+a[6,3]*f3+a[6,4]*f4+a[6,5]*f5));
> #Final Stage
eqqf:=exp(I*w*(x+h))=exp(I*w*x)+h*(b[1]*f1+b[2]*f2+b[3]*f3+b[4]*f4+b[5]*f5+b[6]*f6);
> #Simplification of the last stage
```

```
eqqf:=simplify(eqqf/exp(I*w*x));
> #Real Part
eqqfa:=evalc(Re(op(1,eqqf)))=evalc(Re(op(2,eqqf)));
> #Imaginary Part
eqqfb:=evalc(Im(op(1,eqqf)))=evalc(Im(op(2,eqqf)));
>
>
> #Simplification of the final stage
eqqf:=simplify(eqqf/exp(I*w*x));
> #Real Part
eqqfaa:=evalc(Re(op(1,eqqf)))=evalc(Re(op(2,eqqf)));
> #Imaginary Part
eqqfbb:=evalc(Im(op(1,eqqf)))=evalc(Im(op(2,eqqf)));
>
> #Definitions of the weights of the final stage
> b[1]:=23/192;
> b[2]:=0;
> b[3]:=125/192;
> b[4]:=0;
>
> #Solving the Two Equations of Exponential Fitting
solution[f]:=solve({eqqfaa,eqqfbb},{b[5],[6]});
> #Assigning the solution
assign(solution[f]);
> #Substituting w*h with v
h:=1;
w:=v;
>
> #Presenting the variable coefficients and their Taylor series expansions
> b[5]:=simplify(combine(b[5]));
> b5t:=convert(taylor(b[5],w=0,18),polynom);
> b[6]:=simplify(combine(b[6]));
> b6t:=convert(taylor(b[6],w=0,18),polynom);
>
> #Confirming Algebraic Order
>
> #Right hand sides of conditions
g(1):=1: #1st Order
g(2):=1/2: #2nd Order
g(3):=1/3: g(4):=1/6: #3rd Order
g(5):=1/4: g(6):=1/8: g(7):=1/12: g(8):=1/24: #4th Order
g(9):=1/5:  g(10):=1/10:  g(11):=1/15:  g(12):=1/30:  g(13):=1/20:  g(14):=1/40:  g(15):=1/60:
g(16):=1/120: g(17):=1/20: #5th Order
>
> #Left hand sides of conditions
for i to stages do:
 alg[1]:=alg[1]+b[i]:
 alg[2]:=alg[2]+b[i]*c[i]:
 alg[3]:=alg[3]+b[i]*c[i]^2:
 alg[5]:=alg[5]+b[i]*c[i]^3:
 alg[9]:=alg[9]+b[i]*c[i]^4:

 for j to stages do:
  alg[4]:=alg[4]+b[i]*a[i,j]*c[j]:
```

```
   alg[6]:=alg[6]+b[i]*c[i]*a[i,j]*c[j]:
   alg[7]:=alg[7]+b[i]*a[i,j]*c[j]^2:
   alg[10]:=alg[10]+b[i]*c[i]^2*a[i,j]*c[j]:
   alg[11]:=alg[11]+b[i]*c[i]*a[i,j]*c[j]^2:
   alg[13]:=alg[13]+b[i]*a[i,j]*c[j]^3:

   for k to stages do:
    alg[8]:=alg[8]+b[i]*a[i,j]*a[j,k]*c[k]:
    alg[12]:=alg[12]+b[i]*c[i]*a[i,j]*a[j,k]*c[k]:
    alg[14]:=alg[14]+b[i]*a[i,j]*c[j]*a[j,k]*c[k]:
    alg[15]:=alg[15]+b[i]*a[i,j]*a[j,k]*c[k]^2:
    for l to stages do:
     alg[16]:=alg[16]+b[i]*a[i,j]*a[j,k]*a[k,l]*c[l]:
    od:
   od:
  od:
  for ij to stages do:
   for ik to stages do:
    alg[17]:=alg[17]+b[i]*a[i,ij]*c[ij]*a[i,ik]*c[ik]:
   od:
  od:
od:
>
> #Creating the difference
for i to alg_max do
 alg[i]:=simplify(alg[i]-g(i)):
od:
>
> #Taylor series expansion of the differences
for i to alg_max do
 ttt[i]:=talg[i]=convert(taylor(alg[i],v,19), polynom):
od:
for i to alg_max do
 ttt[i]:=talg[i]=convert(taylor(alg[i],v,10), polynom):
od;
>
```

Second Trigonometrically-Fitted Method

```
> restart;
>
> stages:=6; #Number of stages
> Digits:=50; #Number of decimal digits for floating-point operations
> alg_max:=17; #Number of algebraic order equations
> #Initialization of algebraic order conditions
alg[]:=array[alg_max]:
for i from 1 to alg_max do
 alg[i]:=0;
od:
>
> #Initialization of Runge-Kutta coefficients
for i to stages do;
 for j to stages do;
  a[i,j]:=0;
 od;
od;
```

```
c[1]:=0:
>
> #Coefficients of Explicit Runge-Kutta method of England
> c[2]:=1/3;
> c[3]:=2/5;
> c[4]:=1;
> c[5]:=2/3;
> c[6]:=4/5;
> a[2,1]:=1/3;
> a[3,1]:=4/25;
> a[3,2]:=6/25;
> a[4,1]:=1/4;
> a[4,2]:=-12/4;
> a[4,3]:=15/4;
> a[5,1]:=6/81;
> a[5,2]:=90/81;
> a[5,3]:=-50/81;
> a[5,4]:=8/81;
> a[6,1]:=6/75;
> a[6,2]:=36/75;
> a[6,3]:=10/75;
> a[6,4]:=8/75;
> a[6,5]:=0;
>
> #Exponential Fitting-1st Order
> #1st Stage
f1:=I*w*exp(I*w*x);
> #2nd Stage
f2:=I*w*(exp(I*w*x)+a[2,1]*h*f1);
> #3rd Stage
f3:=I*w*(exp(I*w*x)+h*(a[3,1]*f1+a[3,2]*f2));
> #4th Stage
f4:=I*w*(exp(I*w*x)+h*(a[4,1]*f1+a[4,2]*f2+a[4,3]*f3));
> #5th Stage
f5:=I*w*(exp(I*w*x)+h*(a[5,1]*f1+a[5,2]*f2+a[5,3]*f3+a[5,4]*f4));
> #6th Stage
f6:=I*w*(exp(I*w*x)+h*(a[6,1]*f1+a[6,2]*f2+a[6,3]*f3+a[6,4]*f4+a[6,5]*f5));
> #Final Stage
eqqf:=exp(I*w*(x+h))=exp(I*w*x)+h*(b[1]*f1+b[2]*f2+b[3]*f3+b[4]*f4+b[5]*f5+b[6]*f6);
> #Simplification of the last stage
eqqf:=simplify(eqqf/exp(I*w*x));
> #Real Part
eqqfa:=evalc(Re(op(1,eqqf)))=evalc(Re(op(2,eqqf)));
> #Imaginary Part
eqqfb:=evalc(Im(op(1,eqqf)))=evalc(Im(op(2,eqqf)));
>
> #Exponential Fitting-2nd Order
> #1st Stage
f1:=(1+I*w*x)*exp(I*w*x);
> #2nd Stage
y2:=(x*exp(I*w*x)+a[2,1]*h*f1):
f2:=diff(y2,x$1);
> #3rd Stage
y3:=(x*exp(I*w*x)+h*(a[3,1]*f1+a[3,2]*f2)):
f3:=diff(y3,x$1);
```

```
> #4th Stage
y4:=(x*exp(I*w*x)+h*(a[4,1]*f1+a[4,2]*f2+a[4,3]*f3)):
f4:=diff(y4,x$1);
> #5th Stage
y5:=(x*exp(I*w*x)+h*(a[5,1]*f1+a[5,2]*f2+a[5,3]*f3+a[5,4]*f4)):
f5:=diff(y5,x$1);
> #6th Stage
y6:=(x*exp(I*w*x)+h*(a[6,1]*f1+a[6,2]*f2+a[6,3]*f3+a[6,4]*f4+a[6,5]*f5)):
f6:=diff(y6,x$1);
> #Final Stage
eqqf:=(x+h)*exp(I*w*(x+h))=x*exp(I*w*x)+h*(b[1]*f1+b[2]*f2+b[3]*f3+b[4]*f4+b[5]*f5+b[6]*-
f6);
> #Simplification of the final stage
eqqf:=simplify(eqqf/exp(I*w*x));
> #Real Part
eqqfaa:=evalc(Re(op(1,eqqf)))=evalc(Re(op(2,eqqf)));
> #Imaginary Part
eqqfbb:=evalc(Im(op(1,eqqf)))=evalc(Im(op(2,eqqf)));
>
> #Definitions of the weights of the final stage
> b[1]:=23/192;
> b[2]:=0;
>
> #Solving the Four Equations of Exponential Fitting
solution[f]:=solve({eqqfaa,eqqfbb,eqqfa,eqqfb},{b[3],[4],[5],[6]});
> #Assigning the solution
assign(solution[f]);
> #Substituting w*h with v
h:=1;
w:=v;
>
> #Presenting the variable coefficients and their Taylor series expansions
> b[3]:=simplify(combine(b[3]));
> b3t:=convert(taylor(b[3],w=0,18),polynom);
> b[4]:=simplify(combine(b[4]));
> b4t:=convert(taylor(b[4],w=0,18),polynom);
> b[5]:=simplify(combine(b[5]));
> b5t:=convert(taylor(b[5],w=0,18),polynom);
> b[6]:=simplify(combine(b[6]));
> b6t:=convert(taylor(b[6],w=0,18),polynom);
>
> #Confirming Algebraic Order
>
> #Right hand sides of conditions
g(1):=1: #1st Order
g(2):=1/2: #2nd Order
g(3):=1/3: g(4):=1/6: #3rd Order
g(5):=1/4: g(6):=1/8: g(7):=1/12: g(8):=1/24: #4th Order
g(9):=1/5: g(10):=1/10: g(11):=1/15: g(12):=1/30: g(13):=1/20: g(14):=1/40: g(15):=1/60:
g(16):=1/120: g(17):=1/20: #5th Order
>
> #Left hand sides of conditions
for i to stages do:
 alg[1]:=alg[1]+b[i]:
 alg[2]:=alg[2]+b[i]*c[i]:
```

```
 alg[3]:=alg[3]+b[i]*c[i]^2:
 alg[5]:=alg[5]+b[i]*c[i]^3:
 alg[9]:=alg[9]+b[i]*c[i]^4:
 for j to stages do:
  alg[4]:=alg[4]+b[i]*a[i,j]*c[j]:
  alg[6]:=alg[6]+b[i]*c[i]*a[i,j]*c[j]:
  alg[7]:=alg[7]+b[i]*a[i,j]*c[j]^2:
  alg[10]:=alg[10]+b[i]*c[i]^2*a[i,j]*c[j]:
  alg[11]:=alg[11]+b[i]*c[i]*a[i,j]*c[j]^2:
  alg[13]:=alg[13]+b[i]*a[i,j]*c[j]^3:
  for k to stages do:
   alg[8]:=alg[8]+b[i]*a[i,j]*a[j,k]*c[k]:
   alg[12]:=alg[12]+b[i]*c[i]*a[i,j]*a[j,k]*c[k]:
   alg[14]:=alg[14]+b[i]*a[i,j]*c[j]*a[j,k]*c[k]:
   alg[15]:=alg[15]+b[i]*a[i,j]*a[j,k]*c[k]^2:
   for l to stages do:
    alg[16]:=alg[16]+b[i]*a[i,j]*a[j,k]*a[k,l]*c[l]:
  od:
 od:
od:

for ij to stages do:
 for ik to stages do:
  alg[17]:=alg[17]+b[i]*a[i,ij]*c[ij]*a[i,ik]*c[ik]:
  od:
 od:
od:
>
> #Creating the difference
for i to alg_max do
 alg[i]:=simplify(alg[i]-g(i)):
od:
>
> #Taylor series expansion of the differences
for i to alg_max do
 ttt[i]:=talg[i]=convert(taylor(alg[i],v,19), polynom):
od:
for i to alg_max do
 ttt[i]:=talg[i]=convert(taylor(alg[i],v,10), polynom):
od;
>
```

Appendix G

Maple Program for the development of the New Four-Step P-stable Trigonometrically-Fitted method

```
> restart;
> y[n]:=exp(v*x);
> f[n]:=diff(y[n],x$2);
> y[n+1]:=exp(v*(x+h));
> f[n+1]:=diff(y[n+1],x$2);
```

```
> y[n-1]:=exp(v*(x-h));
> f[n-1]:=diff(y[n-1],x$2);
> y[n+2]:=exp(v*(x+2*h));
> f[n+2]:=diff(y[n+2],x$2);
> y[n-2]:=exp(v*(x-2*h));
> f[n-2]:=diff(y[n-2],x$2);


> final:=y[n+2]-2*y[n+1]+2*y[n]-2*y[n-1]+y[n-2]=h^2*(b[0]*(f[n+2]+f[n-2])+b[1]*(f[n+1]+f[n-
1])+b[2]*f[n]);
> final:=combine(final/exp(v*x));
> final:=expand(final);
> final1:=simplify(convert(final,trig));


> y[n]:=exp(-v*x);
> f[n]:=diff(y[n],x$2);
> y[n+1]:=exp(-v*(x+h));
> f[n+1]:=diff(y[n+1],x$2);
> y[n-1]:=exp(-v*(x-h));
> f[n-1]:=diff(y[n-1],x$2);
> y[n+2]:=exp(-v*(x+2*h));
> f[n+2]:=diff(y[n+2],x$2);
> y[n-2]:=exp(-v*(x-2*h));
> f[n-2]:=diff(y[n-2],x$2);
>  final:=y[n+2]-2*y[n+1]+2*y[n]-2*y[n-1]+y[n-2]=h^2*(b[0]*(f[n+2]+f[n-2])+b[1]*(f[n+1]+f[n-
1])+b[2]*f[n]);;
> final:=combine(final/exp(-v*x));
> final:=expand(final);
> final2:=simplify(convert(final,trig));
> simplify(final1-final2);
> eq1:=final1;
> eq 1i:=subs(v=I*v,eq1);
> eq1:=simplify(eq 1i);
>
>
> y[n]:=x*exp(v*x);
> f[n]:=diff(y[n],x$2);
> y[n+1]:=(x+h)*exp(v*(x+h));
> f[n+1]:=diff(y[n+1],x$2);
> y[n-1]:=(x-h)*exp(v*(x-h));
> f[n-1]:=diff(y[n-1],x$2);
> y[n+2]:=(x+2*h)*exp(v*(x+2*h));
> f[n+2]:=diff(y[n+2],x$2);
> y[n-2]:=(x-2*h)*exp(v*(x-2*h));
> f[n-2]:=diff(y[n-2],x$2);

>  final:=y[n+2]-2*y[n+1]+2*y[n]-2*y[n-1]+y[n-2]=h^2*(b[0]*(f[n+2]+f[n-2])+b[1]*(f[n+1]+f[n-
1])+b[2]*f[n]);
> final:=combine(final/exp(v*x));
> final:=expand(final);
> final1:=simplify(convert(final,trig));

> y[n]:=x*exp(-v*x);
```

```
> f[n]:=diff(y[n],x$2);
> y[n+1]:=(x+h)*exp(-v*(x+h));
> f[n+1]:=diff(y[n+1],x$2);
> y[n-1]:=(x-h)*exp(-v*(x-h));
> f[n-1]:=diff(y[n-1],x$2);
> y[n+2]:=(x+2*h)*exp(-v*(x+2*h));
> f[n+2]:=diff(y[n+2],x$2);
> y[n-2]:=(x-2*h)*exp(-v*(x-2*h));
> f[n-2]:=diff(y[n-2],x$2);
>   final:=y[n+2]-2*y[n+1]+2*y[n]-2*y[n-1]+y[n-2]=h^2*(b[0]*(f[n+2]+f[n-2])+b[1]*(f[n+1]+f[n-
1])+b[2]*f[n]);;
> final:=combine(final/exp(-v*x));
> final:=expand(final);
> final2:=simplify(convert(final,trig));
>
> eq 2a:=subs(v=I*v,final1);
> eq 2b:=subs(v=I*v,final2);
> eq 2a1:=simplify(evalc(Re(eq 2a)));
> eq 2a2:=simplify(evalc(Im(eq 2a)));
> eq 2b1:=simplify(evalc(Re(eq 2b)));
> eq 2b2:=simplify(evalc(Im(eq 2b)));
> eq1;
> eq2:=eq 2a2;
>
>
> y[n]:=x^n;
> f[n]:=diff(y[n],x$2);
> y[n+1]:=(x+h)^n;
> f[n+1]:=diff(y[n+1],x$2);
> y[n-1]:=(x-h)^n;
> f[n-1]:=diff(y[n-1],x$2);
> y[n+2]:=(x+2*h)^n;
> f[n+2]:=diff(y[n+2],x$2);
> y[n-2]:=(x-2*h)^n;
> f[n-2]:=diff(y[n-2],x$2);


>   final:=y[n+2]-2*y[n+1]+2*y[n]-2*y[n-1]+y[n-2]=h^2*(b[0]*(f[n+2]+f[n-2])+b[1]*(f[n+1]+f[n-
1])+b[2]*f[n]);
> n:=0;
> eq3:=simplify(final);
> n:=2;
> eq3:=simplify(final);
> eq3:=simplify(eq3/h^2);
>
> solut:=solve({eq1,eq2,eq3},{b[0],[1],[2]});
> assign(solut);
> h:=1;
> b[0]:=combine(b[0]);
> b[1]:=combine(b[1]);
> b[2]:=combine(b[2]);
> b0t:=convert(taylor(b[0],v=0,22),polynom);
> b1t:=convert(taylor(b[1],v=0,22),polynom);
> b2t:=convert(taylor(b[2],v=0,22),polynom);
```

```
>
>
>
> restart;
> qnp2:=convert(taylor(q(x+2*h),h=0,13),polynom);
> qnp1:=convert(taylor(q(x+h),h=0,13),polynom);
> qnm1:=convert(taylor(q(x-h),h=0,13),polynom);
> qnm2:=convert(taylor(q(x-2*h),h=0,13),polynom);
> snp2:=convert(taylor(diff(q(x+2*h),x$2),h=0,13),polynom);
> sn:=convert(taylor(diff(q(x+h),x$2),h=0,13),polynom);
> snm1:=convert(taylor(diff(q(x-h),x$2),h=0,13),polynom);
> snm2:=convert(taylor(diff(q(x-2*h),x$2),h=0,13),polynom);
> qn:=q(x);
> sn:=diff(q(x),x$2);


>
> b[0]:=subs(v=v*h,3/40+19/3024*v^2+139/259200*v^4+5771/119750400*v^6+271391/
59439744000*v^8+135227/301771008000*v^10+238998847/5335311421440000*v^12);
> b[1]:=subs(v=v*h,13/15-19/756*v^2+113/113400*v^4-293/29937600*v^6-8213/
81729648000*v^8-15563/980755776000*v^10-128309/166728481920000*v^12);
> b[2]:=subs(v=v*h,7/60+19/504*v^2-103/33600*v^4-73/950400*v^6-973199/
108972864000*v^8-188411/217945728000*v^10-11185379/127031224320000*v^12);
> lte[1]:=simplify(a[0]*h^2*(sn1-2*sn+snm1));
> lte[2]:=simplify(a[1]*h^2*(snp2-4*snp1+6*(sn+lte[1])-4*snm1+snm2));
>
> lte:=simplify(qnp2-2*qnp1+2*qn-2*qnm1+qnm2-
h^2*(b[0]*(snp2+snm2)+b[1]*(snp1+snm1)+b[2]*(sn+lte[2])));
> coeff(lte,h,8);
>
> restart;
> ypn:=y[n]+a[0]*H^2*(y[n+1]-2*y[n]+y[n-1]);
> yppn:=y[n]+a[1]*H^2*(y[n+2]-4*y[n+1]+6*ypn-4*y[n-1]+y[n-2]);
> stab:=y[n+2]-2*y[n+1]+2*y[n]-2*y[n-1]+y[n-2]+H^2*(b[0]*(y[n+2]+y[n-
2])+b[1]*(y[n+1]+y[n-1])+b[2]*yppn);
> A[2]:=simplify(coeff(stab,y[n+2]));
> A[1]:=simplify(coeff(stab,y[n+1]));
> A[0]:=simplify(coeff(stab,y[n]));
> chr:=A[2]*(z^4+1)+A[1]*(z^3+z)+A[0]*z^2;
> chr1:=simplify(subs(z=exp(I*H),chr));
> eqs 1:=simplify(evalc(Re(chr1)));
> eqs 2:=simplify(evalc(Im(chr1)));
> chr2:=simplify(subs(z=exp(-I*H),chr));
> eqs 3:=simplify(evalc(Re(chr2)));
> eqs 4:=simplify(evalc(Im(chr2)));
> simplify(eqs 1-eqs 3);
> simplify(eqs 2+eqs 4);
> chr3:=simplify(subs(z=-exp(I*H),chr));
> eqs 5:=simplify(evalc(Re(chr3)));
> eqs 6:=simplify(evalc(Im(chr3)));
> chr4:=simplify(subs(z=-exp(-I*H),chr));
> eqs 7:=simplify(evalc(Re(chr4)));
> eqs 8:=simplify(evalc(Im(chr4)));
> simplify(eqs 7-eqs 5);
> simplify(eqs 8+eqs 6);
```

```
> solut1:=solve({eqs 1,eqs 2,eqs 5,eqs 6},{a[0],[1]});
> assign(solut1);
> a[0]:=combine(a[0]);
>
> a[1]:=combine(a[1]);
> b[0]:=(-14*cos(v)-2*cos(3*v)-5*v*sin(v)+8*cos(2*v)+8-v*sin(3*v)+4*v*sin(2*v)+2*v^3*-
sin(v))/(v^3*sin(3*v)+5*v^3*sin(v)-4*v^3*sin(2*v));
> b[1]:=(2*cos(4*v)-2-4*cos(3*v)+4*cos(v)+2*v*sin(3*v)+10*v*sin(v)-8*v*sin(2*v)-
4*v^3*sin(2*v))/(v^3*sin(3*v)+5*v^3*sin(v)-4*v^3*sin(2*v));
> b[2]:=(2*v^3*sin(3*v)+6*v^3*sin(v)+20*cos(v)+12*cos(3*v)-10*v*sin(v)-2*v*sin(3*v)+8*v*-
sin(2*v)-4*cos(4*v)-16*cos(2*v)-12)/(v^3*sin(3*v)+5*v^3*sin(v)-4*v^3*sin(2*v));
> b[0]:=subs(v=H,b[0]);
> b[1]:=subs(v=H,b[1]);
> b[2]:=subs(v=H,b[2]);
> a1t:=convert(series(a[1],H=0,32),polynom);
> a0t:=convert(taylor(a[0],H=0,26),polynom);
> P[1]:=simplify(2*A[2]-2*A[1]+A[0]);
> P[3]:=simplify(2*A[2]+2*A[1]+A[0]);
> P[2]:=12*A[2]-2*A[0];
> combine(P[2]);
> NH:=simplify(P[2]^2-4*P[1]*P[3]);
> plot([P[1],P[2],P[3],NH], H=0.02..2, style=[line,line,line,line],thickness=[3,3,3,3],symbol=
[box,circle,diamond,cross],symbolsize=[20,20,20,20],title="Stability Polynomials for the New
Method");
>
```

6

Determination of Structure in Electronic Structure Calculations

BY MICHAEL SPRINGBORG
Physical and Theoretical Chemistry, University of Saarland, 66123, Saarbrücken, Germany

1 Introduction

Theoretical studies of materials have developed into an important part of science that contributes with information that often complements experimental studies in a very constructive way. For instance, different scenarios for the chemical interactions between different molecules or for the bonding of one molecule to another or to a surface may be distinguished by comparing experimental and theoretical information. Accordingly, if a molecule, constituting of two parts, A and B, is known to adsorb on the surface of crystalline D, it may not be known whether the A or the B part (or both) is taking part in forming the chemical bond between the molecule and the substrate. By studying all possibilities theoretically and compare with experimental results from, *e.g.*, some kind of spectroscopy, it may be possible to identify which bonding pattern is observed in the experiment.

However, such studies (and many other theoretical studies) rely heavily on some kind of information on the possible structure of the system of interest. Thus, if the molecule mentioned above contains just slightly more than 10 atoms, one usually assumes that at least parts of the structure resemble those of other 'related systems,' where 'related systems' are molecules containing the same types of atoms in the same type of bonding situations.

In many cases this approach may provide accurate and useful results, but there are also situations where it may not be particularly useful. A well-known class of materials that, moreover, constitutes a larger part of the systems that shall be discussed in this presentation, is that of clusters and colloids. These materials are intermediates between smaller molecules and extended, macroscopic solids. Typically, they contain from some 10s till several 100 000s of atoms. Often, they have only some few types of atoms with, in some cases, the exception of the surface, where ligands, that saturate dangling bonds, may occur. Thus, it may be assumed that the structure of these nanoparticles resembles that of small, finite parts of the infinite, crystalline material. However,

Chemical Modelling: Applications and Theory, Volume 4

very often this assumption is not correct (as we also shall see in this presentation), making it necessary to be able to determine the structure in other ways.

Often, experimental studies on such materials may give some information. However, a number of issues makes it extremely difficult to extract definite information on the structure of such nanoparticles from the experimental results. Thus, in many cases they are produced in some solution, meaning that the solvent as well as the presence of other nanoparticles may affect the properties of the individual clusters. Furthermore, in many cases the precise size of the clusters is only known approximately. In addition, in those cases where the size is known, they are often produced in gas phase in so small amounts that statistics make the experimental results imprecise.

Thus, in such cases, as well as for many other types of materials, theoretical studies can not rely on available information on the structure of the system of interest. In that case the theoretical studies should be able to predict the structure without biasing the results through more or less realistic assumptions.

But, if the systems of our interest are not very small we face two serious computational problems. We shall briefly discuss those here. Moreover, we emphasize that although larger parts of the presentation will focus on clusters and colloids, almost all of the ideas and approaches can be, and in many cases also are being, used for many other systems with a larger number of atoms and a low symmetry.

Figure 1 shows a schematic flow chart for a typical electronic-structure calculation. Once the number and types of the atoms for the system that shall be studied are known, some initial structure is chosen and various properties for this structure are calculated. These properties are first of all the total energy but may also be others like the forces acting on the nuclei (*i.e.*, the derivatives of the total energy with respect to the nuclear coordinates). Also available

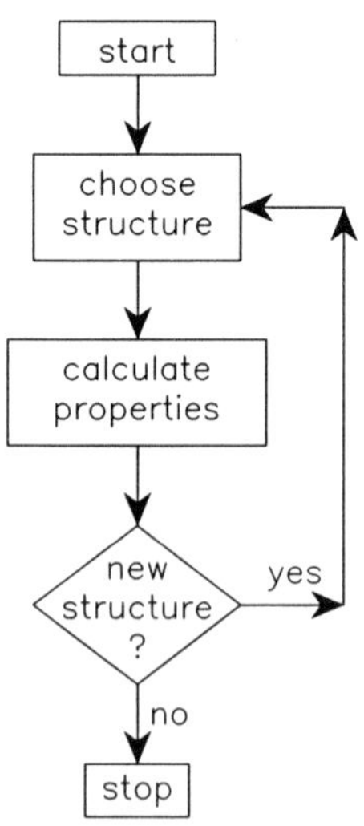

Figure 1 *A flow chart for a typical electronic-structure calculation*

experimental information may be sought calculated. Subsequently, one may ask whether a new structure shall be studied. There can be a number of reasons for choosing to do so, including that the calculated forces are not approximately vanishing, that the calculated properties differ significantly from those obtained in experiment, or that one simply wants to explore further parts of the structure space.

For larger systems this approach is related with two serious problems. First, if the system of interest contains N nuclei and M electrons, the computational costs for passing the loop in Figure 1 just one single time scale as the size of the system to some power, *i.e.*, like M^k or N^k, where k is some power from 2 and upwards. For some of the most popular approaches, like density-functional methods and Hartree-Fock methods (see, *e.g.*, 1), typically $k = 3$, but for more advanced methods k may be larger than 7. Thus, in this case, it should be obvious that for just intermediately large systems, the calculation of the properties for just one single geometry may become computationally very costly.

At this place it shall be mentioned that special methods that scale essentially linearly ($k = 1$) with the system size have been developed (see, *e.g.*, 2–6). These methods may be useful for larger systems, but shall not be discussed further here. Here, we are, instead, concerned with the second problem.

Independently of the scaling of the computational needs as a function of the size of the system, another, complementary, problem causes additional complications. This problem is related to the fact that the number of inequivalent minima on the total-energy surface as a function of structure grows very fast with the size of the system. In fact, it has been shown[7] that the number of local total-energy minima grows faster than any polynomial of the number of atoms in the system or, alternatively expressed, that the determination of the global total-energy minimum is a so-called NP-hard problem.

A model system that is so simple that detailed studies can be performed is that of a cluster of identical atoms for which it is assumed that the total energy can be written as a sum of pair potentials, each one being a simple Lennard-Jones potential. This system was studied by Tsai and Gordon,[8] who found a rapidly increasing number of inequivalent metastable structures as a function of size, cf. Table 1. In Figure 2 the results have been fitted with an exponential, $a \exp(bN)$, and it is seen that the fit follows the calculated results fairly close. The fit gave $a = 0.00341$ and $b = 0.983$, which in turn means that for $N = 55$ the fit predicts that of the order of 10^{21} inequivalent minima exist.

This result points directly to a central issue of this presentation, *i.e.*, how do we determine the ground-state structure for a system with at least some 10s of

Table 1 *Calculated number of (second line) inequivalent total-energy minima and (third line) inequivalent saddle points for Lennard-Jones clusters as a function of the number of atoms N in the cluster. From 8*

N	7	8	9	10	11	12	13
Number	4	8	21	64	152	464	1328
Number	12	40	152	584	911	2803	8453

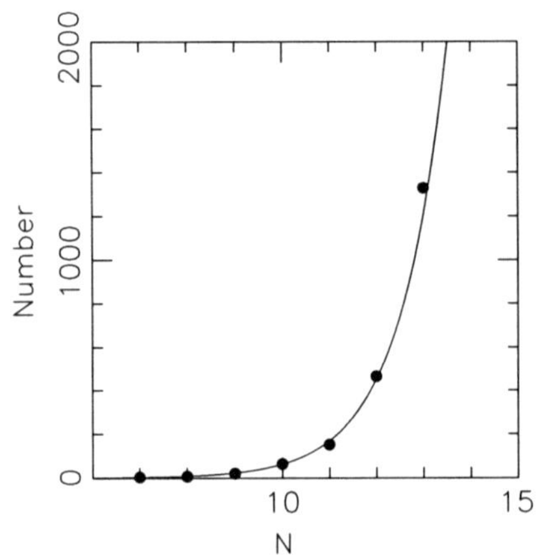

Figure 2 *A schematic representation of number of inequivalent local total-energy minima from the results of Table 1 (the black circles) together with a fit with an exponential (the full curve)*

atoms and for which no, or only very limited, information on the structure can be used. Although the example we have discussed is taken from the class of clusters and colloids with, moreover, interatomic interactions given by Lennard-Jones potentials, the general question remains also for many other classes of materials, including larger, biochemically relevant systems.

In most electronic-structure calculations, the Born-Oppenheimer approximation is invoked. This leads to flow charts like that of Figure 1, where the electronic properties are determined for a given set of nuclear positions. This is done by solving some kind of time-independent electronic Schrödinger equation,

$$\hat{H}_e \Psi_e = E_e \Psi_e, \tag{1}$$

where $\hat{H}_e$ is the Hamilton operator for the electrons in the field of the nuclei [we may also include density-functional approaches in our discussion: also in that case equations like Eq. (1) will have to be solved, but in that case $\hat{H}_e$ is related to the Kohn-Sham operators]. Ψ_e is the electronic wavefunction. The important point is that, although Eq. (1) is complicated, the variational principle allows for systematically approaching the lowest electronic energy $E_e^{(0)}$ from above by approximating

$$\Psi_e \simeq \Phi \tag{2}$$

and using that

$$\frac{\langle \Phi | \hat{H}_e | \Phi \rangle}{\langle \Phi | \Phi \rangle} \geq E_e^{(0)}. \tag{3}$$

This variational principle is the basis for most of the currently used electronic-structure methods. *i.e.*, by systematically improving the approximate function Φ, one may get an, in principle, arbitrarily close approximation to the lowest electronic energy $E_e^{(0)}$ for a given structure.

However, the ground-state *structure* is obtained by minimizing the total energy E, and not only the electronic energy E_e. Employing the Born-Oppenheimer approximation, the nuclei are treated as classical and not quantum particles, and there is no variational principle that can be used in systematically getting arbitrarily close to the true ground-state structure, even when assuming that the total energy can be calculated accurately for any structure.

Instead, most often one has to change the structure 'by hand' until one is confident that the structure of the lowest total energy has been identified. But it shall be stressed that there is absolutely no approach that with absolutely certainty can guarantee that precisely that structure has been found. Trying to prove so would require complete knowledge about the total-energy surface as a function of all internal degrees of freedom. This information is beyond what ever will be available, even if the computers keep on becoming more and more powerful, and the programs more and more efficient.

That this information will never become available, can be seen by, *e.g.*, considering the largest possible computer, *i.e.*, the complete universe. Lloyd[9] estimated that the complete universe, viewed as an enormous quantum computer that has been operating since the Big Bang, could have performed of the order of 10^{120} binary operations. Although this number is very large, it is also finite. Therefore, it means that considering, *e.g.*, a system of 42 atoms, assuming that a total-energy calculation for this system and a given structure is just one single binary operation, and considering just 10 different values for each internal structural degree of freedom, exactly 10^{120} calculations are needed for obtaining this, relatively crude, total-energy hyper-surface. *i.e.*, the calculations would have required the use of the complete universe since the Big Bang!

In total, it is obvious that for any but the absolutely smallest systems it is not possible to explore anything but very limited parts of the total-energy hyper-surface, and, moreover, that it is never possible to be absolutely sure that the true global total-energy minimum has been found. Any total-energy minimum may provide a total energy that is close to that of the global minimum, although the structures of the two may be markedly different, simply due to the very large number of local total-energy minima. It is also obvious that any attempt to identify the global total-energy minimum has to be based on some kind of qualified search in the multi-dimensional structure space.

It is one of the purposes of this presentation to discuss some of the approaches that have been and still are being used in searching for the global total-energy minimum in the structure space. We shall mainly focus on clusters/colloids, partly because these systems are at the heart of our own research, but we stress that the approaches are fairly general and can be — and are being — used for very many other systems with complicated, multi-dimensional structure spaces.

Another purpose is to discuss approaches for studying theoretically the transition from one state to another. Consider, *e.g.*, Figure 3. At some time a

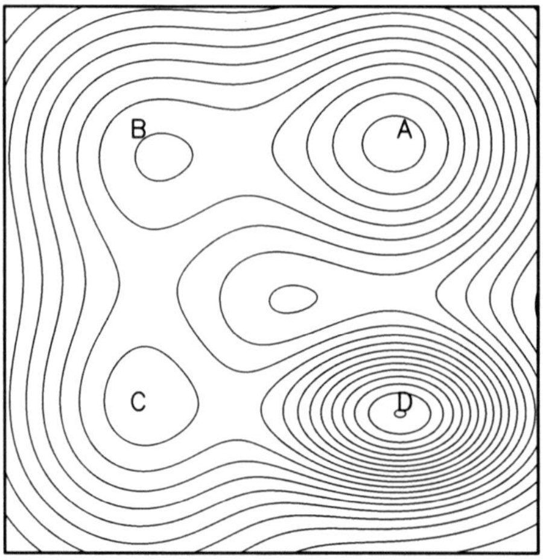

Figure 3 *A hypothetic potential-energy surface for a system with two internal degrees of freedom. A, B, C, and D mark different local total-energy minima*

given system is found in the structure marked with A in the figure. This structure is that of a local total-energy minimum, but the structure of D has a lower total energy, whereas those of B and C have a higher total energy. The structure of A could be that of one isomer of a certain set of atoms, whereas those of B, C, and D correspond to other isomers. In that case, for each of the four isomers all atoms form a molecule (or cluster, colloid, ...). Alternatively, the structure of A may represent that of more, non-interacting molecules that ultimately may react with each other leading to a set of products, whose structures are represented with D. *i.e.*, we consider a reaction like that of

$$R_1 + R_2 + \cdots + R_{N_R} \rightarrow P_1 + P_2 + \cdots + P_{N_P}, \tag{4}$$

where R_1, R_2, ..., R_{N_R} are a set of reacting molecules, in Figure 3 represented through the system A, whereas P_1, P_2, ..., P_{N_P} are the products, represented in Figure 3 through the system D. Whether and how such a reaction will take place is first of all determined through two energy parameters, *i.e.*, the difference in the total energies of the A and the D structures, and in the highest energy (relative to that of A — this energy is the activation energy) that the system has when passing from A via some transition state to D.

Performing total-energy calculations and optimizing the structure will most often lead to the structures at A and D as well as their total energies. However, nothing will in most cases be known about the total-energy surface, and maybe not even the local total-energy-minima structures B and C will be identified through the calculations. The calculations will thus reveal information on the relative total energies of the reactants and the products, but nothing will be known about the activation energy.

However, precise information on the activation energy is important for the understanding of chemical-reaction kinetics and, therefore, much effort is invested in obtaining this information. In some sense the problem of determining the activation energy is similar to that of a person standing in the middle of some valley in a mountainous area who knows that the goal of the tour is another valley behind some mountains but who does not know the easiest way there: One possibility may be to follow more or less the straight line joining the two valleys, although it is not clear whether this is the easiest one. In Figure 3 this will correspond to following the roughly straight line joining A and D.

In the example of Figure 3 we are, however, in the unrealistic situation that we know how the total energy varies as a function of the structure. We can thus see that the barrier height along this part is larger than what is found when following a path from A, via B and C, to D. But to identify such a path when only knowing the structures at A and D, their total energies, as well as, *e.g.*, the first and second derivatives of the total energy as a function of nuclear coordinates at these two structures, is a highly non-trivial task. In this presentation we shall discuss some few approaches that are used for this endeavour.

Ultimately, the systems we are interested in are often larger than very small molecules, and the approaches we shall discuss will then first of all give information on total energies as well as nuclear coordinates for a larger set of atoms. Therefore, a non-trivial task is to extract relevant information from the coordinates of the nuclei and from the total energies. We shall, therefore, also discuss some so-called 'descriptors' that have been introduced for this purpose.

We shall focus on a few classes of materials. For the purpose of discussing the determination of global total-energy minima we shall consider clusters and clusters. We shall in the next section present various methods for the global total-energy minimization for such systems, but stress that the approaches in most cases can be applied for other larger systems, too. The clusters are used mainly for the purpose of providing an adequate set of systems for illustrating the approaches and, therefore, we have not attempted to discuss the very many different studies devoted to cluster-structure optimizations for very many different types of clusters or colloids.

In Section 3 we shall present some of the descriptors we mentioned above, and in Section 4 a subjectively selected set of examples of applications of the approaches is presented and discussed. Subsequently, we shall in Sections 5 and 6 discuss chemical reactions and, in particular, how the activation energies are sought determined theoretically. This is the subject of Section 5, whereas Section 6 is devoted to the discussion of some few selected examples. Once again we emphasize that the examples not have been chosen as being anything but representing applications of the various methods. Finally, we conclude in Section 7. For the sake of completeness we mention the textbooks by Jensen[10] and Wales[11] where more general descriptions of computational methods and of total-energy surfaces are presented.

2 Determining the Global Total-Energy Minima for Clusters

As outlined above, it will never become possible to obtain the complete information on the total energy as a function of structure for systems containing just some few 10s of atoms. Therefore, in order to search for global or local total-energy minima, one has to use methods that search the structure space in some kind of intelligent way and/or to incorporate available (*e.g.*, experimental) information. The danger is, of course, that the calculations become so biased that the true global total-energy structure is not identified. Therefore, such methods could be combined with some degree of randomness, but ultimately all methods are nothing but more or less intelligent tricks devised for identifying the total-energy minima.

In this Section we shall outline the principles behind various approaches for identifying these structures. Besides the individual methods themselves, other methods that use a combination of those are been used.

In the next section we shall briefly discuss how the outcome of such calculations can be analysed, and in Section 4 we shall present some subjectively selected examples of applications of those methods.

2.1 Random *vs.* Selected Structures. – A completely unbiased approach is that of studying randomly generated positions for the N nuclei. Of course, it shall be assured that the atoms are not so far apart that they do not interact with each other and, therefore, the random positions are usually confined to a finite volume, for instance a cube or a sphere. The volume of that could be slightly larger than the sum of the atomic volumes of the N nuclei. On the other hand, it shall also be assured that no two atoms are so close that they feel a very strong interatomic repulsion, so that such configurations will be excluded aforehand, too.

Once such a structure has been generated, a nearby total-energy minimum can be identified, if the forces are available. In standard electronic-structure calculations, the force on the kth nucleus is given by

$$\vec{F}_k = -\vec{\nabla}_k E \tag{5}$$

where $\vec{\nabla}_k$ is the gradient operator for the coordinates of the kth nucleus and where the total energy, when using the Born-Oppenheimer approximation, is split into the electronic term above [Eq. (1)] and a nucleus-nucleus repulsion,

$$E = E_e + E_n \tag{6}$$

with

$$E_n = \frac{1}{2} \sum_{k \neq l = 1}^{N} \frac{Z_k Z_l e^2}{4\pi\varepsilon_0 \left| \vec{R}_k - \vec{R}_l \right|} \tag{7}$$

($\vec{R}_k$ is the position of the kth nucleus; $Z_k e$ its charge).

From Eqs. (5) and (6) it is obvious that the forces in this case contain two contributions, one from the electronic energy and one from the nuclear repulsions. Whereas the latter is simple to calculate, the former can be complicated, in particular when employing accurate electronic-structure methods in the calculation of E_e (see, *e.g.*, 1).

The fact that the calculation of E_e as well as of its derivatives with respect to nuclear coordinates with accurate electronic-structure methods is non-trivial and often time-consuming, makes the use of simpler, parameterized methods attractive, in particular when studying larger systems of low symmetry and/or studying (very) many structures (which easily is the case when attempting to optimize the structure). With these, the total energy is given as some analytical or numerical function of the nuclear coordinates and atom types, *i.e.*,

$$E = E(Z_1, Z_2, \ldots, Z_N, \vec{R}_1, \vec{R}_2, \ldots, \vec{R}_N). \tag{8}$$

A particularly simple case is that of pair potentials,

$$E = \frac{1}{2} \sum_{k \neq l = 1}^{N} E_{kl}\left(\left|\vec{R}_k - \vec{R}_l\right|\right). \tag{9}$$

We shall in Section 4 briefly describe some of those as well as more advanced, but still approximate, methods.

Independent of the theoretical approach that is used in describing the interatomic interactions, the forces, Eq. (5), can be used in identifying the structures of local total-energy minima. The simplest way is the steepest-descent method where new structures are obtained from old ones by using

$$\vec{R}_k \rightarrow \vec{R}_k + \tau \vec{F}_k . \tag{10}$$

Here, τ is some fixed constant, and the procedure is repeated until the forces are close to vanishing. More advanced methods include that of conjugated gradients, with which a smaller number of steps is required before the forces are approximately vanishing.

Ultimately, such calculations lead to a number of different structures of local total-energy minima as well as their total energies, and by starting with sufficiently many randomly generated structures, one may hope that also the structure of the global total-energy minimum is included. In the spirit of Figure 3, this means that the four structures A, B, C, and D all will have been identified, and by comparing, the structure of D is identified as that of the lowest total energy. However, as the figure indicates, some local total-energy minima are found in more 'narrow' parts of structure space, and, as the size of the system increases, the number of local total-energy minima grows essentially exponentially and, therefore, there is no guarantee that the true global total-energy minimum is found.

An absolutely opposite approach is based on using the maximum amount of chemical intuition and experimental information in creating a small set of 'realistic' structures. One may subsequently relax these structures locally, *i.e.*, treat these initial structures absolutely the same way as the randomly generated structures were treated above.

None of these two approaches is optimal. Most often, as we shall see further below, the structures of the global total-energy minima contain some building blocks that show much similarity with 'typical' building blocks from related systems, but also some additional component of disorder, change, or randomness. Therefore, approaches that combine these two extremal methods are often those that most efficiently identify the global total-energy minima.

2.2 Molecular-Dynamics and Monte Carlo Simulations. – Molecular-dynamics and Monte Carlo calculations are based on the idea of simulating experimental situations, where an initial structure (this could, *e.g.*, be a randomly generated structure or one that is considered 'realistic' in the sense above) of the atoms evolve in time until it ultimately reaches the structure of the lowest total energy. Both approaches have some relations to a classical description of particles (*i.e.*, the atoms) moving according to Newton's equation of motion, but this resemblance is more pronounced for the molecular-dynamics calculations than for the Monte Carlo simulations.

Let us assume that at a certain time t the N nuclei have the positions $\vec{R}_1, \vec{R}_2, \ldots, \vec{R}_N$ and the momenta $\vec{P}_1, \vec{P}_2, \ldots, \vec{P}_N$. At a small time step later, *i.e.*, at $t + \Delta t$, Newton's laws may be used in obtaining

$$\vec{P}_k(t+\Delta t) = \vec{P}_k(t) + \Delta t \cdot \vec{F}_k(t), \tag{11}$$

and

$$\vec{R}_k(t+\Delta t) = \vec{R}_k(t) + \frac{\Delta t}{M_k} \cdot \vec{P}_k(t). \tag{12}$$

Here, $\vec{F}_k$ is the force acting on the kth nucleus, which can be calculated as discussed briefly in the previous subsection, and M_k its mass.

By alternatingly using Eqs. (11) and (12) one arrives at the so-called leap-frog algorithms. According to these, starting at a certain set of nuclear coordinates and a certain set of momenta, the forces acting on the nuclei are calculated. These are used in calculating new momenta according to Eq. (11). The new momenta are subsequently used in, from Eq. (12), calculating new positions of the nuclei.

As an alternative one may use the so-called Verlet algorithm. A Taylor expansion gives

$$\vec{R}_k(t+\Delta t) = \vec{R}_k(t) + \frac{\Delta t}{M_k}\vec{P}_k(t) + \frac{(\Delta t)^2}{2M_k}\vec{F}_k(t) + \cdots \tag{13}$$

as well as

$$\vec{R}_k(t-\Delta t) = \vec{R}_k(t) - \frac{\Delta t}{M_k}\vec{P}_k(t) + \frac{(\Delta t)^2}{2M_k}\vec{F}_k(t) + \cdots. \tag{14}$$

Adding the two equations results in

$$\vec{R}_k(t+\Delta t) = 2\,\vec{R}_k(t) - \vec{R}_k(t-\Delta t) + \frac{(\Delta t)^2}{M_k}\vec{F}_k(t) + \cdots, \tag{15}$$

whereby the direct calculation of the momenta is removed, although they still may be determined.

At a given time, the kinetic energy of the atoms can be calculated through

$$E_{\text{kin}}(t) = \sum_{k=1}^{N} \frac{P_k^2}{2M_k}. \tag{16}$$

Moreover, the potential energy is given through the total-energy expressions above, Eqs. (6) and (8), independently of whether accurate, *ab initio* or density-functional, or more approximate model potentials are used,

$$E_{\text{pot}}(t) = E. \tag{17}$$

When the system is isolated, the sum of the potential and kinetic energy is a constant,

$$E_{\text{tot}}(t) = E_{\text{kin}}(t) + E_{\text{pot}}(t) \equiv \text{constant}, \tag{18}$$

(if necessary, this can be forced by, *e.g.*, scaling the momentum so that E_{tot} remains constant) and, using the example of Figure 3 as an illustration, the system will be able to move around on the complete potential-energy surface of the figure, when the kinetic energy is sufficiently large to overcome the barriers between the different local minima.

Ultimately, the kinetic energy defines a temperature,

$$E_{\text{kin}} = N \cdot \frac{3}{2}kT = \frac{3N}{2}kT, \tag{19}$$

or, alternatively,

$$E_{\text{kin}} = \frac{3N-6}{2}kT, \tag{20}$$

(both expressions are being used, although they are not identical), and by gradually reducing the temperature, it may be hoped that the system gets trapped in the structure of the global total-energy minimum (*i.e.*, in D in Figure 3), although there is no guarantee that this indeed is the case. This is the principle of the method of simulated annealing.

Closely related to this approach is that of Monte Carlo optimizations. Starting from a given structure, a random change in the structure is proposed. The total-energy difference between the two structures, ΔE, is calculated, and the change/move/displacement is accepted with probability 1 if $\Delta E < 0$ and otherwise with probability

$$P = \exp\left(-\frac{\Delta E}{kT}\right). \tag{21}$$

In order to determine whether the move in the latter case will be made, one calculates then a random number in the interval [0;1] and from its value, compared to P above, it is decided whether the move is realized. Through this approach it is, as above, possible for the system to move around in the structure space, and by gradually reducing T, it is hoped, as for the simulated annealing and molecular dynamics, that the system ultimately ends in the global total-energy minimum.

Also the random-tunneling algorithm[12] is closely related to Monte Carlo and molecular-dynamics simulations. In this approach the system is allowed to tunnel through energy barriers. Moreover, instead of considering just one single system, an ensemble of several equivalent systems (but with different structures) is studied.

2.3 The Car-Parrinello Method. – In the preceding section we presented methods that were constructed by considering the nuclei as being particles that move according to the laws of classical mechanics. Thereby, the kinetic energy, E_{kin}, and the potential energy, E_{pot}, were introduced as central quantities, whose sum would remain constant when no energy transfer between the surroundings were allowed. Thereby, we obtained a set of equations of motions of the particles, explicitly using the laws of Newton.

An alternative, but equivalent, description can be obtained by considering the difference,

$$L = E_{\text{kin}} - E_{\text{pot}}, \tag{22}$$

i.e., the Lagrangian (see, *e.g.*, 1,13). From this, the same equations of motion as above can be derived.

In the approaches we have discussed so far, the Born-Oppenheimer approximation is employed, *i.e.*, the nuclei are kept at some positions and the electronic structure for this structure is calculated self-consistently, after which the positions of the nuclei may be changed., *e.g.*, by using the forces acting on the nuclei. This corresponds to the representation of Figure 1. As an alternative, Car and Parrinello[14] suggested to determine the electronic distribution and the nuclear coordinates simultaneously. To this end they constructed an artificial Lagrangian,

$$L = E_{\text{kin,n}} + E_{\text{kin,e}} - E_{\text{pot}}, \tag{23}$$

i.e., as a sum of a kinetic-energy term from the nuclei, a similar term from the electrons, and a potential-energy term. For the former they used the expression from a classical (Newtonian) description of the nuclear motion,

$$E_{\text{kin,n}} = \frac{1}{2}\sum_{k=1}^{N} M_k \left|\dot{\vec{R}}_k\right|^2, \tag{24}$$

(where the dot indicates differentiation with respect to time). However, for the kinetic energy of the electrons they used a fictitious expression,

$$E_{\text{kin,e}} = \mu \sum_{i=1}^{N} \left\langle \dot{\psi}_i \middle| \dot{\psi}_i \right\rangle, \tag{25}$$

where μ is a fictitious mass of the electrons (that in the calculations is chosen according to what is convenient), and the sum runs over all occupied orbitals.

Finally, the potential energy is written as the total-energy expression modified with an extra term that shall guarantee that the orbitals are ortho-normal,

$$E_{\text{pot}} = E - \sum_{i,j=1}^{N} \lambda_{i,j}\left(\langle \psi_i | \psi_j \rangle - \delta_{i,j}\right), \tag{26}$$

where the $\lambda_{i,j}$ are Lagrange multipliers for the ortho-normality constraints.

Ultimately this leads to

$$M_k \ddot{\vec{R}}_k = -\vec{\nabla}_{\vec{R}_k} E, \tag{27}$$

which is nothing but the classical equations of motions for the nuclei, as well as

$$\mu \ddot{\phi}_i(\vec{r}, t) = -\frac{\delta E}{\delta \phi_i^*(\vec{r}, t)} + \sum_j \lambda_{i,j} \phi_j(\vec{r}, t). \tag{28}$$

Within Hartree-Fock or density-functional approaches, we have

$$\frac{\delta E}{\delta \phi_i^*(\vec{r}, t)} = \hat{h}(t) \phi_i(\vec{r}, t), \tag{29}$$

where $\hat{h}(t)$ is the Fock or the Kohn-Sham operator for the nuclear positions at time t.

Eqs. (27) and (28) are now sought solved simultaneously, and not, as in most other electronic-structure methods, sequentially, ultimately leading to the determination of the lowest total energy configuration both of the nuclei and of the electrons.

However, just as for the other approaches we have discussed so far, the Car-Parrinello approach is not explicitly devised for the optimization of structure for a given system but may also be used for studying other, most notably dynamical, properties of the system of interest. Accordingly, these methods are more general than just being able to optimize structure, which, on the other hand, may make them less efficient in determining the global total-energy minimum for more complicated systems. This is in contrast to the methods that we now shall discuss and that have been devised explicitly for optimizing structure.

2.4 Eigenmode Methods. – In the previous section we discussed the results of Tsai and Jordan[8] on the number of metastable structures for clusters with atoms that are interacting via Lennard-Jones potentials. We shall repeatedly return to this model system, because it has been, and still is, a system that is excellent for testing methods. However, we shall first discuss the method that Tsai and Jordan used in determining the structures and, in addition, also the saddle points between the metastable structures. Thus, the method will also be relevant below in Section 5.

The eigenmode method goes back to the 1970s but has repeatedly been used and improved since then.[15–23] It is based on a Taylor expansion of the total energy around a stationary point at $\vec{R}_0$,

$$E(\vec{R}) = E_0 + \vec{g} \cdot \vec{h} + \frac{1}{2}\vec{h}^T \cdot \underline{\underline{H}} \cdot \vec{h} + \cdots, \tag{30}$$

where

$$\vec{R} = \vec{R}_0 + \vec{h}, \tag{31}$$

and E_0, $\vec{g}$, and $\underline{\underline{H}}$ are the total energy, the gradient vector, and the second derivative (*i.e.*, Hessian) matrix, respectively, at $\vec{R} = \vec{R}_0$. Finally, $\vec{R}$ represents all internal coordinates of the nuclei.

At a stationary point we have

$$\frac{dE}{d\vec{h}} = \vec{0}, \tag{32}$$

and truncating Eq. (30) after the second-order terms in $\vec{h}$, gives then

$$\vec{h} = -\underline{\underline{H}}^{-1} \cdot \vec{g}. \tag{33}$$

This is the Newton-Raphson method. Starting with some structure, $\vec{R}_0$, a new structure, $\vec{R}$, can be obtained through Eq. (31) where $\vec{h}$ is given by Eq. (33), once the gradient vector $\vec{g}$ and the Hessian $\underline{\underline{H}}$ have been calculated.

By diagonalizing $\underline{\underline{H}}$,

$$\underline{\underline{H}} = \underline{\underline{U}}^\dagger \cdot \underline{\underline{b}} \cdot \underline{\underline{U}}, \tag{34}$$

where $\underline{\underline{b}}$ is a diagonal matrix containing the eigenvalues of $\underline{\underline{H}}$, and $\underline{\underline{U}}$ contains the eigenvectors, $\{\vec{V}_i\}$, we obtain first

$$\underline{\underline{H}}^{-1} = \underline{\underline{U}}^\dagger \cdot \underline{\underline{b}}^{-1} \cdot \underline{\underline{U}}, \tag{35}$$

and subsequently

$$\vec{h} = -\sum_i \frac{F_i}{b_i} \vec{V}_i, \tag{36}$$

where

$$F_i = \vec{g} \cdot \vec{V}_i. \tag{37}$$

Inserting this into Eq. (30), one obtains

$$\Delta E \equiv E(\vec{R}) - E_0 = -\sum_i \frac{F_i^2}{2b_i}. \tag{38}$$

The Newton-Raphson method is accordingly based on performing steps that are decomposed into a sum of orthogonal eigenmode substeps (since the $\{\vec{V}_i\}$ are orthogonal) with the modes with positive eigenvalues, b_i, leading toward lower energy and those with negative eigenvalues leading to higher energy.

Hilderbrandt[15] suggested to modify this approach by replacing b_i by its absolute value,

$$b_i \rightarrow |b_i|, \tag{39}$$

whereby, cf. Eq. (38), convergence to a local total-energy minimum is guaranteed.

This is the eigenmode method. It has been modified, so that also saddle points can be located. We shall return to that in Section 5.

The Hessian, $\underline{\underline{H}}$, can also be used in other ways in optimizing the structure. First, Eq. (33) gives a simple estimate for the optimized structure, once the Hessian has been calculated. Fernández-Serra *et al.*[24] suggested to introduce a Hessian obtained from a simple model Hamiltonian as an efficient way of obtaining reasonable estimates for how to change the structure towards a local total-energy minimum. They showed that this approach could reduce considerably the number of steps before convergence was reached.

2.5 GDIIS. – Another approach based on the Hessian and devised for structure optimization is the GDIIS,[25] *i.e.*, geometry optimization using direct inversion in the iterative subspace. GDIIS is a special version, devised for structure optimization, of DIIS. It has, *e.g.*, been described by Farkas and Schlegel.[26]

One performs a set of total-energy calculations for closely related structures around a certain starting geometry. For each of those, the total energy is expanded according to Eq. (30). Assuming that the sum can be truncated after the quadratic term, for each calculation a difference between the 'true' minimum and the structure that is considered can be calculated. This amounts to the term in Eq. (33),

$$\vec{h}_i = -\underline{\underline{H}}^{-1} \cdot \vec{g}_i, \tag{40}$$

where we will assume that we can use the same $\underline{\underline{H}}$ for all the structures, that furthermore are labeled by i, $i = 1,2,\ldots, K$.

We shall now determine a new structure,

$$\vec{R}^* = \sum_{i=1}^{K} c_i \vec{R}_i, \tag{41}$$

with

$$\sum_{i=1}^{K} c_i = 1. \tag{42}$$

Still assuming the harmonic approximation for the total energy as a function of structure, the error vector at this new structure is just as well a linear combination of the errors of Eq. (40), *i.e.*,

$$\vec{h}^* = \sum_{i=1}^{K} c_i \vec{h}_i . \tag{43}$$

Minimizing $|\vec{h}^*|^2$ under the constraint of Eq. (42) results in a system of $(K+1)$ equations of which the first K ones are for the coefficients c_i and the last is for a Lagrange multiplier for the constraint of Eq. (42). Solving these equations one finds the coefficients $\{c_i\}$ from which a new starting structure can be determined. Subsequently, one can repeat the process from this new starting geometry.

In their presentation of the method, Farkas and Schlegel[26] showed that with this approach the structure of large systems could be found also in cases where other methods would fail, in particular after they included certain improvements that shall not be discussed here.

2.6 Lattice Growth. – Often it is found that the structure of a given cluster with N atoms resembles that of a high-symmetric, compact core containing N_c atoms on whose surface the remaining N–N_c atoms are placed. Therefore, by constructing 'by hand' such structures and, subsequently, letting them relax to their closest total-energy minima, the computational efforts for searching the structure space are reduced significantly. Of course, due to the fixed choice of the structure of the core, the structure optimization is partly biased. This may be an advantage because the dimension of the structure space is drastically reduced making it more likely that the structure of the global total-energy minimum is identified *if* it has a structure that belongs to those that are been studied. On the other hand, a disadvantage is that structural changes as a function of N may not be identified.

Northby[27] discussed in detail the foundations of the lattice-growth approach and applied it in determining the structures of clusters for which the atoms interact with each other via Lennard-Jones potentials. As core structures Northby used structures derived from an icosahedron. He argued that also for cluster sizes for which icosahedra cannot be constructed directly, the icosahedron forms an important structural motif. Based on a single icosahedron, there are various ways of adding additional atoms. One set of structures is formed by the 'MIC' structures, *i.e.*, multi-layer icosahedral structures. Here, the additional atoms are placed at locations of high symmetry on the surface, in particular at tetrahedrally bonded sites, at edge locations with two neighbours in the underlying layer, or at vertices with a single underlying neighbour. Alternatively, also 'PIC' structures, *i.e.*, polyicosahedral structures, may be found. These are structures consisting of more interpenetrating icosahedra, with double icosahedra ('DIC') being the most important structures.

Also Solov'yov *et al.*[28,29] have proposed a method that is based on adding atom by atom to a given core. Here, several different possibilities are explored, *i.e.*, whether the new atom should be added on a surface, inside a surface, or at the center of mass of the cluster.

2.7 Cluster Growth. – The lattice-growth method assumes that, starting with a pre-defined core, for a given cluster there is only a finite set of lattice positions that can be filled with additional atoms. A generalization of this approach is to let additional atoms take any positions as they want. Thus, from one or more structures for the cluster with N atoms, atom number $N + 1$ is placed at various randomly chosen positions. Subsequently, each of those structures is allowed to relax, and the structures of the lowest total energy are kept for the addition of a further atom.

This approach has been discussed in some detail by, *e.g.*, Poteau and Spiegelmann,[30] who used it in optimizing the structures of smaller Na clusters.

By keeping not only the single structure of the lowest total energy for a given size of the cluster (*i.e.*, for a given value of N) one obtains several advantages. First, if more different structures are very close in energy, it may be expected that in an experimental situation all of those may be formed, and in particular if the energy barriers separating them are large, they will all be observed in experiment. Moreover, in this situation it can not be excluded that the energetically lowest structure of the cluster with $N + 1$ atoms is derived from a structure of the cluster with N atoms that is not the energetically lowest one. Finally, the different structures may be derived from fundamentally different structural motifs, so that by keeping more different structures it may be hoped that fundamental structural changes as a function of size may be identified in the calculations.

2.8 Aufbau/Abbau Method. – As an extension of this approach we have proposed a so-called Aufbau/Abbau method.[31–33] The Aufbau part of it is similar to the growth algorithm above, *i.e.*, to one or more given structures of N atoms an additional atom is added at a randomly chosen position. However, in addition an Abbau part is included where atoms are removed from larger clusters. The two parts are carried through independently of each other and only if the results of the two sets of calculations agree it is assumed that the structures of the lowest total energy have been identified.

In detail the calculations are performed as follows:

1) We consider two cluster sizes with N and $N + K$ atoms with $K \simeq 5$–10. For each of those we study a set of randomly generated structures, $N_{\text{ran}} \simeq 1000$. Using the quasi-Newton method the N_{ran} relaxed structures are identified and

the structures of the lowest total energy selected. Each of the N_{ran} starting structures for a cluster with N atoms is generated using a random-number generator for positions within a sphere or a cube of volume $V_{cl} = (p \cdot b_{nn})^3 N$, where b_{nn} is the nearest-neighbour distance of crystalline material and $p = 0.8$, 1.0, 1.2, *i.e.*, we consider slightly compressed, normal, and slightly expanded structures. We include the constraints that the smallest allowed inter-atomic distance is 0.5 b_{nn} and each atom has to interact with at least two others.

2) One by one, each of the N atoms is displaced randomly, and the closest local minima is determined. If the new structure has a lower total energy than the original one, this new one is kept, and the old one discarded. This is repeated approximately 500–1000 times (depending on cluster size).

3) This leaves us with two 'source' clusters of N and $N + K$ atoms with their lowest total energies. One by one an atom is added at a random position to the structure with N atoms (many hundred times for each size), and the structures are relaxed. In parallel, one by one an atom is removed from the structure with $N + K$ atoms — for each intermediate cluster with N' atoms we consider all $N' + 1$ possible configurations, that one can obtain by removing one atom from the cluster with $N' + 1$ atoms. From the two series of structures for $N \leq M \leq N + K$ the structures of the lowest energies are chosen and these are used as seeds for a new set of calculations. First, when no lower total energies are found, it is assumed that the structures of the global-total-energy minima have been identified, and we proceed to larger clusters.

Moreover, by keeping information on not only the single energetically lowest isomer, but more low-lying ones, it is possible to identify more energetically lowest-lying isomers.

2.9 The Basin Hopping Method. – A popular method for unbiased structure optimization is the basin-hopping method.[34] It is closely related to the Monte Carlo method above, but considers a transformed total-energy surface. Thus, for each structure, $\vec{R}$, instead of considering the true total energy E at that structure, one considers the total energy $\tilde{E}$ that the structure would have after having relaxed it to its closest total-energy minimum,

$$\tilde{E}(\vec{R}) = \min\{E(\vec{R})\}, \tag{44}$$

where it is indicated that $\tilde{E}(\vec{R})$ is the lowest total energy that is obtained when starting from the structure $\vec{R}$ but letting it locally relax. Ultimately, this leads to a transformed total-energy surface as schematically shown in Figure 4.

As seen in the figure, the transformed energy surface $\tilde{E}$ is much less structured than E, although the energies at the global and local total-energy minima are identical. The simpler shape of the transformed structure makes it significantly easier to identify the global total-energy minimum, without biasing the calculation in any way.

In a practical calculation, instead of changing the structure according to the Monte Carlo steps outlined above in Secction 2.2 directly, the structure is first locally relaxed using the methods of Section 2.1.

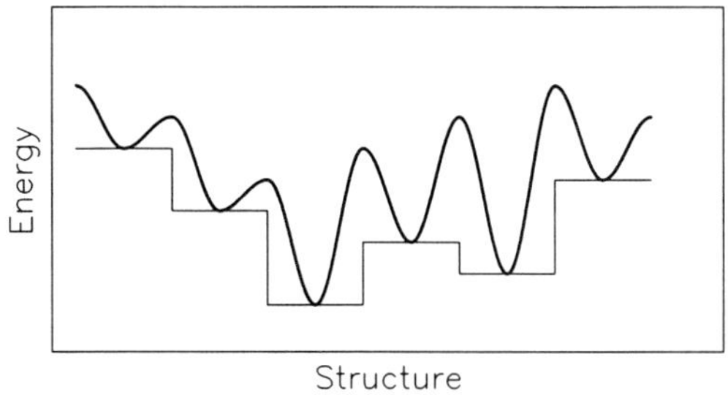

Figure 4 *A schematic representation of the true total energy as a function of structure (the thick curve) as well as the transformed energy (thin curve) as used in the basin-hopping method*

2.10 Genetic Algorithms. – As many of the other methods we have discussed above, also the methods of genetic algorithms are based on a combination of keeping good building blocks and randomness. There are, however, a number of details that makes these methods quite different from the other ones.

Genetic algorithms have been used for a very long time in optimization problems (see. *e.g.*, 35), but first during the first halfpart of the 1990s methods based on this approach were used for optimizing the structure of clusters.[36–41] There are many variants of them, but here we shall just outline the main principles.

Let us assume that we shall optimize the structure of a cluster with N atoms. We create a larger number, K, of initial structures. This can be done, *e.g.*, by using the random structures we discussed in Section 2.1, but some additional structures may also be explicitly constructed, for instance when taking suggestions from other studies or chemical intuition into account. Thus, one may include structures that are thought to be 'realistic' guesses for the structure of the global total-energy minimum, or one may take the structure of $N-1$ atoms and add a single atom to that.

Each of those structures is relaxed locally, and the K (different) structures of the lowest total energy are selected. These form the first generation in the optimization process. Subsequently, operations that are considered to have their origin in natural evolution are applied. First, from these K members, or parents, K children are created. Here, at least two different methods have been proposed. Both are based on taking each of the K parents and cutting them into two (not necessarily equally large) parts, but in a more recently proposed approach,[42] the two parts from one parent are simply interchanged (see Figure 5), whereas in the other, original approach the two parts from two different parents are interchanged (see Figure 6). In particular for clusters with more types of atoms, the first approach is easier to implement. In order to increase the flexibility of the approach, one may include random, rigid orientations of the two parts before joining them.

Each of the children is relaxed to its closest total-energy minimum, giving in total K parents and K children. Out of these $2K$ members, those with the lowest total energy are selected as the K parents for the next generation. This process is

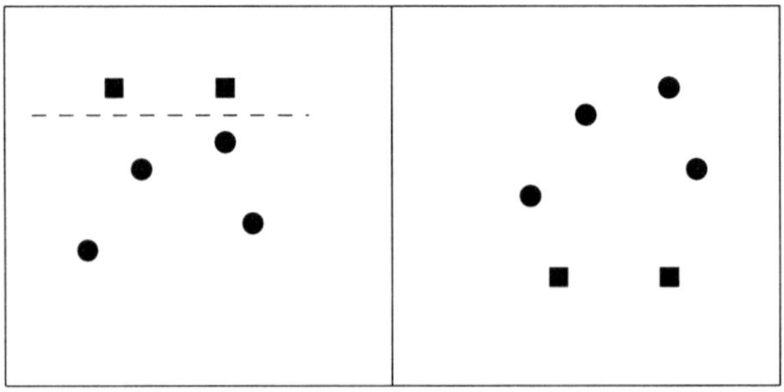

Figure 5 *A schematic representation of the way children structures are created from the parent structures in the genetic algorithms when using a 'one parent → one child' strategy. The parent in the left panel is cut into two halves (made clear through the different symbols for the two different parts as well as through the dashed line that marks the cut) that subsequently are interchanged leading to the structure of the right panel*

then repeated until the lowest total energy stays unchanged for several/many generations, at which point it is assumed that the global total-energy minimum has been identified.

On top of these operations, various additional 'tricks', partly borrowed from biology, have been introduced in order to accelerate convergence. These include mutations where some random perturbation of the cluster regularly (but not for every single cluster) is introduced. This could, *e.g.*, be a random displacement of one of the atoms. Also the possibility of letting pre-selected structures survive although their total energies are not sufficiently low (this could, *e.g.*, be structures with structural motifs that are considered important for the structures of the lowest total energy) can be included in the calculations.

The number of members per generation may be relatively low, from around 10, till fairly large, *i.e.*, of the order of 100, and also the number of generations may vary from some 100s till above 100 000. But as for any other method, ultimately there is identically no guarantee that the structure of the global total-energy minimum has been identified.

2.11 Tabu Search. – Recently, Cheng and Fournier[43] suggested a new algorithm, the tabu search in descriptor space, for unbiased structure optimization. By incorporating extra constraints on the permitted structures, the method was found to be significantly more efficient than, *e.g.*, genetic algorithms, although, as the authors state, it is more complicated and requires more user intervention.

Whereas the methods we have discussed so far focus on minimizing the total energy, Cheng and Fournier consider further descriptors (besides the total energy) of the structures. First, for each atom they calculate the number of neighbours, c_k, within a sphere of a pre-defined radius centered around atom k. From the N values they define an average

$$\bar{c} = \frac{1}{N}\sum_{k=1}^{N} c_k \tag{45}$$

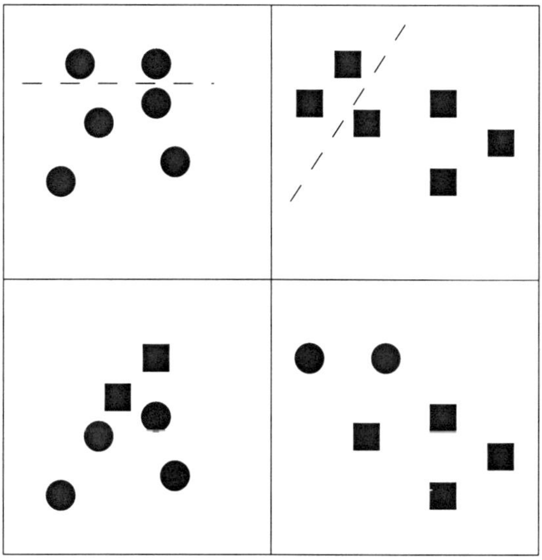

Figure 6 *A schematic representation of the way children structures are created from the parent structures in the genetic algorithms when using a 'two parents → two children' strategy. The parents in the upper panels are each cut into two halves (made clear through the dashed lines that mark the cuts) that subsequently are interchanged leading to the structures of the lower panels. In order to emphasize the interchange, the atoms of the two parent clusters have been given different symbols*

and a root-mean-square of deviations,

$$\Delta c = \left[\frac{1}{N} \sum_{k-1}^{N} (c_k - \bar{c})^2 \right]^{1/2}. \tag{46}$$

As two further descriptors, the largest and smallest values of c_k are used,

$$\begin{aligned} c_- &= \min\{c_k\} \\ c_+ &= \max\{c_k\}. \end{aligned} \tag{47}$$

Finally, two descriptors are derived from the moments of inertia, $I_a \geq I_b \geq I_c$:

$$\begin{aligned} \zeta &= \frac{(I_c - I_b)^2 + (I_b - I_a)^2 + I_a - I_c)^2}{I_a^2 + I_b^2 + I_c^2} \\ \eta &= \frac{2I_b - I_a - I_c}{I_a}. \end{aligned} \tag{48}$$

Initially a set of (randomly) generated structures is relaxed. These may be modified in various ways, but the important thing is that the descriptors above are introduced in order to avoid that the search revisits parts of the structure space that already have been included. Thus, a new set of candidate structures is generated and using the descriptors above their structures are compared with those of the original set. If the descriptors are very similar in value, the new candidate structures are accepted with only a low probability.

The strength of this approach is accordingly that the automatic search in structure space largely avoids returning to the same kind of structures, *i.e.*, a 'tabu' component is introduced. The authors studied subsequently clusters for which the interatomic potentials are given by Lennard-Jones potentials and found that their approach required between 10 and 100 times fewer total-energy evaluations than a good genetic algorithm.

2.12 Combining the Methods. – Ultimately, the approaches for determining the global total-energy minima are all based on some kind of tricks. In this spirit, one may try to devise any method as long as it is hoped that it ultimately will provide the global total-energy minima. Therefore, also combinations of the methods we have discussed so far may be considered useful. Here, we shall briefly mention two recent approaches in this direction.

Lee *et al.*[44] presented a method, the conformational space annealing method, that is based on combining genetic algorithms, molecular dynamics, and simulated annealing. As in the genetic-algorithm approaches, several structures are considered. Moreover, as in simulated annealing, a parameter that plays the role of a temperature is introduced and gradually reduced. However, in this case the parameter D_{cut} is rather quantifying similarities between structures, so that it in some sense is similar to the descriptors discussed in the previous subsection on the Tabu search. If two structures have a difference smaller than D_{cut} (based, *e.g.*, on the descriptors of the preceding subsection, although the authors use some other descriptors), the two structures are considered similar, and only one is kept. When D_{cut} is large (in the beginning of a calculation), the pool of structures contains only structures that are very diverse, assuring that a fairly large part of the structure space is being scanned for the total-energy minima. Finally, as in the so-called Monte Carlo method with minimization, the structures are locally relaxed.

The advantage of this (and of the Tabu) method is particularly easily identified when noticing how a numerical calculation is performed. Then, a structure is considered optimized when the forces acting on the nuclei are below a certain threshold. This means that two structures may be essentially identical, but nevertheless differ slightly due to numerical truncation errors. Through the additional descriptors, it can be avoided that such a pair of almost identical structures is kept in the calculation.

Finally, Goedecker *et al.*[45] combined two different methods for calculating the total energy for a given structure, *i.e.*, a parameterized electronic-structure method and a parameter-free density-functional method. These two methods were combined with a molecular-dynamics-like method for structure optimization, and by combining the two electronic-structure methods it is possible to perform a large structure-space search with the more approximate method before the more exact method is used in fine-tuning the total energies.

3 Descriptors for Cluster Properties

Ultimately, the optimization of the structure of one or more clusters leads to nothing but a larger amount of numbers, including primarily the total energy and the nuclear coordinates. Therefore, a central issue is that of extracting useful information from those, for instance in the case that a whole series of clusters have been considered for which maybe even more isomers for each size have been identified. In this section we shall discuss some of the means that have been developed for the purpose of extracting general trends from the numerical results. In the next section we shall discuss the results of some few selected examples of structure optimization and to that end use the descriptors that will be described here. Also at this place we emphasize that the ideas presented here not are restricted to clusters, but can be used for many other systems with a large number of atoms and a low symmetry, although we here shall focus on clusters.

3.1 Energetics. – Let us assume that we have optimized the structure for a whole class of clusters and that we have more different isomers for each cluster size. Then, we will let $E(N, k)$ be the total energy for the energetically kth lowest isomer for the cluster of N units (where a unit in most cases corresponds to a single atom).

It turns out that the relative total energy per unit, $E(N, k)/N$, in most cases is a very unstructured function that is more or less monotonically decreasing as a function of N. The latter statement is equivalent to stating that any cluster is stabler than two separated fragments. Moreover, the different curves for different k turn out to lie more or less on top of each other.

This means that it in most cases not is possible to identify particularly stable clusters by looking at the relative total energy per unit. Therefore, other descriptors have been introduced.

The quantity

$$\Delta_1 E(N) = -[E(N, 1) - E(N-1, 1)] \tag{49}$$

describes the energy costs related to adding one unit to the most stable isomer of the cluster with $N - 1$ units in order to arrive at the most stable isomer of the cluster with N units. Thus, this function raises abruptly if the cluster of N units is particularly stable compared to the cluster of $N - 1$ units.

Equivalently,

$$\Delta_2 E(N) = E(N+1, 1) + E(N-1, 1) - 2E(N, 1) \tag{50}$$

compares the total energy of the most stable cluster of N units with the most stable structures of $N + 1$ and $N - 1$ units and has therefore peaks when the N-unit cluster is particularly stable. It may also be considered a finite-difference approximation to the second-order derivative of the total energy as a function of N.

In all these cases the most stable cluster of a given size is compared with the most stable clusters of neighbouring sizes. Alternatively, a cluster may also be considered particularly stable if the total-energy difference

$$\Delta_{12}E(N) = E(N, 2) - E(N, 1) \tag{51}$$

is particularly large. In that case the most stable isomer of the cluster of N units has a total energy that is significantly lower than that of the second-most stable isomer.

All these descriptors are accordingly meant to identify particularly stable structures. They do, however, not give any detailed explanation for the reason behind particular stability. To this end, various structural descriptors can be helpful.

3.2 Shape. – In order to analyse the structure of a given structure, it is useful to first define the center of it,

$$\vec{R}_0 = \frac{1}{N}\sum_{k=1}^{N} \vec{R}_k, \tag{52}$$

where the sum runs over all atoms. Subsequently, we may for each atom define an internal position through

$$\vec{r}_k = \vec{R}_k - \vec{R}_0 \,. \tag{53}$$

The quantity

$$r_k = \left|\vec{r}_k\right| \tag{54}$$

may be called the radial distance for the kth atom.

The eigenvalues $I_{\alpha\alpha}$ of the 3×3 matrix with the moments of inertia,

$$I_{st} = \sum_{k=1}^{N} s_k t_k, \tag{55}$$

with $s, t = x, y, z$ being the components of the internal position vectors, can be used in giving information on the overall shape of the cluster.[33] Thus, if they all have the same value, the cluster is essentially spherical (it could, in principle, also form a cube which, however, is highly unlikely), whereas two eigenvalues that are larger than the average suggest a lens-like shape, and two eigenvalues that are smaller than the average suggest a cigar-like shape. The eigenvalues, $I_{\alpha\alpha}$, are related to the descriptors of Eq. (48).

3.3 Atomic Positions. – The internal positions, Eq. (53), describe precisely the structure of the cluster, but is very difficult to depict. By analyzing them, it is possible to identify symmetry properties of the cluster. However, whereas in the mathematical sense it is relatively straightforward to check whether any proposed symmetry operation indeed maps the system on itself, the fact that the nuclear coordinates have been obtained in a numerical optimization makes it necessary to allow for some uncertainty in the nuclear coordinates. This fact can make it less trivial to determine the symmetry properties of larger clusters.

Moreover, it shall be remembered that even a high-symmetric structure that has been slightly disturbed (maybe just by moving one single atom a little, or adding a single atom to a high-symmetric structure) may appear as having essentially no symmetry at all. Therefore, although the symmetry of a cluster does provide some information, it is not exhaustive.

For a highly symmetric cluster, the radial distances of Eq. (53) will take only some few values, whereas a highly distorted system will have a whole distribution of radial distances. Therefore, plotting the radial distances as a function of cluster size can give information on the construction of atomic shells. However, just as above, it shall be remembered that for a highly symmetric cluster with one extra atom, the center, Eq. (52), will be displaced and, therefore, the occurrence of atomic shells will then not be so easily identified.

The pair-correlation functions,

$$g_{\mathrm{AB}}(d) = \sum_{i=1}^{N_{\mathrm{A}}} \sum_{j=1}^{N_{\mathrm{B}}} g_0\left(\left|\vec{R}_i - \vec{R}_j\right| - d\right), \tag{56}$$

where $g_0(x)$ is a normalized function that is sharply peaked at $x = 0$ (in the extreme case, it becomes a Dirac δ function), and where the i and j summations run over all atoms of type A and B, respectively (these types may also be identical), give information on the short- and long-range order. For an infinite, periodic, crystalline system, $g_{\mathrm{AB}}(d)$ consists of regularly spaced peaks and vanishes in between, whereas $g_{\mathrm{AB}}(d)$ becomes increasingly unstructured with increasing d for a disordered system.

3.4 Structural Similarity. – A frequently occurring question is, whether two objects are similar in some sense. This may, *e.g.*, be the case when discussing whether the cluster of N units (or atoms) is similar to the one of $N-1$ units plus one additional unit, *i.e.*, whether the clusters show a regular growth in which unit after unit is added to a given core. Or, it may be discussed whether the cluster is similar to a piece of the infinite, periodic crystal or of some other larger system (for instance a large, high-symmetry object like an icosahedron). Due to structural relaxations, it will hardly ever be found that the two objects are exactly identical, but only more or less approximately. Therefore, it is important to devise descriptors that can quantify the similarities.

We have introduced similarity functions to this purpose.[33] They are based on comparing either the radial distances or the interatomic distances. Let us assume that we want to compare two objects with N_1 and N_2 atoms. For clusters of only one type of atoms the objects could be the cluster of N atoms and the one of $N - 1$ atoms, or it could be the cluster of N atoms and a spherical cut-out of the crystal, also of N atoms.

Let us assume that $N_1 \geq N_2$. First we calculate all the relevant quantities (*i.e.*, radial distances or interatomic distances) for the smaller object. These quantities are sorted and denoted $x_i^{(0)}, i = 1, 2, 3, \ldots, N_l$, where N_l is either N_2 or $\frac{N_2(N_2-1)}{2}$ when using the radial distances or the interatomic distances as structure descriptors, respectively. Subsequently, we consider all $\binom{N_1}{N_2}$ structures that are obtained by removing N_1–N_2 atoms of the larger system and keeping the N_2 remaining atoms at their unchanged positions. For each of those, we calculate and sort the relevant quantities, x_i, $i = 1,2,3,\ldots,N_l$. The two structures can then be compared through the quantity

$$q = \left[\frac{1}{N_l}\sum_{i=1}^{N_l}\left(x_i^{(0)} - x_i\right)^2\right]^{1/2}. \tag{57}$$

Of the $\binom{N_1}{N_2}$ different values of q we choose the smallest one, $q_{\min}$, (this is the one for which the two structures are most similar) and define finally the similarity function

$$S = \frac{1}{1 + q_{\min}/u_l}, \tag{58}$$

where u_l is some length unit. S approaches 1 if the two objects are very similar, and 0 if they are very different.

As we shall see below, we can use the similarity functions in quantifying a number of relevant issues: whether the cluster of N units is similar to that of $N - 1$ unit, whether the cluster is similar to a piece of the crystal, whether the cluster is similar to a piece of a large icosahedron, and whether the structure that has been optimized with one approach is similar to the one that has been optimized with another approach.

Finally, for clusters with more types of atoms appropriate generalizations can easily be constructed so that only the same type of atoms (when studying the radial distances) or of pairs of atoms (when studying the interatomic distances) are compared.

3.5 Structural Motifs. – With the similarity functions introduced above it is possible to obtain a quantitative measure for whether a cluster is more similar to one type of structure or another type of structure. However, in some cases it may be more interesting to obtain information about whether a specific structural motif is found or not. Or, equivalently, which structural motifs are found more frequently. To this end, the common-neighbour analysis[46,47] can be useful.

Let us consider a cluster consisting of only one type of atoms. One defines a cut-off distance, $d_{\rm cut}$, that determines whether two atoms are bonded or not: if the interatomic distance is smaller than $d_{\rm cut}$, the atoms are bonded, otherwise not. Subsequently, one considers all the $\binom{N}{2}$ different pairs of atoms and calculate for each of those, three integers j, k, and l. j is the number of other atoms that are bonded to both the two reference atoms, *i.e.*, the number of common or shared neighbours. k is the number of bonds between the shared atoms (excluding bonds to any of the two selected reference atoms). Finally, l is the number of bonds in the longest sequence of bonds between the shared neighbours.

As simple examples we consider the two two-dimensional, infinite, periodic lattices that are shown in Figures 7 and 8. Since all atoms are equivalent, we can focus on some simple structural motifs. In Figure 7, *e.g.*, most pairs of atoms result in $i = j = k = 0$, which is not interesting. Exceptions are pairs of atoms like the two squares in the figure. For this pair, one atom (the one between the two atoms) is a shared neighbour, resulting in $(i,j,k) = (1,0,0)$. For the pair of atoms marked with open circles, two atoms are shared neighbours, but these are not bonded to each other. Thus, in this case we have $(i,j,k) = (2,0,0)$. It is easy to see that these two sets of indices occur in the same amount.

On the other hand, for the structure of Figure 8 three types of pairs lead to non-vanishing integers. For the pair marked with the squares we have $(i,j,k) = (1,0,0)$, whereas for the pair marked with the open circles we find $(i,j,k) = (2,0,0)$. Finally, the pair marked with the stars gives $(i,j,k) = (2,1,1)$.

For a finite, two-dimensional structure, we will most likely find neither the structure of Figure 7, nor the one of Figure 8, but maybe something in between. By analysing the occurrence of the different sets of integers, we get information on the frequency with which the two types of structures occur. This means that finding the same number of $(i,j,k) = (2,0,0)$ and $(i,j,k) = (2,1,1)$ suggests that the structure of Figure 8 is dominating, whereas a small occurrence of $(i,j,k) = (2,1,1)$ compared to that of $(i,j,k) = (2,0,0)$ suggests that the structure of Figure 7 is dominating.

A related approach has recently been proposed by Cheng *et al.*[48] As above, a cut-off distance is defined that determines the maximum inter-atomic distance

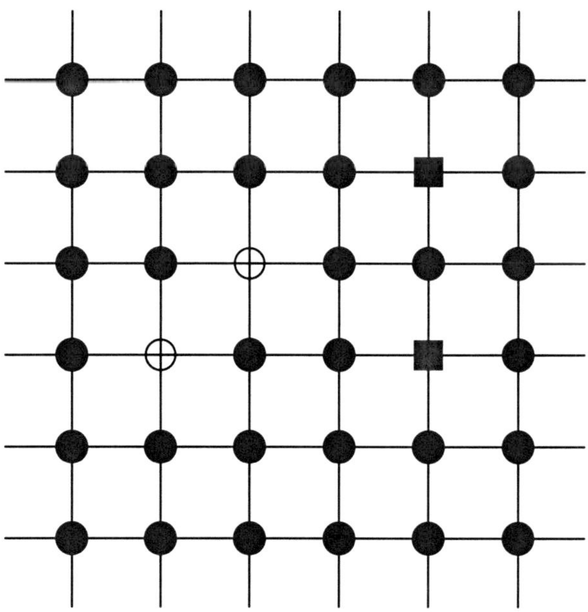

Figure 7 *A schematic representation of an infinite, periodic, two-dimensional lattice. The lines mark interatomic bonds. For details, see the text*

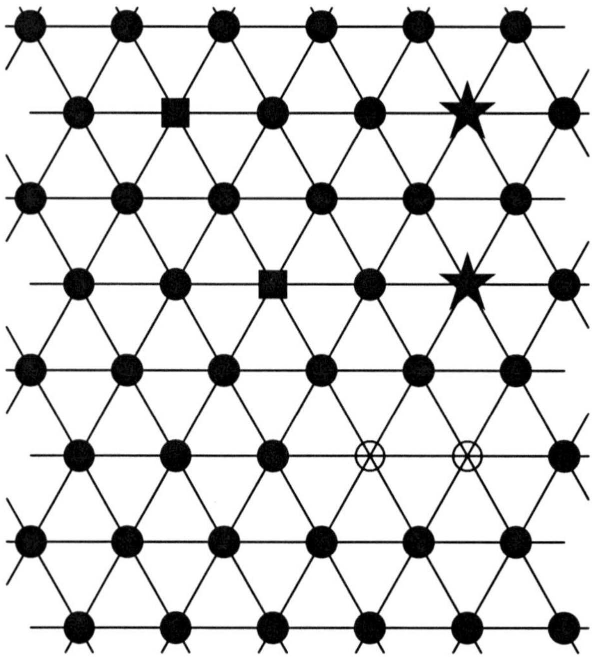

Figure 8 *As Figure 7, but for a different structure*

for atoms that are considered bonded. Subsequently, for each structure a so-called connectivity table is constructed. This is a 12-dimensional array where the pth element is the number of atoms with p nearest neighbours. The approach is best suited for comparing different structures of the same number of atoms.

3.6 Phase Transitions. – In the previous subsections we discussed ways to identify structural motifs for a given cluster. Thus, once the structure has been optimized, it can be analyzed to which extend the structure resembles one or another reference system. In this way changes from predominantly one structure to another as the clusters become large can be identified.

In this subsection we shall discuss means to identify another type of phase transitions, *i.e.*, changes for a fixed size but as a function of temperature. Theoretical studies devoted to these transitions have to include some means of describing temperature, *e.g.*, by being based on molecular-dynamics simulations (cf. Section 2.2). However, when increasing the temperature in a such calculation, first of all the atoms will become less localized and, therefore, it may not be easy to identify temperatures for which the structure changes.

One possibility, used, *e.g.*, by Rey and Gallego,[49] is to use the eigenvalues of the matrix for the moments of inertia, Eq. (55). These quantities will first of all be able to catch transitions from a solid-like to a liquid-like behaviour, where the atoms change from being more or less confined at somewhat fixed positions to becoming much more mobile, so that the cluster may become overall more spherical, at least when averaging the atomic positions over a finite time interval.

Another possibility, also used by Rey and Gallego[49] and put forward by Jellinek *et al.*,[50] is to consider the fluctuations in the inter-atomic distances. This can be done through the parameter (the so-called Lindemann index)

$$\delta = \frac{2}{N(N-1)} \sum_{i<j=1}^{N} \frac{\left(\langle d_{ij}^2 \rangle - \langle d_{ij} \rangle^2\right)^{1/2}}{\langle d_{ij} \rangle}, \tag{59}$$

where $\langle d_{ij} \rangle$ is the time average of the distance between atom i and j.

This parameter may even be resolved so that individual atoms can be analyzed, as done, *e.g.*, by Shimizu *et al.*[51] They defined for each atom (k)

$$\delta_k = \frac{1}{\langle N^{(k)} \rangle} \sum_{i[\neq k)} \frac{(\langle d_{ik}^2 \rangle - \langle d_{ik} \rangle^2)^{1/2}}{\langle d_{ik} \rangle}, \tag{60}$$

where $\langle N^{(k)} \rangle$ is the average of the number of nearest neighbours for the kth atom, and the summation is over only those neighbours.

In their study, Rey and Gallego[49] studied clusters with two types of atoms, A and B, and, therefore, they also introduced a parameter that should monitor a change from an ordered to a segregated phase. From the internal positions for the two types of atoms, Eq. (53), they defined

$$\vec{d} = \frac{1}{N_A} \sum_{i=1}^{N_A} \vec{r}_i^{(A)} - \frac{1}{N_B} \sum_{i=1}^{N_B} \vec{r}_i^{(B)}, \tag{61}$$

where the first summation runs over all A atoms and the second over all B atoms. Finally, the authors study

$$d = \left| \vec{d} \right| \tag{62}$$

as a function of temperature.

The so-called adjacent matrices (see, *e.g.*, 51,52) provide other ways of studying the temporal development of the structure. These $N \times N$ matrices are at any time t defined as

$$A_{ij}(t) = \begin{cases} 1 & \text{if } d_{ij} < d_{\text{cut}} \\ 0 & \text{if } d_{ij} \geq d_{\text{cut}} \end{cases}, \tag{63}$$

where d_{cut} is a pre-defined cut-off radius that describes when two atoms are considered bonded to each other.

Shimizu *et al.*[51] defined from these a distance index for each atom through

$$D_i(t) = \left\{ \sum_{j=1}^{N} \left[A_{ij}(t + \Delta t) - A_{ij}(t) \right] \right\}^{1/2}, \tag{64}$$

whereas Calvo and Spiegelmann[52] compared the adjacent matrix for the cluster with that of a reference structure (*e.g.*, a piece of the solid) through

$$l = \frac{1}{2}\sum_{i,j=1}^{N}\left(A_{ij} - A_{ij}^{(0)}\right)^2, \tag{65}$$

where $A_{ij}^{(0)}$ are the elements of the adjacent matrix of the reference system.

A problem with the adjacent matrices is that through the simple interchange of two equivalent atoms, the quantity l will change, although the structure looks the same.

Finally, from the energy as a function of temperature, $E(T)$, one may calculate heat capacities

$$C_V = \frac{\partial E(T)}{\partial T} \tag{66}$$

that shows peaks at temperatures where a phase transition occurs. It is important to remember that $E(T)$ also contains a contribution from entropy (cf., *e.g.*, 53,54) which for instance can be calculated by means of the partition function, although it shall be emphasized that this is not trivial. $E(T)$ or $C_V(T)$ provide the so-called caloric curves.

4 Examples for Optimizing the Structure of Clusters

After having discussed methods for searching the global total-energy minimum for larger systems, most notably clusters, in Section 2 and presenting, in Section 3, various ways of analyzing and presenting the results, we shall in this section present some few, subjectively chosen, examples that illustrate the results of the two preceding sections. We emphasize that the presentation concentrates on illustrating the concepts we have discussed and not on giving an exhaustive or complete discussion on the systems we shall treat. For more comprehensive discussions the reader is referred to, *e.g.*, the recent review by Baletto and Ferrando[55] as well as the older discussions by Wales and Scheraga,[56] by Doye,[57] and by Hartke.[36] Moreover, the general conclusions, both on the performance of the different computational approaches and on the interpretation of the results in terms of the descriptors, should be valid also when applying the methods on other larger systems of low symmetry.

4.1 One-Component Lennard-Jones Clusters. – Lennard-Jones clusters remain being the reference model system for testing new theoretical developments. For clusters consisting of only one type of atoms, the total energy is written as

$$E = 2\varepsilon \sum_{i\neq j=1}^{N}\left[\left(\frac{\sigma}{d_{ij}}\right)^{12} - \left(\frac{\sigma}{d_{ij}}\right)^{6}\right] \tag{67}$$

with d_{ij} being the interatomic distance between atoms i and j. ε defines an energy scale and σ a length scale. Therefore, independently of the precise values of these two parameters, the same structures and total energies are found (except for this scaling). Exactly this property makes these clusters excellent model systems for testing and comparing various theoretical approaches.

Northby[27] optimized the structures of these clusters for $13 \leq N \leq 147$ using the lattice-growth method of Section 2.6. He presented first of all a discussion of how the structure evolves when atom after atom is added to the cluster. By studying $\Delta_1 E(N)$ of Eq. (49) he identified a number of values for which the clusters were particularly stable.

The purpose of the later study of Tsai and Jordan[8] was different. They used an eigenmode method of Section 2.4 in identifying as many local total-energy minima and saddle points as possible for $7 \leq N \leq 13$. Some of their results were presented above in Figure 2 and Table 1.

The basin-hopping method of Section 2.9 was presented in 1997 by Wales and Doye.[34] As an illustration they applied the method on Lennard-Jones clusters with $2 \leq N \leq 110$. They presented a detailed compilation of the global total-energy minima for all those structures together with references to the work where the structures for the first time was identified. They found that, with only three exceptions, they could reproduce the earlier results, and for the three exceptions they found new structures with lower total energies. Although this does suggest that their method is highly reliable and also efficient, the authors themselves were slightly less optimistic, because, as they state, 'global optimization for Lennard-Jones clusters at most sizes is a relatively easy task.' Nevertheless, those as well as later results for other types of clusters have clearly demonstrated that the basin-hopping method is a powerful method for optimization structures of large systems.

In their presentation of the random-tunneling algorithm method (briefly mentioned in Section 2.2), Jiang *et al.*[12] showed that for smaller Lennard-Jones clusters ($N \leq 38$) their method was computationally highly efficient, requiring one to two orders of magnitude less total-energy calculations than those of some other structure-optimization methods. In addition, they presented the lowest total energies for $21 \leq N \leq 100$. Compared with the results of Wales and Doye[34] discussed above, the results agree in very many cases, but there is at least one exception ($N = 98$) where the total energy as calculated by Jiang *et al.* is lower than the one calculated by Wales and Doye.

In a later work, Doye and Wales[53] studied the properties of a single Lennard-Jones cluster ($N = 38$) as a function of temperature. They found that at low temperatures, the structure is like that of a piece of the fcc crystal structure, but at somewhat elevated temperatures it adopts a structure with significant similarity with an icosahedron, until it at even higher temperatures becomes liquid-like.

Doye and Calvo[54] included entropy effects for large Lennard-Jones clusters and used the results in estimating which structures will be dominating as a function of size and temperature of the cluster. Their main results are reproduced in Figure 9. Most interesting may be that the crystal structure is not found before N is well above 100 000.

Frantz[58] studied the temperature-dependent properties of Lennard-Jones clusters using a Monte Carlo method (Section 2.2). The two parameters ε and σ

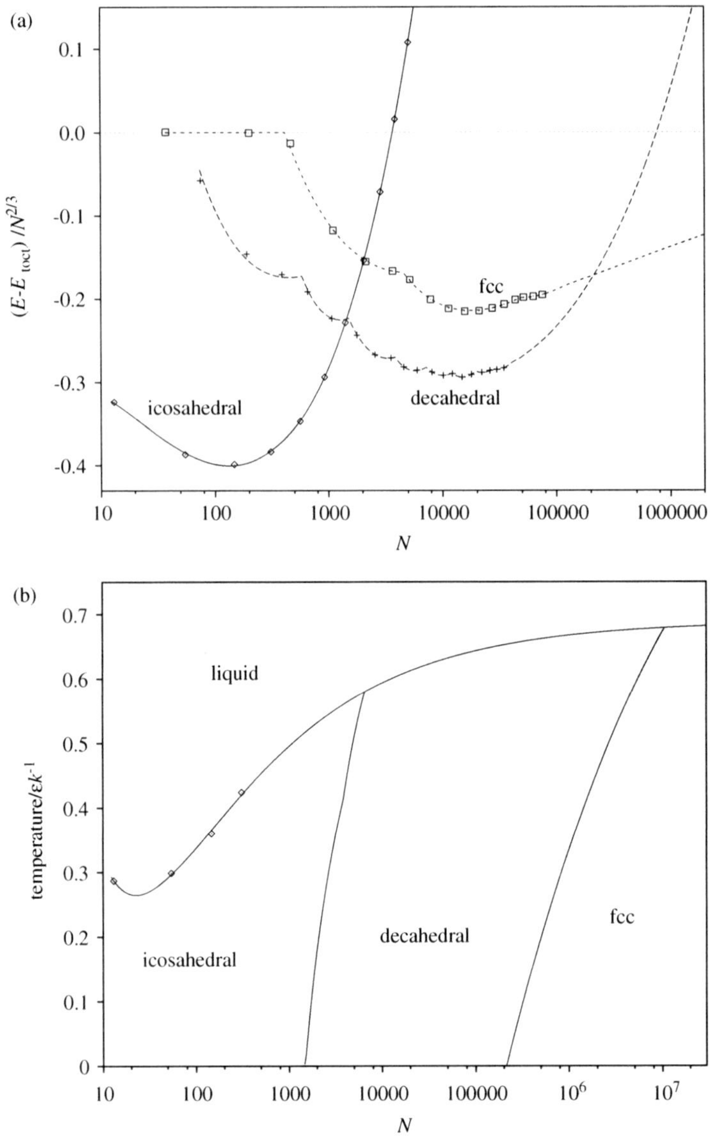

Figure 9 *(a) shows the energies of various structural types of Lennard-Jones clusters. The circles represent calculated values, and the continuous curves are estimates based on fitting the calculated results. (b) shows the phase diagram for the clusters resulting from the results of (a). Reproduced with permission of American Chemical Society from 54*

were fixed at the values that are relevant for Ar systems, *i.e.*, $\varepsilon = 119.4$ K. Figure 10 shows some of the results of the study for $T = 0$. In particular, it is interesting to note that $\Delta_2 E(N)$ and $\Delta_{12} E(N)$ have peaks for essentially the same values of N, implying that a cluster that is particularly stable compared with the two clusters with one atom more or less, also is particularly stable when comparing it with the second-lowest isomer of the same size.

Subsequently, Frantz calculated the total energy as a function of temperature for the different clusters, which resulted in the curves of Figure 11. The curves are seen to posses a change in the slope over a more or less narrow temperature interval, that becomes more narrow when the clusters are larger. These changes in the slope signal phase transitions (see Section 3.6). Ultimately, for the infinite system the slope of the energy will become discontinuous at the temperature of a phase transition, but for the smaller, finite systems, this transition is obviously smeared out.

The derivative, Eq. (66), will have a peak at the phase-transition temperature. In Figure 12 the position and height of this peak is shown. For $30 \leq N \leq 37$, there are two peaks, which also is the case for $N = 27$ and $N = 58$, although the second peak in these two cases is very broad. Comparing Figures 10 and 12 suggests that there is some correlation between high stability (peaks in Figure 10) and sharper transition (high peaks in Figure 12).

In their study, Cheng *et al.*[48] compared different Lennard-Jones clusters with 98 atoms using their proposed connectivity tables. The results for eight isomers are reproduced in Table 2. It is clear that all clusters have comparable total energies as well as number of nearest-neighbour pairs. Therefore, in order to clearly distinguish between the isomers, further descriptors are needed. Their connectivity tables are indeed one way of distinguishing, as the table clearly

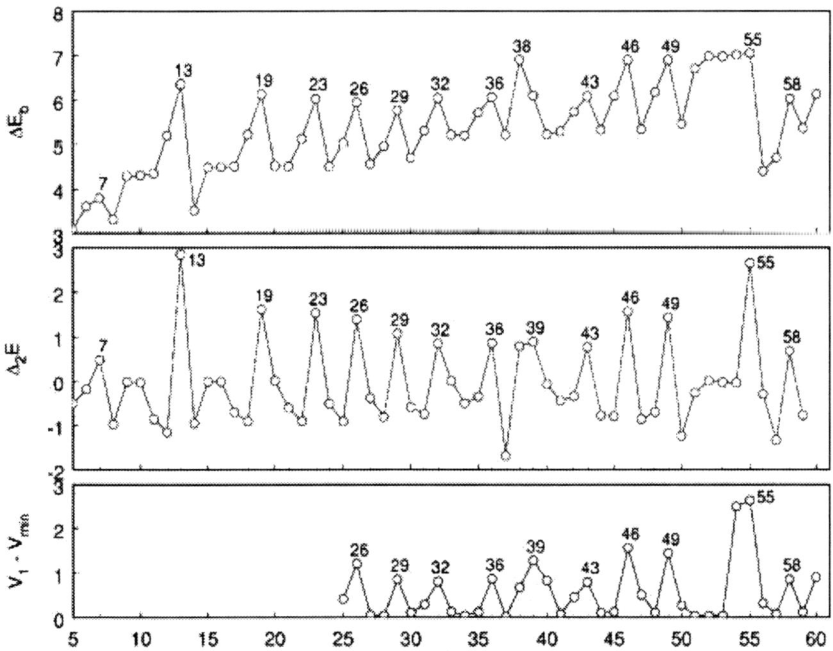

Figure 10 *From top to bottom:* $\triangle_1 E(N)$, $\triangle_2 E(N)$, *and* $\triangle_{12} E(N)$ *for Lennard-Jones clusters. Reproduced with permission of American Institute of Physics from 58*

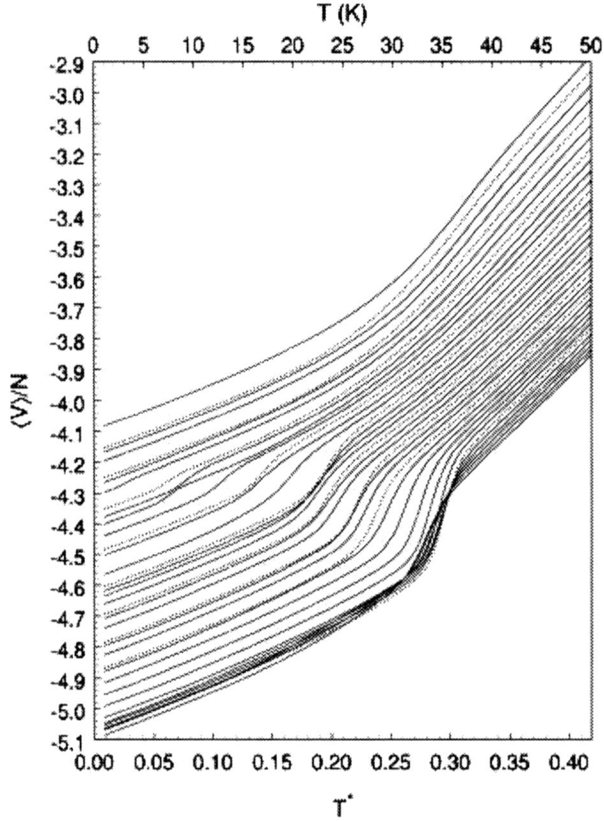

Figure 11 *The total energy per atom as a function of temperature for Lennard-Jones clusters with between 20 (top curve) and 60 atoms (bottom curve). Reproduced with permission of American Institute of Physics from 58*

demonstrates. On the other hand, Cheng *et al.* did unfortunately not compare their descriptors with other ones like, *e.g.*, the common neighbour analysis.

Both Cheng *et al.*[48] and Cheng and Fournier[43] used Lennard-Jones clusters in demonstrating that the computational efforts related to optimizing the structure of clusters in an unbiased way could be drastically reduced through the introduction of some ways of avoiding that the structure search revisits parts of the structure space. In the former work this is made possible by using the connectivity tables as descriptors, in the latter through the descriptors of Eqs. (45)–(48).

4.2 Two-Component Lennard-Jones Clusters. – By generalizing the Lennard-Jones potential to

$$E = 2\sum_{i \neq j=1}^{N} \varepsilon_{ij}\left[\left(\frac{\sigma_{ij}}{d_{ij}}\right)^{12} - \left(\frac{\sigma_{ij}}{d_{ij}}\right)^{6}\right], \tag{68}$$

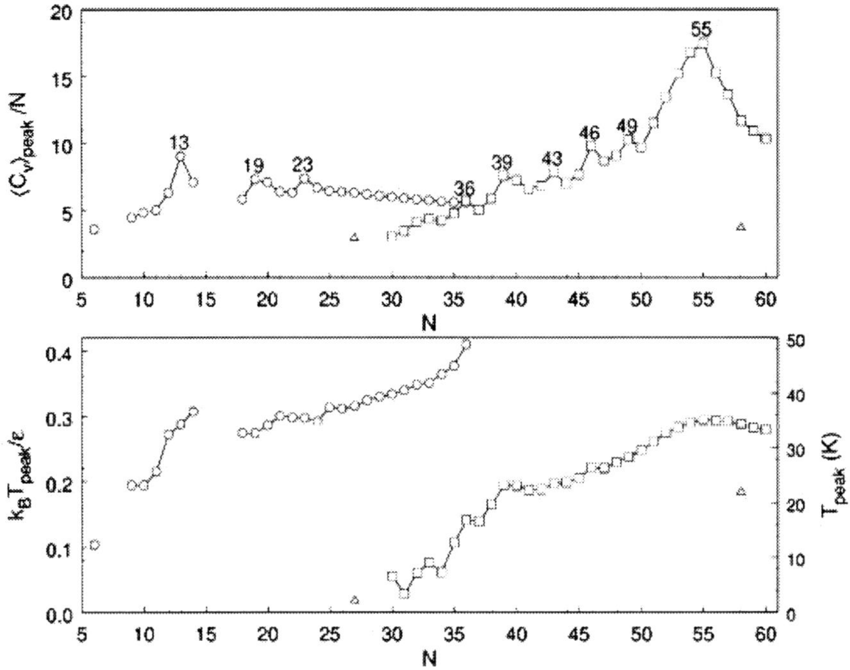

Figure 12 *The upper plot shows the peak value of the heat capacity as a function of N and the lower plot shows the temperature of the peak heat capacity, all for Lennard-Jones clusters. In some cases ($30 \leq N \leq 37$ as well as $N = 27$ and $N = 58$) two peaks are found. Reproduced with permission of American Institute of Physics from 58*

i.e., the parameters σ and ε depend on the (type of) atoms, it is possible to model clusters consisting of more types of atoms.

Both Gregurick *et al.*[37] and Rey and Gallego[49] have presented results of such studies where the clusters consist of two types of atoms. Gregurick *et al.*[37] studied A_N and BA_N clusters, whereas Rey and Gallego[49] studied $A_{13}B_{13}$ clusters. Here, we shall only briefly discuss the second work. In order to reduce the number of parameters Rey and Gallego fixed $\sigma_{AA} = \sigma_{BB}$, $\sigma_{AB} = (1 + \Delta)\sigma_{AA}$, $\varepsilon_{AA} = \varepsilon_{BB}$, and $\varepsilon_{AB} = \alpha\varepsilon_{AA}$. Setting $\alpha = 1$ and varying Δ, the parameter d of Eq. (62) possesses the temperature dependence shown in Figure 13. A separation of the two types of atoms is clearly recognizable, and also the turning-on of a melting at a certain temperature (depending on Δ) can be identified.

4.3 Morse Clusters.

– As another generalization one may consider clusters consisting of atoms that are interacting *via* Morse potentials. Assuming that we have only one type of atoms, this gives

$$E = \frac{\varepsilon}{2} \sum_{i \neq j = 1}^{N} \exp[\alpha(d_0 - d_{ij})] \cdot \left\{ \exp[\alpha(d_0 - d_{ij})] - 2 \right\}. \tag{69}$$

Table 2 *Properties of various isomers of Lennard-Jones clusters with 98 atoms. n_{nn} is the number of nearest-neighbour pairs, E is the total energy, and $1 \leq i \leq 12$ is the number of atoms that have i nearest neighbours. From 48*

n_{nn}	E	1	2	3	4	5	6	7	8	9	10	11	12
432	−543.665	0	0	0	0	0	24	24	0	4	18	0	28
430	−541.406	0	0	0	1	1	24	20	2	5	16	2	27
437	−543.643	0	0	0	0	0	19	9	26	12	0	3	29
437	−543.547	0	0	0	0	0	20	7	27	12	0	3	29
441	−539.720	0	0	0	0	0	29	19	13	3	7	4	32
428	−541.895	0	0	0	0	0	23	22	5	11	9	1	27
430	−541.436	0	0	0	0	0	21	17	12	11	4	2	29
429	−541.870	0	0	0	0	0	24	15	18	3	9	0	29

As for the Lennard-Jones potential, two parameters, d_0 and ε, can be fixed without loss of generality, as they do nothing but define the length and energy scales, respectively. However, the additional parameter α can be varied and different values will lead to qualitatively different results. Thus, the larger α, the more confined is the potential, and smaller values of α give potentials that are more long-ranged.

Doye and Wales[59] as well as Roberts *et al.*[60] studied such clusters for different values of α. As representative results we show in Figure 14 the calculated $\Delta_2E(N)$ for $N \leq 80$ for different values of α from the study of Doye and Wales.[59] The authors noticed that for $\alpha = 6$ the interatomic potential shows strong similarities with the Lennard-Jones potential and, therefore, the particularly stable clusters for Morse potentials with $\alpha = 6$ resemble those that were found for Lennard-Jones potentials. Otherwise, it is readily seen in the figure that the results depend strongly on α. Ultimately, this suggests that those values of N for which the clusters are particularly stable will depend critically on the details of the interatomic potentials.

4.4 Sodium Clusters. – Since the pioneering work of Knight *et al.*[61] more than 20 years ago on mass abundance spectra of sodium clusters, these have been the prototype for metallic clusters. Knight *et al.* measured the relative abundance of sodium clusters with up to somewhat more than 70 atoms. Instead of finding a smooth curve as a function of N, certain sharp peaks were observed, indicating that for certain values of N, particularly large amounts of this cluster were found, *i.e.*, that clusters with these sizes were particularly stable.

In order to explain the occurrence of these so-called magic numbers, Knight *et al.*[61] considered one of the simplest possible models, *i.e.*, the jellium model. For Na, all but the outermost 3*s* electron are tightly bound to the nucleus, whereas the valence 3*s* electron is fairly delocalized. Assuming that this electron is so delocalized that it does not feel the precise position of the core electrons and the nuclei, the charge of the latter can be smeared out (then forming the

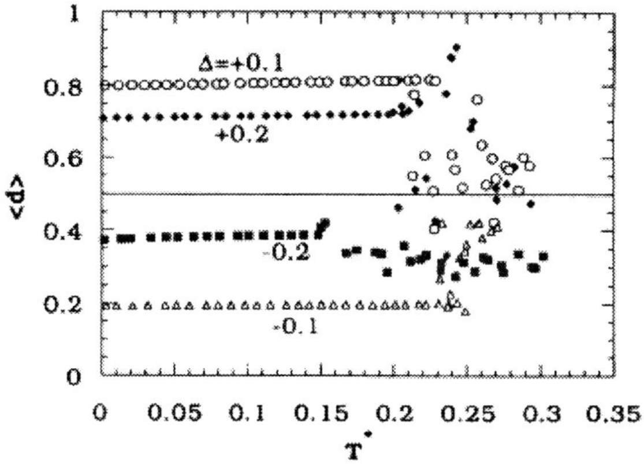

Figure 13 *Temperature dependence of the average separation between the centers of mass of the A and B atoms in the $A_{13}B_{13}$ Lennard-Jones clusters for different values of the parameter Δ. Reproduced with permission of American Physical Society from 49*

jellium) and the properties of the valence electrons can be studied under the assumption that these move in the potential created by the jellium.

In the simplest case, the sodium clusters are spherical. In this case, the valence electrons move in a spherically symmetric potential, and by calculating the total energy as a function of size of the cluster, Knight *et al.*[61] could show that the magic numbers correlate with electronic shell-filling, very similar to what is found for the inert gases in the periodic table. It turned out that $\Delta_2 E(N)$ of Eq. (50) has peaks whenever an electronic shell (or, for larger clusters, a group of nearly-degenerate shells) is filled.

Although the jellium model has been successful in explaining very many properties of clusters of simple metals (see, *e.g.*, 62,63), the clusters are after all made up of discrete atoms and, therefore, a highly relevant question is whether the predictions of the jellium model will hold when explicitly including the positions of the atoms in the calculations.

An early attempt in this direction is the work of Iñiguez *et al.*[64] who used density-functional theory in studying the properties of various metal clusters, including those of sodium. However, in order to simplify the calculations, they approximated the true three-dimensional potential with the spherically symmetric component around the center of mass of the cluster. In that case, the spherical symmetry, that also was used in the jellium studies mentioned above, is recovered. Therefore, it may not surprise that the calculated magic numbers coincide closely with those of the jellium calculations. In fact, Balian and Bloch[65] have shown quite generally, that electronic shells will tend to form group of several nearly-degenerate shells with a pattern that is dictated almost exclusively by the symmetry of the system and largely independent of the details of the potential.

Using a tight-binding model together with a Monte Carlo growth method (see Section 2.7), Poteau and Spiegelmann[30] studied Na_N clusters with $4 \leq N \leq 21$. For the spherical jellium model, magic numbers are found for N

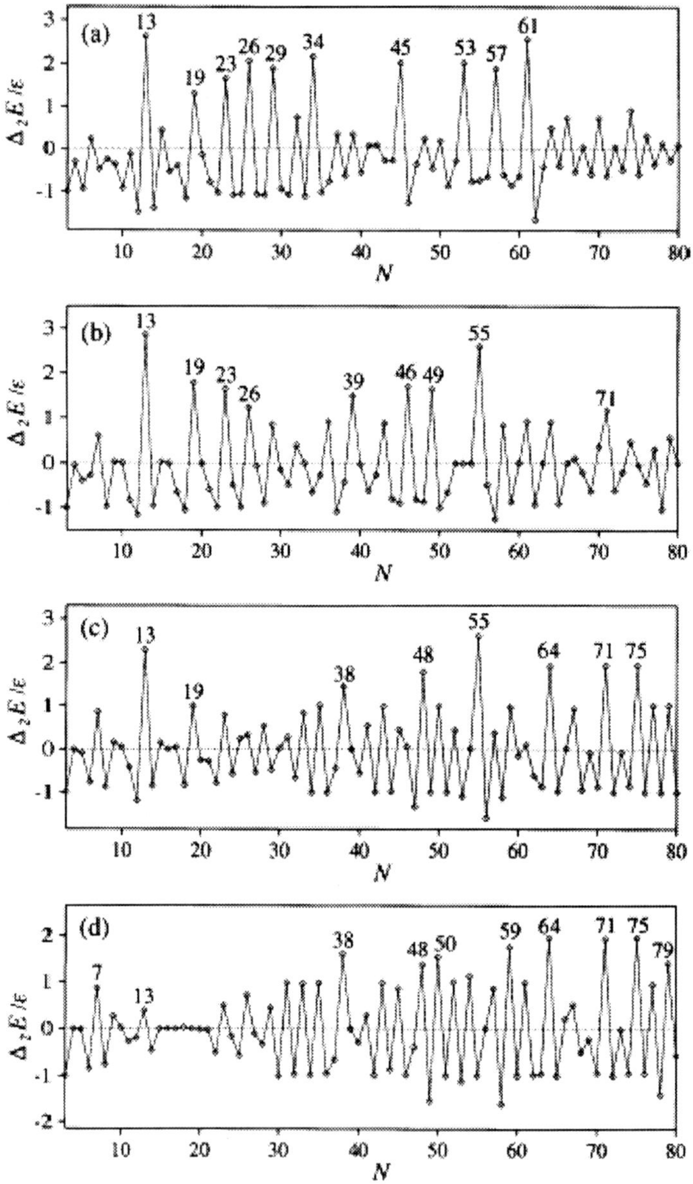

Figure 14 *$\Delta_2 E(N)$ for clusters for which the atoms interact via Morse potentials. The parameter α equals 3, 6, 10, and 14 from the top to the bottom, respectively. Reproduced with permission from 59*

= 2,8,18,20,34,... Studying $\Delta_2 E(N)$, Poteau and Spiegelmann found some indication of particularly high stability for $N = 8$, and to a lesser extend for $N = 18$ and $N = 20$. Much more pronounced was an even-odd oscillation, *i.e.*, for N odd, the clusters were significantly less stable than for N even. A further interesting result in this respect is that only for $N = 8$ and $N = 20$ the clusters were found to have an overall spherical structure, which was found by analysing the quantities of Eq. (55). This finding was recently confirmed by Solov'yov *et al.*[66] who used different parameter-free electronic-structure methods in studying the structural and electronic properties of Na_N clusters with N up to 20. Both for neutral and singly charged clusters, they used the method for structural optimization, that is briefly described in Section 2.6, and calculated subsequently various properties, including total energy per atom, the dipole moment, the moments of inertia, the ionization potential, and the polarizability. The study shall, however, not be discussed here further.

In another work, Poteau *et al.*[67] studied the temperature-dependent properties of the Na clusters using a tight-binding model together with a Monte Carlo thermodynamic method (cf. Section 2.2). By studying the Lindemann index [Eq. (60)] as a function of temperature they could identify phase-transition temperatures, although, as also seen by analysing the heat capacity and the moments of inertia as functions of temperature, these transitions are not sharp.

More recently, the same group[52] extended the study of the temperature dependence of the properties of Na_N clusters to $8 \leq N \leq 147$, although they did not consider all sizes. As above, heat capacity, Lindemann index, and moments of inertia were used in identifying melting temperatures. Moreover, the authors used two different models, *i.e.*, an embedded-atom method (see below) and a tight-binding model, in calculating the total energy for a given structure. Finally, from the heat capacity as a function of T, they also calculated the latent heat,

$$L(N) = \int_{T_{\min}}^{T_{\max}} [C_V(T) - (3N - 6)kT]dT, \tag{70}$$

where the integration interval $[T_{\min}; T_{\max}]$ is over the temperature interval where C_V has its peak due to the phase transition (*i.e.*, melting).

Their results, both for neutral and singly charged clusters, are reproduced in Figure 15 for the two models. It is clear that none of the curves is a smooth function of cluster size, and also that the two models yield quite different results. However, it shall be remembered that only small inaccuracies in the energies easily give inaccuracies in temperatures of the order of several 100 K: 300 K corresponds to only 0.025 eV.

Kronik *et al.*[68] studied the structure and the polarizability of Na_N clusters for $N \leq 20$ using a parameter-free density-functional method. When the clusters are particularly stable due to electronic shell-filling effects, the electrons are hardly able to respond to external potentials like that of an electrostatic field and, accordingly, the polarizability is particularly low. This trend has been seen in experiment (see 68). Kronik *et al.* were able to reproduce this trend, cf. Figure 16. Furthermore, they found that for clusters with more than roughly 10 atoms several isomers were energetically very close, although they have quite different polarizabilities, so that at not very elevated temperatures the measured polarizability becomes a superposition of the ones of more different isomers and, accordingly, a very unstructured function of cluster size.

Finally, in a very recent study Noya *et al.*[69] performed unbiased structure optimizations for maybe the largest cluster range that so far has been considered, *i.e.*, Na_N clusters with $N \leq 380$. They used the basin-hopping method, described in Section 2.9, and compared the results of two model potentials, *i.e.*, the Gupta potential and the Murrell-Mottram potential. Their central result is the magic numbers, which can be extracted from Figure 17. For the smallest clusters (*i.e.*, $N < 100$), the two potentials seem to predict quite similar sequences of magic numbers, but for the larger clusters, deviations are seen. In particular, the magic cluster for $N = 309$ (an icosahedron) which is found with the Murrell-Mottram potential, is not found with the Gupta method.

4.5 Other Metal Clusters. – As we discussed in Section 1, one has to make a compromise between either accurate calculations on few, selected structures of some selected clusters or, alternatively, less accurate calculations using approximate descriptions of the total energy on a large number of structures and/or cluster sizes. Above, we briefly mentioned some of the approximate model potentials and here we shall extend the discussion and apply those for several different types of metals.

For metals the electrons are more or less delocalized and do not participate in the formation of directional, covalent bonds. Instead, metal atoms most often prefer bonding situations characterized by high coordination and, if being able to separate the total energy into atomic components, the binding energy of the individual atoms will be the higher the more nearest neighbours each atom has. This property is the basis for most of the approximate potentials that have been devised for describing the interatomic interactions of metal atoms.

One popular class of potentials is formed by the embedded-atom approaches.[70–75] Here, the basis idea is that each atom is considered as being a guest embedded into the host formed by all the other atoms. The other atoms provide an extra electron density, into which the guest atom is placed. The energy costs related to this is described in forms of a function that depends on the electron density due to the host but at the site of the guest atom. This energy contribution is augmented with pair potentials.

Other potentials are the Gupta[76] and Sutton-Chen[77] ones where the total energy is written as simple functions of all interatomic distances. With few exceptions the studies we shall review in this subsection have been obtained using those potentials.

In a very general study, Doye and Wales[78] optimized all structures for clusters with N up to 80 for more different Sutton-Chen potentials according to which the total energy has the form

$$E = \varepsilon \sum_{i=1}^{N} \left[\frac{1}{2} \sum_{i \neq j=1}^{N} \left(\frac{a}{d_{ij}} \right)^n - c\sqrt{\rho_i} \right], \tag{71}$$

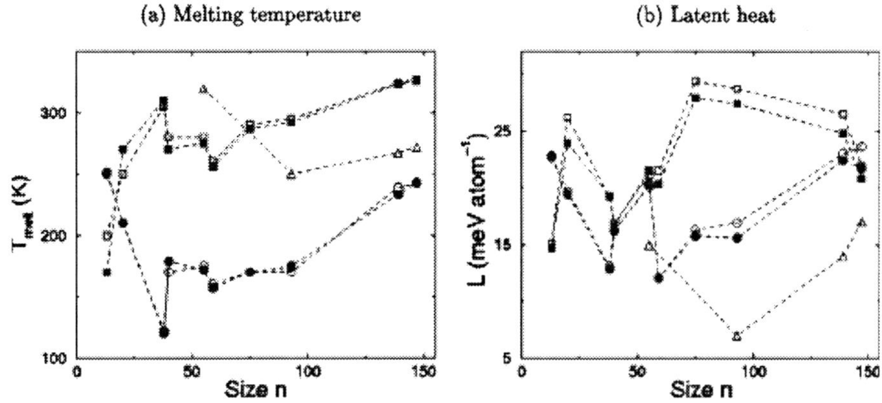

Figure 15 *Melting temperature and latent heat for sodium clusters of different sizes. Open and closed symbols mark results for the singly charged and the neutral clusters, respectively, and the squares and the circles represent results from calculations using the tight-binding and the embedded-atom model, respectively. Finally, experimental results are shown with the triangles. Reproduced with permission of American Institute of Physics from 52*

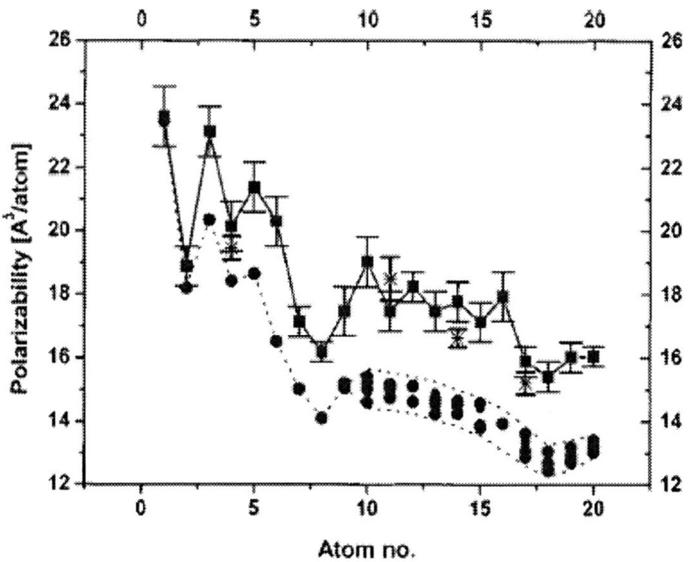

Figure 16 *Polarizability of Na clusters as a function of cluster size. The squares mark experimental results, the circles theoretical results at $T = 0$ K, and the stars theoretical results at $T = 750$ K. The dashed lines are guides to the eye. Reproduced with permission of American Physical Society from 68*

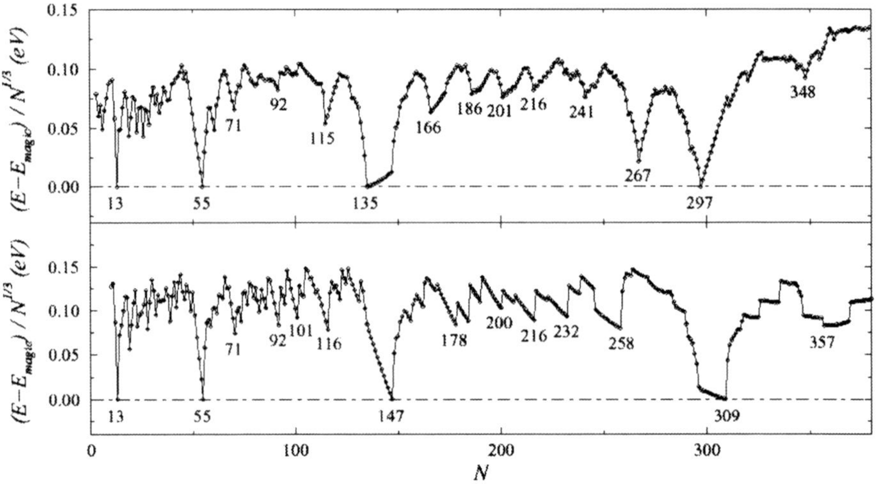

Figure 17 *Variation in the total energy for Na_N clusters as found with (upper panel) the Gupta potential and (lower panel) the Murrell-Mottram potential. Local minima correspond to particularly stable clusters. Reproduced with permission from 69*

where

$$\rho_i = \sum_{i \neq j=1}^{N} \left(\frac{a}{d_{ij}} \right)^m . \tag{72}$$

Different potentials, that can describe different elements, are obtained for different values of (n,m), and Doye and Wales[78] studied such ones [*i.e.*, $(n,m) = (12,6)$, $(n,m) = (9,6)$, and $(n,m) = (10,8)$], whereas c was set equal to values typical for different elements.

In another work, Poteau and Pastor used a phenomenological tight-binding model together with genetic algorithms in optimizing the structures of small metal clusters with one electron per atom.[79]

Ni_N clusters have been the subject of very many theoretical studies (see, *e.g.*, 33 and references therein). Wetzel and DePristo[80] studied Ni_N clusters for $24 \leq N \leq 55$ using the so-called effective-medium potential which is similar to the embedded-atom methods described above. They used a molecular-dynamics approach in optimizing the structure, and identified particularly stable structures through the total-energy difference between the energetically two lowest isomers for a given N, *i.e.*, $\Delta_{12}E(N)$ of Eq. (51).

On the other hand, Michaelian *et al.*[81] studied the structures of some few selected clusters of Ni, Ag, and Au using the Gupta potential.

In our own work,[33] that we here shall discuss briefly, we used the embedded-atom potential together with the Aufbau/Abbau method of Section 2.8 in optimizing the structure. The stability function, $\Delta_2E(N)$, Eq. (50), is shown in Figure 18. It shows a set of peaks that correspond to particularly stable clusters. It is interesting to notice that if one instead studies the total-energy difference

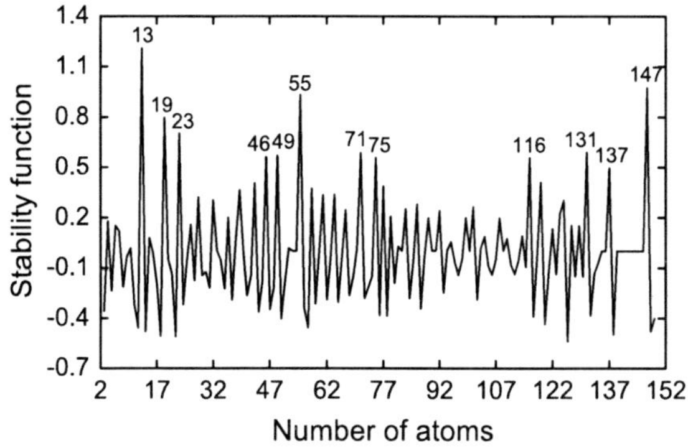

Figure 18 *The stability function $\Delta_2E(N)$ for Ni_N clusters as a function of cluster size*

between the energetically two lowest isomers, $\Delta_{12}E(N)$ of Eq. (51), that is shown in the lowest panel in Figure 19, then the clusters that are stable compared to the clusters with one atom more or less, in most cases also are stable compared with the energetically next higher isomer of the same size. Figure 19 shows also that the total-energy difference between the different isomers in very many case is small, *i.e.*, significantly less than 1 eV, meaning that for not too small clusters and at not very low temperatures, one will expect several isomers to be produced in an experiment. This finding is in accord with what we discussed above for Na clusters.[68]

The shape analysis (cf. Section 3.2 and Figure 20) shows that many of the particularly stable clusters are roughly spherical. Moreover, it is seen that for clusters with $N \geq 50$, the overall shape changes less rapidly as a function of N, and that the different isomers for the same N very often have the same overall shape.

Some similarity functions (cf. Section 3.4) are shown in Figure 21. It is interesting to notice that there are sizes where the structure of the clusters resembles a piece of the fcc crystal structure, but for the largest clusters of this study, the structure is most similar to that of an icosahedron. Thus, the transition to a small crystal will take place for N above 150 — actually, the results of Figure 9 suggest that the transition will take place first for much larger values of N.

Finally, we show in Figure 22 similarity functions that describe whether the energetically lowest isomer of the cluster with N atoms resembles any of the four lowest isomers of the cluster with $N - 1$ atom plus one extra atom. As the two lowest panel shows, this is indeed the case, and most often it is not the energetically lowest isomer of the cluster with $N - 1$ atoms that is most similar to the one with N atoms. This, together with the results of Figure 19, strongly suggests that experimentally produced samples contain several different isomers and that the cluster growth follows in a way where not only one but several isomers are important. Ultimately, this means that theoretical studies, like the ones that are the topic of this presentation, may have to deal with not only one single global total-energy minimum but with several structures.

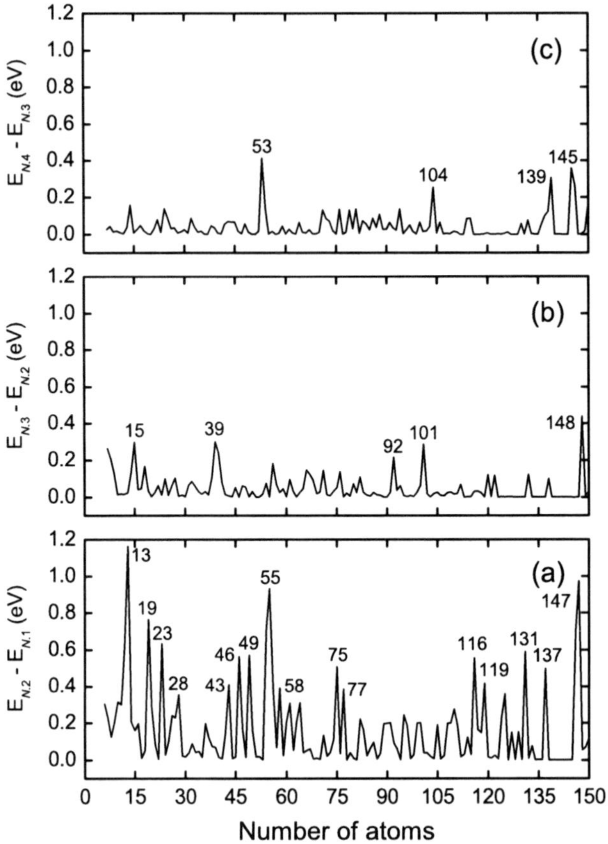

Figure 19 *The total-energy difference for Ni_N clusters between the two energetically neighbouring isomers as a function of cluster size. (a) shows the difference between the second and the first [i.e., $\Delta_{12}E(N)$], (b) between the third and the second and (c) between fourth and the third isomer*

The study on Ni_N clusters was presented in some detail in order to demonstrate what kind of information can be obtained through careful analysis of the results of unbiased structure optimizations for a whole series of clusters. Of course, that study is not the only one, but out of the very many others we shall just pick out some few further ones and discuss them briefly.

The clusters of some elements have been at the center of the scientific interest, whereas those of others have not been so. Maybe the most studied element in this context is gold. However, here various theoretical approaches have problems in obtaining qualitatively the same results. Thus, clusters of gold atoms seem to be a topic just of its own: the results depend very sensitively on the approach used in describing the interatomic interactions, and different theoretical approaches may lead to markedly different predictions. We shall, therefore, not discuss this element further here: the interested reader may consult, *e.g.*, 55,82.

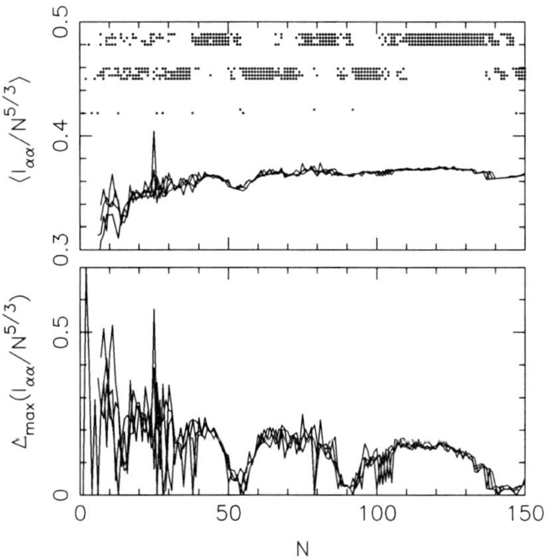

Figure 20 *Different properties of* Ni_N *clusters related to the eigenvalues* $I_{\alpha\alpha}$ *of the matrix with the moments of inertia. In the upper panel we show the average value together with points indicating whether clusters with overall spherical shape (lowest set of rows), overall cigar shape (middle set of rows), or overall lens shape (upper set of rows) are found for a certain size. Moreover, in each set of rows, the lowest row corresponds to the energetically lowest isomer, the second one to the energetically second-lowest isomer, etc. In the lower panel we show the maximum difference of the eigenvalues for the four different isomers*

But also the other coinage elements have been studied intensively, although here the results depend much less sensitively on the approximate or 'exact' description that is used for the interatomic interactions. Among those is Ag. Clusters of Ag were studied by Baletto *et al.*[83] who used a molecular-dynamics method to determine the stable structures for selected cluster sizes with up to almost 600 atoms. In order to identify the crystal-growth modes, they used the common-neighbour analysis (Section 3.5) in distinguishing between icosahedral, decahedral, and fcc-like motifs. From their analysis they were also able to propose a cluster-growth mechanism.

Much smaller clusters were studied by Zhao *et al.*[84] Using a tight-binding model together with genetic algorithms they examined clusters with up to 21 atoms. Particularly stable clusters were found for $N = 8$ and $N = 18$, which is in accordance with the predictions of the jellium model, but also the cluster with $N = 13$ was found to be more stable than the neighbouring clusters. The latter can be explained as being due to packing effects and, accordingly, the authors suggest that both electronic and packing effects have to be taken into account when studying the stability of clusters.

As an example of an older work we mention the study of Jennison *et al.*[85] who studied some selected geometries of Ru, Pd, and Ag clusters with 55, 135, and 140 atoms with different geometries using parameter-free density-functional calculations. The work, just some 10 years old, appeared at a time

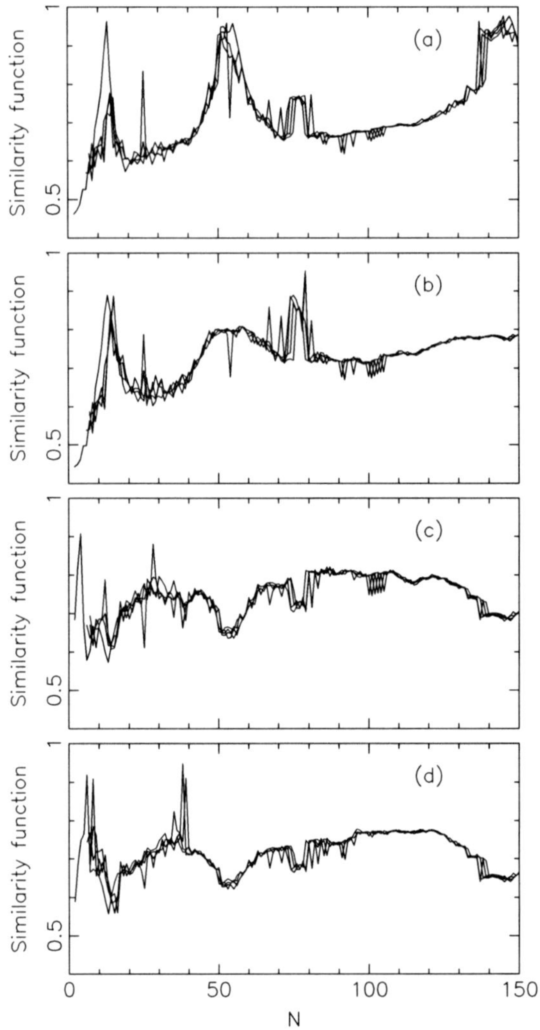

Figure 21 *Each panel shows the similarity function for all four isomers of Ni_N clusters when comparing with (a) an icosahedral cluster, and (b–d) a spherical fragment of the fcc crystal when the center of the fragment is placed at (b) the position of an atom, (c) the middle of a nearest-neighbor bond, and (d) the center of the cube, respectively*

where unbiased structure optimization was just about to emerge and, instead, much computational power was invested in performing accurate calculations on selected structures of larger systems.

Turning to the more recent work, we mention the study of Sebetci and Güvenç[86] on Pt_N clusters with $22 \leq N \leq 56$. They used an embedded-atom approach in describing the interatomic interactions and two different approaches in optimizing the structures, *i.e.*, the basin-hopping algorithm and a molecular-dynamics approach. Their approach allowed for a comparison

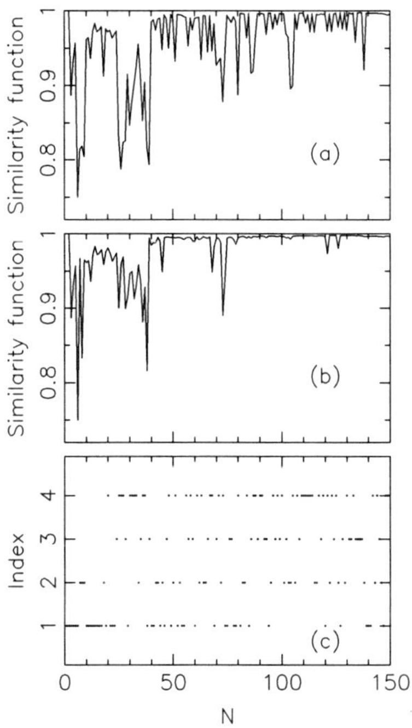

Figure 22 *(a) and (b) show the similarity functions for* Ni_N *clusters that describe whether the cluster with N atoms is similar to that of N − 1 atoms plus an extra atom, when (a) only considering the lowest-energy isomer for the (N − 1)-atom cluster and (b) considering all four isomers for that cluster. (c) shows which isomer in the latter case is most similar to the one of N atoms*

between the two approaches for structure optimization, and they found that the basin-hopping algorithm was more efficient than the molecular-dynamics approach (that was combined with simulated annealing).

In a subsequent work, Sebetci and Güvenç[87] studied clusters of more different metals, *i.e.*, Al, Au, and Pt with up to 80 atoms. An interesting aspect is that for 55 atoms one may construct no less than three highly-symmetric structures, a cuboctahedron, an dodecahedron, and an icosahedron. Sebetci and Güvenç[87] found, however, that for all the three elements they studied, not the cluster with 55, but the one with 54 atoms was particularly stable when using $\Delta_2 E(N)$ in identifying stable clusters. The one with 54 atoms was found to be the 55-atomic icosahedron but without the central atoms.

Quite a different approach was undertaken by Ahlrichs and Elliott[88] who studied Al_N clusters using an accurate density-functional approach. Due to the computational demands of the electronic-structure approach, only the smallest clusters could be studied and for those it was not possible to perform a complete structure optimization. In addition, the authors studied also selected, high-symmetric clusters for larger *N*. Among others, they compared their results with the predictions of the jellium model and found only a very marginal correspondence. Due to the partly directional, covalent bonds of Al, this finding may not surprise.

In another study, Lai *et al.*[89] studied the structures of clusters of various metals. In describing the interatomic interactions they used the Gupta potential.[76] According to this, the total energy is written as

$$E = \sum_{i=1}^{N} [V_r(i) + V_d(i)], \tag{73}$$

where the second term is due to interactions between the atomic orbitals and is approximated as

$$V_d(i) = -\left\{ \sum_{i \neq j=1}^{N} \eta^2 \exp[-2q(d_{ij}/d_0 - 1)] \right\}^{1/2} \tag{74}$$

and the first term represents repulsive interactions,

$$V_r(i) = \zeta \sum_{i \neq j=1}^{N} \exp[-p(d_{ij}/d_0 - 1)]. \tag{75}$$

Lai *et al.*[89] optimized the structures for the clusters using both the basin-hopping method and genetic algorithms. They found excellent agreement between the two approaches. Furthermore, it is interesting to notice that the elements with just one single valence *s* electron per atom, *i.e.*, Na, K, Rb, and Cs, possess very similar properties. This is exemplified in Figure 23 where we show the quantities $\Delta_1 E(N)$ and $\Delta_2 E(N)$ for those clusters. Comparing with the similar results for clusters of Pb, Figure 24, it is clear that the latter is markedly different from the former.

In their study on Fe_N clusters, Bobadova-Parvanova *et al.*[42] used the 'single-parent' genetic-algorithm approach (see Section 2.10), where each member of the new generation is created by interchanging the two parts of the same parent. In calculating the total energy and the electronic properties, they used a parameterized, tight-binding, density-functional method. Their results for clusters with 4, 7, 10, and 19 atoms indicated that the clusters have structures very similar to those of optimized Lennard-Jones clusters. Therefore, they extended the study to the whole range of $2 \leq N \leq 26$ by using the structures of the optimized Lennard-Jones clusters as starting points for calculations where a steepest-descent method (see Section 2.1) is used in identifying the next local total-energy minimum. Thus, this study combined completely unbiased structure optimizations with calculations where the results for 'related systems' (*i.e.*, the clusters that were optimized as well as Lennard-Jones clusters) is used in guessing realistic structures.

In all studies we have discussed in this section, approximate methods are used in calculating the electronic properties and the total energy for a given structure. This implies that the results will never be better than the approximations in the methods. An important question is therefore, whether the approximate methods yield reliable results. Here, in particular small clusters provide a useful test ground for this issue, simply because the computational needs for accurate calculations for the small systems are not overwhelming.

Rogan *et al.*[90] considered therefore Pd_N clusters with $2 \leq N \leq 13$ using a whole range of accurate and approximate methods in combination with genetic algorithms for the structure optimization. Besides the two different types of embedded-atom approximations (*i.e.*, the one due to Foiles, Baskes, and Daw[70–72] and the one due to Voter and Chen[73–75]), they also used the Gupta potential,[76] the potential due to Sutton and Chen,[77] and the potential of Murrell and Mottran.[91,92] As parameter-free, *ab initio* methods they used both a Car-Parrinello approach and the SIESTA code.[93–95]

Figures 25–27 show the results of their study. It is obvious that the various methods do not agree in many details, although, for instance the fact that the clusters are particularly stable for $N = 4$, $N = 6$, and $N = 13$ is found by all approaches. Moreover, with just a single exception, all methods yield the same symmetries for the clusters for all N. Thus, trends but not details seem to be accessible with most methods. It should also be mentioned that the two parameter-free methods show deviations in their predictions that are comparable with those of the parameterized methods, suggesting that the use of parameter-free methods may not necessarily give results that are particularly accurate.

4.6 Non-Metal Clusters. – For clusters of metal atoms, the electrons are largely delocalized over the complete cluster and directional bonds between the atoms do hardly exist. For other types of atoms, this is not the case, and one has to treat the electronic interactions explicitly in order to obtain accurate descriptions of the systems. This means that simplified potentials as the ones we discussed in the preceding section are not sufficiently accurate. However, when the electronic interactions are included explicitly, the computational demands on the calculations grow, which easily can put severe limitations on the possibilities of theoretical studies. Nevertheless, many studies have been devoted to those systems and here we shall briefly review some few of those, concentrating on just three elements.

Silicon is the element in semiconductor industry. Due to the miniaturization of all semiconductor devices, it is also technological relevant to study the properties of very small structures based on only Si. Therefore, very many studies have been devoted to Si clusters. This also means that it is natural to apply new theoretical developments to Si clusters. Accordingly, Ballone *et al.*[96] used the Car-Parrinello method, shortly after its introduction, to the study of the structural properties of Si_N clusters with N up to 10.

In a much more recent study, Sieck *et al.*[97] used molecular-dynamics calculations in studying some few, selected Si_N clusters, *i.e.*, for $N = 25$, 29, 35, 71, and 239, in connection with a parameterized density-functional method. The fact that they found several isomers for clusters of these sizes should not surprise, but is rather a confirmation of the finding for Lennard-Jones clusters

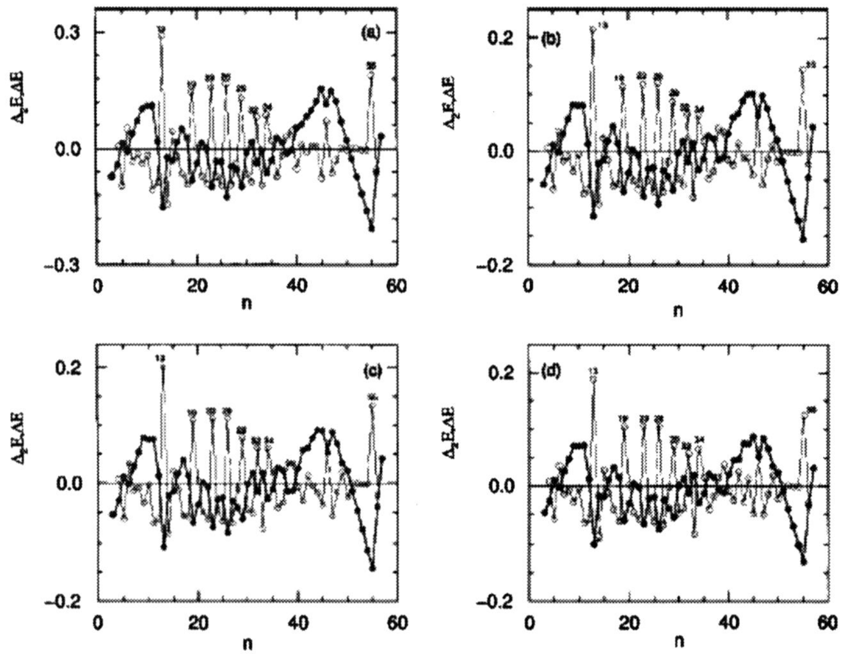

Figure 23 *$\Delta_1E(N)$ (closed circles) and $\Delta_2E(N)$ (open circles) for clusters of (a) Na, (b) K, (c) Rb, and (d) Cs. Reproduced with permission of the American Institute of Physics from 89*

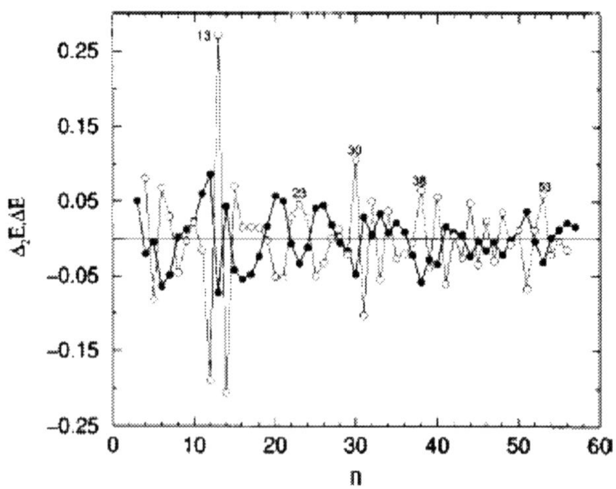

Figure 24 *As Figure 23, but for Pb clusters. Reproduced with permission of the American Institute of Physics from 89*

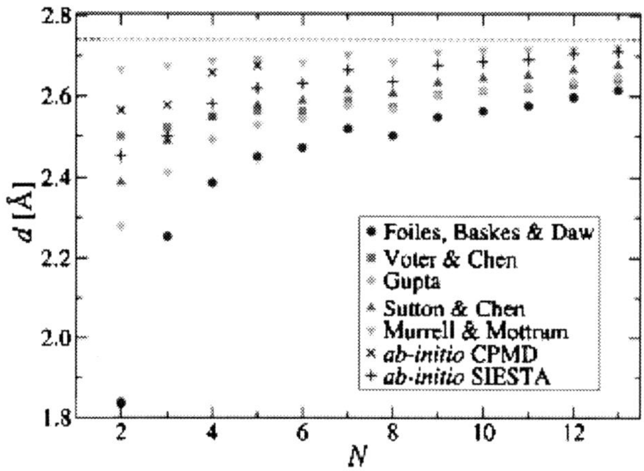

Figure 25 *The average bond length for* Pd_N *clusters as a function of N as obtained with different theoretical methods. The dashed line represents the value for the infinite crystal. Reproduced with permission of the American Physical Society from 90*

as we discussed in Section 1: that clusters with just some few atoms have (very) many local total-energy minima.

Finally, Goedecker *et al.*[45] demonstrated the feasibility of their structure-optimization method (mentioned in Section 2.12) by calculating the structures of more different, smaller Si_N clusters.

Also Ge_N clusters have been the subject of several theoretical studies. Wang *et al.*[98] combined parameter-free density-functional studies with more structure-optimization methods, including molecular-dynamics and simulated-annealing approaches, as well as genetic algorithms, in studying Ge_N clusters with $2 \leq N \leq 25$. $\Delta_2 E(N)$, shown in Figure 28, was used in identifying the particularly stable clusters. This curve is markedly different from the ones we have found and discussed for the metal clusters in the preceding subsections, which is a clear indication of the very different type of interatomic interactions for those clusters.

Parameter-free density-functional calculations on Ge_{20} clusters were performed by Li *et al.*[99] In order to optimize the structures they used a molecular-dynamics approach. They compared the results with those of similar calculations on Si_{20} clusters and found a remarkable similarity between the two elements. Moreover, many of the structures for Ge_{20} could be interpreted as being formed by two interacting Ge_{10} units.

Finally, Hohl *et al.*[100,101] performed some of the first unbiased structure optimizations of clusters using the parameter-free Car-Parrinello method in combination with density-functional methods. They studied the elements S and Se which are known for having a whole wealth of crystal structures of which many are based on finite Se_N or S_N units.

4.7 Metal Clusters with More Types of Atoms. – Above we have seen that for not too small clusters the properties depend critically on the number of atoms,

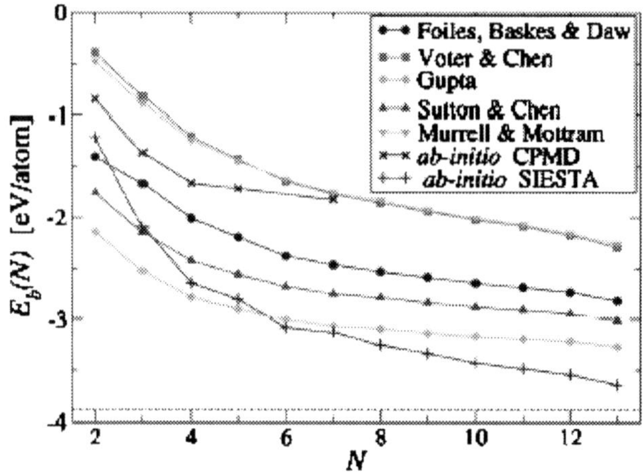

Figure 26 *The binding energy per atom for* Pd_N *clusters as a function of N as obtained with different theoretical methods. The dashed line represents the value for the infinite crystal. Reproduced with permission of the American Physical Society from 90*

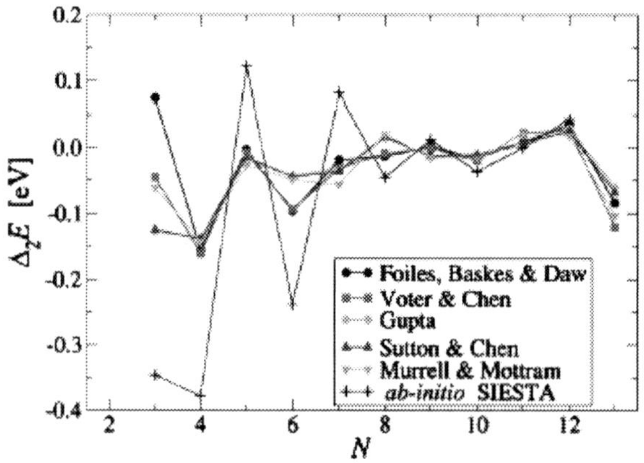

Figure 27 $\triangle_2 E(N) = 2E_b(N) - E_b(N-1) - E_b(N+1)$*, with* $E_b(M)$ *being the binding energy for the* Pd_M *cluster, for* Pd_N *clusters as a function of N as obtained with different theoretical methods. Notice that this definition differs from the one used in the remaining parts of this paper, and that particularly stable clusters are those with particularly large and negative values of* $\Delta_2 E(N)$*. Reproduced with permission of the American Physical Society from 90*

i.e., the cluster with 100 atoms is not simply a small part of the system with 10^{23} atoms. Additional complications/degrees of freedom arise when the system consists of more different types of atoms. One may, as an example, imagine two special cases:

Considering, *e.g.*, a metal cluster Me_N with N atoms and largely undirectional (isentropic) binding interactions. If a certain fraction of the metal atoms is replaced by another type of atoms, one has, in principle, $\binom{N}{N_a}$ possible ways

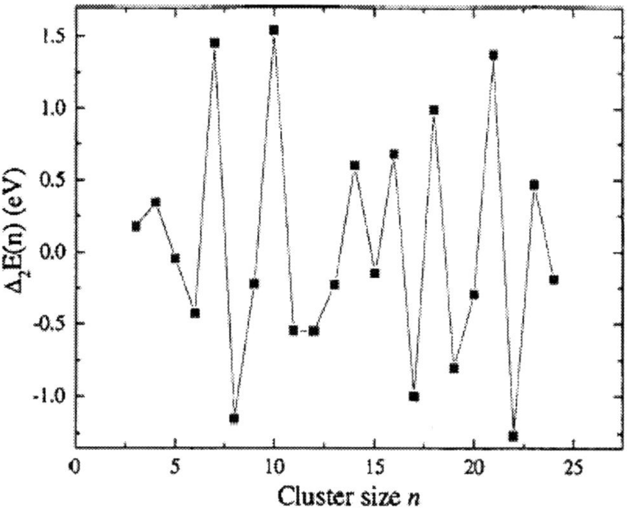

Figure 28 *$\Delta_2E(N)$ for Ge_N clusters as a function of N. Reproduced with permission of the American Physical Society from 98*

of replacing N_a of the N metal atoms with another type of metal atoms, as long as it can be assumed that the structure of the binary system is as that of the original system, except for replacing some of the atoms with others. For N_a and N not too small this leads to a large increase in the computational demands for studying those systems.

As the other extreme case we consider the situation where the units (that above were atoms) are, *e.g.*, molecules. These could for instance be water molecules. Then, the interactions between the units become directional, and, therefore, the determination of the optimum structure is less trivial, because each unit has now three additional structural degrees of freedom (*i.e.*, also the orientation and not only the position).

In this and the next section we shall consider both types of examples as well as some examples that lie in between.

Montejano *et al.*[102] used the embedded-atom method in calculating the total energy for a given structure in studying Cu–Ni and Cu–Pd clusters. They considered only high-symmetric structures like clusters with 55 atoms. They found a clear difference between the two systems. For Cu–Ni there is a partial segregation with Cu on the surface and Ni in the interior, whereas for Cu–Pd the situation is more complex with a competition between two effects: that Pd atoms tend to segregate to the surface, and that this binary systems tends to form ordered structures. Ultimately, this led to a complicated behaviour of these clusters.

Christensen *et al.*[103] studied bimetallic clusters using a Monte-Carlo method together with the effective-medium approximation for the description of the interatomic interactions (this method is closely related to the embedded-atom method). As initial structures they use ones that were derived from the fcc crystal structure. Focusing on the Cu–Ni system, they used the results in predicting the critical size of the clusters of several different bimetallic systems above which the atoms will segregate. They found that this size lies in the range of around 100 (for Pd–Ru) to well above 2000 (for Cu–Ni).

On the other hand, López *et al.*[104] used molecular-dynamics simulations in systematically studying $Cu_{n-x}Au_x$ clusters with $n = 13$ or 14, and x ranging from 0 to n. Due to the small size of the clusters (13 or 14 atoms) it was possible to obtain a detailed description of the structure of all stoichiometries. The interatomic interactions were described with an approximate Gupta-like potential. They found that for $n = 13$ all clusters have an icosahedral-like structure. On the other hand, for $n = 14$ the pure Au and Cu clusters have different structures, and in this case it was found that all clusters, except for the pure Au one, prefers the same type of structure. This finding indicates that it is not trivial to guess the structure of binary compounds from those of the pure system.

A similar study was undertaken by Rey *et al.*[105] but for $Ni_{n-x}Al_x$ clusters with $n = 13$, 19, and 55, and with the embedded-atom approach for describing the interatomic interactions. As for Cu and Au, the pure clusters for these sizes are icosahedral-like, and also the binary clusters have this structure. Moreover, they observed a general tendency for Al to segregate to the surface for those systems, although this behaviour was competing with a tendency towards ordering.

Shimizu *et al.*[51] used the Morse potential in describing the interatomic interactions between the atoms of two-dimensional binary metal clusters, that were considered as model systems for extracting general information. Moreover, they used molecular-dynamics simulations in determining the structures of the clusters at different temperatures, in particular focusing on spontaneous alloying. By studying the Lindemann-like index $\delta_k(t)$, Eq. (61), at different temperatures they could identify a melting of the surface at considerably lower temperatures as those where the rest of the cluster would melt.

In two recent papers Ferrando and coworkers[106,107] studied a whole series of bimetallic clusters using a parameterized potential for describing the interatomic interactions and genetic algorithms in optimizing the structures. In particular they studied clusters with 34 and 38 atoms and found that the structures of those bear no resemblance to those of the corresponding pure clusters of the same size.

The same group has also studied the formation of a shell of one type of metals atoms on the core of another type.[108]. They used the same form for the interatomic interactions but optimized the structures using a molecular-dynamics approach. In order to study the growth process, various scenarios for diffusion processes of the shell-type of atoms on the core were examined. The study revealed that for the two systems of their interest, *i.e.*, a Pd or a Cu shell on an Ag core, the behaviour with temperature is different. Thus, whereas a single-layer shell in both cases easily can be grown on the core, for Ag–Pd more regular shells are obtained at low temperatures, whereas for Ag–Cu the lowest number of defects is found at somewhat elevated temperatures. As a further extension of this study, the authors also used molecular-dynamics simulations in studying the formation of three-shell bimetallic clusters, *i.e.*, structures with a shell of B atoms covering a core with A atoms, for which the shell of B atoms is further covered by a shell of A atoms.[109] In both cases, the radial distributions of the atoms show that the core/shell separation is not perfect. Finally, the group of Ferrando performed unbiased structure optimizations using genetic algorithms with a simple potential for the interatomic interactions on Ag–Ni and Ag–Cu nanoparticles.[110] Subsequently, the results in some few cases were controlled using parameter-free density-functional calculations. They found a very good agreement between the two computational methods. Moreover, more interesting is it that they predicted the existence of particularly stable bimetallic clusters consisting of a high-symmetric core of one metal covered completely with a shell of the other metal. For the two systems of their study, the core consists of Ni or Pd, whereas the shell consists of Ag.

As a very recent example of theoretical studies of metal clusters with more than one type of atoms we mention the work of Chui and Chan.[111] They studied $Pt_{n-x}Co_x$ clusters with $n = 500$ and $x = 125$, 250, and 375 and used the semiempirical Sutton-Chen potential together with molecular-dynamics simulation in obtaining the structure. For all systems they could see some effects towards segregation where Pt atoms would form the major parts of the surface region. However, details of the results were strongly dependent on the details of the calculations and of the stoichiometry.

Finally, Weigend *et al.*[112] suggested that the energy change related to the replacement of an atom of type A by one of type B at site $\vec{R}_k$ can calculated to first order through

$$\Delta_k^{\mathrm{A}} - \left.\frac{\partial E^{\mathrm{A}}}{\partial Z}\right|_{\vec{R}_k} \Delta Z + \left.\frac{\partial E^{A}}{\partial e}\right|_{\vec{R}_k} \Delta e. \tag{76}$$

Here, ΔZ is the change in nuclear charge and Δe the change in the number of electrons due to the substitution. Assuming that $\Delta Z = \Delta e$, they find

$$\Delta_k^{\mathrm{A}} = \left[\sum_{k\neq i=1}^{N} \frac{Z_{\mathrm{A}}}{\left|\vec{R}_i - \vec{R}_k\right|} - \int \frac{\rho(\vec{r})}{\left|\vec{R}_k - \vec{r}\right|}\, d\vec{r}\right] \Delta Z. \tag{77}$$

Finally, when more atoms are being substituted one has to add the contributions from the individual substitutions.

The strength of this approach is that it allows for efficient calculation of energy differences, as long as it can be assumed that the structure of the cluster does not change significantly. In that case one may also apply computationally heavier approaches, which is what the authors did in testing their ideas. They considered $Pt_{13-x}Ir_x$ clusters and used a parameter-free density-functional program and found indeed a good agreement between exact calculations and those obtained using the estimates above.

4.8 Non-Metal Clusters with More Types of Atoms. – When the atoms are bonded together via directed, covalent bonds, the computational demands increase. Therefore, as above for the non-metal clusters with just one type of bonds, the theoretical studies of such systems with more types of atoms are not easy. Nevertheless, some studies in this direction have been reported, and here we shall review some few of those.

In two works, Flikkema and Bromley[113,114] used a specifically tailored semiempirical potential for describing the interatomic interactions of $(SiO_2)_N$ clusters. In order to optimize the structure, they used the basin-hopping algorithm. Subsequently, parameter-free density-functional calculations were carried through for the optimized structures.

They predicted a transition from a column-like to a disk-like structure for $N = 23$. Their calculated structures around those sizes are shown in Figure 29. Besides illustrating the transition between the two types of structures, the figure shows also that these structures are everything else but closed packed. Accordingly, for these, covalent and not metallic bonding is important.

Whereas the SiO_2 clusters possessed very open structures, as also is the case for the solids, other oxides have more compact structures, mainly due to ionic bonding. MgO is one example of a such system. This was studied by Roberts and Johnston[115] who used a simple model potential for describing the interatomic interactions,

$$E = \sum_{i>j=1}^{N} \left[\frac{q_i q_j e^2}{4\pi\varepsilon_0 \left| \vec{R}_i - \vec{R}_j \right|} + B_{ij} \exp\left(- \frac{\left| \vec{R}_i - \vec{R}_j \right|}{\rho_{ij}} \right) \right]. \tag{78}$$

The first term is the classical interaction between point charges and the second term is a repulsive Born-Mayer potential.

Roberts and Johnston considered stoichiometric clusters $(MgO)_n$ and used a genetic-algorithms approach in optimizing the structures. In this case, the cutting and mating processes have to be performed with care: by simply cutting two clusters and interchanging the halves, the stoichiometry may not be kept. Roberts and Johnston devised, however, a method with which the children clusters and the parent clusters have the same stoichiometry.

Roberts and Johnston studied the effects of the formal charge of the atoms, *i.e.*, whether the clusters formally are to be described as $(Mg^{+}O^{-})_n$ or $(Mg^{2+}O^{2-})_n$ or something in between. Thereby, the charges q_i of Eq. (78) are modified, and by simultaneously modifying the parameters B_{ij} and ρ_{ij} it is in principle possible to adjust the potential so that certain fundamental properties (like bond length of the diatomic molecule or of the crystal) are reproduced correctly, although Roberts and Johnston did not do so.

The authors found that for clusters with $q_i = \pm 1$ the calculated structures for small values of n agree with those of more accurate calculations and that for larger values of n clusters that look like smaller pieces of the rocksalt structure are found. On the other hand, for $q_i = \pm 2$ many structures were cage-like, in contrast to other theoretical or experimental findings for these systems. This does suggest that $(Mg^{+}O^{-})_n$ provides a better description of the clusters than $(Mg^{2+}O^{2-})_n$ does. An example of their findings is shown in Figure 30.

In a series of recent papers, Head and coworkers (see, *e.g.*, 116) studied the structures of silicon clusters for which dangling bonds at the surface have been saturated with hydrogen atoms. In optimizing the structure they used genetic algorithms combined with different more or less accurate parameterized potentials for the description of the interatomic interactions. They found that their results depend very sensitively on the quality of the approximate potential and that they ultimately had to devise their own approximate potential in order to get reliable results. Also the genetic algorithms were not always able to find the correct ground-state structure. In total, their results point very clearly to the two general problems of such studies: the results may be biased by inaccurate

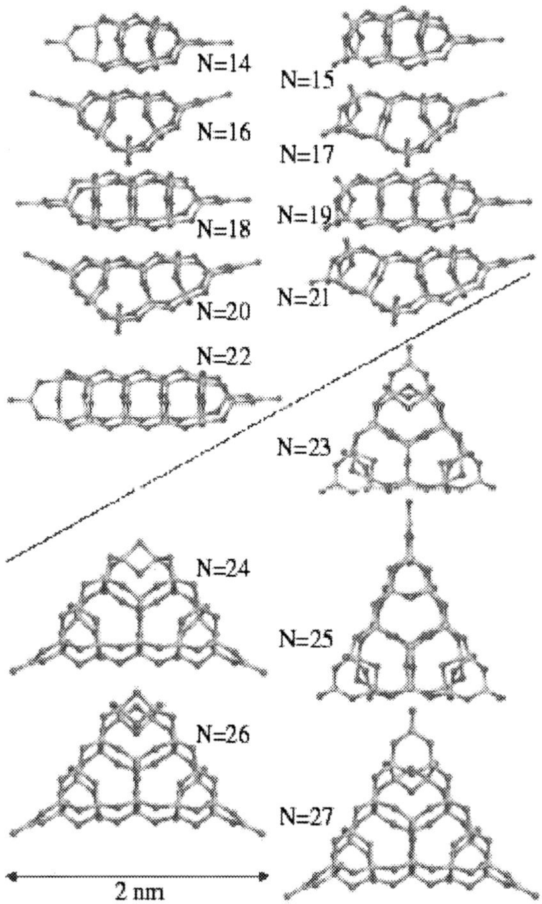

Figure 29 *Lowest-energy structures for* $(SiO_2)_N$ *clusters. The dashed line separates columnar and disklike clusters. Reproduced with permission of the American Physical Society from 114*

approximate potentials and there is absolutely no guarantee that a structure search will give the correct ground-state structure.

III-V and II-VI semiconductors crystallize most often in the zincblende or the wurtzite structure. Then, all atoms are fourfold coordinated, and the total energy for the two crystal structures are very close. Finite clusters of those semiconductors may therefore have a structure related to that of the infinite crystals. In our own work[117] we have made this assumption and, subsequently, studied the structural and electronic properties of such clusters. However, there is absolutely no guarantee that this assumption is justified. Burnin *et al.*[118] studied recently both experimentally and theoretically one example of those systems, *i.e.*, stoichiometric Zn_nS_n clusters. They used a simplified interaction potential together with genetic algorithms in determining the structure of the clusters. Finally, for the optimized structures, parameter-free density-functional calculations were carried through. The experimental results indicate that the clusters for n = 3, 6, and 13 are particularly stable. Moreover, the calculations predict that the most stable structures are planar for small n and become closed-cage polyhedra with all atoms in three-fold coordinated states, cf. Figure 31. Thus, the results show marked differences to the assumption that the structures are related to those of the infinite crystals.

HAlO is a nanostructured material. It is, moreover, stoichiometric, but little is known about its structure. Thus, it is an example of a material where calculations can be of help. Using a parameterized density-functional method together with two different unbiased structure optimization methods (*i.e.*, the Aufbau part of the Aufbau/Abbau method described in Section 2.8 as well as the single-parent genetic-algorithm method described in Section 2.10) we studied $(HAlO)_n$ clusters.[119] The results gave that for the clusters that were considered (*i.e.*, up to $n = 26$), the structure consists of a core of Al and O atoms (with essentially only heteroatomic bonds) covered by a layer of H atoms. The latter can be seen in the radial distances for the individual types of atoms, cf. Figure 32. Moreover, further calculations with a parameter-free density-functional method confirmed that the H atoms would prefer to be bonded to the Al atoms, and not to the O atoms, *i.e.*, the formation of hydroxy groups was not preferred. Finally, it was found that when bringing two clusters together the hydrogen atoms would still prefer to stay on the surface on a core formed by the Al and O atoms. This study clearly demonstrates how theoretical studies can be useful in getting information that is not directly accessible with experimental methods, although it shall be stressed that the theoretical study would not have been possible without valuable information and feedback from experiment.

As the last example we shall study clusters of water molecules. Here, we have clusters formed by only weakly interacting units, but for which the units have an internal structure and, consequently, the interactions are directional. For water clusters the TIP potentials[120–122] are very popular and have, therefore, been used in optimizing the structure of water clusters. One of the first studies in this direction is due to Tsai and Jordan[8] who used their eigenmode method (see Section 2.4) in optimizing the structure for clusters with up to 5 units. Later

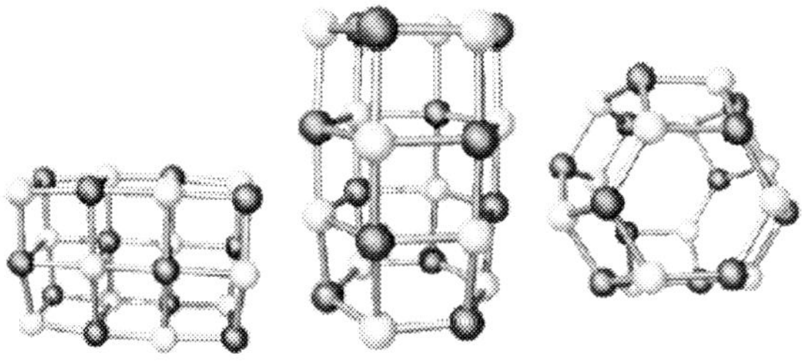

Figure 30 *Lowest-energy structures for $(Mg^{q+}O^{q-})_{12}$ clusters for different values of q. Reproduced with permission from 115*

clusters with more than 20 molecules have been studied (see 36) and it has been found that for the smallest clusters (with up to 5 units), the structure is essentially planar, after which three-dimensional structures develop. These findings have been confirmed by accurate density-functional calculations with the Car-Parrinello method. Subsequently, also structures for water clusters containing various ions (Na^+, K^+, Cs^+) have been studied.

4.9 Clusters on Surfaces. – In many experimental studies clusters deposited or grown on some surface are studied instead of free, isolated clusters in the gas phase. Of course, the presence of a substrate may modify the structure of the clusters in unknown ways and, therefore, theoretical studies of such systems are highly relevant.

In principle, the substrate provides an external medium on which the cluster is either grown or deposited. In the first case, the methods we have discussed in Section 2 can still be applied with the modification that the presence of the medium shall be included first of all as providing an external (structured) potential in which the cluster grows. The structure of the substrate may be optimized, too, but without searching for its global total-energy minimum. When the cluster is deposited on the substrate, some molecular-dynamics approach is most useful.

Wright *et al.*[123] studied the structure of small Pt, Pd, and Ni clusters on the Pt(001) surface using the embedded-atom method for describing the interatomic interactions. Since the systems of their study were sufficiently small, they could search the structure space and found a competition between the formation of islands and that of chains for clusters with 9 atoms.

Later, Shluger *et al.*[124] studied $(NaCl)_n$ clusters (with $1 \leq n \leq 48$) on the MgO(100) surface. They used simple pair potentials in describing the interactions between the atoms. First, they studied a number of selected geometries in order to obtain information on the relative importance of the different type of interactions. Subsequently, they constructed different models for the structure of the cluster on the substrate and relaxed these structures. Thus, in this case the very complex structure is sought understood by reducing it to simpler fragments that each is studied in greater detail.

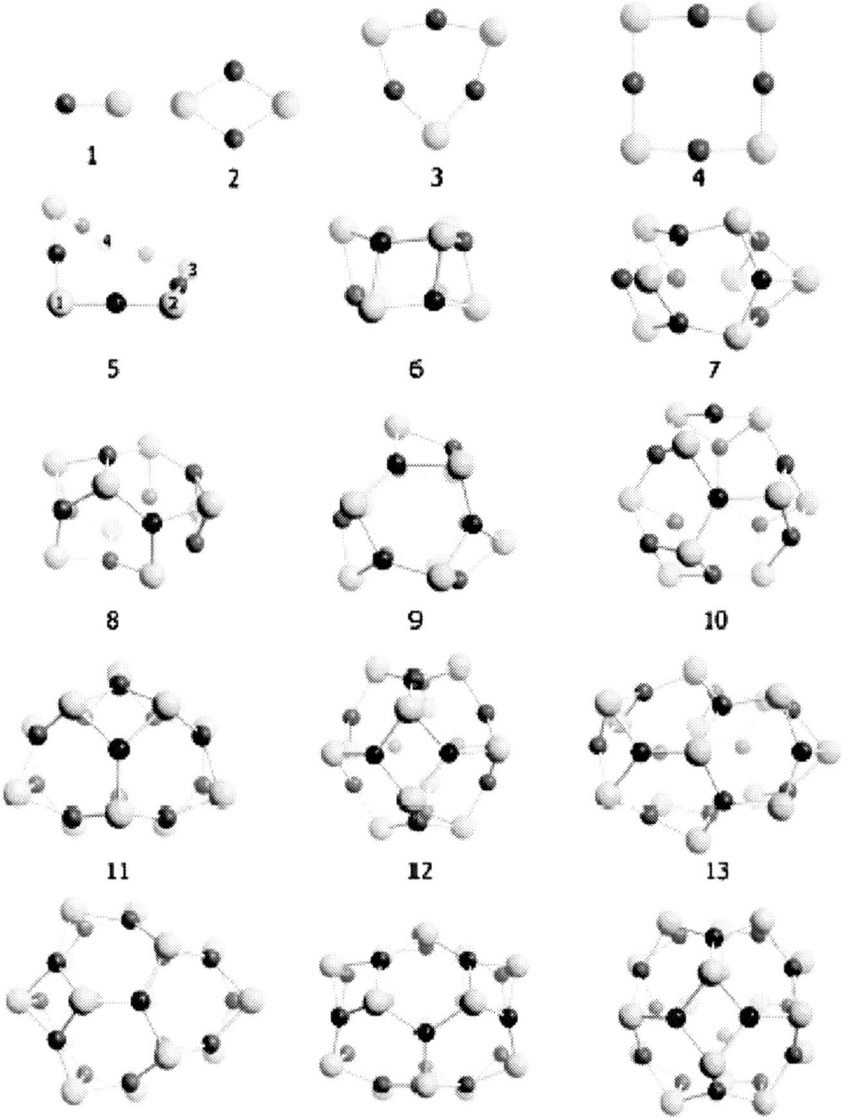

Figure 31 *Lowest-energy structures for neutral Zn_nS_n clusters for different values of n. Dark (light) spheres mark zinc (sulphur) atoms. Reproduced with permission of the American Chemical Society from 118*

5 Determining Saddle Points and Reaction Paths

In some sense one may consider the problem of finding a structure of a local total-energy minimum easier than that of finding a saddle point. Referring to Figure 3, starting essentially anywhere in the structure space, a steepest-descent method will bring you to one of the four local total-energy minima. Trying sufficiently many times, one may ultimately identify the structure of the global total-energy minimum. However, this approach will hardly be able to identify any of the saddle points between the local total-energy minima. Accordingly, neither reaction paths will be identified. It is the purpose of this section to discuss approaches for determining the saddle points. For further information the reader is referred to, *e.g.*, 125–127.

Staying with the example of Figure 3, even if information on the total-energy surface is available it is not obvious how a system will evolve from one structure to another. Imagine that the structure A in the figure represents the structure of some reacting molecules and that the structure D represents the products. Thus, somehow the system evolves from structure A to structure D. From the figure it is obvious that there is a saddle point on the total energy surface roughly in the middle between the two structures A and D, and one may try to calculate the total-energy variations when the system changes from A, via the above-mentioned saddle point, to D. However, the total-energy surface of Figure 3 suggests that lower energy-barriers are encountered if the system evolves from A, via B and C, to D.

One question we shall address in this section is, how one calculates the total-energy path when the start and end points (*i.e.*, A and D in the above example) are known, but nothing else is known about the total-energy surface. An even more complicated question is that of studying how a system will change when only the starting point, *i.e.*, A, is known. This issue shall hardly be discussed here, except for referring to a single work in the next section.

5.1 Interpolation. – The simplest approach (that still often is used) is to define some fictitious reaction coordinate δ that equals 0 (1) for the initial (final) structure and then let all internal coordinates (*i.e.*, bond lengths, bond angles, dihedral angles, *etc.*) vary linearly between the two end values. It is very unlikely that this approach indeed will pass through a saddle point and, therefore, it will usually lead to an upper bound for the transition energy.

5.2 Eigenmode Methods. – In Section 2.4 we saw how the eigenvalues of the Hessian matrix could be used in forcing a structure to automatically change towards a local total-energy minimum. Thereby, the change of Eq. (39) was necessary.

Once a local total-energy minimum has been located, one may follow the eigenvectors of the Hessian. However, the Hessian is a function of structure, so when calculating the Hessian at the local total-energy minimum, one gets a set of eigenvectors of which one is chosen. The system is changed a little according to this eigenvector, and the Hessian at this new structure is calculated and, subsequently, its eigenvectors are determined. These are most likely not identical to those at the minimum, so one chooses the eigenvector at the new structures which is most similar to (has the largest overlap with) the previously used eigenvector. By repeating this step, one will ultimately arrive at a

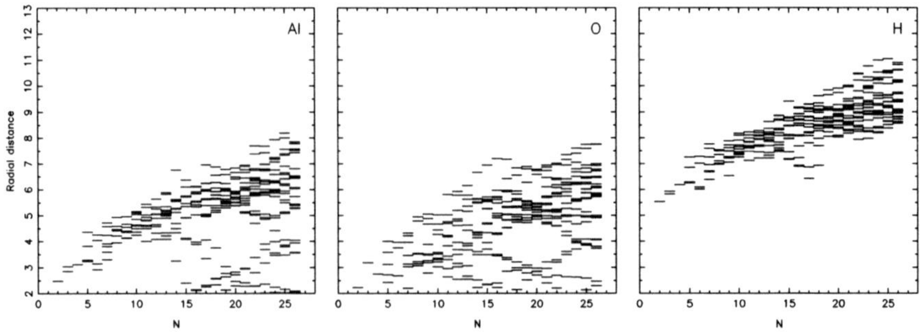

Figure 32 *The radial distances (in a.u.) for Al, O, and H atoms, separately, as a function of the size of the cluster n for* $(HAlO)_n$ *clusters. In each panel, a small horizontal line shows that at least one atom of the corresponding type has that distance to the center of the cluster for a given value of n*

saddle point. This is the approach that was suggested originally by Hildebrandt.[15]

There are several ways of modifying this approach (see, *e.g.*, 8) that, however, all are based on using the eigenvectors of the Hessian.

5.3 The Intrinsic Reaction Path. – The intrinsic reaction path was originally proposed by Fukui.[128,129] In this case structure is changed according to

$$d\vec{R}_k = -\frac{1}{\sqrt{M_k}}\vec{\nabla}_{\vec{R}_k} E \cdot dt, \tag{79}$$

i.e., a steepest-descent-like method is used. The only difference is that the gradients contain mass-depending weights.

Ultimately, the intrinsic reaction path defines a unique way in which the structure of a system is allowed to evolve. Thus, in a molecular-dynamics calculation the system is no longer allowed to change structure completely freely, but has to change along the intrinsic reaction path. This leads to an extra constraint on the possible structural degrees of freedom that has to be included in a practical molecular-dynamics calculation. This was done recently by Michalak and Ziegler[130] who showed how the approach could be combined with the Car-Parrinello method that we discussed in Section 2.3.

5.4 Changing the Fitness Function. – Determining the structures of local total-energy minima is a 'simple' minimization problem, which not is the case for the determination of saddle points. However, it may be possible to study another function than the total energy, *i.e.*, a function that has extrema where the total energy has saddle points. This idea was used by Chaudhury *et al.*,[131] who suggested to consider the quantity

$$f = e^{-F} \tag{80}$$

with

$$F = (E - E_L)^2 + \beta \sum_{i=1}^{N} \left| \vec{\nabla}_{\vec{R}_i} E \right|^2 + \sum_{i=1}^{3N} \eta_i e^{p_i \lambda_i}. \tag{81}$$

In this equation, E_L is introduced in order to tune the search to locate a critical point around the energy E_L. Moreover, $\beta \neq 0$ makes the search for local extrema more efficient. Finally, λ_i are the eigenvalues of the Hessian which all are positive (negative) at local minima (maxima) of the total energy except for six that have the value 0 (and, therefore, should not be included in the summation). On the other hand, at a saddle point one eigenvalue is negative and the others are positive, so that by choosing the quantities p_i and η_i appropriately, it is possible to push the calculation towards the determination of this saddle point.

The approach of Chaudhury *et al.*[131] can in principle be combined with any of the search methods we have discussed in Section 2 (as long as the Hessian can be calculated relatively easy — this seems to be the critical point of the approach), but Chaudhury *et al.*[131] combined it with a genetic-algorithm search. Subsequently they demonstrated its validity for clusters of Lennard-Jones atoms.

5.5 Chain-of-States Methods. – The chain-of-states methods (see 132) are methods that are devised for calculating a set of structures between two local total-energy minima. Referring to Figure 3, the idea is to start out from two well-known structures, *e.g.*, A and D in the figure. Subsequently, a set of intermediate structures shall be determined that shall fulfill the following criteria:

(i) the 0th and *P*th structure shall represent the start and end structure, respectively;
(ii) adjacent structures shall be similar;
(iii) all clusters shall form a chain of structures between the 0th and the *P*th structure;
(iv) it shall be avoided that a total-energy minimization gives only the 0th and the *P*th structure.

For each of the structures we have *N* vectors describing the nuclear coordinates. We now construct the following quantity

$$\begin{aligned} &S(\{\vec{R}_i^{(0)}\}, \{\vec{R}_i^{(1)}\}, \ldots, \{\vec{R}_i^{(P-1)}\}, \{\vec{R}_i^{(P)}\}) \\ &= \sum_{p=0}^{P} E(\{\vec{R}_i^{(p)}\}) + \frac{k}{2} \sum_{p=1}^{P} \sum_{i=1}^{N} \left| \vec{R}_i^{(p)} - \vec{R}_i^{(p-1)} \right|^2. \end{aligned} \tag{82}$$

S depends accordingly on all the $P + 1$ structures along the chain, each described by $\{\vec{R}_i^{(p)}\}$, where *i* denotes the nucleus and *p* the structure along the chain. The first term on the right-hand side is the sum of the total energy for each of these structures that in principle shall be minimized. However, in order to avoid that all clusters take one out of the two structures at the ends (*i.e.*, at $p = 0$ or at $p = P$), the second term is included that becomes very large if the structure suddenly changes abruptly along the chain.

Ultimately, S is sought minimized under the constraint that the structures $\{\vec{R}_i^{(0)}\}$ and $\{\vec{R}_i^{(P)}\}$ are known and fixed. This leads to a set of structures that, hopefully, will represent a good approximation to those along a reaction path.

5.6 Nudged Elastic-Band Methods. – Jónsson *et al.*[132] demonstrated that the chain-of-states method suffered from a severe shortcoming. Independent of the choice of the value of the (unphysical) spring constant k, the system tends to avoid the saddle-point region. Although the energy at the saddle point is the lowest barrier that shall be overcome along the transition from initial to final structure, the second term in Eq. (82) may lead to a preference for structures with a higher total energy than at the saddle point, but with a larger similarity with the structures along the path. Thus, the method minimizes the second term of Eq. (82) on the cost of the first term. Reducing the force constant k makes this effect less drastic, but instead no structures near the saddle point may be found. Hence, in particular closest to the saddle point, which often is the most interesting region, the method may fail.

As a simple cure, Jónsson *et al.*[132] showed that one may modify the chain-of-states method slightly, whereby the nudged elastic-band method results. Once again taking Figure 3 as reference and assuming that the initial and the final structures are those at A and D, respectively, the chain-of-states method corresponds to having a sequence of structures between A and D. For each of those, one may calculate a force that describes how the structure will change if allowed to. The force is given as the gradient of S of Eq. (82) and contains accordingly two terms: one term from the true physical interactions and one term from the unphysical springs. The second term is the one that may force the system to stay away from a local total-energy minimum. However, the unphysical spring forces are also those that can drive the system away from a saddle point if using the chain-of-states method. Therefore, in order to force the system to pass through the saddle point, Jónsson *et al.*[132] suggested to modify the spring forces so that only the components along the physical forces are kept.

Since its introduction almost 10 years ago, several improvements have been proposed (see, *e.g.*, 133–135), but the main ideas remain. We shall not discuss the details of the improvements further here, but instead below present some examples of its applications.

5.7 String Methods. – A recently proposed approach that is closely related to the nudged elastic-band method of the preceding section is the string method of E *et al.*[136] (see also 137). We shall here briefly outline the basics of the method and, thereby, also draw the connection to the nudged elastic-band method, following the presentation of Kanai *et al.*[137]

The total energy for the system of interest is a function of all nuclear coordinates

$$\vec{R} \equiv (\vec{R}_1, \vec{R}_2, \vec{R}_3, \ldots, \vec{R}_N). \tag{83}$$

Out of many paths, denoted ϕ, that connect two local total-energy minima, we seek one ϕ^* that passes through the saddle point. This path will then be a minimum-energy-path. Such a path will satisfy

$$(\vec{\nabla} E[\phi^*])^{\perp} = 0. \tag{84}$$

Here, E is the total energy as a function of $\vec{R}$, and $(\vec{\nabla} E[\phi^*])^{\perp}$ is the component of

$$\vec{\nabla} E = (\vec{\nabla}_{\vec{R}_1} E, \vec{\nabla}_{\vec{R}_2} E, \vec{\nabla}_{\vec{R}_3} E, \ldots, \vec{\nabla}_{\vec{R}_N} E) \tag{85}$$

that is perpendicular to ϕ^*. Notice that ϕ is a short-hand notation for a continuous curve in the space of $\vec{R}$ of Eq. (83), *i.e.*, ϕ is a path in a $3N$ dimensional space.

Just as indicated in the introduction to this section, we introduce a parameter δ whose value is 0 and 1 for the initial and final states, respectively. Then, ϕ becomes a function of this parameter,

$$\phi = \phi(\delta). \tag{86}$$

For the nudged elastic-band method, $\phi(\delta)$ represents the system images along the path joining initial and final states. Moreover, in that case one seeks only a discrete set of those, *i.e.*, only the structures for a finite, discrete set of δ values.

In order to identify the special path ϕ^*, a chosen path ϕ is modified until it satisfies the criterion (84). However, the parameterization (86) shall stay valid during the deformation of the path, which can be expressed through the constraint

$$\frac{d}{d\delta}\left|\frac{d\phi(\delta)}{d\delta}\cdot\frac{d\phi(\delta)}{d\delta}\right|^{1/2} = 0 \tag{87}$$

This equation gives upon integration

$$\left|\frac{d\phi(\delta)}{d\delta}\right|\cdot\frac{d\phi(\delta)}{d\delta} = c, \tag{88}$$

where c is some constant, or upon integration from the initial ($\delta = 0$) to the final ($\delta = 1$) state,

$$\int_0^1 \left|\frac{d\phi(\delta)}{d\delta}\right|^2 d\delta = c. \tag{89}$$

One may now combine Eqs. (86) and (87) or (88) together with Eq. (84) in obtaining

$$(\vec{\nabla} E[\phi^*(\delta)])^{\perp} - \lambda(\delta)\,\hat{\tau}^*(\delta) = 0. \tag{90}$$

Here,

$$\hat{\tau}^*(\delta) = \frac{1}{\left|\frac{d\phi^*(\delta)}{d\delta}\right|}\frac{d\phi^*(\delta)}{d\delta} \tag{91}$$

is a unit tangent vector to the path, $\lambda(\delta)$ is a Lagrange multiplier for the constraint (87) or (88), and

$$[\vec{\nabla}E]^{\perp} = \vec{\nabla}E - (\vec{\nabla}E \cdot \hat{\tau})\hat{\tau}. \tag{92}$$

In order to solve this equation one may perform a steepest-descent dynamics in the path space, *i.e.*, introduce a fictitious time variable t and let the path ϕ depend on t. Then, starting from some initial path, one may let it evolve in time according to

$$\dot{\phi}(\delta;t) = -(\vec{\nabla}E[\phi(\delta;t)])^{\perp} + \lambda(\delta;t)\hat{\tau}(\delta;t). \tag{93}$$

Alternatively, applying the same steepest-descent approach within the nudged elastic-band method leads to

$$\dot{\phi}(\delta;t) = -(\vec{\nabla}E[\phi(\delta;t)])^{\perp} + k\left[\frac{\partial^2\phi(\delta;t)}{\partial\delta^2}\hat{\tau}(\delta;t)\right]\hat{\tau}(\delta;t). \tag{94}$$

Thus, the two approaches are very similar, although not identical.

In the original approach, E *et al.*[136] combined the string method with a molecular-mechanics method for the calculation of the total energy for a given structure, whereas Kanai *et al.*[137] combined it with the parameter-free density-functional Car-Parrinello method. Below we shall present results from the latter study.

5.8 Approximating the Total-Energy Surface. – Another approach is based on approximating the total energy around the initial and final states with some adequate analytical expressions of the structural coordinates. This could, *e.g.*, be a second- or higher-order Taylor expansion. Having two such expansions, one around each local total-energy minimum, the two expressions can be set equal to each other and that solution with the lowest total energy will then provide an estimate for the transition state. One may either use this directly or try to refine it using other methods; for further discussion, see, *e.g.*, 138.

6 Examples for Saddle-Point and Reaction-Path Calculations

There exists very many theoretical studies devoted to the calculation of reaction paths (and accompanying variations in the total energy) for chemical reactions and to the determination of saddle points. It is not possible to present an adequate discussion of just a very small fraction of those, and, therefore, we shall here just randomly discuss briefly some, mainly recent, studies.

In their discussion of the eigenmode method, Tsai and Jordan[8] illustrated the approach through two simple systems. One of those was Lennard-Jones clusters, and the other was small clusters of water molecules. For both they tried to identify as many local total-energy minima and saddle points as possible. The numbers for the Lennard-Jones clusters are reproduced in Table 1 and it is remarkable to see that the number of transition states exceeds by far the number of local total-energy minima. This was also the case for the clusters of water molecules.

As mentioned above, Chaudhury *et al.*[131] suggested to consider a different fitness function, Eqs. (80) and (81). As test system they, too, considered Lennard-Jones clusters and demonstrated the feasibility of their method. However, it shall be stressed that Lennard-Jones clusters are special: except for scalings, only one type of interactions (including strength and range) is treated, and, moreover, it is trivial to calculate any derivatives of the total energy with respect to nuclear coordinates.

Michalak and Ziegler[130] studied the total-energy variations along the intrinsic reaction paths (Section 5.3) using a parameter-free, density-functional, molecular-dynamics method. They studied some simple reactions, *i.e.*, HCN→CNH, the conrotatory ring opening of cyclobutene, the SN2 reaction $Cl^- + CH_3Cl \rightarrow ClCH_3 + Cl^-$, and the chloropropene isomerization $Cl{-}CH_2{-}CH{=}CH_2 \rightarrow CH_2{=}CH_2Cl$. As an example of their study we show in Figure 33 results from their molecular-dynamics calculations on the reaction HCN→CNH at different temperatures and with different constraints. Only in the panels d) and f) of that figure, a different constraint than the intrinsic-reaction constraint was enforced.

Maragakis *et al.*[133] studied the performance of a modified version of the nudged elastic-band approach for several more or less complex systems, ranging from smaller organic molecules, via large bio-molecules, to solids. Their basic idea for improving the nudged elastic-band method is to modify it so that extra images are inserted automatically into the sequence of structures at regions where rapid structural changes occur. They combined the method with several, parameterized and parameter-free, total-energy methods and found that their method is powerful. Subsequently, they applied the method for, *e.g.*, a double proton transfer process in an adenine-thymine base pair, see Figure 34. The final state was found to be roughly 8 kcal/mol above the initial state, and the transition state about 15 kcal/mol above the initial state.

Also Chu *et al.*[134] presented results of an improved version of the nudged elastic-band method. They applied the method on more different systems, including an α-helix to π-helix transition of an alanine decapeptide and the oxidation of dimethyl sulfide.

Another improvement was proposed by Trygubenko and Wales[135] who subsequently applied it to Lennard-Jones clusters with 7, 38, and 75 atoms.

Sarkar *et al.*[139] used the nudged elastic-band method in studying the stability of core/shell nanoparticles consisting of a core of one semiconductor covered with a shell of another semiconductor. In this case the two semiconductors were CdS and CdSe, and it was investigated whether processes, where a S and a Se

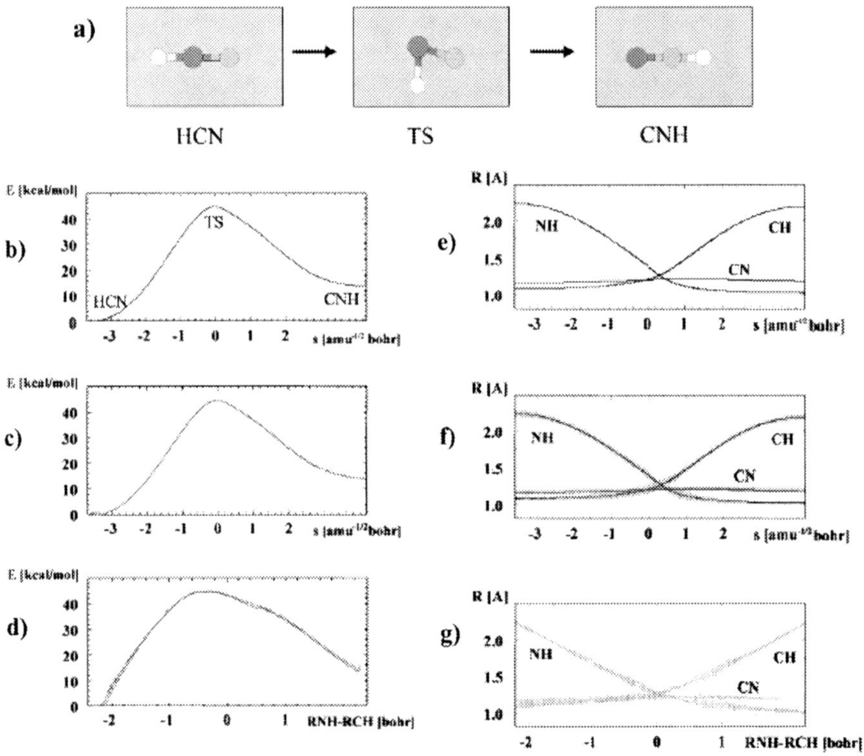

Figure 33 *Results of a study of the reaction HCN → CNH. a) shows the initial and final state together with the transition state, whereas b), c), and d) show the total-energy variations. Interatomic distances are shown in e), f), and g). The calculations of b) and d) were done at 0 K, those of c) and f) at 300 K. In d) and g) it was assumed that the bond-length difference between N–H and C–H bonds is a constant, and the calculations were done at 300 K. Reproduced with permission of American Chemical Society from 130*

atom from the two parts could be interchanged, were possible. Some examples of the results are shown in Figure 35, where it is seen that the total energy may be lowered upon the interchange, but in all cases the barrier is so high, that the processes are not realistic.

In their presentation of the string method, Kanai *et al.*[137] applied the method to the adsorption of hydrogen on the Si(100) surface, both at low and at large hydrogen coverage, in particular for the processes where the H_2 molecule change adsorption site.

In a recent paper, Olsen *et al.*[127] compared a whole class of methods for identifying saddle points in the case that only the initial but not the final state was known. As test system they studied a seven-atom Pt island on a Pt(111) surface and they allowed either one atom, the 7 atoms of the island, or all 175 atoms of their system to move. In order to reduce the computational needs, the interatomic interactions were approximated through Morse potentials. Through this simplification, detailed studies of the performance of the different

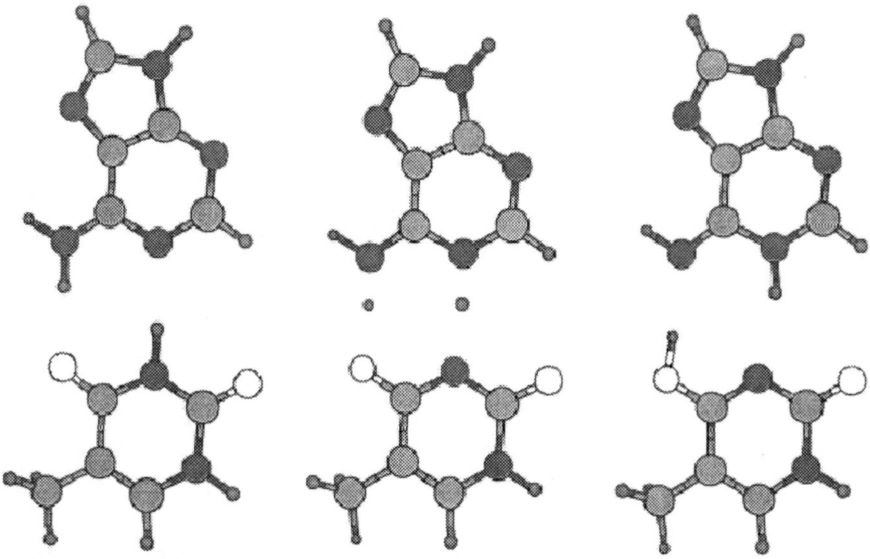

Figure 34 *From left to right: initial, transition, and final state for the double proton transfer process in an adenine-thymine base pair. Reproduced with permission of American Institute of Physics from 133*

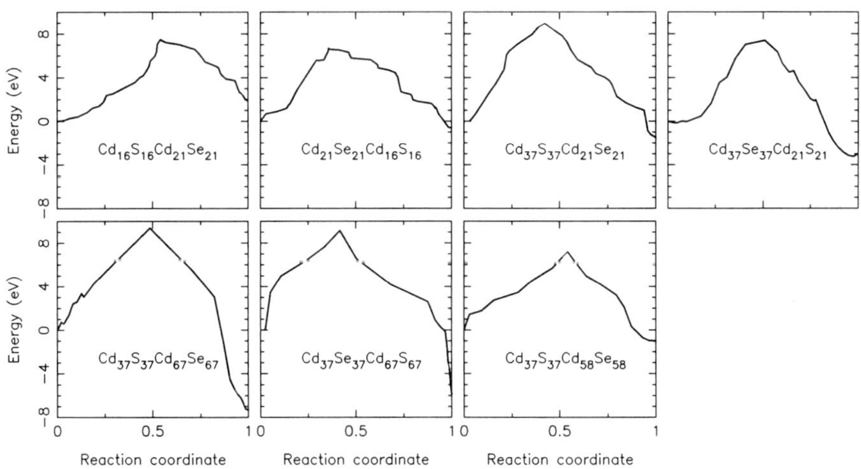

Figure 35 *The total-energy variation when interchanging a close S–Se pair between the core and the shell of semiconductor core/shell nanoparticles. The reaction coordinate equals 0 before and 1 after the interchange, and the $A_aB_bC_cD_d$ notation means an A_aB_b core covered by a C_cD_d shell before the interchange*

methods were made possible. It turned out that the success of the calculations depends sensitively on carefully adjusting internal parameters, which does make the approaches difficult to use as 'black boxes'. However, we shall here not discuss their conclusions further; instead the interested reader is referred to

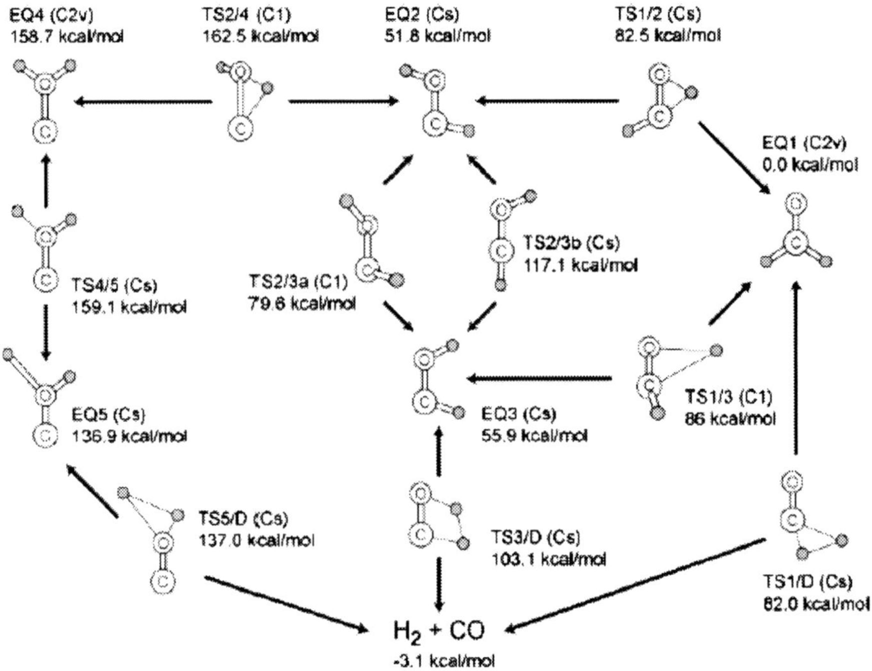

Figure 36 *Different (meta-)stable structures and transition states for the formaldehyde molecule, COH_2. Reproduced with permission of American Chemical Society from 138*

the original paper both for a description of the methods and for a detailed discussion of the results.

In the last work we shall discuss here, Maeda and Ohno[138] used their scaled hypersphere search method, related to approximating the total-energy surface with some analytical form (see Section 5.8) for a couple of smaller systems. Their results on formaldehyde, summarized in Figure 36, represent the best way of closing the presentation of the problem of determining the structure in electronic-structure calculations. This molecule has only four atoms but nevertheless, the total-energy surface is complicated and full of local minima and saddle points!

7 Conclusions

The total energy as a function of structure has been the central issue of this presentation. We have focused on two different aspects in this connection, both related to the situation where information (for instance the total energy as well as a number of its derivatives with respect to nuclear coordinates) can be obtained for any discrete set of structures, but further information on the total energy as a function of structure is not available. The two issues differed, however, in fundamental aspects.

First, we studied methods devised to identify the structure of the global total-energy minimum, when essentially nothing except for the number and types of atoms were known. Most methods were based on the principle that once some structure was suggested, a locally relaxed structure could 'easily' be found by using calculated information on the forces acting on the nuclei. The non-trivial part was to search the structure space efficiently but still largely unbiased. The fact that chemical interactions (*i.e.*, bonds) usually are fairly short-ranged made it possible to apply methods based on the keeping of good structural elements. Growth methods, our own Aufbau/Abbau method, and genetic algorithms are all based on this basic principle. Nevertheless, no method can guarantee that the true global total-energy minimum will be found and it is always possible that certain important structures will be overseen. However, we could hopefully show that methods that are explicitly devised for optimizing the structure in the above-mentioned situation usually perform better than more general methods (like, *e.g.*, simulated annealing or molecular dynamics) that can be used for other purposes, too.

In the examples we emphasized that one has to cope with one further fundamental problem, *i.e.*, accurate electronic-structure methods will usually become too costly when applied together with unbiased structure-optimization methods for not too small systems, partly because the computational costs for a single total-energy calculation grow rapidly with system size, and partly because the number of total-energy calculations easily becomes very large when searching the global total-energy minimum in an unbiased way (well above 10^7 for systems with of the order of 20 atoms). Therefore, very many studies employ parameterized methods, which, of course, does introduce a further degree of uncertainty into the results.

Finally, we shall therefore stress that the results of such calculations never can be considered providing the final truth for the systems of interest. The results should always be checked with those of other methods and with results of experimental studies. Furthermore, in the latter not only energetic, but also kinetic effects may be of ultimate importance, that are beyond most of the theoretical methods we have discussed here.

In the second, shorter, part we discussed a somewhat different situation, *i.e.*, the situation where we know an initial state that will evolve (*e.g.*, due to temperature) towards another, final state and thereby pass a saddle-point on the total-energy surface. The energy of the saddle point relative to that of the initial state is of ultimate importance in the determination of the rate constant (*e.g.*, through the Arrhenius equation). Here, two different situations may be considered: either both initial and final states are known and only the minimum-energy path connecting them in the structure space is sought, or only the initial state is known and one wants to explore the development of the system of interest with time.

For the first situation, the last decade or so has seen the development of accurate and reliable methods based on chains-of-states methods or extensions hereof (most notably, the nudged elastic-band method). On the other hand, the second situation remains to be a challenge, and also here it shall be remembered that the experimental situation may be different from the one simulated on the computer, *i.e.*, also here kinetic effects could be dominating.

But in total, many specifically developed methods have emerged over the last years, that are surprisingly reliable in providing accurate results also for large, complex systems. This is encouraging in particular when taking into account the limited computer capacity at our disposal and the complexity of the problem at hand, as we discussed in the introduction.

Finally, although major parts of the presentation have focused on systems that are at the heart of our own research, *i.e.*, clusters and colloids, very many of the ideas, both concerning computational methods for identifying global total-energy minima and saddle points and concerning ways of extracting essential information from the results (using 'descriptors') are applicable for essentially any other system for which only little information on the structure is available and which has a larger number of atoms and (maybe) a low symmetry.

References

1. M. Springborg, *Methods of Electronic-Structure Calculations*, Wiley, Chichester, UK, 2000.
2. G. Galli, *Phys. Stat. Sol. b*, 2000, **217**, 231.
3. D.R. Bowler, T. Miyazaki and M.J. Gillan, *J. Phys. Cond. Matt.*, 2002, **14**, 2781.
4. C. Ochsenfeld, J. Kussmann and F. Koziol, *Angew. Chem. Int. Ed.*, 2004, **43**, 4485.
5. W. Liang, Y. Shao, C. Ochsenfeld, A.T. Bell and M. Head-Gordon, *Chem. Phys. Lett.*, 2002, **43**, 4485.
6. S. Li, W. Li and T. Fang, *J. Am. Chem. Soc.*, 2005, **127**, 7215.
7. L.T. Wille and J. Vennik, *J. Phys. A*, 1985, **18**, L419.
8. C.J. Tsai and K.D. Jordan, *J. Phys. Chem.*, 1993, **97**, 11227.
9. S. Lloyd, *Phys. Rev. Lett.*, 2002, **88**, 237901.
10. F. Jensen, *Introduction to Computational Chemistry*, Wiley, Chichester, UK, 1999.
11. D.J. Wales, *Energy Landscapes*, Cambridge University Press, Cambridge, UK, 2003.
12. H. Jiang, W. Cai and X. Shao, *Phys. Chem. Chem. Phys.*, 2002, **4**, 4782.
13. H. Goldstein, *Classical Mechanics*, Addison-Wesley, Reading, Massachusetts, 1950.
14. R. Car and M. Parrinello, *Phys. Rev. Lett.*, 1985, **55**, 2471.
15. R.L. Hilderbrandt, *Comput. Chem.*, 1977, **1**, 179.
16. C.J. Cerjan and W.H. Miller, *J. Chem. Phys.*, 1981, **75**, 2800.
17. J. Simons, P. Jørgensen, H. Taylor and J. Ozment, *J. Phys. Chem.*, 1983, **87**, 2745.
18. A. Banerjee, N. Adams, J. Simons and R. Shepard, *J. Phys. Chem.*, 1985, **89**, 52.
19. J. Nichols, H. Taylor, P. Schmidt and J. Simons, *J. Chem. Phys.*, 1990, **92**, 340.
20. D.J. Wales, *J. Chem. Soc., Faraday Trans.*, 1990, **86**, 3505.
21. J. Baker, *J. Comput. Chem.*, 1986, **7**, 385.
22. D.J. Wales, *J. Chem. Phys.*, 1989, **91**, 7002.
23. D.J. Wales and R.S. Berry, *J. Chem. Phys.*, 1990, **92**, 4283.
24. M.V. Fernández-Serra, E. Artacho and J.M. Soler, *Phys. Rev. B*, 2003, **67**, 100101.
25. P. Császár and P. Pulay, *J. Mol. Struct.*, 1984, **114**, 31.
26. Ö. Farkas and H.B. Schlegel, *Phys. Chem. Chem. Phys.*, 2002, **4**, 11.
27. J.A. Northby, *J. Chem. Phys.*, 1987, **87**, 6166.

28. I.A. Solov'yov, A.V. Solov'yov, W. Greiner, A. Koshelev and A. Shutovich, *Phys. Rev. Lett.*, 2003, **90**, 053401.
29. I.A. Solov'yov, A.V. Solov'yov and W. Greiner, *Int. J. Mod. Phys. E*, 2004, **13**, 697.
30. R. Poteau and F. Spiegelmann, *J. Chem. Phys.*, 1993, **98**, 6540.
31. V.G. Grigoryan and M. Springborg, *Phys. Chem. Chem. Phys.*, 2001, **3**, 5125.
32. V.G. Grigoryan and M. Springborg, *Chem. Phys. Lett.*, 2003, **375**, 219.
33. V.G. Grigoryan and M. Springborg, *Phys. Rev. B*, 2004, **70**, 205415.
34. D.J. Wales and J.P.K. Doye, *J. Phys. Chem. A*, 1997, **101**, 5111.
35. D.E. Goldberg, *Genetic Algorithms in Search, Optimization and Machine Learning*, Addison-Wesley, Boston, Massachusetts, 1989.
36. B. Hartke, *Angew. Chem.*, 2002, **14**, 1534.
37. S.K. Gregurick, M.H. Alexander and B. Hartke, *J. Chem. Phys.*, 1996, **104**, 2684.
38. B. Hartke, *Chem. Phys. Lett.*, 1996, **258**, 144.
39. D.M. Deaven and K.M. Ho, *Phys. Rev. Lett.*, 1995, **75**, 288.
40. B. Hartke, *J. Phys. Chem.*, 1993, **97**, 9973.
41. Y. Xiao and D.E. Williams, *Chem. Phys. Lett.*, 1993, **215**, 17.
42. P. Bobadova-Parvanova, K.A. Jackson, S. Srinivas, M. Horoi, C. Köhler and G. Seifert, *J. Chem. Phys.*, 2002, **116**, 3576.
43. J. Cheng and R. Fournier, *Theor. Chem. Acc.*, 2004, **112**, 7.
44. J. Lee, I.-H. Lee and J. Lee, *Phys. Rev. Lett.*, 2003, **91**, 080201.
45. S. Goedecker, W. Hellmann and T. Lenosky, *Phys. Rev. Lett.*, 2005, **95**, 055501.
46. J.D. Honeycutt and H.C. Andersen, *J. Phys. Chem.*, 1987, **91**, 4950.
47. D. Faken and H. Jónsson, *Comp. Mat. Sci.*, 1994, **2**, 279.
48. L. Cheng, W. Cai and X. Shao, *Chem. Phys. Lett.*, 2004, **389**, 309.
49. C. Rey and L.J. Gallego, *Phys. Rev. B*, 1995, **51**, 13691.
50. J. Jellinek, T.L. Beck and R.S. Berry, *J. Chem. Phys.*, 1986, **84**, 2783.
51. Y. Shimizu, K.S. Ikeda and S. Sawada, *Phys. Rev. B*, 2001, **64**, 075412.
52. F. Calvo and F. Spiegelmann, *J. Chem. Phys.*, 2000, **112**, 2888.
53. J.P.K. Doye and D.J. Wales, *Phys. Rev. Lett.*, 1998, **80**, 1357.
54. J.P.K. Doye and F. Calvo, *J. Chem. Phys.*, 2002, **116**, 8307.
55. F. Baletto and R. Ferrando, *Rev. Mod. Phys.*, 2005, **77**, 371.
56. D.J. Wales and H.A. Scheraga, *Science*, 1999, **285**, 1368.
57. J.P.K. Doye, in *Global Optimization — Selected Case Studies*, ed. J.D. Pinter, Kluwer, Dordrecht, 2006.
58. D.D. Frantz, *J. Chem. Phys.*, 2001, **115**, 6136.
59. J.P.K. Doye and D.J. Wales, *J. Chem. Soc. Faraday Trans.*, 1997, **93**, 4233.
60. C. Roberts, R.L. Johnston and N.T. Wilson, *Theor. Chem. Acc.*, 2000, **104**, 123.
61. W.D. Knight, K. Clemenger, W.A. de Heer, W.A. Saunders, M.Y. Chou and M.L. Cohen, *Phys. Rev. Lett.*, 1984, **52**, 2141.
62. M. Brack, *Rev. Mod. Phys.*, 1993, **65**, 677.
63. W.A. de Heer, *Rev. Mod. Phys.*, 1993, **65**, 611.
64. M.P. Iñiguez, M.J. Lopez, J.A. Alonso and J.M. Soler, *Z. Phys. D*, 1989, **11**, 163.
65. R. Balian and C. Bloch, *Ann. Phys.*, 1972, **69**, 76.
66. I.A. Solov'yov, A.V. Solov'yov and W. Greiner, *Phys. Rev. A*, 2002, **65**, 053203.
67. R. Poteau, F. Spiegelmann and P. Labastie, *Z. Phys. D*, 1994, **30**, 57.
68. L. Kronik, I. Vasiliev and J.R. Chelikowsky, *Phys. Rev. B*, 2000, **62**, 9992.
69. E.G. Noya, J.P.K. Doye and D.J. Wales, cond-mat/0506329.
70. M.S. Daw and M.I. Baskes, *Phys. Rev. Lett.*, 1983, **50**, 1285.
71. M.S. Daw and M.I. Baskes, *Phys. Rev. B*, 1984, **29**, 6443.

72. S.M. Foiles, M.I. Baskes and M.S. Daw, *Phys. Rev. B*, 1986, **33**, 7983.
73. A.F. Voter and S.P. Chen, in *Characterization of Defects in Materials*, ed. R.W. Siegal, J.R. Weertman and R. Sinclair, MRS Symposia Proceedings No. 82, Materials Research Society, Pittsburgh, 1987.
74. A. Voter, Los Alamos Unclassified Technical Report No LA-UR 93-3901 (1993).
75. A.F. Voter, in *Intermetallic Compounds*, ed. J.H. Westbrook and R.L. Fleischer, John Wiley, New York, 1995.
76. R.P. Gupta, *Phys. Rev. B*, 1981, **23**, 6265.
77. A.P. Sutton and J. Chen, *Philos. Mag. Lett.*, 1990, **61**, 139.
78. J.P.K. Doye and D.J. Wales, *New J. Chem.*, 1998, **1998**, 733.
79. R. Poteau and G.M. Pastor, *Eur. Phys. J. D*, 1999, **9**, 235.
80. T.L. Wetzel and A.E. DePristo, *J. Chem. Phys.*, 1996, **105**, 572.
81. K. Michaelian, N. Rendón and I.L. Garzón, *Phys. Rev. B*, 1999, **60**, 2000.
82. W. Fa, C. Luo and J. Dong, *Phys. Rev. B*, 2005, **72**, 205428.
83. F. Baletto, C. Mottet and R. Ferrando, *Phys. Rev. B*, 2001, **63**, 155408.
84. J. Zhao, Y. Luo and G. Wang, *Eur. Phys. J. D*, 2001, **14**, 309.
85. D.R. Jennison, P.A. Schultz and M.P. Sears, *J. Chem. Phys.*, 1997, **106**, 1856.
86. A. Sebetci and Z.B. Güvenç, *Eur. Phys. J. D*, 2004, **30**, 71.
87. A. Sebetci and Z.B. Güvenç, *Model. Simul. Mater. Sci. Eng.*, 2005, **13**, 683.
88. R. Ahlrichs and S.D. Elliott, *Phys. Chem. Chem. Phys.*, 1999, **1**, 13.
89. S.K. Lai, P.J. Hsu, K.L. Wu, W.K. Liu and M. Iwamatsu, *J. Chem. Phys.*, 2002, **117**, 10715.
90. J. Rogan, G. García, J.A. Valdivia, W. Orellana, A.H. Romero, R. Ramírez and M. Kiwi, *Phys. Rev. B*, 2005, **72**, 115421.
91. J.N. Murrell and R.E. Mottram, *Mol. Phys.*, 1990, **69**, 571.
92. H. Cox, R.L. Johnston and J. Murrell, *J. Solid State Chem.*, 1999, **145**, 517.
93. P. Ordejón, E. Artacho and J.M. Soler, *Phys. Rev. B*, 1996, **53**, 10441.
94. J.M. Soler, E. Artacho, J.D. Gale, A. García, J. Junquera, P. Ordejón and D. Sánchez-Portal, *J. Phys. Cond. Matt.*, 2002, **14**, 2745.
95. J.M. Soler, M.R. Beltrán, K. Michaelian, I.L. Garzón, P. Ordejón, D. Sánchez-Portal and E. Artacho, *Phys. Rev. B*, 2000, **61**, 5771.
96. P. Ballone, W. Andreoni, R. Car and M. Parrinello, *Phys. Rev. Lett.*, 1988, **60**, 271.
97. A. Sieck, Th. Frauenheim and K.A. Jackson, *Phys. Stat. Sol. b*, 2003, **240**, 537.
98. J. Wang, G. Wang and J. Zhao, *Phys. Rev. B*, 2001, **64**, 205411.
99. B.-X. Li, P.-L. Cao and F.-S. Liang, *Physica B*, 2003, **337**, 69.
100. D. Hohl, R.O. Jones, R. Car and M. Parrinello, *Chem. Phys. Lett.*, 1987, **139**, 540.
101. D. Hohl, R.O. Jones, R. Car and M. Parrinello, *J. Chem. Phys.*, 1988, **89**, 6823.
102. J.M. Montejano-Carrizales, M.P. Iñiguez and J.A. Alonso, *Phys. Rev. B*, 1994, **49**, 16649.
103. A. Christensen, P. Stoltze and J.K. Nørskov, *J. Phys. Cond. Matt.*, 1995, **7**, 1047.
104. M.J. López, P.A. Marcos and J.A. Alonso, *J. Chem. Phys.*, 1996, **104**, 1056.
105. C. Rey, J. García-Rodeja and L.J. Gallego, *Phys. Rev. B*, 1996, **54**, 2942.
106. A. Rapallo, G. Rossi, R. Ferrando, A. Fortunelli, B.C. Curley, L.D. Lloyd, G.M. Tarbuck and R.L. Johnston, *J. Chem. Phys.*, 2005, **122**, 194308.
107. G. Rossi, R. Ferrando, A. Rapallo, A. Furtunelli, B.C. Curley, L.D. Lloyd and R.L. Johnston, *J. Chem. Phys.*, 2005, **122**, 194309.
108. F. Baletto, C. Mottet and R. Ferrando, *Phys. Rev. B*, 2002, **66**, 155420.
109. F. Baletto, C. Mottet and R. Ferrando, *Phys. Rev. Lett.*, 2003, **90**, 135504.
110. G. Rossi, A. Rapallo, C. Mottet, A. Fortunelli, F. Baletto and R. Ferrando, *Phys. Rev. Lett.*, 2004, **93**, 105503.

111. Y.H. Chui and K.-Y. Chan, *Chem. Phys. Lett.*, 2005, **408**, 49.
112. F. Weigend, C. Schrodt and R. Ahlrichs, *J. Chem. Phys.*, 2004, **121**, 10380.
113. E. Flikkema and S.T. Bromley, *J. Phys. Chem. B*, 2004, **108**, 9638.
114. S.T. Bromley and E. Flikkema, *Phys. Rev. Lett.*, 2005, **95**, 185505.
115. C. Roberts and R.L. Johnston, *Phys. Chem. Chem. Phys.*, 2001, **3**, 5024.
116. Y. Ge and J.D. Head, *Chem. Phys. Lett.*, 2004, **398**, 107.
117. J.-O. Joswig, S. Roy, P. Sarkar and M. Springborg, *Chem. Phys. Lett.*, 2002, **365**, 75.
118. A. Burnin, E. Sanville and J.J. BelBruno, *J. Phys. Chem. A*, 2005, **109**, 5026.
119. Y. Dong, M. Burkhart, M. Veith and M. Springborg, *J. Phys. Chem. B*, 2005, **109**, 22820.
120. W.L. Jorgensen, J. Chandrasekhar, J.D. Madura, R.W. Impey and M.L. Klein, *J. Chem. Phys.*, 1983, **79**, 926.
121. W.L. Jorgensen and J.D. Madura, *Mol. Phys.*, 1985, **56**, 1381.
122. W.L. Jorgensen and J. Tirado-Rives, *J. Am. Chem. Soc.*, 1988, **110**, 1657.
123. A.F. Wright, M.S. Daw and C.Y. Fong, *Phys. Rev. B*, 1990, **42**, 9409.
124. A.L. Shluger, A.L. Rohl and D.H. Gay, *Phys. Rev. B*, 1995, **51**, 13631.
125. T. Ziegler and J. Autschbach, *Chem. Rev.*, 2005, **105**, 2695.
126. K. Fukui, *J. Phys. Chem.*, 1970, **74**, 4161.
127. R.A. Olsen, G.J. Kroes, G. Henkelman, A. Arnaldsson and H. Jónsson, *J. Chem. Phys.*, 2004, **121**, 9776.
128. K. Fukui, *Acc. Chem. Res.*, 1981, **14**, 363.
129. A. Tachibana and K. Fukui, *Theor. Chim. Acta*, 1980, **57**, 81.
130. A. Michalak and T. Ziegler, *J. Phys. Chem. A*, 2001, **105**, 4333.
131. P. Chaudhury, S.P. Bhattacharyya and W. Quapp, *Chem. Phys.*, 2000, **253**, 295.
132. H. Jónsson, G. Mills and K.W. Jacobsen, in *Classical and Quantum Simulations in Condensed Phase Simulations*, ed. B.J. Berne, G. Ciccoti and D.F. Coker, World Scientific, Singapore, 1998.
133. P. Maragakis, S.A. Andreev, Y. Brumer, D.R. Reichman and E. Kaxiras, *J. Chem. Phys.*, 2002, **117**, 4651.
134. J.-W. Chu, B.L. Trout and B.R. Brooks, *J. Chem. Phys.*, 2003, **119**, 12708.
135. S.A. Trygubenko and D.J. Wales, *J. Chem. Phys.*, 2004, **120**, 2082.
136. W.E.W. Ren and E. Vanden-Eijnden, *Phys. Rev. B*, 2002, **66**, 052301.
137. Y. Kanai, A. Tilocca, A. Selloni and R. Car, *J. Chem. Phys.*, 2004, **121**, 3359.
138. S. Maeda and K. Ohno, *J. Phys. Chem. A*, 2005, **109**, 5742.
139. P. Sarkar, M. Springborg and G. Seifert, *Chem. Phys. Lett.*, 2005, **405**, 103.

7

Simulation of Liquids

BY B.D. TODD[a] AND D.J. SEARLES[b]

[a] *Centre for Molecular Simulation, Swinburne University of Technology, P.O. Box 218, Hawthorn, Victoria 3122, Australia*

[b] *Nanoscale Science and Technology Centre and School of Science, Griffith University, Brisbane, Qld 4111, Australia*

1 Introduction

After spending a number of months reviewing the literature in this field, two inevitable and immediate conclusions were drawn: first, the field is growing at a phenomenal rate, with contributions ranging from fundamental studies of the liquid state through to highly applied and industrially focussed research; and second – a corollary to the first – that, because of this huge growth, it is simply impossible to adequately cover all topics in a general review of simulation of the liquid state. The growth in this field has also been largely driven by a further two factors: the economic imperatives of developing smarter, smaller, more energy efficient technologies based on the controlled manipulation of inorganic and organic molecules, and the steady and (currently) predictable doubling of computing power approximately every 18 months. The former driver provides the impetus for studying more and more ambitious physical and biological systems, whereas the latter provides the computational power to successfully accomplish such studies.

Bearing this in mind we have deliberately targeted subjects that we feel are of overall general interest to the simulation community. We also focus now and then on topics that are of particular interest to us, where we feel that they will also be viewed favourably by others. We therefore structure the review in essentially two unequal halves: the first half dealing with the techniques and algorithms of molecular simulation, and the second half dealing more generally with the properties that can be determined from such simulations. In Sections 2–5 we consider recent developments in classical simulation techniques (including systems at thermodynamic equilibrium and nonequilibrium states), potential energy interaction potentials, quantum mechanical procedures, and microscopic chaos. Sections 6–10 cover simulation results and span thermodynamics and transport, phase diagrams and transitions, complex fluids, confined fluids and water and aqueous systems. We offer some brief concluding remarks in Section 11.

Chemical Modelling: Applications and Theory, Volume 4

At the onset, we apologise to readers for not covering the following topics: recent developments in dissipative particle dynamics (DPD), smooth particle hydrodynamics (SPH), Lattice Boltzmann methods, liquid crystals, biomolecular systems in general (though some developments and simulations are discussed) and glasses in general (though again, some discussion of several papers on supercooled liquids and glasses is presented). Nor are we able to cover topics in any depth. Our goal is to provide a flavour of what has been done in the past two years in this field but not to be detailed or comprehensive in our discussion (at times, we need to probe back further in time to give greater perspective). Indeed, it is impossible to do this in such a general review. As such, we have most probably missed some important work, and we apologise in advance for any such unintentional neglect. Nevertheless, we have endeavoured to bring workers in the field up-to-date with the broader simulation literature and hope that they will find it a useful tool for their own research.

2 Classical Simulation Techniques

2.1 Statistical Mechanical Ensembles and Equilibrium Techniques. – Thermostats continue to play an important role in the development of practical statistical mechanical ensembles for use in computer simulation of fluids. Branka *et al.*[1] propose a modified Nosé-Hoover thermostat in which the thermostat multiplier that couples the momenta to the force, ζ, is an odd power of some integer n, *i.e.* $\varsigma \rightarrow \varsigma^{2n-1}$. Apart from this modification, the same Nosé-Hoover equations of motion apply. They find that at optimal values of $n = 2$ or 3 the system generates more chaotic dynamics than the standard Nosé-Hoover thermostat. Their thermostat is able to thermalise small and stiff systems more efficiently than the traditional Nosé-Hoover thermostat. Bright *et al.*[2] point out that there are an infinite number of ways of thermostatting a system, namely by fixing any of the sum of the momenta raised to any arbitrary power, *i.e.* $\Sigma|p_i|^{\mu+1} = constant$. However, they demonstrate that there is only one unique value of μ, namely $\mu = 1$, which makes the traditional Gaussian thermostat minimize the phase space compression. It is also the only value of μ for which the conjugate pairing rule[3] remains valid and the phase space distribution function is preserved by field-free equations of motion. All other $\mu \neq 1$ thermostats actually perform work on the system and only a $\mu = 1$ thermostat allows for an equilibrium state.

Powles *et al.*[4] provide a nice review of the various definitions of temperature, including the thermodynamic, kinetic, Monte Carlo (MC), Brownian dynamics (BD), Rugh and configurational temperatures. They point out that the configurational temperature can actually be derived from a suitably formulated hypervirial introduced by Hirschfelder[5] in 1960. They then studied and characterized the configurational temperature for the soft sphere fluid potential of the form $\phi(r) \sim r^{-n}$ over a wide range of densities and potential steepness, n. Braga and Travis[6] have derived a new form of the configurational temperature thermostat based on the Nosé-Hoover equations of motion. Their thermostat is an improvement of the one proposed by Delhommelle and Evans[7] in that it does not contain a term proportional to the spatial gradient of the temperature, which can lead to stiff equations of motion, and also because it generates the canonical ensemble. Their thermostat is shown to be more robust than that of Delhommelle and Evans and may be useful for systems involving large changes in temperature over small time scales, and possibly in nonequilibrium molecular dynamics (NEMD) simulations. Work is in progress to adapt this new thermostat to the isothermal-isobaric ensemble, NEMD systems and systems with holonomic constraints, such as molecular fluids. In the case of stochastic thermostats, Uberuaga *et al.*[8] have observed, as have previous researchers, that initially independent molecular dynamics (MD) trajectories become synchronized in time even if the same random number seeds are used and the systems are in the same potential energy basin. For the first time, they derive an expression for the time dependence of the synchronization for the harmonic potential well. This observation is then used to enhance various aspects of the simulation, such as efficient thermalisation of a system of atoms. Their study is only relevant for stochastic thermostats, *not* for deterministic thermostats such as Nosé-Hoover or Gaussian.

In an attempt to enable Gaussian isokinetic simulations that ergodically sample the available phase space, Minary *et al.*[9] developed an "extended system" formulation of the isokinetic equations of motion. In this system they are able to generate an equilibrium distribution function that is canonical. However this is accomplished at the sacrifice of a non-conserved kinetic energy. Instead the so-called extended system "kinetic energy" is conserved, in which a particle's momentum is coupled to an external momentum bath. Gauss's principle is then used on the extended system to generate the appropriate constraint equations. The method was extended by incorporating holonomic constraints for biomolecular simulation studies that were demonstrated to have enhanced equilibration and sampling efficiency. In a companion publication, the method was also applied to *ab initio* Car-Parrinello MD (CPMD)[10] and was successfully used to study liquid aluminium, solid silicon, reconstruction of the Si(100) surface and the Diels-Alder addition of 1,3-butadiene to the reconstructed Si(100) surface.

Boinepalli and Attard[11] have developed a modification to the hybrid grand canonical-MD (GCMD) ensemble that follows the extended system approach of previous workers. In their variant, the particle number is a continuous variable with a fractional part that represents the coupling of solute to solvent. Whereas previous workers evolve this variable in time deterministically via appropriate equations of motion, Boinepalli and Attard evolve them stochastically. Insertion and deletion of particles are therefore replaced by growing and shrinking. The result of this procedure is that the equilibrium grand canonical distribution is maintained. Their results for the diffusion coefficient, pressure, density and chemical potential are in good agreement with previously published values. Attard also derives statistical mechanical theory[12] for what he terms are steady-state systems, though the theory is only really developed for systems with temperature gradients and so can not be universally applied, *e.g.* to a system driven by a velocity gradient, *etc.* In one of his papers[12] he derives a probability distribution, which is the foundation of the theory. The validity of this approach is questionable since it seems to be founded on the principle of equal *a-priori* probability, a cornerstone of equilibrium statistical mechanics, but which has no justification for extension into nonequilibrium systems in general. As discussed previously by Evans *et al.*[13] (see discussion on g-SLLOD in Section 2.2) a truly general nonequilibrium *steady-state* probability distribution function is fractal in nature with a non-vanishing phase-space compressibility factor. Though Attard derives a probability distribution that, supposedly, can be used in MC simulations,[14] it is clear that this distribution is not the same as the Kawasaki nonequilibrium distribution function. The theory may have some value for systems close to equilibrium but can not remain valid in the nonlinear regime, as indeed is noted by Attard himself.[12]

An additional tool for Langevin dynamics in the constant pressure constant temperature ensemble (*NpT*) has been developed by Quigley and Probert.[15] Their system is based upon an extended Hamiltonian framework developed by Hoover[16] but an extension is also made to incorporate the Parrinello and Rahman scheme.[17,18] They furthermore develop a suitable integrator based upon Liouville time evolution operators. Their scheme provides good sampling of the available phase space and generates the required probability distribution.

We now comment on two new schemes for simulating chemically reactive fluids. Dahler and Qin[19] formulated a statistical mechanical treatment of chemical reactions, in which the reaction rate is expressed as a convolution integral of a correlation function of reaction rate densities with the chemical affinity. Identifying the affinity as the driving force of the reaction enables a direct analogy between this formulation and standard response theory formulations for the transport of mass, momentum and energy. The theory is not yet tested against simulation or experimental data. Lisal *et al.*[20] formulate a hybrid scheme, termed "dual control cell reaction ensemble MD" (DCC-R×MD), to compute diffusion coefficients, permeability and the mass flux of fluids in porous media. The scheme couples reaction ensemble MC (REMC) with the dual control volume grand canonical MD method (DCV-GCMD). The method thus allows simultaneous simulations of reaction dynamics and adsorption processes and could have important applications in membrane separation studies.

Accelerated MD methods continue to attract interest due to their promise in studying large molecular systems such as biomolecules with multiple relaxation timescales. Minary *et al.*[21] note that while multiple timestep integrators go some way towards solving this bottleneck, they are still limited by resonance phenomena that restrict integration timesteps to <8 fs in biological systems. They propose resonance-free equations of motion that generate the canonical distribution in configurational space and test their algorithm on several model systems. They find they are able to integrate their equations of motion with order of magnitude improvement compared to typical multiple timestep methods. An alternative approach by Hamelberg *et al.*,[22] based upon the accelerated MD scheme of Voter,[23,24] applies a bias potential to the true potential in regions of potential energy minima, where the system can be trapped for long periods of time and hence limit the ergodicity of the system relative to the available simulation time. The application of this bias potential raises the energy minima, allowing the system to escape these trapped regions in a fraction of the time they would otherwise take, while still being able to sample the available conformational space. In a similar vein, Miron and Fichthorn[25] consolidate smaller energy barriers in the potential energy surface into a "collective" larger state, thus speeding up the motion between energy barriers via "hyperdynamics". Their algorithm is done "on the fly" such that necessary slow dynamics is faithfully observed and the simulation can be tuned to follow these "slow" events that are of particular interest. Though not in the spirit of accelerated MD, Zheng *et al.*[26] seek to escape the limitations of timescale bottlenecks for electrostatic interactions via a cell multipole method, allowing for a linear dependence on the number of particles, compared to the quadratic dependence of conventional simulation methods. The method is particularly efficient for programming on massively parallel computers, which is increasingly the way that supercomputing is heading, and is applied to both MD and grand canonical MC simulations. In all such schemes the question of suitable integrators is invariably linked to the scheme, and most authors describe the type and justification of integrator used, or else derive the integrator suitable for their needs. A recent example of this is for the case of translational and rotational motion of rigid molecules in the microcanonical, canonical and isothermal-isobaric ensembles. For this, Kamberaj *et al.*[27] derive time reversible integrators based on quaternions that are suitable for rigid body simulations, such as Gay-Berne molecules. Like most other schemes, this one starts with the Liouville equation and uses Trotter factorisation to obtain the integrators, which, in this case, turn out to be similar to the velocity Verlet integrator and demonstrates good stability compared to comparable standard methods.

Finally, we comment on a method, in some ways analogous to the method-of-planes techniques (referred to in Section 9: see also Daivis *et al.*[28]), of calculating unbiased estimators for the radial distribution function. In this method, proposed by Adib and Jarzynski,[29] the usual method of histogramming $g(r)$ in bins of width Δr is re-considered. The authors derive an exact statistical mechanical expression in which $g(r)$ is expressed as a simple bulk average over some spherical region between the radius of the solute and the radius of the largest sphere that can fit inside the simulation cell. In this way the approximate and necessarily biased nature of binning is removed entirely.

2.2 Nonequilibrium MD Simulations and Hybrid Atomistic-Continuum Schemes. – There have been relatively few new NEMD techniques to emerge. Most of these are based essentially on existing techniques, such as the SLLOD equations of motion[30] for simulating homogeneous shear or elongational flows. Pan *et al.*[31] revised an operator splitting algorithm to perform isokinetic NEMD simulations using the SLLOD algorithm. Their scheme, based upon the work of Zhang *et al.*[32,33] and Martinya and co-workers,[34–36] conserves the kinetic energy using a Gaussian thermostat. They demonstrated that the algorithm could accurately reproduce energies, pressures and viscosities for a simple WCA fluid using integrator timesteps an order of magnitude greater than commonly used in NEMD simulations. However, it was found to fail for low shear rates and so was unable to probe the Newtonian regime.

Petravic[37] extended earlier studies on the time dependence of phase variables under steady shear flow. This periodic dependence is due to the non-autonomous nature of the periodic boundary conditions.[38] Using linear response theory Petravic was able to study the density dependence of such systems and found that the time dependence is a nonlinear effect at low densities. Petravic and Harrowell[39] derived linear response theory expressions for the shear viscosity and thermal conductivity of fluids driven by thermostatted boundary conditions. These expressions are likely to be useful in nanofluidics, in which explicit modelling of boundaries and the rheology of fluid/wall interfaces is important. Terao and Müller-Plathe[40] devised an NEMD scheme for computing thermal conductivities in which the heat sink is localized but the heat source is uniformly distributed in the system. Their algorithm makes use of numerical noise as a source for the heat and does not require the computation of energy fluxes. Their algorithm is tested to reproduce the known thermal conductivity of the SPC/E model for water. Its application or usefulness for systems under flow or non-uniform density profiles remains untested.

Ge *et al.*[41] have used power-law scaling exponents of the shear dependent isotropic pressure to predict the solid-liquid phase coexistence for a simple Lennard-Jones fluid. The novel aspect of this work is that it uses standard NEMD SLLOD techniques for a non-equilibrium system to predict the *equilibrium* phase diagram at the liquid-solid interface. The method is seen to be computationally efficient and self-starting, requiring no knowledge of a starting phase, such as is required for the highly useful Gibbs-Duhem method. Whether the technique can be useful for more complex fluids remains unanswered for the time being.

NEMD simulations of elongational flow continue to attract a cautious interest. The cautiousness is associated with confusion in the simulation community over what exactly are the correct equations of motion to use when simulating homogeneous elongational flow. The first implementations of indefinite steady-state planar elongational flow[42–44] were performed with the SLLOD algorithm using periodic boundaries based upon the reproducible lattices of Kraynik and Reinelt,[45] but the use of SLLOD was earlier questioned by Tuckerman *et al.*[46] who devised their own variant, known as the g-SLLOD equations. The g-SLLOD algorithm was criticized for a number of reasons[13] and a sense of uncertainty has lingered over how to simulate elongation. This debate has been re-opened recently by Edwards and colleagues, who maintain that g-SLLOD is correct.[47–49] However, their re-derivation of g-SLLOD contains the same problems as in the original derivation by Tuckerman *et al.*,[46] as well as others. It has been recently shown[50] that g-SLLOD predicts the wrong energy dissipation rate for a viscous fluid under steady elongation, whereas SLLOD predicts the correct dissipation. Daivis and Todd[51] have, from first principles, re-derived the SLLOD equations of motion for generalized flows.

They show that for elongational flow the SLLOD equations are identical to Newton's equations of motion with the inclusion of an additional external force that must exist in order to sustain a steady elongational flow. Their derivation shows that SLLOD is the correct set of equations to use when performing NEMD simulations of elongational flow. No doubt the issue will continue to be debated in the literature for some time to come.

Controversy aside, Matin *et al.*[52] devised a cell neighbour list method for planar elongational flow that is compatible with the Kraynik-Reinelt scheme of boundary conditions. The scheme is highly efficient for large system sizes and is suitable for chain molecules. Hunt and Todd[53] made the observation that the Kraynik-Reinelt scheme of periodic boundary conditions is in fact equivalent to the Arnold cat map, a famous mapping of the torus onto itself, which has been highly instrumental in the study of low dimensional chaos. They re-derived the scheme of periodic boundary conditions and showed that use of the cat-map formalism considerably simplifies the mathematics and increases the intuitive understanding of how these boundary conditions work. It is the first connection made between a dynamical map and a scheme of boundary conditions useful for MD simulations. This discovery linked the cat map to the inherent chaos in the convective part of elongational flow and it is suggested[50] that this may be a source of inherent instability in the microscopic dynamics of elongational flow, something that was observed in the earlier work of Todd and Daivis.[54]

As computing power increases in accord with Moore's Law, the prospects for applying molecular simulation to important problems in molecular biology and nanofluidics continue to tantalize and attract researchers. While most of these studies on biological systems involve equilibrium structure and energy minimization calculations, a growing number are seeking to devise new, or use existing, techniques to study large molecular systems under nonequilibrium conditions. Work on nanofluidics is progressing rapidly. Delgado-Buscalioni and Coveney[55,56] devised a hybrid continuum-particle scheme for simulating fluid flows of moderate to high densities. Their scheme couples the mass, energy and momentum fluxes between two regions, one a particulate region operating under Newton's particle dynamics, the other a continuum region described by continuum fluid dynamics. Their scheme has been tested for both steady and unsteady flows and they find that the typical hydrodynamic flow variables evolve as would be expected. The goal of such a scheme is to enable longer real time simulations of flows that take into account the microscopic details of the molecular components, without the need for doing brute-force MD. They apply their algorithm to the test case of a single polymer chain tethered to a wall that itself is immersed in an explicit solvent and compare their results to a brute-force MD simulation of the same system.[57]They find very good agreement in the properties computed by both methods, with the advantage that their algorithm consumes only some 6% of the CPU time of the MD simulation.

Hybrid atomistic-continuum methods have also been developed and explored by other authors. For example, Werder *et al.*[58] propose a similar scheme in which the atomistic part of the simulation evolves via MD, whereas the continuum part evolves under classical Navier-Stokes equations. The main technical problem in all such approaches is to carefully match the interface boundary conditions between the atomistic and continuum regions, and these challenges are addressed. Particle insertion into the atomistic section is undertaken by the USHER algorithm [59] in which insertions occur where the potential energy corresponds to the mean potential energy of the system. The whole simulation process occurs in an iterative manner, in which boundary conditions from the continuum (the continuum velocity field) are fed into the atomistic component of the simulation. In turn, the generated atomistic velocity field is then fed back into the continuum as the updated continuum boundary condition, and so on. In this way the combined system is constantly updating the flow profile. The advantage of this scheme over others that involve coupling of fluxes between atomistic and continuum regions is that only velocity boundary conditions are required, which are easier to extract than fluxes. Their scheme was tested against a full MD simulation of liquid argon around a carbon nanotube, and good agreement was found between the two flow fields generated. Future work will extend the system to more complex molecules, including aqueous systems.

Nie *et al.*[60] also use a hybrid scheme to study cavity flow that is driven by a moving wall. In this case both the stresses and velocity fields are computed. They find that slip behaviour is important to quantify but this can not be done by a purely Navier-Stokes boundary condition. Comparison of the hybrid scheme with a full MD simulation demonstrates good agreement between flow velocities and shear stresses, including around corners. The Navier-Stokes solution is also in good agreement with both MD and hybrid solutions provided that the scale over which the continuum model fails is smaller than the inertial length, defined as the ratio of the kinematic viscosity to the flow velocity, and is typically only a few molecular diameters. Once again the hybrid scheme allows for the simulation of much larger systems with speed-ups of several orders of magnitude compared to MD. It is this speed-up factor that makes such schemes attractive. What is however lacking is a detailed statistical mechanical analysis of the validity of these schemes, particularly the coupling of fluxes and flow fields in the interface between the atomic and continuum boundaries, and a link between this and generalized hydrodynamics. Some progress along these lines has been made recently by Ripoll and Ernst[61] but in terms of dissipative particle dynamics and not coupled hybrid systems (for a review of the links between microscopic and macroscopic fluid mechanics and smooth particle methods, see Hoover and Hoover[62]). This would be a worthy exercise, in much the same way as the development of statistical mechanics for thermostatted particle systems in the 1980s and 1990s allowed for greater insights into the dynamical properties of such systems and a theoretical justification for the use of thermostats.

In an attempt to enhance coarse-graining towards biomolecular modelling applications, Ayton *et al.*[63] devise their so-called "BLOB/BLOBS" methodology, in which a "ghost" particle that represents a volume of coarse-grained fluid element is first embedded in a sea of atomistic particles. This ghost particle, or "blob", gathers dynamical information without disturbing the surrounding microstructure. Then, in a second stage of the simulation, the "intelligent" blob is surrounded by "like-minded" blobs and the coarse-grained mesoscopic level of the simulation is implemented via equations of motion analogous to the SLLOD equations for NEMD. The mesoscale simulation effectively speeds up the simulation by a factor of 40 compared to MD. There is again no attempt made to justify the methodology via either statistical mechanics or hydrodynamics. Interestingly, Okumura and Heyes[64] show that classical Navier-Stokes hydrodynamics for nonstationary conditions is valid down to nanometer length scales, breaking down only at very narrow channel widths. This result is in agreement with the results of Travis *et al* [65]earlier for nonequilibrium steady-state (stationary) flows. The blob idea is considered further by Chao *et al.*,[66] though quite distinct and independent of Ayton *et al.'s* method. Chao *et al.'s* blob model takes into account stereochemical information. In this way a particular material (*e.g.* molecule) is separated into rigid blobs and suitable coarse-grain blob inter- and intra-molecular potentials are calculated for them, allowing for faster implementation compared to molecularly detailed MD. Blobs of both spherical and ellipsoidal geometry are considered. Good agreement with atomistic MD simulation is found for the radial distribution functions for some trial molecular systems.

3 Potential Energy Hypersurfaces for Liquid State Simulations

For simple fluids composed of small molecules, the current equilibrium simulation techniques are highly accurate and the computational tools exist for carrying out long enough simulations of a sufficiently large system to obtain experimental accuracy. Thus, for these systems the accuracy of the potential energy surface is normally the factor that will restrict the accuracy that can be achieved for the determination of their properties.

Potential energy surfaces can be obtained empirically, or using quantum chemical calculations (or a combination of these). Empirical methods tend to be accurate, provided the system studied does not differ too much from the one(s) for which it was parameterised, and the properties that are considered are similar to those considered in the parameterisation. Jorgensen and Tirado-Rives discuss various potential energy functions in their review on simulations of water and organic biomolecular systems.[67] Quantum chemical methods, in contrast, should apply widely, but are limited by the accuracy of the quantum chemical methods employed. The sophistication (and therefore accuracy) that is computationally feasible usually rapidly deteriorates as the system's size increases, and errors become relatively large when weakly bound systems are considered.

Potentials can be developed that are transferable (*i.e.* the potentials are of a specific form that are applicable to a range of systems and/or for systems in a wide range of thermodynamic states), or so that they are as accurate as possible for a single system. Transferable potentials have a fixed analytic form, usually constructed based on physical and chemical properties of the system, and with parameters that are determined by the atoms or groups in the system of interest. There has been considerable recent work in this area in the past few years – particularly in the incorporation of both experimental and quantum chemical information in construction of the potential (see for example[68–72]), however here we will focus on developments in the construction of potentials where transferability to different systems is not the main motivation.

Since it is the forces between molecules that are important for dynamics in liquids state simulations, rather than developing potential energy surfaces, effective force fields can be directly obtained. One interesting approach to development of these force fields is to use forces determined during *ab initio* MD simulations as data for parameterisation of effective force fields. This is referred to as a Force-Matching procedure. This approach is advantageous in that it ensures that physically important regions within the force field are well sampled, however it suffers from the possibility that it will not be highly transferable. Once an adequate force field has been developed, long simulations can be carried out much more quickly. Promising results for liquid water[73] and HF[74] have recently been obtained over a range of state points where the force-matching procedure involves use of CPMD simulations.

The interaction energy of a system of N particles, $\Delta V(\mathbf{r}_1, \mathbf{r}_2, \mathbf{r}_3, \ldots)$ can be expressed as a sum of interaction potentials of increasing order:

$$\begin{aligned}\Delta V(\mathbf{r}_1,\mathbf{r}_2,\mathbf{r}_3,\ldots) &= V(\mathbf{r}_1,\mathbf{r}_2,\mathbf{r}_3,\ldots) - \sum_i V(\mathbf{r}_i) \\ &= \sum_{i,j>i} V(\mathbf{r}_i,\mathbf{r}_j) + \sum_{i,j>i,k>j} V(\mathbf{r}_i,\mathbf{r}_j,\mathbf{r}_k) + \ldots\end{aligned} \tag{1}$$

where $V(\mathbf{r}_1, \mathbf{r}_2, \mathbf{r}_3, \ldots)$ is the total energy of the system, $V(\mathbf{r}_i)$ is the monomer potential energy, $V(\mathbf{r}_i, \mathbf{r}_j)$ represents the pair-potential interaction and $V(\mathbf{r}_i, \mathbf{r}_j, \mathbf{r}_k)$ is the triplet potential interaction. Usually the dominant term is the pair-interaction, $V(\mathbf{r}_i, \mathbf{r}_j)$ and the remaining terms $(\sum_{i,j>i,k>j} V(\mathbf{r}_i,\mathbf{r}_j,\mathbf{r}_k) + \ldots)$ are referred to as many-body terms. Some interaction potentials, such as that of neon, are well represented by pair-potentials.

To construct an *ab initio* pair-potential, quantum mechanical calculations are carried out at a range of bond-lengths and a curve is fitted through the data. In many cases, even the error resulting from the assumption of pair-additivity is small compared to the errors in the pair-potentials, so development of techniques to determine accurate pair potentials is a field in which important research is being carried out.

Recently, considerable effort has been devoted to obtaining quantum mechanical intermolecular potentials suitable for fluid simulations. For some examples see 75 for $(HF)_2$; 76 for $(N_2)_2$; 77–80 for $(CO)_2$; 81 for the benzene dimer; 82 for the SiH_4 dimer; 83 for water-argon and water-methane potentials; 84 for the formamide dimer and N,N-dimethylformamide; 85 for lithium iodide in dimethylsulfoxide; and 86 for Ni^{2+} in aqueous solution. Rather than discuss each of these studies, here we will focus on a few important developments that we anticipate could alter the capacity or approach to development of interaction potentials in the future. In the following we will consider studies on quantum mechanically determined potentials for weak intermolecular interactions such as those between rare gas atoms; the current status of work incorporating three-body interactions in quantum mechanical potentials; and development of potentials for interactions between fluid molecules in confined systems.

3.1 Quantum Mechanical Interaction Potentials for Weak Interactions. – Development of quantum mechanical intermolecular potentials is challenging because of the weakness of the interactions, and because the dominant forces are often dispersion forces which are much more difficult to determine accurately than other interactions such as electrostatic interactions.

Most developmental work has been on rare gases. The interactions in these systems are particularly weak, but this fact is balanced by the fact that they are monatomic, and therefore the number of degrees of freedom and the number of electrons are reduced. The status of studies on potential energy surfaces of rare gases, and the structural, thermodynamic and transport properties of their fluids was reviewed up to mid-2003 by Searles and Huber.[87] It was indicated that experimental accuracy (or better) can be obtained for most properties of liquid neon, which is generally well represented by a pair-potential. For argon, the pair-potential is already highly accurate, but three-body interactions need to be determined more accurately for some properties. For krypton and radon even the pair-potential is unsatisfactory. While the potential energy surface for helium should be adequate, it is necessary to incorporate quantum effects in order to give accurate results, so few liquid state properties had been determined.

The accuracy of the interaction potential is limited by the size of the calculations required, and while these results are for specific systems, they give an indication of the current limitations for interactions dominated by dispersion forces. Although computer speed will continue to increase, if *ab initio* calculations are to be practical for calculations of interaction potentials of a wide range of molecules, it will be necessary to employ faster schemes. At the time of the review,[87] high level wavefunction-based quantum mechanical methods were required to obtain the accurate pair-potentials for neon and argon. For neon the best potential that was available was by Gdanitz[88] (employed in simulations by Muthusamy *et al.*[89]) which was based on coupled cluster calculations at the CCSD(T) level (with corrections for missing correlations, basis set incompleteness, missing clusters and relativistic effects), and the most accurate quantum mechanical pair-potential for argon was by Cybulski and Toczylowski[90] which was also obtained using the CCSD(T) method. Since this time, further work has been carried out to determine accurate rare gas potentials (see 77,91–93 and references therein). The best results are still generally based on coupled cluster (CCSD(T)) methods with corrections, except for helium where CCSD(T) and a number of other high level *ab initio* calculations are able to produce similar results.[94]

The development of density functional theory (DFT) methods has revolutionised computational quantum chemistry, allowing much larger systems to be studied accurately. However, until very recently this method has not assisted in the study of intermolecular interactions of rare gases or other weakly interacting systems. Although there is no fundamental reason why it cannot be applied to these systems, there has been no functional that adequately describes the dispersive forces responsible for the intermolecular interactions in these systems.[95,96] Recently, however the symmetry-adapted perturbation theory (SAPT) of intermolecular forces has been used in conjunction with DFT to give accurate intermolecular forces.[97,98] This method was originally proposed in 2001,[99] but due to the way in which it was implemented, the results were not particularly accurate. However recent advances seem to have produced a robust and efficient scheme, and similar accuracy to results from the most accurate wavefunction based methods (CCSD(T)) has been obtained. Importantly the SAPT(DFT) methods are valid at all separations of interacting molecules, they are memory efficient, and they are insensitive to the density functional that is selected (in contrast to conventional DFT methods[97]).

Figure 1 compares the widely accepted empirical pair potential for argon of Aziz[100] with a CCSD(T) potential[77] which is one of the best *ab initio* quantum chemical potentials currently available, the argon van der Waals potential with a dispersion term[485] and the SAPT(DFT) results of Podeszwa and Szalewicz.[97] Clearly the SAPT(DFT) potential performs well, and it is reported to require only a fraction of the computational effort of the CCSD(T) potential. This approach is still new, and its performance for a wide variety of systems needs to be investigated. However it promises to be of great use in the efficient determination of accurate interaction energies for use in the simulations of liquids.

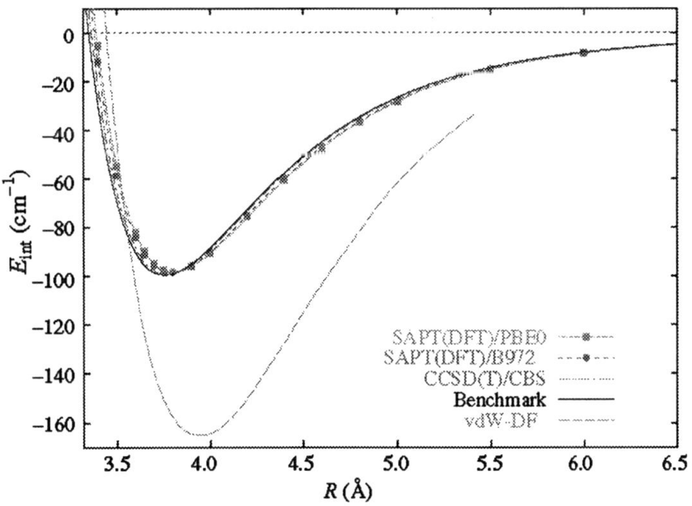

Figure 1 *Comparison of argon interaction-potentials. The solid line is the empirical pair potential;[100] the dashed line is the van der Waals + dispersion model of Dion et al.;[485] the dot-dashed line is a CCSD(T) ab initio potential;[77] and the dotted lines with circle and square symbols are the SAPT(DFT) potentials of Podeszwa and Szalewicz[97] with different functionals. Reprinted from, 97, Copyright (2005) with permission from Elsevier*

In particular, the *ab initio* MD methods such as the CPMD method or AMPD simulations which rely on calculation of the total interaction potential 'on the fly', and therefore currently relies on DFT methods, might be able to be applied to a wider range of liquids.

An alternative approach to obtaining accurate *ab initio* energies in a computationally efficient manner is the Hybrid Methods for Interaction Energies (HM-IE) method that has been developed by Sandler and coworkers.[102,103] In this approach it is assumed that the effects of electron correlation and basis set size are additive in order to reproduce the results of CCSD(T) calculations using a large basis set, with considerably less computational effort. The method is similar to the 'hybrid' or 'compound' quantum mechanical methods such as the widely used G3 method, and those of Dunning and Peterson.[104] The HM-IE method has been used to determine the interaction potentials for Ne_2, $(C_2H_2)_2$, N_2-benzene[102] and the acetylene dimmer[103] which have been shown to approximate the large basis set CCSD(T) results very well. The CCSD(T) calculations with a small basis set are corrected for the basis set error using Møller-Plesset (MP) calculations. It is found in 102 that use of an average of MP2 and MP3 level calculations to estimate this error produces remarkably good results over a wide range of intermolecular separations and molecular orientations, and this was also the case in subsequent work.[103] Studies on a wider range of systems will show how widely this method can be used reliably, and therefore its potential as a predictive approach.

3.2 Three-Body Interactions. – While the three-body interaction term (see equation (1)) can be neglected in some systems, it can be very important in others. In addition, the importance of the inclusion of three-body terms varies with the property considered.[87] In most classical liquid state simulations today, two-body effective empirical potentials are used. Since many of the parameters used in these potentials are determined using liquid state data, it is presumed that some of the effects due to the three-body terms will be accounted for. Because determination of forces between molecules is the most computationally intensive part of a simulation, and inclusion of three-body interactions increases the time by a factor of order N, it is often assumed that effective pair potentials provide adequate representations of the system.

If three-body interactions are considered explicitly in simulations, the Axilrod-Teller-Muto[78,79] (ATM) empirical three-body interaction term is usually employed. This term represents the three-body dipole-dipole-dipole dispersion energy, and while it has been shown to improve simulation results in a number of cases, it was more recently shown that the success was due to the complete cancellation of other many-body terms in these cases[105] so this term should be utilised with caution. Nevertheless, some useful information on the importance of three-body terms can be obtained using the ATM terms (see, for example the study on the importance of three-body terms in clusters and films[106]), and treatment of these systems with more accurate representations of this term would be of interest.

Much work has recently been carried out to quantify the three-body and many-body interactions in small clusters (mostly rare gases and water), which have implications on the liquid state properties, however here we consider some studies that have directly determined their influence on the bulk fluid properties. In reference[87] the significant influence of three-body interactions on properties of rare gas fluids is discussed, and a recent manuscript by Szalewicz *et al.*[107] thoroughly reviews the importance of many-body forces in general. Here we just summarise some important recent results.

The SAPT method has been applied to determine the three-body potential of water at the SCF level by Mas *et al.*[108] and this potential has been used in conjunction with a MP4 SAPT pair potential, and SAPT(DFT) pair potential[107] in molecular simulations by these authors.[107,108] In this work the water molecule is assumed to be rigid and quantum effects are not included, and therefore as pointed out by the authors, complete agreement with experiment cannot be expected. Nevertheless they were able to demonstrate the importance of inclusion of three-body terms in these systems, showing that hydrogen bonding is suppressed if they are excluded. As well as using the three-body potentials, simulations were carried out with four and higher-body interactions approximated using a polarization model. The simulations demonstrate that inclusion of three-body terms is crucial to obtaining accurate structural and energetic results for water, and suggest that higher-order many-body contributions improve the O–O distribution function and might be important for the accurate modelling of liquid water.

A review on the water trimer by Keutsch *et al.*[109] points out the importance of three-body interactions in bulk water, and provides a comprehensive discussion on approaches taken to incorporate these effects into the potential models. These include the use of effective potentials with polarisable potentials.

The importance of many-body interactions for accurate prediction of phase diagrams of rare gases has been demonstrated.[87] A study on liquid mercury that uses an *ab initio* pair potential and a semiempirical many-body correction to produce an effective three-body potential that was able to predict the liquid-vapour phase diagram well, [110] suggests that the main deficiency of the classical pair potential is the neglect of many-body effects.

Moghaddam *et al.*[111] discuss the consequences of three-body hydrophobic interactions on properties of a system consisting of methane-like nonpolar solutes in water. The implications of their findings on understanding protein folding are considered in detail.

3.3 Potential Energy Functions for Confined Fluids. – The recent demand for knowledge about behaviour at the nanoscale has enhanced interest in fluid-wall interactions, due to their relative importance in these (and smaller) systems. When large pores are considered, and wall atoms are small, it is reasonable to use simple, computational efficient models of the walls where the details of the atomic structure of the wall are averaged out (*e.g.* such as the Steele 10-4-3 model for slit pores[112,113]). Such models are still important, and new model potentials have recently been developed. These include a potential model for

Lennard-Jones cylindrical walls and fluids[114] that gives the same energies as the potential of Petersen *et al.*[115], but has an analytic form. New, effective, structureless potentials have also been proposed for the adsorption of polar molecules on graphite.[116] Since the carbon atoms of graphite have a non-zero quadrupole moment, electrostatic interactions between the absorbed molecules can occur, and in this paper it is shown that these interactions are significant when the adsorbate is strongly polar. The difference between single-site and three-site models of carbon dioxide in carbon slit pores has been examined by Bhatia *et al.*[117] who show that the packing in these models is very different, highlighting the importance of multi-site models.

The structureless wall-fluid interaction potentials described above require input of parameters (e.g. the Lennard-Jones parameters σ and ε[101]) that are associated with the wall and fluid atoms or molecules involved in the interactions. These will also be required if a model potential such as the Lennard-Jones potential[101] is used. Often, empirically determined bulk parameters are used for the fluid. In small pores and/or low fluid densities this may become problematic due to the fact that bulk fluid many-body interactions that are implicitly influencing these parameters may be quite different in the pore.[118] In fact, in these cases three-body interactions involving a fluid molecule and two wall atoms will often be more likely to occur. Parameters should ideally be determined using empirical data for the fluid wall interaction, or using *ab initio* calculations of these. A number of studies have used *ab initio* calculations to examine the interaction of molecules with systems that are expected to resemble carbon nanotubes and graphene sheets, such as polyaromatic hydrocarbons [(119–124)]. Some studies have even determined potential energy curves (for example 120–122), however accurate *ab initio* multidimensional surfaces for many-atom adsorbates have not yet been obtained. Due to the size of adequate representations of graphene, as well as the weakness of interactions, such calculations are challenging and a number of approaches have been developed. Bauschlicher and coworkers use a combined quantum mechanical/molecular mechanics (QM/MM) method (the ONIOM method) to treat hydrogen on a carbon nanotube.[125] Cunning *et al.*[118] use the concept of basis sets of high local quality to look at interactions between neon and large polyaromatic hydrocarbons as models of graphene. Ghorai *et al.*[126] used *ab initio* data to provide new values for the Lennard-Jones CH_4-zeolite interaction parameters, σ and ε. They show that the reparameterised potentials give much better adsorption isotherms and self-diffusion coefficients than the earlier work.

Klauda *et al.*[127] use HM-IE *ab initio* calculations (see Section 3.1) to determine the effect of curvature of a carbon wall on its interactions with N_2 and O_2. Interactions with graphene, fullerene (C_{60}) and schwarzite (C_{168}) are compared and curvature is seen to have a significant effect – particularly for the N_2-carbon surface interactions.

In even more realistic studies of fluids in porous systems, the details of the long and short range ordering of the pores have been examined. Gubbins and co-workers and Snook and co-workers have carried out a substantial amount of work in this direction for both ordered and amorphous systems. Snook and co-workers have recently use *ab initio* electronic structure calculations in order to assist in the accurate modelling of graphene layers and graphene walls.[128] They find that curved sheets of graphene are likely to occur. They point out that the DFT methods available for use in their structure calculations is not able to model dispersion interactions well, and indicate that further work in this direction is therefore required.

4 Quantum Mechanical Considerations

The consideration of quantum effects in the simulations of liquids can take on many forms depending on the system and property of interest.

Although interactions between molecules *must* be treated using empirical or quantum mechanical methods, for simple fluids it is a good approximation to consider the dynamics of simple liquids using classical mechanics except in the smallest molecules or at low temperatures. In a purely theoretical calculation of liquid state properties, either (i) a potential energy surface is calculated; or (ii) the forces on each particle are calculated 'on the fly' by methods such as the Born-Oppenheimer (BO) MD simulations, *ab initio* CPMD simulations and Atom-Centred Density Matrix Propagation (ADMP) schemes.[129] Approaches (i) and (ii) both have advantages and disadvantages. In the first case, high-level quantum calculations are possible, but some interpolation error may result, and it is usual to assume that the potential energy surface is pair-wise additive for intermolecular interactions. Furthermore chemical reactions cannot be properly treated unless multi-state empirical valence bond (MS-EVB) type surfaces are used.[130] In the second case, the energy is calculated only at the geometries that are sampled and all molecules are included in the *ab initio* calculation, so many-body interactions are properly incorporated. However due to the enormous number of configurations sampled in a typical simulation, the latter method has severe restrictions on the sophistication of the *ab initio* potential energy calculations and the size of the simulation box that can be treated.

4.1 Born-Oppenheimer, Car-Parrinello and Atom-Centred Density Matrix Propagation Methods. – Despite the computational demands of these approaches, with the increase in computing power and development of more sophisticated algorithms, BOMD, CPMD and ADMP simulations are becoming feasible and are especially important in the study of molecules where the number of degrees of freedom and significance of many-body interactions or the occurrence of reactions make development of accurate potential energy surfaces problematic. Recent applications of these methods are expansive and we mention only a few of the numerous interesting applications. Bühl *et al.*[131] studied the coordination environment of the aqueous uranyl (VI) ion, which is important in nuclear waste processing. They use CPMD methods with the gradient-corrected B3LYP functional, a pseudopotential for the U atom, and periodic boundary conditions. Zhuang and Dellago[132] study HCl dissociation in liquid glycerol, demonstrating the cooperative hydrogen bonding effects required in order to produce HCl ionisation.

Recently use of localised Wannier functions instead of delocalised Bloch states (CP orbitals) in CPMD simulations has proved an efficient and effective approach for study of fluids such as water[133] and DMSO/water mixture.[134] Use of localised Wannier functions also has the advantage that they allow electrons to be assigned to bonds, making visualisation of bonding and structure of molecules easy and facilitating comparisons with standard chemical bonding models.

Blumberger, Sprik and co-workers have made considerable advances in the simulation of redox reactions in solution (see for example 135–138). They have treated various redox reactions and gained important insight into the structure and solvation of ions using the extension to CPMD simulations that allows consideration of changes in the number of electrons in a system.[139] A grand-canonical approach is used where the (integer) number of electrons is allowed to vary. To avoid difficulties associated with periodic boundary conditions of charged system, finite systems can be considered which are large enough that boundary effects can be considered small, or boundaries can be imposed that result in neutral overall charge. In the latter case, use of periodic boundary conditions will still be substantial, and then only differences in properties where these effects cancel can be considered.

4.2 Hybrid Methods. – For molecules that are immersed in a solvent, hybrid methods have been developed where the solute, and possibly molecules closely associated with it, is treated using *ab initio* quantum chemical calculations and molecular mechanics methods are employed for the other molecules. Alternatively, for large molecules, part of the molecule might be treated using *ab initio* quantum mechanical methods whereas the rest of the molecule and the solvent can be treated in a simpler manner. This greatly reduces the computation time, but ensures that the molecules or interactions of interest are treated accurately. It is necessary that the interface between the subsystems dealt with using different schemes is treated carefully, and various procedures have been introduced to address this. Reference 140 reviews QM/MM methods for studying enzymatic reactions and includes a discussion on ways in which this interface can be treated. In 141 Sebastiani and Röthlisberger show that using sufficiently large solvation cells, QM/MM methods can give NMR chemical shifts comparable to those obtained using full QM calculations but at a fraction of the computation time. Now QM/MM methods have been made available in commercial and publicly accessible packages, and these methods are becoming commonly used, with perhaps the ONIOM method being the most well known.

Jorgensen and coworkers have used QM/MM MC simulations to elucidate transition states and reaction mechanisms of common and important organic reactions. For example, they studied the nucleophilic aromatic substitution reaction between an azide ion and 4-fluoronitobenzne in protic and dipolar aprotic solvents[142] and were able to correctly predict the solvent-induced changes in the activation energies or reaction rates of this process. A step in the mechanism of a macrophomate synthase catalysis of 2-pyrone derivatives to benzoate analogues[143] was also elucidated using QM/MM MC calculations.

Importantly, the simulations allow transition structures, including the solvent if significant, to be visualised. This work demonstrates that simulations of reactions of simple molecules can now be investigated using liquid state simulations where the solvent is treated explicitly.

QM/MM methods are often used in the study of biological systems where long-range electrostatic interactions are likely due to polar solvents and presence of counterions. Furthermore, although the simulation cells are often quite small, electrostatic cut-offs are often applied. This can lead to artefacts in reaction free energies and barriers. An efficient linear-scaling Ewald method has recently been developed which extends the commonly used Ewald summation method so that it can be applied to QM/MM calculations.[144] It has been applied to a range of systems and shown to give reliable results.

A new hybrid approach that promises to be very useful for the study of chemical reactions and systems containing charged species has been developed by Rega *et al.*[145] It combines the ONIOM scheme with an ADMP approach that is shown to be significantly faster than hybrid BOMD, and because it is an atom centred approach it can treat systems with periodic boundaries and long-ranged interactions.

4.3 Cluster Calculations. – While properties such as the pressure, temperature, shear stress, *etc.* can be considered to depend on the positions and momentum of the nuclei only, other properties, such as vibrational frequencies, electric field gradient and chemical shifts require the full quantum mechanical structure of the molecules. However, many 'quantum properties' can be readily calculated from averaging static quantum mechanical results for the fluid-state configurations of clusters of molecules once the ensemble of liquid state configurations are generated. Such calculations are often referred to as 'cluster calculations' and the configurations can be generated using MC or MD simulations (classical, BOMD, CPMD *etc.*). The accuracy of the results depends on the adequate sampling using the MD or MC simulation methods, and the accuracy of the quantum chemical calculations, which in turn will depend on the size of the clusters of molecules that are required. Recently vibrational calculations on aqueous Li^+,[146] and N-methylacetamide,[147] and NMR property calculations for liquid water[148,149] and other hydrogen-bonding fluids[150] have been published that demonstrate the utility of cluster calculations using classical and CPMD methods. A new and different approach is used by Müller *et al.*[151] where a quantum chemical property surface is calculated (analogously to the potential energy surface) and used to determine the property at each timestep of a MD simulation. A comparison of recent results for the calculation of the ^{17}O and D quadrupole coupling constants (QCC) are given in Table 1. Various methods of calculation of NMR properties of liquids have recently been discussed in reviews.[152,153]

4.4 Dynamical Quantum Effects. – When dynamical quantum effects become important, corrections must be made. At the simplest level, quantum corrections can be made following Hirschfelder *et al.*[157] allowing corrections for

Table 1 *Comparison of ^{17}O and D quadrupole coupling constants determined in simulations with the experimental values*

Reference	*D QCC/ kHz*	*^{17}O QCC/ MHz*	*Details of method*
Pennanen *et al.* (2004) [149]	186	8.56	CPMD Cluster Calculation; 400 clusters; HF calcs HF/aug-cc-pVDZ on central molecule; 32 molecules in the liquid state unit cell
Pennanen *et al.* (2004) [149]	184	7.77	CPMD Cluster Calculation; 400 clusters; DFT calcs at B3LYP/aug-cc-pVDZ on central molecule; 32 molecules in the liquid state unit cell
Müller *et al.* (2004) [151]	252 ± 1		Empirical potential; quantum mechanical efg surface; 500 molecules in liquid state unit cell; 'quasi-experimental' determination of QCC
Müller *et al.* (2004) [151]	267.3 ± 0.1		Empirical potential; quantum mechanical efg surface; 500 molecules in liquid state unit cell; direct calculation
Eggenberger *et al.* (1993) [154]		8.9 ± 0.3	Cluster calculation; 42 clusters; Empirical potential; HF efg calculations; best estimate from clusters of 5–9 molecules; 125 molecules in liquid state unit cell.
Experimental	247		from 155
Experimental		8.96	from 155
		8.0 ± 0.2	from 156

quantum effects on the second virial coefficient, radial distribution function *etc.* Wigner-Kirkwood perturbation theory can also be applied. However, in other cases (such as the light rare gas liquids) methods such as Centroid MD (CMD), Path Integral MC (PIMC) or Path Integral MD (PIMD) may be necessary. PIMD and PIMC simulations express the partition function of a quantum mechanical system in terms of a path integral. The path integral is equivalent to a configuration integral of a system of (artificial) classical 'ring polymers', and therefore it is possible to obtain equilibrium quantum statistical properties using modified classical simulations of the ring polymers. Due to its importance, many studies on the consequences of treating nuclei using a quantum mechanical approach have been on liquid water. Recently PIMD methods[158] and CMD approaches[159,160] have been used to look at water or rigid liquid water. In these studies the isotope effects on dynamical properties are found to be accurately predicted by the quantum calculations, and the quantum corrections become more significant as the temperature is reduced, consistent with physical expectations. PIMD simulations of liquid hydrogen fluid[161] have also been carried out which indicate that inclusion of nuclear quantum effect strengthens hydrogen bonding.

In classical MD simulations it is often assumed that the nuclei move on a potential energy surface. Implicit in this assumption is that the Born-Oppenheimer approximation applies: that electronic and nuclear motion can be separated and that an adiabatic potential energy surface exists. In many processes, including charge-transfer processes, photo-induced reactions and some recombination processes, this is not the case. The system moves between electronic states and due to the quantum nature of these states, a quantum mechanical treatment of the processes is required. Several approaches have been taken to treatment of these systems and are summarised in a recent review.[162]

Röthlisberger and co-workers have considered numerous problems of biological importance by combining QM/MM calculations with CPMD DFT quantum and classical simulations. In reference[163] the retinal chromophore of rhodopsin is investigated using classical and CPMD approaches, and although agreement between results from the two methods depends on the property considered, the structural properties and dynamics are very similar. They also use time-dependent DFT approaches to study photo-induced processes.[164,165]

5 Lyapunov Exponents

The chaoticity of a fluid can be measured in terms of the Lyapunov exponents, the set of which are referred to as the Lyapunov spectrum.[166] At a molecular level, where the degrees of freedom of a system are treated as the momenta and positions of all the particles in the system, simple fluids are chaotic and mixing. In fact, at equilibrium fluids are usually more chaotic than nonequilibrium fluids under the same conditions, which can be considered to be due to the development of organised behaviour often associated with the application of a field. Chaoticity leads to the decorrelation of properties with time. Over several decades, methods have been developed for accurate determination of these exponents, and exact relationships have been derived which relate the properties to physical properties, such as transport coefficients.[167–169]

Simulation studies on the Lyapunov exponents of fluids over the past few years have been mainly associated with finding insights into the physical behaviour of a system from examination of Lyapunov exponents, their vectors and their spectra.[170,171] Analyses of which particles or modes contribute to particular Lyapunov exponents have been carried out, and investigations on how trends in the exponents can be linked to changes in the fluid properties have been carried out. A number of groups have been looking at the exponents that have small magnitudes – the so-called 'Lyapunov modes' or 'Posch modes' that have been identified as Goldstone modes and linked to the hydrodynamic behaviour of a system.[170,172] These modes have now been observed in fluids of 'soft' particles.[173,174]

Changes in the largest Lyapunov exponent have also been used to signal phase changes, however work in this area has mostly been at the fluid dynamics level, using integral equations.

De Wijn[175] has used cylindrical scatters as a representation of a system of many hard spherical particles and has examined their Lyapunov spectrum as part of a study on the interaction of cylindrical scatters with high-dimensional billiards. Comparisons of the results with those of hard disks and the high-dimensional Lorentz gas are made.

Recently, the Lyapunov instability of fluids in pores has also been examined using simulation,[176] highlighting the fact that singularities are observed when the pore width is twice the width of the particle size. However it is expected that this effect will be weaker in soft spheres.

Frascoli *et al.*[177] have examined the Lyapunov spectrum for a bulk system of WCA particles undergoing elongational flow. In the limit of large system size the system is shown to satisfy the conjugate pairing rule, which is consistent with the fact that the underlying equations of motion are symplectic.[178] From a practical point, satisfaction of the conjugate pairing rule means that the transport properties of a fluid can be calculated in an efficient manner from the Lyapunov exponents. The results are compared with those obtained using the SLLOD equations of motion for Couette flow which does not have a symplectic structure, and numerical evidence suggests that pairing in this system might only occur in the weak field limit. This result is also consistent with earlier studies.[178,179]

The chaoticity of a fluid is also often treated at a fluid dynamics level, rather than modelling all components at the molecular level, and in this case it is important to distinguish between Lagrangian chaos, in which chaos exists within a fluid volume moving with the flow, and Eulerian chaos, which is related to the onset of turbulence. In order to examine Lagrangian chaos in these systems, test particles can be considered. Recently calculation of Lyapunov exponents of Lagrangian particles within a fluid have been used to examine phenomena such as the mixing of fluids,[180,181] phase changes in dilute polymers[182] and the effect of polymer additives on turbulence.[183]

We briefly mention a couple of interesting papers that are directly related to chaos and transport. Ihm *et al.*[184] demonstrate that the Kolmogorov entropy η_{KS} is related to the self-diffusion coefficient (D) in simple liquids by the relationship

$$h_{KS}/\nu \propto \left(D/\sigma^2\nu\right)^\eta \tag{2}$$

where v is the average collision frequency, σ is the effective atomic radius and η is an exponent that depends on the form of the potential interaction but is independent of density and temperature. Based on their work on Helfand moments, Viscardy and Gaspard[185] compute the shear viscosity by the escape rate method, which has previously been developed and used by Gaspard and colleagues to compute diffusion and reaction-diffusion processes.

6 Thermodynamic and Transport Properties

6.1 Thermodynamic Properties. – The drive towards a consistent theory of thermodynamics far from equilibrium continues to motivate a number of researchers and we consider some of these in this section. A major and long-lasting controversy in formulations of irreversible thermodynamics is due to the definition and/or existence of generalized thermodynamic variables and the lack of experimental tests to differentiate between the models proposed. As such, molecular simulation continues to be a favoured approach in helping to validate these models. Several fine reviews exist in the literature, such as those by Casas-Vazquez and Jou[186] and Kröger[187] and the reader is referred to these and others for a more thorough exposé.

Daivis and Matin[188] propose an exact theory of steady-state thermodynamics for a shearing fluid that is valid in the linear regime. Their work is an extension of previous formulations by Evans and Hanley[189–191] but goes further by deriving a simple expression for the change in free energy due to a change in strain rate and thermodynamically evaluating the entropy difference between an equilibrium state and a shearing steady-state. The novelty arises in splitting the work into two components, one a viscous and the other an elastic component. The former is identified with the irreversible work while the latter is identified with the reversible work. This, they find, leads to the determination of the steady-state entropy, which, for a simple atomic fluid in the weak-field limit, is found to vary as the square of the strain rate. Their theory is used to compute thermodynamic quantities of a shearing Lennard-Jones fluid by NEMD simulation and is demonstrated to be consistent with expectations.

An alternative approach to describe steady-state thermodynamics for shear flow was formulated by Taniguchi and Morriss.[192] Their method involves the development of a canonical distribution for shear flow by a Lagrangian formalism of classical mechanics. They then derive the Evans-Hanley thermodynamics, *i.e.*

$$dU = TdS + \zeta d\dot{\gamma} \tag{3}$$

where U, T, S, $\dot{\gamma}$ and ζ are the internal energy, temperature, entropy, strain rate and a coefficient conjugate to the strain rate, respectively, from their distribution. Once again they validate their formalism by NEMD simulations of a simple fluid. Their approach is based more heavily on nonequilibrium statistical mechanics, whereas the approach by Daivis and Matin is a purely phenomenological theory. It would be a useful and constructive exercise to compare both theories for consistency checks and for underpinning theoretical connections (or differences) between them, but this remains to be done.

Another approach towards a thermodynamics of steady-state systems is presented by Santamaría-Holek *et al.*[193] In this formulation a local thermodynamic equilibrium is assumed to exist. The probability density and associated conjugate chemical potential are interpreted as mesoscopic thermodynamic variables from which the Fokker-Planck equation is derived. Nonequilibrium equations of state are derived for a gas of shearing Brownian particles in both dilute and dense states. It is found that for low shear rates the first normal stress difference is quadratic in strain rate and the viscosity is given as a simple power law in the strain rate, in contrast to standard mode-coupling theory predictions (see Section 6.3).

The approaches of Daivis and Matin and Taniguchi and Morriss are essentially exact, in that they are applicable generally to simple shearing fluids in the linear regime. However, in order to cross over the many timescales of complex fluids Kröger and Öttinger[194] develop a scheme known as "beyond-equilibrium" MD (BEMD), which is based on the GENERIC algorithm.[195] In effect these schemes integrate out the fast degrees of freedom from a system, evolving only the slower experimentally measurable variables of interest. This reduces the computational effort when compared to NEMD methods, in which all degrees of freedom are evolved. It starts with a generalized canonical ensemble, compatible with the GENERIC equation of motion for the time evolution of the macroscopic variable $\mathbf{x}$ (where $\mathbf{x}$ is a vector, whose components are density, velocity and temperature). The so-called friction matrix appearing in the equation of motion for $\mathbf{x}(t)$ is evaluated along short trajectories of length τ_s and thus the equation of motion is solvable for the relevant thermodynamic variables. Their method is applied to a rarefied gas under shear flow and compared with traditional NEMD simulations. BEMD is seen to be computationally more efficient than NEMD at weak field strengths ($\dot{\gamma}<0.1$) but fails at higher strain rates. It cannot therefore be used universally and it is not yet clear how useful it will be for complex fluids (*e.g.*, polymer melts and solutions) in which nonlinearities appear at weaker field strengths (in principle, this is where it should be useful).

Ansumali *et al.*[196] propose that the grand potential, written in the Eulerian coordinate system as

$$\Omega_E = \left(-p + \frac{(\rho u)^2}{2\rho}\right)\delta V \tag{4}$$

is the natural thermodynamic potential to study the onset of incompressibility in fluid flows. Here p is the pressure, ρ is the density, ρu is the momentum density and V is the volume. They show that use of the grand potential instead of the entropy significantly simplifies the hydrodynamic equations for low Mach number flows. Reduced compressible Navier-Stokes (RCNS) equations are thereby derivable, while the incompressible Navier-Stokes (INS) equations are a quasi-stationary solution of the RCNS. Simon *et al.*[197] study the thermal flux through a surface of *n*-octane at the liquid-vapour interface via nonequilibrium MD (driven by a temperature gradient) and nonequilibrium thermodynamics methods. The agreement between both methods is supportive of the basic assumptions used in nonequilibrium thermodynamics (primarily that of local equilibrium in a temperature gradient) and allows for linear flux-force transport expressions with surface tension dependent transport coefficients.

We now briefly consider another important aspect of nonequilibrium thermodynamics, namely phase transformations and how they are modelled. Galenko and Jou[198] develop a thermodynamic formalism for rapid phase transformations within a diffuse interface of a binary system in which the system is in a state of local nonequilibrium. The phase-field method, in which the phase-"field" variable Φ varies smoothly and continuously between one pure phase (in which $\Phi = +1$) and another (in which $\Phi = -1$), is used to derive defining equations that are compatible with both extended irreversible thermodynamics (macroscopic) and the fluctuation-dissipation theorem (microscopic). The value of $\Phi = 0$ defines the interface between the two phases. Because the method is compatible with extended irreversible thermodynamics (in contrast to classical irreversible thermodynamics, in which local equilibrium is assumed) the local nonequilibrium nature of the system means that rapid phase transitions are able to be modelled.

In a sequence of consecutive papers, Reguera and Rubí[199,200] formulate a thermodynamic description of homogeneous nucleation of clusters in *inhomogeneous* media, namely systems that are both spatially inhomogeneous and under the influence of either (a) a temperature gradient[199] or (b) a steady-state shear flow.[200] The study of nucleation is relevant and important because it forms the first stage in many phase transitions. They extend the mesoscopic nonequilibrium thermodynamics formalism (MNET) of Rubí and colleagues (see earlier references contained in 199) for homogeneous media so that it is extended to inhomogeneous media. This is accomplished by dividing the system volume into smaller volume elements that are small enough to be approximately homogeneous but large enough to be thermodynamically valid. Thus, thermodynamic gradients must be sufficiently larger than the length of the volume elements, which themselves must be larger than typical molecular length scales. Local equilibrium is assumed to hold and the Fokker-Planck equation is derived from the MNET formalism. This equation describes the dynamics of the nucleation phenomenon and provides a mesoscopic description. Macroscopic evolution is then obtained by averaging the relevant probability densities obtained in the Fokker-Planck equation and constructing hydrodynamic equations from them. Nucleation rates are computed and condensation and crystallisation under a temperature gradient and mechanical stress are discussed. Not surprisingly, they find that shear flow influences transport and the evolution of cluster formation, this in turn affecting the nucleation and growth rates. Temperature gradients are seen to be particularly important in the case of polymer crystallisation, in which thermal diffusion plays a crucial role.

6.2 Free Energies and Entropy Production. – In the past few years, development of new theories have lead to completely new ways of determining free energy changes. Traditionally, free energy changes or differences ($\Delta A_{1\to2}$) of two equilibrium states can be obtained directly from the statistical mechanical definition of the free energy, A, in terms of the partition function, $A = -k_BT\ln\int d\mathbf{\Gamma}e^{-\beta H(\mathbf{\Gamma})} = -k_BT\ln Z$, where k_B is Boltzmann's constant, $H(\mathbf{\Gamma})$ is the Hamiltonian at $\mathbf{\Gamma}$, T is the temperature of the system, and $\beta = 1/(k_BT)$. That is,

$$\begin{aligned}\Delta A_{1\to2} =& k_BT\ln(Z_1) - k_BT\ln(Z_2)\\ =& -k_BT\ln\left\langle e^{-\beta(H_2-H_1)}\right\rangle_1\end{aligned} \tag{5}$$

where the initial state has Hamiltonian H_1 and the final state has Hamiltonian H_2, and where the ensemble average in the second line is carried out using the equilibrium distribution of state 1. Due to numerical issues related to determination of the final expression directly when the free energy difference is large, it is usual to break the procedure into tractable (often non-physical) stages that take the system step-wise from state 1 to state 2, and then to use thermodynamic integration to obtain an overall result. Bias potentials or reweighting schemes (*e.g.* umbrella sampling methods) have been used to improve the sampling. Alternatively, if the change from state 1 to state 2 is carried out so slowly that the process can be considered 'reversible' the well know relationship between the change in free energy and the reversible work done in transforming from state 1 to 2, $w_{1\rightarrow 2;rev}$, applies:

$$\Delta A_{1\rightarrow 2} = w_{1\rightarrow 2;rev} \tag{6}$$

In simulations, the change from one state to another is achieved by slowly and continuously varying a control parameter that produces state 1 initially and state 2 finally. A completely reversible situation can only be approached, and the procedure is often referred to as a 'slow switching' approach.

Development of the Jarzynski equality (also known as the Work Relation or the Nonequilibrium Work Relation) in 1997, and the Crooks identity (or Crooks Fluctuation Relation/Theorem) have provided a new approach to the calculation of free energy differences, enabling their determination by following irreversible (and therefore possibly fast) pathways. The approach has not yet resulted in a more efficient algorithm than the conventional approaches, but as they are new methods, this may still be possible. These methods are often referred to as a 'fast switching' method. Here we firstly consider developments that are still being made in the more traditional approaches, and then describe the newer methods, citing papers where the different approaches have been used and compared. A special issue of Molecular Simulation "Challenges in Free Energy Calculations" published in Jan-Feb 2002[201] includes articles on the different approaches to free energy calculations, including the Jarzynski approach.[202]

A number of papers have discussed the theoretical and practical aspects of calculation of free energy changes from 'end-point' methods (first equality in equation (5)), where the free energy is calculated from the logarithm of the partition function of the final and initial states, particularly in the studies of ligand-receptor binding free energies.[203–205] It therefore requires accurate knowledge of these partition functions, which can be problematic – especially in systems with large, flexible molecules in a solvent. Woo and Roux[203] describe a new scheme based on determination of the potential of mean force where ligands are constrained in their binding configuration in their unbound state, and which carefully included solvent effects. Swanson *et al.*[204] discuss the appropriateness of approximations made in end-point methods, emphasising the importance of including translation/orientational effects and presenting a MD simulation method of doing so. Swendsen *et al.*[205] present a new adaptive method of calculating free energies from the partition function.

Anwar and Heyes[206] present a new approach for calculation of free-energies in charged molecular systems when particles need to be created or annihilated during the process. This procedure is expected to be particularly useful in determination of free energies and hence study of reactions and phase diagrams in systems with large and complex molecules. Escobedo and coworkers[207,208] have developed and used a new importance weighting method that uses multicanonical sampling and is useful for multicomponent systems. Neimark and Vishnyakov[209] have developed a method useful for study of small, confined, inhomogeneous systems when particle creation and annihilation is required.

The Jarzynski equality states,

$$\left\langle e^{-\beta w_{1\to 2}} \right\rangle_1 = e^{-\beta \Delta A_{1\to 2}} \tag{7}$$

where the work is that done in transforming a system from state 1 to state 2 along any nonequilibrium pathway. In contrast to relation (6), the path does not have to be reversible, so the system can be out of equilibrium during the process. Equation (6) is a direct consequence of (7) obtained by taking the infinitely slow path. A recent review on the Jarzynski equality and related theorems discusses the method and its application.[210]

Lechner *et al.*[211] have used equation (7) in free energy calculations, taking advantage of the fact that the procedure does not need to sample physically realistic trajectories to replace solution of the Newtonian dynamics by an appropriate mapping. This improves the efficiency of the usual implementation of (7) where Newtonian trajectories are used in the transformation from state 1 to state 2. They study a system where a particle is moved through a dense fluid and find an order of magnitude improvement in the results compared to the standard fast-switching technique.

Various groups have developed reweighting and optimisation schemes to improve the converges of the Jarzynski equality (see for examples,[212–216]). Gore *et al.*[217] examine the finite sampling error for the Jarzynski equality in the near-equilibrium regime, and develop a bias-corrected estimation.

The Crooks identity can be expressed as,

$$\frac{P(w_{1\to 2} = a)}{P(w_{2\to 1} = -a)} = e^{-\beta(\Delta A_{1\to 2} - a)} \tag{8}$$

where $P(w_{1\to 2} = a)$ is the probability that the work done along a path from state 1 to 2 is equal to a, and $\Delta A_{1\to 2}$ is the free energy difference between the initial and final states. This relationship has been verified experimentally[218] and used to obtain free energy changes.

A number of papers have presented alternative derivations or generalisations of the Jarzynski equality and fluctuation relations, including their application to quantum systems.[219–222]

Liquid simulation studies have been essential in assessing the applicability of various fluctuation relations to real physical systems. These are important relations in nonequilibrium statistical mechanics that are valid far from equilibrium and can be used to derive Green-Kubo relations for transport coefficients.[223,224] They show how thermodynamic irreversibility emerges from reversible equations of motion. When applied to nonequilibrium steady states, the relationships take on the form,

$$\frac{\mathrm{P}(\Sigma_t = a)}{\mathrm{P}(\Sigma_t = -a)} = e^a \tag{9}$$

where $\mathrm{P}(\Sigma_t = a)$ is the probability that the extensive entropy production, Σ, takes on a value a over a period, t. It can be used to provide a simple proof that the entropy production must be positive, as well as providing information on the distribution of the entropy production. Simulations have been carried out on a wide range of systems and they have also been verified experimentally.[225–227]

In derivation of the steady state relations, artificial, reversible thermostats such as the Gaussian thermostat or Nosé-Hoover thermostat have been used. However, Williams *et al.*[228] have recently used MD simulations to verify assumptions made in their derivation that shows that the FR is insensitive to the details of the thermostatting mechanism.

6.3 Transport Properties. – The behaviour of non-Newtonian simple liquids – in particular the dependence of thermodynamic and transport properties on strain rate – remains to this day an interesting and as yet unresolved issue. Since the early formulations of mode-coupling theory [229] that predicts non-analytic dependencies of pressure, energy and viscosity on strain rate (in particular, pressure and energy scale as $\dot{\gamma}^{3/2}$ whereas viscosity scales as $\dot{\gamma}^{1/2}$), computer simulations have been an invaluable aid in testing these predictions. Until recently most computer simulations (particularly NEMD simulation) of simple liquids confirmed these predictions. This was questioned in 2001 by Marcelli *et al.*[230] who performed very accurate simulations of liquid argon using 2 + 3-body interaction forces and found quadratic dependencies of both energy and pressure. In a later study Ge *et al.*[231,232] found that the mode-coupling predictions were valid near the triple point, but broke down elsewhere. Most recently Ge *et al.*[233] and Todd[234] have fully characterized this behaviour and find a remarkably simple linear relationship that describes the scaling exponent as a function of density and temperature. In particular, if X represents any of energy, pressure or viscosity, then its non-Newtonian behaviour may be characterized by a power law of the form

$$X \sim x\dot{\gamma}^{\alpha} \tag{10}$$

where x is a constant, and the power-law exponent α is simply expressible as

$$\alpha = A + BT - C\rho \tag{11}$$

where A, B and C are positive constants. It is of interest that these power-law dependencies are qualitatively consistent with the theoretical predictions of Santamaría-Holek *et al.*,[193] discussed in Section 6.1.

Complementing simulation results are some recent theoretical developments. Wada and Sasa[235] consider long-range momentum correlations in a simple fluid and find the standard mode-coupling prediction for the pressure holds, *i.e.*

pressure scales as $\dot{\gamma}^{3/2}$. However, we point out that their model is not directly comparable with the simulation results of Ge *et al.* and Todd because of the larger strain rates necessary to obtain reasonable simulation statistics. It is clear that what is really necessary to allow direct comparisons between theory and simulation is for either the development of a theory valid at high strain rates, or else perform highly accurate simulations at sufficiently weak field strengths. Both remain difficult technical challenges.

Das and Bhattacharjee[236] derive the frequency and shear dependent viscosity of a simple fluid at the critical point and find good agreement with recent experimental measurements of Berg *et al.*[237] Ernst[238] calculates universal power law tails for single and multi-particle time correlation functions and finds that the collisional transfer component of the stress autocorrelation function in a classical dense fluid has the same long-time behaviour as the velocity autocorrelation function for the Lorentz gas, *i.e.*

$$C(t) \sim t^{-(1+d/2)} \tag{12}$$

where d is the dimension. Computer simulations of such time correlation functions are difficult to perform for weak fields, in the sense that highly accurate statistical data are difficult to obtain. Ge[239] performed some of these via the transient-time correlation function (TTCF) formalism and found that exponential decay was a better fit to the data rather than power-law decay, but the results are not extensive. Hess *et al.*[240] compute an effective viscosity based on equilibrium stress fluctuations rather than the usual stress autocorrelations. By analysing the cross-over between short-time elastic and long-time viscous processes they are able to estimate the zero-shear viscosity as an alternative to standard NEMD extrapolation methods. Viscardy and Gaspard[241] propose a new formulation for computing the viscosity by Helfand moments by incorporating the minimum image convention used in MD simulations with periodic boundary conditions. In this formulation the viscosity is computed via Einstein-like relations, where

$$\eta = \lim_{t\to\infty} \frac{\beta}{2Vt} \left\langle \left[G_{xy}(t) - G_{xy}(0)\right]^2 \right\rangle \tag{13}$$

and the Helfand moments are defined as

$$G_{ij}(t) = G_{ij}(0) + \int_0^t J_{ij}(\tau)\, d\tau \tag{14}$$

where J_{ij} is the relevant microscopic current appearing in the corresponding Green-Kubo expression. The Helfand-moment method is equivalent to the Green-Kubo method, obeys the central limit theorems and is simply computed.

The effects of thermostats on the transport properties of simple fluids continue to be explored. In particular the configurational thermostat is studied in NEMD simulations to address a number of shortcomings of previous work that used profile biased kinetic thermostats. Delhommelle *et al.*[242] perform

NEMD simulations of steady and oscillatory shear with a configurational thermostat. Shear thickening is observed for both steady shear and low frequency oscillatory shear as shear rate increases. Furthermore, at higher frequencies shear induced ordering commences at higher strain rates. These results are consistent with observations of colloidal suspensions. The geometry of the fluid is also examined under these conditions, showing hexagonal plane structures either perpendicular (low strain) or parallel (high strain) to the flow velocity. These simulations suggest that these non-Newtonian effects, usually attributed to a solvent, can occur in a pure fluid. In simulations this observation is made because use of a configurational thermostat no longer constrains the fluid to the desired Newtonian linear form.

Since Erpenbeck[243] observed so-called string phases in NEMD simulations of a simple fluid (*i.e.* an alignment of particles along the direction of flow mimicking solid-like ordering), this phenomenon has been a subject of considerable controversy. The most potent criticism, supported by carefully performed NEMD simulations (see for example the study by Travis *et al.*[244]), has been that these strings were artefacts of the thermostatting mechanism. In particular, an incorrect assumption on the form of the streaming velocity profile leads to an enhanced ordering in the fluid due to the way the equations of motion account for the dissipation of heat. To support this explanation, Delhommelle *et al.* have repeated these simulations, but by incorporating a configurational thermostat rather than a kinetic thermostat, thereby eliminating any assumption about the form of the streaming velocity profile. They find that NEMD simulations performed with the configurational thermostat shows no evidence of string formation[245] and furthermore that use of profile biased thermostats can give unphysical steady state flow profiles and that the actual temperature can be significantly lower than that expected even at relatively low shear rates.[246]

In a series of papers, Petravic has examined the effects of strain, stress relaxation and cooperative effects in nonequilibrium and equilibrium simple fluids. In the first of these Petravic[247] studied the transport, structure and thermodynamics of the dense Lennard-Jones liquid/solid. The diffusion, viscosity and thermal conductivity coefficients are characterized in the vicinity of the liquid-solid phase transition, as is the shear stress. In the second paper[248] a small dense liquid is found to maintain its liquid structure and diffusion coefficient when an imposed constant strain is applied, but does not relax in shear stress. This effect is greater as the system size decreases but its effect is lessened with increasing temperature, and the origin of this effect is purely configurational. Finally, a systematic study[249] is made of the cooperative effects in shear relaxation using equilibrium MD. Stress relaxation is found to be cooperative with a correlation length ξ that increases with decreasing temperature at constant pressure. For $\xi > L$, where L is the simulation box size, the shear stress depends upon the strain induced by the periodic boundary condition, but this is not the case for the transport properties.

Another systematic study of transport for simple fluids, this time the study of "hard" and "soft" potential fluids, has been undertaken Heyes and colleagues.

This follows from previous work of these authors that studied the transport of fluids interacting via a steeply repulsive potential of the form

$$\phi(r) = \varepsilon(\sigma/r)^n \tag{15}$$

where n is large. In the first of these new studies, Rickayzen *et al.*[250] consider the infinite frequency moduli and time correlation functions of steeply repulsive fluids with an interaction potential of the form

$$\phi(r) = k_B T \exp[-\alpha f(r)] \tag{16}$$

where α is a measure of the potential steepness. They compare these quantities with equivalent quantities computed via the r^{-n} potential and find that, by-and-large, the properties studied are independent of the specific form of the potential. Instead the infinite frequency moduli are proportional to α while, in the short-time limit, the time correlation functions are of the form

$$C(t) + 1 - T(\alpha t)^2 + O\left[(\alpha t)^4\right] \tag{17}$$

where temperature (T) and time (t) are in reduced units. They conclude that for hard-sphere type potentials, mechanical and dynamical properties are insensitive to the precise analytical form of the potential, unlike the case for soft-sphere potentials. Their theoretical treatment could therefore be useful in the study of colloidal suspensions. Extending their work on steeply repulsive fluids further, Rickayzen and Heyes[251] investigate the memory function for such fluids, finding a closed expression that is simple to evaluate. In the limit of hard spheres this expression reduces to the Enskog expression. The significance of this result is that their expression is a generalization of the Enskog expression that is valid for softer potentials. In another study, Branka and Heyes[252] extend the theoretical analysis of Dufty,[253] which showed that the stress time autocorrelation function for hard-sphere systems consists of a singular (at $t = 0$) and non-singular part (which tends to the Enskog solution at long time). This singular part is of the form sec$h(a_n nt)$, where a_n is a coefficient and n is the potential stiffness. Branka and Heyes extend Dufty's treatment to study the bulk viscosity and thermal conductivity. They find that the corresponding bulk viscosity time correlation functions (involving pressure) can not be expressed simply as the Dufty form, but is well represented by

$$C(t) = \sec h(a_n nt) + w(a_n nt) \tag{18}$$

where the function w satisfies the limits $w(t) \to 0$ when $t \to 0, \infty$. This form satisfies the requirements to reproduce the Enskog bulk viscosity and the correct short-time decay. A similar conclusion is drawn for the thermal conductivity correlation function (involving heat flux). In other papers[254] these authors study the effects of potential softness on the transport coefficients, particularly shear (η_s) and bulk viscosities (η_b), self-diffusion (D) and thermal conductivity (λ), comparing these to semi-empirical expressions with various degrees of success (D being best and λ being worst). While all these previous simulations were performed by MD methods, Branka and Heyes[255] also extend the work to more realistic colloidal systems by BD simulations.

In an interesting study of collective dynamics, Schoen and Porcheron[256] show by MD simulation that the dynamic structure factor $S(k_{\|}, \omega)$ (measured in light scattering experiments) may be usefully employed to observe phase transitions in confined fluids. For thermodynamically stable states, $S(k_{\|}, \omega)$ displays three peaks, whereas as one approaches the phase transition the three peaks are replaced by one that becomes a single delta-function peak exactly at the spinodal, as seen in Figure 2. Experimental measurement of $S(k_{\|}, \omega)$ should therefore provide significant insight into the collective dynamics of phase behaviour in both confined and bulk fluids.

The study of supercooled liquids continues to attract significant theoretical and simulation study. Miyazaki *et al.*[257] study supercooled 2-dimensional liquids under shear using mode-coupling theory and SLLOD NEMD methods. They find significant reduction of the shear viscosity and structural relaxation time due to shear, where both quantities decrease as $\dot{\gamma}^{-\nu}$ where $\nu \leq 1$, and that the relaxation is almost isotropic, despite the strong anisotropy in the direction of flow. Limitations of their model and directions for future study are also identified. The decoupling of transport coefficients in supercooled liquids is studied by Jung *et al.*[258] In particular, they study the breakdown of the Stokes-Einstein formula, $D \propto T/\eta$, for a supercooled liquid, finding this breakdown to result from fluctuations in the dynamics of low temperature glass formers. Their results differ from mode-coupling theory and predicts a dynamic scaling of the form $\tau\ (l) \sim l^{z}$, where $\tau\ (l)$ is a structural relaxation time for events at length scale l and z is a scaling exponent. Following up on this work, Berthier *et al.*[259] study the relationship between self-diffusion and spatio-temporal excitation lines by examining the incoherent scattering functions. They determine a

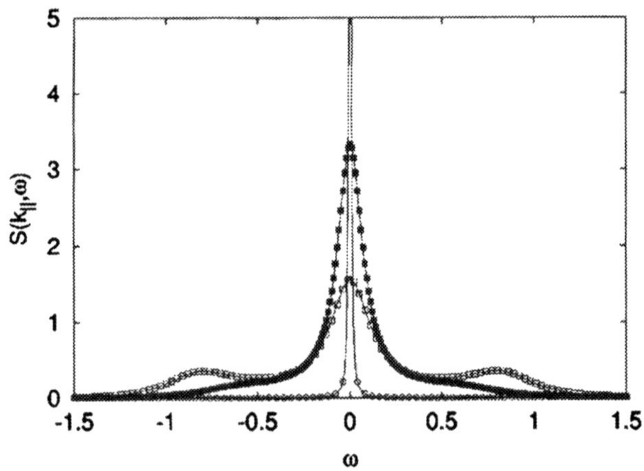

Figure 2 *Change in dynamic structure factor as the density of the confined fluid is decreased. The change from three peaks to a singularity at $\omega = 0$ is clearly visible at the spinodal. Reprinted figure with permission from 256. Copyright 2003 by the American Physical Society*

length scale for the onset of Fickian diffusion, characterize its temperature dependence and give analytic approximations to the scattering functions. In trying to understand how structure and dynamic heterogeneity are related in supercooled liquids, Widmer-Cooper *et al.*[260] ask the question: "what aspect of the dynamic heterogeneity actually arises from the structure?" To help answer this question they simulate a 2-dimensional binary liquid interacting by a purely repulsive potential by MD in the isothermal-isobaric ensemble. They run a number of trajectories with the same initial configurations but different random momenta and define the ensemble mean of the squared displacement of particle i ($\langle \Delta r_i^2 \rangle$) to be the "propensity for motion" of that particle and then examine the spatial distribution of these propensities as well as their scatter. They show that there is significant variation in the evolution of a specific particle configuration. Therefore not all particle dynamics can be correlated with an initial configuration for a supercooled liquid. Their method establishes that the spatial variation in the propensity for particle motion, rather than particle motion itself, is what is determined by the initial configuration. This description of supercooling is an alternative to that of potential energy surfaces.

To close this Section we comment on two papers that do not fit under any "neat" heading. The first of these is by Xiao *et al.*,[261] who study the final stages of the collapse of an unstable bubble or cavity using MD simulations of an equilibrated Lennard-Jones fluid from which a sphere of molecules has been removed. They find that the temperature inside this bubble can reach up to an equivalent of 6000 K for water. It is at these temperatures that sonoluminescence is observed experimentally. The mechanism of bubble collapse is found to be oscillatory in time, in agreement with classical hydrodynamics predictions and experimental observation. The second paper, by Lue,[262] studies the collision statistics of hard hypersphere fluids by MD in 3, 4 and 5 dimensions. Equations of state, self-diffusion coefficients, shear viscosities and thermal conductivities are determined as functions of density. Exact expressions for the mean-free path in terms of the average collision time and the compressibility factor in terms of collision rate are also derived. Work such as this, abstract as it may appear, may be valuable in the development of microscopic theories of fluid transport as well as provide insight into transport processes in general.

7 Phase Diagrams and Phase Transitions

7.1 Bulk Fluids. – Here we focus on liquid-vapour transitions and liquid-solid transitions, and briefly mention some recent deposition studies.

Methods for simulation of the liquid-vapour coexistence are well developed and were reviewed by Panagiotopoulos[263] however in some cases these have been shown to be sensitive to the potential energy surface and factors such as many-body interactions, and therefore new results continue to be obtained to investigate these issues and to study new systems (see for example,[110] for mercury,[264] for methane, and[265] for water). The Gibbs ensemble approaches and grand canonical and isothermal-isobaric MC simulations with histogram reweighting are usually employed. However, new methods continue to be developed for more efficient determination of vapour-liquid coexistence curves. Recently a multibaric-multithermal ensemble simulation method for efficient determination of the full phase coexistence curve of a Lennard-Jones fluid[266] was proposed and was shown to be a useful approach. McGrath *et al.*[267] have developed a program for the efficient determination of vapour-liquid phase diagrams using *ab initio* methods. A theoretical paper on the origin of condensation and phase separations highlights the fact that contributions of gradients and dividing surfaces to the system entropy need to be considered in studies of phase separation in small systems.[268] A new MD method, the superposition state MD method of Venkatnathan and Voth[269] which is useful for study of systems with multiple minima, might also be applicable in this context. Pamies *et al.*[270] compare the canonical MD method and the Gibbs ensemble MC method for determination of the coexistence line of methane and propane.

MC methods for the determination of phase diagrams of binary and ternary systems have recently been investigated. The flat-histogram method, which is a histogram reweighting method that allows a system to visit macrostates with a uniform distribution, has been combined with transition-matrix MC methods and has been found useful for the determination phase diagrams for binary systems.[271] A flat-histogram method for determination of coexistence points that can be applied to ternary as well as binary mixtures has also be developed and used to study Lennard-Jones mixtures.[272] In a similar approach, Escobedo and coworkers have used multi-canonical sampling to simulate along a macrostate path connecting two phases and thus determine free-energy profiles of binary systems.[207,208]

An interesting MD simulation study on how water boils has been published by Zahn.[273] A path sampling method is used, and shows that the boiling process is initiated by formation of cavities in the liquid that merge. Insight into the process is obtained by visualising the formation and merging of these cavities as boiling proceeds.

Development of efficient and reliable methods of determining melting curves has progressed in recent years. Experience with the rare gases[87] suggests that melting points are not particularly sensitive to details of the potential energy surfaces (although one should be cautious that this is not due to a cancellation of errors). However their determination by simulation has proved challenging since it is difficult to model sufficiently realistic melting using finite, periodic systems. Furthermore, standard Gibbs ensemble methods are generally not suitable due to the difficulty of carrying out particle insertions in dense systems (with the exception of hard sphere systems).

There are now several widely used approaches to determination of melting curves and solid-liquid coexistence lines: gradual heating of a perfect crystal, gradual heating of a solid containing voids or defects;[274–277] simulation of coexisting liquid and solid phases; and approaches where the differences in the Gibbs free energy of the solid and liquid phases are determined. In addition, a novel approach where NEMD simulations are used to predict equilibrium melting points, has been applied to determine the melting curve for Lennard-Jones fluids.[278]

It is understood that a perfect periodic crystal superheats and will melt at a temperature significantly higher than the thermodynamic melting temperature. The introduction of voids or defects prevent this, and it has been shown to be an efficient and reliable approach to determination of melting points and solid-liquid coexistence curves of a wide range of atomic and molecular systems (1-ethyl-3-methylimidazoium hexafluorophosphate;[279] ammonium nitrate;[280] nitromethane;[281] ammonium dinitramide[282] and rare gases[276–277,283]). This method is also advantageous in that it allows the melting process to be visualised and for solid state phase transitions that may occur to be identified. Phase coexistence studies have also been carried out for water.[284]

An alternative approach is to measure the free energies of the liquid and solid phases as a function of temperature, until they coincide. This is a thermodynamically rigorous method, and has been applied to a range of systems (for example: NaCl;[285] hard spheres;[286] model protein;[287] Lennard-Jones fluids;[288] water[289]). A number of variations have been adopted to determine the free energies: some involving relation to a reference state [285–287] (often using a multiple histogram reweighting method[263]); and others where the liquid is converted to a solid using reversible pathways.[288] In the latter method, the pathways are generally unrealistic, and as simple as possible. This is an attractive approach in that it avoids the need to compare the liquid and solid states to some reference state.

Eike *et al.*[290] propose a general free energy method for determination of the liquid-solid coexistence line. Firstly they apply isobaric-isothermal temperature increases to identify an approximate melting point, and then free energy difference methods are used to determine a single coexistence point. The free energy as a function of temperature is also determined using the multiple histogram reweighting method, and finally the Clapeyron equation is integrated to determine a full coexistence curve. Their approach has been tested on NaCl and the Lennard-Jones fluid, with good results.

There is still debate as to the best approach, with the gradual heating with voids approach currently being more flexible and efficient, and having the advantage that it gives some insight into the processes involved in melting, while the determination of the difference in free energies using reversible pathways is attractive from the view of thermodynamic rigor. Nevertheless, provided sufficient computer time is available it appears that both methods are able to predict melting points accurately and are able to be used as predictive and interpretive tools. In fact Aguado and Madden[291] use simulations of MgO to understand discrepancies between MD simulation melting points and coexistence curves, and the experimental results. They suggest that it is caused by a solid-solid phase transition that occurs just before melting.

The silicon melting line has been determined using a coexistence simulation[292] and is able to predict a decrease in T_m with increase in pressure, consistent with experiment. Vega *et al.*[289] carry out Gibbs-Duhem integration to determine the coexistence line of water, and the change in the coexistence line with potential model, and use this to predict the melting temperatures of a range of common models of water. The freezing of water from a salt solution has also been investigated with visualisation of the rejection of brine from the solution giving valuable insight into this process.[293]

Wang *et al.*[294] have studied the crystallization and melting of icosahedral gold nanoclusters, by gradually heating and cooling the system. Changes in the structure of the clusters and diffusion within the clusters are measured using various parameters, and are visualised. They find that melting does not occur below the melting point, but that surfaces soften. Alavi and Thompson[295] also observe surface softening in simulations of nitromethane nanoclusters. They find that the melting temperature increases with nanoparticle size, but has almost converged to the bulk phase value when clusters of 480 molecules are considered. Since the minimum systems size considered by Wang *et al.*[294] is 600 atoms, these results are not inconsistent.

A new phase has been suggested in hard spheres using a Gibbs-ensemble simulation algorithm[296–298] from consideration of compressibility data. This new phase has a zigzag pattern in configuration space and appears between the metastable fluid and solid phases but this has been questioned as an artefact of the simulation. More work is required to resolve the issue. A substantial range of compressibility data also been presented for use in development of hard sphere equations of state.[299] Giovambattista *et al.*[300] have simulated phase transitions in amorphous ice using simulations. Branka and Heyes have determined the equation of state for a fluid with inverse power intermolecular interactions using theory and computer simulations.[301]

Some interesting simulations have been used to investigate crystallization and deposition processes, in order to obtain a fundamental understanding of how these processes can be inhibited.[302,303] The inhibition of formation of waxes (mixtures of normal alkanes that form lamellar structures) was examined.[302] Gas hydrate formation has been investigated, and it was demonstrated that inhibition of methane hydrate by a octomer of polyvinylpyrrolidone was able to be simulated,[302] and simulations were used to assist in developing a new class of inhibitors of gas-hydrate formation.[303]

7.2 Phase Transitions in Confined Systems. – Studies on confined systems show that even relatively simple systems display complex phase behaviour, which is sensitive to changes in the potential interactions, structure of the pore and models for the pore walls. Despite these factors, some simulations have been able to model experimentally observed behaviour accurately.

Sałamacha and coworkers[304–306] have carried out a series of studies on Lennard-Jones fluids confined to nanoscopic slit pores made from parallel planes of face centred cubic crystals. Grand canonical and canonical ensemble MC simulations have been used to determine the structure and phase behaviour as the width of the pore and the strength of the fluid-wall interactions were varied. The pore widths were small: accommodating 2 to 5 layers of fluid molecules.[304,305] The strength of the fluid-wall interaction is linked to the degree of corrugation of the surface, and it is found that the structure of the solid phases formed at low temperatures have a structure resembling the solid when this interaction is strong. The pore width also dramatically influences the structure of the phases. The complex phase diagrams of these systems were calculated. The way in which the pores fill is examined and is also found to be sensitive to the fluid-wall interaction strength.[304,306]

Wongkoblap *et al.*[307] study Lennard-Jones fluids in finite pores, and compare their results with Grand canonical ensemble simulations of infinite pores. Slit pores of 3 finite layers of hexagonally arranged carbon atoms were constructed. They compare the efficiency of Gibbs ensemble simulations (where only the pore is modelled) with Canonical ensemble simulations where the pore is situated in a cubic cell with the bulk fluid, and find that while the results are mostly the same, the Gibbs ensemble method is more efficient. However, the meniscus is only able to be modelled in the canonical ensemble.

Hung and coworkers have studied the freezing and melting of Lennard-Jones carbon tetrachloride within carbon nanotubes.[308,309] In accord with the studies on carbon slit pores, the phase behaviour is complex and sensitive to conditions. Multiple melting transitions temperatures are observed and formation of the usual solid structures is prevented by the geometry of the wall, and concentric layers of molecules are seen to form at low temperatures. A parallel tempering scheme and free energy calculations are used to reliably determine the phase transition temperatures. These results are in agreement with experimental studies on carbon tetrachloride; and in agreement with the Gibbs-Thomson equation, the melting point of the fluid is depressed in the nanotubes and the depression becomes larger as the nanotube diameter decreases.

Cámara and Bresme[310] also studied crystallization of a Lennard-Jones fluid under confinement. They looked at the system in the vicinity of the triple point where the solid-fluid and fluid-fluid interactions were the same. The paper examines the forces exerted on the crystal as the melt crystallizes. They also find that the solid phase is stabilised by confinement, and that crystallisation is sensitive to the width of the pore and the lattice spacing of the solid.

Coasne *et al.*[311] studied freezing and melting of a binary mixture of argon and krypton in a structureless slit pore. Comparison of the results with the bulk mixture are made. Interestingly it is found that the melting point increases in these systems compared to the bulk, in qualitative agreement with experiment.

A study on torsion-induced phase transitions in fluids is carried out by Sacquin-Mora *et al.*[312] using grand canonical equilibrium MC simulations. A slit pore that has a repulsive interaction with the fluid molecules is 'chemically decorated' with attractive regions on both surfaces. These are positioned so that fluid molecules can form bridges across the pore. Such 'bridge phases' have been studied computationally and experimentally and are anticipated to play a role in nanofluidics. If the regions are misaligned, and the bridge is subject to a strain due to rotation of the substrates, it is found that the confined fluid will undergo phase transitions, and these are examined in this paper. Nanofluidics in pores with patterned, striped wall surfaces has been studied using MC simulations by Schneemilch and Quirke.[313] They carefully examine the wall-vapour interfaces perpendicular and parallel to the wall and their dependence on the wavelength of the stripes, and test the applicability of free energy and surface tension models to the results.

Lattice models have also been applied to the study of confined fluids. Water in a hydrophobic tube was examined by Maibaum and Chandler[314] who find that water will flow in pulses under certain conditions, and that density fluctuations at the entrances to the tube influence this behaviour. Rother *et al.*[315] use lattice models to examine the effect of confinement on the preferential adsorption of a component from a binary liquid system near liquid/liquid phase separation.

Neimark and Vishnyakov[316,317] have carried out an interesting study into the formation of a bubble in a Lennard-Jones fluid confined to a spherical pore at a metastable state. Various simulation techniques are used and compared. The Lennard-Jones parameters are selected to resemble nitrogen and the results are compared with experimental results, with qualitative agreement obtained.

Dzubiella and Hansen[318] investigate the competition of hydrophobic and Coulombic interactions between nanosized solutes, that can ultimately lead to phase separation when the concentration of a solute becomes sufficiently high. Spherical solutes of identical radii but various surface charge patterns are considered in constant pressure MD simulations. They examine how the competition between hydrophobicity (which would lead to aggregation of hydrophobic particles) and electrostatics (leading to separation of charged solutes) determines the structure and behaviour of a solution. They also examine how the surface charge pattern influences the results. Images, density profiles, orientational ordering, solvation free energies and mean force calculations are used in the interpretation of the simulation data. The results also compared with those obtained with a continuous solvent, and an important conclusion of this work is that it is essential to treat solvents at a molecule level if even the correct qualitative behaviour is to be reproduced, and charge distributions on surfaces of solutes should be realistically modelled. This clearly has implications in the accurate modelling of protein aggregation and interactions.

Toxvaerd[319] has studied nucleation in homogeneous systems, in systems with a carrier gas, and in confined systems. It is shown how confinement catalyses nucleation due to the formation of dense phases near the wall.

8 Complex Fluids

Enormous effort is spent on studying complex fluids, more-so than any of the previous topics reviewed above. These fluids include polymer solutions and melts, alkanes, colloidal systems, electrolytes, liquid crystals, micelles, surfactants, dendrimers and, increasingly, biological systems such as DNA and proteins in solution. There are therefore many specialist areas and it is impossible to review them all here. As such, we sample only a select few areas that reflect our own personal interests, and apologise to readers who have specific interests elsewhere. First, we briefly look over some simulations on colloidal systems, alkanes, dendrimers, biomolecular systems, *etc*, and will then concentrate most discussion on polymer solutions and melts, as the latter are overwhelmingly the subject of intense study by molecular simulation techniques. This is probably because (a) the increased power of computing now makes the simulation of long-chain molecules possible, which can therefore be used to test a variety of polymer theories and (b) polymers are an extremely important and omnipresent material in modern society, ranging in scale from nanocomposite materials for aerospace applications, to plastic chairs and cling-wrap, right down to miniaturized components in microfluidic and microelectronic devices. Hence there is a very strong commercial interest in being able to design tailor-made polymers for specific products. For an excellent review on the use of molecular simulation in relation to polymer materials, see Theodorou.[320] A thorough and comprehensive review of simulation and theory for complex nonequilibrium fluids is given by Kröger.[187]

8.1 Colloids, Dendrimers, Alkanes, Biomolecular Systems, etc. – Colloidal systems are of interest to study by simulation because of their technological relevance and also because they can be studied experimentally by techniques such as static or dynamic light scattering. Simulation can then be a powerful tool in understanding the molecular origins of experimentally observed phenomena. Nielaba *et al.*[321] study the structure, phase transitions and elastic properties of model colloids by MC methods. In addition to static properties, their phase diagrams under an oscillatory field are studied in order to mimic the interaction of colloids in external laser beams. Colloids are modelled as either hard or soft disks. Systems near walls, with impurities and binary mixtures in 2 and 3 dimensions are studied. Of particular interest is the observation, consistent with experiment, that hard disks show a re-entrant liquid phase over a broad section of thermodynamic state space. This is similar to the experimental observation that stable charged colloids under the influence of a laser field can undergo an initial freezing, followed by re-melting as the amplitude of the laser field is increased. These observed features are consistent with a recent dislocation unbinding theory of laser induced melting by Nelson and colleagues.[322,323] Addition of nonadsorbing polymer into a colloidal suspension can induce phase separation characterized by colloid rich (polymer depleted) and colloid depleted (polymer rich) phases.

The colloid-polymer model of Asakura and Oosawa[324] and Vrij[325] has been examined by Vink and Horbach[326,327] by grand canonical MC simulation, in which the critical point for the model is determined and the 3D Ising universality class is established. Vink *et al.*[328] further study the critical behaviour in greater detail using finite scaling methods to extrapolate the critical behaviour in the thermodynamic limit. They find that the critical exponent of the interfacial tension σ (where $\sigma = \sigma_0 d^{2\nu}$, d is a distance measure from the critical point and ν the exponent) is close to the expected 3D Ising value of $2\nu \approx 1.26$, which is also consistent with experimental observation.

Colloids under shear as studied by either BD or NEMD continue to grow in popularity. Once again these studies are motivated by the processing requirements of many colloidal systems. But they are also being studied because they are akin to soft-glassy systems and are therefore interesting from a fundamental science perspective. An example of theoretical interest is the origin of the violation of the fluctuation-dissipation theorem (FDT) for self-diffusion in a sheared colloidal suspension, examined by Szamel.[329] He does this by computing the self-diffusion and mobility tensors and shows that Einstein's relation between the two is violated (at equilibrium the relation is valid, as it must be). Next, he shows that by using transverse components of the self-diffusion and mobility tensors one can define an effective shear-rate temperature in terms of the relevant diffusion tensor component force autocorrelation functions. This simple expression is then a measure of FDT violation that can be used in BD simulations. Finally, by using mode-coupling theory to compute the friction tensor and effective temperature, Szamel shows that FDT violation is due to the nonequilibrium nature of the stationary sheared state. These theoretical predictions now need to be tested by numerical simulation.

BD simulations have also been carried out to study crystal nucleation of weakly shearing colloidal suspensions. Blaak *et al.*[330,331] find that the rate of crystal nucleation decreases as shear rate increases (found earlier by Butler and Harrowell[332]) but the size of the nucleation zone increases. A simple phenomenological theory is shown to be sufficient to account for these observations. Furthermore, the nuclei orient with respect to the flow velocity such that the smallest eigenvalue of the moment of inertia tensor of a nucleus (corresponding to the largest dimension of the nucleus) is in the direction of the vorticity, whereas the largest eigenvalue (smallest dimension) is perpendicular to it and slightly tilted with respect to the shear gradient.

Banchio and Brady[333] develop a new Stokesian dynamics algorithm for Brownian suspensions, based on the accelerated Stokesian dynamics formulation of Sierou and Brady.[334] In this formalism, the many-body long range hydrodynamic interaction is computed using fast Fourier transforms, and the algorithm is found to scale as $O(N^{1.25} \log N)$. A suitable and accurate approximation brings this scaling down to $O(N \log N)$ for nonequilibrium BD simulations and $O(N)$ for equilibrium simulations and makes the procedure attractive for systems of the order of 500 particles. Based on the accelerated Stokesian dynamics algorithm, Sierou and Brady[334] apply the technique to study shear induced self-diffusion in a non-colloidal suspension of up to 1000 particles and find good agreement with experimental self-diffusivities.

NEMD simulations of shearing colloidal suspensions have also been attempted by Delhommelle[335] and Delhommelle and Petravic.[336] Delhommelle shows[335] that the formation of "lanes" of particles in conventional NEMD or NEBD simulations is an artefact of using a kinetic thermostat. By using instead a configurational thermostat for NEMD simulations, these lanes disappear; equivalently use of hydrodynamic interactions in NEBD also fixes this problem. Following from this, Delhommelle and Petravic[336] show that shear thickening (the increase in shear viscosity as a function of strain rate) in a colloidal suspension results from the expulsion of solvent from the interstice between jammed colloidal particles, thus increasing resistance to flow. The relationship between this mechanism and the normal phenomenon of shear dilatancy for a pure liquid under shear is not discussed however, but would be a useful and insightful study.

The properties of molten alkali halides have recently been studied by both equilibrium MD and NEMD. Molten alkali halides are of industrial interest because they are used as electrolytes in high temperature carbonate fuel cells. At such high temperatures it is very difficult to experimentally measure important transport properties, such as viscosities and thermal conductivities. Molecular simulation therefore becomes a feasible and attractive alternative. Galamba *et al.*[337] compute the zero-shear viscosity of molten NaCl and KCl by Green-Kubo calculations via equilibrium MD simulations in the microcanonical ensemble (*NVE*). Comparison to available experimental data shows that use of the BMHTF interionic potential (see relevant references for this potential in 337) results in overestimating the shear viscosity by 10–15%. Next, these authors compute the thermal conductivity for these systems, again by Green-Kubo EMD simulations.[338] Comparison with existing experimental data shows again that the BMHTF potential overestimates the thermal conductivity by 10–20%. It is of note that both these studies used the Ewald method to compute electrostatic interactions. NEMD simulations of molten NaCl were also performed by Delhommelle and Petravic[339–341] but their aim was to study the relative merits of kinetic and configurational thermostats for such a system. It was shown here that a kinetic thermostat can significantly alter the microscopic mechanisms that control the conductivity of a molten salt. Galamba *et al.*[342] also performed NEMD simulations to compute the strain-rate dependent shear viscosity for the BMHTF and MWGKL potentials (see interaction potential energy references therein). Extrapolations to the zero-shear viscosity were made by fitting with a Carreau equation and a mode-coupling$\left(\dot{\gamma}^{1/2}\right)$relationship, but they were not in agreement when extrapolating to the zero-shear viscosity. Recent NEMD simulations of Todd[234] demonstrate that the mode-coupling relationship is unlikely to be valid at all thermodynamic state points at typical NEMD strain rates.

Bulk alkanes under shear have been studied by NEMD techniques for many years now, where their technological importance as lubricants is highly valued. Zhang and Ely[343] use the Buckingham exponential-6 potential to compute the shear viscosity of a variety of alkanes, alcohols and mixtures and find good agreement with experimental values for *n*-alkane and alcohol systems. McCabe *et al.*[344] compare the explicit-atom model of Borodin *et al.* for perfluoroalkanes with those of the united atom model of Cui *et al.* by EMD and NEMD simulation (see references in 344 for these potentials). The Newtonian shear viscosity for the former model is found to be in excellent agreement with experiment, whereas the united-atom model predicts significantly lower values. Finally, hydrogen bonding in shearing ethanol is studied by Petravic and Delhommelle[345] via NEMD simulation. They find hydrogen bonding dissociations overwhelmingly take place at low strain rates, but shear thinning is not observed until shear-induced alignment of hydrogen bonds with the direction of shear takes place.

To date there are few molecular simulation studies of dendrimers, though this is likely to increase rapidly in the next few years. Dendrimers are a particularly interesting class of synthetic polymer that is highly branched, having a fractal-like tree structure. They have been touted as possible rheology modifiers and drug delivery agents. Dendrimers are formed around a central core with relatively short chains with multifunctional groups built up one layer per generation number. Therefore, their structure is characterized by: (1) the number of generations (g); (2) the functionality of the end groups (f) (*i.e.*, the number of linear arms emanating from each node); and (3) the number of monomers per unit chain (b). Their structure is represented diagrammatically in Figure 3. The total number of monomers, N, grows exponentially with generation number as

$$N = fb[(f-1)^{g+1} - 1]/(f-2) + 1 \tag{19}$$

Because the rate of growth is faster than the available volume ($\sim g^3$), the combination of increasing density and excluded volume results in unusual bulk and solution properties compared to traditional polymers. Giupponi and Buzza[346] study by lattice MC dendrimers of generation 1-6 in solvents of variable quality. They find that the exact scaling of the radius of gyration (R_g) is not $R_g \sim N^{1/3}$, but rather conforms to the Flory theory predictions of $R_g \sim N^{1/5} [(g+1)m]^{2/5}$ for athermal solvents (m is the length between monomers) and $R_g \sim N^{1/4}[(g+1)m]^{1/4}$ in θ solvents. However, for all solvents the dendrimers have a dense core and terminal groups are distributed throughout the interior. This is in conformity with previous experimental observations,[347] computer simulations[348] and the theoretical analysis (revised de Gennes model) of Zook and Pickett.[349] In a subsequent study, Giupponi and Buzza[350] perform lattice MC simulations of amphiphilic dendrimers, such that terminal monomers interact differently with the solvent compared to internal monomers. The motivation behind this study is to control dendrimer conformations by amphiphilic interactions, with the long-term aim of using such dendrimers as "smart" materials that can respond to their environment. Depending on the types of interactions considered, dendrimers could form micelles of various conformations.

In the first NEMD simulation of dense dendrimer melts undergoing planar shear flow, Bosko *et al.*[351–353] study the conformations and viscoelastic behaviour of dendrimers of generation 1–4. The radius of gyration follows a simple power law scaling as a function of strain rate, whereas the fractal dimension is a maximum at equilibrium ($d_f \sim 3$) which decreases monotonically as a function of strain rate to ~ 2.6 at reduced strain rates of 0.1. Average eigenvalues of the tensor of gyration are also computed. This analysis shows that the onset of shear thinning is accompanied by significant molecular deformation (ellipsoidal deformation), whereas molecular alignment with the flow field has a less significant role. The distribution of terminal end groups also showed significant penetration into the interior of the dendrimer at all strain rates, again in conformity to experiment, simulation and theory. Of interest is that the dendrimers studied displayed neither Rouse nor reptation-like behaviour.

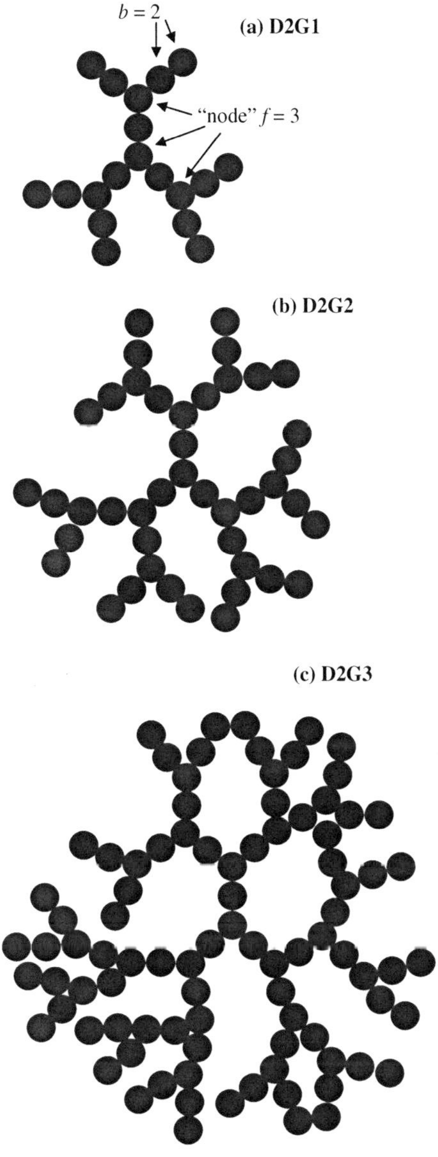

Figure 3 *Two dimensional schematic representation of dendrimers of generation (a) 1, (b) 2 and (c) 3 with functionality $f = 3$ and spacers $b = 2$*

Instead the zero-shear viscosity scaled as $\eta_0 \sim M^{0.65}$ (see Figure 4), where M is the molecular mass. Back-folding of end groups reduces molecular entanglements, thereby reducing the shear viscosity at weak fields compared to linear chains of equal molecular weight. At higher strain rates, superior molecular alignment of linear chains in relation to dendrimers of equal molecular weight results in the opposite feature: the viscosity of linear chains is lower than that of

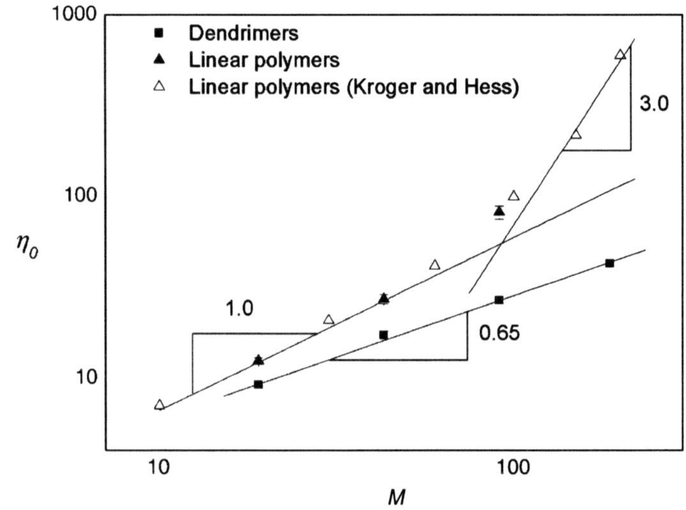

Figure 4 *Zero shear viscosity as a function of molecular mass for dendrimer and linear polymer models from Bosko et al.*[352] *Also shown is data for linear polymers taken from Kröger and Hess.*[388] *Reused with permission from 352. Copyright 2004, American Institute of Physics*

dendrimers. Calculation of birefringence extinction angles and angular velocities also shows that dendrimers rotate more freely than linear chains of equal molecular weight due to their compact structure. The previous simulations were performed in the (*NVT*) ensemble, but simulations were also performed[353] in the (*NpT*) ensemble and comparisons to (*NVT*) simulations were made. Additionally, a study of blends of linear chains and dendrimers was performed, as well as mixtures of dendrimer in solvent by explicit NEMD. These studies are the first to show how dendrimers may be used as viscosity modifiers, and viscosity curves were computed for various concentration ratios.

While bio-molecular simulation is not the subject of this review, we mention several works that may be useful, particularly in terms of model and algorithm development. Sunthar and Prakash[354] perform BD simulations of single chain model DNA in solution under steady extensional flow. DNA is coarse-grained as an ensemble of non-interacting bead-spring chains, with both excluded volume and hydrodynamic interactions accounted for. These simulations essentially model the experiments of Smith and Chu,[355] in which DNA subject to an extensional flow was observed to evolve from coil-like to fully elongated states. Essentially two regimes are observed. First, a universal regime in which the chain is coiled and parameter-free predictions of properties such as the expansion ratio (ratio of the extension of the molecule in the expanding flow direction to its value at equilibrium) is possible. These quantities are thus independent of the specific values of hydrodynamic and excluded-volume interaction parameters and depend instead only on the radius of gyration, solvent quality and Weissenberg number. In the other, non-universal regime, the chain unravels into a stretched state and is finite-size limited. These simulations go someway towards more universal models of DNA, as is common in conventional polymer statistics.

How water layers in or around the walls of cells is of fundamental interest in molecular biology. Layering near cell walls will profoundly affect the transport of proteins and nutrients to and from cells, so is very important. Pandit *et al.*[356] use MD to study the layering of water near the interface of a phospholipids membrane wall. They find four layered regions of water, where each layer has distinct orientational properties. Zhou *et al.*[357] perform the first MD study of NO diffusion in water. NO is considered an important neuromodulator in the brain and its transport under *in vivo* conditions is relatively unknown. Motivated by this, Zhou *et al.* compute the self and mutual diffusion coefficients of dilute NO in solution in the bulk. They find that under *in vivo* thermodynamic conditions, NO does not significantly alter the structure of water or the tendency of water to form H bonds between water molecules; nor in fact do H bonds form between water hydrogens and either the nitrogen or oxygen atoms of NO. At these low concentrations, NO exists as free molecules that do not form complexes with water. However, these studies were done in bulk conditions. An interesting extension would be to perform such simulations in the vicinity of cell walls.

A coarse-grain molecular model, in which molecules are represented as bead-springs for easy implementation into MD algorithms, is presented for double-helix molecules by Tepper and Voth.[358] By incorporating hydrophobic/hydrophilic interactions between base pairs, backbone and solvent, accounting for phosphate-phosphate repulsion along the backbone, and correct stacking energies for base pairs, they show that an initially linear chain spontaneous assembles into a double-helix structure. The evolution of the initial linear chain into a double-helix is shown in Figure 5. That the double-helix structure can spontaneously arise from such a coarse-grain molecular model could prove useful in more accurate studies of DNA by molecular simulation.

Motivated by a desire to usefully apply molecular simulation to the study of virology, Rapaport[359] uses MD and specifically designed molecular potentials to study the assembly process of capsids. A capsid is a protein shell that provides the "packaging" for either spherical or polyhedral viruses. In the spirit of the simplifying nature of coarse-graining, the model is devoid of biochemical complexity, but provides a suitable representation of the important molecular components. The specifics of applying MD techniques to these unusual molecular structures are described and shell growth and statistics monitored and discussed. This work is not only relevant for molecular simulation studies of viruses but also of self-assembly in nano-components for nanotechnology applications, likely to become one of the leading technological thrusts of this century.

8.2 Polymers. – *8.2.1 Algorithms and Coarse-Graining.* Overwhelmingly, the majority of simulation algorithms over the past decade for polymers in melt or solution have been based on MC methods. This is because MC methods, while complicated for high molecular weight chains, are considerably faster than MD

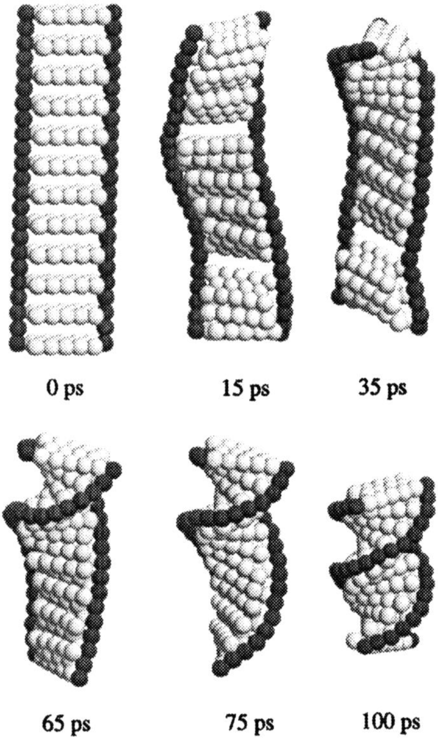

Figure 5 *Snapshots over evolution time of spontaneous helix formation via the coarse-grained model of Tepper and Voth. Reused with permission from 358. Copyright 2005, American Institute of Physics*

simulations, sometimes by many orders of magnitude. These simulations are of course limited to equilibrium or weak field flows. One of the most popular algorithms is the configurational-bias (CB) scheme, which originally worked by re-growing chain ends stepwise. Modifications to the CB algorithm have allowed interior segment growth as well, which is particularly advantageous for high molecular weight polymers. Extending this further, Rane and Mattice[360] perform regrowth of interior segments by MC on the second-nearest-neighbour-diamond lattice (hexagonal packing structure), where internal rearrangements of molecules are determined from a set of predetermined bond vectors. Significant speed up for relaxation processes were found for both linear polyethylene (PE) and polyethylene-oxide. An alternative to the CB scheme is the double-rebridging schemes of Theodorou, Uhlherr and colleagues, in which bond lengths and angles are rigid. In this method the connectivity of the inner segments of a molecule is exchanged between two chains in one MC move. This is particularly efficient for high MW entangled melts and improves the relaxation of long-range correlations. Banaszak and de Pablo[361] propose a variable connectivity double-rebridging algorithm, in which bond lengths and angles are now flexible, based upon the single rebridging scheme of Chen and Escobedo.[362] They apply their algorithm to PE C_{200} – C_{1000} chains and find good agreement with alternative MC and MD methods. In particular, the end-to-end autocorrelation function relaxation is significantly faster. Indeed, the goal of transcending the limitations of chain entanglements has been the subject of much study and Auhl *et al.*[363] perform a systematic study of the equilibration of long chains by double-rebridging and suitable pre-packing algorithms. While these simulations were performed on linear chains, it should also be suitable for branched and star polymers. Work by Uhlherr[364] has also demonstrated how large scale MC can be effectively domain decomposed for parallel computation, based on his "Collingwood" method[365] (named after the striped pattern of the Guernsey worn by players of the Collingwood Australian Rules football club in Melbourne, Australia!). While the method was demonstrated for a simple atomic fluid, it is also applicable for heterogeneous networks, such as polymer melts[365,366] and affords considerable computational speed-up.

While many MC simulations for melts are for monodisperse systems, polydispersity can be simulated, based on the work of Pant and Theodorou.[367] While the original work set up a relationship between the distribution of chemical potentials and the distribution of chain lengths for homogeneous bulk systems, Daoulas *et al.*[368] now extend the work such that a relationship between these two distributions is achieved for inhomogeneous systems; in effect a set of three simultaneous equations is derived, which when solved leads to the desired relationship. Their distributions are verified by end-bridging MC simulations of two polydisperse PE melts grafted onto a surface. The methods described above are not however suitable for single chain studies. In such cases so called "static" MC may be preferable. Inda and Frenkel[486] use static MC coupled with a multiple histogram method to study the stress-strain curve of single linear chains of length $N = 100$ to $N = 600$. Good agreement is found with scaling theory predictions.

Topological constraints are an extremely important property that affects the dynamics and conformations of polymer melts. Everaers and co-workers[370,371] provide an approximate solution to the problem of computing the shortest multiple disconnected path of topologically constrained chains. Their solution involved MD simulation of bead-spring chains, disabling the intrachain excluded volume interactions (but retaining interchain excluded volume interactions), and minimizing the energy by cooling the system to zero temperature. This way the minimum total energy of all primitive paths is obtained. Length and energy minimization methods to obtain primitive paths are analysed by Zhou and Larson[372] via MD simulation. Length, rather than energy, minimization yields superior agreement with the Doi-Edwards model predictions, with a quadratic primitive path potential with prefactor ~ 1.5. Instead of using MD simulation, Kröger[373] uses a geometrical algorithm to find model independent approximate solutions to the problem in two and three dimensions. The polymer contour is approximated by a line with kinks and segments. A systematic geometrical minimization of all primitive paths is performed until a final minimum is obtained. Melnik *et al.*[374] present distance geometry algorithms to generate initial configurations of dense polymer systems, thereby increasing the efficiency of either MD or MC algorithms.

Efforts to introduce new coarse-graining schemes are always advancing. More effective coarse-graining of molecules means that longer length and time scales are accessible to molecular simulation and more accurate results obtainable. Some recent reviews of coarse-graining schemes have been provided by Baschnagel *et al.*[375] and Müller-Plathe.[376] With the aim of faithfully reproducing equilibrium structure and thermodynamic properties of phenyl-ring chains, Zacharopoulos *et al.*[377] perform coarse-graining by first computing atomistic interaction potentials and then tabulating these for use in an actual MC simulation. Linear interpolation is used to approximate the potential at any given relative molecular separation, $\mathbf{r}_{ij}$. The scheme is tested against a fully atomistic liquid benzene model, simulated in an MD ensemble and excellent agreement is found for pressure, molecular and atomic pair distribution function and orientational correlation with up to 15 times the computational efficiency of fully atomistic MD. Ismail *et al.* approach the problem via application of wavelet transformations of freely-jointed chains[378] and self-avoiding chains.[379] In this approach successive averaging of the molecular topology is achieved. For a freely-jointed chain, if the vector $\mathbf{R} = \{\mathbf{r}_1, \ldots, \mathbf{r}_N\}$ represents the sites of the defining beads of a chain, then a wavelet pair is defined as

$$\mathbf{r}_n^{(k)} = 1/2\left[\mathbf{r}_{2n-1}^{(k-1)} + \mathbf{r}_{2n}^{(k-1)}\right] \tag{20}$$

and

$$\mathbf{w}_n^{(k)} = 1/2\left[\mathbf{r}_{2n-1}^{(k-1)} - \mathbf{r}_{2n}^{(k-1)}\right] \tag{21}$$

where **r** and **w** are sets of averages and differences, respectively. By repeating this procedure as many times as required for the degree of coarse-graining required, one obtains a set of $2N$ transformed variables. By keeping only the odd or even numbered averages and differences one can obtain the required N data points. Therefore, although the total number of variables remains in-tact, the sets of variables are different, and these correspond to the degree of coarse-graining desired. A similar scheme is implemented for the self-avoiding walk chain.[379] Reasonable scaling behaviour is found for the squared radius of gyration as a function of chain length for both types of chains.

Öttinger considers a different approach to coarse-graining for wormlike chains.[380] His procedure is to remove the structural details between the persistence length (length between statistically correlated chain segments) and the entanglement length (distance between entanglements) to produce a more smooth and persistent chain with Kuhn length equivalent to the entanglement length. This is done by considering first a Hookean bead-spring chain of N beans, with vectors spanning $\mathbf{R} = \{\mathbf{r}_1, \ldots, \mathbf{r}_N\}$, and spring constant k. Each bead is in turn "connected" to the corresponding atom of the finer detailed atomistic chain by another Hookean spring with spring constant k'. In the limit of large k', the two chains coincide; if the reverse is true then the coarse-grained chain is smooth. A suitable choice of the ratio of k/k' thus provides a measure of chain smoothness and the coarse-grained bead positions can be obtained by minimizing the total spring energy. An example of such a coarse-grained chain is shown in Figure 6 for an atomistic chain of $N = 85$, and spring constant ratio of 25. Thus the resulting chains, called "tapeworms" can be constructed, each of desired width. Optimised values of k/k' were obtained for polyethylene, polystyrene and polypropylene.

In a study of force-extension behaviour in typical bead-spring chain models, Underhill and Doyle[381] come up with a method termed the Polymer Ensemble Transformation (PET) for coarse-graining. PET involves determining an appropriate spring force law to model polymers at a given set of reference points, simultaneously smoothing out any unwanted details of the polymer between these references points. The segment of polymer to be modelled is placed in the so-called constant extension ensemble, the partition function of which is

$$\Omega(\mathbf{r}) = \int_V \exp(-U/k_B T)\,\delta(\mathbf{r} - \mathbf{R}_{tot})d\mathbf{V} \tag{22}$$

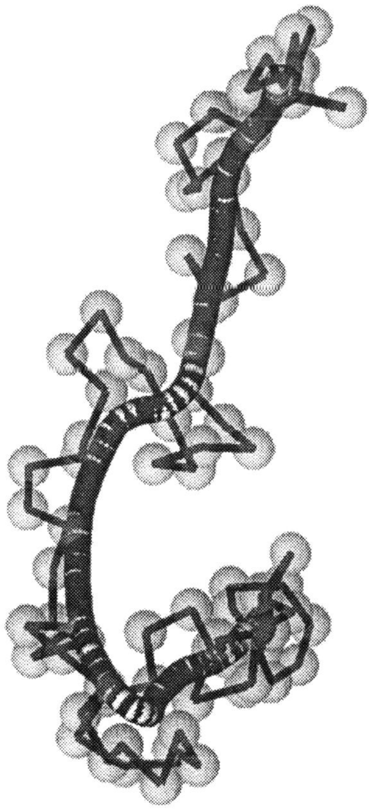

Figure 6 *Primitive path construction from an atomistic chain of N = 85 and k/k′ = 25. Reprinted from 380. Copyright (2004), with permission from Elsevier*

where $\mathbf{R}_{tot}$ is the coordinate of the chain end. The average force required to keep the chain at fixed extension is computed, from which the spring force is set to be equal to this average force. For some models, such as the freely-jointed chain, the polymer force-extension behaviour is faithfully represented by this procedure, but this is not true for other models, such as the wormlike chain.

8.2.2 Polymer Solutions. Successive fine graining, a technique developed by Prabhakar *et al.*[382] is a new theoretical technique devised to obtain realistic predictions of the rheological properties of polymer solutions. In this method one starts with a coarse-grained bead-spring chain and increase the number of beads (termed "fine graining"). The requisite number of Kuhn steps required to accurately model the chain is obtained from BD simulation. If N_S is the number of springs in the coarse-grained polymer and N_K is the Kuhn length, then successive BD simulations of the chain may be performed (with hydrodynamic interaction included), each with increasing numbers of $N_S = N_K$. By extrapolating to the limit $N_S \rightarrow N_K$ one can predict quantities of interest. The model is found to give very good extrapolated extensional viscosities in agreement with experimental measurements.

As is known, hydrodynamic interactions are important in simulating polymer solutions and can not generally be ignored. Kikuchi *et al.*[487] study the role of hydrodynamic interactions by a hybrid MD scheme, in which the polymer is evolved under MD equations of motion, whereas the solvent is modelled using a particle-based mesoscopic model for simulating fluctuating hydrodynamics. By ignoring the molecular details of the solvent, the polymer is modelled within a "hydrodynamic heat bath". They find that polymer collapse is accelerated by the hydrodynamic interactions and present a scaling theory for collapse times that agree well with the simulation data. For some model proteins, however, hydrodynamics does not affect the kinetics significantly.

In one of only a few NEMD simulations of polymers in solution, Kairn *et al.*[488] study the concentration dependence of the conformational and rheological behaviour of short chain polymer solutions under shear. NEMD has the great advantage over BD simulations in that no assumptions about frictional drag, hydrodynamic interactions, *etc*, are needed. The system is exactly solvable, the only approximations being the nature of the intermolecular interactions and molecular models employed. Radii of gyration, end-to-end distance, site statistics, shear viscosity, first and second normal stress differences are all computed as functions of strain rate and concentration and tested against commonly cited theoretical models. They find that, contrary to accepted mean free and scaling theories, something other than screening causes the concentration dependence of the radius of gyration in short chain polymer solutions. Simulations such as this are very useful in the test of existing or emerging theories of polymer solutions, such as a more recent mean-field theory for the concentration dependent shear induced anisotropy of Kröger and De Angelis.[383]

Linear polymers in the presence of spherical colloids with radius comparable to or smaller than the polymer radius of gyration have also been studied by

Doxastakis *et al.*[384] By off-lattice MC simulation they find that at low densities chain ends prefer to locate close to colloid surfaces and tend to align parallel to them. At higher concentrations, chain ends still prefer to locate near colloidal surfaces, but now polymer alignment is layered parallel and perpendicular to them. Packing effects are also studied, as are size and orientation of chains. The colloidal filler is seen to significantly alter the chain conformations near its vicinity, which in turn will affect material properties particularly at high polymer concentrations.

8.2.3 Polymer Melts. Polymer melts are an extremely important field of study for molecular simulation researchers. Melts play an important role in the processing of polymers into products; it is only in the melt phase that they can be suitably shaped. The processing of melts also involves a number of different types of flow geometries, often combining both shear and elongational flow. We briefly report on some studies of melts in equilibrium, under shear and elongational flow.

Macromolecules, because of their length, are able to form knots, either intra or inter-molecular. The former knot is commonly found in bio-molecules (*e.g.* DNA), whereas the latter predominantly in polymer melts. However, both types of knots can occur in both types of systems. Knots or entanglements form the most important type of topological constraint for large macromolecules and their study by molecular simulation is very valuable in understanding the mechanisms behind their formation and destruction. In such a study, Kim and Klein[385] perform equilibrium MD simulations of linear polyethylene of chains ranging from $N = 53$ to $N = 293$. Knots unravel by a sequence of expansions, contractions and migration. Interestingly the knot size grows as $\langle M_{knot} \rangle \sim M^{0.4}$, whereas the unentanglement time scales as $(M - \langle M_{knot} \rangle)^{5/2}$. Padding and Briels[386] investigate time-integrated, as opposed to time-resolved, measurements of the entanglement mass. Time resolved measurements (either from experiment or MD simulation) can give estimates of the Doi-Edwards reptation entanglement mass from either the dynamic structure factor or the zero-shear relaxation modulus, whereas time-integrated measurements estimate the so-called "rubbery" entanglement mass, obtained through the plateau of the shear relaxation modulus (which is a time integrated quantity), giving a value that is roughly 1.5 times smaller than the reptation entanglement mass. They conclude that because the various methods of obtaining the reptation entanglement mass are consistent to within error bars, this value is to be preferred. Harmandaris *et al.*[387] study the cross-over from Rouse to reptation dynamics by long-time (up to 300 ns) atomistic MD simulation of polyethylene melts of length C_{78} to C_{250}. From a plot of self-diffusion coefficient as a function of molecular weight, they find the transition from Rouse to reptation dynamics occurs at roughly C_{156}. Estimates of the monomer friction factor, reptation tube diameter and zero shear viscosity were obtained by a suitable mapping scheme and results were consistent with experiment, as shown for the zero-shear viscosity in Figure 7. Sen *et al.*[489] perform equilibrium MD simulations of a bead-spring model polymer and compute the zero shear viscosity by integrating the stress

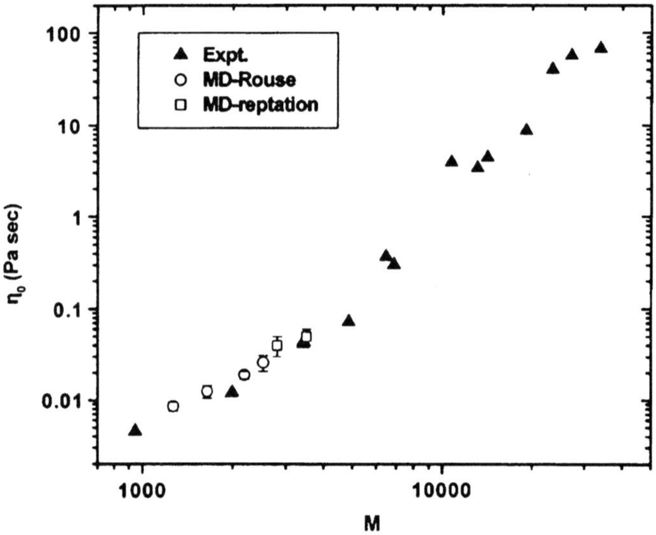

Figure 7 *Zero shear viscosity as a function of molecular mass. Mapping of atomistic MD data onto Rouse and reptation models are shown, as well as experimental values. Reprinted with permission from 387. Copyright 2003 American Chemical Society*

autocorrelation time function. They find reptation dynamics kicks in at about $N \sim 100$, in agreement with NEMD estimates of Kröger and Hess.[388] From the plateau shear modulus they find the "rubbery" entanglement length to be $\sim$28, close to the value found by Kremer and Grest,[389] but contradicting NEMD estimates of $\sim$80. The matter of just what the entanglement length is thus still appears a little nebulous at this point in time. Note that reptation/entanglement lengths/masses will all depend on the type of molecular model used, so direct comparisons between different results need to be treated cautiously and appropriately mapped between different models.

Rheological experiments on shearing melts are overwhelmingly performed by oscillatory shear flow, in which a sample of fluid is subjected to small amplitude oscillatory shear. From the dynamic moduli (G', G'') one can obtain information about the liquid-like viscous behaviour (loss modulus, G'') or the solid-like elastic behaviour (storage modulus, G'). NEMD can be a powerful simulation tool to study viscoelasticity and directly relate microscopic behaviour to macroscopic observables. Indeed, this has been done for simple fluids under oscillatory elongation in the past,[390,391] but for the first time Cifre *et al.*[392] perform an NEMD simulation of low molecular weight polymer melts ($N = 5$ to 70) under oscillatory shear. These simulations are in good agreement with theoretical expectations, such as the Cox-Merz rule, which states

$$|\eta^*(\omega)| = \eta(\omega) \tag{23}$$

where $\eta^*(\omega)$ is the dynamic (complex) viscosity, defined as

$$\eta^*(\omega) = -i(G'(\omega) + iG''(\omega))/\omega \tag{24}$$

and $\eta(\omega)$ is the frequency dependent (or, equivalently, strain rate dependent) shear viscosity.

Yamamoto and Onuki[393] perform steady-state NEMD simulations of shearing bead-spring chains. Once more they find the entanglement length to be of order $N \sim 100$. The novel aspect of their work is their visualization of the entanglement process. By monitoring the sudden enhancements in interaction energies near entanglement points they find that at high shear rates the chains take on bent conformations. The chains stretch as a network, but as the bends approach chain ends they disentangle and lead to experimentally observable stress overshoots. Daivis *et al.*[394] study the nonlinear shear and elongational rheology of freely-jointed (tangent Lennard-Jones) chains by NEMD. For the first time steady-state planar elongational viscosities of chains up to $N = 50$ are obtained by use of the spatio-temporal periodic boundary conditions devised by Kraynik and Reinelt[45] and implemented by Todd and Daivis.[42–43,52,54] In addition to steady-state shear and elongational viscosities, first and second normal stresses are also computed. The viscometric data are analysed via the third order retarded motion expansion (RME). Good agreement to RME predictions in the limit as $\dot{\gamma}$, $\dot{\varepsilon} \rightarrow 0$ is found for the shear viscosity and first and second elongational viscosities. Lowest order nonlinear RME coefficients are in agreement with simulation data for shear and elongation for low N, but not for higher order coefficients, probably due to insufficient data at low strain rates for higher N. Simulations are performed in both the NVT and NpT ensembles. A qualitative comparison of the elongational viscosities in the NpT ensemble bears close agreement with experimentally determined elongational viscosities obtained by Laun and Münsted.[395] Figure 8 shows this qualitative comparison. As both confidence and computing power grows, more ambitious MD simulations are being performed on shearing and elongating polymer melts, and even networked cross-linked "rubbery" polymers, such as the study by Svaneborg *et al.*[396] Such studies, coupled with further theoretical developments in molecular models of melt rheology (see for example the recent extension of the tube model by Marrucci and Ianniruberto[397]) means that far greater insights into the underlying dynamics and structure of macromolecules will be gained in the coming years.

There are, as has been mentioned before, too many studies to be able to faithfully describe them all, or to describe them in the detail that they deserve. A review such as this can only sample a cross section of the literature to give the reader a flavour of what is being done; it is up to the reader to determine what particular studies are of personal interest to pursue further. Tremendously detailed simulations by Theodorou, Mavrantzas and co-workers for specifically detailed polymers deserve mention, and interested readers are referred to the following references:[229,398–401]. Branched and star polymers will also be increasingly studied,[402,403] as will supercooled polymers and glasses,[404,405] di-block co-polymers,[406–408] and the all-important behaviour of polymer melts near surfaces,[409–412] so important for emerging nanotechnology (confined polymer melts are reviewed in Section 9). Nor do we discuss here recent developments in polymer theory, such as those of Doi and colleagues for inhomogeneous

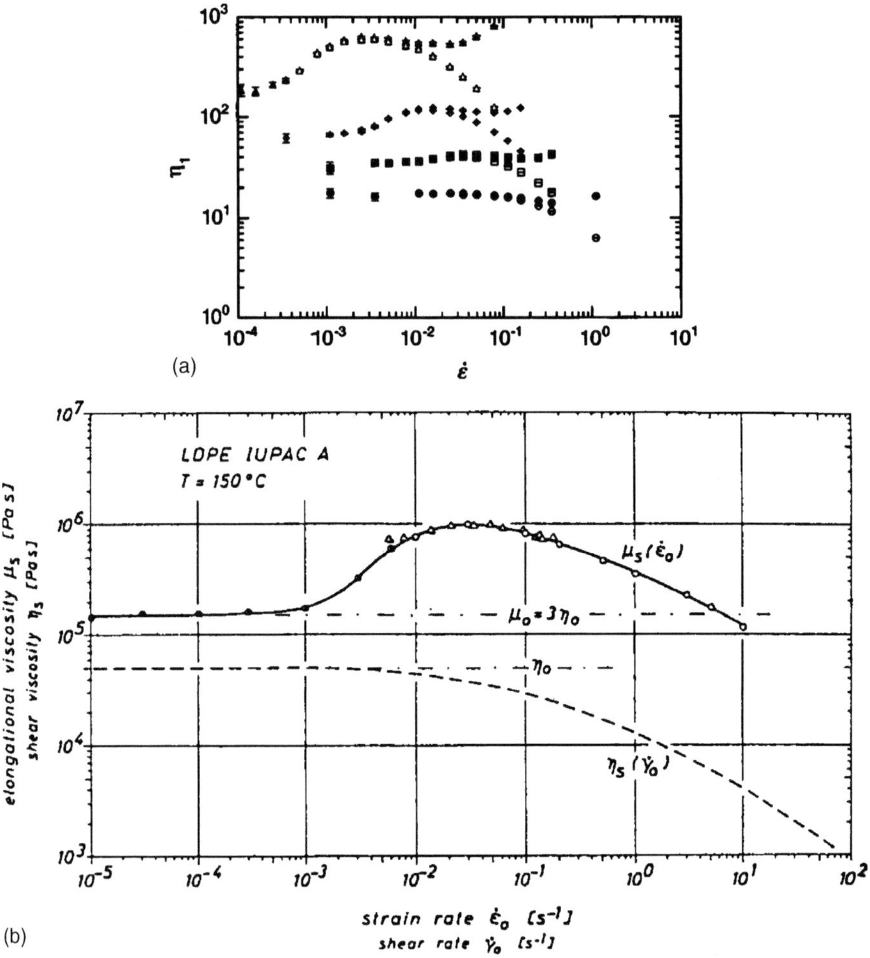

Figure 8 *(a) NEMD simulated elongational viscosities for the Lennard-Jones freely-jointed tangent chain. Chain lengths are $N = 4$ (circles), 10 (squares), 20 (diamonds) and 50 (triangles). Filled symbols are constant volume and open symbols are constant normal pressure simulations. Reprinted from 394 Copyright (2003), with permission from Elsevier. (b) Experimental shear and elongational viscosities of linear polyethylene, reprinted with permission from Laun and Münsted.[395] Qualitative agreement is found as N increases for constant pressure simulations, as these more faithfully represent experimental conditions*

dense polymers and block copolymer melts and blends.[413,414] Space and time restrictions prevent an adequate description of these studies and more.

9 Confined Fluids

The study of highly confined fluids by molecular simulation techniques continues to grow in popularity. While there are many individual and widely diverse studies, most tend to fall within the boundaries of nanofluidics, friction (including stick-slip boundary conditions), transport and structure, complex fluids and phase equilibria.

9.1 Nanofluidics, Friction, Stick-Slip Boundary Conditions, Transport and Structure. – The development of models that go beyond classical Navier-Stokes (NS) hydrodynamics is crucial for the prediction and understanding of flows in highly confined geometries of the order of molecular length scales (nanofluidics). The NS equations and continuum models have been shown to be remarkably accurate for simple fluids confined to narrow channels up to a few atomic diameters in width but break down dramatically at smaller confinement dimensions. This has been demonstrated by both simulation[65,415] and experiment (see for example Becker and Mugele[416]). The exact level of approximation for which NS hydrodynamics remains valid will depend on a number of factors such as the degree of confinement, rate of flow, fluid density and very importantly the nature of the fluid-fluid and fluid-wall interaction potentials. The nature and range of the interaction potentials will determine not only the degree of stick-slip but also the degree to which the fluid's transport can be described by "local" hydrodynamics.

A study of the breakdown of continuum hydrodynamics was recently performed by Luan and Robbins.[417] In particular they examined the validity of using the popular continuum "contact mechanics" for estimating the forces between interfaces. The model was checked against MD simulations. Not surprisingly, away from the interfacial regions the continuum model agreed well with MD simulation results, but broke down significantly (up to an order of magnitude for friction) near the interfaces. In a first serious attempt to study the non-local viscosity of confined fluids, Zhang *et al.*[418,419] model the flow by a non-local constitutive equation first proposed by Evans and Morriss.[30] If one assumes steady-state flow then the generalization of Newton's law of viscosity is

$$\sigma_{xy}(y) = \int_{-\infty}^{\infty} dy' \, \eta(y, y - y') \, \dot{\gamma}(y') \tag{25}$$

where $\sigma_{xy}(y)$ is the shear stress at some point y in the fluid (here the geometry is such that flow occurs in the x-direction and the channel is separated in the y-direction) and $\dot{\gamma}$ is the strain rate. η is a viscosity kernel that is a non-local function of position. In other words, the stress at y is not just proportional to the strain rate at y (as is assumed by Newton's law of viscosity), but rather depends on the entire strain rate distribution across the fluid. The problem is how to extract the viscosity kernel from the available stress and strain rate simulation data (in their case, generated by NEMD simulation of planar Poiseuille flow). Zhang *et al.* did this by, firstly approximating the kernel by an effective kernel, such that $\eta(y, y - y') \approx \eta(y - y')$, and then by working in Fourier space to obtain the transformed $\tilde{\eta}(k_y)$, after which $\eta(y)$ could be obtained via inverse transform. Their studies showed that the method was reasonable for relatively wide channel widths, but the approximation breaks down dramatically at lower channel widths and high fluid densities. A full theoretical analysis of the methodology has been performed by Cadusch *et al.*,[420] including a thorough discussion of the limitations of the model. Based on this work, it is clear that a better alternative method needs to be applied to compute the exact kernel, $\eta(y, y - y')$. The potential of this formulation (which is in the spirit of generalized hydrodynamics) lays in its ability to accurately predict the flow fields of fluids highly confined in arbitrary geometries. If successful, the method could be an extremely powerful and useful tool in the study of nanofluidics.

Once again in the spirit of generalized hydrodynamics, Al-Ghoul and Eu[421] develop a continuum hydrodynamic model for micro-confined fluids. Their constitutive equations are both nonlinear (unlike Newton's law of viscosity or Fourier's law of thermal conductivity) and thermodynamically consistent. Boundary conditions are implemented by a Langmuir adsorption model. The governing generalized hydrodynamic equations, incorporating both boundary conditions and nonlinear constitutive equations, are then solved to predict flow and stress profiles in good agreement with available published data. Their formulation is valid for high Knudsen number flows with moderate aspect ratios (width/length). The current formulation is unlikely to be successful for high density highly confined (nano-scale confinement) flows, but further modifications (including accounting for non-local phenomenology) could make it so.

An alternative to generalized hydrodynamics is kinetic theory, based on the Enskog theory. Kinetic theories for confined fluids were pioneered in the 1980s by Davis and colleagues and made more suitable for dense inhomogeneous fluids by Pozhar and colleagues in the 1990s. However, their application still relies on a number of approximations and in general the governing equations are not easy to work with. Nevertheless they can be useful and provide a basis for further theoretical development. In an attempt to simplify the governing kinetic equation, Guo *et al.*[422] simplify the equation of motion for the one-particle distribution function. They do this by splitting the Enskog collision operator into a Boltzmann collision part and an excess part, both of which can be approximated. The relaxation time that results from the simplification of the Boltzmann collision part is then estimated by the local average density model (LADM) that is often used for kinetic theory models. Once these quantities are determined, the distribution function can be solved, allowing for the computation of density and velocity profiles, both of which are seen to be in reasonable agreement with MD and MC simulation results.

A comparison of mesoscopic simulation methods with MD simulations has been performed by Denniston and Robbins.[423] They study a binary mixture of simple Lennard-Jones fluids and "map" out the required parameters of the mesoscopic model from their MD simulation data. Their mapping scheme is more complete than those of previous workers because in addition to accounting for the interfacial order parameter and density profiles, they also consider the stress. Their mapping consists of using MD simulations to parameterise the popular mesoscale Lattice Boltzmann simulation technique and find that a number of assumptions of these mesoscale models break down at the interface. By suitable mapping (or "tailoring") of mesoscale parameters by using accurate MD simulation data, many of these defects can be minimized, thus improving the reliability of these schemes at interfaces.

Jepps and Bhatia[490] have extended the so-called method-of-planes (MOP) technique[424] for exactly computing the pressure tensor for a confined nonequilibrium fluid in cylindrical geometry. They term their derivation the 'method-of-circles' for obvious reasons and find excellent agreement with a direct integration of the momentum continuity equation for a fluid under the influence of an external driving field. Zhang and Todd[425] extended MOP for both the pressure tensor and heat flux vector to include the contribution of 3-body forces. Their exact statistical mechanical calculations were compared to exact hydrodynamic calculations by directly integrating the continuity equations for momentum and energy for the special case of planar Poiseuille flow of a highly confined fluid. Both methods were found to be in excellent agreement. For the simple fluid studied, they found that 3-body effects have a small influence on either the shear stress (and by implication, shear viscosity) and heat flux, but does have a significant effect on the normal pressures. The technique should be useful for computing many-body contributions to the transport properties of fluids where these are important, *e.g.* liquid metals. Ikeshoji *et al.*[369] compute a coarse-grained pressure for inhomogeneous fluids by following the method established by Schofield and Henderson.[426] In this case the configurational contribution to the pressure tensor between atoms i and j is given as

$$P_{ij,\alpha\beta}(\mathbf{r}) = -\int_{C_{ij}} \hat{\boldsymbol{\alpha}}.\mathbf{F}_{ij}\,\delta(\mathbf{r}-\boldsymbol{\lambda})d\lambda_\beta \tag{26}$$

where $\hat{\boldsymbol{\alpha}}.\mathbf{F}_{ij}$ is the α-component of the pair-wise force between the atoms ($\hat{\boldsymbol{\alpha}}$ is the unit vector in the α-direction), C_{ij} is an integration contour from $\mathbf{r}_i$ to $\mathbf{r}_j$, and $d\lambda_\beta$ is the projection of the contour in the β-direction. By integrating over a local volume they obtain coarse-grain pressures in both planar and spherical geometries. Their expressions compare well with MOP calculations for planar geometry and display comparable or slightly improved statistics. Han and Lee[427] go one step further still by generalizing this approach to arbitrary flow geometries. MOP is powerful but limited to the computation of fluxes that are normal to the confinement surface. Han and Lee remove this limitation by writing the O_{ij} tensor (which is a differential operator resulting from an expansion of the difference of delta function terms in atomic positions between particles i and j) in integral, rather than differential, form. In fact, this method and the one by Ikeshoji *et al.* are similar to an earlier procedure devised by Monaghan and Morriss[428] in which momentum and energy fluxes at any arbitrary point $\mathbf{r}$ in an inhomogeneous fluid can be computed exactly, and all these methods are predated by the Schofield and Henderson work. The formulation by Han and Lee allows for the computation of fluxes, averaged across finite areas (as with MOP), that do not need to be restricted to being normal to the confinement surface. By doing this they find a significant heat flux parallel to the streaming

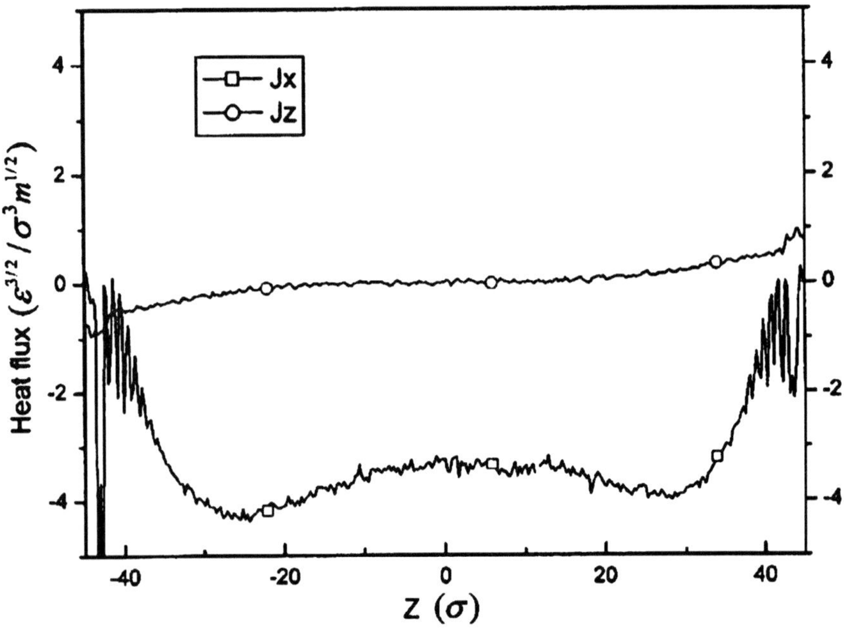

Figure 9 *Heat fluxes transverse (Jz) and parallel (Jx) to the flow for an atomic fluid undergoing planar Poiseuille flow in the x-direction. Solid atomistic walls are separated in the z-direction. Reprinted figure with permission from 427. Copyright 2004 by the American Physical Society*

velocity (see Figure 9), something which MOP was unable to observe. Such parallel heat fluxes have in fact been theoretically predicted and observed in previous simulation studies by Cordero and Risso[429] and demonstrated to exist theoretically by Daivis and Coelho.[430] Nevertheless, the generalization of MOP to arbitrary geometries is a significant step forward in studying transport processes in inhomogeneous nonequilibrium fluids.

Studies of interfacial friction and stick-slip boundary conditions by molecular simulation techniques continue to attract attention. Such studies are likely to increase, not only because of their relevance in nanotechnology, but also because a deeper molecular understanding of stick-slip conditions is essential in earthquake prediction, a science that has come to the fore of public consciousness due to the massive Asian tsunami in December 2004. While there is much literature on the subject of solid-solid friction and stick-slip, as well as recent experimental observations of liquid-solid stick-slip, here we only report on liquid-solid interfaces that are studied by molecular simulation methods. Zhang and Kwok[431] study slip in a confined fluid by a mean field Lattice Boltzmann method. Their simulations show that macroscopic slip (defined by the slip length, which is determined by extrapolation of the streaming velocity to zero) can be observed, even though there is no actual microscopic slip. This disparity is due to the lack of microscopic fine scale resolution in making a macroscopic extrapolation. They further study the solid-fluid interaction strength and contact angles to the degree of slip. The degree of macroscopic slip (even though there is no microscopic slip) is larger for stronger fluid-solid interactions, in agreement with MD simulations. The advantage of using a no-slip boundary condition Lattice Boltzmann method rather than standard MD is that it allows one to probe mesoscale dynamics that are currently beyond the length and time scales of MD.

Hendy *et al.*,[432] motivated by the technological desirability of increasing mixing rates in microfluidic devices, consider patterned wettability to induce variations in slip. They do this both theoretically and by MD simulation. They model a pressure driven flow by using the Navier-Stokes equation for a viscous incompressible fluid with slip boundary conditions, and allow the slip length to vary in the direction of flow (this models the patterned wettability). They then solve the resulting equation for the flow profile by a perturbation method. MD simulations are performed for a simple Lennard-Jones fluid confined between atomistic walls. The variable wettability is modelled via a fluid-wall interaction potential that is just the standard 6–12 Lennard-Jones potential, modified to include a variable multiplicative factor c_{fs} ($0 < c_{fs} \leq 1$) on the attractive part of the 6–12 potential. The MD simulation results are in reasonable qualitative agreement with theory, though it is clear that modelling of nano-scale phenomena via modified Navier-Stokes treatments is unlikely to give very good agreement with MD simulation results due to a host of approximations inherent in the Navier-Stokes formulations for highly confined fluids (such as assumptions of constant viscosity and Newton's constitutive laws, discussed previously, *etc*). However, their motivation in this study was to demonstrate that changes in slip length (caused by patterned wettability) can induce flow transverse to the walls and hence enhance mixing, which is something they successfully achieved. Similarly, Priezjev *et al.*[433] study slip length for a fluid undergoing planar shear flow. Patterned wettability is introduced by having a 'bottom' patterned striped substrate that in effect represent regions of no shear and finite or no slip. The 'top' substrate is homogeneous and moves with constant velocity. Both transverse (stripes perpendicular to flow direction) and longitudinal (stripes parallel to flow direction) flows are considered. Velocity fields are computed for the confined fluid by both MD simulation and hydrodynamic solution to the Stokes equation. Good agreement for slip lengths are found between both methods when the width of the patterned stripes is of the order of the molecular diameter or greater. Roughening features are observed for transverse but not longitudinal flows. Enhanced slip is found by MD simulation for longitudinal flows for smaller strip widths, compared to the hydrodynamic prediction. Reduction in molecular ordering within the first fluid layer accounts for this enhanced slip length.

Sokhan and Quirke[434] devise a method of computing interfacial friction and the Maxwell slip coefficients (α) by equilibrium MD, in which α is computed from the relaxation time, which itself can be estimated by an exponential fit to the collective velocity autocorrelation function. Their equilibrium method is compared to an NEMD method devised by the authors previously and excellent agreement is found. They study the density dependence of the slip coefficient, finding it to be sensitive to the surface corrugation of the wall and a non-monotonous function of fluid density, though effectively constant over a large density range.

Jabbarzadeh *et al.*[435] study by NEMD simulation a six-layer dodecane liquid film confined between mica walls. Their simulations show a significantly reduced friction state and they argue that this low friction state is not due to wall slip but rather to layer sliding throughout the fluid. While their results are in conformity with recent experimental observations on low friction states of OMCTS and squalene, we note that the authors thermostatted the fluid. It has previously been pointed out[436] that thermostatting a highly confined inhomogeneous fluid can lead to a number of spurious unphysical results, including the generation of artificially ordered fluid layers (*i.e.*, string phases). While the authors say that they used a profile unbiased thermostat, nevertheless even the use of such a thermostat is not theoretically sound unless the thermostat is formulated to be inhomogeneous and hence able to account for all non-zero wavevector temperature components (see for example the thermostat of Baranyai *et al.*[437]), and this is unlikely to have been the case. Therefore though the results are interesting, one may still question whether they are physically meaningful until the thermostatting issue is correctly resolved. An alternative and physically more realistic approach would be to dispense with a thermostat on the fluid altogether and just remove excess heat by a wall thermostat.

Butler and Harrowell[438] perform NEMD simulations of planar Couette flow to study the interface between a shearing liquid and the confining strained crystal. They find increasing velocity field penetration into the crystal as the shear stress increases along the coexistence line. Alternating slip-stick is found, where slip occurs with significant fluid disordering at the interface. The authors pay closer attention to thermostatting the fluid, noting that the use of either a wall thermostat or a homogeneous fluid thermostat does not affect the position of the nonequilibrium solid-liquid coexistence line, provided that the interfacial temperature is used as the system temperature for the wall thermostat.

As discussed in Section 7.2, Schneemilch and Quirke[313] perform MC simulations to study contact angles of a wetting fluid at chemically striped nanopatterned surfaces. Lateral stress is observed at wall-vapour interfaces that are released at the unbending transition. The contact angle varies sinusoidally across the surface at the liquid-vapour interface parallel to the stripe. Sacquin-Mora *et al.*[439] have studied stress-strain relationships in liquid bridges between confining walls by grand canonical equilibrium MC simulation. They find qualitative similarities between the torsional stress curve for such a system and a confined sheared system. Càmara and Bresme[440] perform MD simulations of a simple Lennard-Jones fluid confined in a wedge shaped pore. The density, energy and normal pressure profiles are found to be non-uniform along the direction of the pore. The origin of the pressure oscillations is due to the creation of new fluid layers and the maxima in the pressure are highly correlated with corresponding maxima in the density and minima in the energy. Solid phases are also seen to occur under high degrees of confinement and states close to the triple point.

It has been known for some time now that the viscosity of highly confined organic liquids (film thicknesses less than 5–8 molecular layers) can be orders of magnitude larger their viscosity in the bulk. However, recent experiments have shown that this is not necessarily the case for highly confined electrolyte solutions (thickness< 1 nm), in which the fluidity remains in tact. Leng and Cummings[441] study the origin of this phenomenon by NEMD simulations of an electrolyte solution (potassium ions in water) confined by two mica clay surfaces that move relative to each other. They find Newtonian plateaus of the shear viscosity comparable to the bulk value for thicknesses between 0.92–2.44 nm, shown in Figure 10. They associate the origin of this persistent liquidity to the rotational dynamics of the water molecules. The rotational correlation times are significantly larger than those in the bulk due to the high degree of confinement whereas they are comparable to the diffusion times in the bulk. Shear thinning is found to be dominated by rotational dynamics. These simulations, coupled with the experimental observations that motivated them, suggest that water can be used as an efficient lubricant for nanotribology applications.

Dellago and Naor[442] have studied the dipole moment of chains of water in molecules in narrow pores using *ab initio* MD simulations. The interest in these systems is due to the fact that proton transport through water in membrane pores can be considered as a one-dimensional proton-hopping process where the proton 'hops' from one site to another. However, defects in the chain produced by hydrogen bonding to other sources can prevent this process due to the unavailability of a site for easy binding. The dipole moment of the molecules in the chain will also affect the process. Since the dipole moment of water in the liquid state is very different to that in the gas phase, it is anticipated to be different again in a membrane environment. CPMD methods were used and a significant lowering of the dipole moment compared to liquid water is observed.

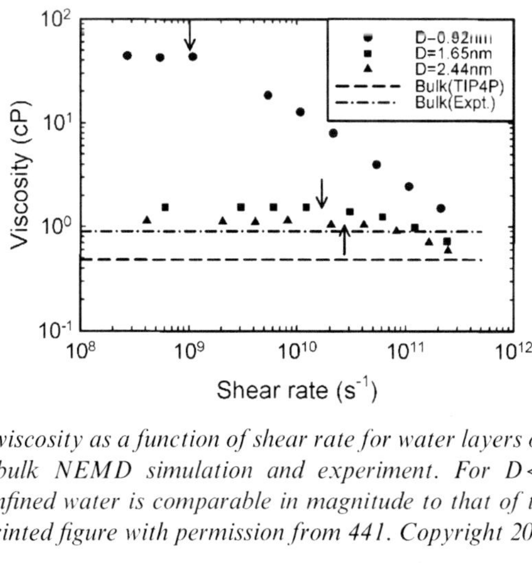

Figure 10 *Shear viscosity as a function of shear rate for water layers of various thickness, D, compared to bulk NEMD simulation and experiment. For D<0.92 the effective viscosity of the confined water is comparable in magnitude to that of the bulk at all shear rates studied. Reprinted figure with permission from 441. Copyright 2005 by the American Physical Society*

Hansen and coworkers[443] have also studied water in nanopores, with particular interest in the filling of pores, and the effect on this process of application of electric fields produced by ion concentration gradients. In reference[443] a hydrophobic, cylindrical structureless pore is considered. In studies of the water permeation, the cylindrical pore is initially empty and the water occupancy is measured as a function of time for different pore widths. A intermittent filling and emptying of the pore that has been reported previously in these systems (for example, see 314), was also observed here. Application of electric fields was found to have significant effects on the behaviour. Simulations with a steady charge gradient are carried out by making the pores impermeable to the charged ions. Application of the field is then found to enable filling of the pore by water molecules at radii smaller than the minimum possible when no charge gradient exists (R_m). Free energy barriers to permeation of the pore by different ions are determined. In simulations that allow the transport of ions through the pore, if the radius is set less than R_m under conditions where water can fill the pore, the passage of the ions eventually reduces the charge gradient to a point where the water can no longer fill the tube.

9.2 Confined Complex Fluids. – Because of their important technological relevance, the study of alkanes and polymers under extreme confinement continues to gain popularity in the simulation community. The difficulty in making experimental measurements for nano-confined systems, and the lack of confirmed theoretical models for such systems, makes molecular simulation the ideal tool to explore thermodynamics, structure and transport at such scales.

Supple and Quirke[444,445] perform MD simulations of oil imbibition in carbon nanotubes. They find that the smallest nanotubes imbibe quickly along the surface of the inner tube and that the penetration length is a linear function of time and derive simple and tractable expressions for this and the velocity of the oil (decane) and relate these to interfacial friction and surface tensions. Expressions are derived for the density profiles of the oil as functions of adsorption energy and pore surface friction. The molecular conformations of decane were also studied as functions of tube radius and pore position. By and large, decane molecules are aligned with the tube axis, though for the larger tubes studied there were also some molecules that have a preference for wrapping around the tube wall in a direction perpendicular to the tube axis. The linear dependence of penetration length at short times is related to the low friction state of graphitic surfaces. It is this low friction of carbon nanotubes that make them of particular technological interest. Motivated by experimental measurements by Granick *et al.*[446,447] that showed enhanced shear thinning of highly confined *n*-dodecane, Cui *et al.*[448] performed NEMD simulations of *n*-dodecane confined by planar mica surfaces that slide over the fluid in equal and opposite directions to mimic planar Couette flow. Their simulation results are comparable to the experimental results of Granick *et al.* The onset of shear thinning occurs at strain rates orders of magnitude smaller than a corresponding bulk fluid. They find that the microscopic origin of this phenomenon is the "freezing-out" of the rotational degrees of freedom and a power-law diffusional relaxation. This increases the largest relaxation times by over seven orders of magnitude. Since the onset of shear thinning occurs at times that correspond to the reciprocal of the largest relaxation times, the net result is an enhancement of the shear thinning process at lower strain rates compared to those in the bulk fluid.

Polymer blends are an important material in a host of industries ranging from traditional plastics to lubricants, anti-corrosion coatings, adhesives and electrical insulators. Blends under confinement are particularly complex systems that involve interplays between enthalpic and entropic forces, finite thickness and wetting effects and phase separation, *etc*. To try to better understand the latter phenomenon, Cavallo *et al.*[449] study the phase separation of a symmetric binary polymer blend confined by flat, structureless walls. Simulations are performed on coarse-grained polymers by equilibrium lattice MC methods. The geometrical structure of chains shows that even for high density blends correlation hole effects exist, particularly for the smallest confinement separations. The scaling behaviour of typical quantities such as the radius of gyration and end-to-end vector are also studied. Deviations from the predictions of the Flory-Huggins theory of "unmixing" (otherwise known as the random phase approximation) are found when (a) the length of the chains is larger than the confinement length and (b) the critical behaviour is in the universality class of the 2-D Ising model.

The stability of thin polymer films on solid substrates is very important for a host of manufacturing processes such as paints, coatings, adhesives and photographic films. As many of these processes involve polymers in solution, the study of solvents on thin film stability is a useful quest. Lin *et al.*[450] study the wetting behaviour of a mixture of short chain molecules and solvent by both MC simulation and self-consistent field calculations. Both methods are in good qualitative agreement. They find that solvent accumulates at the interfaces and reduce the liquid-vapour interface tension. Wetting can be inhibited by solvent molecules penetrating between polymer and substrate, thereby displacing the former. The phase behaviour of the system is also studied and the shapes of the chains covering the substrate are found to be either spread out or have "cap" like structures, depending on the relative strengths of the substrate-solvent and substrate-polymer interactions. As with the previously described study by Schneemilch and Quirke,[313] Yaneva *et al.*[451] are motivated to study the structure of polymer nano-drops adsorbed on a chemically "striped" surface, which has technological significance to processes such as photolithography and microphase separation in diblock copolymer films, amongst numerous others. For such patterned surfaces liquid bridges have been both experimentally observed and theoretically predicted. Yaneva *et al.* study these by the aid of MD simulation of coarse-grained polymer chains of length $N = 32$. Droplet shape, size, contact angle and density profiles are all studied and reasonable qualitative agreement with theoretical predictions is obtained.

MC simulations of linear polymer chains confined between flat slits of thickness comparable to the transverse diameter of the chains have been performed by Vacatello.[452] He finds that the force between the slits immersed in the polymeric liquid is a result of chain layering. In particular, when the slit width is of the order of the transverse diameter of the chains, polymer mediated attractions and repulsions occur between the plates, with an oscillatory behaviour similar to those observed experimentally. Maury-Evertsz *et al.*[453] also perform MC simulation of confined single-chain homopolymers and compute their equilibrium thermodynamic properties, including the coil-to-globule transition. In contrast to the relatively large number of studies of linear chain molecules confined by flat surfaces, Lamura *et al.*[454] study helical polymers either confined by a cylindrical wall or wrapped around a cylindrical rod. The motivations behind this study are obviously biological, where helical structures are very common for biological macromolecules (*e.g.* DNA, F-actin, microtubuli and peptides in solution), and bio-nanotechnology will increasingly manipulate the flow of biological molecules through and around highly confined geometries. Their simulation algorithm involves replacing a polymer by a curve of fixed contour length. The rotation of a vector that correspond to the tangent vector and the principal axes of the polymer cross section evolves in time by generalized Frenet differential equations that are discretised and solved numerically. In the limit of large persistence length in comparison to confinement diameter and helix radius, they find wormlike chain behaviour. While the "simulation" is not molecular simulation in the conventional sense, it would be useful to test these results against either MC or MD coarse-grain simulations.

Sikorski and Romiszowski[455] study confined branched star polymers by on-lattice MC simulation. Attractive forces are excluded and only excluded volume accounted for, thus making the simulations relevant for chains in a good solvent. Contrary to expectation, they find that the diffusion constant is very similar for either moderate or highly confined chains and scales approximately as N^{-1}, though a more accurate representation is suggested by

$$D \propto N^{-x} \exp(-\alpha N / R_e) \tag{27}$$

where x may be 1, 2 or 3 (depending on the theory) and R_e is the entanglement length. Based on what we know about the transport of highly confined systems, further simulations, preferably by MD rather than MC, should be able to give more accurate measurements of diffusion and structure for these topologically interesting molecules. What is clear is that chain topology greatly influences the structure and transport properties of polymer fluids and that these systems will be increasingly studied by molecular simulation methods, both in the bulk and in confined geometries. Maury-Evertsz *et al.*[456] follow up their previous study of linear homopolymers by performing a similar MC study for confined star-branched polymers, paying particular attention to the transition from coil-to-globule and liquid-like to solid-like behaviour as a function of degree of confinement.

The structure and dynamics of supercooled polymer melts in confined planar geometries has also been studied by Varnik *et al.*[457] and Meyer and Baschnagel[458] by equilibrium MD simulation. Studying confined supercooled systems could give evidence for the existence of a glass correlation length, which is the length scale over which cooperative rearrangements of the molecules need to take place for relaxation. In the former study, equilibrium MD simulations of a coarse-grained polymer melt confined to film thicknesses between 3-55 times the bulk radius of gyration are carried out with the intention of gaining insight into the dynamics, and is an extension of the previous work of these authors. Intra-chain Rouse mode relaxations are found to be faster in the films compared to the bulk, in contrast to previous studies. This is consistent with an enhancement of the cage dynamics (*i.e.* faster relaxation of the local density fluctuations in the film). This acceleration increases with decreasing film thickness. In the latter study, equilibrium MD simulations are performed to examine the structure of a supercooled melt in confined geometry. As the temperature is lowered systematically, crystalline order emanates from the walls (where chains align parallel to them) and propagates into the film, seen in Figure 11. This occurs whether or not the walls are smooth or have greater structure. Further study of the effect of wall roughness on the relaxation dynamics has been conducted by Scheidler *et al.*,[459] again by MD simulation. In short, increasing the roughness of the walls can slow down the dynamics near the wall by orders of magnitude and increase the glass transition temperature as the confinement length decreases. This is the case even if wall structure has little effect on the static structure of the melt, though the dynamics, being highly cooperative in nature, does propagate further away from the walls.

The study of translocation of polymers by molecular simulation is likely to increase because of its relevance in molecular biology (*e.g.* transport of DNA across pores into the cell nucleus, channel protein transport in membranes, *etc*) and biotechnology (*e.g.* drug delivery). However, the mechanism behind translocation is still not certain and various models have been proposed. Milchev *et al.*[460] consider the suggestion by Park and Sung[461] that translocation may be adsorption-induced. They do this by performing MC simulations of coarse-grained polymers of chain lengths between $N = 32$ to $N = 256$ monomers per chain that pass through a narrow pore in a membrane. The

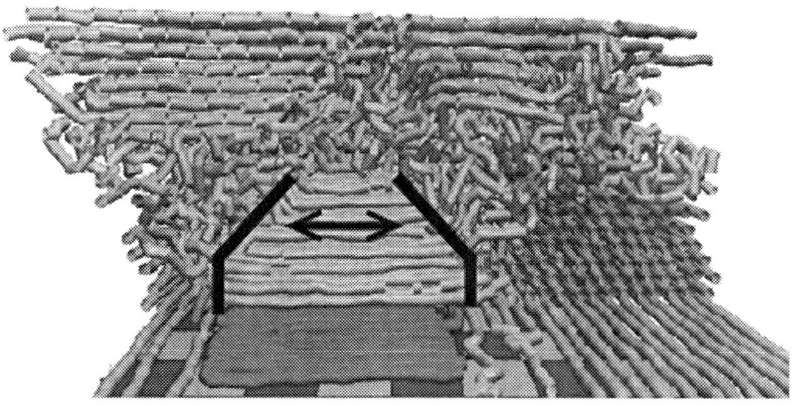

Figure 11 *Growth front (highlighted by thick lines) of polymer chains aligning parallel to the walls, extending into the interior. Figure reproduced with permission from 458*

degree of adsorption can be varied by the strength of the wall-polymer interaction energy, where one side is purely attractive and the other purely repulsive. MC simulations that follow Rouse-like dynamics are performed. They find that the translocation time (the time taken for the chain to pass through the pore) scales as $N^{2.23\pm0.04}$, in agreement with the work of Chaung *et al.*[462] that predicts it should scale proportionally to the Rouse time, *i.e.* $N^{2\nu+1}$, where $\upsilon = 0.59$ in three dimensions. This disagrees with the earlier conclusions of Park and Sung ($\propto N^3$) and Muthukumar[463] ($\propto N^2$). Another interesting study of biological membranes was undertaken by Loison *et al.*[464] who perform large-scale MD simulations of lipid bilayers and observe the spontaneous formation of pores. Such pores are very important for the transport of molecules across membranes. Their aim is to compare their simulation results with simple mesoscopic theories based upon the idea by Lister[465] that a pore carries a hydrophilic edge with an energy E given by

$$E = E_0 + \lambda c - \gamma a \tag{28}$$

where E_0 is the core energy, λ is the line tension of the pore edge, c is the contour length, a is the area of the pore and γ is the surface tension of the bilayer. Their simulations show that such a simple mesoscopic model can be a good representation for molecular processes, in particular for tensionless membranes where pores spontaneously appear due to thermal fluctuations. They study the spatial distribution of pores (see Figure 12), as well as their size and shape distributions, and find good agreement with the mesoscopic model. Pores are found not to be circular, but rather have a fractal structure. However,

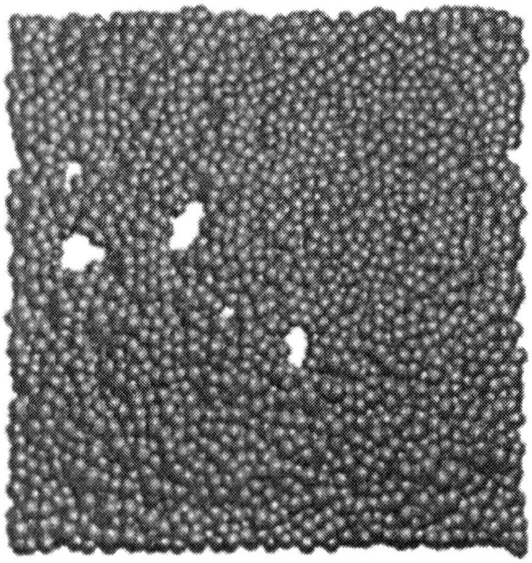

Figure 12 *Snapshot (top view) of one membrane, clearly showing the spontaneous formation of pores. Reused with permission from 464. Copyright 2004, American Institute of Physics*

their interaction potentials were essentially short-ranged Lennard-Jones type. Realistic biological systems (membranes composed of amphiphile-solvent mixtures) will have long range and electrostatic interactions. An important follow-on study would be to consider the use of more realistic potentials and see whether these simulation results are still in agreement with mesoscopic models.

MD simulations of single-stranded DNA in aqueous saline solution confined to a pore of width 2–3 nm has been performed by Cui.[466] DNA strands of 10 and 20 bases were simulated but again long range electrostatic interactions were not considered for the sake of computational expediency. This was justified by reference to similar work in the biomolecular simulation community, though it would seem reasonable that long range electrostatics will have important consequences and more realistic simulations will be attempted as computational resources increase in the coming years. Nevertheless, Cui performs an extensive set of computations investigating the DNA conformations as functions of pore width and chain length, adsorption of ions to DNA, distribution of water and ions (ions tend towards the pore centre, while Na^+ ions have an oscillatory density distribution perpendicular to the walls that is correlated with the distribution of H_2O in that peaks in the Na^+ density correspond to troughs in the H_2O distribution) and the orientation of DNA with respect to the confinement walls. Micro-confined DNA (confinement lengths 10–100 μm) under flow conditions has been studied by coarse-grained Langevin (Brownian) dynamics by Jendrejack *et al.*[491] Chains move towards the centre of the channel at higher velocities as molecular weight increases, in agreement with experimental observation. The inclusion of the hydrodynamic interaction is essential to reproduce this experimental observation, without which the opposite unphysical process is observed. A simple kinetic theory is found to predict these observations well. Comparison of such a simulation system with single molecule experiments is reported by Chen *et al.*[467] and good agreement is found between experiment, BD simulation and scaling theory predictions.

9.3 Simple Models. – Simulations are usually used for the direct calculation of properties or as an aid in the understanding of physical or chemical phenomena. However, they are also often carried as an aid in the development of simple models for future studies. This is particularly evident in the study of adsorption and flow in microporous systems, where standard hydrodynamic theories are inadequate but can in some cases be extended to treat the effects due to the confinement. Typically simulations of nanosystems need to be on longer timescales than those in the bulk due to the inhomogeneity of the system. Thus development of efficient models is important, and there has therefore been much activity in this field in recent years. Theories such as density functional theories have been extended and verified using simulation methods and simple statistical mechanical models have also been developed.

Dubbeldam *et al.*[468] use random walk theory and extended dynamically corrected transition state theory in order to compute the self-diffusivity of adsorbed molecules in nanoporous confined systems at nonzero loading. The results are compared with MD simulation results with good agreement.

Moreover, it is noted that this method can be applied to studies of slow diffusion, inaccessible in MD simulations. The approach seems very flexible in that it is applicable to a wide range of pore structures and fluids, provided the free-energy barriers are sufficiently high for transition state theory to be valid. The method therefore will fail at sufficiently high temperatures. Studies on diffusion of methane, ethane, and propane in LTL- and LTA-type zeolites were considered.

Bhatia and coworkers have developed simple statistical mechanical models to study diffusion in pores of various structures. Lennard-Jones fluids in slit[469,470] and cylindrical pores[471–473] have been considered. For systems at low densities (methane in a slit pore at 0.1–1 nm^{-3}), the oscillator model, a molecular flow model that ignores interactions between fluid molecules but considers interactions between the fluids and wall are found to be accurate. The oscillator model can therefore be considered to be an extension of the theory of Knudsen flow which takes into account the interaction of the particle with the wall. At higher densities viscous effects need to be considered and slip flow models have been developed.[470,473] The transport diffusion constants as a function of pore width and density, are compared with NEMD and MD simulation results in each case. The choice of optimum model for each condition is analysed, and very good results are obtained, provided the density is not too high. While initial studies were carried out with simple slit and cylindrical pores, these restrictions are not necessary and the extension to systems consisting of a series of slit pores of differing wall widths has also been considered.[474]

Kotdawala *et al.*[475] develop a mean-field perturbation theory for the study of polar molecules in slit pores and validate their model by comparison with MC simulations. Their theory incorporates the electrostatic interactions, and allows prediction of adsorption isotherms for molecules confined between two parallel plates. The model is used to study sorption of water molecules, and results of this model are compared with others in the literature.

Nguyen and Bhatia[476,477] have developed new (molecular) non-local density functional theories that account for the finite thickness of pore walls in porous systems. It has been demonstrated using experimental data that nanoporous carbons such as activated carbons and coal chars have pore walls that typically only consist of one or two layers. The extension of the DFT methods to treat such systems was therefore essential if accurate modelling is to be carried out. These models have been used to determine the capacity of CO_2 in carbon slit pores, and the results are compared with simulation.[478]

Often when structureless pore walls are treated, a parameter is introduced that modifies the specular reflection so that it becomes partly diffuse. This serves to model the surface roughness and introduce momentum transfer with the wall. Bhatia *et al.*[479] have recently studied the diffusion of light gases in single walled carbon nanotubes, analysing the diffusive and viscous contributions to this process. This allowed them to obtain an indication of the characteristics of the fluid that will influence the degree of diffuse reflection that occurs. They find that in single-walled carbon nanotubes, the reflection is close to being specular. They find that it is more diffuse for smaller fluids, and that the degree of diffuse reflection increases with density. Furthermore, they find that specular reflection leads to a dominance of diffusive transport. This is consistent with simulations of atomically detailed carbon nanotubes.[480]

10 Water

Numerous studies on water have been mentioned above. For liquid water, see 67, 73, 108, 109 for potential energy surfaces; 40,107–109,133,141,148,149,158–160 for simulations and properties; and 265, 273, 284, 289 for phase diagrams and transitions. For aqueous solutions, see 83, 86 for potential energy surfaces; and 111, 134, 293, 315, 357 for simulations. Water and solutions in pores and membranes is discussed in 314, 356, 441–443, 475. However, due to the huge amount of interest in liquid water and aqueous solutions there is much that has not been covered so far. Work has been carried out on water due to its importance, but also because it is a challenging system from theoretical, simulation and experimental perspectives. Many new potential energy surfaces for water appear in the literature each year, and it is usually one of the first systems on which new simulations are tested. Here we mention some of the important studies that have not been considered above.

CPMD simulations have been carried out on water to critically assess how accurately thermodynamic and dynamic properties can be calculated using this method with a range of density functionals.[481] A simulation cell size of just 32 molecules is found to give reasonable system-size convergence and the properties are not strongly influenced by the fictitious electron mass used in the CP Lagrangian, provided it remains within a specified range. However some results are quite poor. The diffusion coefficient is 10 times too small and the radial distribution functions are overstructured. It is suggested that a quantum treatment of the hydrogen atoms might improve the results. However this work indicates that even for a relatively simple system as water, it is important to continue to seek new simulation techniques.

Vande Vondele *et al.*[482] also examine the ability of water to be modelled by CPMD methods and carry out a careful analysis of the ability of different DFT functionals to produce reliable results. The effect of temperature on the results is also examined and agreement with experimental results is obtained. Different functionals are shown to produce different qualitative results, and the authors highlight the importance of testing the functional on the fluid system of interest before using it as a predictive tool.

A new empirical potential for water has been developed using spectroscopic data, which is able to model condensed water with good accuracy.[483] The potential is referred to as the VRT(ASP-W)III potential (the third fitting of the Anisotropic Site Potential with Woemer dispersion to Vibration-Rotation Tunnelling data). It give excellent results for vibrational properties of water clusters up to $(H_2O)_6$, but unlike earlier spectroscopically derived potentials also models the liquid state well. MC simulations are used to study the liquid state properties. It is noted that this potential only partly accounts for many-body interactions (the induction term) and the simulations do not include dynamical quantum effects, nevertheless it is found that this is sufficient to give good accuracy in general.

Water adsorption in carbon-slit nanopores has been studied in detail by Striolo *et al.*[484] using GCMC calculations. This is one of the few studies that has considered water in atomically structured pores. The adsorption isotherms are calculated at various pore widths, and hysteresis is observed in adsorption/desorption. Using their results they propose that for fluid separation or gas storage, narrow pores in materials with uniform pore distribution size should be designed.

11 Conclusions

Based on the number, diversity and depth of the studies undertaken over the past two years we are confident that the future of molecular simulation is very healthy. In terms of the volume of work undertaken, clearly the growth areas are in complex fluids and confined fluids. The simulation of biomolecular fluids in particular (DNA, proteins in solution, *etc*) will attract considerable effort in the years to come. New theories continue to be proposed and simulation will play a key part in assessing their validity, particularly when experimental verification is unavailable or even unfeasible at the present time. New results from simulation will also demand a re-working of existing theories, and the validity of new simulation techniques themselves will be judged against available experimental data. There is much to do, and without a doubt much will be reported when the next review of the liquid state will be conducted in two years time.

References

1. A.C. Branka, M. Kowalik and K.W. Wojciechowski, *J. Chem. Phys.*, 2003, **119**, 1929.
2. J.N. Bright, D.J. Evans and D.J. Searles, *J. Chem. Phys.*, 2005, **122**, 194106.
3. S. Sarman, D.J. Evans and G.P. Morriss, *Phys. Rev. A*, 1992, **45**, 2233.
4. J.G. Powles, G. Rickayzen and D.M. Heyes, *Mol. Phys.*, 2005, **103**, 1361.
5. J.O. Hirschfelder, *J. Chem. Phys.*, 1960, **33**, 1462.
6. C. Braga and K.P. Travis, *J. Chem. Phys.*, 2005, **123**, 134101.
7. J. Delhommelle and D.J. Evans, *Mol. Phys.*, 2001, **99**, 1825.
8. B.P. Uberuaga, M. Anghel and A.F. Voter, *J. Chem. Phys.*, 2004, **120**, 6363.
9. P. Minary, G.J. Martyna and M.E. Tuckerman, *J. Chem. Phys.*, 2003, **118**, 2510.
10. P. Minary, G.J. Martyna and M.E. Tuckerman, *J. Chem. Phys.*, 2003, **118**, 2527.
11. S. Boinepalli and P. Attard, *J. Chem. Phys.*, 2003, **119**, 12769.
12. P. Attard, *J. Chem. Phys.*, 2004, **121**, 7076.
13. D.J. Evans, D.J. Searles, W.G. Hoover, B.L. Holian, H.A. Posch and G.P. Morriss, *J. Chem. Phys.*, 1998, **108**, 4351.
14. P. Attard, *J. Chem. Phys.*, 2005, **122**, 244105.
15. D. Quigley and M.I.J. Probert, *J. Chem. Phys.*, 2004, **120**, 11432.
16. W.G. Hoover, *Phys. Rev. A*, 1986, **34**, 2499.
17. M. Parrinello and A. Rahman, *Phys. Rev. Lett.*, 1980, **23**, 1196.

18. M. Parrinello and A. Rahman, *J. Appl. Phys.*, 1981, **52**, 7182.
19. J.S. Dahler and L. Qin, *J. Chem. Phys.*, 2003, **118**, 8396.
20. M. Lisal, J.K.S. Brennan, W.R. Smith and F.R. Siperstein, *J. Chem. Phys.*, 2004, **121**, 4901.
21. P. Minary, M.E. Tuckerman and G.J. Martyna, *Phys. Rev. Lett.*, 2004, **93**, 150201.
22. D. Hamelberg, J. Mongan and J.A. McCammon, *J. Chem. Phys.*, 2004, **120**, 11919.
23. A.F. Voter, *Phys. Rev. Lett.*, 1997, **78**, 3908.
24. A.F. Voter, *J. Chem. Phys.*, 1997, **106**, 4665.
25. R.A. Miron and K.A. Fichthorn, *Phys. Rev. Lett.*, 2004, **93**, 128301.
26. J. Zheng, R. Balasundaram, S.H. Gehrke, G.S. Heffelfinger, W.A. Goddard III and S. Jiang, *J. Chem. Phys.*, 2003, **118**, 5347.
27. H. Kamberaj, R.J. Low and M.P. Neal, *J. Chem. Phys.*, 2005, **122**, 224114.
28. P.J. Daivis, K.P. Travis and B.D. Todd, *J. Chem. Phys.*, 1996, **104**, 9651.
29. A.B. Adib and C. Jarzynski, *J. Chem. Phys.*, 2005, **122**, 04114.
30. D.J. Evans and G.P. Morriss, *Statistical Mechanics of Nonequilibrium Liquids, Academic Press*, London, 1990.
31. G. Pan, J.F. Ely, C. McCabe and D.J. Isbister, *J. Chem. Phys.*, 2005, **122**, 094114.
32. F. Zhang, *J. Chem. Phys.*, 1997, **106**, 6102.
33. F. Zhang, D.J. Searles, D.J. Evans, J.S.d.T. Hansen and D.J. Isbister, *J. Chem. Phys.*, 1999, **111**, 18.
34. G.J. Martinya, M.E. Tuckerman, D.J. Tobias and M.L. Klein, *Mol. Phys.*, 1996, **87**, 1117.
35. M.E. Tuckerman, C.J. Mundy and G.J. Martinya, *Europhys. Lett.*, 1999, **45**, 149.
36. M.E. Tuckerman, Y. Liu, G. Ciccotti and G.J. Martinya, *J. Chem. Phys.*, 2001, **115**, 1678.
37. J. Petravic, *Phys. Rev. E*, 2005, **71**, 011202.
38. J. Petravic and D.J. Evans, *Mol. Phys.*, 1998, **95**, 219.
39. J. Petravic and P. Harrowell, *Phys. Rev. E*, 2005, **71**, 061201.
40. T. Terao and F. Müller-Plathe, *J. Chem. Phys.*, 2005, **122**, 081103.
41. J. Ge, G.-W. Wu, B.D. Todd and R.J. Sadus, *J. Chem. Phys.*, 2003, **119**, 11017.
42. B.D. Todd and P.J. Daivis, *Phys. Rev. Lett.*, 1998, **81**, 1118.
43. B.D. Todd and P.J. Daivis, *Comput. Phys. Commun.*, 1999, **117**, 191.
44. A. Baranyai and P.T. Cummings, *J. Chem. Phys.*, 1999, **110**, 42.
45. A.M. Kraynik and D.A. Reinelt, *Int. J. Multiphase Flow*, 1992, **18**, 1045.
46. M.E. Tuckerman, C.J. Mundy, S. Balasubramanian and M.L. Klein, *J. Chem. Phys.*, 1997, **106**, 5615.
47. B.J. Edwards and M. Dressler, *J. Non-Newtonian Fluid Mech.*, 2001, **96**, 163.
48. C. Baig, B.J. Edwards, D.J. Keffer and H.D. Cochran, *J. Chem. Phys.*, 2005, **122**, 114103.
49. C. Baig, B.J. Edwards, D.J. Keffer and H.D. Cochran, *J. Chem. Phys.*, 2005, **122**, 184906.
50. B.D. Todd, *Mol. Sim.*, 2005, **31**, 411.
51. P.J. Daivis and B.D. Todd, *in press J. Chem. Phys.*, 2005.
52. M.L. Matin, P.J. Daivis and B.D. Todd, *Comput. Phys. Commun.*, 2003 **151**, 35.
53. T.A. Hunt and B.D. Todd, *Mol. Phys.*, 2003, **101**, 3445.
54. B.D. Todd and P.J. Daivis, *J. Chem. Phys.*, 2000, **112**, 40.
55. R. Delgado-Buscalioni and P.V. Coveney, *Phys. Rev. E*, 2003, **67**, 046704.
56. R. Delgado-Buscalioni and P.V. Coveney, *Phil. Trans. Royal Soc. Lond. A*, 2004, **362**, 1639.

57. R. Barsky, R. Delgado-Buscalioni and P.V. Coveney, *J. Chem. Phys.*, 2004, **121**, 2403.
58. T. Werder, J.H. Walther and P. Koumoutsakos, *J. Comput. Phys.*, 2005, **205**, 373.
59. R. Delgado-Buscalioni and P.V. Coveney, *J. Chem. Phys.*, 2003, **119**, 978.
60. X. Nie, S. Chen and M.O. Robbins, *Phys. Fluids*, 2004, **16**, 3579.
61. M. Ripoll and M.H. Ernst, *Phys. Rev. E*, 2004, **71**, 041104.
62. W.G. Hoover and C.G. Hoover, *Mol. Phys.*, 2003, **101**, 1559.
63. G.S. Ayton, H.L. Tepper, D.T. Mirijanian and G.A. Voth, *J. Chem. Phys.*, 2004, **120**, 4074.
64. H. Okumura and D.M. Heyes, *Phys. Rev. E*, 2004, **70**, 061206.
65. K.P. Travis, B.D. Todd and D.J. Evans, *Phys. Rev. E*, 1997, **55**, 4288.
66. S.D. Chao, J.D. Kress and A. Redondo, *J. Chem. Phys.*, 2005, **122**, 234912.
67. W.L. Jorgensen and J. Tirado-Rives, *PNAS*, 2005, **102**, 6665.
68. E. Bourasseau, J.-B. Maillet, L. Mondelain and P.-M. Anglade, *Mol. Sim.*, 2005, **31**, 705.
69. Y. Boutard, P. Ungerer, J.M. Teuler, M.G. Ahunbay, S.F. Sabater, J. Pérez-Pellitero, A.D. Mackie and E. Bourasseau, *Fluid Phase Equilib.*, 2005, **236**, 25.
70. J.S. Smith, O. Borodin and G.D. Smith, *J. Phys. Chem. B*, 2004, **108**, 20340.
71. A. Grossfield, P. Ren and J.W. Ponder, *J. Am. Chem. Soc.*, 2003, **125**, 15671.
72. E. Harder, B. Kim, R.A. Friesner and B.J. Berne, *J. Chem. Theory and Comp.*, 2005, **1**, 169.
73. S. Izvekov, M. Parrinello, C.J. Burnham and G.A. Voth, *J. Chem. Phys.*, 2004, **120**, 10896.
74. S. Izvekov and G.A. Voth, *J. Phys. Chem. B*, 2005, **109**, 6573.
75. S.J. Wierzchowski and D.A. Kofke, *J. Chem. Phys.*, 2003, **119**, 6092.
76. M.H.K. Jafari, A. Maghari and S. Shahbazian, *Chem. Phys.*, 2005, **314**, 249.
77. K. Patkowski, G. Murdachaew, C.-M. Fou and K. Szalewicz, *Mol. Phys.*, 2005, **103**, 2031.
78. Y. Muto, *Proc. Phys. Math. Soc. Japan*, 1943, **17**, 629.
79. B.M. Axilrod and E. Teller, *J. Chem. Phys.*, 1943, **11**, 299.
80. G.W.M. Vissers, A. Heßelmann, G. Jansen, P.E.S. Wormer and A. van der Avoird, *J. Chem. Phys.*, 2005, **122**, 054306.
81. I. Cacelli, G. Cinacchi, G. Prampolini and A. Tani, *J. Am. Chem. Soc.*, 2004, **126**, 14278.
82. Y. Sakiyama, S. Takagi and Y. Matsumoto, *J. Chem. Phys.*, 2005, **122**, 234501.
83. B.J. Anderson, J.W. Tester and B.L. Trout, *J. Phys. Chem. B*, 2004, **108**, 18705.
84. Y.P. Puhovski, L.P. Safonova and B.M. Rode, *J. Mol. Liquids*, 2003, **103–104**, 15.
85. U. Onthong, T. Megyes, I. Bakó, T. Radnai, K. Hermansson and M. Probst, *Chem. Phys. Lett.*, 2005, **401**, 217.
86. S. Iuchi, A. Morita and S. Kato, *J. Chem. Phys.*, 2004, **121**, 8446.
87. D.J. Searles and H. Huber, in *Electronic Excitations in Liquified Rare Gases*, ed. W. F. Schmidt and E. Illenberger, American Scientific Publishers, California, 2005, 1.
88. R.J. Gdanitz, *Chem. Phys. Lett.*, 2001, **348**, 67.
89. V. Muthusamy, M.G. Müller, H. Huber and R.J. Gdanitz, *Collect. Czech. Chem. Commun.*, 2003, **68**, 627.
90. S.M. Cybulski and R.R. Toczylowski, *J. Chem. Phys.*, 1999, **111**, 10520.
91. A.E. Nasrabad, R. Laghaei and U.K. Deiters, *J. Chem. Phys.*, 2004, **121**, 6423.
92. A.E. Nasrabad and U.K. Deiters, *J. Chem. Phys.*, 2003, **119**, 947.
93. V. Muthusamy, C. Bratschi, H. Huber and R.J. Gdanitz, *Fluid Phase Equilib.*, 2004, **218**, 285.

94. J.B. Anderson, *J. Chem. Phys.*, 2004, **120**, 9886.
95. T. van Mourik and R.J. Gdanitz, *J. Chem. Phys.*, 2002, **116**, 9620.
96. S.M. Cybulski and C.E. Seversen, *J. Chem. Phys.*, 2005, **122**, 014117.
97. R. Podeszwa and K. Szalewicz, *Chem. Phys. Lett.*, 2005, **412**, 488.
98. R. Bukowski, R. Podeszwa and K. Szalewicz, *Chem. Phys. Lett.*, 2005, **414**, 111.
99. H.L. Williams and C.F. Chabalowski, *J. Phys. Chem. A*, 2001, **105**, 646.
100. R.A. Aziz, *J. Chem. Phys.*, 1993, **99**, 4518.
101. M.P. Allen and D.J. Tildesley, *Computer Simulation of Liquids*. Clarendon Press, Oxford, 1987.
102. J.B. Klauda, S.L. Garrison, J. Jiang, G. Arora and S.I. Sandler, *J. Phys. Chem. A.*, 2004, **108**, 107.
103. S.L. Garrison and S.I. Sandler, *J. Phys. Chem. B*, 2004, **108**, 18972.
104. T.H. Dunning and K.A. Petersen, *J. Chem. Phys.*, 2000, **113**, 7799.
105. V.F. Lotrich and K. Szalewicz, *J. Chem. Phys.*, 1997, **106**, 9688.
106. S.M. Gatica, M.M. Calbi, M.W. Cole and D. Velegol, *Phys. Rev. B*, 2003, **68**, 205409.
107. K. Szalewicz, R. Bukowski and B. Jeziorski, in *Theory and Applications of Computational Chemistry The First Forty Years*, ed. C. Dykstra, G. Frenking, K. Kim and G. Scuseria, Elsevier, 2005, 919.
108. E.M. Mas, R. Bukowski and K. Szalewicz, *J. Chem. Phys.*, 2003, **118**, 4386.
109. F.N. Keutsch, J.D. Cruzan and R.J. Saykally, *Chem. Rev.*, 2003, **103**, 2533.
110. G. Raabe and R.J. Sadus, *J. Chem. Phys.*, 2003, **119**, 6691.
111. M.S. Moghaddam, S. Shimizu and H.S. Chan, *J. Am. Chem. Soc.*, 2005, **127**, 202.
112. W.A. Steele, *J. Phys. Chem.*, 1978, **83**, 817.
113. W.A. Steele, *Surface Science*, 1973, **36**, 317.
114. X. Zhang, W. Wang and G. Jiang, *Fluid Phase Equilib.*, 2004, **218**, 239.
115. B.K. Petersen, J.P.R.B. Walton and K.E. Gubbins, *J. Chem. Soc., Faraday Trans. 2*, 1986, **82**, 1789.
116. X. Zhao and J. K. Johnson, *Mol. Sim.*, 2005, **31**, 1.
117. S.K. Bhatia, K. Tran, T.X. Nguyen and D. Nicholson, *Langmuir*, 2004, **20**, 9612.
118. B. Cunning, D.J. Searles and S. Bhatia, *in preparation*, 2005.
119. O. Hübner, A. Glöss, M. Fichtner and W. Klopper, *J. Phys. Chem. A*, 2004, **108**, 3019.
120. T. Heine, L. Zhechkov and G. Seifert, *Phys. Chem. Chem. Phys.*, 2004, **6**, 980.
121. Y. Miura, H. Kasai, W.A. Diño, H. Nakanishi and T. Sugimoto, *J. Phys. Soc. Jpn.*, 2003, **72**, 995.
122. W.A. Diño, Y. Miura, H. Nakanishi, H. Kasai and T. Sugimoto, *J. Phys. Soc. Jpn.*, 2003, **72**, 1867.
123. M. Cinke, J. Li, C.W. Bauschlicher Jr, A. Ricca and M. Meyyappan, *Chem. Phys. Lett.*, 2003, **376**, 761.
124. C.W. Bauschlicher Jr. and A. Ricca, *Phys. Rev. B*, 2004, **70**, 115409.
125. C.W. Bauschlicher Jr., *Nano Letters*, 2001, **1**, 223.
126. P.K. Ghorai, M. Sluiter, S. Yashonath and Y. Kawazoe, *J. Am. Chem. Soc.*, 2003, **125**, 16192.
127. J.B. Klauda, J. Jiang and S.I. Sandler, *J. Phys. Chem. B*, 2004, **108**, 9842.
128. I. Snook, A. Barnard, S. Russo, R. Springal and J. Srbinovsky, *Mol. Sim.*, 2005, **31**, 495.
129. H.B. Schlegel, S.S. Iyengar, X. Li, J.M. Millam, G.A. Voth, G.E. Scuseria and M.J. Frisch, *J. Chem. Phys.*, 2002, **117**, 8694.
130. S.S. Iyengar, C.J. Burnham, M.K. Petersen and G.A. Voth, *Computing in Science and Engineering*, 2003, **5**, 31.
131. M. Bühl, R. Diss and G. Wipff, *J. Chem. Phys.*, 2005, **127**, 13506.

132. W. Zhuang and C. Dellago, *J. Phys. Chem. B*, 2004, **108**, 19647.
133. M. Sharma, Y. Wu and R. Car, *Int. J. Quantum Chem.*, 2003, **95**, 821.
134. B. Kirchner and J. Hutter, *J. Chem. Phys.*, 2004, **121**, 5133.
135. J. Blumberger and M. Sprik, *J. Phys. Chem. B*, 2004, **108**, 6529.
136. J. Blumberger and M. Sprik, *J. Phys. Chem. B*, 2005, **109**, 6793.
137. J. Blumberger, L. Bernasconi, I. Tavernelli, R. Vuilleumier and M. Sprik, *J. Am. Chem. Soc.*, 2004, **126**, 3928.
138. J. Blumberger, Y. Tateyama and M. Sprik, *Comput. Phys. Commun.*, 2005, **169**, 256.
139. I. Tavernelli, R. Vuilleumier and M. Sprik, *Phys. Rev. Lett.*, 2002, **88**, 213002.
140. R. A. Friesner and V. Guallar, *Annu. Rev. Phys. Chem.*, 2005, **56**, 389.
141. D. Sebastiani and U. Rothlisberger, *J. Phys. Chem. B*, 2004, **108**, 2807.
142. O. Acevedo and W.L. Jorgensen, *Organic Letters*, 2004, **6**, 2881.
143. C.R.W. Guimarães, M. Udier-Blgovic and W.L. Jorgensen, *J. Am. Chem. Soc.*, 2005, **127**, 3577.
144. K. Nam, J. Gao and D.M. York, *J. Chem. Theory and Comp.*, 2005, **1**, 2.
145. N. Rega, S.S. Iyengar, G.A. Voth, H.B. Schlegel, T. Vreven and M.J. Frisch, *J. Phys. Chem. B*, 2004, **108**, 4210.
146. L. Pejov, D. Spangberg and K. Hermansson, *J. Phys. Chem. A*, 2005, **109**, 5144.
147. M.P. Gaigeot, R. Vuilleumier, M. Sprik and D. Borgis, *J. Chem. Theory and Comp.*, 2005, **1**, 772.
148. E.H. Hardy, M.G. Müller, P.S. Vogt, C. Bratschi, B. Kirchner, H. Huber and D.J. Searles, *J. Chem. Phys.*, 2003, **119**, 6184.
149. T.S. Pennanen, J. Vaara, P. Lantto, A.J. Sillanpää, K. Laasonen and J. Jokisaari, *J. Am. Chem. Soc.*, 2004, **126**, 11093.
150. J. Schmidt and D. Sebastiani, *J. Chem. Phys.*, 2005, **123**, 074501.
151. M.G. Müller, E.H. Hardy, P.S. Vogt, C. Bratschi, B. Kirchner, H. Huber and D.J. Searles, *J. Am. Chem. Soc.*, 2004, **126**, 4704.
152. D.J. Searles and H. Huber, in *Calculation of NMR and EPR Parameters Theory and Applications*, ed. M. Kaupp, M. Bühl and V.G. Malkin, Wiley-VCH, Weinheim, 2005, 175.
153. D.J. Searles and H. Huber, in *Encyclopedia of Nuclear Magnetic Resonance*, ed. D.M. Grant and R. K. Harris, John Wiley & Sons, Chichester, 2002, Vol. 9.
154. R. Eggenberger, S. Gerber, H. Huber, D. Searles and M. Welker, *Mol. Phys.*, 1993, **80**, 1177.
155. J.A. Ropp, C. Lawrence, T.C. Farrar and J.L. Skinner, *J. Am. Chem. Soc.*, 2001, **123**, 8047.
156. J.R.C. van der Maarel, D. Lankhorst, J. de Bleijser and J.C. Leyte, *Chem. Phys. Lett.*, 1985, **122**, 541.
157. J.O. Hirschfelder, C.F. Curtiss and R.B. Bird, *Molecular Theory of Gases and Liquids. Wiley*, New York, 1954.
158. B. Chen, I. Ivanov, M.L. Klein and M. Parrinello, *Phys. Rev. Lett.*, 2003, **91**, 215503.
159. L. Hernández de la Peña and P.G. Kusalik, *J. Am. Chem. Soc.*, 2005, **127**, 5246.
160. L. Hernández de la Peña and P.G. Kusalik, *J. Chem. Phys.*, 2004, **121**, 5992.
161. S. Raugei and M.L. Klein, *J. Am. Chem. Soc.*, 2003, **125**, 8992.
162. A.W. Jasper, S. Nangia, C. Zhu and D.G. Truhlar, *Acc. Chem. Res.*, 2005, in press.
163. U.F. Rohrig, L. Guidoni and U. Röthlisberger, *Chem. Phys. Chem.*, 2005, **6**, 1836.
164. I. Tavernelli, U.F. Rohrig and U. Röthlisberger, *Mol. Phys.*, 2005, **103**, 963.
165. M.E. Moret, E. Tapavicza, L. Guidoni, U.F. Rohrig, M. Sulpizi, I. Tavernelli and U. Röthlisberger, *Chimia*, 2005, **59**, 493.

166. J.-P. Eckmann and D. Ruelle, *Rev. Mod. Phys.*, 1985, **57**, 617.
167. D.J. Evans, E.G.D. Cohen and G.P. Morriss, *Phys. Rev. A*, 1990, **42**, 5990.
168. P. Gaspard and G. Nicolis, *Phys. Rev. Lett.*, 1990, **65**, 1963.
169. P. Gaspard and J.R. Dorfman, *Phys. Rev. E*, 1995, **52**, 3525.
170. A.S. de Wijn and H. van Beijeren, *Phys. Rev. E*, 2004, **70**, 16207.
171. T. Taniguchi and G.P. Morriss, *Phys. Rev. Lett.*, 2005, **94**, 154101.
172. J.-P. Eckmann, C. Forster, H.A. Posch and E. Zabey, *J. Stat. Phys.*, 2005, **118**, 795.
173. C. Forster and H.A. Posch, *New Journal of Physics*, 2005, **7**, 32.
174. H.-l. Yang and R. Gunter, *Phys. Rev. E*, 2005, **71**, 036211.
175. A.S. de Wijn, *Phys. Rev. E.*, 2005, **72**, 026216.
176. C. Forster, D. Mukamel and H.A. Posch, *Phys. Rev. E*, 2004, **69**, 066124.
177. F. Frascoli, D.J. Searles and B.D. Todd, in press, 2005, *Phys. Rev. E.*
178. D.J. Searles, D.J. Evans and D.J. Isbister, *Chaos*, 1998, **8**, 337.
179. P. Debabrata and R. van Zon, *Phys. Rev. E*, 2002, **66**, 021101.
180. C.A. Florek and C.L.I. Tucker, *Physics of Fluids*, 2005, **17**, 053101.
181. F. Schoenfeld, V. Hessel and C. Hofmann, *Lab on a Chip*, 2004, **4**, 65.
182. T. Burghelea, E. Segra and V. Seteinberg, *Europhys. Lett.*, 2004, **68**, 529.
183. G. Boffetta, C. Antonio and S. Musacchio, *Phys. Rev. Lett.*, 2003, **91**, 034501.
184. D. Ihm, Y.-H. Shin, J.-W. Lee and E.K. Lee, *Phys. Rev. E*, 2003, **67**, 027205.
185. S. Viscardy and P. Gaspard, *Phys. Rev. E*, 2003, **68**, 041205.
186. J. Casas-Vazquez and D. Jou, *Rep. Prog. Phys.*, 2003, **66**, 1937.
187. M. Kröger, *Phys. Rep.*, 2004, **390**, 453.
188. P.J. Daivis and M.L. Matin, *J. Chem. Phys.*, 2003, **118**, 11111.
189. D.J. Evans and H.J.M. Hanley, *Phys. Lett.*, 1980, **80A**, 175.
190. H.J.M. Hanley and D.J. Evans, *J. Chem. Phys.*, 1982, **83**, 3297.
191. D.J. Evans, *J. Chem. Phys.*, 1983, **83**, 3297.
192. T. Taniguchi and G.P. Morriss, *Phys. Rev. E*, 2004, **70**, 056124.
193. I. Santamaría-Holek, J.M. Rubí and A. Pérez-Madrid, *New Journal of Physics*, 2004, **7**, 35.
194. M. Kröger and H.C. Öttinger, *J. Non-Newtonian Fluid Mech.*, 2004, **120**, 175.
195. H.C. Öttinger and M. Grmela, *Phys. Rev. E*, 1997, **56**, 6633.
196. S. Ansumali, I.V. Karlin and H.C. Öttinger, *Phys. Rev. Lett.*, 2005, **94**, 080602.
197. J.-M. Simon, S. Kjelstrup, D. Bedeaux and B. Hafskjold, *J. Phys. Chem. B.*, 2004, **108**, 7186.
198. P. Galenko and D. Jou, *Phys. Rev. E*, 2005, **71**, 046125.
199. D. Reguera and J.M. Rubí, *J. Chem. Phys.*, 2003, **119**, 9877.
200. D. Reguera and J.M. Rubí, *J. Chem. Phys.*, 2003, **119**, 9888.
201. *Molecular Simulation*, edited by N. Quirke, K. E. Gubbins, J. N. Onuchic and A. H. Fuchs (Taylor & Francis, London, 2002), Vol. 28.
202. G. Hummer, *Mol. Sim.*, 2002, **28**, 81.
203. H.-J. Woo and B. Roux, *PNAS*, 2005, **102**, 6825.
204. J.M.J. Swanson, R.H. Henchman and J.A. McCammon, *Biophys. J.*, 2004, **86**, 67.
205. R.H. Swendsen, M. Fasnacht and J.M. Rosenberg, *Comput. Phys. Commun.*, 2005, **169**, 274.
206. J. Anwar and D.M. Heyes, *J. Chem. Phys.*, 2005, **122**, 224117.
207. M.K. Fenwick and F.A. Escobedo, *J. Chem. Phys.*, 2004, **120**, 3066.
208. I.D. Gospodinov and F.A. Escobedo, *J. Chem. Phys.*, 2005, **122**, 164103.
209. A.V. Neimark and A. Vishnyakov, *J. Chem. Phys.*, 2005, **122**, 234108.
210. A. E. Roitberg, *Annual Reports in Computational Chemistry*, 2005, **1**, 103.

211. W. Lechner, H. Oberhofer, C. Dellago and P.L. Geissler, *J. Chem. Phys.*, 2005, submitted.
212. J. MacFadyen and I. Andricioaei, *J. Chem. Phys.*, 2005, **123**, 074107.
213. D. Wu and D.A. Kofke, *J. Chem. Phys.*, 2005, **122**, 204104.
214. H. Oberhofer, C. Dellago and P.L. Geissler, *J. Phys. Chem. B.*, 2005, **109**, 6902.
215. E. Atilgan and S.X. Sun, *J. Chem. Phys.*, 2004, **121**, 10392.
216. F.M. Ytreberg and D.M. Zuckerman, *J. Comput. Chem.*, 2004, **25**, 1749.
217. J. Gore, F. Ritort and C. Bustamante, *PNAS*, 2003, **100**, 12564.
218. D. Collin, F. Ritort, C. Jarzynski, S.B. Smith, I. Tinoco Jr and C. Bustamante, *Nature*, 2005, **437**, 231.
219. D.J. Evans, *Mol. Phys.*, 2003, **101**, 1551.
220. T. Monnai, *Phys. Rev. E.*, 2005, **72**, 027102.
221. J.C. Reid, E.M. Sevick and D.J. Evans, *Europhys. Lett.*, 2005, in press.
222. C. Jarzynski, *J. Stat. Mech. Theory and Expt.*, 2004, P09005.
223. D.J. Searles and D.J. Evans, *Aust. J. Chem.*, 2004, **57**, 1119.
224. D.J. Evans and D.J. Searles, *Adv. Phys.*, 2002, **51**, 1529.
225. G.M. Wang, E.M. Sevick, E. Mittag, D.J. Searles and D.J. Evans, *Phys. Rev. Lett.*, 2002, **89**, 050601.
226. D.M. Carberry, J.C. Reid, G.M. Wang, E.M. Sevick, D.J. Searles and D.J. Evans, *Phys. Rev. Lett.*, 2004, **92**, 140601.
227. G.M. Wang, J.C. Reid, D.M. Carberry, D.R.M. Williams, E.M. Sevick and D. J. Evans, *Phys. Rev. E*, 2005, **71**, 046142.
228. S.R. Williams, D.J. Searles and D.J. Evans, *Phys. Rev. A*, 2004, **70**, 066113.
229. N.C. Karayiannis, V.G. Mavrantzas and D.N. Theodorou, *Macromolecules*, 2004, **37**, 2978.
230. G. Marcelli, B.D. Todd and R.J. Sadus, *Phys. Rev. E*, 2001, **63**, 021204.
231. J. Ge, G. Marcelli, B.D. Todd and R.J. Sadus, *Phys. Rev. E*, 2001, **64**, 021201.
232. J. Ge, G. Marcelli, B.D. Todd and R.J. Sadus, *Phys. Rev. E.*, 2002, **65**, 069901(E).
233. J. Ge, B.D. Todd and R.J. Sadus, *Phys. Rev. E.*, 2003, **67**, 061201.
234. B.D. Todd, *Phys. Rev. E*, 2005, **72**, 041204.
235. H. Wada and S. Sasa, *Phys. Rev. E.*, 2003, **67**, 065302(R).
236. P. Das and J.K. Bhattacharjee, *Phys. Rev. E*, 2005, **71**, 036145.
237. R.F. Berg, M.R. Moldover and G.A. Zimmerli, *Phys. Rev. E*, 1999, **60**, 4079.
238. M.H. Ernst, *Phys. Rev. E*, 2005, **71**, 030101.
239. J. Ge, PhD dissertation, Swinburne University of Technology, 2004.
240. S. Hess, M. Kröger and D.J. Evans, *Phys. Rev. E*, 2003, **67**, 042201.
241. S. Viscardy and P. Gaspard, *Phys. Rev. E*, 2003, **68**, 041204.
242. J. Delhommelle, J. Petravic and D.J. Evans, *J. Chem. Phys.*, 2004, **120**, 6117.
243. J.J. Erpenbeck, *Phys. Rev. Lett.*, 1984, **52**, 1333.
244. K.P. Travis, P.J. Daivis and D.J. Evans, *J. Chem. Phys.*, 1996, **105**, 3893.
245. J. Delhommelle, J. Petravic and D.J. Evans, *Phys. Rev. E*, 2003, **68**, 031201.
246. J. Delhommelle, J. Petravic and D.J. Evans, *J. Chem. Phys.*, 2003, **119**, 11005.
247. J. Petravic, *J. Chem. Phys.*, 2004, **120**, 7041.
248. J. Petravic, *J. Chem. Phys.*, 2004, **120**, 10188.
249. J. Petravic, *J. Chem. Phys.*, 2004, **121**, 11202.
250. G. Rickayzen, J.G. Powles and D.M. Heyes, *J. Chem. Phys.*, 2003, **118**, 11048.
251. G. Rickayzen and D.M. Heyes, *Phys. Rev. E*, 2005, **71**, 061204.
252. A.C. Branka and D.M. Heyes, *Phys. Rev. E*, 2004, **69**, 021202.
253. J.W. Dufty, *Mol. Phys.*, 2002, **100**, 2331.
254. D.M. Heyes and A.C. Branka, *Phys. Chem. Chem. Phys.*, 2005, **7**, 1220.

255. A.C. Branka and D.M. Heyes, *Mol. Phys.*, 2005, **103**, 2359.
256. M. Schoen and F. Porcheron, *Phys. Rev. E*, 2003, **67**, 051202.
257. K. Miyazaki, D.R. Reichman and R. Yamamoto, *Phys. Rev. E*, 2004, **70**, 011501.
258. Y. Jung, J.P. Garrahan and D. Chandler, *Phys. Rev. E*, 2004, **69**, 061205.
259. L. Berthier, D. Chandler and J.P. Garrahan, *Europhys. Lett.*, 2005, **69**, 320.
260. A. Widmer-Cooper, P. Harrowell and H. Fynewever, *Phys. Rev. Lett.*, 2004, **93**, 135701.
261. C. Xiao, D.M. Heyes and J.G. Powles, *Physica Status Solidi B: Basic Solid State Physics*, 2005, **242**, 749.
262. L. Lue, *J. Chem. Phys.*, 2005, **122**, 044513.
263. A.Z. Panagiotopoulos, *J. Phys.: Condens. Matter*, 2000, **12**, R25.
264. I. Skarmoutsos, L.I. Kampanakis and J. Samios, *J. Mol. Liquids*, 2005, **117**, 33.
265. E. Johansson, K. Bolton and P. Ahlstrom, *J. Chem. Phys.*, 2005, **123**, 024504.
266. H. Okumura and Y. Okamoto, *Comput. Phys. Commun.*, 2005, **169**, 317.
267. M.J. McGrath, J.I. Siepmann, I.-F.W. Kuo, C.J. Mundy, J.V. Vondele, M. Sprik, J. Hutter, F. Mohamed, M. Krack and M. Parrinello, *Comput. Phys. Commun.*, 2005, **169**, 289.
268. D.H.E. Gross and J.F. Kenney, *J. Chem. Phys.*, 2005, **122**, 224111.
269. A. Venkatnathan and G.A. Voth, *J. Chem. Theory and Comp.*, 2005, **1**, 36.
270. J.C. Pamies, C. McCabe, P.T. Cummings and L.F. Vega, *Mol. Sim.*, 2003, **29**, 463.
271. V.K. Shen and J.R. Errington, *J. Chem. Phys.*, 2005, **122**, 064508.
272. J.R. Errington and V.K. Shen, *J. Chem. Phys.*, 2005, **123**, 164102.
273. D. Zahn, *Phys. Rev. Lett.*, 2004, **93**, 227801.
274. J.F. Lutsko, D. Wolf, S.R. Phillpot and S. Yip, *Phys. Rev. B*, 1989, **40**, 2841.
275. S.R. Phillpot, J.F. Lutsko, D. Wolf and S. Yip, *Phys. Rev. B*, 1989, **40**, 2831.
276. J. Solca, A.J. Dyson, G. Steinebrunner, B. Kirchner and H. Huber, *Chem. Phys.*, 1997, **224**, 253.
277. J. Solca, A.J. Dyson, G. Steinebrunner, B. Kirchner and H. Huber, *J. Chem. Phys.*, 1998, **108**, 4107.
278. J. Ge, G.-W. Wu, B.D. Todd and R.J. Sadus, *J. Chem. Phys.*, 2003, **119**, 11017.
279. S. Alavi and D.L. Thompson, *J. Chem. Phys.*, 2005, **122**, 154704.
280. G.F. Velardez, S. Alavi and D.L. Thompson, *J. Chem. Phys.*, 2004, **120**, 9151.
281. P.M. Agrawal, B.M. Rice and D.L. Thompson, *J. Chem. Phys.*, 2003, **119**, 9671.
282. G.F. Velardez, S. Alavi and D.L. Thompson, *J. Chem. Phys.*, 2003, **119**, 6698.
283. P.M. Agrawal, B.M. Rice and D.L. Thompson, *J. Chem. Phys.*, 2003, **118**, 9680.
284. T. Bryk and A.D.J. Haymet, *Mol. Sim.*, 2004, **30**, 131.
285. J. Anwar, D. Frenkel and M.G. Noro, *J. Chem. Phys.*, 2003, **118**, 728.
286. J. Chang and S. I. Sandler, *J. Chem. Phys.*, 2003, **118**, 8390.
287. J. Chang and A.M. Lenhoff, *J. Chem. Phys.*, 2004, **120**, 3003.
288. G. Grochola, *J. Chem. Phys.*, 2004, **120**, 2122.
289. C. Vega, E. Sanz and J.L.F. Abascal, *J. Chem. Phys.*, 2005, **122**, 114507.
290. D. M. Eike, J.F. Brennecke and E.J. Maginn, *J. Chem. Phys.*, 2005, **122**, 014115.
291. A. Aguado and P.A. Madden, *Phys. Rev. Lett.*, 2005, **94**, 068501.
292. S. Yoo, X.C. Zeng and J.R. Morris, *J. Chem. Phys.*, 2004, **120**, 1654.
293. L. Vrbka and P. Jungwirth, *Phys. Rev. Lett.*, 2005, **95**, 148501.
294. Y. Wang, S. Teitel and C. Dellago, *J. Chem. Phys.*, 2005, **122**, 214722.
295. S. Alavi and D.L. Thompson, *J. Chem. Phys.*, 2004, **120**, 10231.
296. G.-W. Wu and R.J. Sadus, *J. Chem. Phys.*, 2004, **121**, 12117.
297. G.-W. Wu and R.J. Sadus, *J. Chem. Phys.*, 2004, **120**, 11686.
298. R. Blaak, H. Löwen and J.J. Barrat, *J. Chem. Phys.*, 2004, **121**, 12115.

299. G.-W. Wu and R.J. Sadus, *AIChE*, 2005, **51**, 309.
300. N. Giovambattista, H.E. Stanley and F. Sciortino, *Phys. Rev. Lett.*, 2003, **91**, 115504.
301. A.C. Branka and D.M. Heyes, *Mol. Phys.*, 2004, **102**, 2049.
302. D.M. Duffy, C. Moon and P.M. Rodger, *Mol. Phys.*, 2004, **102**, 203.
303. M.T. Storr, P.C. Taylor, J.-P. Monfort and P.M. Rodger, *J. Am. Chem. Soc.*, 2004, **126**, 1569.
304. L. Sałamacha, A. Patrykiejew and S. Sokołowski, *J. Chem. Phys.*, 2005, **122**, 074703.
305. L. Sałamacha, A. Patrykiejew, S. Sokołowski and K. Binder, *J. Chem. Phys.*, 2004, 120.
306. L. Sałamacha, A. Patrykiejew, S. Sokołowski and K. Binder, *Eur. Phys. J.*, 2004, **13**, 261.
307. A. Wongkoblap, S. Junpirom and D.D. Do, *Adsorption Science & Technology*, 2005, **23**, 1.
308. F.R. Hung, G. Dudziak, M. Sliwinska-Bartkowiak and K.E. Gubbins, *Mol. Phys.*, 2004, **102**, 223.
309. F.R. Hung, B. Coasne, E.E. Santiso, K.E. Gubbins, F.R. Siperstein and M. Sliwinska-Bartkowiak, *J. Chem. Phys.*, 2005, **122**, 144706.
310. L.G. Cámara and F. Bresme, *J. Chem. Phys.*, 2003, **119**, 2792.
311. B. Coasne, J. Czwartos, K.E. Gubbins, F.R. Hung and M. Sliwinska-Bartkowiak, *Mol. Phys.*, 2004, **102**, 2149.
312. S. Sacquin-Mora, A.H. Fuchs and M. Schoen, *J. Chem. Phys.*, 2004, **121**, 9077.
313. M. Schneemilch and N. Quirke, *Mol. Sim.*, 2003, **29**, 685.
314. L. Maibaum and D. Chandler, *J. Phys. Chem. B*, 2003, **107**, 1189.
315. G. Rother, D. Woywod, M. Schoen and G.H. Findenegg, *J. Chem. Phys.*, 2004, **120**, 11864.
316. A.V. Neimark and A. Vishnyakov, *J. Chem. Phys.*, 2005, **122**, 189903.
317. A.V. Neimark and A. Vishnyakov, *J. Chem. Phys.*, 2005, **122**, 054707.
318. J. Dzubiella and J.-P. Hansen, *J. Chem. Phys.*, 2004, **121**, 5514.
319. S. Toxvaerd, *Mol. Sim.*, 2004, **30**, 179.
320. D.N. Theodorou, *Mol. Phys.*, 2004, **20**, 147.
321. P. Nielaba, K. Binder, D. Chaudhuri, P. Henseler, M. Lohrer, A. Ricca, S. Senguptu and W. Strepp, *J. Phys. Condens. Matter*, 2004, **16**, S4115.
322. E. Frey, D.R. Nelson and L. Radzihovsky, *Phys. Rev. Lett.*, 1999, **83**, 2977.
323. L. Radzihovsky, E. Frey and D.R. Nelson, *Phys. Rev. E*, 2001, **63**, 2977.
324. S. Asakura and F. Oosawa, *J. Chem. Phys.*, 1954, **22**, 1255.
325. A. Vrij, *Pure Appl. Chem.*, 1976, **48**, 471.
326. R.L.C. Vink and J. Horbach, *J. Phys. Condens. Matter*, 2004, **16**, 3807.
327. R.L.C. Vink and J. Horbach, *J. Chem. Phys.*, 2004, **121**, 3253.
328. R.L.C. Vink, J. Horbach and K. Binder, *Phys. Rev. E*, 2005, **71**, 011401.
329. G. Szamel, *Phys. Rev. Lett.*, 2004, **93**, 178301.
330. R. Blaak, S. Auer, D. Frenkel and H. Löwen, *Phys. Rev. Lett.*, 2004, **93**, 068303.
331. R. Blaak, S. Auer, D. Frenkel and H. Löwen, *J. Phys.: Condens. Matter*, 2004, **16**, S3873.
332. S. Butler and P. Harrowell, *Phys. Rev. E*, 1995, **42**, 6424.
333. A.J. Banchio and J.F. Brady, *J. Chem. Phys.*, 2003, **118**, 10323.
334. A. Sieroi and J.F. Brady, *J. Fluid Mech.*, 2001, **448**, 115.
335. J. Delhommelle, *Phys. Rev. E*, 2005, **71**, 016705.
336. J. Delhommelle and J. Petravic, *J. Chem. Phys.*, 2005, **123**, 074707.
337. N. Galamba, C.A. Nieto de Castro and J.F. Ely, *J. Phys. Chem. B*, 2004, **108**, 3658.

338. N. Galamba, C.A. Nieto de Castro and J.F. Ely, *J. Chem. Phys.*, 2004, **120**, 8676.
339. J. Delhommelle and J. Petravic, *J. Chem. Phys.*, 2003, **118**, 2783.
340. J. Petravic and J. Delhommelle, *J. Chem. Phys.*, 2003, **118**, 7477.
341. J. Petravic and J. Delhommelle, *J. Chem. Phys.*, 2003, **119**, 8511.
342. N. Galamba, C.A. Nieto de Castro and J.F. Ely, *J. Chem. Phys.*, 2005, **122**, 224501.
343. H. Zhang and J.F. Ely, *Fluid Phase Equilib.*, 2004, **217**, 111.
344. C. McCabe, D. Bedrov, O. Borodin, G.D. Smith and P.T. Cummings, *Ind. Eng. Chem. Res*, 2003, **42**, 6956.
345. J. Petravic and J. Delhommelle, *J. Chem. Phys.*, 2005, **122**, 234509.
346. G. Giupponi and D.M.A. Buzza, *J. Chem. Phys.*, 2004, **120**, 10290.
347. R. Scherrenberg, B. Coussens, P. Van Vliet, G. Edouard, J. Brackman and E. De Brabander, *Macromolecules*, 1998, **31**, 456.
348. E.G. Timoshenko, Y.A. Kuznetsov and R. Connolly, *J. Chem. Phys.*, 2002, **117**, 9050.
349. T.C. Zook and G.T. Pickerr, *Phys. Rev. Lett.*, 2003, **90**, 015502.
350. G. Giupponi and D.M.A. Buzza, *J. Chem. Phys.*, 2005, **122**, 194903.
351. J.T. Bosko, B.D. Todd and R.J. Sadus, *J. Chem. Phys.*, 2004, **121**, 1091.
352. J.T. Bosko, B.D. Todd and R.J. Sadus, *J. Chem. Phys.*, 2004, **121**, 12050.
353. J.T. Bosko, B.D. Todd and R.J. Sadus, *J. Chem. Phys.*, 2005, **123**, 034905.
354. P. Sunthar and J.R. Prakash, *Macromolecules*, 2005, **38**, 617.
355. D.E. Smith and S. Chu, *Science*, 1998, **281**, 1335.
356. S.A. Pandit, D. Bostick and M.L. Berkowitz, *J. Chem. Phys.*, 2003, **119**, 2199.
357. Z.W. Zhou, B.D. Todd, K.P. Travis and R.J. Sadus, *J. Chem. Phys.*, 2005, **123**, 054505.
358. H.L. Tepper and G.A. Voth, *J. Chem. Phys.*, 2005, **122**, 124906.
359. D.C. Rapaport, *Phys. Rev. E.*, 2004, **70**, 051905.
360. S.S. Rane and W.L. Mattice, *J. Chem. Phys.*, 2005, **122**, 234913.
361. B.J. Banaszak and J.J. de Pablo, *J. Chem. Phys.*, 2003, **119**, 2456.
362. Z. Chen and F.A. Escobedo, *J. Chem. Phys.*, 2000, **113**, 11382.
363. R. Auhl, R. Everaers, G.S. Grest, K. Kremer and S.J. Plimpton, *J. Chem. Phys.*, 2003, **119**, 12718.
364. A. Uhlherr, *Comput. Phys. Commun.*, 2003, **155**, 31.
365. A. Uhlherr, S.J. Leak, N.E. Adam, P.E. Nyberg, M. Doxastakis, V.G. Mavrantzas and D.N. Theodorou, *Comput. Phys. Commun.*, 2002, **144**, 1.
366. A. Uhlherr, M. Doxastakis, V.G. Mavrantzas, D.N. Theodorou, S.J. Leak, N.E. Adam and P.E. Nyberg, *Europhys. Lett.*, 2002, **57**, 506.
367. P.V.K. Pant and D.N. Theodorou, *Macromolecules*, 1995, **28**, 7224.
368. K.C. Daoulas, A.F. Terzis and V.G. Mavrantzas, *Macromolecules*, 2003, **36**, 6674.
369. T. Ikeshoji, B. Hafskjold and H. Furuholt, *Mol. Sim.*, 2003, **29**, 101.
370. R. Everaers, S.K. Sukumaran, G.S. Grest, C. Svaneborg, A. Sivasubramanian and K. Kremer, *Science*, 2004, **303**, 823.
371. S.K. Sukumaran, G.S. Grest, K. Kremer and R. Everaers, *J. Polym. Sci. B*, 2005, **43**, 917.
372. Q. Zhou and R.G. Larson, *Macromolecules*, 2005, **38**, 5761.
373. M. Kröger, *Comput. Phys. Commun.*, 2005, **168**, 209.
374. R.V.N. Melnik, A. Uhlherr, J. Hodgkin and F. De Hoog, *Computers and Mathematics with Applications*, 2003, **45**, 515.
375. J. Baschnagel, K. Binder, P. Doruker, A.A. Gusev, O. Hahn, K. Kremer, W.L. Mattice, F. Müller-Plathe, M. Murat, W. Paul, S. Santos, U.W. Suter and V. Tries, *Adv. Polym. Sci.*, 2000, **152**, 41.

376. F. Müller-Plathe, *Soft Mater.*, 2003, **1**, 1.
377. N. Zacharopoulos, N. Vergadou and D.N. Theodorou, *J. Chem. Phys.*, 2005, **122**, 244111.
378. A.E. Ismail, G.C. Rutledge and G. Stephanopoulos, *J. Chem. Phys.*, 2005, **122**, 234901.
379. A.E. Ismail, G. Stephanopoulos and G.C. Rutledge, *J. Chem. Phys.*, 2005, **122**, 234902.
380. H.C. Öttinger, *J. Non-Newtonian Fluid Mech.*, 2004, **120**, 207.
381. P.T. Underhill and P.S. Doyle, *J. Non-Newtonian Fluid Mech.*, 2004, **122**, 3.
382. R. Prabhakar, J.R. Prakash and T. Sridhar, *J. Rheol.*, 2004, **48**, 1251.
383. M. Kröger and E. De Angelis, *J. Non-Newtonian Fluid Mech.*, 2005, **125**, 87.
384. M. Doxastakis, Y.-L. Chen, O. Guzmán and J.J. de Pablo, *J. Chem. Phys.*, 2004, **120**, 9335.
385. E.-G. Kim and M.L. Klein, *Macromolecules*, 2004, **37**, 1674.
386. J.T. Padding and W.J. Briels, *J. Chem. Phys.*, 2004, **120**, 2996.
387. V.A. Harmandaris, V.G. Mavrantzas, D.N. Theodorou, M. Kröger, J. Ramirez, H.C. Öttinger and D. Vlassopoulos, *Macromolecules*, 2003, **36**, 1376.
388. M. Kröger and S. Hess, *Phys. Rev. Lett.*, 2000, **85**, 1128.
389. K. Kremer and G.S. Grest, *J. Chem. Phys.*, 1990, **92**, 5057.
390. B.D. Todd and P.J. Daivis, *J. Chem. Phys.*, 1997, **107**, 1617.
391. P.J. Daivis and B.D. Todd, *Int. J. Thermophys.*, 1998, **19**, 1063.
392. J.G.H. Cifre, S. Hess and M. Kröger, *Macromol. Theory Simul.*, 2004, **13**, 748.
393. R. Yamamoto and A. Onuki, *Phys. Rev. E*, 2004, **70**, 041801.
394. P.J. Daivis, M.L. Matin and B.D. Todd, *J. Non-Newtonian Fluid Mech.*, 2003, **111**, 1.
395. H.M. Laun and H. Münsted, *Rheol. Acta*, 1978, **17**, 415.
396. C. Svaneborg, G.S. Grest and R. Everaers, *Phys. Rev. Lett.*, 2004, **93**, 257801.
397. G. Marrucci and G. Ianniruberto, *Macromolecules*, 2004, **37**, 3934.
398. M. Doxastakis, D.N. Theodorou, G. Fytas, F. Kremer, R. Faller, F. Muller-Plathe and N. Hadjichristidis, *J. Chem. Phys.*, 2003, **119**, 6883.
399. P. Gestoso, E. Nicol, M. Doxastakis and D.N. Theodorou, *Macromolecules*, 2003, **36**, 6925.
400. C.D. Wick and D.N. Theodorou, *Macromolecules*, 2004, **37**, 7026.
401. G. Tsolou, V.G. Mavrantzas and D.N. Theodorou, *Macromolecules*, 2005, **38**, 1478.
402. R. Connolly, G. Bellesia, E.G. Timoshenko, Y.A. Kuznetsov, S. Elli and F. Ganazzoli, *Macromolecules*, 2005, **38**, 5288.
403. L.D. Peristeras, I.G. Economou and D.N. Theodorou, *Macromolecules*, 2005, **38**, 386.
404. M. Aichele, Y. Gebremichael, F.W. Starr, J. Baschnagel and S.C. Glotzer, *J. Chem. Phys.*, 2003, **119**, 5290.
405. T.S. Jain and J.J. de Pablo, *Phys. Rev. Lett.*, 2004, **92**, 155505.
406. Q. Wang, P.F. Nealey and J.J. de Pablo, *J. Chem. Phys.*, 2003, **118**, 11278.
407. G. Srinivas and M.L. Klein, *Mol. Phys.*, 2004, **102**, 883.
408. G. Srinivas, D.E. Discher and M.L. Klein, *Nature Materials*, 2004, **3**, 638.
409. G. Sun, M. Kappl, T. Pakula, K. Kremer and H.-J. Butt, *Langmuir*, 2004, **20**, 8030.
410. K.C. Daoulas, V.A. Harmandaris and V.G. Mavrantzas, *Macromolecules*, 2005, **38**, 5780.
411. K.C. Daoulas, D.N. Theodorou, V.A. Harmandaris, N.C. Karayiannis and V.G. Mavrantzas, *Macromolecules*, 2005, **38**, 7134.

412. V.A. Harmandaris, K.C. Daoulas and V.G. Mavrantzas, *Macromolecules*, 2005, **38**, 5796.
413. T. Shima, H. Kuni, Y. Okabe, M. Doi, X.-F. Yuan and T. Kawakatsu, *Macromolecules*, 2003, **36**, 9199.
414. T. Uneyama and M. Doi, *Macromolecules*, 2005, **38**, 196.
415. K.P. Travis and K.E. Gubbins, *J. Chem. Phys.*, 2000, **112**, 1984.
416. T. Becker and F. Mugele, *Phys. Rev. Lett.*, 2003, **91**, 166104.
417. B. Luan and M.O. Robbins, *Nature*, 2005, **435**, 929.
418. J. Zhang, B.D. Todd and K.P. Travis, *J. Chem. Phys.*, 2005, **122**, 219901.
419. J. Zhang, B.D. Todd and K.P. Travis, *J. Chem. Phys.*, 2004, **121**, 10778.
420. P.J. Cadusch, B.D. Todd, J. Zhang and P.J. Daivis, in preparation.
421. M. Al-Ghoul and B.C. Eu, *Phys. Rev. E*, 2004, **70**, 016301.
422. Z. Guo, T.S. Zhao and Y. Shi, *Phys. Rev. E*, 2005, **71**, 035301(R).
423. C. Denniston and M.O. Robbins, *Phys. Rev. E*, 2004, **69**, 021505.
424. B.D. Todd, D.J. Evans and P.J. Daivis, *Phys. Rev. E*, 1995, **52**, 1627.
425. J. Zhang and B.D. Todd, *Phys. Rev. E*, 2004, **69**, 031111.
426. P. Schofield and J. R. Henderson, *Proc. R. Soc. London, A.*, 1982, **379**, 231.
427. M. Han and J. S. Lee, *Phys. Rev. E*, 2004, **70**, 061205.
428. D.R.J. Monaghan and G.P. Morriss, *Phys. Rev. E*, 1997, **56**, 476.
429. P. Cordero and D. Risso, *Physica A*, 1998, **257**, 36.
430. P.J. Daivis and J.L.K. Coelho, *Phys. Rev. E*, 2000, **61**, 6003.
431. J. Zhang and D.Y. Kwok, *Phys. Rev. E*, 2004, **70**, 056701.
432. S.C. Hendy, M. Jasperse and J. Burnell, *Phys. Rev. E*, 2005, **72**, 016303.
433. N.V. Priezjev, A.A. Darhuber and S.M. Troian, *Phys. Rev. E*, 2005, **71**, 041608.
434. V.P. Sokhan and N. Quirke, *Mol. Sim.*, 2004, **30**, 217.
435. A. Jabbarzadeh, P. Harrowell and R.I. Tanner, *Phys. Rev. Lett.*, 2005, **94**, 126103.
436. B.D. Todd, D.J. Evans, K.P. Travis and P.J. Daivis, *J. Chem. Phys.*, 1999, **111**, 10730.
437. A. Baranyai, D.J. Evans and P.J. Daivis, *Phys. Rev. A*, 1992, **46**, 7593.
438. S. Butler and P. Harrowell, *Phys. Rev. E*, 2003, **67**, 051503.
439. S. Sacquin-Mora, A.H. Fuchs and M. Schoen, *Phys. Rev. E*, 2003, **68**, 066103.
440. L.G. Cámara and F. Bresme, *J. Chem. Phys.*, 2004, **120**, 11355.
441. Y. Leng and P.T. Cummings, *Phys. Rev. Lett.*, 2005, **94**, 026101.
442. C. Dellago and M.M. Naor, *Comput. Phys. Commun.*, 2005, **169**, 36.
443. J. Dzubiella and J.-P. Hansen, *J. Chem. Phys.*, 2005, **122**, 234706.
444. S. Supple and N. Quirke, *J. Chem. Phys.*, 2004, **121**, 8571.
445. S. Supple and N. Quirke, *J. Chem. Phys.*, 2005, **122**, 104706.
446. S. Granick, *Science*, 1991, **253**, 1374.
447. H.W. Hu, G.A. Carson and S. Granick, *Phys. Rev. Lett.*, 1991, **66**, 2758.
448. S.T. Cui, C. McCabe, P.T. Cummings and H.D. Cochran, *J. Chem. Phys.*, 2003, **118**, 8941.
449. A. Cavallo, H. Müller and K. Binder, *J. Phys. Chem. B.*, 2005, **109**, 6544.
450. Y.-C. Lin, H. Müller and K. Binder, *J. Chem. Phys.*, 2004, **121**, 3816.
451. J. Yaneva, A. Milchev and K. Binder, *J. Chem. Phys.*, 2004, **121**, 12632.
452. M. Vacatello, *Macromol. Theory Simul.*, 2004, **13**, 30.
453. J.R. Maury-Evertsz, L.A. Estévez and G.E. López, *J. Chem. Phys.*, 2003, **119**, 9925.
454. A. Lamura, T.W. Burkhardt and G. Gompper, *Phys. Rev. E*, 2004, **70**, 051804.
455. A. Sikorski and P. Romiszowski, *J. Chem. Phys.*, 2004, **120**, 7206.
456. J.R. Maury-Evertsz, L.A. Estévez and G.E. López, *J. Chem. Phys.*, 2004, **121**, 8651.

457. F. Varnik, J. Baschnagel, K. Binder and M. Mareschal, *Eur. Phys. J. E*, 2003, **12**, 167.
458. H. Meyer and J. Baschnagel, *Eur. Phys. J. E.*, 2003, **12**, 147.
459. P. Scheidler, W. Kob and K. Binder, *Eur. Phys. J. E*, 2003, **12**, 5.
460. A. Milchev, K. Binder and A. Bhattacharya, *J. Chem. Phys.*, 2004, **121**, 6042.
461. P.J. Park and W. Sung, *J. Chem. Phys.*, 1998, **108**, 3013.
462. J. Chaung, Y. Kantor and M. Kardar, *Phys. Rev. E*, 2001, **65**, 011802.
463. Muthukumar, *J. Chem. Phys.*, 1999, **111**, 10371.
464. C. Loison, M. Mareschal and F. Schmid, *J. Chem. Phys.*, 2004, **121**, 1890.
465. J.D. Lister, *Phys. Lett.*, 1975, **53A**, 193.
466. S.T. Cui, *Mol. Phys.*, 2004, **102**, 139.
467. Y.-L. Chen, M.D. Graham, J.J. de Pablo, G.C. Randall, M. Gupta and P.S. Doyle, *Phys. Rev. E*, 2004, **70**, 060901(R).
468. D. Dubbeldam, E. Beerdsen, T.J.H. Vlugt and B. Smit, *J. Chem. Phys.*, 2005, **122**, 224712.
469. O.G. Jepps, S.K. Bhatia and D.J. Searles, *Phys. Rev. Lett.*, 2003, **91**, 0126102.
470. O.G. Jepps, S.K. Bhatia and D.J. Searles, *J. Chem. Phys.*, 2004, **120**, 5396.
471. S.K. Bhatia and D. Nicholson, *Phys. Rev. Lett.*, 2003, **90**, 016105.
472. S.K. Bhatia and D. Nicholson, *J. Chem. Phys.*, 2003, **119**, 1719.
473. S.K. Bhatia, O.G. Jepps and D. Nicholson, *J. Chem. Phys.*, 2004, **120**, 4472.
474. O.G. Jepps, S.K. Bhatia and D.J. Searles, *Langmuir*, 2005, **21**, 229.
475. R.R. Kotdawala, N. Kazantzis and R.W. Thompson, *J. Math. Chem.*, 2005, **38**, 325.
476. T.X. Nguyen and S.K. Bhatia, *Langmuir*, 2004, **20**, 3532.
477. T.X. Nguyen and S.K. Bhatia, *J. Phys. Chem. B*, 2004, **108**, 14032.
478. S.K. Bhatia, K. Tran, T.X. Nguyen and D. Nicholson, *Langmuir*, 2004, **20**, 9612.
479. S.K. Bhatia, H. Chen and D.S. Sholl, *Mol. Sim.*, 2005, **31**, 643.
480. D.M. Ackerman, A.I. Skoulidas, D.S. Sholl and J.K. Johnson, *Mol. Sim.*, 2003, **29**, 677.
481. J.C. Grossman, E. Schwegler, E.W. Draeger, F. Gygi and G. Galli, *J. Chem. Phys.*, 2004, **120**, 300.
482. J. Vande Vondele, F. Mohamed, M. Krack, J. Hutter, M. Sprik and M. Parrinello, *J. Chem. Phys.*, 2005, **122**, 014515.
483. N. Goldman, C. Leforestier and R.J. Saykally, *Phil. Trans. R. Soc. A*, 2005, **363**, 493.
484. A. Striolo, A.A. Chialvo, P.T. Cummings and K.E. Gubbins, *Langmuir*, 2003, **19**, 8583.
485. M. Dion, H. Rydberg, E. Schröder, D.C. Langreth and B.I. Lundqvist, *Phys. Rev. Lett.*, 2004, **92**, 246401.
486. M.A. Inda and D. Frenkel, *Macromol. Theory Simul.*, 2004, **13**, 36.
487. N. Kikuchi, J.F. Ryder, C.M. Pooley and J.M. Yeomans, *Phys. Rev. E*, 2005, **71**, 061804.
488. T. Kairn, P.J. Daivis, M.L. Matin and I.K. Snook, *Polymer*, 2004, **45**, 2453.
489. S. Sen, S.K. Kumar and P. Keblinski, *Macromolecules*, 2005, **38**, 650.
490. O.G. Jepps and S.K. Bhatia, *Phys. Rev. E*, 2003, **67**, 041206.
491. R.M. Jendrejack, D.C. Schwartz, J.J. de Pablo and M.D. Graham, *J. Chem. Phys.*, 2004, **120**, 2513.

8

Combinatorial Enumeration in Chemistry

BY A. MILIČEVIĆ[a] AND N. TRINAJSTIĆ[b]
[a] *The Institute of Medical Research and Occupational Health, P.O.B. 291, HR-10002 Zagreb, Croatia*
[b] *The Rugjer Bošković Institute, P.O.B. 180, HR-10002 Zagreb, Croatia*

1 Introduction

In the present Report we review the literature on combinatorial enumeration in chemistry published in the last two years – from June 2003 to May 2005. This represents a continuation of our reporting on advances in counting objects in chemistry since June 1999. The first Report was entitled *Enumeration in Chemistry*[1] in which we reviewed historically important enumerations, enumeration methods and the literature published between June 1999 to May 2001. The second Report was entitled *Combinatorial Enumeration in Chemistry*[2] in which we reviewed the literature published between June 2001 and May 2003 and some past achievements related to the discussed results.

As before, we will review current results and will also present the related past results to help the reader to see how the certain area of combinatorial enumeration evolved.

2 Current Results

2.1 Isomer Enumeration. – In the past two years (2003–2005) there were somewhat less reports on the enumeration of isomeric structures unlike the periods between 1999–2001 and 2001–2003.

2.1.1 Chiral Isomers of Alkanes. In a continuation of his enumerative works,[3–5] Lukovits[6] reported constructive enumeration of chiral alkanes. It should be noted that the *constructive enumeration* means counting objects by constructing each object separately using a suitable code.[7,8] In his work, Lukovits represented alkanes by chemical trees. *Chemical trees* are graph-theoretical trees[9] that represent the carbon skeleton of alkanes and in which *no* vertex has a degree greater than 4.[10] The degree of a vertex is the number of edges meeting at this vertex.[11]

Chemical Modelling: Applications and Theory, Volume 4

Lukovits' approach[6] is based on the *lowest degree first* (LDF) tree and the concept of canonical labeling.[12] A LDF tree is a Morgan tree (and consequently a physical tree, as well). A *physical tree*[13] is a tree obtained by assigning labels to its vertices consecutively and each vertex to be labeled must be adjacent to an already labeled vertex. Therefore, each vertex, except the vertex labeled 1, has exactly one neighbor with a lower label. Such labeling results in the vertex-adjacency matrix[10] A of a physical tree that contains only one non-zero element in each column of its upper triangle.[14] This fact allows the vertex-adjacency matrix to be replaced by the condensed representation, called the *compressed adjacency matrix* and denoted by CAM. Each digit in CAM denotes the row in the upper-half of the vertex-adjacency matrix in which digit 1 is placed. Thus CAM is a linear representation of the vertex-adjacency matrix composed of $N-1$ entries and it can be set up directly by inspecting the labeling of a physical tree. It can be also used for coding trees. Physical trees with no vertex-degree greater than 4 have been generated and enumerated using the N-tuple code.[13,15,16] An illustrative example of a physical tree representing the carbon skeleton of 3-methyl-4-ethylhexane, the upper-half of the corresponding vertex-adjacency matrix and the related CAM code is given in Figure 1.

A *Morgan-tree*, so named by Lukovits[3] in honour of H.L. Morgan, who years ago proposed an algorithm for unique labeling of chemical compounds,[17] is a physical tree in which the vertex-labeling must obey an additional restriction with respect to physical trees:[3,18] Each new label must be attached to a vertex adjacent to the vertex labeled with the lowest ordinal. The structure of the vertex-adjacency matrix belonging to a Morgan tree also allows it to be condensed to CAM. The number of Morgan trees is a small fraction of the number of physical trees. Figure 2 displays the Morgan tree representing the carbon skeleton of 3-methyl-4-ethylhexane, the upper-half of the corresponding vertex-adjacency matrix and the related CAM code.

A *LDF tree* is a tree in which a new label is attached to the vertex that has a first neighbour labeled with the lowest ordinal k and which in addition has the lowest degree of all vertices attached to the vertex labeled k. Labeling in LDF trees must start at an end-vertex. The structure of the vertex-adjacency matrix belonging to a LDF tree also allows it to be condensed to CAM. Labeling of a LDF tree is exemplified in Figure 3. This figure displays the LDF tree representing the carbon skeleton of 3-methyl-4-ethylhexane, the upper-half of the corresponding vertex-adjacency matrix and the related CAM code.

If the CAM codes of physical tree A, Morgan tree B and LDF tree C are lexicographically ordered, the obtained order is CAM(A) $<$ CAM (B) $<$ CAM (C). LDF trees, therefore, have the maximal lexicographical value of all CAM codes related to the same molecule; hence they are unique.

Enumeration of chiral alkanes is carried out in the following way. The computer program performing the generation procedure,[3,5,18] starts calculations by generating the LDF trees with N vertices, and then sorts out all chiral trees. To determine whether the LDF code represents a chiral tree, all vertices of degrees three or four need to be inspected. If two branches attached to these vertices are identical, the tree is *not* a chiral tree. To accomplish this task, two

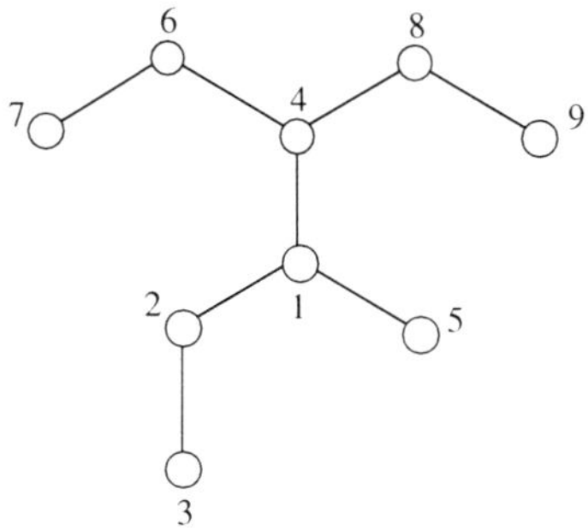

A

$$
\begin{array}{ccccccccc}
0 & 1 & 0 & 1 & 1 & 0 & 0 & 0 & 0 \\
 & 0 & 1 & 0 & 0 & 0 & 0 & 0 & 0 \\
 & & 0 & 0 & 0 & 0 & 0 & 0 & 0 \\
 & & & 0 & 0 & 1 & 0 & 1 & 0 \\
 & & & & 0 & 0 & 0 & 0 & 0 \\
 & & & & & 0 & 1 & 0 & 0 \\
 & & & & & & 0 & 0 & 0 \\
 & & & & & & & 0 & 1 \\
 & & & & & & & & 0
\end{array}
$$

12114648

Figure 1 *Physical tree depicting the carbon skeleton of 3-methyl-4-ethylhexane, the upper-half of the corresponding vertex-adjacency matrix and the CAM code*

rules are used to delete non-chiral structures and to determine whether two branches are identical or not. For example, among the 9 heptane trees, 18 octane trees, 35 nonane trees and 75 decane trees there are 2, 5, 15 and 40 trees, respectively, possessing at least one chiral vertex. As an illustration, all chiral octane trees are given in Figure 4.

2.1.2 Counting Disconnected Structures. A great deal of research has been done on counting connected structures, *e.g.*[1,2] However, Petkovšek and Pisanski[19] presented a method, using the graph-theoretical language,[9–11] for enumerating disconnected graphs (structures) with a given number vertices if the number of connected graphs is known. They have also shown that the method can be used in reverse direction, that is, if the numbers of *all* structures are known, the number of connected structures can be readily determined. Their method can be summarized as follows. If $F(x)$ is the generating function of the

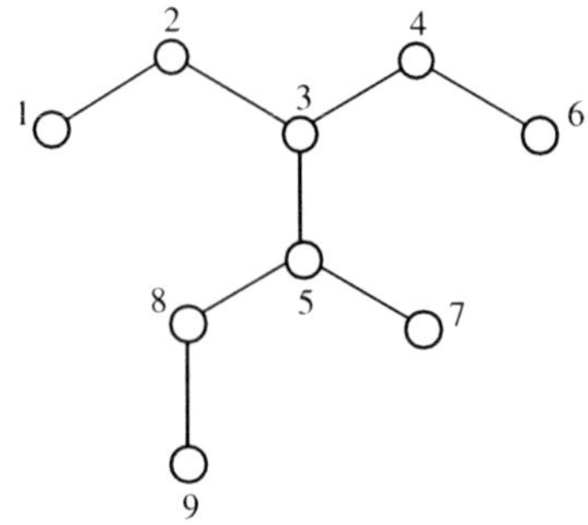

B

$$
\begin{array}{ccccccccc}
0 & 1 & 0 & 0 & 0 & 0 & 0 & 0 & 0 \\
 & 0 & 1 & 0 & 0 & 0 & 0 & 0 & 0 \\
 & & 0 & 1 & 1 & 0 & 0 & 0 & 0 \\
 & & & 0 & 0 & 1 & 0 & 0 & 0 \\
 & & & & 0 & 0 & 1 & 1 & 0 \\
 & & & & & 0 & 0 & 0 & 0 \\
 & & & & & & 0 & 0 & 0 \\
 & & & & & & & 0 & 1 \\
 & & & & & & & & 0
\end{array}
$$

12334558

Figure 2 *Morgan tree depicting the carbon skeleton of 3-methyl-4-ethylhexane, the upper-half of the corresponding vertex-adjacency matrix and the CAM code*

counting sequence of connected graphs, then the corresponding generation function $G(x)$ of the counting sequence of all graphs is given by:[20]

$$G(x) = \exp \sum_{k \geq 1} F(x^k)/k \tag{1}$$

Applying a variant of the Möbius inversion to equation (1), it is possible to express $F(x)$ in terms of $G(x)$:

$$F(x) = \sum_{k \geq 1} [\mu(x)/k] \log G(x^k) \tag{2}$$

where μ stands for the Möbius function.[20]

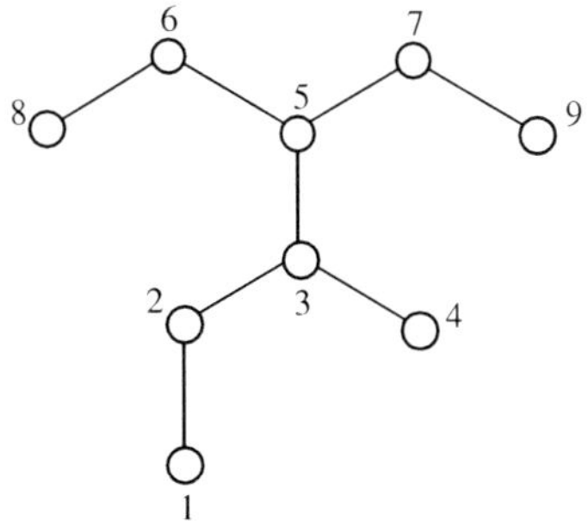

C

0	1	0	0	0	0	0	0	0
	0	1	0	0	0	0	0	0
		0	1	1	0	0	0	0
			0	0	0	0	0	0
				0	1	1	0	0
					0	0	1	0
						0	0	1
							0	0
								0

12335567

Figure 3 *LDF tree depicting the carbon skeleton of 3-methyl-4-ethylhexane, the upper-half of the corresponding vertex-adjacency matrix and the CAM code*

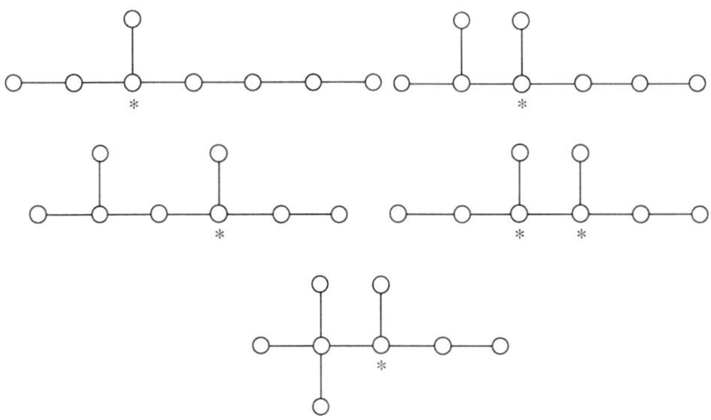

Figure 4 *All chiral octane trees. Asterisks denote chiral vertices*

An illustrative example is given below. The number of connected simple graphs (that is, graphs without loops and multiple edges)[11] with N vertices is given by:

$$F(x) = x + x^2 + 2x^3 + 6x^4 + 21x^5 + 112x^6 + 853x^7 + 11117x^8 + 261080x^9 + 11716571x^{10} + 1006700565x^{11} + \dots \quad (3)$$

and the number of all simple graphs with N vertices is given by:

$$G(x) = x + 2x^2 + 4x^3 + 11x^4 + 34x^5 + 156x^6 + 1044x^7 + 12346x^8 + 274668x^9 + 12005168x^{10} + 1018997864x^{11} + \dots \quad (4)$$

while the corresponding sequence of disconnected simple graphs is:

$$0, 1, 2, 5, 13, 44, 191, 1229, 13588, 288597, 12297299, \dots \quad (5)$$

Figure 5 gives as an example *all* simple graphs with *five* vertices.

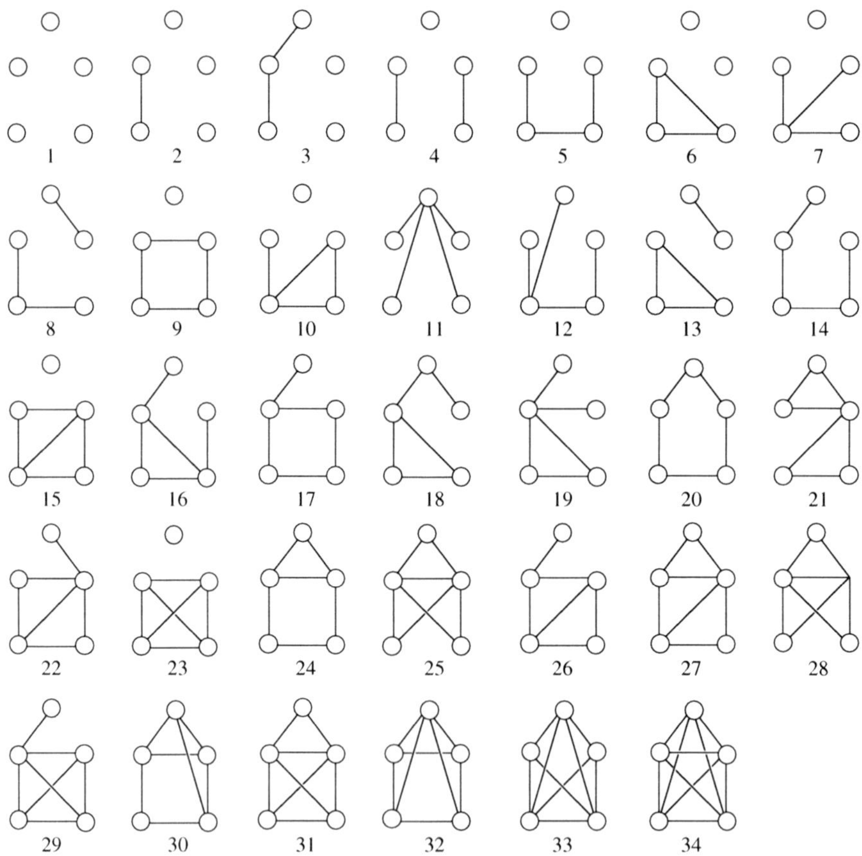

Figure 5 *All simple graphs with five vertices*

Numbers given above in (3) and (4) can be checked in the *On-Line Encyclopedia of Integer Sequences* – see sequences A001349 (3) and A000088 (4). This *Encyclopedia* is a marvellous, free of charge, source of a great many integer sequences published electronically by Neil J.A. Sloane in 2005 at the address http://www.research.att.com/njas/sequence/. It should be appreciated by all those who do combinatorial enumeration or any other research in which there is need for integer sequences. The number of all connected and all simple graphs with up to nine vertices can be also found in the book *Graphical Enumeration* by Harary and Palmer.[21] However, the number of all simple graphs with nine vertices (308708) in Harary's masterpiece *Graph Theory*[9] must be a misprint since the correct number is 274668.

2.1.3 Isomeric Fullerenes. Fullerenes C_N are carbon cages built from fused pentagons and hexagons. With increasing the value of N, the number of possible fullerenes increases considerably. The smallest fullerene is C_{20} (there was a recent report on the existence of this fullerene)[22] and C_{22} fullerene is not possible. In the mentioned *On-Line Encyclopedia of Integer Sequences*, the following numbers of isomeric fullerenes are given starting with C_{20} and increasing the subscript by 2 – see sequence A007894:

$$1, 0, 1, 1, 2, 3, 6, 6, 15, 17, 40, 45, 89, 116, 199, 271, 437, 580, 924, 1205, 1812, 2385, \ldots \quad (6)$$

These numbers have been known for some time, *e.g.*[23–28] Nevertheless, enumeration of fullerenes, notably higher fullerenes with $N > 100$ is still being pursued. Thus, Livshits and Lozovik[29] introduced the *cut-and-unfold* approach to fullerene enumeration. Their approach combines the formalism of a fullerene graph cut-and-fold onto a planar triangular lattice and a topological description of closed quasi-2D clusters. The authors have tested their approach against the available data[27] and enumerated fullerenes for $100 \le N \le 110$ (see Table 1) and fullerenes that satisfy the isolated pentagon rule (IPR)[30,31] for $102 \le N \le 150$ (see Table 2). The IPR predicts fullerenes with all the pentagons isolated by hexagons to be more stable than those with abutted pentagons.

Table 1 *Fullerene-numbers (FN) for $100 \le N \le 110$ obtained by the cut-and-unfold approach. These results are taken from ref. 29*

N	*FN*
100	285 914[a]
102	341 658[a]
104	419 013[a]
106	497 529
108	604 217
110	713 319

[a] Numbers labeled by *a* are also given in the *On-Line Encyclopedia of Integer Sequences*.

Table 2 *Numbers of isolated-pentagon fullerenes (NIPF) for $102 \leq N \leq 150$ obtained by the cut-and-unfold approach. These results are taken from ref. 29*

N	*NIPF*
102	616
104	823
106	1233
108	1799
110	2355
112	3342
114	4468
116	6063
118	8148
120	10774
122	13977
124	18769
126	23589
128	30683
130	39393
132	49878
134	62372
136	79362
138	98541
140	12354
142	151201
144	186611
146	225245
148	277930
150	335569

Buckminsterfullerene, the C_{60} isomer,[32] is the smallest possible fullerene that obeys the IPR and the most stable one among the 1812 cages with 60 carbon atoms.[33,34]

2.1.4 Mathematical Models of Isomerism. Papers in this area are highly mathematical and often difficult to read. Here we report on publications by Iliev, Fujita and Balasubramanian that have appeared in the last few years.

Lunn and Senior[35] proposed many years ago a mathematical model of isomerism in organic chemistry which Iliev generalized in recent years.[36–38] Lunn and Senior considered three types of isomerism: (i) univalent substitution isomerism (positional isomerism), (ii) skeletal (structural) isomerism and (iii) enantiomorphism. *Univalent substitution isomerism* is the relationship existing between any two compounds A_1 and A_2 with the same empirical formula in case that structural formula of A_1 can be converted into that of A_2 by a permutation of the univalent substituents without disturbing the skeleton. Then the molecules A_1 and A_2 are said to be *univalent substitution isomers*. For example, 1-chloropropane and 2-chloropropane are univalent substitution (positional) isomers. *Skeletal isomers* are any two compounds A_1 and A_2 with the same empirical formula that differ in the connectivity (Lunn and Senior used the older term connexity) relations within their respective structures. Or in other words, *skeletal isomerism* is the relationship that exists between compound A_1 and the isomeric compound A_2 if the skeleton of A_1 cannot be converted into that of A_2 without disturbing its skeleton. Then the molecules A_1 and A_2 are said to be skeletal isomers. For example, butane and isobutane are skeletal isomers. *Enantiomorphism* or optical isomerism is a special form of stereoisomerism. A pair of molecules is defined as enantiomeric if one molecule in the pair is not a superimposable mirror image of the other.

Substitution isomerism, structural isomerism and stereoisomerism define three equivalence relations on the set T_λ of all structural formulas $A = (A_1, A_2, \ldots)$ with a fixed empirical formula, skeleton and with d univalent substituents. On the other hand, the symmetric group S_d and any of its subgroups acts naturally on the set T_λ. The basic idea of the Lunn-Senior model is that one can obtain these equivalence relations by means of the action of three permutation groups $G, G', G'' \leq S_d$, respectively, and G is not generally the 3-dimensional symmetry group of the molecule. Thus, Lunn and Senior[35] supposed that the number of orbits constitutes the upper boundaries of the number of isomers obtained by the experiment (though the computed and experimental values often coincide). In his first paper (part I),[36] Iliev defined precisely the set where the above groups act (structural formulas are identified with the tabloids with d nodes) and introduced a partial order that models the substitution reactions between isomers. This partial order can be used for prediction of certain substitution reactions and for proving the impossibility of such reactions. For any permutation group $W \leq S_d$ (W could be, for example, any of G, G' or G'') and for any one-dimensional character χ of W, Iliev defined W-orbits with a special maximum property, which in the case $W = G'$ and χ is the two-valued character with kernel $G \leq G'$, specializes to the chirality. Thereby Iliev was also able to get a type property[35] of the molecule under consideration. Note that the type properties of the molecule do not depend on the nature of the univalent substituents. Moreover, Iliev established a formula for counting these special W-orbits. Thus, in particular, one obtains the well-known inventory of the enantiomorphous pairs. In case that χ is the unit character of W, one gets a formula for counting all W-orbits. Iliev also generalized the formula obtained by Ruch *et al.*[39] for the number of isomers corresponding to a given partition of d. In addition, the results of Ruch *et al.*[39] which relates the dominance order between partitions and the existence of chiral pairs is reached as a consequence of a much more general statement.

Iliev devoted the second part[38] of his generalization of the Lunn-Senior mathematical model of isomerism to identification of the isomers of a given molecule with their structural formulas on the basis of their substitution reactions. A classical prototype of this identification is the Körner scheme of substitution reactions[40] between the three di-substituted and the three tri-substituted homogeneous products of benzene, which allows all these products to be identified as ortho, meta, para, vicinal, asymmetric, symmetric, respectively, in agreement with the combinatorics of the substituent positions on the six-membered ring.

In his pioneering work, Körner was concerned with the dibromo- and tribromo-benzenes.[40] A few words about now almost forgotten Körner are in order. Wilhelm Guglielmo Körner (1839–1925) was a student of Friedrich August Kekulé (1824–1896) and professor of chemistry at the University of Milan. He devised one of the chemical proofs of the equivalence of the six hydrogen atoms of benzene and was the first to propose the accepted formula for pyridine. It should be noted that the problem of assigning structures to isomeric benzene derivatives was known as orientation in the benzene nucleus. This was solved in two ways; one way is known as the Körner absolute method (the Körner scheme of substitution reactions) and the other as the interconversion method.[41] Illiev developed the algebraic approach within the framework of the Lunn-Senior model that allows for all couples of isomers that can and that cannot be distinguished by means of substitution reactions to be found. He also introduced distinguishability by means of characters, proved that the enantiomers cannot be distinguish either by substitution reactions or by characters and discussed the examples of ethylene, benzene and cyclopropane.

In a related paper, Iliev[37] considered organic compounds with one mono-substituted and at least three di-substituted homogeneous derivatives. Examples of molecules studied were benzene and cyclopropane. The Lunn–Senior permutation groups of substitutional isomerism of these compounds were described and upper bounds of the numbers of their di-substituted and tri-substituted homogeneous derivatives were found. Lists of the possible simple substitution reactions among di-substituted homogeneous derivatives, di-substituted heterogeneous and tri-substituted homogeneous derivatives were given. Substitution reactions were found that allowed some derivatives to be identified by their structural formulas.

Very active in devising mathematical models of isomerism is Shinsaku Fujita. His aim is to restructure stereochemistry by clarifying ambiguous concepts. Fujita's efforts in the past were directed toward defining precisely the basic stereochemical concepts: *chirality* and *stereogenicity*,[42,43] since in many textbooks the difference between these two fundamental concepts was not fully demonstrated, *e.g.*[44,45] Stereochemistry is concerned with at least two aspects of organic molecules: One is the *geometric* aspect known as chirality and the other is the *nomenclature* aspect called stereogenicity. The geometrical aspect was studied by point groups which are well-suited to treat continuous objects. But point groups are unsuitable for treating those stereochemical problems which are characterized by a discrete nature, such as isomer classification and combinatorial enumeration. Stereogenicity was related to the *RS*-nomeclature,[46–48] where the permutability of ligands determines the *R*- and *S*-descriptors. Such permutability was treated by permutation groups, which appear to be insufficient to treat chirality.[49] In order to discuss chirality and stereogenicity from a common point, Fujita[43] introduced the concept of stereoisomeric groups, which is a combination between the coset representation of point groups and permutation groups. He then proposed the concepts of *holantimers* and *stereoisograms* and concluded that *RS*-stereogenicity is more useful than stereogenicity in discussing the *R,S*-nomeclature.[50] Above we stated when two molecules make an enantiomeric pair. Additionally, two molecules can be defined to be enantiomeric as follows. When a molecule is regarded as a skeleton with ligands, its mirror image comprises the corresponding mirror skeleton and the corresponding ligands of opposite chirality. On the other hand, two molecules based on the same skeleton are defined as being *holantimeric* if the chirality of each ligand in one molecule of the pair is changed into opposite chirality to give the other molecule based on that same skeleton. These two molecules are then said to be holantimers. Roots of the terms *holantimers* and *holantimeric*, proposed by Fujita, are Greek words *holo* or *hol* meaning *whole* or *entire*, and *anti* meaning *opposite to*. Fujita introduced the concept of *stereoisogram* in order to correlate a set of stereoisomers based on holantimeric and enantiomeric relationships. The term *stereoisogram* results from combining the terms *stereoisomer* and *isogram*. Stereogenicity, characterized by a stereoisogram and called *RS*-stereogenicity, has been identified as a property that should be correlated with chirality.

The above concepts were applied to tetrahedral organic molecules and to square-planar inorganic complexes.[50,51] Fujita earlier[52] enumerated tetrahedral molecules with various substituents with respect to their molecular formulas and their point-group symmetries. He also classified them in terms of permutation-group symmetries in addition to point-group symmetries.[49] Using the concepts of holantimers and stereoisograms, Fujita[50] examined all the enumerated tetrahedral molecules and found that their stereoisograms can be classified into five stereogenecity types: Type I (chiral/*RS*-stereogenic), Type II (chiral/*RS*-astereogenic), Type III (chiral/*RS*-stereogenic), Type IV (achiral/*RS*-astereogenic) and Type V (achiral/*RS*-stereogenic). The stereoisograms of square-planar complexes were classified into two types: Type II and Type IV. These are called *RS*-stereogenicity types since each depends on whether it can be specified by the *RS*-nomeclature or not. Later Fujita[53] was able to give a proof for the existence of five stereogenecity types that was based on the existence of five factor groups derived from *RS*-stereoisomeric groups. In the same paper, Fujita also put the concepts of holantimers and stereoisograms on the firm mathematical basis.

However, further questions emerged from this work on the relationship between stereogenicity and chirality,[50,51] *i.e.*, how stereogenicity and *RS*-stereogenicity (or stereoisomeric groups and *R,S*-stereoisomeric groups) differ and what happens if they are identical. Since these questions remained after his approach was applied to tetrahedral molecules, Fujita applied his approach to other types of molecules. In a two-part series of papers[54,55] he studied allene derivatives and square-planar complexes in more detail. In the first part,[54] Fujita, like in the case of tetrahedral molecules, discussed chirality, *R,S*-stereogenicity, stereogenicity and skeletal isomerism of allene derivatives using his terminology for groups (point groups, *R,S*-permutation groups, *R,S*-stereoisomeric groups and isoskeletal groups) and for isomers (enantiomers, holantimers, *R,S*-diastereomers, diastereomers and isoskeletal isomers). In the case of allene derivatives, Fujita established that *R,S*-stereoisomeric groups coincide with stereoisomeric groups so that diastereomers are identical with

R,S-diastereomers. In this first paper of the series,[54] Fujita also proposed the concept of *stereoisogram set* to discuss the relationship between (R,S-)diastereomers and isoskeletal isomers. Each stereoisogram set for allene derivatives contains three stereoisograms to represent isoskeletal isomerism and each of these stereoisograms was classified into five stereogenicity types (Types I–V). Combination of these stereogenicity types can be used to characterize the isomerism of allene derivatives. Fujita[54] also discussed the case of the tetrahedral skeleton, for which he found the following hierarchy of groups: point groups $\subset$ *R,S*-stereoisomeric groups = stereoisomeric groups = isoskeletal groups.

In the second part,[55] Fujita found the following hierarchy of groups for square-planar complexes: point groups = *R,S*-stereoisomeric groups $\subset$ stereoisomeric groups = isoskeletal groups. It appears that the *R,S*-nomenclature is not applicable to square-planar complexes because *R,S*-stereoisomeric groups coincide with point groups, so that all square-planar complexes are determined to be *R,S*-astereogenic. Further, there exist no isoskeletal isomers for square-planar complexes, because isoskeletal groups coincide with stereoisomeric groups. In order to discuss stereogenicity, Fujita proposed the concept of *extended stereoisogram*, which contains three degenerate stereoisograms. Applying this concept, square-planar complexes were classified into types (Types II-II-II, IV-IV-IV, *etc.*) on the basis of relevant stereoisograms (Types I–V). The meaning of the number *three* of degenerate stereoisograms in an extended stereoisogram is that square-planar complexes cannot be named by a dichromous nomenclature like *R,S*-nomenclature or *E*/*Z*-nomeclature.

Fujita[56] also discussed the *E*/*Z*-nomenclature for ethylene derivatives using group-theoretical arguments. He comprehensively discussed chirality, *R,S*-stereogenicity, stereogenicity and isoskeletal isomerism for ethylene derivatives by considering point groups, *R,S*-stereoisomeric groups, stereoisomeric groups and isoskeletal groups. He selected ethylene derivatives as a typical example of general cases in which stereoisomeric groups should be considered in addition to *R,S*-stereoisomeric groups. He established that the *R,S*-stereoisomeric groups for ethylene derivatives do not coincide with their stereoisomeric groups, so that diastereomers (*E*/*Z*-isomers) are not identical with *R,S*-diastereomers. In order to consider the relationships amongst *R,S*-diastereomers, *m*-diastereomers (cis/trans-isomers) and isoskeletal isomers, Fujita utilized the concepts of extended stereoisograms and extended stereoisogram sets. Using these concepts, Fujita classified ethylene derivatives into Types II-II/II-II/II-II, IV-IV/IV-IV/IV-IV, *etc.* on the basis of relevant stereoisograms (Types I-V). Fujita concluded that the stereoisomerism of ethylene derivatives should be treated in terms of *m*-diastereomers characterized by *E*/*Z*-nomeclature and not to be treated in terms of *R,S*-diastereomers characterized by *R,S*-nomenclature.

In a separate paper, Fujita jointly with Sherif El-Basil[57] used the concept of doubly-colored graphs to visualize the abstract concepts such as subductions of coset representations, double cosets, and unit-subduced-cycle-indices,[58] which had been mathematically formulated in the framework of coset algebraic theory developed by Fujita.[42]

In another two-part series,[59,60] Fujita presented the stereochemical extension of the Pólya theorem.[61] In the first part,[59] he points out that the Pólya theorem is concerned with graphs and not with molecules although the key paper of Pólya[61] contains also chemical compounds in the title: *Kombinatorische Anzahlbestimmungen für Gruppen, Graphen and chemische Verbindungen* (translated later into English as *Combinatorial Enumeration of Groups, Graphs, and Chemical Compounds*). The basic assumption was that chemical compounds can be regarded as kinds of graphs. Although this assumption works in many cases, there are problems when the Pólya theorem is applied to chiral/achiral ligands. Fujita set a task to correct this situation. In the past, several approaches were proposed to treat the chirality/achirality of ligands. Hässelbarth and Ruch[39,62–64] reported an approach based on double cosets of permutation groups, whilst Fujita[42] proposed the unit-subduced-cycle-index approach. In this approach, the author has shown that the sphericity of an orbit is important to stereochemical discussion when taking account of chirality/achirality of ligands[49,65] and that the chirality fittingness due to sphericity produces the unit-subduced-cycle-index with the chirality fittingness that can be applied to combinatorial enumerations.[66]

In this work,[59] Fujita reexamined the original definition of the sphericity concept in terms of conjugacy classes so that the Pólya theorem could be substantially extended in combination with the sphericity concept. To take account of chiral ligands together with achiral ligands, permutations of the coset representations for cyclic subgroups have been classified into proper and improper elements. This has resulted in classifying the k-cycle contained in each permutation as an enantiospheric, homospheric or hermispheric cycle. Sphericity indices were then defined according to the enantiospheric, homospheric or hermispheric character of each k-cycle. Now, using sphericity indices, cycle indices with chirality fittingness have replaced the Pólya cycle indices. The use of cycle indices with chirality fittingness to enumerate stereoisomers with chiral and achiral ligands has been illustrated on allene derivatives.

In the second part of the series,[60] Fujita proposed a proligand approach to enumerate nonrigid isomers. The proligand approach is based on the concepts of proligand and promolecule[52] and it takes into consideration both chiral and achiral (pro)ligands. Fujita used the extended sphericity indices of k-cycles, which were also been defined according to the enantiospheric, homospheric or hermispheric character of each k-cycle. Then, as before, the cycle indices with chirality fittingness have been defined so as to enumerate nonrigid stereoisomers with chiral and achiral ligands. The proligand approach was applied to tetramethylmethane and compared to the Pólya approach with the conclusion that the Pólya theorem is applicable to graphs and his (Fujita's) approach to nonrigid molecules.

Balasubramanian[67] reported all irreducible representations of the octahedral (cubic) symmetry for vertex-, face- and edge-colorings of a pair of closely related Platonic bodies, the octahedron and its dual the cube using multinomial combinatorics. Combinatorial tables for all irreducible representations and all multinomial partitions were constructed and visualized by the Young tableaux.

These enumerations constitute multinomial expansions of character-based cycle index polynomials and grow in combinatorial complexity as a function of vertex- or edge-coloring partitions. It should be noted that the Pólya combinatorial theory of enumeration[61] provides the generating function for multinomial expansion in terms of the ordinary cycle index of a group which is the sum of all orbits of permutation of a group when it acts on a given set divided by the order of the group. Several authors have independently generalized the cycle index, *e.g.*[68–70] Balasubramanian has not only generalized the cycle index, which he called the generalized character cycle index, but also provided a physical and geometric interpretation of the numbers generated.[67,71,72]

Balasubramanian[73,74] also carried out combinatorial and group-theoretical analyses of a $C_{48}N_{12}$ dodecaazafullerene. The IUPAC name of this compound is rather unwieldy (8,13,18,23,26,29,32,35,40,45,50)-dodeca[60-S_6]fullerene. Balasubramanian[75] found earlier that there are 233,227,974,475 possible isomers of this compound. Many of these isomers are not expected to have much of the life-time, since if two nitrogen atoms are adjacent in an isomer, there will be nitrogen-nitrogen repulsion that would destabilize the structure. Balasubramanian[73] presented the correlation table of the rotational levels of $C_{48}N_{12}$ into the S_6 group and also its induction into the I_h parent group. He predicted the properties of the ^3A_u, ^3E_g, and ^3E_u excited states of $C_{48}N_{12}$ that lie 1.9 eV above the ^1A_g ground state and also predicted that the ^3E_g and ^3E_u states would undergo the Jahn-Teller distortion into chiral structures with no symmetry and an achiral structure with C_1 symmetry. In a related paper, Balasubramanian[74] presented the nuclear spin statistics of $C_{48}N_{12}$: the nuclear spin multiplets and statistical weights of ^{14}N spin-1 bosons, and the ^{13}C nuclear spin species and spin statistical weights of $^{13}C_{48}N_{12}$.

Balasubramanian[76] also considered combinatorial and group-theoretical analysis of three $C_{60}H_{36}$ isomers. The first paper on $C_{60}H_{36}$ appeared in 1990, in which Haufler *et al.*[77] demonstrated the formation of $C_{60}H_{36}$ through low-temperature Birch reduction of C_{60} and since that time $C_{60}H_{36}$ has been studied by a number of authors. Balasubramanian[75] predicted in 1991 that there were over 600 trillion isomers including chiral isomers of $C_{60}H_{36}$ (the exact number of isomers being 600,873,146,368,170). Among them, there are only four T isomers without chirality (or eight isomers with chirality), 31280 C_3 isomers[78] whilst the majority of $C_{60}H_{36}$ isomers belong to the C_1 group. The current report by Balasubramanian was stimulated by a recent paper of Gakh *et al.*[79] in which the authors presented NMR spectra of three isomers of $C_{60}H_{36}$ obtained using the high temperature hydrogenation reaction of C_{60}. Balasubramanian obtained the generating functions for nuclear spin species of protons and ^{13}C nuclei. Special cases of generating functions produce the NMR spectral patterns. The predicted NMR results for the three $C_{60}H_{36}$ isomers with C_1, C_3 and T symmetries agreed with the experimental results. Balasubramanian also predicted the ESR hyperfine patterns of radicals obtained from the three $C_{60}H_{36}$ isomers and gave complete tables of the nuclear spin species of these isomers.

Balasubramanian[80] additionally prepared a nice and very informative author's review on the applications of enumerative combinatorics to chemistry and spectroscopy for a special issue of *Advances in Quantum Chemistry*, dedicated to Chemical Graph Theory (guest editor: D.J. Klein). The author first presented the essence of enumerative combinatorics by means of the Pólya theorem,[61] generalization of the Pólya theorem to all irreducible representations,[81–83] the de Bruijn[84,85] extension of the Pólya theorem and the Balasubramanian generalization of the de Bruijn extension to all characters of any finite permutation group.[86] He then showed how these techniques can produce full multinomial combinatorics of all irreducible representations of any group. As an illustration, he used octahedral (cubic) and icosahedral symmetries as a function of partitions for vertex-, face- and edge-coloring. He has also given full combinatorial tables for all irreducible representations and all multinomial representations. Application was illustrated on fullerenes. This Balasubramanian's review may be regarded as a sequel to his 1985 review[87] which followed the path set by Rouvray's early review on enumeration in chemistry.[88]

2.1.5 Stereoisomers of Oligoinositols. In our last Report,[2] we presented enumerative works on oligoinositols by Dolhaine *et al.*[89,90] and by Dolhaine and Hönig.[91,92] The problem of counting oligoinositols was started by the question posed by Hudlicky *et al.*[93] These authors wondered how many stereoisomers exist of diinositols (**b**) and triinositols (**c**) existed. See the considered oligoinositols in their Haworth projections in Figure 6.

Stereoisomerism of **a**–**e** is due to tetrahedral stereocenters (chiral centers). The number of stereoisomers depends on the possibilities of distributing C–O bonds above and below the average planes of the cyclohexane rings. Because of the rapid interconversion of conformers, each cyclohexane ring can be schematized by a planar hexagon (see Figure 6). Since the considered oligoinositols exhibit high constitutional symmetry, many conformers are identical; thus the number of distinct stereoisomers is much less than 2^c, where c is the number of chiral centers.

Hudlicky *et al.*[93] gave two answers to their question for **b**: $2^{12} = 4096$ stereoisomers (theoretical result) and 990 stereoisomers (true result). The authors did not explain how they reached the latter number. Dolhaine *et al.*[89] obtained 528 as the number of stereoisomers of **b** by applying their program ISOMERS. In one of the two subsequent papers, Dolhaine and Hönig[92] published tables of oligoinositol isomers up to tetramers. But, they did *not* answer the question regarding the number of stereoisomers of **c**. They produced, however, the number of all stereoisomeric linear triinositols, *i.e.*, the sum of stereoisomers enumerations for **c** and its constitutional isomers **d** and ***e***. They also asked about the number of achiral stereoisomers among all stereoisomeric diinositols and triinositols and treated this problem by inspecting many thousands of stereoisomers.

Rücker *et al.*[94] provided the answer. These authors used three methods to count the stereoisomers of oligoinositols: (i) Manual exhaustive construction of all stereoisomers, (ii) Counting without construction by using the

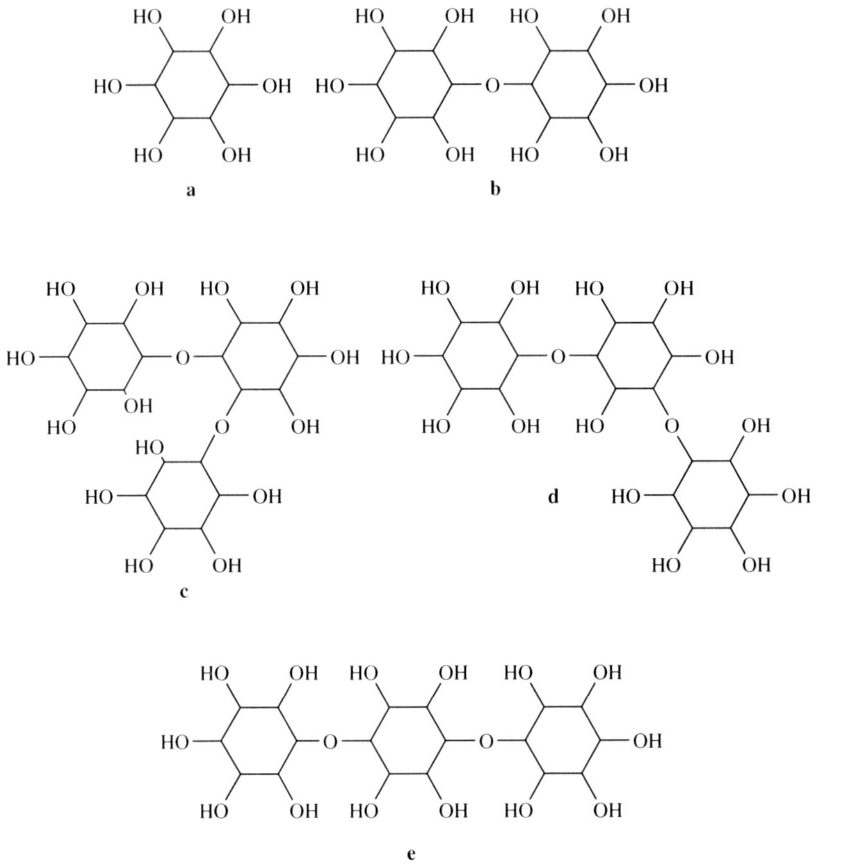

Figure 6 *Considered diinositols and triinositols in their Haworth projections*

Cauchy-Frobenius lemma (or the Burnside lemma)[95] and (iii) Application of the computer program MOLGEN 3.5.[96–99] Actually, the paper of Rücker *et al.*[94] aimed to show the potential of constructive enumeration for problems with up to a certain level of complexity, especially if it can be helped by symmetry arguments. The other two methods are used to support and check the results obtained by manual enumeration. Rücker *et al.*[94] obtained the following results: (1) There are 9 monoinositols (**a**) and 7 are achiral, (2) there are 528 diinositols (**b**) and 48 are achiral (Dolhaine *et al.*[89] give 48 in their 1999 paper whilst in the second[92] of their 2002 papers Dolhaine and Hönig give 46 as the number of achiral isomers of **b**) and (3) there are 82176 triinositols (**c**–**e**) with only a small fraction of isomers being achiral (768). All three counting methods used produced the same stereoisomer counts, but the differentiation between chiral and achiral stereisomers was not implemented in the version of MOLGEN employed in the paper.

2.2 Kekulé Structures. – Kekulé structures are still attracting attention of chemists and mathematicians. In the language of mathematics, Kekulé structures are often called *perfect matchings*.[100–103] Matching theory is a well-developed part of graph theory,[100] but it is still of interest,[104–107] especially in the theory of benzenoid hydrocarbons, *e.g.*[108–112] The importance of Kekulé structures, among other concepts, in the graph-theoretically formulated electronic-structure theory was recently discussed by Klein.[113] Kekulé structures are essential in the conjugated-circuit model of aromaticity[114–116] and in the Simpson-Herndon variant[117,118] of the valence-bond resonance theory.[119] It should be also noted that a widely used topological index,[120,121] called the *Hosoya index*,[10] is based on matchings in a molecular graph.[122] Similarly, the key concept in the topological resonance energy (TRE) is the *matching polynomial*.[123–126] The TRE is still used in assessing the aromatic character of conjugated systems, in recent years especially of fullerenes, *e.g.*[127] It should also be noted that the absolute value of the constant coefficient in the matching polynomial yields the number of Kekulé structures of a given conjugated system. Balasubramanian,[128] for example, calculated matching polynomials of fullerenes C_{20}–C_{50} and thus also produced their *K*-numbers. His results were, for example, utilized by Torrens in his work on structural properties of fullerenes.[129–134]

2.2.1 Benzenoid Hydrocarbons. Došlić[135] reported counting Kekulé structures in benzenoid parallelograms and triangular benzenoids using an approach based on the theory of lattice animals and lattice paths. A *lattice animal* is a 1-connected collection of congruent regular polygons arranged in a plane such that two polygons are either completely disjoint or have a common edge. A benzenoid in the parallelogram-like shape, called the *benzenoid parallelogram*, $B_{m,n}$, consists of $m \times n$ hexagons, arranged in m rows, each row containing n hexagons, shifted by half a hexagon to the right from the row immediately below. A *triangular benzenoid* T_m consists of m rows with the number of hexagons in each row decreasing by one from the lowest to one hexagon in the uppermost row, each row shifted by one and a half hexagon to the right from the row immediately below.

Došlić[135] proved that there is a one-to-one correspondence between the number of Kekulé structures in a benzenoid parallelogram and the number of all square-lattice paths from (0,0) to (n,m) with steps (1,0) and (0,1). As illustrative examples we give all Kekulé structures of anthanthrene in Figure 7 and the corresponding square-lattice paths in Figure 8.

A consequence of the above correspondence is the exact formula for counting the number of Kekulé structures K of benzenoid parallelograms:

$$K(B_{m,n}) = K_{m,n} = \binom{m+n}{n} = \binom{m+n}{m} = \frac{(m+n)!}{m!n!} \tag{7}$$

The same result was obtained by Gordon and Davison[136] in 1952, but it was derived in a less formal way than by Došlić. The application of equation (7) is shown in Figure 9.

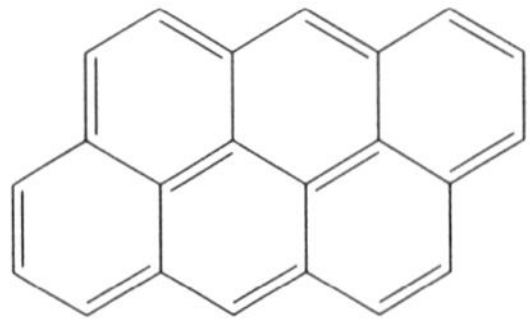

1

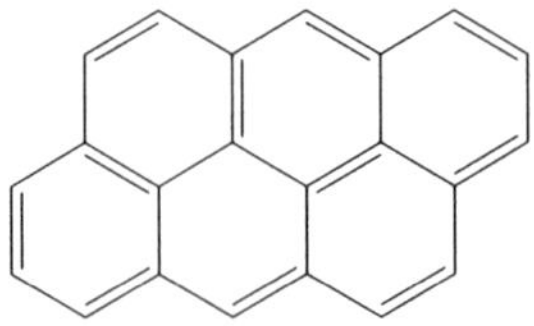

2

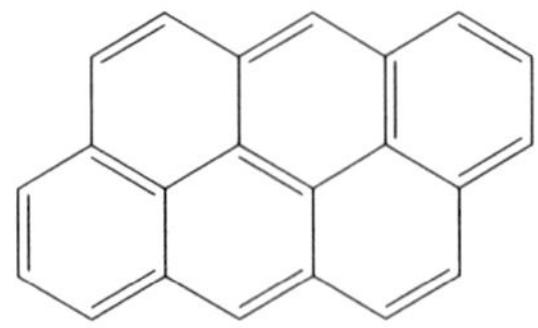

3

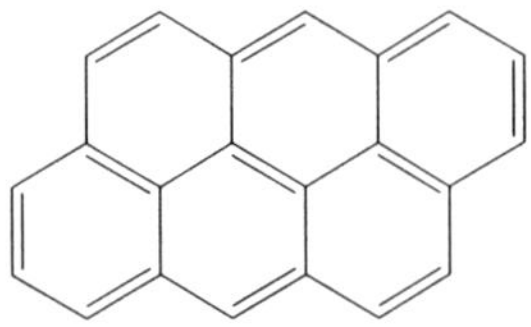

4

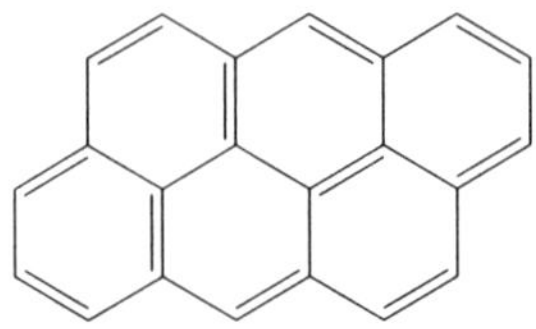

5

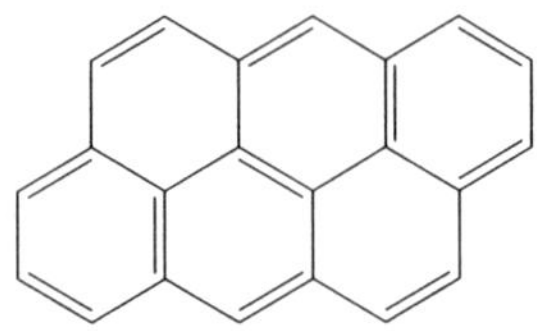

6

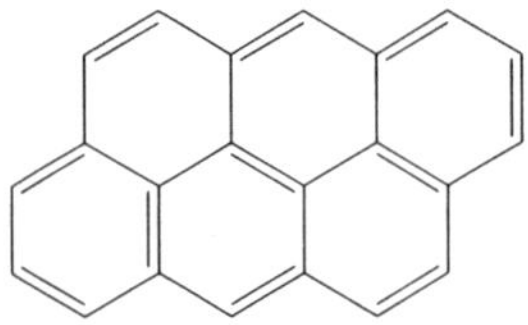

7

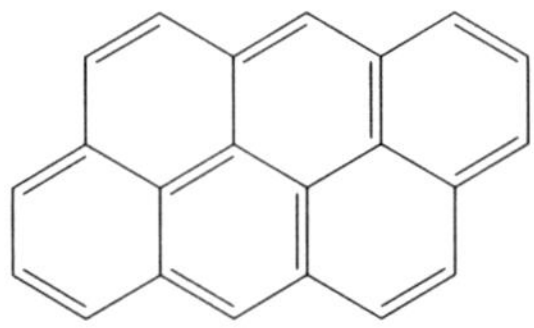

8

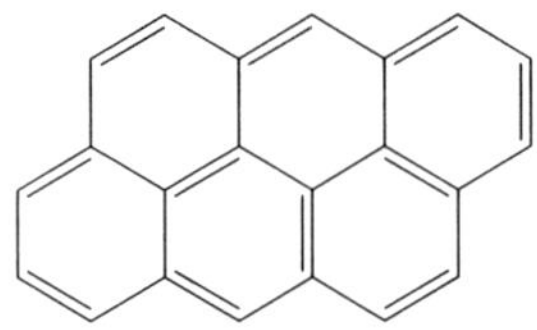

9

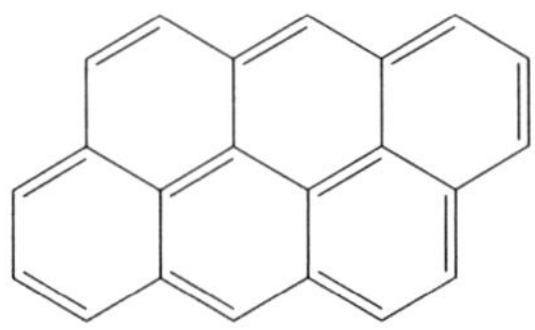

10

Figure 7 *Kekulé structures of anthanthrene*

The same formula as (7) was also given by Yen[137] in 1971. In his paper, Yen considered the parallelogram-shaped, the symmetric-circular-shaped, the rectangular-shaped and the skew-strip-shaped benzenoid hydrocarbons. Došlić did not refer to Yen's results.

Gordon and Davison[136] initiated the path counting method. In the *path counting method* the number of Kekulé structures K of a benzenoid B is equal to the number of the mutually self-avoiding directed peak-to-valley paths:[138,139]

$$K = \det |\boldsymbol{P}| \tag{8}$$

where $\boldsymbol{P}$ is a matrix whose elements $(\boldsymbol{P})_{ij}$ represent counts of self-avoiding paths in B starting at peak(s) and ending at valley(s). A *peak* is a vertex on the upper perimeter of B that lies above its adjacent vertices, whilst a valley is a vertex lying below its nearest neighbors on the lower perimeter of B. The peaks and valleys must match if the K-number is to be non-zero. Identification of peaks

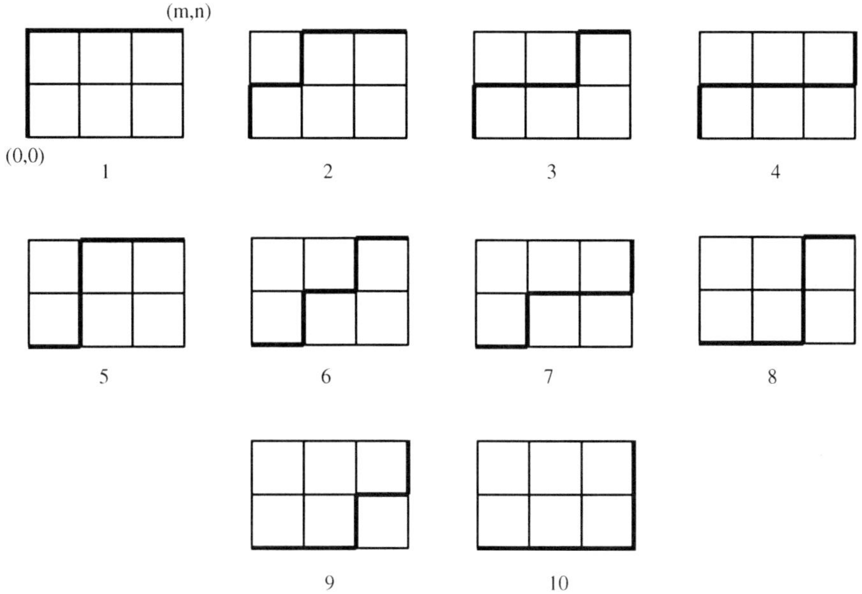

Figure 8 *Paths on the square-lattice corresponding to Kekulé structures of anthanthrene*

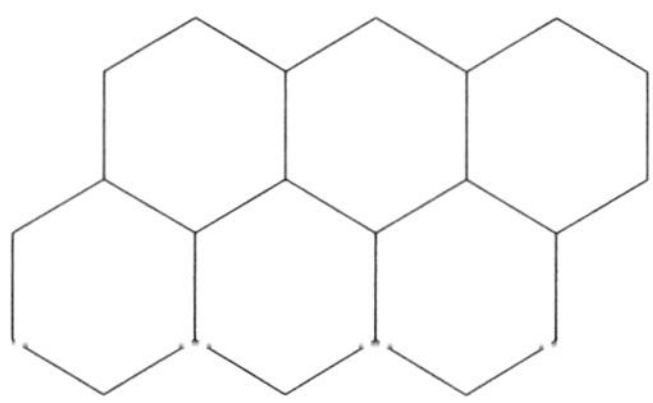

B_{23}

$$K_{2,3} = \binom{2+3}{3} = 10$$

Figure 9 *Counting Kekulé structures of anthanthrene by using equation (7)*

and valleys depends on the orientation of a benzenoid. For convenience, the north-south orientation[140] is used. The K-number is, of course, invariant to the orientation of the benzenoid. For example, anthanthrene considered in Figures 6 and 8, has three peaks and three valleys.

The use of equation (8) depends on the enumeration of paths connecting peaks and valleys. A very efficient and elegant enumeration method is founded on the *Pascal recurrence algorithm* based on the Pascal triangle.[141] As an illustrative example, we give in Figure 10 the application of the Pascal recurrence algorithm and the use of equation (8).

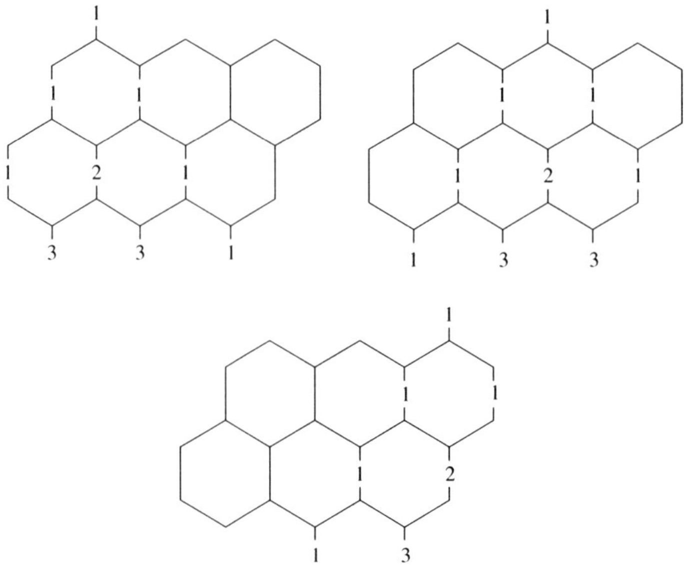

$$K = \det\begin{bmatrix} 3 & 3 & 1 \\ 1 & 3 & 3 \\ 0 & 1 & 3 \end{bmatrix} = 10$$

Figure 10 *Counting Kekulé structures of anthanthrene by applying the Pascal recurrence algorithm*[141]

Therefore, both approaches, the Pascal recurrence algorithm and the approach based on counting the square-lattice paths are similar. However, the Pascal recurrence algorithm is applicable to a wider range of benzenoids and not only to benzenoid parallelograms.

The counting formula for the Kekulé structures of triangular benzenoids is rather simple:[135]

$$K(T_m) = K_m = C_{m+1} \quad (9)$$

where C_{m+1} is the $(m+1)$ *Catalan number*.[142,143] An illustrative example is shown in Figure 11.

The Catalan numbers C_N (rarely called *Segner numbers* after Johann Andreas von Segner (1704–1777))[144] can also be found in the *On-Line Encyclopedia of Integer Sequences*-see sequence A000108: 1, 1, 2, 5, 14, 42, 132, 429, 1430, … or they can be generated using the following formula:

$$C_N = 2N!/N!\,(N+1)! \quad (10)$$

The Catalan numbers could have also been called the Euler numbers because they were discovered by Leonhard Euler (1707–1783) for counting

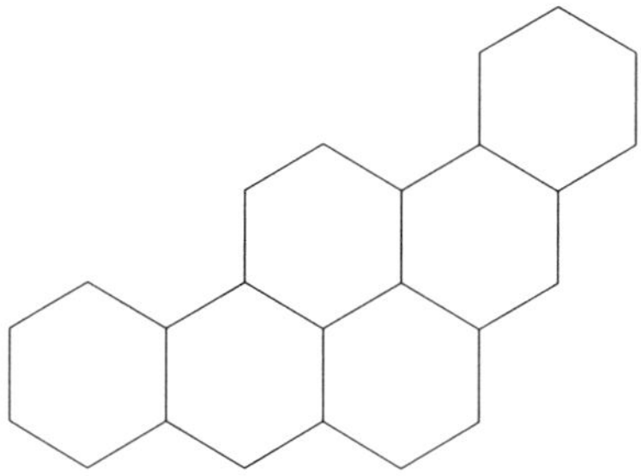

$K_3 = C_4 = 14$

Figure 11 *Counting Kekulé structures of dibenzo[a,i]pyrene by using equation (9)*

triangulations. A triangulation of an N-gon is a division of its inside into triangles. For example, a pentagon has five different triangulations. Von Segner was also involved in counting triangulations. Nevertheless, Catalan numbers are named after Eugene Charles Catalan (1814–1894), who rediscovered these numbers in connection with a different problem, that is, he solved the question: In how many ways can the product $x_1x_2 \ldots x_N$ be parenthesized? The term Segner number is ocassionally encountered. *e.g.*, in the *On-Line Encyclopedia of Integer Sequences*.

The Pascal recurrence algorithm is also applicable to triangular benzenoids, see Figure 12.

Karimi *et al.*[145] also proposed a counting method based on peaks and valleys. These authors transformed benzenoid parallelograms into square-meshes (tilted by 45 degrees square-lattices) and established that there is a one-to-one correspondence between Kekulé structures and descending paths from the top to the bottom of a square-mesh. They also generalized their counting approach to non-parallelogram benzenoids using the idea of *diminished square-mesh* which is obtained as a result of deleting one or more submeshes from a given square-mesh. In their paper, Karimi *et al.*[145] did not refer to the Pascal recurrence algorithm[141] which represents a simpler way of counting peak-to-valley paths.

A variant of the path-counting method, called the reduced graph approach,[146] was based on the idea of transforming the hexagonal network into the trigonal network and counting paths connecting peaks and valleys (though these terms were not used).[147,148] This approach was neglected by both Došlić[135] and Karimi *et al.*[145] though it preceded their reports by more than twenty years.

Došlić[135] also overlooked a more recent paper by Lukovits[149] on the resonance energy in graphite. In his paper, Lukovits calculated the number of Kekulé structures $K_{m,n}$ in the (m,n) parallelogram-shaped graphite sheets using the following recursive equation:

$$K_{m,n} = K_{m-1,n} + K_{m,n-1} \tag{11}$$

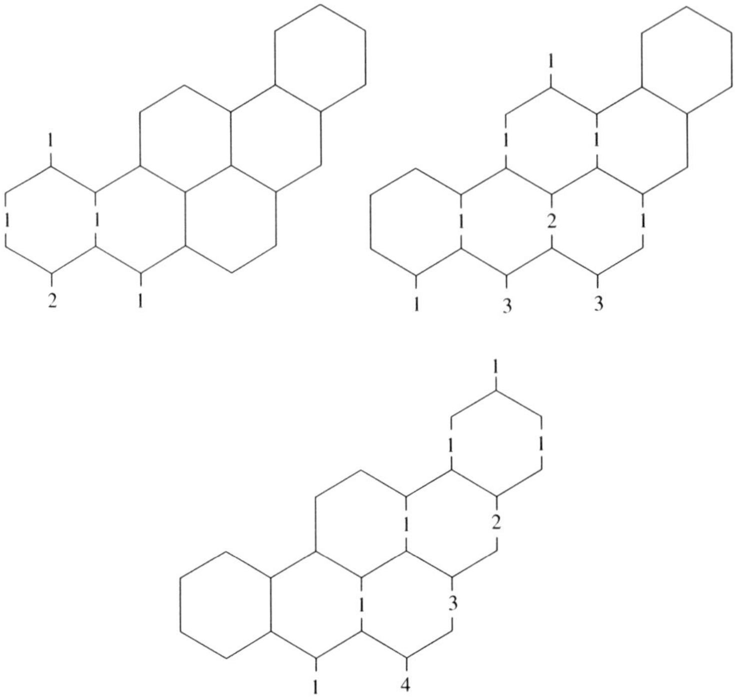

$$K(T_3) = \begin{bmatrix} 2 & 1 & 0 \\ 1 & 3 & 3 \\ 0 & 1 & 4 \end{bmatrix} = 14$$

Figure 12 *Counting Kekulé structures of dibenzo[a,i]pyrene by applying the Pascal recurrence algorithm*[141]

Lukovits has also shown that equation (11) is just another way of expressing equation (7).

2.2.2 Algebraic Kekulé Structures. Randić[115,150] introduced a novel description of Kekulé structures by replacing their standard representation, which he called *geometric* Kekulé structures, by what he called *algebraic* (numerical) representation. The algebraic Kekulé structures are constructed in the following way: Each double bond in a geometric Kekulé structure gets weight 2 for two π–electrons making up the double bond and if the double bond is shared by two rings, it gets the weight of only 1. Then the numbers assigned to bonds in each ring of a polycyclic conjugated system are added up. Hence, every geometric Kekulé structure gives rise to a numerical Kekulé structure that can simply be encoded by a linear sequence of numbers. The algebraic Kekulé structure can be generated for any class of conjugated systems, *i.e.*, conjugated (alternant, nonalternant, coronoids, *etc.*) hydrocarbons or heterosystems possessing geometric Kekulé structures.[151–158] As an example, the geometric Kekulé structures and the corresponding algebraic Kekulé structures of chrysene are given in Figures 13 and 14.

Geometric Kekulé structures provide a linear code for Kekulé structures. For example, we give below the codes for chrysene Kekulé structures. They are listed in the lexicographic order:

4 6 2 6
4 6 3 5
5 3 6 4
5 4 3 6
5 4 4 5
6 2 6 4
6 3 3 6
6 3 4 5

Gutman *et al.*[159] proved that cata-condensed benzenoids (catafusenes) have a one-to-one correspondence between geometric and algebraic Kekulé structures and that peri-condensed benzenoids (perifusenes) and corona-condensed benzenoids (coronafusenes) do not satisfy this rule. Vukičević *et al.*[160] proved the theorem on the necessary and sufficient conditions for the algebraic Kekulé structure of a perifusene or coronafusene to correspond to more than one geometric Kekulé structure. Pyrene is the simplest example of a perifusene in which two distinct geometric Kekulé structures correspond to the same algebraic Kekulé structure 5 3 3 5 (this coding of algebraic Kekulé structures of perifusene is based on the Wiswesser coding system for benzenoids).[16,153,161]

2.2.3 Fullerenes. In a series of papers,[129–134] Torrens used a set of structural parameters of fullerenes reflecting the presence of contiguous pentagons to study their periodic properties. The parameters he considered are the number of edges common to two pentagons p, the number of vertices common to three pentagons q, the number of pairs of nonadjacent pentagon edges shared between two other pentagons r and ratios q/p and r/p. He correlated $ln\ K$, $ln(per\ \mathbf{A})$ and $ln(per\ \mathbf{A})/ln\ K$ with parameters p, q, r, q/p and r/p. In doing this, Torrens really continued earlier efforts of Cash.[162,163] The values of K, $per\ \mathbf{A}$ and $ln(per\ \mathbf{A})/ln\ K$, p, q, r, q/p and r/p for fullerenes with a given symmetry from C_{20} to C_{44} are presented in Table 3. For obvious reasons, the IPR fullerenes have zero values for parameters p, q, and r. For example, buckminsterfullerene possesses the following values $K = 12500$, $per\ \mathbf{A} = 395{,}974{,}320$, $ln(per\ \mathbf{A})/ln\ K = 2.0986$, $p = 0$, $q = 0$ and $r = 0$.

Torrens also listed the numbers of Kekulé structures of some larger IPR fullerenes with icosahedral symmetry, *e.g.*, C_{80}: 140,625; C_{180}: 1,389,029,765,625; C_{240}: 21,587,074,966,666,816.

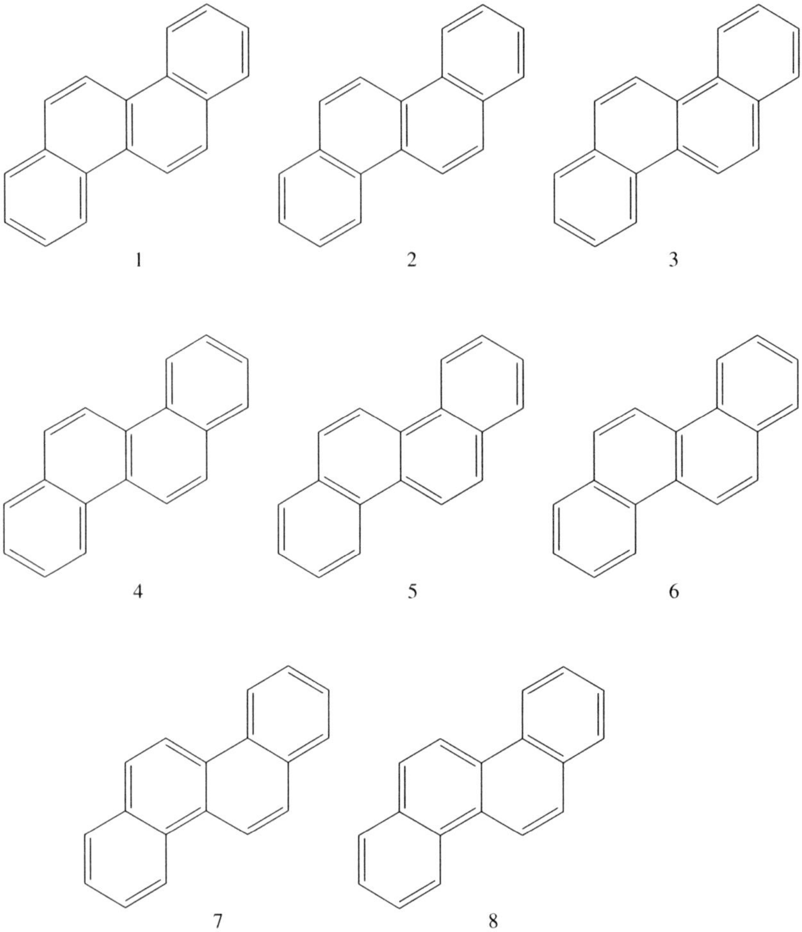

Figure 13 *Geometric Kekulé structures of chrysene*

Torrens reported several correlations between ln *K*, *ln*(*per* ***A***) and *ln*(*per* ***A***)/*ln* *K* and *p*, *q*, *r*, q/p and r/p. We selected the following:

$$\begin{gathered} ln\ K = 10.2 - 0.407\,p + 0.238\,q + 0.792\,q/p + 0.320\,r/p \\ n = 28 \qquad R = 0.931 \qquad S = 0.339 \qquad F = 28.7 \end{gathered} \tag{12}$$

$$\begin{gathered} ln(per\ \boldsymbol{A}) = 20.2 - 0.660\,p + 0.383\,q \\ n = 29 \qquad R = 0.949 \qquad S = 0.757 \qquad F = 118.5 \end{gathered} \tag{13}$$

$$\begin{gathered} ln(per\ 2A)/ln\ K = 1.88 + 0.0361\,p - 0.0490\,q + 0.00953\,r + 0.0497\,q/p \\ - 0.2.53\,r/p \\ n = 28 \qquad R = 0.941 \qquad S = 0.019 \qquad F = 34.2 \end{gathered} \tag{14}$$

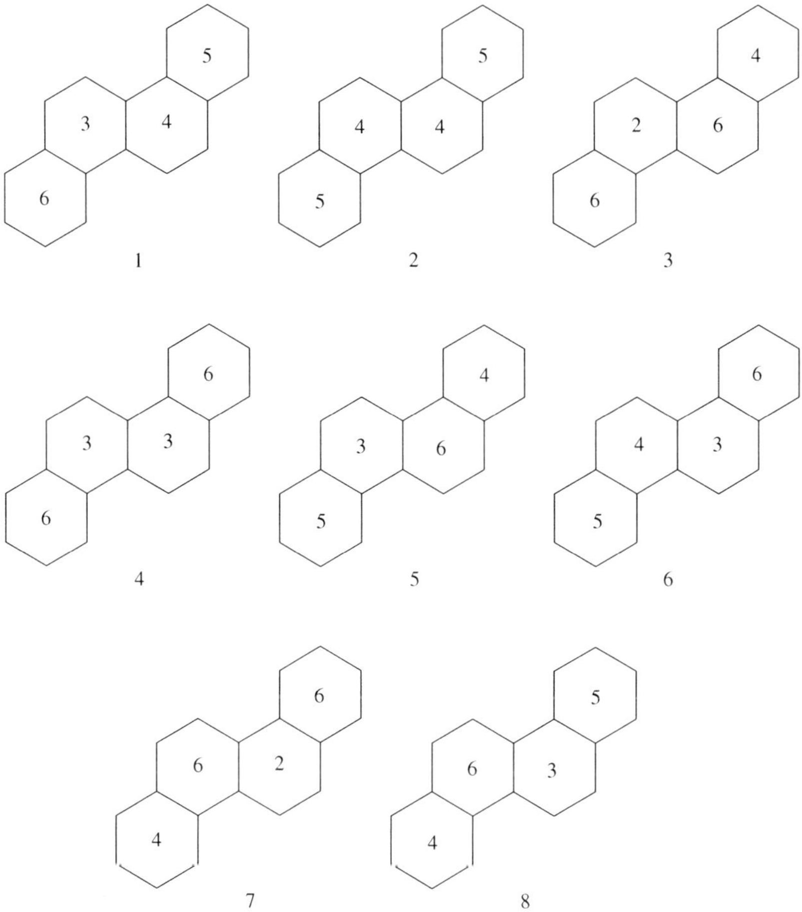

Figure 14 *Algebraic Kekulé structures of chrysene*

where n is the number of fullerenes considered, R is the correlation coefficient, S is the standard error of estimate and F is the result of Fisher's test. In the case of equation (13), the parameters of C_{60} were included in the correlation. We repeated the correlation between $ln(per\ A)$ for C_{20}–C_{44}, without including C_{60}, and their structural parameters, using the CROMRsel procedure.[164–167] CROMRsel is a multivariate procedure that has been designed to select the best possible model among the set of models obtained for a given number of descriptors, the criterion being the standard error of estimate. The quality of the CROMRsel models is expressed by fitted (descriptive) statistical parameters: the correlation coefficient (R_{fit}), the standard error of estimate (S_{fit}) and Fisher's test (F). The models are also cross(internally)-validated by a leave-one-out procedure. Statistical parameters for the cross-validated models are symbolized by R_{cv} and S_{cv}, where subscript cv denotes the cross-validation. The

Table 3 *The values of K, per(**A**), ln[per(**A**)]/ln K, p, q, and r parameters for fullerenes from C_{20} to C_{44} of a given symmetry. These results are taken from refs. 131–134*

Fullerene	K^a	*per* (**A**)^a	*ln[per(**A**)]/ ln K*	*p*	*q*	*r*	*q/p*	*r/p*
C_{20} (I_h)	36	1392	2.0199	30	20	30	0.6667	1.0000
C_{24} (D_{6d})	54	4692	2.1192	24	12	36	0.5000	1.5000
C_{26} (D_{3h})	63	8553	2.1853	21	8	30	0.3810	1.4286
C_{28} (T_d)	75	15705	2.2378	18	4	24	0.2222	1.3333
C_{28} (D_2)	90	16196	2.1540	20	8	24	0.4000	1.2000
C_{30} (C_{2v}) I	107	29621	2.2034	17	4	20	0.2353	1.1765
C_{30} (C_{2v}) II	117	30053	2.1651	18	6	20	0.3333	1.1111
C_{30} (D_{5h})	151	31945	2.0672	20	10	20	0.5000	1.0000
C_{32} (D_3)	144	55140	2.1968	15	2	18	0.1333	1.2000
C_{32} (C_2) I	151	55705	2.1780	16	4	16	0.2500	1.0000
C_{32} (C_2) II	168	57092	2.1375	17	6	16	0.3529	0.9412
C_{32} (D_2)	184	58384	2.1045	18	8	15	0.4444	0.8333
C_{34} (C_{3v})	195	103665	2.1902	15	3	15	0.2000	1.0000
C_{34} (C_s)	196	104484	2.1896	15	3	16	0.2000	1.0667
C_{34} (C_2) I	204	103544	2.1714	14	2	14	0.1429	1.0000
C_{34} (C_2) II	212	107720	2.1632	17	6	16	0.3529	0.9412
C_{36} (D_{6h})	272	192528	2.1706	12	0	12	0.0000	1.0000
C_{36} (D_{2d})	288	192720	2.1489	12	0	12	0.0000	1.0000
C_{36} (C_{2v})	312	197340	2.1231	13	2	10	0.1538	0.7692
C_{36} (D_{3h})	364	207924	2.0764	15	6	6	0.4000	0.4000
C_{38} (C_{2v})	360	366820	2.1768	14	2	14	0.1429	1.0000
C_{38} (C_{3v})	378	363300	2.1572	12	1	9	0.0833	0.7500
C_{38} (D_{3h})	456	411768	2.1116	18	8	18	0.4444	1.0000
C_{40} (D_{5d}) I	562	515781	2.0775	10	0	10	0.0000	1.0000
C_{40} (T_d)	576	704640	2.1185	12	4	0	0.3333	0.0000
C_{40} (D_{5d}) II	701	803177	2.0750	20	10	20	0.5000	1.0000
C_{44} (T)	864	2478744	2.1775	12	4	0	0.3333	0.0000
C_{44} (D_{3h})	960	2436480	2.1416	9	2	0	0.2222	0.0000

[a] Torrens, according to his e-mail to N.T. of April 19, 2005, calculated *per A* with programs provided by Dr. Gordon G. Cash (Office of Pollution Prevention and Toxics Risk Assessment Division, U.S. Environmental Protection Agency, Washington, USA) – the latest program is described in G. Delic and G.G. Cash, *Comput. Phys. Commun.*, 2000, **124**, 315. He translated these programs from C-language into Fortran. In the e-mail of April 25, 2005, Torrens stated that he took the fullerene *K*-numbers from the literature, *e.g.*, K. Balasubramanian, *J. Chem. Inf. Comput. Sci.*, 1994, **34**, 421 and H.-Y. Zhu and D.J. Klein, *J. Mol. Struct.* (*Theochem*), 1995, *338*, 11.

obtained regression equation is:

$$\begin{gathered} ln(per\ A) = 20.9(\pm 1.0) - 0.722(\pm 0.089)\,p + 0.437(\pm 0.091)\,q \\ n = 28 \quad R_{fit} = 0.911 \quad R_{cv} = 0.880 \\ S_{fit}(n - I - 1) = 0.76 \quad S_{cv}(n - I - 1) = 0.87 \quad F = 60.6 \end{gathered} \tag{15}$$

where I is the number of descriptors used in the regression. We did this because it is not clear why Torrens included buckminsterfullerene in getting the regression equation (13), unless he wanted to obtain the statistical parameters comparable to equations (12) and (14), which appears to be the case. When we used C_{60} parameters in the regression, we reproduced his results for fitted statistical parameters (he made no cross-validation of his models): $n = 29$, $R_{fit} = 0.948$, $R_{cv} = 0.932$, $S_{fit}\ (n{-}I{-}1) = 0.757$, $S_{cv}(n{-}I{-}1) = 0.873$, $F = 118.5$.

The principal component analysis of the structural parameters, supported by the cluster analysis, classified fullerenes into five groups. The periodic table of fullerenes was built on the structural parameters, the principal component analysis and the cluster analysis. The conclusion reached by Torrens is that the preriodicity of fullerene properties is *not* general. Torrens[168–171] also studied the periodic properties of carbon nanotubes as well.

Jiang *et al.*[172] studied the relationship between the permanent of the vertex-adjacency matrix of fullerenes and their structural parameters *p*, *q* and *r*, but also introduced a set of additional structural parameters because, as stated above, *p*, *q* and *r* are zero for the IPR fullerenes. The new structural parameters that count contiguous hexagons are denoted by *u*, *v* and *w*. The *u* and *v* enumerate, respectively, the number of edges common to two hexagons and the number of vertices common to three hexagons. The third parameter *w* enumerates the number of pairs of nonadjacent hexagon edges shared with other two hexagons. Using the stepwise regression, the best correlation between *ln*(*per A*) and parameters *p*, *q*, *r*, *u*, *v* and *w* for the same set of fullerenes, including C_{60} used by Torrens, is obtained as follows:

$$\begin{gathered} ln(per\ \boldsymbol{A}) = 13.28 - 0.20\,p + 0.22\,u \\ n = 29 \quad R = 1.000 \quad S = 0.016 \quad F = 28300 \end{gathered} \tag{16}$$

where *per A* was calculated with their own algorithm.[173] This is a very good correlation, which indicates that only two parameters, *p* and *u*, can give almost full information on *ln*(*per A*). We repeated as above the correlation between *ln*(*per A*) for C_{20}–C_{44}, without including C_{60}, and the same set of structural parameters, using the CROMRsel procedure.[164–167] The obtained regression equation is a very good one:

$$\begin{gathered} ln(per\ 2A) = 13.207(\pm 0.054) - 0.1971(\pm 0.0027)\,p + 0.2165(\pm 0.0022)\,u \\ n = 28 \quad R_{fit} = 0.9996 \quad R_{cv} = 0.9994 \\ S_{fit}(n - I - 1) = 0.054 \quad S_{cv}(n - I - 1) = 0.062 \quad F = 14255 \end{gathered} \tag{17}$$

We also repeated their calculations for C_{20}–C_{44} including C_{60} and statistical parameters, but did not reproduce results of Jiang *et al.*[172]: $n = 29$, $R_{fit} = 0.9997$, $R_{cv} = 0.9997$, $S_{fit}\ (n{-}I{-}1) = 0.055$, $S_{cv}(n{-}I{-}1) = 0.062$, $F = 24792$.

2.2.4 Buckminsterfullerene. Vukičević and Randić[174] examined the 12,500 Kekulé structures of buckminsterfullerene (this number was first obtained by Klein *et al.*[175] using the transfer-matrix method) and were able to classify them according to the different innate degree of freedom that they possess. The *innate degree of freedom* or, for short, the *degree of freedom* (*df*) of a Kekulé structure is the smallest possible number of carbon-carbon double bonds that, when selected, determine the locations of all the remaining carbon-carbon double bonds.[176,177] Vukičević and Randić[174] developed a computational algorithm for finding degrees of freedom and applied it to 12,500 Kekulé structures of buckminsterfullerene. The results they obtained are given in Table 4.

Kekulé structures of buckminsterfullerene belong to six classes. Only one Kekulé structures of buckminsterfullerene possesses the highest degree of freedom ($df = 10$). This is the Kekulé structure in which the carbon-carbon double bonds are exocyclic to pentagons.[178] There are two distinct, that is symmetry non-equivalent, Kekulé structures with $df = 9$ – one appearing 20 times and the other 60 times. The base numbers in the last column of Table 4 indicate how often distinct Kekulé structures occur – their numbers are given as exponents. For example, 23 over 120 (see the first base number and its exponent in the third row of the last column) means that 23 of 47 distinct Kekulé structures with $df = 7$ occur 120 times. The summation of exponents gives the total number of distinct Kekulé structures whilst the multiplication of exponents by base numbers gives the total number of structures with a given degree of freedom. The sum of the numbers in the second column in Table 4 gives 158 distinct Kekulé structures. This number agrees with earlier findings of Babić and Trinajstić,[179] who reported (in 1994) the frequency of occurrence of these 158 distinct Kekulé structures among 12,500 Kekulé structures of buckminsterfullerene, which is identical with the decomposition presented in the last row and the last column of Table 4. These authors considered the combinatorial assembling of fullerenes from identical fragments using symmetry arguments.

In a subsequent paper, Vukičević and Randić in collaboration with Sir Harold W. Kroto[180] catalogued all 158 distinct Kekulé structures producing the *Atlas of Kekulé Structures of Buckminsterfullerene*. Everything they did with

Table 4 *The degrees of freedom (df) of distinct Kekulé structures (KS) of buckminsterfullerene, their frequence of occurrence and their partition. These results are taken from ref. 174*

df	*Distinct KS*	*Total number of KS*	*Partition of KS and their multiplicity*
5	36	3170	$120^{19} + 60^{13} + 40 + 30 + 20^2$
6	39	3116	$120^{18} + 60^{13} + 30^5 + 10^2 + 6$
7	47	4060	$120^{23} + 60^{19} + 40^2 + 30^2 + 20$
8	33	2073	$120^{10} + 60^9 + 40^4 + 30^3 + 20^2 + 12 + 10^2 + 6 + 5$
9	2	80	$60 + 20$
10	1	1	1
Total	158	12500	$120^{70} + 60^{55} + 40^7 + 30^{11} + 20^6 + 12 + 10^4 + 6^2 + 5 + 1$

the Kekulé structures of buckminsterfullerene, they also collected in an author's review.[181] In that review, they discussed the hidden treasures of Kekulé structures (*e.g.*, degrees of freedom, conjugated circuits, geometric *versus* algebraic Kekulé structures), Clar structures of buckminsterfullerene, enumeration of conjugated circuits, selected mathematical properties of Kekulé structures and in the Appendix they presented the computational algorithm in pseudocode.

2.2.5 Nanotubes. Lukovits and Janežić[182] reported the number of Kekulé structures in polyphenanthrenes (the zigzag shaped benzenoid chains), $(1,1)_n$ armchair-type carbon nanotubes and extended $(1,1)_n$ nanotubes. The lower index denotes the number of hexagons layers in a nanotube. A $(n,n)_1$ nanotube that contains a single-hexagon-layer is called the nanoribbon. They enumerated the Kekulé structures in polyphenanthrenes in terms of the number of hexagons h using the Gordon-Davison method,[136] which can be summarized by a simple counting formula:[183]

$$K(h) = K(h-1) + \mathrm{K}(h-2) \tag{18}$$

with $\mathrm{K}(-2) = 0$ and $\mathrm{K}(-1) = 1$(by definition).

Lukovits and Janežič[182] generated the number of Kekulé structures of $(1,1)_n$ armchair-type carbon nanotubes using the following recurrence equation:[184]

$$K(1,1)_n = 3\ K(1,1)_{n-1} - K(1,1)_{n-2} \tag{19}$$

with $K(1,1)_{-1} = 1$ and $K(1,1)_0 = 2$, whilst $n =$ the naphthalene unit. They also generated the number of Kekulé structures of extended $(1,1)_n$ nanotubes by means of:

$$K^*(1,1)_n = K(1,1)_n - K^*(1,1)_{n-1} \tag{20}$$

with $K^*(1,1)_{-1} = 1$ and $K^*(1,1)_0 = 3$. The numbers of Kekulé structures in polyphenanthrenes, $(1,1)_n$ armchair-type carbon nanotubes and extended $(1,1)_n$ nanotubes are given in Table 5.

From Table 5, one can see that the Kekulé structures in polyphenanthrenes, $(1,1)_n$ armchair-type carbon nanotubes and in extended $(1,1)_n$ nanotubes are related by the following equalities:

$$K(2h = n) = K^*(1,1)_{n-1} \tag{21}$$

$$K(2h + 1 = n + 1) = K(1,1)_n \tag{22}$$

The K-numbers are used to generate the conjugated circuits (see later text) that were used to calculate the resonance energies of nanotubes.

Equation (18) is a *Fibonacci-type* recurrence relationship[185,186] and the K-numbers for polyphenanthrenes are the *Fibonacci numbers.*[142,143,187] The Fibonacci numbers can also be found in the *On-Line Encyclopedia of Integer Sequences* – see sequence A000045: 0, 1, 1, 2, 3, 5, 8, 13, 21, 34, 55, 89, 144, 233, 377, 610, ... The Fibonacci numbers were named after the Italian

Table 5 *The number of Kekulé structures in polyphenanthrenes K(h) in terms of the number of hexagons h, the number of Kekulé structures in $(1,1)_n$ nanotubes $K(1,1)_n$ and in extended $(1,1)_n$ nanotubes $K^*(1,1)_n$ in terms of napthalene units n (n = 2h). These results are taken from ref. 182*

h(n)	*K(h)*	$K(1,1)_n$	$K^*(1,1)_n$
−2 (−1)	0	1	5
−1	1		
0 (0)	1	2	3
1	2		
2 (1)	3	5	8
3	5		
4 (2)	8	13	21
5	13		
6 (3)	21	34	55
7	34		
8 (4)	55	49	144
9	89		
10 (5)	144	233	377
11	233		
12 (6)	377	324	987
13	610		
14 (7)	987	1597	2584
15	1597		
16 (8)	2584	4181	6785
17	4181		
18 (9)	6785	10946	17711
19	10946		
20 (10)	17711	28657	46368
21	28657		
22 (11)	46368	75025	121393
23	75025		
24 (12)	121393	196418	317811

mathematician Fibonacci (c.1175–c.1250), also known as Leonardo da (of) Pisa or Leonardo Pisano. The use of the subriquet Fibonacci, a contraction of filius Bonacci, son of Bonaccio, for Leonardo Pisano was introduced in 1838 by the mathematical historian Guillaume Libri. There is no evidence that Leonardo referred to himself as Fibonacci or was ever called so by his contemporaries. He actually called himself Leonardo Pisano Bigollo. No one has been able to trace the origin of Bigollo in his name, but it has been speculated that he called himself Bigollo (ignorant peasant) to differentiate himself, who never attended university, from the pretentious ignorant academics of those days. Additionally, his father's name was not Bonaccio: His name was Guglielmo. The problem that gave rise to the Fibonacci series was specified by Leonardo Pisano in his book *Liber Abaci* (*The Book of Abacus*, published in 1202), there serving as a model for breeding rabbits. The Fibonacci rabbit population model is perhaps the first study in mathematical biology, many centuries before the emergence of mathematical biology as a part of biological science.

2.2.6 Enumeration of Kekulé and Longer Range Resonance Valence Bond Structures. Cai *et al.*[188] proposed a novel procedure for counting Kekulé structures and longer range resonance valence bond (VB) structures (excited VB structures) that find use in the VB theory. Kekulé structures and their numbers are denoted by K, whilst two kinds of longer range resonance VB structures and their numbers are denoted by S_3 (singly-excited VB structures) and $D_{3,3}$ (doubly-excited VB structures), where the suffixes indicate the number of bonds between two through the ring(s) connected sites. This notation is used by Flocke and Schmalz[189] and they also found, in their VB calculations on the singlet ground states of fullerenes from C_{20} to C_{60}, that the most important excited structures in this case are S_3 and $D_{3,3}$. As examples, we give one Kekulé structure and one S_3 structure of benzene and one $D_{3,3}$ structure of naphthalene in Figure 15.

The procedure for counting S_3 and $D_{3,3}$ is based on the *extended vertex-adjacency matrix* and the *adjacency bonding array*. The extended vertex-adjacency matrix of a Kekulé structure contains elements 1 or 2 depending on the single or double bond between carbon atoms in the Kekulé structure and it is called *extended* because of the added column and row in which elements are set to 1 or 0 depending whether the carbon atom in the Kekulé structure is connected to the hydrogen bond or not. For example, the extended vertex-adjacency matrix for the Kekulé structure of benzene (see Figure 15) is:

$$
\begin{array}{c} j/i \\ 1 \\ 2 \\ 3 \\ 4 \\ 5 \\ 6 \\ 7 \end{array}
\begin{array}{c} \begin{array}{ccccccc} 1 & 2 & 3 & 4 & 5 & 6 & 7 \end{array} \\
\begin{bmatrix}
0 & 1 & 0 & 0 & 0 & 2 & 1 \\
1 & 0 & 2 & 0 & 0 & 0 & 1 \\
0 & 2 & 0 & 1 & 0 & 0 & 1 \\
0 & 0 & 1 & 0 & 2 & 0 & 1 \\
0 & 0 & 0 & 2 & 0 & 1 & 1 \\
2 & 0 & 0 & 0 & 1 & 0 & 1 \\
1 & 1 & 1 & 1 & 1 & 1 & 0
\end{bmatrix} \end{array}
$$

Since this matrix is a sparse matrix, it is not economical to be explicitly used in computer programming. In order to save computer memory, the authors used the technique of the *bond list*, which encodes the position and the value of the non-zero elements in the upper-triangle (or lower-triangle) of the matrix. Thus, each member in the bond list is coded by $(i, j, [A]_{ij})$, with i and j referring

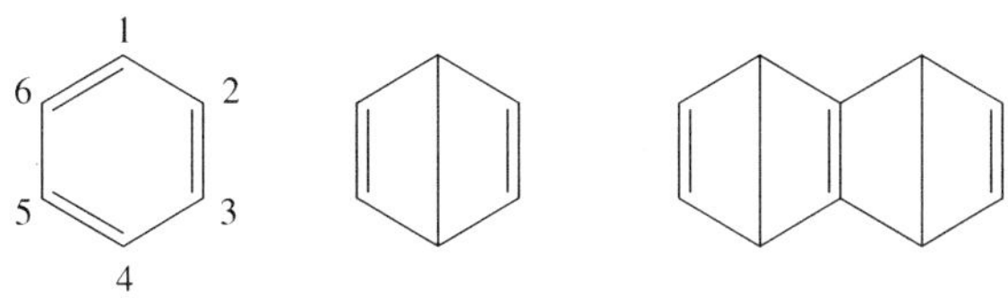

K S_3 $D_{3,3}$

Figure 15 *Labeled Kekulé structure K, S_3 structure of benzene, and $D_{3,3}$ structure of naphthalene*

to the position and $[A]_{ij}$ to the value of the non-zero matrix element. The bond list for the Kekulé structure of benzene, whose extended vertex-adjacency matrix is given above, is (1,2,1), (1,6,2), (1,7,1), (2,3,2), (2,7,1), (3,4,1), (3,7,1), (4,5,2), (4,7,1), (5,6,1), (5,7,1), (6,7,1).

The next step was the introduction of the adjacency bonding array. The adjacency bonding array is based on the concept of quadrivalence of carbon. The skeleton (hydrogen atoms included) of a conjugated molecule provides 3 valences and a Kekulé structure or a higher VB structure contributes the fourth valence. Then, a 2-dimensional array $a[N][3]$ (in the C-language) is used to store the atom connectivity of the molecular skeleton, which the authors call the skeleton adjacency bonding array: $a[i][k]$, $k = 0,1,2$ registers adjacent atoms (0 for a hydrogen atom) of the i-th carbon atom C_i. An additional array $K[N]$, which is referred to as a Kekulé or a higher VB adjacency bonding array, is required for notation of each Kekulé or higher VB structure. $K[i]$ records the atom paired with C_i in the Kekulé structure or higher VB structures. The Kekulé structure of benzene from Figure 15 is denoted by:

$$a[6][3] = \begin{bmatrix} 0 & 2 & 6 \\ 0 & 1 & 3 \\ 0 & 2 & 4 \\ 0 & 3 & 5 \\ 0 & 4 & 6 \\ 0 & 5 & 1 \end{bmatrix} \quad \text{combined with } K[6] = \begin{bmatrix} 6 \\ 3 \\ 2 \\ 5 \\ 4 \\ 1 \end{bmatrix} \tag{23}$$

The arrays $a[N][3]$ and $K[N]$ are also sometimes called connection matrices and are defined as

$$C(i,c) = \begin{cases} i_c \equiv c^{\text{th}} \text{ site to which site } i \text{ is attached}, \ c \leq d, \ c = 1,2,3 \\ 0 \quad \text{when } c < d_i \leq 3 \end{cases} \tag{24}$$

$$K(i) \equiv \text{j for } \{i,j\} \in \text{Kekulé structure under consideration} \tag{25}$$

Since the skeleton adjacency array is the same for all the Kekulé and higher resonance VB structures in one molecule, it is only needed to save a one-dimensional array $K[N]$ for each structure.

Based on the adjacency bonding arrays, Cai *et al.*[188] developed very efficient algorithms for the systematic search for K, S_3 and $D_{3,3}$ structures. These authors listed in their paper K, $K + S_3$ and $K + S_3 + D_{3,3}$ numbers for 72 selected conjugated molecules.

In Table 6, we present the K, $K + S_3$ and $K + S_3 + D_{3,3}$ numbers of several conjugated molecules taken from Table 3 from the paper by Cai *et al.*[188]

2.3 Walks. – The definitions of walks and random walks are given in a number of mathematical texts, *e.g.*[9,11,103,190,191] Here we briefly repeat these definitions. A *walk* in a (molecular) graph G is an alternating sequence of vertices and edges of G, such that each edge e begins and ends with the vertices

Table 6 *K, S_3 and $D_{3,3}$ numbers for several conjugated molecules taken from Cai et al.*[188]

Conjugated molecule	K	$K + S_3$	$K + S_3 + D_{3,3}$
Benzene	2	5	
Naphthalene	3	17	39
Pyrene	6	62	301
Triphenylene	9	84	411
Perylene	9	117	722
Chrysene	9	121	747
Anthanthrene	10	162	1209
Benzo[*ghi*]perylene	14	204	1449
Benzo[*e*]anthanthrene	19	333	2827
Coronene	20	332	2690
Dibenzo[*bc,ef*]coronene	30	702	7711
Ovalene	50	1210	14181
Quaterrylene	81	2241	30492
Buckminsterfullerene	1250	722300	20633840

immediately preceding and following e in the sequence. The *length* of a walk is the number of edges in it. Repetition of vertices and edges is allowed in a walk.

Simple random walks probabilistically grown step by step on a graph are distinguished from walk enumerations and associated equipoise random walks. A *random walk* in a (molecular) graph is naturally designated by a probability measure, of which there are at least two natural intrinsic ones. The first is the probability measure that entails starting walks from each vertex with equal probability and subsequent steps are such that each neighboring vertex is stepped to with equal probability, so that the probability of stepping from vertex i to vertex j is $1/di$, where di is the *degree* of vertex i. The walks of the consequently generated distribution are referred to[192] as *simple random* walks. The second is the probability measure that takes each possible walk of a given length as equally probable. These probability measures are generally quite different, though if a graph is regular (*i.e.*, having all vertices of the same degree), then these two probability measures are equivalent. The walks generated by this second probability measure are called *equipoise random* walks. Simple random walks are not much used in chemical graph theory, whereas the equipoise random walks have been used fairly frequently.[193–208] Equipoise or simple random walks have often been called just *random*, probably without recognition of the alternative type – or perhaps due to confusion of the two possibilities. Simple random walks have been used to obtain probabilistic solutions to mathematical, physical, computational or chemical problems, *e.g.*[192,209–215]

2.3.1 Enumeration of Random Walks. Counting simple random walks was reported by Klein *et al.*[216] In parallel to the generation of walks from the powers of the adjacency matrix (see, for example, our Report in ref. 2) that may be viewed as an identification of the distribution for equipoise random walks, Klein *et al.*[216] generated the distribution for simple random walks by powers of a *Markov matrix* $\boldsymbol{M}$ with elements that are probabilities for associated individual steps:

$$(\boldsymbol{M})_{ij} = \begin{cases} 1/d_j & i \neq j \\ 0 & \text{otherwise} \end{cases} \tag{26}$$

The $(\boldsymbol{M}^l)_{ij}$ is the probability for an l-step random walk beginning at vertex j to end at vertex i.

The *random-walk count* of the length l $(rwc)_l$ at vertex i is given by:

$$(rwc)_l(i) = \sum_{j=1}^{N} (2M^l)_{ij} \tag{27}$$

If we consider only diagonal elements of matrix $\boldsymbol{M}$, $(\boldsymbol{M}^l)_{ii}$, the self-returning l-step random walk $(srrw)_l$ at vertex i is:

$$(srrw)_l\ (i) = (\boldsymbol{M}^l)_{ii} \tag{28}$$

which, of course, is the probability for a simple random walk starting at vertex i to return to i after l steps. These vertex characteristics were earlier discussed by Klein[217] in the context of the general *sum rules* for *resistance distances*,[218] though $(srrw)_l(i)$ was in his paper denoted $w_l(i)$.

Klein *et al.*[216] have also shown that the $(rwc)_l$ and $(srrw)_l$ can be computed *via* eingevalues of matrix $\boldsymbol{M}$. However, matrix $\boldsymbol{M}$ is nonsymmetric unless the graph is regular. But in the general case the eigenvalues of $\boldsymbol{M}$ are related to those of a symmetric matrix, since $\boldsymbol{M}$ can be given as:

$$\boldsymbol{M} = \boldsymbol{A}\ \boldsymbol{\Delta}^{-1} \tag{29}$$

where $\boldsymbol{A}$ is the vertex-adjacency matrix, Δ is the diagonal matrix with elements $(\Delta)_{ii} = d_i$ and Δ^{-1} is its inverse. Matrix $\boldsymbol{M}$ can be related to an auxiliary matrix $\boldsymbol{H}$:

$$\boldsymbol{H} = \boldsymbol{\Delta}^{-1/2}\ \boldsymbol{M}\ \boldsymbol{\Delta}^{1/2} \tag{30}$$

Matrices $\boldsymbol{H}$ and $\boldsymbol{M}$ are related by a similarity transformation, so that $\boldsymbol{H}$ has the same eigenvalues as $\boldsymbol{M}$ and eigenvectors c'_r, which are simply related to those of c_r. Hence:

$$c'_{\mathrm{r}} = \boldsymbol{\Delta}^{-1/2}\ c_r \tag{31}$$

$$c_{\mathrm{r}} = \boldsymbol{\Delta}^{1/2}\ c'_r \tag{32}$$

If $\boldsymbol{A}\ \boldsymbol{\Delta}^{-1}$ is substituted for $\boldsymbol{M}$ in equation (30):

$$\boldsymbol{H} = \boldsymbol{\Delta}^{-1/2}\ \boldsymbol{A}\ \boldsymbol{\Delta}^{-1/2} \tag{33}$$

one sees that $\boldsymbol{H}$ is symmetric, whence one can use standard symmetric-matrix diagonalization routines to obtain the c_r and associated eigenvalues λ. Then:

$$(rwc)_l(i) = \sum_{r}^{N} (c_{ri} d_i^{-1/2}) \lambda_r^l \left(\sum_{j=1}^{N} c_{rj} d_j^{-1/2} \right) \tag{34}$$

where the c_{ri} are components of c_i. For $(srrw)_f(i)$, a similar expression applies with the j-sum eliminated and j set equal to i. Analogous formulas for equipose walks were reported by Rücker and Rücker[198,202–204,206,208] and have been reviewed by us in our last Report.[2] The $\boldsymbol{H}$ matrix is also related to the *connectivity index* of Randić.[219,220] The connectivity index χ is given as the half-sum of the off-diagonal elements of $\boldsymbol{H}$:

$$\chi = (1/2)\sum_{i \neq j} (2H)_{ij} \tag{35}$$

Klein *et al.*[216] have also shown that the *combinatorial Laplacian matrix L* (often called just the *Laplacian matrix*[221,222]) is related to matrix $\boldsymbol{H}$:

$$\boldsymbol{L} = \boldsymbol{\Delta} - \boldsymbol{A} = \boldsymbol{\Delta}^{-1/2}\,(\boldsymbol{I} - \boldsymbol{H})\,\boldsymbol{\Delta}^{1/2} \tag{36}$$

The eigenvalues λ of $\boldsymbol{L}$ give the *Wiener index W* of acyclic structures:[223,224]

$$W = N\sum_{\lambda=2}^{N} 1/\lambda \tag{37}$$

When the above formula is extended to cycle-containing structures, a variant of the Wiener index[225] is obtained, which is named the *quasi-Wiener index* or *Kirchhoff index* and is usually denoted by W^*.[223,226–229]

The difference matrix $\boldsymbol{I} - \boldsymbol{H}$ is called the *normalized Laplacian matrix* $\boldsymbol{L}_{norm}$ (also sometimes called just the *Laplacian matrix*) of G and there is much theory about it.[230] The matrix $\boldsymbol{L}_{norm}$ is clearly also related to the connectivity index:

$$\chi = (1/2)\sum_{i \neq j} (L)_{norm} \tag{38}$$

2.3.2 *Walks on Plerographs and Kenographs*. In his paper on the mathematical theory of isomers (published in 1847), Arthur Cayley (1821–1895), a distinguished 19th century English mathematician, considered two types of molecular graphs, which he named *plerograms* and *kenograms*.[231] In modern chemical graph theory, plerograms (P) are molecular graphs in which *all* atoms are represented by vertices whilst kenograms (K) are referred to as *hydrogen-suppressed* or *hydrogen-depleted* molecular graphs.[232] In their book *Mathematical Concepts in Organic Chemistry*, Gutman and Polansky[101] used the terms *complete molecular graphs* and *skeleton graphs* for plerograms and kenograms, respectively. The term *thorn* or *thorny* graps for plerograms is also found in the literature.[233,234]

Vukičević *et al.*[235] introduced the terms *plerographs* and *kenographs* for plerograms and kenograms, respectively, because plerograms and kenograms are *graphs* rather than types of *diagrams*, and these authors also wanted to preserve the roots of Cayley's terms: *plero* (from the Greek word *pleres* = full) and *keno* (from the Greek word *kenos* = empty). Therefore, Vukičević *et al.*[235] substituted *grams* in Cayley's terms by *graphs*. Cayley could not do this because the name *graph* was not yet adopted in 1847. Introduction of the term *graph* happened later, when one-page paper by Sylvester[236] appeared in *Nature* (in 1877), in which he introduced this term, stimulated as he stated by *chemicographs* in *Lectures notes for chemical students* (London, 1866) by Edward Frankland (1825–1899). Frankland used in his *Notes* the term *graphic-like symbolic formulae*. If *plerographs* and *kenographs* are used to represent hydrocarbons, then one may think of them as CH-graphs and C-graphs, respectively. In Figure 16, we give as an illustration the plerograph and the kenograph representing 2,3-dimethylhexane.

Various molecular descriptors have been computed for plerographs usually in terms of the corresponding descriptors for kenographs.[237–239] Vukičević *et al.*[235] calculated walks for plerographs and kenographs representing alkanes. In the case of alkanes, all vertices in the related kenographs correspond to carbon atoms, while in plerographs the mono-valent vertices correspond to hydrogen atoms while four-valent vertices correspond to carbon atoms.

Vukičević *et al.*[235] obtained the number of walks by the matrix multiplication, since it is well-established that the number of walks of length λ beginning at vertex i and ending at vertex j is given by the element $(A^{\lambda})_{ij}$, the ij-element, in the λ-th power of the vertex-adjacency matrix A (see details in our previous Report[2]). They calculated walks with up to length 7 on plerographs and kenographs for octanes using a *C++* program. The obtained relationship between the number of walks in octane-plerographs $w(P)$ and the number of walks in octane-kenographs $w(K)$ is approximately linear:

$$w(P) = 26678.9 + 2.1\ w(K) \tag{39}$$

with the correlation coefficient $R = 0.995$. Therefore, if the walks are used, for example, in the structure-property-activity modeling[121] of acyclic structures, then it will be sufficient in most cases to use kenographs for representing the molecules under study. The use of plerographs would be justified only in case they provide new insights, though their use will be more time-consuming since they are more complex structures than kenographs as can be seen in Figure 16. It should also be pointed out that plerographs correspond more closely to molecular structures than kenographs.

2.3.3 Walk Counts and Zagreb Indices. Zagreb indices M_1 and M_2 have been introduced more than 30 years ago.[240] They are simply defined in terms of vertex-degrees:

$$M_1 = \sum_{\text{vertices}} d_i\, d_i \tag{40}$$

$$M_2 = \sum_{\text{edges}} d_i\, d_j \tag{41}$$

In recent years these indices have been modified[241] and their mathematical properties investigated.[242–248] Although they have initially been used to study the molecular branching,[249] they soon afterwards found application in the

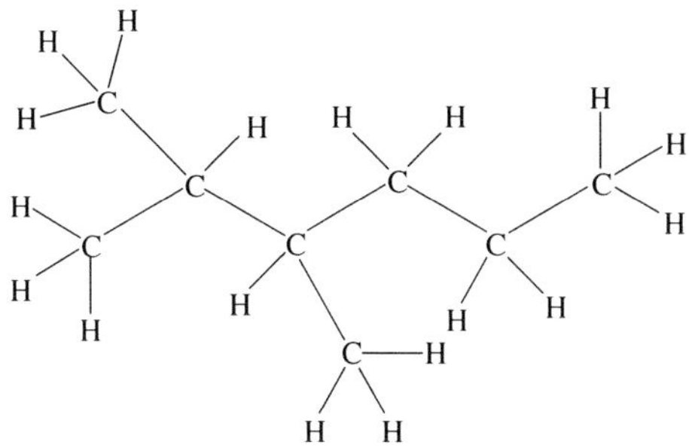

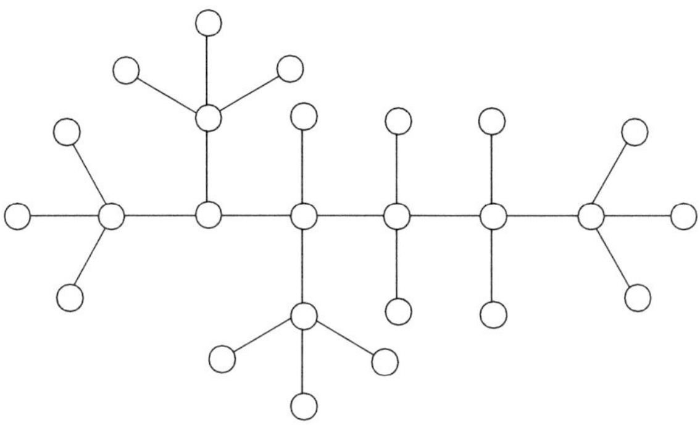

P

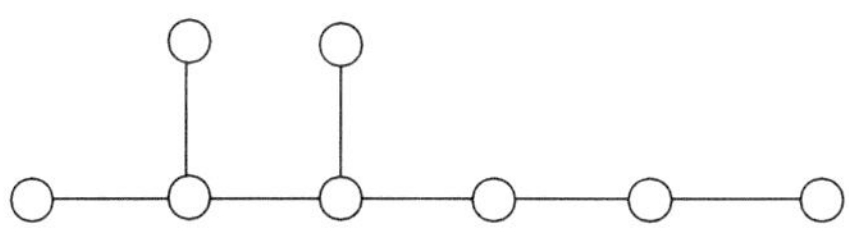

K

Figure 16 *Two graph-teoretical representation of 2,3-dimethylhexane: plerograph P and kenograph K*

structure-property-activity modeling, *e.g.*[121] A few years ago Gutman *et al.*[250] established the relationship between the two-step walks over all vertices in a molecular graph G, denoted by mwc_2 (*mwc* stands for *molecular walk count*) and the sum of the squares of the vertex-degrees (that is, the first Zagreb index M_1):

$$mwc_2 = \sum_{\text{vertices}} d_i\, d_i = M_1 \tag{42}$$

Now Braun *et al.*[251] established the equivalence of molecular walk count of length 3 mwc_3 and the second Zagreb index M_2 and also confirmed the above result of Gutman *et al.*[250] Using the definition of molecular walk count[252,253] (see also our second Report[2]), Braun *et al.*[251] derived the following formula:

$$mwc_3 = 2\sum_{\text{edges}} d_i\, d_j = 2M_2 \tag{43}$$

Thus, equations (42) and (43) show that both Zagreb indices may be regarded as combinatorial quantities related to simple walks on molecular graphs.

2.4 Structural Complexity. – Several papers reporting studies on the complexity of Platonic solids, Archimedean solids, fullerenes and nanotubes have appeared in the last couple of years. These are attractive structures and are of current chemical interest.

2.4.1 Platonic Solids. Because of their high symmetry-based beauty, Platonic solids attract attention not only of scientists, but also of artists.[254,255] *Platonic solids* are regular convex polyhedra and there are five of them: tetrahedron (T), cube (C), octahedron (O), icosahedron (I) and dodecahedron (D) (see their shapes in Figure 17). Note that a convex polyhedron is *regular* if all of its faces are regular polygons. A polyhedron is *convex* if every dihedral angle is less than 180°. The dihedral angle is the angle formed by two polygons joined along a common edge.

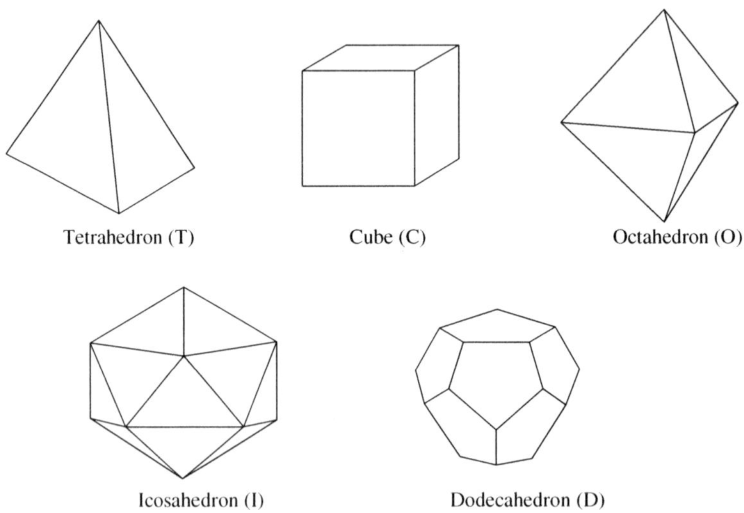

Figure 17 *The shapes of Platonic solids*

Regular convex polyhedra are called Platonic solids because they play an important role in Plato's philosophy. Plato (*ca.* 428–347), however, was not the first to write about regular polyhedra – the mathematician Theatetus (*ca.* 380 B.C.), a friend of Plato's and pupil of Socrates (*ca.* 470–399), discovered the octahedron and the icosahedron and was also the first to write about the five regular polyhedra.[256]

In chemistry, Platonic solids serve as versatile models for molecular shapes.[257–259] Among them, the tetrahedron is especially important in organic chemistry since more than 130 year ago Le Bel[260] and van't Hoff[261] used it independently to model the 3-dimensional structure of carbon compounds. The tetrahedron, the cube and the dodecahedrane are used to model three strained organic molecules: tetrahedrane C_4H_4 (some of its derivatives are known[262]), cubane C_8H_8 (prepared in 1964 by Eaton and Cole[263]) and dodecahedrane $C_{20}H_{20}$ (prepared in 1982 by Paquette *et al.*[264]). The C_{20} fullerene (still unprepared) is the unsaturated analogue of dodecahedrane.[22] Hydrocarbons modeled by the octahedron and icosahedron are not known. However, they model well inorganic compounds.[265] Thus, there are known octahedral molecules, *e.g.*, SF_6, octahedral complexes, *e.g.*, $Co(NH_3)_6{}^{+3}IrCl_6{}^{-3}$, whilst the icosahedron models boron compounds, *e.g.*, $B_{12}H_{12}{}^{2-}$.[266,267]

Bonchev[268] discussed the global, relative and local complexity of Schlegel graphs[269] representing Platonic solids by using several combinatorial measures of complexity: the total subgraph count,[270,271] overall connectivity,[272,273] overall Wiener indices[274] and total walk count.[275] We also discussed these complexity measures in our second Report.[2] Bonchev derived equations for the first several orders of global, relative and local complexity indices as functions of the number of vertices and vertex-degrees and used them to assess the complexity of the Platonic solids. The relative complexity index was defined as the ratio of the complexity index selected and its value for the complete graph having the same number of vertices as the respective Platonic solid. A simple graph is called a *complete graph* if each pair of its vertices is adjacent.[9–11] The global complexity index measures the complexity of the whole structure and the local complexity index is defined *per* vertex or *per* edge.

Bonchev[268] obtained the following *relative* complexity order of the Platonic solids in their Schlegel representation: T > O > I > C > D. Judging by this order the larger Schlegel graphs of a Platonic solid, the lesser chance for it to be close to the related complete graph. This is violated only by icosahedron, presumably because of its high local symmetry. Both kinds of *local* complexity indices (*per* vertex and *per* edge) produced the following order: I > O > D ≥ C ≥ T. The order obtained by the *global* complexity indices is: I > D > O > C > T, indicating the icosahedron as the most complex Platonic solid. This global ordering concides with orderings produced[276] by using the Bertz index[277] and the Estrada second-order edge-connectivity index.[278] The cyclomatic number also gives the same complexity order.[279] The *cyclomatic number* (or the circuit rank) μ of a polycyclic graph G[10,11] is the minimum number of edges that have to be removed from G in order to convert it into an acyclic graph. It is given by the expression: $\mu = E - V - 1$, where E is the number of edges and V is the number of vertices of G.

In a related paper, Balaban and Bonchev[280] produced a complexity order of Platonic solids according to their sphericity expressed in terms of the solid angles of their vertices. *Sphericity* determines how closely a polyhedron, viewed as a geometrical object, approaches a sphere. One measure of sphericity is provided by the solid angle (usually given in steradians or in degrees).[281] The solid angle θ of Platonic solids may be computed as follows:

$$\sin(\delta/2) = \cos(\pi/a)/\cos(\alpha/2) \quad (44)$$

$$\theta = a\,\delta - (a-2)\pi \quad (45)$$

where δ is the dihedral angle and a is the number of edges belonging to n-gons meeting at a vertex forming equal planar angles α. For example, θ for the icosahedron (a = 5, α = 60°, δ = 138° 57′) is 2.635 steradians or 150° 57′. The obtained complexity order based on the solid angle is D > I > C > O > T. This order is in agreement with complexity orders of Platonic solids based on the indices derived from the resistance-distance matrix.[276,279,282]

2.4.2 Archimedean Solids. Archimedean solids[283] (also known as Archimedean bodies[284] and Archimedean polyhedra[255]) were until recently much less studied than the Platonic solids. They came into the focus of research interest after the discovery of buckminsterfullerene, a pure carbon molecule consisting of 60 atoms, and assignment of its structure as that of the truncated icosahedron.[32] As far as we know, buckminsterfullerene is the only real molecule whose structure is modeled by an Archimedean solid. Nevertheless, the Archimedean solids are occasionally studied, usually in conjunction with the Platonic solids and other polyhedra, *e.g.*[285–287]

Archimedean solids are semiregular convex polyhedra.[269] A *semiregular polyhedron* is a polyhedron whose faces are regular polygons, although not all the same, and each of whose vertices is symmetrically equivalent to every other vertex (thus, they are vertex-transitive structures). There are 13 Archimedean solids: truncated tetrahedron (TT), truncated cube (TC), truncated octahedron (TO), truncated icosahedron (TI), truncated dodecahedron (TD), cuboctahedron (CO), truncated cuboctahedron (TCO), snub cuboctahedron (SCO), rhombicuboctahedron (RCO), icosidodecahedron (ID), truncated icosidodecahedron (TID), snub icosidodecahedron (SID) and rhombicosidodecahedron (RID). Their shapes are shown in Figure 18.

The first surviving description of the Archimedean solids is that of the Greek geometer Pappus of Alexandria who lived in the fourth century (around 320).[288,289] Pappus of Alexandria attributed the invention of semiregular convex polyhedra to Archimedes (287–212 B.C.); hence, the name *Archimedean solids*. The painter-mathematicians of the Renaissance were interested in the Golden Cut and in its appearance in the Platonic solids and Archimedean solids.[254] In 1492, the Archimedean solids were rediscovered by the painter and

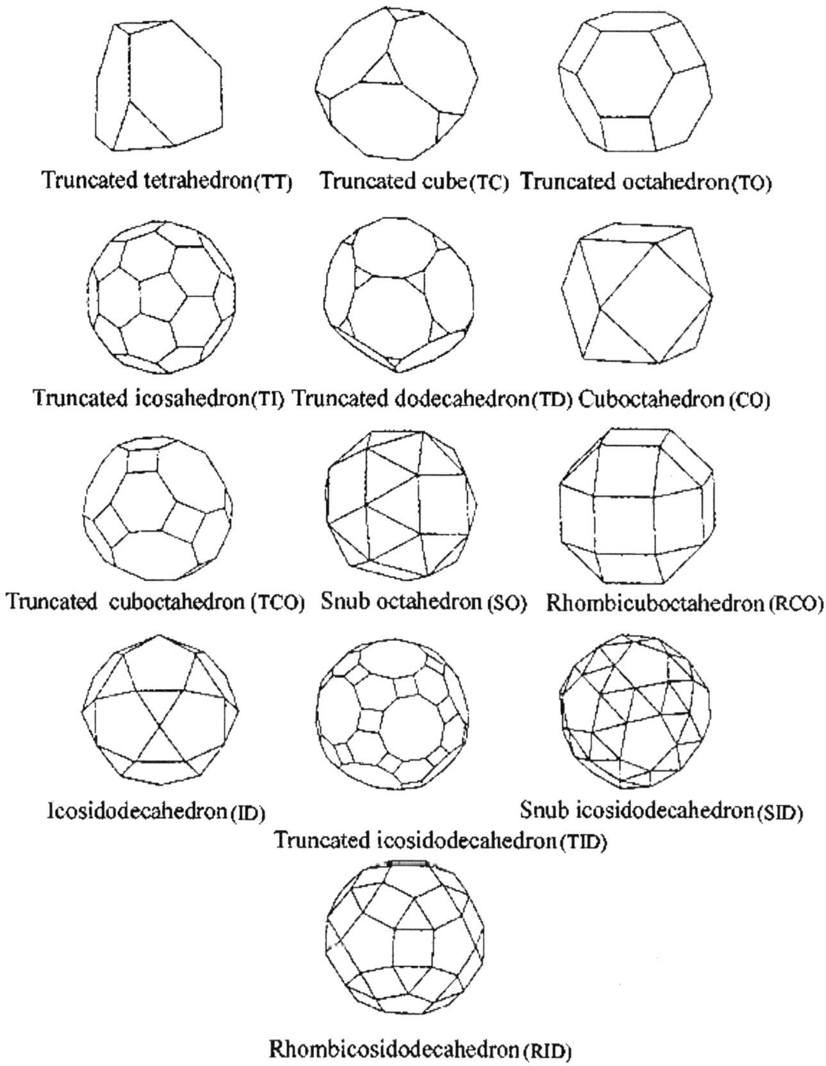

Figure 18 *The shapes of Archimedean solids*

mathematician Piero della Francesca.[290] In his masterpiece *De Divina Proportione* (1509, reproduced in 1956), Luca Pacioli examined, besides the Platonic solids, some of the Archimedean solids, in particular the cubooctahedron. But the Renaissance author who was perhaps most interested in the Archimedean solids was Daniel Barbaro, as can be seen in his book *Prattica de la Perspectiva* (1569). It was, however, Johannes Keppler (1571–1630) who catalogued the 13 Archimedean solids in 1619 and gave them their now generally accepted names.

Rajtmajer *et al.*[291] considered the complexity of the Archimedean solids by means of four complexity indices: the index based on the concept of the augmented vertex-degree, and three indices based on the total number of walks, the total number of paths and the total number of trails. Note, a walk is a *trail* if all the edges are distinct and a *path* if all vertices are distinct.[9,11] The complexity index, based on the concept of the *augmented vertex-degree*, was recently introduced by Randić and Plavšić[292–294] and reviewed by us in our second Report on combinatorial enumeration[2] and by several other groups.[206,295] This concept is based on the notion of partial additivity of vertex-degrees. The augmented degree of a given vertex is obtained by adding to its degree, the degrees of vertices with the weight that depends on their distances from this vertex. The complexity index is then given as the sum of augmented degrees of all vertices in a structure *not* equivalent by symmetry. The greater value of the index, the greater is the complexity of the structure. Since the Archimedean solids are vertex-transitive structures, the complexity index ξ for them is identical with the augmented degree of a single vertex:

$$\xi = (AVD)_i = \sum_{i=1}^{\lambda_{\max}} d_i/2^{\lambda(\mathrm{ij})} \tag{46}$$

where d_i is the degree of vertex i, $\lambda(ij)$ is the shortest distance in terms of the number of edges between vertices i and j, and λ_{max} is the value of the maximal shortest distance.

All the Schlegel graphs representing Archimedean solids are *regular* graphs, that is, all vertices have the same degree d. The number of walks of length q, $w(q)$, in these graphs is equal to $V\,d^q$. Note that the first vertex can be chosen in V ways and its each successor in the walk in d ways. Total walk counts, $twc(q)$, were calculated by summing up all walks up to a certain length q, $w(q)$:

$$twc(q) = \sum_q w(q) \tag{47}$$

Rajtmajer *et al.*[291] calculated the total numbers of paths and trails using their own computer program. They obtained the following ordering of the Archimedean solids by the four complexity indices:

(i) The augmented vertex-degree
TT < TC < TO < TD < TCO < TID < TI < CO < RCO < ID < RID < SCO < SID

(ii) The total number of walks with up to the length of 8
TT < TC = TO < TCO < TI = TD < CO < TID < RCO < ID < RID < SCO < SID

(iii) The total number of paths with up to the length of 8
TT < TC < TO < TD < TCO < CO < TI < TID < RCO < ID < RID < SCO < SID

(iv) The total number of trails with up to the length of 8
TT < TC < TO < TD< TCO < TI < TID < CO < RCO < ID < RID < SCO < SID

All four indices predict the truncated tetrahedron (TT), the truncated cube (TC) and the truncated octahedron (TO) as the least complex structures, and the rhombicuboctahedron (RCO), the icosidodecahedron (ID), the rhombicosidodecahedron (RID), the snub cuboctahedron (SCO) and the snub icosidodecahedron (SID) as the most complex structures among the Archimedean solids. The ordering of TI, TD, CO, TCO and TID differs from index to index. The above can be visualized by the Hasse diagram based on the partial order of the four complexity indices (see Figure 19). It should be noted that some authors consider the complexity as a partially-ordered quantity.[296–300] The Hasse diagram reflecting the partial order of indices appears to be a very useful device for appraising the structural complexity of molecules and graphs.

The diagram in Figure 19 is such that in going downward along a path from structure X to structure Y, all four indices have a smaller value for X than for Y. Furthermore, the two structures X and Y are directly linked by an edge downward from X to Y if and only if no third structure is placed by this partial ordering between X and Y.

The complexity order of Archimedean solids in terms of the solid angle of their vertices is[280] TT < CO < TC < TO < RCO < ID < TCO < TD < TCO < RID < TID. The two chiral Archimedean solids (snub octahedron, snub icosidodecahedron) were not considered. This order disagree with all four complexity given above, except in the case of the truncated tetrahedron which is predicted to be the least complex of all Archimedean solids. This discrepancy is perhaps due to different bases of the compared complexity orders; the above orders being the result of 2D representation and the Balaban-Bonchev order of 3D structure of Archimedean solids.

2.4.3 Fullerenes and Nanotubes. Randić *et al.*[301] considered the problem of evaluating the complexities of 19 lower fullerenes having from 20 to 50 carbon atoms and of buckminsterfullerene C_{60}. They selected this set of fullerenes because they had been already studied by Laidboeur *et al.*[302] who depicted all these fullerenes by means of their Schlegel graphs and denoted their symmetries. Schlegel graphs, named after the 19th century German mathematician Victor Schlegel,[303] are convenient graph-theoretical representations of fullerenes – they are planar 3-connected graphs.[269,304] In Figure 20, we give, as an example, the Schlegel graph representing buckminsterfullerene.

Randić *et al.*[301] studied the complexity of lower fullerenes by using the distance degree sequences *DDS*s for all symmetry non-equivalent vertices, the multiplicity μ of the *DDS*s, which are defined as the cardinality of each equivalence class, the augmented degrees for all vertices of different equivalence class ξ (see text above),[292–294] which represent a measure of the local

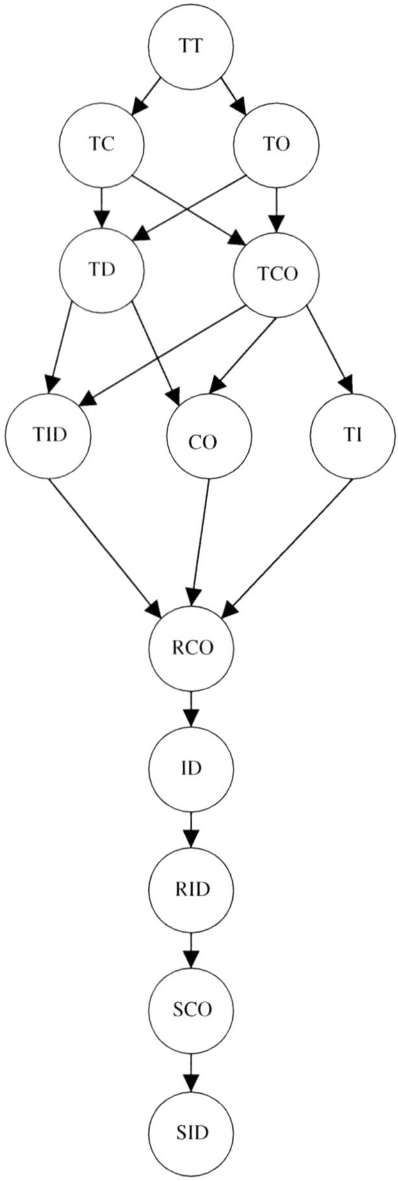

Figure 19 *The partial order by means of the Hasse diagram for Archimedean solids based on the four complexity indices*

complexity, the average value of the augmented vertex-degrees ξ_{av} and the sum of all augmented vertex-degrees as the overall complexity index $V\xi_{av}$. For illustration, we give below the values of these complexity measures for buckminsterfullerene (since we gave in Figure 20 the Schlegel graph of C_{60}, these

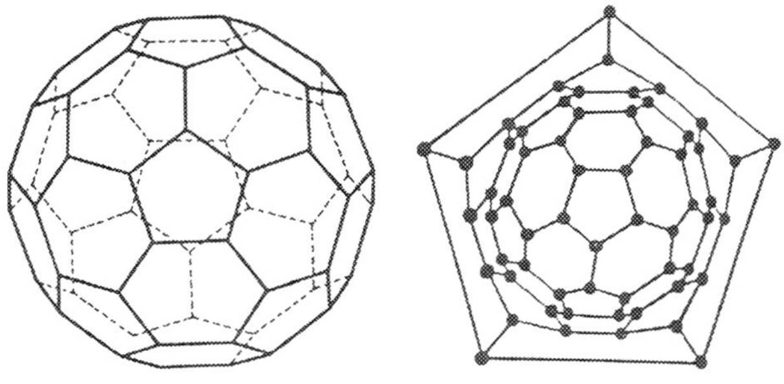

C_{60} Schlegel graph of C_{60}

Figure 20 *The truncated icosahedron structural model of buckminsterfullerene C_{60} and the corresponding Schlegel graph*

measures can be checked easily):

$$\begin{array}{ll} DDS: & 3, 9, 18, 24, 30, 30, 24, 9, 3 \\ \mu: & 60 \\ \xi = \xi & 18.5098 \\ V\xi_{av} & 1110.5880 \end{array}$$

Both ξ_{av} and $V\xi_{av}$ increase with the size of the fullerene. However, the authors considered two isomers of C_{38} (D_{3h} and C_{3v}), three isomers of C_{40} (T_d, C_{3v} and D_{5d}) and two isomers of C_{44} (T and D_{3h}) Since the augmented vertex-degree depends on the symmetry characteristics of a structure, it is expected that the indices ξ_{av} and $V\xi_{av}$ would have higher values for a less symmetric structure. Thus, C_{38} (C_{3v}) and C_{44} (D_{3h}) are predicted to be more complex than C_{38} (D_{3h}) and C_{44} (T). However, in the case of three C_{40} isomers, two isomers possessing different symmetries (T_d and C_{3v}) have identical values of ξ_{av} and $V\xi_{av}$. This is probably a coincidence, possibly as a result of the average *DDS* being the same for both fullerenes: 3.00, 9.00, 18.00, 22.50, 25.50, 22.50, 12.00, 5.40.

In the same article, Randić *et al.*[301] also considered the complexity of single-wall carbon nanotubes. These nanotubes can be constructed by rolling a long strip of a planar graphene sheet. Their properties depend on how the ends of the strip become connected. In this paper, the authors used the concept introduced by Klein *et al.*[305]: The geometry of a nanotube if defined in terms of two parameters – the *twist* t_+ and the *countertwist* t_- under the condition that t_+ must be an integer higher than, or equal to, 3, whilst t_- can have any integer value between 0 and t_+, that is, $t_- \leq t_+$. The numbers t_- and t_+ indicate how many linearly condensed hexagons in two direction diverging from one hexagon by 120° in the nanotube are involved in the repeating pattern on the nanotube until they overlap when rolling the strip. When $t_- = 0$, the nanotube is achiral and is called a zigzag nanotube whose diameter in Ångströms is approximately $0.78t_+$. When $t_+ = t_-$, the nanotube is also achiral and is called an armchair nanotube with the diameter $0.78t_+\sqrt{3}$. When $t_+ \neq t_-$, the nanotube is chiral with the diameter $0.78(t_-^2 + t_+^2 + t_-t_+)^{1/2}$. The twist t_+ and the countertwist t_- parameters determine the helicity of nanotubes by $\sin(1 - t_-/t_+)\pi$.

The complexity of a nanotube depends on its length, diameter and helicity. The values of the ratio t_-/t_+ range from zero for the zigzag nanotubes through 0.5 for the skew nanotubes to 1 for the armchair nanotubes. Therefore, the helicity of nanotubes ranges from 0 for the zigzag and armchair nanotubes to 1 for the skew nanotubes. Hence, the skew nanotubes are more complex than either zigzag or armchair nanotubes. The key parameters for studying complexity of nanotubes are the twist t_+ and the countertwist t_- since they determine the diameter and helicity of nanotubes.

2.5 Other Enumerations. – In this section, we present combinatorial enumerations that do not fit in any of the above sections, though some are related to discussions above or in our previous Reports.[1,2]

2.5.1 The Number of Spanning Trees. Kirby *et al.*[306] put forward a theorem for counting spanning trees in general molecular graphs (that is, non-planar graphs with loops and multiple edges) with particular application to toroidal fullerenes. A spanning tree is a connected subgraph of graph G containing all vertices of G.[11] The problem of counting spanning trees goes back to the work by Kirchhoff[307] and subsequent mathematical formulations.[190] Spanning trees are used to study the complexity of reaction mechanisms,[308] of molecular graphs,[206,309,310] in calculating the magnetic properties of conjugated polycyclic molecules by means of the ring-current model within the framework of the π-electron molecular orbital theory.[311–313]

An important contribution to the enumeration of spanning trees of molecular *planar* graphs was the theorem by Gutman *et al*:[314]

$$t(G) = det\ \mathbf{D} \tag{48}$$

where $t(G)$ is the number of spanning trees and $\boldsymbol{D}$ is the difference matrix defined as follows:

$$\boldsymbol{D} = \boldsymbol{\Delta}^* - \boldsymbol{A}^* \tag{49}$$

where $\boldsymbol{A}^*$ is the vertex-adjacency matrix of an inner dual G^* of G and $\boldsymbol{\Delta}^*$ is a diagonal matrix of G^* with elements $\lambda^*(k)$, where $\lambda^*(k)$ is the size of cycle k (called the patch cycle by Kirby *et al.*[306]) in G. Matrix $\boldsymbol{D}$ may also be regarded as a *Laplacian matrix*[315–317] of the vertex-weighted inner dual G^*, the weights of vertices in G^* being equal to the sizes of cycles in the parent graph G.

An inner dual G^* of a planar polycyclic graph G is obtained by placing a vertex in each cycle of G and pairs of vertices in G* are connected if the corresponding rings in G have common edges. Examples of a graph and its inner dual are given in Figure 21.

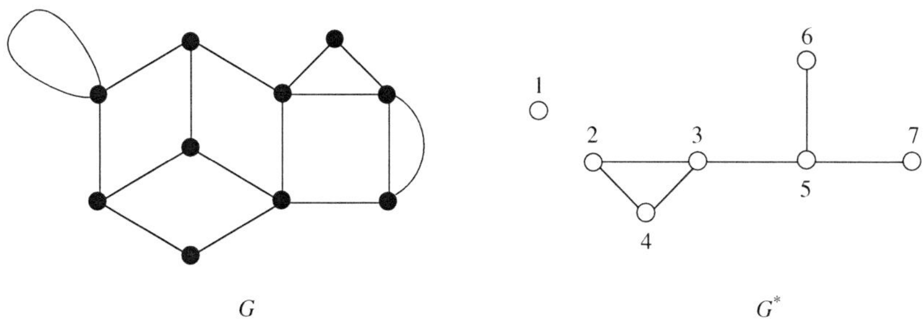

Figure 21 *Graph G and its labelled disconnected inner dual G**

The difference matrix $\boldsymbol{D}$ belonging to the inner dual G^* from Figure 21 and the corresponding number of spanning trees are given below.

$$\begin{matrix} 1 & 0 & 0 & 0 & 0 & 0 & 0 \\ 0 & 4 & -1 & -1 & 0 & 0 & 0 \\ 0 & -1 & 4 & -1 & -1 & 0 & 0 \\ 0 & -1 & -1 & 4 & 0 & 0 & 0 \\ 0 & 0 & -1 & 0 & 4 & -1 & -1 \\ 0 & 0 & 0 & 0 & -1 & 3 & 0 \\ 0 & 0 & 0 & 0 & -1 & 0 & 2 \end{matrix}$$

$$t(G) = 860$$

The theorem stated as equation (48) is a special case of the theorem derived by Kirby *et al.*:[306]

$$t(G) = \det \boldsymbol{V}/(\det \boldsymbol{U})^2 \tag{50}$$

where the $\boldsymbol{V}$-matrix is defined as:

$$\boldsymbol{V} = \boldsymbol{Z}\boldsymbol{Z}^{\mathrm{T}} \tag{51}$$

and the $\boldsymbol{U}$-matrix is a non-singular matrix to be chosen in such a way that its determinant is 1 or some small integer. The $\boldsymbol{Z}$-matrix is the cycle-edge incidence matrix.[317,318]

The cycle-edge incidence matrix $\boldsymbol{Z}$ is a $c \times e$ matrix (c and e being the numbers of cycles and edges, respectively), which is determined by the incidences of cycles and edges in G:

$$[\boldsymbol{Z}]_{\mathrm{ij}} = \begin{cases} 1 & \text{if a cycle } i \text{ is incident with an edge } j \\ 0 & \text{otherwise} \end{cases} \tag{52}$$

Kirby *et al.*[306] considered three kinds of cycles: independent cycles, fundamental cycles and patch cycles and showed that they all produce the same number of spanning trees. However, *det* $\boldsymbol{U}$ is equal to 1 only for the fundamental set of cycles of any graph and for patch cycles of any planargraph.

If equations (49) and (51) are applied to planar graphs and use is made of patch cycles, then:

$$\boldsymbol{V} = \boldsymbol{Z}\boldsymbol{Z}^{\mathrm{T}} = \Delta^{*} - A^{*} = \boldsymbol{D} \tag{53}$$

and consequently:

$$t(G) = \det \boldsymbol{V}/(\det \boldsymbol{U})^{2} = det\ \boldsymbol{D} \tag{54}$$

2.5.2 The Number of Conjugated Circuits in Graphite. Lukovits[149] calculated the number of conjugated circuits in (m,n) parallelogram-shaped graphite sheets and used them to obtain the resonance energies per electron (REPE) of graphite sheets. He computed the benzene-like, the naphthalene-like and the anthracene-like conjugated circuits $[6]_{m,n}$, $[10]_{m,n}$ and $[14]_{m,n}$ for several values of m and n ($m,n \leq 6$) using the following equations:

(i) The benzene-like conjugated circuits

$$[6]_{m,n} = 2(m+n-1)!/(m-1)!\,(n-1)! \tag{55}$$

or

$$[6]_{m,n} = [6]_{m,n-1} + [6]_{m-1,n} + [6]_{m-1,n}/(\mathrm{m}-1) \tag{56}$$

(ii) The naphthalene-like conjugated circuits

$$[10]_{m,n} = [6]_{m-1,n} + [6]_{m,n-1} \tag{57}$$

From equations (56) and (57) it follows that if m and n are sufficiently large, then $[6]_{m,n} = [10]_{m,n}$ in (m,n) parallelogram-shaped graphite sheets. By setting $m = n$, from equations (55), (56) and (57), Lukovits[149] obtained:

$$[10]_{n,n} = 2\{(2n-2)!\,(2\mathrm{n}-2)/[(n-1)!]^{2}\} \tag{58}$$

(ii) The anthracene-like conjugated circuits

$$[14]_{m,n} = [6]_{m-2,n} + [6]_{m,n-2} \tag{59}$$

In any conjugated hydrocarbon G made up of hexagons subunits, the REPE is given by:[200,319,320]

$$\mathrm{REPE}_{G} = \{[6]_{G}\, R_{6} + [10]_{G}\, R_{10} + [14]_{G}\, R_{14} + \ \ldots\ \}/V\, K_{G} \tag{60}$$

where V is the number of carbon atoms in G. For (n,n) parallelogram-shaped graphite sheets, V is equal to:

$$V = 2(n^{2} + 2n) \tag{61}$$

Randić,[114,115] who introduced the concept of conjugated circuits, used R_{4n+2} to denote the energy contributions of the $(4n+2)$-membered conjugated circuits to the resonance energy.

Lukovits[149] calculated the resonance energy per electron (REPE) of graphite sheets using the expression that he obtained from equations (11), (55), (58), (60) and (61):

$$[\text{REPE}]_{n,n} = [n\,R_6/(2n^2 + 4n)] + \{[n^2 - n)R_{10}/[(2n - 1)\,(n^2 - 2n)]\} \tag{62}$$

Lukovits has taken the values of the parameters $R_6 = 0.8220$ eV and $R_{10} =$ 0.3355 eV from Randić *et al.*[321] The value of parameter R_{14} is rather small (0.0580 eV) and its inclusion in the REPE expression depends on the number of the anthracene-like conjugated circuits. Since Lukovits established that the number of $[14]_{m,n}$ circuits is much smaller than the number of $[10]_{m,n}$ circuits, he did not include them in the REPE calculation of graphite sheets.

2.5.3 The Number of Conjugated Circuits in Nanotubes. Lukovits and Janežič[182] calculated the number of conjugated circuits in armchair $(1,1)_n$ carbon nanotubes and $(n,n)_1$ nanoribbons and used them to obtain the resonance energies per electron (REPE) of these structures. For any $(1,1)_n$, nanotube, they derived the following formulae for 6- and 10-membered circuits in terms of the corresponding Kekulé structures:

$$\begin{aligned}[6](1,1)_n =& 4(K^*(1,1)_{n-2}K(1,1)_{-1} + K^*(1,1)_{n-3}K(1,1)_0 \\ &+ K^*(1,1)_{n-4}K(1,1)_1 + \ldots + K^*(1,1)_{-1}K(1,1)_{n-2}+\end{aligned} \tag{63}$$

$$[10](1,1)_n = [6](1,1)_{n-1} \tag{64}$$

Using data from Table 5, one can easily obtain the number of two smallest conjugated circuits of $(1,1)_n$ nanotubes.

For nanoribbons, the counting formulas for 6- and 10-membered circuits are given in terms of the *K*-numbers of polyphenanthrenes (note that $K(2h) = K(n)_{\text{ph}}$)

$$[6](n,n)_1 = 4n\;K(2n-3)_{\text{ph}} \tag{65}$$

$$[10](n,n)_1 = 4n\;K(2n-4)_{\text{ph}} \tag{66}$$

Lukovits and Janežič[182] calculated the resonance energies of armchair $(1,1)_n$ carbon nanotubes and $(n,n)_1$ nanoribbons for a given n using equation (64) and found for n very large REPE of $(1,1)_n$ nanotubes is 0.160 eV and for nanoribbons 0.142 eV. Since these values are a bit larger than REPE for buckminsterfullerene (0.12 eV),[303] Lukovits and Janežič[182] concluded that carbon nanotubes belong to the most aromatic structures.

2.5.4 Combinatorial Matrices. There are a number of combinatorial matrices available in the literature, usually used to generate graph invariants.[120,121,317] We have already mentioned some of these matrices, such as the random-walk

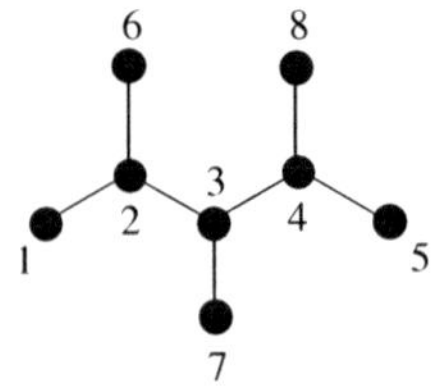

(i) Edge-graphical matrix $^{e}\boldsymbol{G}$

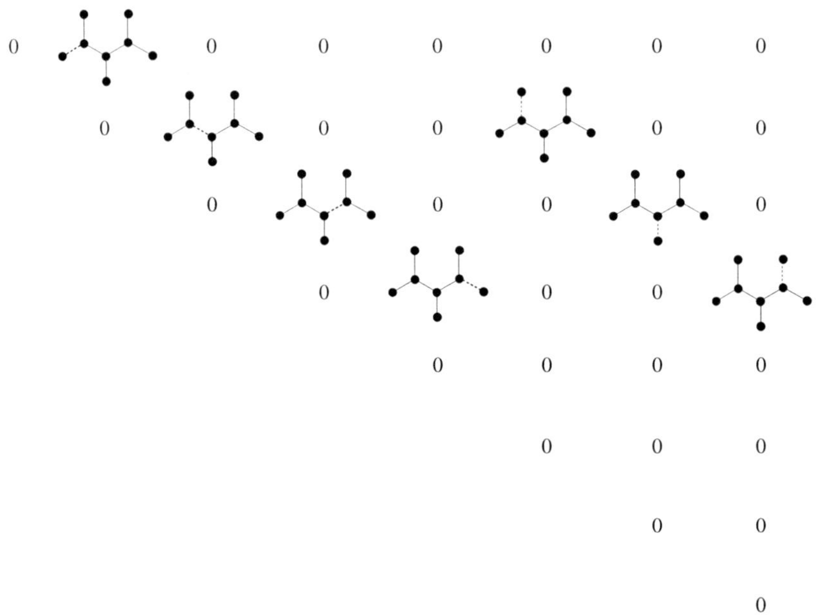

Figure 22 *Four types of graphical matrices for a tree representing the carbon skeleton of 2,3,4-trimethylpentane*

Markov matrix (section 2.3.1), the Laplacian or difference matrix (sections 2.3.1 and 2.5.1) and the incidence matrix (section 2.5.1).

Randić *et al.*[322] proposed a novel type of combinatorial matrices, which they called *graphical matrices*. They used these matrices to generate the distance-based molecular descriptors. Graphical matrices are combinatorial matrices whose elements are subgraphs of the graph rather than numbers – hence their name. Thus far, very little work has been reported on these matrices.[323] However, many combinatorial matrices,[120] such as the Wiener matrices[324–326] and the Hosoya matrices,[327] may be regarded as the numerical realizations of the corresponding graphical matrices.[317]

The advantage of a graphical matrix lies in the fact that it allows for a great many possibilities of numerical realizations. In order to obtain a numerical form of a graphical matrix, one needs to select a graph invariant

(ii) Sparse vertex-graphical matrix ${}^{sij}\boldsymbol{G}$

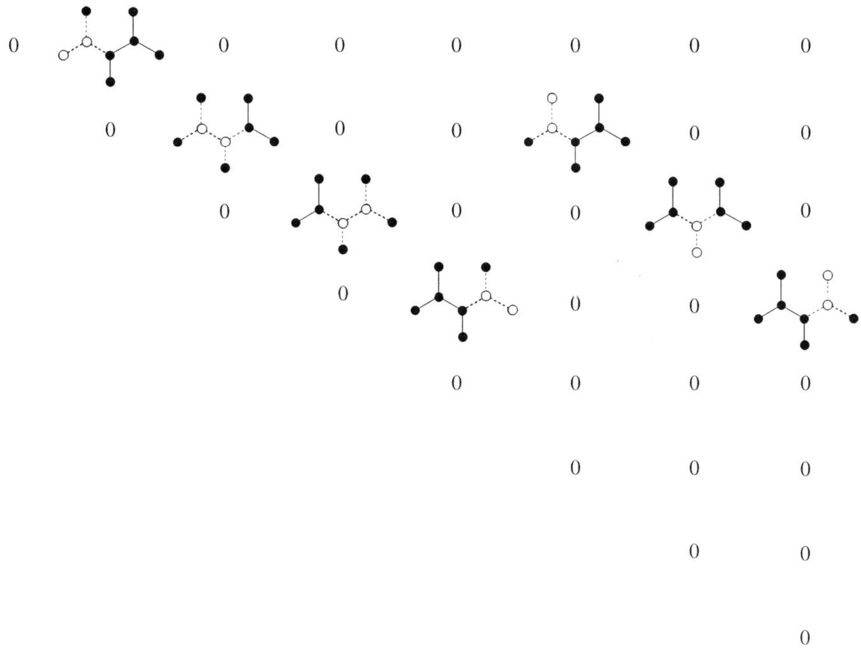

Figure 22 (continued)

and replace all the graphical elements (subgraphs of some form) by the corresponding numerical values of the selected invariant. In this way, the numerical form of the graphical matrix is established and one can select another or the same type of invariant – this time an invariant of the numerical matrix. Graph invariants generated in this way are *double invariants*[323] in view of the fact that two invariants are used in constructing a given molecular descriptor.

Nikolić *et al.*[328] classified graphical matrices $\boldsymbol{G}$ into four types-two being *sparse* matrices: (i) the *edge-graphical* matrices ${}^{e}\boldsymbol{G}$, (ii) the *vertex-graphical* matrices ${}^{sij}\boldsymbol{G}$, and two being *dense* matrices: (iii) the *path-graphical* matrices ${}^{p}\boldsymbol{G}$ and (iv) the *vertex-graphical* matrices ${}^{dij}\boldsymbol{G}$. The elements of the edge-graphical matrix ${}^{e}\boldsymbol{G}$ are obtained by consecutive removal of *edges* connecting vertices i and j from graph $\boldsymbol{G}$. The elements of the sparse vertex-graphical matrix ${}^{sij}\boldsymbol{G}$ are obtained by the consecutive removal of *adjacent* vertices i and j, and the incident edges from $\boldsymbol{G}$. The elements of the path-graphical matrices ${}^{p}\boldsymbol{G}$ are obtained by consecutive removal of the *paths* joining vertices i and j from G. The elements of the dense vertex-graphical matrices ${}^{dij}\boldsymbol{G}$ are obtained by

(iii) Path-graphical matrix ${}^{p}\boldsymbol{G}$

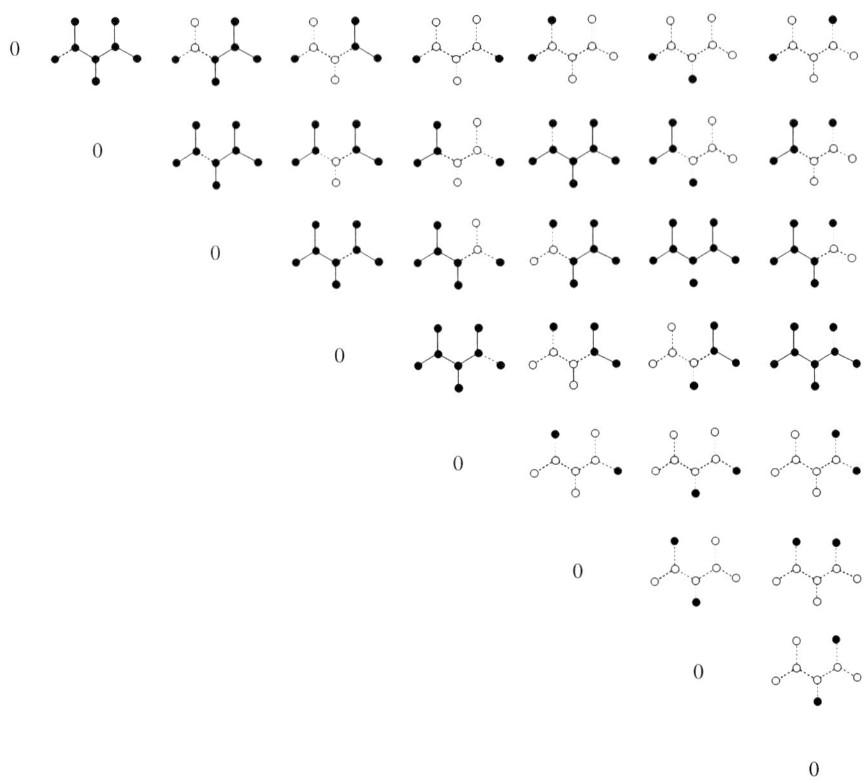

Figure 22 (continued)

removing pairs of vertices *i* and j at increasing distances from *G*. In their paper, Randić *et al.*[322] discussed only the dense vertex-graphical matrix ${}^{dij}\boldsymbol{G}$. In Figure 22, we give all four types of graphical matrices for 2,3,4-trimethylpentane. Since the graphical matrices are square symmetrical matrices, we give only their upper triangles.

The next step is to replace (sub)graphs with the invariants of choice. For this purpose, Randić *et al.*[322] used the Wiener index[225,329,330] and Nikolić *et al.*[328] used the Wiener index and the Hosoya index.[122] The numbers that replace the subgraphs in the graphical matrices are obtained by *summing up* their Wiener indices or their Hosoya indices. The numerical forms of four graphical matrices presented in Figure 22, obtained using the Wiener number, are given in Table 7. The summation of the matrix-elements in each matrix-triangle gives the following descriptors of 2,3,4-trimethylpentane: the *edge-Wiener-Wiener index* ${}^{e}WW$, the sparse *vertex-Wiener-Wiener index* ${}^{suv}WW$, the *path-Wiener-Wiener number* ${}^{p}WW$ and the dense *vertex-Wiener-Wiener index* ${}^{duv}WW$. These numbers are also given in Table 7.

(iv) Dense vertex-graphical matrix $^{dij}\boldsymbol{G}$

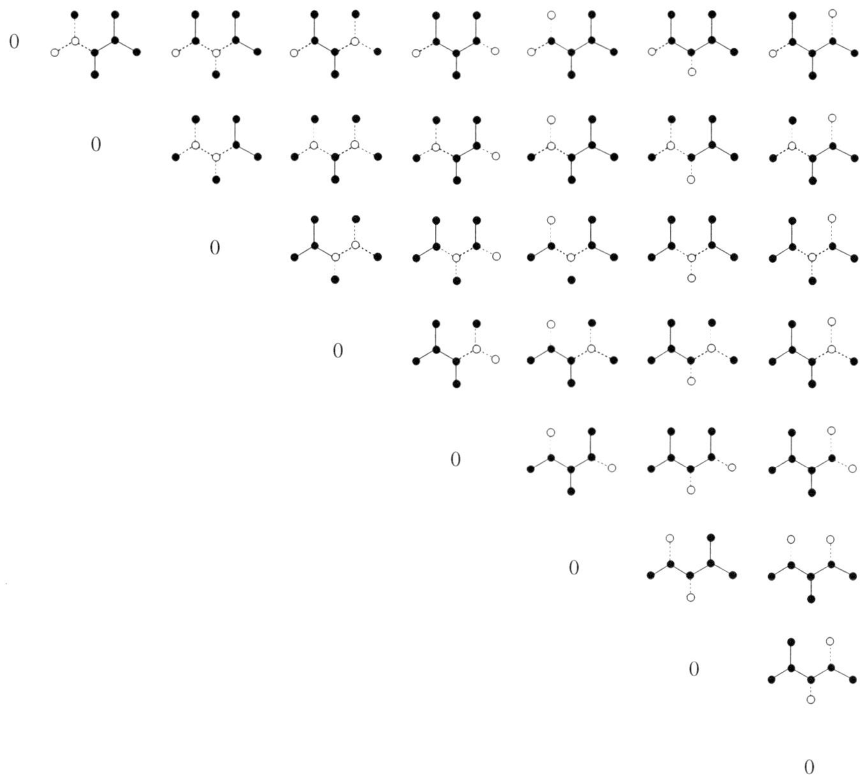

Figure 22 (continued)

It should be noted that if we replace subgraphs in the edge-Wiener matrix by *products* of the number of vertices in each subgraph, the obtained numerical matrix, called the *Wiener matrix* $\boldsymbol{W}$,[324–326] also called the edge-Wiener matrix $^{e}\boldsymbol{W}$,[120,317] gives the standard Wiener index W. Similarly, if we replace each non-zero element in the edge-Wiener matrix with its inverse value, the obtained modified edge-Wiener matrix ^{m}W gives the modified Wiener index $^{m}\boldsymbol{W}$.[331] Finally, if we replace subgraphs in the path-Wiener matrix by *products* of the numbers of vertices in each subgraph, the obtained numerical matrix $^{h}\boldsymbol{W}$ gives the *hyper-Wiener index*.[317,324,325,332] All the three results are valid *only* for acyclic graphs. Examples of the edge-Wiener matrix, the modified edge-Wiener matrix and the hyper-Wiener matrix and the corresponding indices are presented in Table 8.

3 Conclusion

In the last six years we reported on a number of novel results achieved in the combinatorial enumeration in chemistry. In the period of 1999–2001, we

Table 7 *Four Wiener-like matrices (only upper triangles are shown) of 2,3,4-trimethylpentane and their corresponding Wiener-Wiener indices*

(i) Edge-Wiener matrix $^{e}\mathbf{W}$							
0	46	0	0	0	0	0	0
	0	22	0	0	46	0	0
		0	22	0	0	0	48
			0	46	0	0	46
				0	0	0	0
					0	0	0
						0	0
							0

$^{e}WW = 276$

(ii) Sparse vertex-Wiener matrix $^{sij}\mathbf{W}$							
0	18	0	0	0	0	0	0
	0	4	0	0	18	0	0
		0	4	0	0	8	0
			0	18	0	0	18
				0	0	0	0
					0	0	0
						0	0
							0

$^{sij}WW = 88$

(iii) Path-Wiener matrix $^{p}\mathbf{W}$							
0	46	18	4	0	0	0	0
	0	22	8	4	46	8	4
		0	22	18	18	48	18
			0	46	4	4	46
				0	0	0	0
					0	0	0
						0	0
							0

$^{p}WW = 384$

(iv) Dense vertex-Wiener matrix $^{dij}\mathbf{W}$							
0	18	4	10	31	29	32	21
	0	4	1	10	18	9	31
		0	4	4	5	8	5
			0	18	10	9	18
				0	31	32	29
					0	32	31
						0	32
							0

$^{dij}WW = 496$

Table 8 *The edge-Wiener matrix $^{e}\mathbf{W}$, the modified edge-Wiener matrix $^{m}\mathbf{W}$ and the hyper-Wiener matrix $^{h}\mathbf{W}$ and the corresponding indices W, ^{m}W and ^{h}W of 2,3,4-trimethylpentane*

0	7	0	0	0	0	0	0
	0	15	0	0	7	0	0
		0	15	0	0	0	7
			0	7	0	0	7
				0	0	0	0
					0	0	0
						0	0
							0

$W = 65$

0	1/7	0	0	0	0	0	0
	0	1/15	0	0	1/7	0	0
		0	1/15	0	0	0	1/7
			0	1/7	0	0	1/7
				0	0	0	0
					0	0	0
						0	0
							0

$^{m}W = 0.848$

0	7	5	3	1	1	1	1
	0	15	9	3	7	6	3
		0	15	5	5	7	5
			0	7	3	3	7
				0	1	1	1
					0	1	1
						0	0
							0

$^{h}W = 122$

considered 96 papers published during these two years that were concerned with combinatorial enumerations. The next two years (2001–2003), there has been even a greater output – we considered 100 papers that have been reporting topics from the area of combinatorial enumerations in chemistry. Finally, in the last two years (2003–2005), there were 89 papers that we included in our report. But for the each period mentioned, we included in our reports a much greater number of references, since we have mentioned older results and related results from other areas of chemistry that find useful chemical enumerations. We also considered journals that published most of combinatorial results in each period. In all three periods (1999–2001, 2001–2003, 2003–2005), the first medium for publishing combinatorial enumerations in chemistry was the *Journal of Chemical Information and Computer Sciences* (JCICS), published by the American Chemical Society. This journal since the beginning of 2005 has split into two journals – the *Journal of Chemical Information and Modeling* and the *Journal of Chemical Theory and Computation*. It is expected that the first of these two journals will inherit the tradition of JCICS and will continue to publish papers about chemical combinatorial enumerations. The next important journals for chemical combinatorics appear to be the *Journal of Mathematical Chemistry* and *MATCH Communications in Mathematical and Computer Chemistry*. Additionally, two more journals with frequent contribution to chemical enumerations were *Chemical Physics Letters* and *Croatica Chemica Acta*, published by the Croatian Chemical Society. In the period 1999–2001, a number of chemical combinatorial papers appeared in the *Bulletin of Chemical Society of Japan* and in the period just reviewed in this report (2003–2005), in the *Internet Electronic Journal of Molecular Design* (www.biochempress.com). Important contribution over the years appeared in *Nature*, the *Journal of the American Chemical Society*, the *Journal of Organic Chemistry*, I*nternational Journal of Quantum Chemistry* and *Theretical Chemistry Accounts* (formerly *Theoretica Chimica Acta*).

Acknowledgment

We thank the Ministry of Science, Education and Sports of the Republic of Croatia for the support through the Grant No. 0098034.

References

1. D.J. Klein, D. Babić and N. Trinajstić, in: *Chemical Modelling – Applications and Theory*, ed. A. Hinchliffe, Specialist Periodical Reports, The Royal Society of Chemistry, Cambridge, UK, 2002, Vol. 2, pp. 56–95.
2. D. Babić, D.J. Klein, J. von Knop and N. Trinajstić, in: *Chemical Modelling – Applications and Theory*, ed. A. Hinchliffe, Specialist Periodical Reports, The Royal Society of Chemistry, Cambridge, UK, 2004, Vol. 3, pp. 126–170.
3. I. Lukovits, *J. Chem. Inf. Comput. Sci.*, 1999, **39**, 563.
4. I. Lukovits, *J. Chem. Inf. Comput. Sci.*, 2000, **40**, 361.
5. I. Lukovits, in: *Topology in Chemistry-Discrete Mathematics of Molecules*, eds. D.H. Rouvray and R.B. King, Horwood Publishing Ltd., Chichester, 2002, pp. 327–337.
6. I. Lukovits, *Croat. Chem. Acta*, 2004, **77**, 295.
7. J. von Knop, W.R. Müller, Ž. Jeričević and N. Trinajstić, *J. Chem. Inf. Comput. Sci.*, 1981, **21**, 91.
8. V. Kvasnička and J. Pospichal, *J. Chem. Inf. Comput. Chem.*, 1990, **30**, 99.
9. F. Harary, *Graph Theory*, Addison-Wesley, Reading, MA, 2nd printing, 1971.

10. N. Trinajstić, *Chemical Graph Theory*, CRC, Boca Raton, Fl, 2nd revised Edn., 1992.
11. R.J. Wilson, *Introduction to Graph Theory*, Oliver & Boyd, Edinburgh, 1972.
12. R.C. Read, in: *Chemical Applications of Graph Theory*, ed. A.T. Balaban, Academic Press, London, 1976, pp. 25–61.
13. J. von Knop, K. Szymanski, W.R. Müller, H.W. Kroto and N. Trinajstić, *J. Comput. Chem.*, 1987, **8**, 549–554.
14. I. Gutman, W. Linert, I. Lukovits and A.A. Dobrynin, *J. Chem. Inf. Comput. Chem.*, 1997, **37**, 349.
15. J. von Knop, W.R. Müller, K. Szymanski and N. Trinajstić, *Computer Generation of Certain Classes of Molecules*, SKTH, Zagreb, 1985.
16. N. Trinajstić, S. Nikolić, J. von Knop, W.R. Müller and K. Szymanski, *Computational Chemical Graph Theory: Characterization, Enumeration and Generation of Chemical Structures by Computer Methods*, Ellis Horwood Ltd., Simon & Schuster, Chichester, 1991.
17. H.L. Morgan, *J. Chem. Doc.*, 1965, **5**, 107.
18. I. Lukovits and I. Gutman, *Croat. Chem. Acta*, 2002, **75**, 563.
19. M. Petkovšek and T. Pisanski, *Croat. Chem. Acta*, 2005, **78**, 563.
20. F. Bergeron, G. Labelle and P. Leroux, *Combinatorial Species and Tree-like Structures*, University Press, Cambridge, 1998.
21. F. Harary and E.M. Palmer, *Graphical Enumeration*, Academic Press, London, 1973, p. 240 (all simple graphs); p. 242 (all connected simple graphs).
22. H. Prinzbach, A. Weiler, P. Landenberger, F. Wahl, J. Wörth, L.T. Scott, M. Gelmont, D. Olevano and B. Von Issendorff, *Nature*, 2000, **407**, 60.
23. X. Liu, D.J. Klein, T.G. Schmalz and W.A. Seitz, *J. Comput. Chem.*, 1991, **12**, 1252.
24. D.E. Manolopoulos, J.C. May and S.E. Down, *Chem. Phys. Lett.*, 1991, **181**, 105.
25. P.W. Fowler, D.E. Manolopoulos and R.C. Batten, *J. Chem. Soc. Faraday Trans.*, 1991, **87**, 3103.
26. D. Babić and N. Trinajstić, *Comput. Chem.*, 1993, **17**, 271.
27. P.W. Fowler and D.E. Manolopoulos, *An Atlas of Fullerenes*, Oxford University Press, New York, 1995.
28. G. Brinkmann and A.W.M. Dress, *J. Algorithms*, 1997, **23**, 345.
29. A.M. Livshits and Y.E. Lozovik, *J. Chem. Inf. Comput. Sci.*, 2004, **44**, 1517.
30. T.G. Schmalz, W.A. Seitz, D.J. Klein and G.E. Hite, *Chem. Phys. Lett.*, 1986, **130**, 203.
31. H.W. Kroto, *Nature*, 1987, **329**, 529.
32. H.W. Kroto, J.R. Heath, S.C. O'Brien, R.F. Curl and R.E. Smalley, *Nature*, 1985, **318**, 162.
33. S.J. Austin, P.W. Fowler, P. Hansen, D.E. Manolopoulos and M. Zheng, *Chem. Phys. Lett.*, 1994, **228**, 478.
34. S.J. Austin, P.W. Fowler, D.E. Manolopoulos, G. Orlandi and F. Zerbetto, *J. Phys. Chem.*, 1995, **99**, 8076.
35. A.C. Lunn and J.K. Senior, *J. Phys. Chem.*, 1929, **33**, 1027.
36. V.V. Iliev, *MATCH Commun. Math. Comput. Chem.*, 1999, **40**, 153.
37. V.V. Iliev, *J. Math. Chem.*, 2003, **34**, 137.
38. V.V. Iliev, *MATCH Commun. Math. Comput. Chem.*, 2004, **50**, 39.
39. E. Ruch, W. Hässelbarth and B. Richter, *Theoret. Chim. Acta*, 1970, **19**, 288.
40. W.G. Körner, *Gazz. Chim. Ital.*, 1874, **4**, 305.

41. *e.g.*, C.R. Noller, *Chemistry of Organic Compounds*, Saunders, Philadelphia, reprinted, 1952.
42. S. Fujita, *Symmetry and Combinatorial Enumeration in Chemistry*, Springer, Berlin, 1991.
43. S. Fujita, *J. Math. Chem.*, 2003, **33**, 81.
44. E. Eliel and S.H. Wilen, *Stereochemistry of Organic Compounds*, Wiley, New York, 1994.
45. S.R. Buxton and S.M. Roberts, *Guide to Organic Stereochemistry*, Addison-Wesley Longman, New York, 1996.
46. R.S. Cahn, C.K. Ingold and V. Prelog, *Angew. Chem. Int. Ed. Engl.*, 1966, **5**, 385.
47. V. Prelog, *Science*, 1976, **193**, 17.
48. V. Prelog and G. Helmchen, *Angew. Chem. Int. Ed. Engl.*, 1982, **21**, 567.
49. S. Fujita, *Bull. Chem. Soc. Jpn.*, 2001, **74**, 1585.
50. S. Fujita, *J. Org. Chem.*, 2004, **69**, 3158.
51. S. Fujita, *J. Math. Chem.*, 2004, **35**, 265.
52. S. Fujita, *Tetrahedron*, 1991, **47**, 31.
53. S. Fujita, *MATCH Commun. Math. Comput. Chem.*, 2005, **54**, 39.
54. S. Fujita, *MATCH Commun. Math. Comput. Chem.*, 2004, **52**, 3.
55. S. Fujita, *MATCH Commun. Math. Comput. Chem.*, 2005, **53**, 147.
56. S. Fujita, *J. Chem. Inf. Comput. Sci.*, 2004, **44**, 1719.
57. S. Fujita and S. El-Basil, *J. Math. Chem.*, 2004, **36**, 211.
58. S. Fujita, *Theoret. Chim. Acta*, 1989, **76**, 247.
59. S. Fujita, *Theoret. Chem. Acc*, 2005, **113**, 73.
60. S. Fujita, *Theoret. Chem. Acc*, 2005, **113**, 80.
61. G. Pólya, *Acta Math.*, 1937, **68**, 145. This paper George Pólya (1887–1985), a Hungarian mathematician who emigrated in 1940 to the USA, has written in German. A translation of this classical paper of combinatorial theory appeared on the 100th anniversary of his birth and is to be found in G. Pólya and R.C. Read, *Combinatorial Enumeration of Groups, Graphs, and Chemical Compounds*, Springer, New York, 1987.
62. W. Hässelbarth and E. Ruch, *Israel J. Chem.*, 1976/1977, **15**, 112.
63. W. Hässelbarth and E. Ruch, *Theoret. Chim. Acta*, 1985, **67**, 339.
64. W. Hässelbarth and E. Ruch, *Theoret. Chim. Acta*, 1985, **67**, 427.
65. S. Fujita, *J. Am. Chem. Soc.*, 1990, **112**, 3390.
66. S. Fujita, *J. Math. Chem.*, 1990, **5**, 121.
67. K. Balasubramanian, *J. Math. Chem.*, 2004, **35**, 345.
68. S.G. Williamson, *J. Comb. Theory A*, 1970, **8**, 162.
69. S.G. Williamson, *J. Comb. Theory A*, 1971, **11**, 122.
70. R. Merris, *Linear Algebra Appl.*, 1980, **29**, 255.
71. K. Balasubramanian, *Theoret. Chim. Acta*, 1981, **59**, 237.
72. K. Balasubramanian, *J. Chem. Phys.*, 1981, **74**, 6824.
73. K. Balasubramanian, *Chem. Phys. Lett.*, 2004, **391**, 64.
74. K. Balasubramanian, *Chem. Phys. Lett.*, 2004, **391**, 69.
75. K. Balasubramanian, *Chem. Phys. Lett.*, 1991, **182**, 257.
76. K. Balasubramanian, *Chem. Phys. Lett.*, 2004, **400**, 78.
77. R.E. Haufler, J. Conceicao, L.P.F. Chibante, Y. Chai, N.E. Byrne, S. Flanagan, M.M. Haley, S.C. O'Brien, C. Pan, Z. Xiao, W.E. Billups, M.A. Ciufolini, R.H. Hauge, J.L. Margrave, L.J. Wilson, R.F. Curl and R.E. Smalley, *J. Phys. Chem.*, 1990, **94**, 8634.
78. S. Fujita, *Bull. Chem. Soc. Jpn.*, 1991, **64**, 3215.

79. A.A. Gakh, A.Y. Romanovich and A. Bax, *J. Am. Chem. Soc.*, 2003, **125**, 7902.
80. K. Balasubramanian, *Advances in Quantum Chemistry*, 2006, in press.
81. K. Balasubramanian, *J. Chem. Phys.*, 1981, **74**, 4572.
82. K. Balasubramanian, *Int. J. Quantum Chem.*, 1981, **20**, 1255.
83. K. Balasubramanian, *J. Chem. Phys.*, 1981, **74**, 6824.
84. N.G. de Bruijn, *J. Comb. Theory*, 1967, **2**, 418.
85. N.G. de Bruijn, *J. Comb. Theory A*, 1972, **12**, 14.
86. K. Balasubramanian, *J. Math. Chem.*, 1993, **14**, 113.
87. K. Balasubramanian, *Chem. Rev.*, 1985, **85**, 599.
88. D.H. Rouvray, *Chem. Soc. Rev.*, 1974, **3**, 355.
89. H. Dolhaine, H. Hönig and M. van Almsick, *MATCH Commun. Math. Comput. Chem.*, 1999, **39**, 21.
90. M. van Almsick, H. Dolhaine and H. Hönig, *MATCH Commun. Math. Comput. Chem.*, 2001, **43**, 153.
91. H. Dolhaine and H. Hönig, *MATCH Commun. Math. Comput. Chem.*, 2002, **46**, 71.
92. H. Dolhaine and H. Hönig, *MATCH Commun. Math. Comput. Chem.*, 2002, **46**, 91.
93. T. Hudlicky, K.A. Abboud, D.A. Entwistle, R. Fan, R. Maurya, A.J. Thorpe, J. Bolonick and B. Myers, *Synthesis*, 1996, 897.
94. C. Rücker, R. Gugisch and A. Kerber, *J. Chem. Inf. Comput. Sci.*, 2004, **44**, 1654.
95. A. Kerber, *Applied Finite Group Actions*, Springer, Berlin, 1998; A. Kerber, *Applied Finite Group Actions*, 2nd edition, Springer, Berlin, 1999.
96. R. Grund, A. Kerber and R. Laue, *MATCH Commun. Math. Comput. Chem.*, 1992, **27**, 87.
97. T. Wieland, A. Kerber and R. Laue, *J. Chem. Inf. Comput. Sci.*, 1996, **36**, 413.
98. C. Benecke, T. Grüner, A. Kerber, R. Laue and T. Wieland, *J. Anal. Chem.*, 1997, **359**, 23.
99. A. Kerber, R. Laue, T. Grüner and M. Meringer, *MATCH Commun. Math. Comput. Chem.*, 1998, **37**, 205.
100. L. Lovasz and M.D. Plummer, *Matching Theory*, Akadémiai Kiadó, Budapest, 1986.
101. I. Gutman and O.E. Polansky, *Mathematical Concepts in Organic Chemistry*, Springer, Berlin, 1986.
102. H. Hosoya, *Comp. Math. Appl.*, 1986, **12B**, 271.
103. D. Veljan, *Combinatorics and Discrete Mathematics*, Algorithm, Zagreb, 2001.
104. T. Bocchi, M. Lewinter and S. Karimi, *J. Math. Chem.*, 2004, **35**, 339.
105. S. Fajtlowicz, P.E. John and H. Sachs, *Croat. Chem. Acta*, 2005, **78**, 195.
106. G.G. Cash, *J. Math. Chem.*, 2005, **37**, 117.
107. A. Graovac, D. Vukičević, D. Ježek and J. Žerovnik, *Croat. Chem. Acta*, 2005, **78**, 283.
108. A.T. Balaban and I. Tomescu, *Dicrete Appl. Math.*, 1988, **19**, 5.
109. I. Gutman and S.J. Cyvin, *Introduction to the Theory of Benzenoid Hydrocarbons*, Springer, Berlin, 1989.
110. P.C.B. Lam, W.C. Shiu and H. Zhang, *MATCH Commun. Math. Comput. Chem.*, 2003, **49**, 127.
111. K. Salem, *MATCH Commun. Math. Comput. Chem.*, 2005, **53**, 419.
112. K. Salem, *MATCH Commun. Math. Comput. Chem.*, 2005, **53**, 427.
113. D.J. Klein, *Internet Electronic J. Mol. Design*, 2003, 2, 814. http://www.biochempress.com.

114. M. Randić, *Chem. Phys. Lett.*, 1976, **38**, 68.
115. M. Randić, *Chem. Rev.*, 2003, **103**, 3449.
116. D.J. Klein, *J. Chem. Educ.*, 1992, **69**, 691.
117. W.T. Simpson, *J. Am. Chem. Soc.*, 1953, **75**, 597.
118. W.C. Herndon, *J. Am. Chem. Soc.*, 1973, **95**, 2404.
119. D.J. Klein and N. Trinajstić, *Pure Appl. Chem.*, 1989, **61**, 2107.
120. *Topological Indices and Related Descriptors inQSAR and QSPR*, eds. J. Devillers and A.T. Balaban, Gordon & Breach, Amsterdam, 1999.
121. R. Todeschini and V. Consonni, *Handbook of Molecular Descriptors*, Wiley-VCH, Weinheim, 2000.
122. H. Hosoya, *Bull. Chem. Soc. Jpn*, 1972, **45**, 3415.
123. I. Gutman, M. Milun and N. Trinajstić, *MATCH Commun. Math. Comput. Chem.*, 1975, **1**, 171.
124. J.-i. Aihara, *J. Am. Chem. Soc.*, 1976, **98**, 2750.
125. I. Gutman, M. Milun and N. Trinajstić, *J. Am. Chem. Soc.*, 1977, **99**, 1692.
126. V.I. Minkin, M.N. Glukhovtsev and B.Y. Simkin, *Aromaticity and Antiaromaticity – Electronic and Structural Aspects*, Wiley, New York, 1994, pp. 14–19.
127. J.-i. Aihara and H. Kanno, *J. Mol. Struct. (Theochem)*, 2005, **722**, 111.
128. K. Balasubramanian, *J. Chem. Inf. Comput. Sci.*, 1994, **34**, 421.
129. F. Torrens, *Internet Electronic J. Mol. Design*, 2002, 1, 351. http://www.biochempress.com.
130. F. Torrens, *Internet Electronic J. Mol. Design*, 2003, 2, 96. http://www.biochempress.com.
131. F. Torrens, *Theoret. Chem. Acc.*, 2003, **110**, 371.
132. F. Torrens, *Internet Electronic J. Mol. Design*, 2003, 2, 546. http://www.biochempress.com.
133. F. Torrens, *J. Chem. Inf. Comput. Sci.*, 2004, **44**, 60.
134. F. Torrens, *J. Mol. Struct. (Theochem)*, 2004, **709**, 135.
135. T. Došlić, *Croat. Chem. Acta*, 2005, **78**, 251.
136. M. Gordon and W.H.T. Davison, *J. Chem. Phys.*, 1952, **20**, 428.
137. T.F. Yen, *Theoret. Chim. Acta*, 1971, **20**, 399.
138. P. John and H. Sachs, *Topics Curr. Chem.*, 1990, **153**, 145.
139. Wc. He and Wj. He, *Topics Curr. Chem.*, 1990, **153**, 195.
140. J.R. Dias, *Handbook of Polycyclic Hydrocarbons. Parft A. Benzenoid Hydrocarbons*, Elsevier, Amsterdam, 1987.
141. D.J. Klein and N. Trinajstić, *J. Mol. Struct. (Theochem)*, 1990, **206**, 135.
142. D.I.A. Cohen, *Basic Techniques of Combinatorial Theory*, Wiley, New York, 1978, pp. 121–132.
143. J.H. Conway and R.K. Guy, *The Book of Numbers*, Springer, New York, 1995, p. 101.
144. J.A. von Segner, *Novi Comm. Acad. Scient. Imper. Petropolitanae*, 1758/1759, **7**, 203.
145. S. Karimi, M. Lewinter and M. Stauffer, *J. Math. Chem.*, 2003, **34**, 297.
146. B. Džonova-Jerman-Blažič and N. Trinajstić, *Comput. Chem.*, 1982, **6**, 121.
147. N. Trinajstić and P. Křivka, in: *Mathematics and Computational Concepts in Chemistry*, ed. N. Trinajstić, Horwood/Wiley, Chichester, 1986, pp. 328–345.
148. P. Křivka, S. Nikolić and N. Trinajstić, *Croat. Chem. Acta*, 1986, **59**, 659.
149. I. Lukovits, *J. Chem. Inf. Comput. Sci.*, 2004, **44**, 1565.
150. M. Randić, *J. Chem. Inf. Comput. Sci.*, 2004, **44**, 365.
151. M. Randić and A.T. Balaban, *Polycyclic Aromat. Compnds.*, 2004, **24**, 173.

152. A.T. Balaban and M. Randić, *J. Chem. Inf. Comput. Sci.*, 2004, **44**, 50.
153. A. Miličević, S. Nikolić and N. Trinajstić, *J. Chem. Inf. Comput. Sci.*, 2004, **44**, 415.
154. A.T. Balaban and M. Randić, *New J. Chem.*, 2004, **28**, 800.
155. A.T. Balaban and M. Randić, *J. Chem. Inf. Comput. Sci.*, 2004, **44**, 1701.
156. D. Vukičević and D.J. Klein, *J. Math. Chem.*, 2005, **37**, 163.
157. A.T. Balaban and M. Randić, *J. Math. Chem.*, 2005, **37**, 443.
158. I. Gutman and N. Turković, *Monat. Chem.*, 2005, **136**, 713.
159. I. Gutman, D. Vukičević, A. Graovac and M. Randić, *J. Chem. Inf. Comput. Sci.*, 2004, **44**, 296.
160. D. Vukičević, M. Randić and A.T. Balaban, *J. Math. Chem.*, 2004, **36**, 271.
161. R.A. Henson, K.J. Windlinx and W.J. Wiwesser, *Comput. Biomed. Res.*, 1975, **8**, 53.
162. G.G. Cash, *J. Phys. Chem. A*, 1997, **101**, 8094.
163. G.G. Cash, *Polycyclic Aromat. Compds.*, 1997, **12**, 61.
164. B. Lučić and N. Trinajstić, *J. Chem. Inf. Comput. Sci.*, 1999, **39**, 121.
165. B. Lučić, N. Trinajstić, S. Sild, M. Karelson and A.R. Katritzky, *J. Chem. Inf. Comput. Sci.*, 1999, **39**, 610.
166. B. Lučić, D. Amić and N. Trinajstić, *J. Chem. Inf. Comput. Sci.*, 2000, **40**, 403.
167. A. Miličević and S. Nikolić, *Croat. Chem. Acta*, 2004, **77**, 97.
168. F. Torrens, *Internet Electronic J. Mol. Design*, 2004, 3, 514. http://www.biochempress.com.
169. F. Torrens, *Internet Electronic J. Mol. Design*, 2005, 4, 59. http://www.biochempress.com.
170. F. Torrens, *Nanotechnology*, 2005, **16**, S181–S189.
171. F. Torrens, *Frontiers in Drug Design & Discovery*, 2005, **1**, 231–266.
172. S. Jiang, H. Liang and F. Bai, *New Structural Parameters and Permanents of Adjacency Matrices of Fullerenes*, to be published.
173. H. Liang and F. Bai, *Comput. Phys. Commun.*, 2004, **163**, 79.
174. D. Vukičević and M. Randić, *Chem. Phys. Lett.*, 2005, **401**, 446.
175. D.J. Klein, T.G. Schmalz, G.E. Hite and W.A. Seitz, *J. Am. Chem. Soc.*, 1986, **108**, 1301.
176. M. Randić and D.J. Klein, in: *Mathematics and Computational Concepts in Chemistry*, ed. N. Trinajstić, Horwood/Wiley, Chichester, 1986, pp. 274–282.
177. M. Randić and D.J. Klein, *J. Comput. Chem.*, 1987, **8**, 516.
178. M. Randić, S. Nikolić and N. Trinajstić, *Croat. Chem. Acta*, 1987, **60**, 595.
179. D. Babić and N. Trinajstić, *Fullerene Sci. Technol.*, 1994, **2**, 343.
180. D. Vukičević, H.W. Kroto and M. Randić, *Croat. Chem. Acta*, 2005, **78**, 223.
181. M. Randić, H.W. Kroto and D. Vukičević, *Advances in Quantum Chemistry*, 2006, in press.
182. I. Lukovits and D. Janežić, *J. Chem. Inf. Comput. Sci.*, 2004, **44**, 410.
183. N. Trinajstić, T.G. Schmalz, T.P. Živković, S. Nikolić, G.E. Hite, D.J. Klein and W.A. Seitz, *New J. Chem.*, 1991, **15**, 27.
184. I. Lukovits, A. Graovac, E. Kálmán, G. Kaptay, P. Nagy, S. Nikolić, J. Sytchev and N. Trinajstić, *J. Chem. Inf. Comput. Sci.*, 2003, **43**, 609.
185. S. El-Basil and D.J. Klein, *J. Math. Chem.*, 1989, **3**, 1.
186. L. Pogliani, M. Randić and N. Trinajstić, *Int. J. Math. Educ. Sci. Technol.*, 1998, **29**, 729.
187. D. Wells, *The Penguin Dictionary of Curious and Interesting Numbers*, Penguin, London, revised Ed., 1997, pp. 42–48, 216.
188. F. Cai, H. Shao, C. Liu and Y. Jiang, *J. Chem. Inf. Model.*, 2005, **45**, 371.

189. N. Flocke and T.G. Schmalz, *Chem. Phys. Lett.*, 1998, **298**, 71.
190. D.M. Cvetković, M. Doob and H. Sachs, *Spectra of Graphs-Theory and Applications*, J.A. Barth, Heidelberg, 3rd revised and enlarged ed., 1995.
191. D. Aldous and J. Fill, *Reversible Markov Chains and Random Walks on Graphs*. www.stat.berkeley.edu/users/aldous/.
192. P.G. Doyle and J.L. Snell, *Random Walks and Electric Networks*, Mathematical Association of America, Washington, DC, 1984.
193. R.A. Marcus, *J. Chem. Phys.*, 1965, **43**, 2643.
194. M. Randić, *J. Comput. Chem.*, 1980, **1**, 386.
195. M. Randić, W.L. Woodworth and A. Graovac, *Int. J. Quantum Chem.*, 1983, **24**, 435.
196. M. Razinger, *Theor. Chim. Acta*, 1986, **70**, 365.
197. D. Bonchev, X. Liu and D.J. Klein, *Croat. Chem. Acta*, 1993, **66**, 141.
198. G. Rücker and C. Rücker, *J. Chem. Inf. Comput. Sci.*, 1993, **33**, 683.
199. M.V. Diudea, *J. Chem. Inf. Comp. Sci.*, 1994, **34**, 1064.
200. M. Randić, *Theor. Chim. Acta*, 1995, **92**, 97.
201. M.V. Diudea, *J. Chem. Inf. Comp. Sci.*, 1997, **37**, 300.
202. G. Rücker and C. Rücker, *J. Chem. Inf. Comput. Sci.*, 1999, **39**, 788.
203. I. Gutman, G. Rücker and C. Rücker, *J. Chem. Inf. Comput. Sci.*, 2001, **41**, 739.
204. G. Rücker and C. Rücker, *J. Chem. Inf. Comput. Sci.*, 2001, **41**, 1457.
205. I. Lukovits, A. Miličević, S. Nikolić and N. Trinajstić, *Internet Electronic J. Mol. Design*, 2002, 1, 388. http//:www.biochempress.com.
206. S. Nikolić, N. Trinajstić, I.M. Tolić, G. Rücker and C. Rücker, in: *Complexity-Introduction and Fundamentals*, eds. D. Bonchev and D.H. Rouvray, Taylor & Francis, London, 2003, pp. 29–89.
207. I. Lukovits and N. Trinajstić, *J. Chem. Inf. Comput. Sci.*, 2003, **43**, 1110.
208. G. Rücker and C. Rücker, *J. Chem. Inf. Comput. Sci.*, 2003, **43**, 1115.
209. G. Polya, *Math. Ann.*, 1921, **84**, 149.
210. P.J. Flory, *Principles of Polymer Chemistry*, University Press, Cornell, Ithaca, 1953.
211. E.W. Montroll, *J. Soc. Indust. Appl. Math.*, 1956, **4**, 241.
212. E.W. Montroll and M.F. Schlesinger, in: *Nonequilibriuum Phenomena II*, eds. J.L. Lebowitz and E.W. Montroll, North-Holland, Amsterdam, 1984, pp. 1.
213. R. Motwani and P. Raghavan, *Randomized Algorithms*, University Press, Cambridge, UK, 1995.
214. P.F. Stadler, *J. Math. Chem.*, 1996, **20**, 1.
215. P.F. Stadler and R. Happel, *J. Math. Biol.*, 1999, **38**, 435.
216. D.J. Klein, J.L. Palacios, M. Randić and N. Trinajstić, *J. Chem. Inf. Comput. Sci.*, 2004, **44**, 1521.
217. D.J. Klein, *Croat. Chem. Acta*, 2002, **75**, 633.
218. D.J. Klein and M. Randić, *J. Math. Chem.*, 1993, **12**, 81.
219. M. Randić, *J. Am. Chem. Soc.*, 1975, **97**, 6609.
220. M. Randić, *J. Mol. Graphics Modell.*, 2001, **20**, 19.
221. B. Mohar, in: *Graph Theory, Combintorics, and Applications*, eds. Y. Alavi, C. Chartrand, O.R. Oilermann and A. Schwenk, Wiley, NY, 1991, p. 871–898.
222. R. Merris, *Lin. Alg. Appl.*, 1994, **197/198**, 143.
223. B. Mohar, D. Babić and N. Trinajstić, *J. Chem. Inf. Comput. Sci.*, 1993, **33**, 153.
224. N. Trinajstić, D. Babić, S. Nikolić, D. Plavšić, D. Amić and Z. Mihalić, *J. Chem. Inf. Comput. Sci.*, 1994, **34**, 368.
225. H. Wiener, *J. Chem. Phys.*, 1947, **69**, 17.

226. D. Bonchev, A.T. Balaban, X. Liu and D.J. Klein, *Int. J. Quantum Chem.*, 1994, **50**, 1.
227. I. Gutman, S.-L. Lee, C.-H. Chu and Y.-L. Luo, *Ind. J. Chem.*, 1994, **33A**, 03.
228. I. Gutman and B. Mohar, *J. Chem. Inf. Comput. Sci.*, 1996, **36**, 982.
229. D. Babić, D.J. Klein, I. Lukovits, S. Nikolić and N. Trinajstić, *Int. J. Quantum Chem.*, 2002, **90**, 166.
230. *e.g.*, F.R.K. Chung, *Spectral Graph Theory*, AMM, Providence, Rhode Island, 1997.
231. A. Cayley, *Phil. Mag.*, 1847, **47**, 444.
232. *e.g.*, A.T. Balaban and O. Ivanciuc, in: *Topological Indices and Related Descriptors in QSAR and QSPR*, eds. J. Devillers and A.T. Balaban, Gordon & Breach, Amsterdam, 1999, pp. 21–57.
233. I. Gutman, *Publ. Inst. Math. (Beograd)*, 1998, **63**, 31.
234. L. Bytautas, D. Bonchev and D.J. Klein, *MATCH Commun. Math. Comput. Chem.*, 2001, **44**, 31–40.
235. D. Vukičević, A. Miličević, S. Nikolić, J. Sedlar and N. Trinajstić, *Arkivoc*, 2005 (x), 33. http://www.arkat-usa.org.
236. J.J. Sylvester, *Nature*, 1877, **17**, 284.
237. I. Gutman, D. Vidović and Lj. Petrović, *J. Chem. Soc. Faraday Trans.*, 1997, **94**, 857.
238. D. Bonchev and D.J. Klein, *Croat. Chem. Acta*, 2002, **75**, 613.
239. D. Vukičević and A. Graovac, *Croat. Chem. Acta*, 2005, **78**, 203.
240. I. Gutman and N. Trinajstić, *Chem. Phys. Lett.*, 1972, **17**, 535.
241. S. Nikolić, G. Kovačević, A. Miličević and N. Trinajstić, *Croat. Chem. Acta*, 2003, **76**, 113.
242. E. Estrada, *Chem. Phys. Lett.*, 2001, **336**, 248.
243. D. Vukičević and N. Trinajstić, *Croat. Chem. Acta*, 2003, **76**, 183.
244. I. Gutman and K.C. Das, *MATCH Commun. Math. Comput. Chem.*, 2004, **50**, 83.
245. K.C. Das and I. Gutman, *MATCH Commun. Math. Comput. Chem.*, 2004, **52**, 103.
246. A. Miličević, S. Nikolić and N. Trinajstić, *Molecular Diversity*, 2004, **8**, 393.
247. D. Vukičević and N. Trinajstić, *MATCH Commun. Math. Comput.Chem.*, 2005, **53**, 111.
248. B. Zhou and I. Gutman, *MATCH Commun. Math. Comput. Chem.*, 2005, **54**, 233.
249. I. Gutman, B. Ruščić, N. Trinajstić and C.F. Wilcox Jr., *J. Chem. Phys.*, 1975, **62**, 3399.
250. I. Gutman, C. Rücker and G. Rücker, *J. Chem. Inf. Comput. Sci.*, 2001, **41**, 739.
251. J. Braun, A. Kerber, M. Meringer and C. Rücker, *MATCH Commun. Math. Comput.Chem.*, 2005, **54**, 163.
252. C. Rücker and G. Rücker, *J. Chem. Inf. Comput. Sci.*, 1993, **33**, 683.
253. C. Rücker and G. Rücker, *J. Chem. Inf. Comput. Sci.*, 2000, **40**, 99.
254. H.E. Huntley, *The Divine Proportion*, Dover Publications, New York, 1970.
255. M. Ghyka, *The Geometry of Art and Life*, Dover Publications, New York, 1977.
256. G. Sarton, *Introduction to the History of Science*, Carnegie Institution, Washington, D.C., 1968, reprint, Vol. 1, p. 116.
257. A. Nickon and E.F. Silversmith, *Organic Chemistry: The Name Game*, Pergamon Press, New York, 1987.
258. M.A. Kuznecov, B.L. Milman and S.M. Sevchenko, *The Shapes of Molecules*, Khimia, St. Petersburg, 1989.
259. I. Hargittai and M. Hargittai, *Symmetry through the Eyes of a Chemist*, Plenum Press, New York.

260. J.-A. Le Bel, *Bull. Chem. Soc. France*, 1874, **22**, 337.
261. J.H. van't Hoff, *Archives Né'erlandaises des Sciences Exactes et Naturelles*, 1874, **9**, 445.
262. *e.g.*, S. Maier, S. Pfriem, U. Schäfer and R. Matusch, *Angew. Chem. Int. Ed. Engl.*, 1978, **17**, 520.
263. P.E. Eaton and T.W. Cole Jr., *J. Am. Chem. Soc.*, 1964, **86**, 3157.
264. L.A. Paquette, R.J. Ternansky and D.W. Balogh, *J. Am. Chem. Soc.*, 1982, **104**, 4502.
265. F.A. Cotton and G. Wilkinson, *Advanced Inorganic Chemistry*, Wiey, New York, 1980.
266. W. Lipscomb, *Boron Hydrides*, Benjamin, New York, 1963.
267. D.B. Sullenger and C.H.L. Kennard, *Sci. Amer.*, 1966, **215**(#1), 96.
268. D. Bonchev, *Croat. Chem. Acta*, 2004, **77**, 167.
269. H.S.M. Coxeter, *Regular Polytopes*, Dover, New York, 1973.
270. S.H. Bertz and T.J. Sommer, *Chem. Comm.*, 1997, 2409.
271. D. Bonchev, *SAR QSAR Environ. Res.*, 1997, **7**, 23.
272. D. Bonchev, in: *Topological Indices and Related Descriptors in QSAR and QSPR*, eds. J. Devillers and A.T. Balaban, Gordon & Breach, Amsterdam, 1999, pp. 361–401.
273. D. Bonchev, *J. Chem. Inf. Comput. Sci.*, 2000, **40**, 934.
274. D. Bonchev, *J. Chem. Inf. Comput. Sci.*, 2001, **41**, 582.
275. G. Rücker and C. Rücker, *J. Chem. Inf. Comput. Sci.*, 2000, **40**, 99.
276. I. Lukovits, S. Nikolić and N. Trinajstić, *Croat. Chem. Acta*, 2000, **73**, 957.
277. S.H. Bertz, *J. Am. Chem. Soc.*, 1981, **103**, 3599.
278. E. Estrada, *J. Chem. Inf. Comput. Sci.*, 1995, **35**, 31.
279. D. Babić, D.J. Klein, I. Lukovits, S. Nikolić and N. Trinajstić, *Int. J. Quantum Chem.*, 2002, **90**, 166.
280. A.T. Balaban and D. Bonchev, *MATCH Commun. Math. Comput. Chem.*, 2005, **54**, 137.
281. A.T. Balaban, *Comput. Math. Appls.*, 1986, **12B**, 999.
282. I. Lukovits, S. Nikolić and N. Trinajstić, *Int. J. Quantum Chem.*, 1999, **71**, 217.
283. R. Pearce and S. Pearce, *Polyhedra Primer*, Nostrand-Rheinhold, New York, 1978.
284. D'Arcy Thompson, *On Growth and Form*, reprint-the first edition 1917, University Press, Cambridge, 1995, p. 121.
285. N. Trinajstić, S. Nikolić, D. Babić and Z. Mihalić, *Bull. Chem. Technol. Macedonia*, 1997, **16**, 43.
286. B. De La Vaissière, P.W. Fowler and M. Deza, *J. Chem. Inf. Comput. Sci.*, 2001, **41**, 376.
287. G.F. Swiegers and T.J. Malafetse, *Chem. Eur. J.*, 2001, **7**, 3637.
288. C.B. Boyer, *A History of Mathematics*, Wiley, New York, 1968, p. 149.
289. E. Laxoras, *Am. Scientist*, 1993, **81**, 208.
290. F. Chung and S. Sternberg, *Am. Scientist*, 1993, **81**, 56.
291. S.M. Rajtmajer, A. Miličević, N. Trinajstić, M. Randić and D. Vukičević, *J. Math. Chem.*, 2006, **39**, 119.
292. M. Randić, *Croat. Chem. Acta*, 2001, **74**, 683.
293. M. Randić and D. Plavšić, *Croat. Chem. Acta*, 2002, **75**, 107.
294. M. Randić and D. Plavšić, *Int. J. Quantum Chem.*, 2003, **91**, 20.
295. P.W. Fowler, *MATCH Commun. Math. Comput. Chem.*, 2003, **48**, 87.
296. D.J. Klein, *J. Math. Chem.*, 1995, **18**, 321.
297. D. Babić and N. Trinajstić, *Discrete Appl. Math.*, 1996, **67**, 5.

298. D.J. Klein and D. Babić, *J. Chem. Inf. Comput. Sci.*, 1997, **37**, 656.
299. D.J. Klein and J. Brickmann (eds.), Partial Orderings in Chemistry, *MATCH Commun. Math-. Comput. Chem.*, 2000, **42**, 7–290.
300. See footnote 27 by D.J. Klein in S. Nikolić, I.M. Tolić, N. Trinajstić and I. Baučić, *Croat. Chem. Acta*, 2000, 73, 909.
301. M. Randić, X. Guo, D. Plavšić and A.T. Balaban, in: *Complexity-Applications*, eds. D. Bonchev and D.H. Rouvray, Taylor & Francis, London, 2005, in press.
302. T. Laidboeur, D. Cabrol-Bass and O. Ivanciuc, *J. Chem. Inf. Comput. Sci.*, 1996, **36**, 811.
303. N. Trinajstić, M. Randić, D.J. Klein, D. Babić and Z. Mihalić, *Croat. Chem. Acta*, 1995, **68**, 241.
304. B. Grünbaum, *Convex Polytopes*, Wiley, New York, 1973.
305. D.J. Klein, W.A. Seitz and T.G. Schmalz, *J. Phys. Chem.*, 1993, **97**, 1231.
306. E.C. Kirby, D.J. Klein, R.B. Mallion, P. Pollak and H. Sachs, *Croat. Chem. Acta*, 2004, **77**, 263.
307. G. Kirchhoff, *Ann. Phys. Chem.*, 1847, **72**, 497.
308. O.N. Temkin, A.V. Zigarnik and D. Bonchev, *Chemical Reaction Networks: A Graph-Theoretical Approach*, CRC Press, Boca Raton, FL, 1996.
309. R.B. Mallion and N. Trinajstić, *MATCH Commun. Math. Comput. Chem.*, 2003, **48**, 97.
310. S. Nikolić, N. Trinajstić, A. Jurić, Z. Mihalić and G. Krilov, *Croat. Chem. Acta*, 1996, **69**, 883.
311. R.B. Mallion, *Proc. Roy. Soc. (London)*, 1975, **A341**, 429.
312. C.W. Haigh and R.B. Mallion, *Croat. Chem. Acta*, 1989, **62**, 1.
313. J.A.N.F. Gomes and R.B. Mallion, *Chem. Rev.*, 2001, **101**, 1349.
314. I. Gutman, R.B. Mallion and J.W. Essam, *Mol. Phys.*, 1983, **50**, 859.
315. B. Mohar, in: *MATH/CHEM/COMP 1988*, ed. A. Graovac, Elsevier, Amsterdam, 1989, pp. 1–8.
316. N. Trinajstić, D. Babić, S. Nikolić, D. Plavšić, D. Amić and Z. Mihalić, *J. Chem. Inf. Comput. Sci.*, 1994, **34**, 368.
317. A. Miličević, S. Nikolić, N. Trinajstić and D. Janešić, *Advances Quantum. Chem.*, 2006, in press.
318. D.J. Klein, A. Miličević and N. Trinajstić, *Incidence, Dissidence, and Comparator Matrices*, to be published.
319. S. Nikolić, M. Randić, D.J. Klein, D. Plavšić and N. Trinajstić, *J. Mol. Struct.*, 1989, **198**, 223.
320. S. Nikolić, N. Trinajstić and D.J. Klein, *Comput. Chem.*, 1990, **14**, 313.
321. M. Randić, D.J. Klein, S. El-Basil and P. Calkins, *Croat. Chem. Acta*, 1996, **69**, 1639.
322. M. Randić, N. Basak and D. Plavšić, *Croat. Chem. Acta*, 2004, **77**, 251.
323. M. Randić, D. Plavšić and M. Razinger, *MATCH Commun. Math. Comput. Chem.*, 1997, **35**, 243.
324. M. Randić, *Chem. Phys. Lett.*, 1993, **211**, 478.
325. M. Randić, X. Guo, T. Oxley and H. Krishnapryan, *J. Chem. Inf. Comput. Sci.*, 1993, **33**, 709.
326. M. Randić, X. Guo, T. Oxley, H. Krishnapryan and L. Naylor, *J. Chem. Inf. Comput. Sci.*, 1994, **34**, 361.
327. M. Randić, *Croat. Chem. Acta*, 1994, **67**, 415.
328. S. Nikolić, A. Miličević and N. Trinajstić, *Croat. Chem. Acta*, 2005, **78**, 241.
329. D.H. Rouvray, *Sci. Am.*, 1986, **254**(9), 40.

330. S. Nikolić, N. Trinajstić and Z. Mihalić, *Croat. Chem. Acta*, 1995, **68**, 105.
331. S. Nikolić, N. Trinajstić and M. Randić, *Chem. Phys. Lett.*, 2001, **333**, 319.
332. O. Ivanciuc and T. Ivanciuc, in: *Topological Indices and Related Descriptors in QSAR and QSPR*, eds. J. Devillers and A.T. Balaban, Gordon & Breach, Amsterdam, 1999, pp. 221–277.

9

Many-Body Perturbation Theory and its Application to the Molecular Structure Problem

BY S. WILSON
Rutherford Appleton Laboratory, Chilton, Oxfordshire

1 Introduction

This report covers developments in the theory and application of many-body perturbation theory to molecular systems during the period June 2003 through to May 2005. It thus continues three earlier reviews in this series[1–3] which, in turn, built on my report, written about twenty-five years ago, for a previous Specialist Periodical Reports series published in 1981.[4]

Applications of many-body perturbation theory to the molecular electronic structure problem have been published during the reporting period in an ever increasing range of scientific areas. In particular, in its second order form, many-body perturbation theory continues to be the most widely used *ab initio* quantum chemical method for describing the effects of electron correlation. A review of the numerous applications reported during the period under consideration is given in Section 4.

Writing in 2005, the centenary of the three famous publications by Albert Einstein which marked a turning point in early twentieth century science, it seemed appropriate in this review to consider some of the major trends in early twenty-first century molecular science and the contribution of many-body perturbation theory to their study.

In the introduction to his *Lectures on Physics*,[5] Richard P. Feynman asserts that the atomic theory of matter is the most important of scientific theories since it underpins our explanation of the material world. He wrote

> *"If, in some cataclysm, all scientific knowledge were to be destroyed, and only one sentence passed on to the next generation of creatures, what statement would contain the most information in the fewest words? I believe it is the atomic hypothesis (or atomic fact, or whatever you wish to call it) that all things are made of atoms little particles that move around in perpetual motion, attracting each other when they are a little distance apart, but repelling upon being squeezed into one another. In that one sentence you*

Chemical Modelling: Applications and Theory, Volume 4

will see an enormous amount of information about the world, if just a little imagination and thinking are applied."

The atomic theory was the work of the nineteenth century beginning with the original publication of John Dalton[6] in Manchester and undisputed acceptance with the Nobel-prizing winning explanation of Brownian motion[†] published by Albert Einstein[8] one hundred years ago in 1905.[‡]

The theoretical explanation of atomic and molecular structure rests on theories discovered in the first half of the twentieth century, quantum mechanics and, for systems containing heavy atoms, relativistic quantum mechanics and quantum electrodynamics. But, as Dirac[9] famously pointed out, these theories lead to equations which are much to complicated to be solved for any system of interest, *i.e.* for chemists, molecules. The first half of the twentieth century saw the development of models, theoretical constructs which attempted to simplify the basic equations of physics by capturing chemical (and physical) insights. Thus, the Born-Oppenheimer approximation[10] separated the electronic and nuclear motion and in doing so realized the concept of molecular structure, which was central to much of the classical chemistry developed during the nineteenth century and early twentieth century, within the framework of quantum mechanics. Both molecular orbital theory and valence bond theory decoupled the many-body problem, which is the problem of molecular electronic structure, to realize a model in which each electron moves in a quasi-independent fashion. Both of these theories, when systematically refined, lead to the "solution" of the Schrödinger equation. With increasingly powerful computers and supercomputers dramatic progress has been made in the second half of the twentieth century. Progress that was recognized in the award of the 1998 Nobel Prize for Chemistry to Sir John Pople FRS. Computation and supercomputation is one of the two themes that I want to identify in this review, concentrating, of course, on recent work.

As the twenty-first century unfolds, it is becoming clear that there has been a paradigm shift in the molecular sciences, in both theoretical and experimental chemistry, the chemical sciences and, indeed, the allied sciences of physics, especially atomic and molecular physics and condensed matter physics, and biology and biochemistry. Attention is increasingly focussed on molecular systems of increasing complexity. The term "complexity" has come to mean a system which displays emergent properties, that is properties which are not shared by the component parts. A crystal of common salt, for example, is not complex. If it is broken in half, it remains common salt with all its properties. Theoretically common salt is a regular lattice, which, by imposing periodic boundary conditions, can be subjected to computations only marginally more difficult than those for a small molecule. Complexity is not synonymous with largeness. Polyvinylchloride is a large molecule it is no more complex than one of its basic monomers. There is no known limit to the degree of complexity of

---

[†] First observed in 1827[7] and named after the English botanist Robert Brown.

[‡] Einstein's paper[8] was entitled "Über die von der molekularkinetischen Theorie der Wärme geforderte Bewegung von in ruhenden Flüssigkeiten suspendierten Teilchen."

molecular systems. The component parts of CuO-based high temperature superconductors are insulators. Superconductivity is believed to be an emergent property of these layered compounds. The theoretical description of complex molecular systems is a problem that requires computation and supercomputation. But computation and supercomputation alone is not enough; theoretical apparatus capable of describing complex molecules need to be developed. Some of this apparatus was developed during the second half of the twentieth century beginning with the seminal paper by Brueckner[11] published fifty years ago at the time of writing which established the many-body expansion for the energy with terms scaling linearly with the number of particles in the target system. This is the second theme of this review, the development of many-body techniques and, in particular, many-body perturbation theory in the theoretical description of molecular systems of increasing complexity.

This report is organized as follows:- In Sections 2 and 3, we consider two of the major trends in early twentieth century science and the role that many-body methods, in general, and many-body perturbation theory, in particular, may play in addressing the problems which the trends suggest. The use of computation and supercomputation is ubiquitous in much of modern science and yet their introduction is relatively recent. Computation has a history of less than fifty years; supercomputation a fraction of that. Their role will be considered in Section 2. In Section 3 attention is turned to complexity and, in particular, to the study of increasingly complex molecular systems. Complexity in the theoretical study of molecules can arise in a number of ways - through the study of larger, more extensive, systems, through the use of relativistic quantum mechanics, in the study of systems requiring more sophisticated many-body formulations and, in particular, multireference techniques, and, through the use of multicomponent formulations. In Section 4, a review of applications of many-body perturbation theory in its simplest form during the reporting period (June 2003-May 2005) is given. This review serves not only to illustrate the ever widening range of problems being addressed by means of many-body perturbation theory but also to demonstrate the prospects for the trends discussed in Sections 2 and 3 - computation and supercomputation, and the study of increasingly complex molecular systems. A concluding Section, Section 5, contains a summary and briefly considers future prospects.

2 Computation and Supercomputation

That computers have revolutionize science and society is now a matter of history, for as Roberts[12] writes

> *"The introduction of electronics-based information technology to western culture (it can be approximately dated from the 1940s) announced the latest of those wave of innovation provoking 'industrial revolutions' since the eighteenth century. Since then, it has swept through societies worldwide.*

Information technology has made anything that requires the understanding of large and complex systems vulnerable to exploration as never before."

Many processes - traffic flows, telecommunications networks, ecological systems -[12]

"All these can be 'modelled' and simulated as never before ..."

as can the molecular structure, properties and processes that form the basis of chemistry and the chemical sciences. According to Lykos,[13] writing in 1982

"The is no forseeable limit to the complexity of the systems that chemists can model."

and therefore

"There is no forseeable scientific computer that chemistry modellers cannot exploit fully."

Computational science is concerned with the effective exploitation of computer hardware in modelling and simulating a wide variety of processes. Computational science is therefore concerned with the development of computer software based on appropriate theoretical models. This software is then executed on a given hardware platform thereby providing information about the studied systems and new insights into their behaviour.

The plan of this section is as follows: We begin by considering the role of computation in general terms both in science as a whole and in applications in the molecular sciences. We then consider what has been termed supercomputational science, the science that can be carried out on computing machines of the highest performance in each generation. Today such machines are sometimes termed *'leadership class'*. We emphasise the importance of software in the effective exploitation of different computer architectures and this leads us to discuss *'literate programming'* which we suggest should be an essential ingredient of computational science and engineering, and, for our present purposes, computational modelling of molecular systems. We conclude this section by giving an explicit example of literate programming in practical many-body perturbation theory by presenting an algorithm for the evaluation of the correlation energy component corresponding to the third order 'ring' diagram.

2.1 The Role of Computation. – The important role that computation was to play in science was recognised at an early stage. The report of a working party[§] on *"Future Facilities for Advanced Research Computing"* published in 1985 begins

"There are many problems that cannot be readily studied by experiment or are not amenable to a theoretical solution by mathematical analysis because of their complexity. It is possible, however, to approach a solution by

[§]The report of a Joint Working Party, June 1985, under the chairmanship of Professor A.J. Forty[14].

numerical computation, given sufficient insight of the problem, a knowledge of the parameters involved and a sufficiently powerful computer."

The report continues

> *"The recognition of the potential of numerical computation as a research technique dates from the vision of von Neumann and others in the 1940s of the future use of digital computers as mathematical tools. However, the real emergence of computational science stems from the 1960s when powerful computers such as the Ferranti Atlas, a supercomputer of its day, became generally available."*

In 1929, not long after the birth of quantum mechanics, Dirac wrote[9] that the

> *"... laws necessary for the mathematical treatment of a large part of physics and the whole of chemistry are thus completely known, and the difficulty is only that the exact application of these laws leads to equations much too complicated to be soluble."*

He continued[9]

> *"It therefore becomes desirable that approximate methods of applying quantum mechanics should be developed, which can lead to an explanation of the main features of complex atomic systems without too much computation."*

The development of approximate methods of applying quantum mechanics to molecules and molecular systems leads immediately to the introduction of models. Seldom is discussion of the properties of a given molecule or molecular complex based on the direct solution of the molecular Schrödinger equation (or its relativistic generalization). Instead, approximations are invoked which not only make the computation tractable but also introduce concepts and models upon which our interpretation of the properties of the molecule are based.

For example, the Born-Oppenheimer approximation is ubiquitous. The separation of the electronic and nuclear motion is most often an excellent approximation. However, it is also fundamental to the concept of molecular structure. The model of fixed nuclei surrounded by electrons which accommodate almost instantly any change in the nuclear positions is basic to qualitative and quantitative discussions of molecular structure.

Orbital models are almost invariably the first approximation for the electronic structure of molecules within the Born-Oppenheimer approximation. By considering the motion of each electron in the averaged field of the remaining electrons in the systems in the self-consistent field theory of Hartree and Fock, a model is obtained in which each electron is described by an spin-orbital, a description which is familiar to all chemists.

Of course, orbital models, such as the widely used Hartree-Fock approximation neglect the effects of electron correlation. One approximation which forms the basis of a computationally tractable approach to the electron correlation problem in atoms and molecules is the many-body perturbation theory (and closely related cluster expansions) which provides a simple diagrammatic representation of the various effects which contribute to correlation effects. The use of diagrammatic perturbation theory remains unfamiliar territory to the majority of chemists, although it is entirely equivalent to the Møller-Plesset perturbation expansion, which in lowest order remains the most widely used *ab initio* approach to the correlation energy problem in molecules.

2.2 Supercomputational Science. – In a volume entitled Supercomputational Science, which records the lectures given at a Summer School held in Abingdon near Oxford in September 1989, Evans and Wilson write[15]

> *"It is widely recognized that computation is now an indispensable alternative to the traditional alternative to the traditional methods of scientific investigation - theoretical analysis and laboratory experiment. Over the last four hundred years, science has progressed by a symbiotic interaction between theory and experiment. Theory has suggested new experiments and then interpreted their results. Equally, experiment has tested existing theories and suggested new theories. Now theory provides the basic equations for the computational approach. The situation is summarized in Figure 1."*

They continue by emphasizing the benefits of the computational approach to problem solving:[15]

> *"The computational approach can frequently afford solutions to scientific problems which are theoretically intractable because of their complexity. The computer can not only provide a route to information which is not*

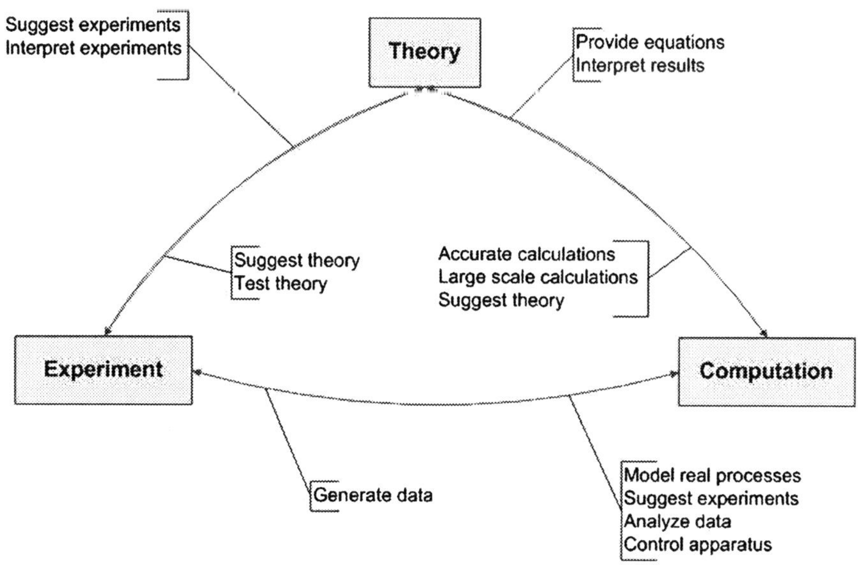

Figure 1 *The relation between theory, computation and experiment*

> *available from laboratory experiments but can also afford additional insight into the problems being studied. Moreover, because it is often more efficient than the alternatives, the computational approach can increasingly be justified in economic terms."*

According to the report *"Future Facilities for Advanced Research Computing"*

> *"The term supercomputer is generally applied to machines which by virtue of their design yield better than expected performance for the technology of their day. They generally deliver this perfomance advantage over a limited but significant set of problems. Clearly, in the early 1960s the Ferranti Atlas would have qualified for this description. Subsequently the mantle passed to the CDC 7600 and (for floating point arithmetic) the IBM 360/195, which in turn have succumbed to machines like the Cray and Cyber 205 since the late 1970s."*

In his discussion of the benefits of supercomputing in 1989, Davies[16] writes

> *"... today's supercomputers can indeed deliver high performance ... on real codes, but that in general the user has to do a significant amount of work to attain it because of the complexity of the supercomputer hardware.."*

Davies[16] writes

> *"A new hardware generation might occur every year or two; new software generations occur nearly every decade or two."*

The cost of developing and maintaining high quality software which can effectively exploit a given hardware platform often exceeds the cost of the hardware itself. Documentation is a key element of high quality software since[17]

> *"The ability to comprehend a program written by other individuals is becoming increasingly important in software development. Given that the general cost of program maintainance may reach 60 percent of the total software costs associated with a certain product, a real demand exists for systems that can present the program text in a readable, understandable format."*

In the following section, we show that not only does literate programming provide high quality documentation of programs and algorithms but also facilitates their publication.

2.3 Literate Programming. – In his 1968 publication *Public Knowledge: the Social Dimension of Science*, John Ziman answers the question *"What is Science?"*[18] by asserting that *"Science is Public Knowledge"*.¶ He continues[20]

¶ In his more recent book *Real Science: What it is,and what it means*,[19] Ziman asserts that "The function of science is to produce knowledge".

> *"Science is not merely published knowledge or information. ... Scientific knowledge is more than this. Its facts and theories must survive a period of critical study and testing by other competent and disinterested individuals, and must have been found so persuasive that they are almost universally accepted."*

In considering Science as Public Knowledge, Ziman emphasizes the restrictions imposed by the traditional scientific journal when he writes[21]

> *"Our experience of the form that such knowledge can take is so limited to verbalization and formulae - to written and printed words and symbols - that we instinctively frame our definitions of Science within the framework of such forms. Our theories and explanations have to pass through this filter ..."*

Even in 1968, Ziman[21] recognized the limitations of the printed medium in that

> *"... physicists and engineers will sometimes make available to their colleagues the tapes of instructions for computer programs that they have devised for some particular purpose, such as for the solution of some difficult equation, of for the reduction and analysis of certain types of data. These tapes are collected in 'libraries' at computer centres, and are used directly to generate further knowledge. It seems to me that these are not just tools of research; they embody information, and play just the same role as would mathematical formulae published in books, or tables of physical data in scientific papers."*

We are concerned here with the effective communication of the ideas and information embodied in quantum chemistry computer software. In 2000, Ziman writes[22]

> *"Research results do not count as scientific unless they are reported, disseminated, shared, and eventually transformed into communal property by being formally published."*

He continues[23]

> *"... what is distinctive about formal scientific communication is neither the medium nor the message: it is that it is published."*

and

> *"... it is fully and freely available for open criticism and constructive use."*

Much contemporary quantum chemistry software does not satisfy these requirements.

An early attempt to formalize the distribution of quantum chemistry software was the Quantum Chemistry Program Exchange (*QCPE*)[24] which was created in 1962 by H. Shull and offered a small collection of some 33 pieces of software. According to the current *QCPE* web-site[24]

> *"QCPE promotes the practice of computational chemistry by gathering and distributing computer software for chemistry and related fields. QCPE holds approximately 772 computational chemistry systems, programs, and routines. Most programs are available in source code with test data sets. QCPE holds many of its programs in several versions for different computer platforms. These programs can be ordered from QCPE for a nominal distribution fee."*

QCPE distributes, for example, an early version of the GAUSSIAN code, GAUSSIAN70.[26]

Another attempt to formalize the distribution of software for atomic and molecular physics is the journal *Computer Physics Communications* which published both scientific papers and computer programs. The first issue of this journal appeared in 1969 with an article by Roberts[25] on *"The publication of scientific FORTRAN programs"*, which aimed to encourage the exchange of ideas embedded in computer codes. In spite of the continued success of this journal, it has nevertheless has to be observed that none of the quantum chemical packages mentioned in our preamble or indeed similar packages have been published via this mechanism.

Literate programming provides a mechanism for introducing higher standards of code documentation and thus greater levels of collaboration in quantum chemistry code development and distribution. Literate programming combines the theoretical development of a particular model with the associated computer code. In 1984, D.E. Knuth published his seminal paper[27] entitled *"Literate programming"* in *The Computer Journal*. Knuth proposed a system of programming, which he termed the web,[‖] for the generation of structured and documented programs. The philosophy of the WEB system is described by Knuth as follows:-

> *"I believe that the time is ripe for significantly better documentation of programs, and that we can best achieve this by considering programs to be works of literature. Hence, my title: Literate Programming".*

Knuth advocates a radical shift of emphasis in the writing of computer programs. He makes this point as follows:-

> *"Let us change our traditional attitude to the construction of programs: Instead of imagining that our main task is to instruct a computer what to do, let us concentrate rather on explaining to human beings what we want a computer to do."*

Knuth made these observations twenty years ago. Yet to date they have had surprising little impact in computational quantum chemistry or, indeed, computational chemistry in general. There is no mention of literate programming methods in, for example, the *Encyclopedia of Computational Chemistry*,[28] a major reference work in the field published in 1998. (A notable exception is the

[‖] This is **not** to be confused with the World Wide Web which had not been proposed at the time Knuth first published his idea.

work of D.B. Cook[29] described in his *Handbook of Computational Quantum Chemistry*[**]). Indeed, to the authors' knowledge, literate programming techniques have not been widely adopted in any of the computational sciences.

Knuth recognizes that the task facing a literate programmer extends beyond that of a computer programmer. The literate programmer strives not only to create correct and efficient computer code but also a description of the theoretical concepts that lie behind the code. Knuth[30] explains the literate programmers task as follows:-

> *"The practitioner of literate programming can be regarded as an essayist, whose main concern is with exposition and excellence of style. Such an author, with thesaurus in hand, chooses the names of variables carefully and explains what each variable means. He or she strives for a program that is comprehensible because its concepts have been introduced in an order that is best for human understanding, using a mixture of formal and informal methods that reinforce each other."*

For many years quantum chemistry has been one of the primary areas of application of computers in the scientific research. The Schrödinger equation for any molecular system can be easily written down. In principle, the solution of this equation yield the structure and properties of a molecule, but in practice this can lead to severe computational demands which may, in fact, render calculation for particular properties of particular systems intractable. It is important that quantum chemistry software be efficient. Thorough documentation of code is an essential ingredient of efficient software.

By combining documentation, including the theoretical foundations of the model being employed, with the associated computer code, literate programming techniques offer, in order of importance, the follow benefits to the quantum chemistry community:-

(*i*) **structure:** Knuth suggests that we should consider "programs to be works of literature". One would not write a book without dividing it into parts and chapters and, for a scientific treatise, the chapters are invariably divided into sections and subsections, and supplemented by appendices. This "literate" approach inevitably leads to modularity of the code. Literate programming will thus naturally lead to a structured, highly structured, and modular programming style.

(*ii*) **collaboration:** it facilitates collaborative working especially between geographically distributed sites. This is the benefit of literate programming upon which the present paper is particularly focussed and which will be discussed in the following section.

(*iii*) **code integrity:** Computer code which is documented to a high standard can be usefully published. At present, most quantum chemical publications describing new methodology stop short of including the relevant

[**] Cook describes the application *a posteriori* of literate programming methods to the last public release of the well known GAUSSIAN package, GAUSSIAN70, for molecular electronic structure calculations.

code. Literate programming allows due credit to be given for coding developments. In this way more science is placed in the public domain and this in turn will foster further development in the field.

(*iv*) **economy/efficiency:** It is widely recognized that well-documented computer codes are cost effective. For many years, it has been recognized that software development costs far exceed computer hardware costs when proper account is taken of the time and effort involved. A well-documented code is far more easily re-engineered than a code which is poorly documented or, indeed, not documented at all. Literate programming affords documentation of the highest quality.

(*v*) **education:** Students and researchers new to quantum chemistry can gain valuable insight into the different methodologies and their practical realization through codes which employ literate programming techniques. For the new entrant to the field, quantum chemistry is made more difficult by the fact that code is 'hidden' by lack of publication and/or poor documentation of computer code.

We have explored the pedagogical benefits of the literate programming approach to quantum chemistry in a recent publication[31] and we are preparing a monograph[32] which will elaborate this aspect. We have also discussed the benefits of literate programming in collaborative projects and in scientific publications.[33]

Literate programs combine text and code in a single file called a WEB file. In general, the literate programming approach does not restrict the language used to generate the text, although TEX[††] or LATEX are most commonly used, and the code may be written in FORTRAN, C or even, as in Knuth original paper[27], PASCAL. In this work, we shall combine LATEX and C. It should be emphasized that the WEB is not employed directly. It is processed by one of two commands called `weave` and `tangle`.The `weave` command generates a LATEX file which can be processed to produce a document in the usual manner.[34] So, for example, the file `example.web` would be processed as follows:-

```
weave example.web > example.tex
latex example.tex > example.dvi
dvips example.dvi > example.ps
```

where the first line extracts a LATEX file with extension `.tex` from the WEB file the LATEX compiler is executed in the second line giving a with extension `.dvi` (a device independent and in the third line the `dvi` is processed using `dvips` leading to a PostScript[‡‡] file `example.ps` which can be printed in the usual way on a printer which understands PostScript. The command `tangle` generates a `c` which can be processed to produce an executable program. So, for

[††] TEX is a trademark of the American Mathematical Society.
[‡‡] PostScript is a trademark of Adobe Systems Incorporated.

```
@
LaTeX source code describing the C code which follows below
...
@
C source code corresponding to the descriptive text above
...
@
LaTeX source code describing the C code which follows below
...
@
C source code corresponding to the descriptive text above
...
@
LaTeX source code describing the C code which follows below
...
@
C source code corresponding to the descriptive text above
...
@
LaTeX source code describing the C code which follows below
...
...
```

Figure 2 *Structure of a* WEB *file after Knuth (The Computer Journal, 1984,* **27:2**, *97). The* WEB *consists of alternating fragments of* LaTeX *source code, which may generate Tables and Figures as well as text and equations, and C source code corresponding to the* LaTeX

example, the same `example.web`, used above would be processed as follows:-

```
tangle example.web > example.c
gcc example.c > example
example < example.in > example.out
```

where the `tangle` command extracts Ccode from the `WEB` in the line, in the second line this C code is compiled using a standard C compiler, `gcc`, to give the executable `example`, and in the line `example` is executed using the input and output `example.in` and `example.out`, respectively.

A number of schemes for formatting the `web` can be envisaged. Knuth[27] suggests using the character @ to delimit different parts of the WEB file and that then the WEB file be structured as illustrated in Figure 2. The text (LaTeX) and code (C) are treated even-handedly. Each fragment of text is associated with the corresponding fragment of C code. Each fragment of C code is paired with a fragment of text.

Literate programming methods have been explored for a number of languages[§§]. We are not advocating a particular method here, but rather the concept of literate programming in practical quantum chemistry.

[§§] e.g., CWEB[38] uses TeX and c, FWEB[39] uses TeX and FORTRAN.

The literate programmer needs a knowledge of the text processing language used (LaTeX in the present work), a knowledge of the programming language used (C in this work) together with a knowledge of literate programming methodology. Although many quantum chemists are familiar with both LaTeX and C, we suggest that it is the need to handle two languages simultaneously together with complex literate programming conventions[¶¶] which has inhibited the widespread adoption of this approach in the computational quantum chemistry literature.

2.4 A Literate Program for Many-Body Perturbation Theory. – Here the application of literate programming methods in many-body perturbation theory will be illustrated by considering the computation of the third order "ring" diagram in the correlation energy expansion for a closed-shell, singlet system. The FORTRAN code for calculating this component of the correlation energy was published by the author[40] in 1978. The original code was written for an IBM360/91 computer. It required 200 kbytes of high speed storage and was overlayed with 6 segments and 2 nodes. In the following we shall ignore the overlay structure. The combined program and test deck consists of 739 lines.

The code presented here was published as part of a program package by D.M. Silver and the present author which calculated the components of the correlation energy corresponding to all second and third order diagrammatic terms. The whole package was originally published in *Computer Physics Communications* in 1978.[41,42]

The particular energy component that will be studied here is the third order 'ring' energy also referred to as the third order hole-particle energy, $E_3(hp)$. The corresponding Brandow diagram (a diagram with antisymmetrized vertices) is shown in Figure 3. The associated Goldstone diagrams are shown in Figure 4, including those in which the spin labels differ.

The algebraic expression corresponding to the Brandow diagram shown in Figure 3 takes the form

$$E_3(hp) = -\sum_{ijk}\sum_{abc}\frac{\langle ij|\mathcal{O}|ab\rangle\langle ak|\mathcal{O}|ic\rangle\langle bc|\mathcal{O}|jk\rangle}{\mathcal{D}_{ijab}\mathcal{D}_{jkbc}}$$

where $\langle pq|\mathcal{O}|rs\rangle$ denotes a two-electron integral with 'antisymmetrized' interaction

$$\mathcal{O} = \frac{1 - P_{12}}{r_{12}}$$

and $\mathcal{D}_{pqrs}$ is a denominator factor which depends on the choice of zero-order hamiltonian for the perturbation expansion. The indices i, j and k label hole states whilst the indices a, b and c label particle states.

The following section contains a literate program for evaluating the third order 'ring' energy in many-body perturbation theory. The corresponding WEB

¶¶ See, for example, N. Ramsey, in[35] and *Literate-Programming Can Be Simple and Extensible*, http://www.literateprogramming.com/farticles.html.

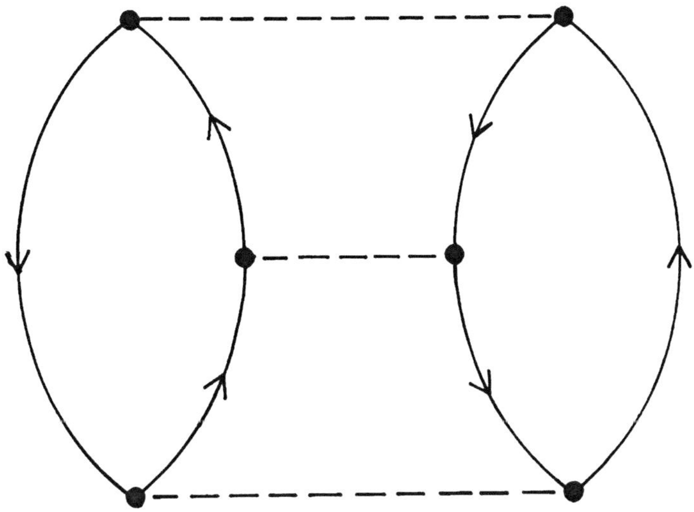

Figure 3 *Brandow diagram for the third order 'ring' energy. This is the 'parent diagram' for a set of eight Goldstone diagrams which can be generated by electron exchange, although only six of these diagrams are unique*

file is written in LaTeX and FORTRAN. The first few lines of this WEB file are shown in Figure 5. The following section was generated by executing the `weave` command. The resulting LaTeX code was incorporated in the file containing the rest of this report.

2.4.1 A Literate Program for Third Order Many-Body Perturbation Theory 'ring' Diagram Components. **Controlling routine:** `sw`- The controlling routine for the computation of $E_3(hp)$ is called `sw` and begins with the following call, declarations and assignments:

```
subroutine sw  (io,ktf,ktr,kts,iprnt)
implicit real*8 (a-h,o-z)
dimension gijk(2,1100),ktf(10),ktr(10),dint(35,35)
nocctm=10
norbtm=35
nvirtm=25
```

In the arguments of `sw`, `io` is the print dataset, `ktf (10)` datasets for labelled lists of integrals, `ktr (10)` work datasets, `kts` is an internal output dataset, and `iprnt` is a print control parameter such that

`iprnt=1`: print out of intermediate results
`iprnt=0`: no print out of intermediate results

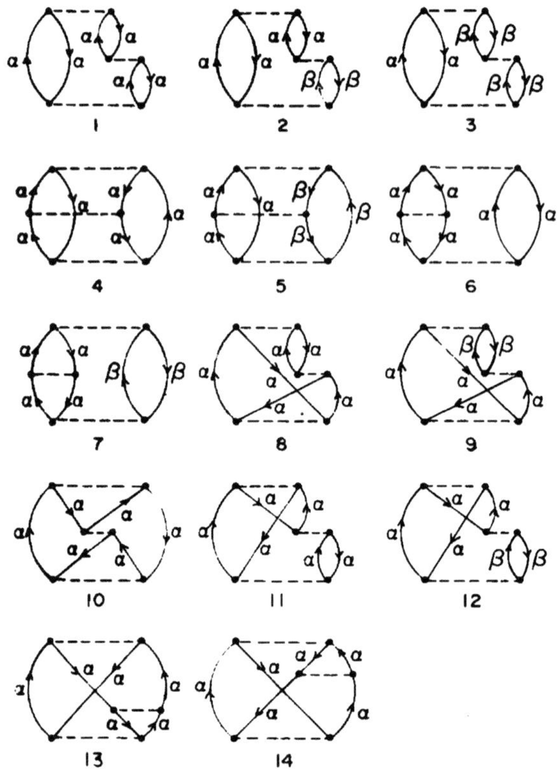

Figure 4 *All Goldstone diagrams for the third order 'ring' energy, including those in which the spin labels differ. Taken from S. Wilson, Comput. Phys. Commun.* **14**, *91 (1978)*

The three variables `nocctm` (n_{occ}^{max}), `norbtm` (n_{orb}^{max}) and `nvirtm` (n_{virt}^{max}) define the maximum number of occupied orbitals, orbitals and virtual orbitals, respectively, that can be consider with the array dimension settings indicated. The array `gijk` is used to store the energy components corresponding to a given *i*, *j*, *k* combination

$$g_{ijk} = \sum_{abc} \frac{\langle ij|\mathcal{O}|ab\rangle\langle ak|\mathcal{O}|ic\rangle\langle bc|\mathcal{O}|jk\rangle}{\mathcal{D}_{ijab}\mathcal{D}_{jkbc}}$$

and is of dimension $2 \times n_g^{max}$, where

$$n_g^{max} = 2n_{occ}^{max}\left(\left(n_{occ}^{max}\right)^2 + n_{occ}^{max}\right)$$

The controlling routine `sw` calls three subroutines-`inpt`:

```
call inpt(io,ktf,ktr,nocctm,norbtm,nvirtm,dint,iprnt)
```

which initialises certain arrays and controls the subroutines `symt` and `ordr`. `symt` handles degenerate symmetry species for linear systems, whilst `ordr`

```
\subsubsection{A literate program for third order many-body perturbation
theory ‘ring’ diagram components}

\paragraph{Controlling routine: \texttt{sw}}

The controlling routine for the computation of $E_3(hp)$ is called \texttt{sw%
} and begins with the following call, declarations and assignments:
@
      subroutine sw  (io,ktf,ktr,kts,iprnt)
      implicit real*8 (a-h,o-z)
      dimension gijk(2,1100),ktf(10),ktr(10),dint(35,35)
      nocctm=10
      norbtm=35
      nvirtm=25
@

\noindent In the arguments of \texttt{sw}, \texttt{io} is the print dataset,
\texttt{ktf(10)} datasets for labelled lists of integrals, \texttt{ktr(10)}
work datasets, \texttt{kts} is an internal output dataset, and \texttt{iprnt}
is a print control parameter such that
\[
\begin{tabular}{ll}
\texttt{iprnt=1}: & print out of intermediate results \\
\texttt{iprnt=0}: & no print out of intermediate results
\end{tabular}
\]
The three variables \texttt{nocctm }($n_{occ}^{\max }$), \texttt{norbtm }($%
n_{orb}^{\max }$) and \texttt{nvirtm }($n_{virt}^{\max }$) define the
maximum number of occupied orbitals, orbitals and virtual orbitals,
respectively, that can be consider with the array dimension settings
indicated. The array \texttt{gijk} is used to store the energy components
corresponding to a given $i$, $j$, $k$ combination
\[
g_{ijk}=\sum_{abc}\frac{\left\langle ij\right| \mathcal{O}\left|
ab\right\rangle \left\langle ak\right| \mathcal{O}\left| ic\right\rangle
\left\langle bc\right| \mathcal{O}\left| jk\right\rangle }{\mathcal{D}_{ijab}%
\mathcal{D}_{jkbc}}
\]
and is of dimension $2\times n_g^{\max }$, where
\[
n_g^{\max }=2n_{occ}^{\max }\left( \left( n_{occ}^{\max }\right)
^2+n_{occ}^{\max }\right)
\]
The controlling routine \texttt{sw} calls three subroutines \texttt{inpt}:
@
      call inpt(io,ktf,ktr,nocctm,norbtm,nvirtm,dint,iprnt)
@
```

Figure 5 *The first few lines of the* WEB *file from which the following section was generated by executing the* weave *command. The descriptive text is written in* LaTeX *and the code in* FORTRAN. *Text and code is separated by the symbol "@"*

arranges the two-electron integrals of the type $\langle ij|\mathcal{O}|ab\rangle$ into separate datasets in a manner that will be descibed in more detail below; `ring`:

```
call ring(gijk,io,ktf(5),ktr,dint)
```

which is the routine where the required 'ring' correlation energy components are computed; `gwrt`:

```
call gwrt(io,kts,gijk,iprnt)
```

which controls the printing of intermediate data where it is required. Control is then returned to the calling routine:

```
return
end
```

and this completes execution of the controlling routine `sw` for the 'ring' diagram energy components. Control is passed back to the calling routine, which is not considered explicitly here, where the 'ring' diagram energy is added to other components of the correlation energy.

Intialization of arrays: `inpt`- The subroutine `inpt` intializes certain arrays and also controls the subroutine `symt`, which handles degenerate symmetry species for linear systems, and `ordr`, which arranges the two-electron integrals of the type $\langle ij|\mathcal{O}|ab\rangle$. All of the quantities appearing in the argument list for `inpt` have been defined above:

```
subroutine inpt(io,ktf,ktr,nocctm,norbtm,nvirtm,dint,iprnt)
```

The declarations and common blocks are as follows:

```
implicit real*8 (a-h,o-z)
dimension ktf(10),ktr(10),label(20),dint(35,35)
logical lsym1(550),lsym2(550)
common/ptind/ ind(60),norb,nnorb,nocc,nnocc,nvirt,nnvirt
common/ptsym/ lsym1,lsym2,invlab(60)
common/ptres/ etwo(2),etotal(2),ediag,eorb(60)
```

The three common blocks store: *(i)* various indices (`ptind`), *(ii)* symmetry data (`ptsym`), *(iii)* results (`ptres`).

Execution begins with the reading of a `label`, `nocc` (n_{occ}) – the number of occupied orbitals, `norb` (n_{orb}) – the number of orbitals, `eorb` (ε_p) - theorbital energies, `invlab` - a symmetry label, and `dint` - the denominator shift integrals. These quantities are read from the dataset `ktf(10)`. `invlab` is stored in the common block `ptsym` along with the logical arrays `lsym1` and `lsym2`.

```
      kt=ktf(10)
      read(kt) label,nocc,norb,eorb,invlab
      read (kt) dint
      rewind kt
```

If intermediate print out is switched 'on', the title "Third-order hole-particle ring diagram" is printed:

```
      if(iprnt.ne.0) write (io,6000)
6000 format(39h1Third-order hole-particle ring diagram/)
```

The number of virtual orbitals, n_{virt}, is determined and the conditions

$$n_{occ} \leq n_{occ}^{\max}; \quad n_{orb} \leq n_{orb}^{\max}; \quad n_{virt} \leq n_{virt}^{\max}$$

checked. If any of these inequalities are not satisfied then an error message is written and execution terminated.

```
      nvirt=norb-nocc
      if(nocc.le.nocctm.and.norb.le.norbtm.and.nvirt.le.nvirtm) go to 3
      write(io,6007) nocctm,nvirtm,norbtm
6007 format(50h0Dimension problem for arrays in ring program sw  //10x
    $36hNumber of doubly occupied orbitals =,i2,5x,
    $31hNumber of unoccupied orbitals =,i3,5x,
    $20hNumber of orbitals =,i3)
      stop
   3 continue
```

Various indices stored in the common block `ptind` are set up. `ind`(i) contains values of

$$\frac{1}{2}(i(i-1))$$

This index is required to determine the position of the elements of a symmetric matrix whose elements have been stored in a one-dimensional arrays as shown in Figure 6. `nnocc` ($\bar{n}_{occ}$), `nnorb` ($\bar{n}_{orb}$), and `nnvirt` ($\bar{n}_{virt}$) are given by

$$\begin{aligned}\bar{n}_{occ} &= \frac{1}{2}(n_{occ}(n_{occ}+1));\\ \bar{n}_{orb} &= \frac{1}{2}(n_{orb}(n_{orb}+1));\\ \bar{n}_{virt} &= \frac{1}{2}(n_{virt}(n_{virt}+1))\end{aligned}$$

Symmetric matrix

$$
\begin{array}{cc}
 & i \\
 & \begin{array}{ccccc} 1 & 2 & 3 & 4 & 5 \end{array} \\
j \begin{array}{c} 1 \\ 2 \\ 3 \\ 4 \\ 5 \end{array} & \begin{bmatrix} 1 & 2 & 4 & 7 & 11 \\ & 3 & 5 & 8 & 12 \\ & & 6 & 9 & 13 \\ & & & 10 & 14 \\ & & & & 15 \end{bmatrix}
\end{array}
$$

One-dimensional array

$$\tfrac{1}{2}\left(i\left(i-1\right)\right)+j$$

$$[\ 1\ \ 2\ \ 3\ \ 4\ \ 5\ \ 6\ \ 7\ \ 8\ \ 9\ \ 10\ \ 11\ \ 12\ \ 13\ \ 14\ \ 15\ ]$$

Values of $\frac{1}{2}(i(i-1))$ for low i values.

i	$\frac{1}{2}(i(i-1))$
1	0
2	1
3	3
4	6
5	10
6	15

Figure 6 *Values of the index used to access elements of a symmetric matrix stored as a one-dimensional array*

```
      j=0
      do 1 i=1,norb
      ind(i)=j
      j=j+i
    1 continue
      nnorb=(norb*(norb+1))/2
      nnocc=ind(nocc+1)
      nnvirt=ind(nvirt+1)
```

Arrays in which the calculated energy components will be stored are set to zero. `etotal` will contain the total energy whilst `etwo` will contain the two-body component. `iz` $=$ `1` corresponds to the Hartree-Fock model zero order hamiltonian, that is the Møller-Plesset expansion whereas `iz` $=$ `2` identifies the shifted denominator scheme which uses the Epstein-Nesbet zero order hamiltonian. `ediag` will be used to store the diagonal components. These energies are stored in the common block `ptres` together with the orbital energy (`eorb(60)`).

```
      do 5 iz=1,2
      etotal(iz)=0.0d+00
      etwo(iz)=0.0d+00
    5 continue
      ediag=0.0d+00
```

Control this then passed to the subroutine `symt` and then subroutine `ordr` before it is returned to the controlling routine `sw`.

```
      call symt
      call ordr(ktf(6),ktr)
      return
      end
```

Handling of degenerate symmetry species for linear molecules: `symt`- Symmetry can be exploited to improve the efficiency of *ab initio* quantum chemical programs. This program recognizes degenerate symmetry species for linear molecules.

```
      subroutine symt
      implicit real*8 (a-h,o-z)
      logical l1,l2,l3,l4,l5,l6,l7,l8,l9,lik,lsym1(550),lsym2(550)
      common/ptind/ ind(60),norb,nnorb,nocc,nnocc,nvirt,nnvirt
      common/ptsym/ lsym1,lsym2,invlab(60)
```

The ith element of the array `invlab` identifies the symmetry of the orbital i. This index is used for handling the degenerate symmetry species that occur with linear molecules. All elements of `invlab` are set equal to 1 if the molecule is not linear or if real spherical harmonics are not used. For linear molecules, the elements of `invlab` are assigned as follows:

$$\begin{array}{ccc} \sigma & \rightarrow & 1 \\ \pi & \rightarrow & 2 \\ \bar{\pi} & \rightarrow & 3 \end{array}$$

If a π orbital is occupied then the corresponding $\bar{\pi}$ orbital must also be occupied and must follow the π orbital in consecutive order. The logical arrays `lsym1` and `lsym2` are initialized by setting all elements to `.false.`.

```
      do 1 i=1,550
      lsym1(i)=.false.
      lsym2(i)=.false.
    1 continue
```

By considering all sets of indices $(i, j, k)(i \geq k)$ all cases which will have a non-zero contribution because of symmetry considerations can be identified

```
      icount=0
      do 10 i=1,nocc
      isym=invlab(i)
      do 10 k=1,i
      lik=(i.ne.(k+1))
      ksym=invlab(k)
      do 10 j=1,nocc
      jsym=invlab(j)
      icount=icount+1
      l1=(isym.eq.1)
      l2=(isym.eq.2)
```

```
      l3=(isym.eq.3)
      l4=(jsym.eq.1)
      l5=(jsym.eq.2)
      l6=(jsym.eq.3)
      l7=(ksym.eq.1)
      l8=(ksym.eq.2)
      l9=(ksym.eq.3)
      if(l1.and.l6.and.l7) lsym1(icount)=.true.
      if(l3.and.l4.and.l7) lsym1(icount)=.true.
      if(l1.and.l4.and.l9) lsym1(icount)=.true.
      if(l3.and.l6.and.l7) lsym1(icount)=.true.
      if(l3.and.l4.and.l9) lsym1(icount)=.true.
      if(l1.and.l6.and.l9) lsym1(icount)=.true.
      if(l3.and.l5.and.l7) lsym1(icount)=.true.
      if(l3.and.l4.and.l8.and.lik) lsym1(icount)=.true.
      if(l1.and.l6.and.l8) lsym1(icount)=.true.
      if(l3.and.l6.and.l9) lsym1(icount)=.true.
      if(l3.and.l6.and.l8.and.lik) lsym1(icount)=.true.
      if(l3.and.l5.and.l9) lsym1(icount)=.true.
      if(l3.and.l5.and.l8.and.lik) lsym1(icount)=.true.
      if(l1.and.l5.and.l7) lsym2(icount)=.true.
      if(l2.and.l4.and.l7) lsym2(icount)=.true.
      if(l1.and.l4.and.l8) lsym2(icount)=.true.
      if(l2.and.l5.and.l7) lsym2(icount)=.true.
      if(l2.and.l4.and.l8) lsym2(icount)=.true.
      if(l1.and.l5.and.l8) lsym2(icount)=.true.
      if(l2.and.l6.and.l7) lsym2(icount)=.true.
      if(l2.and.l4.and.l9) lsym2(icount)=.true.
      if(l1.and.l5.and.l9) lsym2(icount)=.true.
      if(l2.and.l5.and.l8) lsym2(icount)=.true.
      if(l2.and.l5.and.l9) lsym2(icount)=.true.
      if(l2.and.l6.and.l8) lsym2(icount)=.true.
      if(l2.and.l6.and.l9) lsym2(icount)=.true.
   10 continue
      return
      end
```

Processing of the two-electron integrals of the type $\langle ij|\mathcal{O}|ab\rangle$: ordr- The subroutine ordr reads the integrals $\langle ij|\mathcal{O}|ab\rangle$ from the dataset jt and creates n_{occ} datasets ktr(i), i = 1, n_{occ}, where dataset ktr(i) contains $\langle ij|\mathcal{O}|ab\rangle$ for a given i arranged in blocks labelled by j. The disposition of the integrals $\langle ij|\mathcal{O}|ab\rangle$ in the datasets ktr(i), i = 1, n_{occ} is illustrated in Figure 7.

The argument list consists of jt and ktr:

```
      subroutine ordr(jt,ktr)
```

	ktr(1): $i = 1$	ktr(2): $i = 2$	...	ktr(nocc): $i = n_{occ}$
$j = 1$	$j = 1$, all a, b	$j = 1$, all a, b	...	$j = 1$, all a, b
$j = 2$	$j = 2$, all a, b	$j = 2$, all a, b	...	$j = 2$, all a, b
...	...	...		...
$j = n_{occ}$	$j = n_{occ}$, all a, b	$j = n_{occ}$, all a, b	...	$j = n_{occ}$, all a, b

Figure 7 *Organization of the integrals $\langle ij|\mathcal{O}|ab\rangle$*

The declarations, common block and equivalence statements are

```
implicit real*8 (a-h,o-z)
dimension a(625),ktr(10),indx(625),lblds(20),na(1)
common/ptind/ ind(60),norb,nnorb,nocc,nnocc,nvirt,nnvirt
equivalence (na(1),a(1))
```

Here `a(625)` is an array used to store values of the integral $\langle ij|\mathcal{O}|ab\rangle$ for all a and b foragiven i and j. It has the dimensions n_{virt}^2. The array `indx(625)` is used to store the corresponding integral labels. The various indices in the common block `ptind` were set up by the subroutine `ordr`. Note the equivalence of the arrays `na` and `a`, which is used for efficient reading and writing of integral lists.

```
    read (jt) lblds
    ies=1
    is=0
    js=0
    do 7 i=1,nocc
    do 7 j=1,i
    n=0
    if(ies.eq.0) go to 20
 10 continue
    if(is-i) 12,11,20
 11 continue
    if(js-j) 12,13,20
 12 continue
    read (jt,end=19) is,js,ns
    go to 10
 13 continue
    n=ns
    go to 20
 19 continue
    ies=0
 20 continue
    if(n.eq.0) go to 5
```

Now read the integral labels into the array `indx` and the integrals into the array `a`, which is related to the array `na` by an equvalence

```
      m=n+n
      call rfst(indx,n,jt)
      call rfst(na,m,jt)
    5 continue
```

Write the labels and integrals to the dataset `ktr(i)`

```
      kt=ktr(i)
      write(kt) i,j,n
      if(n.eq.0) go to 6
      call wfst(indx,n,kt)
      call wfst(na,m,kt)
    6 continue
```

If $i \neq j$ write the labels and integrals to the dataset `ktr(j)`

```
      if(i.eq.j) goto 7
      kt=ktr(j)
      write(kt) i,j,n
      if(n.eq.0) go to 7
      call wfst(indx,n,kt)
      call wfst(na,m,kt)
    7 continue
```

Execution of the subroutine `ordr` concludes by rewinding the datasets `jt` and `ktr(i)`, `i` $= 1, n_{occ}$.

```
    3 continue
      rewind jt
      do 4 i=1,nocc
      kt=ktr(i)
    4 rewind kt
      return
      end
```

Evaluation of the third order 'ring' diagram energy components: `ring- ring` is the subroutine in which the third order 'ring' diagram energy components are actually evaluated. The computation proceeds in two main steps.

In the first step, the integrals $\langle ij|\mathcal{O}|ab\rangle$ and $\langle bc|\mathcal{O}|jk\rangle$ together with the corresponding denominator factors $\mathcal{D}_{ijab}$ and $\mathcal{D}_{jkbc}$ are combined to form an intermediate f_{ijkac} by summing over the index b

$$f_{ijkac} = \sum_{b} \frac{\langle ij|\mathcal{O}|ab\rangle\langle bc|\mathcal{O}|jk\rangle}{\mathcal{D}_{ijab}\mathcal{D}_{jkbc}}$$

The summation shown here is over spin orbitals, but in practice it is carried out over spatial orbitals for which upper case indices are employed. The different spin cases which can arise are summarized in Table 1. The difference spin cases are distinguished by the index μ.

The second step involves the contraction of the integrals of the type $\langle ak|\mathcal{O}|ic\rangle$ by summation over the indices a and c to give

Table 1 *Spin cases which arise in the calculation of the intermediate* F^{μ}_{IJKAC}

μ	$\langle IJ\|\frac{1}{r_{12}}\|AB\rangle$	$\langle IJ\|\frac{1}{r_{12}}\|BA\rangle$	$\langle JK\|\frac{1}{r_{12}}\|BC\rangle$	$\langle JK\|\frac{1}{r_{12}}\|CB\rangle$	F^{μ}_{IJKAC}
1	$(\alpha\alpha\alpha\alpha)$	$(\alpha\alpha\alpha\alpha)$	$(\alpha\alpha\alpha\alpha)$	$(\alpha\alpha\alpha\alpha)$	$(\alpha\alpha\alpha\alpha\alpha)$
2	$(\alpha\alpha\alpha\alpha)$	$(\alpha\alpha\alpha\alpha)$	$(\alpha\beta\alpha\beta)$	$(-)$	$(\alpha\alpha\beta\alpha\beta)$
3	$(\alpha\beta\alpha\beta)$	$(-)$	$(\beta\alpha\beta\alpha)$	$(-)$	$(\alpha\beta\alpha\alpha\alpha)$
4	$(\alpha\beta\alpha\beta)$	$(-)$	$(\beta\beta\beta\beta)$	$(\beta\beta\beta\beta)$	$(\alpha\beta\beta\alpha\beta)$
5	$(-)$	$(\alpha\beta\alpha\beta)$	$(-)$	$(\beta\alpha\beta\alpha)$	$(\alpha\beta\alpha\beta\beta)$

Table 2 *Spin cases which arise in the calculation of the intermediate* G^{μ}_{IJK}

μ	$\langle IC\|\mathcal{O}\|AK\rangle$	F^{μ}_{IJKAC}	G^{μ}_{IJK}
1	$(\alpha\alpha\alpha\alpha)$	$(\alpha\alpha\alpha\alpha\alpha)$	$(\alpha\alpha\alpha)$
2	$(\alpha\alpha\alpha\alpha)$	$(\alpha\beta\alpha\alpha\alpha)$	$(\alpha\beta\alpha)$
3	$(\alpha\beta\beta\alpha)$	$(\alpha\beta\alpha\beta\beta)$	
4	$(\alpha\beta\alpha\beta)$	$(\alpha\alpha\beta\alpha\beta)$	$(\alpha\alpha\beta)$
5	$(\alpha\beta\alpha\beta)$	$(\alpha\beta\beta\alpha\beta)$	$(\alpha\beta\beta)$

$$g_{ijk} = \sum_{ac} f_{ijkac}\langle ak|\mathcal{O}|ic\rangle.$$

In practice, the summation is carried out over spatial orbitals. The different spin cases are again distinguished by the index μ. The spin case which arise are dein Table 2.

The `ring` subroutine calling arguments are as follows:

```
subroutine ring(gijk,io,jt,ktr,dint)
```

where

gijk:	energy components for a given i, j, k
io:	printed output
jt:	dataset containing $\langle ic\|\mathcal{O}\|ak\rangle$
ktr:	datasets containing $\langle ij\|\mathcal{O}\|ab\rangle$
dint:	denominator integrals

The declarations, data, common block and equivalence statements are:

```
 implicit real*8 (a-h,o-z)
 dimension g1(2),g2(2),g3(2),g4(2),f1(2,2),f2(2,2),f3(2,2),f4(2,2),
$f5(2,2),d1(2),d2(2),d3(2),d4(2),d5(2),vijab(625),vjkbc(625),
$gijk(2,1),ijab(625),jkbc(625),ktr(10),lblds(20),dint(35,35),
$itypea(625),itypec(625),locab(325),locbc(325),rint(950),icak(950),
$nint(1),nijab(1),njkbc(1),iar(950),icr(950),ityper(950)
 logical j1,j2,j3,j4,j5,l1,l2,l3,l4,l5,l6,l7,l8,l9,l10,lsym1(550),
```

```
 $lsym2(550)
  data t1/0.0d+00/,t2/0.0d+00/,t3/0.0d+00/,t4/0.0d+00/,t5/0.0d+00/
  common/ptind/ ind(60),norb,nnorb,nocc,nnocc,nvirt,nnvirt
  common/ptsym/ lsym1,lsym2,invlab(60)
  common/ptres/ etwo(2),etotal(2),ediag,eorb(60)
  equivalence (nijab(1),vijab(1)),(njkbc(1),vjkbc(1))
 1,           (nint(1),rint(1))
  limit=nvirt*nvirt+nnvirt
  nd=2
  read(jt) lblds
  isym0=0
  indx=0
```

Start a loop over i and assign the dataset `it` containing the $\langle ij|\mathcal{O}|bc\rangle$ for a given i.

```
do 1 i=1,nocc
it=ktr(i)
```

and then start a loop over k, assign `isym`, test for the case $i = k$, and assign the dataset `kt` containing the $\langle ij|\mathcal{O}|bc\rangle$ for a given k.

```
do 2 k=1,i
isym=isym0
l9=(i.eq.k)
kt=ktr(k)
```

The integrals of the type $\langle ic|\mathcal{O}|ak\rangle$ are now read from the dataset `jt`. First the labels i (`ir`) and k (`kr`) are read together with the number of integrals in the block n_r (`nr`).

```
read(jt,end=900) ir,kr,nr
```

If an error is encountered control passes to statement `900`. Check that the indices `ir` and `kr` correspond to the required block of integrals:

```
if(ir.ne.i.and.kr.ne.k) goto 901
```

If they do not then handle the error by passing control to statement `901`. Also check that the number of integrals in the block, `nr`, is not greater than the program limitations imposed by array dimensions.

```
if(nr.gt.limit) goto 903
```

If this limit is exceeded, handle this error by passing control to statement `903`. If the number of integrals in the (i,k) block is 0 then the processing of this block can be skipped.

```
if(nr.eq.0) go to 2
```

The integral labels and the integrals $\langle ic|\mathcal{O}|ak\rangle$ themselves are read from the dataset `jt`.

```
      nnr=nr+nr
      call  rfst(icak,nr,jt)
      call  rfst(nint,nnr,jt)
```

The integral labels for the (i,k) block of integrals are now unpacked. In this code, `iar (iz)` contains the index a, `icr(iz)` the label c, and `ityper(iz)` contains an index which defines the integral type.

```
      do 3 iz=1,nr
      label=icak(iz)
      laba=label/1200
      iar(iz)=laba-(laba/60)*60-nocc
      labc=label/20
      icr(iz)=labc-(labc/60)*60-nocc
      ityper(iz)=label-(label/20)*20
    3 continue
```

The integrals $\langle ic|\mathcal{O}|ak\rangle$ are now available for all (a,c) for a given (i,k). The program now loops over the third occupied orbital index j.

```
      do 7 j=1,nocc
      isym=isym+1

      read(it,end=900) mi,mj,mn
```

If an error is encountered control passes to statement `900`. Check that the indices `mi` and `mj` correspond to the required block of integrals:

```
      if((mi.ne.i.or.mj.ne.j).and.(mi.ne.j.or.mj.ne.i)) goto 901
```

If they do not then handle the error by passing control to statement `901`. Also check that the number of integrals in the block, `mn`, is not greater than the program limitations imposed by array dimensions.

```
      if(mn.gt.limit) goto 903
```

If they do not then handle the error by passing control to statement `903`. Also check that the number of integrals in the block, `mn`, is not greater than the program limitations imposed by array dimensions.

```
      if(mn.eq.0) go to 5
```

The integral labels and the integrals $\langle ij|\mathcal{O}|ab\rangle$ themselves are read from the dataset it.

```
      mnn=mn+mn
      call  rfst(ijab,mn,it)
      call  rfst(nijab,mnn,it)
    5 continue
```

If $i = k$, $\langle jk|\mathcal{O}|bc\rangle$ the block is equivalent to the $\langle ij|\mathcal{O}|ab\rangle$ block and so the former do not have to be read. Control therefore passes to `8`.

```
      if(l9) goto 8
```

The integrals of the type $\langle jk|\mathcal{O}|bc\rangle$ are now read from the dataset `kt`. First the labels j (`ni`) and k (`nj`) are read together with the number of integrals in the block n_n (nn)

```
read(kt,end=900) ni,nj,nn
```

If an error is encountered control passes to statement `900`. Check that the indices `ni` and `nj` correspond to the required block of integrals:

```
if((ni.ne.j.or.nj.ne.k).and.(ni.ne.k.or.nj.ne.j)) goto 901
```

If they do not then handle the error by passing control to statement `901`. Also check that the number of integrals in the block, `nn`, is not greater than the program limitations imposed by array dimensions.

```
if(nn.gt.limit) goto 903
```

If they do not then handle the error by passing control to statement `903`. If the number of integrals in the (j,k) block is 0 then the processing of this block can be skipped.

```
if(nn.eq.0) go to 7
```

The integral labels and the integrals $\langle jk|\mathcal{O}|bc\rangle$ are read from the dataset `kt`.

```
nnn=nn+nn
call  rfst(jkbc,nn,kt)
call  rfst(njkbc,nnn,kt)
```

If the number of integrals in the $\langle jk|\mathcal{O}|bc\rangle$ block is 0 then processing of this block can be skipped.

```
if(mn.eq.0) go to 7
goto 9
```

The following code is only executed if $i = k$. The block of integrals $\langle jk|\mathcal{O}|bc\rangle$ is obtained from the block $\langle ij|\mathcal{O}|ab\rangle$.

```
 8 continue
   if(mn.eq.0) go to 7
   nn=mn
   do 10 iz=1,nn
   vjkbc(iz)=vijab(iz)
   jkbc(iz)=ijab(iz)
10 continue
 9 continue
   if(lsym1(isym)) goto 7
   l6=lsym2(isym)
```

Set the elements of the arrays `g1`, `g2`, `g3` and `g4` corresponding to the intermediate G^{μ}_{IJK}, μ = 1, 2, 3, 4, to 0. The index `iz` = `1`, `nd` distinguished different denominator factors.

```
      do 4 iz=1,nd
      g1(iz)=0.0d+00
      g2(iz)=0.0d+00
      g3(iz)=0.0d+00
      g4(iz)=0.0d+00
 4    continue
      l7=(i.gt.j)
      l8=(j.gt.k)
```

The arrays pointing to the integrals $\langle ij|\mathcal{O}|ab\rangle$ and $\langle jk|\mathcal{O}|bc\rangle$ in the integral arrays `vijab` and `vjkbc`, respectively, are now assigned. Set the elements of the arrays `locab` and `locbc` to 0.

```
      do 20 ix=1,nnvirt
      locab(ix)=0
      locbc(ix)=0
20    continue
```

The following code assigns the values of `locab`:

```
      iz=0
21    continue
      iz=iz+1
      label=ijab(iz)
      laba=label/1200
      laba=laba-(laba/60)*60
      laba=laba-nocc
      labb=label/20
      labb=labb-(labb/60)*60
      labb=labb-nocc
      itypea(iz)=label-(label/20)*20
      indab=ind(laba)+labb
      if(locab(indab).eq.0) locab(indab)=iz
      if(locab(indab).eq.iz-1) locab(indab)=1-iz
      if(iz.lt.mn) goto 21
```

Labels are unpacked by exploiting integer arithmetic. Execution reaches this point when `iz` = `mn` and then continues by assigning values of `locbc`:

```
      iz=0
22    continue
      iz=iz+1
      label=jkbc(iz)
      labb=label/1200
```

```
      labb=labb-(labb/60)*60
      labb=labb-nocc
      labc=label/20
      labc=labc-(labc/60)*60
      labc=labc-nocc
      indbc=ind(labb)+labc
      itypec(iz)=label-(label/20)*20
      if(locbc(indbc).eq.0) locbc(indbc)=iz
      if(locbc(indbc).eq.iz-1) locbc(indbc)=1-iz
      if(iz.lt.nn) goto 22
```

Execution reaches this point when `iz` = `nn` and all the non-zero elements of `locbc` are assigned.

The index ir counts the 'ring' integrals. The program now processes the 'ring' integrals $\langle ic|\mathcal{O}|ak\rangle$ to form the intermediates F^{μ}_{IJKAC} and then the G^{μ}_{IJK}

```
      ir=0
   25 continue
      ir=ir+1
      ia=iar(ir)
      ic=icr(ir)
      l10=(ia.eq.ic)
      itype=ityper(ir)
```

Assign the three 'ring' integrals to the scalars `vr1`, `vr2`, `vr3`:

```
      vr1=0.0d+00
      vr2=0.0d+00
      vr3=0.0d+00
      if(itype.ne.1.and.itype.ne.3.and.itype.ne.5.and.itype.ne.7)
     $ goto 50
      vr1=rint(ir)
      if(ir.ge.nr) go to 52
      iaa=iar(ir+1)
      icc=icr(ir+1)
      if(iaa.ne.ia.or.icc.ne.ic) goto 52
      ir=ir+1
      itype=ityper(ir)
   50 continue
      if(itype.ne.2.and.itype.ne.4.and.itype.ne.6.and.itype.ne.8)
     $ goto 51
      vr2=rint(ir)
      if(itype.ne.8) go to 52
      if(ir.ge.nr) go to 52
      iaa=iar(ir+1)
      icc=icr(ir+1)
```

```
      if(iaa.ne.ia.or.icc.ne.ic) goto 52
      ir=ir+1
      itype=ityper(ir)
   51 continue
      if(itype.ne.9) goto 901
      vr3=rint(ir)
   52 continue
```

Set the arrays used for storing the intermediates F^{μ}_{IJKAC} to 0. `f1`, `f2`, `f3`, `f4`, `f5` correspond to the different spin cases: $\mu = 1, 2, 3, 4, 5$, shown in Table 1. `ix` = `1,2` distinguish F^{μ}_{IJKAC} and F^{μ}_{IJKCA}. `iz` = `1,2` allows calculations for different denominator factors to be carried out at the same time.

```
      do 42 ix=1,2
      do 42 iz=1,nd
      f1(ix,iz)=0.0d+00
      f2(ix,iz)=0.0d+00
      f3(ix,iz)=0.0d+00
      f4(ix,iz)=0.0d+00
      f5(ix,iz)=0.0d+00
   42 continue
```

Now the summation over b begins:

```
      do 40 ib=1,nvirt
      kb=ib+nocc
```

and the two cases F^{μ}_{IJKAC} and F^{μ}_{IJKCA} considered.

```
      do 41 ix=1,2
      ka=ia+nocc
      kc=ic+nocc
      l1=(ia.ge.ib)
      l2=(ib.ge.ic)
      if(l1) iab=ind(ia)+ib
      if(l2) ibc=ind(ib)+ic
      if(.not.l1) iab=ind(ib)+ia
      if(.not.l2) ibc=ind(ic)+ib
      jab=locab(iab)
      jbc=locbc(ibc)
      if(jab.eq.0.or.jbc.eq.0) goto 43
      v1=0.0d+00
      v2=0.0d+00
      v3=0.0d+00
      v4=0.0d+00
      if(jab.lt.0) goto 54
      itype=itypea(jab)
      if(itype.lt.5) v1=vijab(jab)
```

```
   if(itype.eq.5) v3=vijab(jab)
   goto 55
54 continue
   jab=iabs(jab)
   itype=itypea(jab)
   v1=vijab(jab)
   v3=vijab(jab+1)
55 continue
   if(jbc.lt.0) goto 56
   jtype=itypec(jbc)
   if(jtype.lt.5) v2=vjkbc(jbc)
   if(jtype.eq.5) v4=vjkbc(jbc)
   goto 57
56 continue
   jbc=iabs(jbc)
   jtype=itypec(jbc)
   v2=vjkbc(jbc)
   v4=vjkbc(jbc+1)
57 continue
```

The denominator factors are now required. They are stored in the arrays `d1`, `d2`, `d3`, `d4`, `d5` corresponding to the different spin cases labelled by $\mu = 1, 2, 3, 4, 5$. This program handles both the Møller-Plesset and the Epstein-Nesbet perturbation series. For the Møller-Plesset expansion, the denominators do not depend on the spin case and are given by

$$\mathscr{D}_{ijab} = \varepsilon_i + \varepsilon_j - \varepsilon_a - \varepsilon_b$$

and

$$\mathscr{D}_{jkbc} = \varepsilon_j + \varepsilon_k - \varepsilon_b - \varepsilon_c$$

The product of these denominator factors are assigned in the following code:

```
   dijab=eorb(i)+eorb(j)-eorb(ka)-eorb(kb)
   djkbc=eorb(j)+eorb(k)-eorb(kb)-eorb(kc)
   d=dijab*djkbc
   if(d.lt.1.0d-10) goto 902
   d=1.0d+00/d
   iz=1
   d1(iz)=d
   d2(iz)=d
   d3(iz)=d
   d4(iz)=d
   d5(iz)=d
```

which includes a check for vanishing denominator factors which would cause overflow. The detection of such factor provokes an error condition via a `goto 902`.

The third order 'ring' energy component for the perturbation series corresponding to the Epstein-Nesbet zero-order hamiltonian is given by

$$E_3(hp) = -\sum_{ijk}\sum_{abc} \Theta_{ik,ac} \frac{\langle ij|\mathcal{O}|ab\rangle\langle ak|\mathcal{O}|ic\rangle\langle bc|\mathcal{O}|jk\rangle}{\left(\mathcal{D}_{ijab} - d_{ijab}\right)\left(\mathcal{D}_{jkbc} - d_{jkbc}\right)}$$

where

$$\Theta_{ik,ac} = 2^{-\gamma_{pq}\gamma_{rs}}\left(\gamma_{pq} + \gamma_{rs}\right)$$

in which

$$\gamma_{pq} = \begin{cases} 0, & \text{if } p = q \\ 1, & \text{if } p \neq q \end{cases}$$

and

$$\begin{aligned} d_{pqrs} = &\left(\left\langle pq\middle|\tfrac{1}{r_{12}}\middle|pq\right\rangle - \left\langle pq\middle|\tfrac{1}{r_{12}}\middle|qp\right\rangle\right) + \left(\left\langle rs\middle|\tfrac{1}{r_{12}}\middle|rs\right\rangle - \left\langle rs\middle|\tfrac{1}{r_{12}}\middle|sr\right\rangle\right) \\ &+ \left(\left\langle pr\middle|\tfrac{1}{r_{12}}\middle|pr\right\rangle - \left\langle pr\middle|\tfrac{1}{r_{12}}\middle|rp\right\rangle\right) + \left(\left\langle ps\middle|\tfrac{1}{r_{12}}\middle|sp\right\rangle - \left\langle ps\middle|\tfrac{1}{r_{12}}\middle|ps\right\rangle\right) \\ &+ \left(\left\langle qr\middle|\tfrac{1}{r_{12}}\middle|rq\right\rangle - \left\langle qr\middle|\tfrac{1}{r_{12}}\middle|qr\right\rangle\right) + \left(\left\langle qs\middle|\tfrac{1}{r_{12}}\middle|sq\right\rangle - \left\langle qs\middle|\tfrac{1}{r_{12}}\middle|qs\right\rangle\right). \end{aligned}$$

The following code sets up the products of these denominator factors.

```
if(l1) x1=dint(ka,kb)
if(l1) x2=dint(kb,ka)
if(.not.l1) x1=dint(kb,ka)
if(.not.l1) x2=dint(ka,kb)
if(l7) x3=dint(i,j)
if(l7) x4=dint(j,i)
if(.not.l7) x3=dint(j,i)
if(.not.l7) x4=dint(i,j)
x5=dint(ka,i)
x6=dint(i,ka)
x7=dint(kb,i)
x8=dint(i,kb)
x9=dint(ka,j)
x10=dint(j,ka)
x11=dint(kb,j)
x12=dint(j,kb)
y=x5-x6+x7-x8+x9-x10+x11-x12
if(i.ne.j.and.ia.ne.ib) s11=-x1+x2-x3+x4+y
if(i.ne.j.and.ia.eq.ib) s11=-x3+x4+y
if(i.eq.j.and.ia.ne.ib) s11=-x1+x2+y
if(i.eq.j.and.ia.eq.ib) s11=y
s12=-x1-x3+x5-x6+x7+x9+x11-x12
s13=-x1-x3+x5+x7-x8+x9-x10+x11
if(l2) x1=dint(kb,kc)
if(l2) x2=dint(kc,kb)
```

```
if(.not.l2) x1=dint(kc,kb)
if(.not.l2) x2=dint(kb,kc)
if(l8) x3=dint(j,k)
if(l8) x4=dint(k,j)
if(.not.l8) x3=dint(k,j)
if(.not.l8) x4=dint(j,k)
x5=dint(kb,j)
x6=dint(j,kb)
x7=dint(kc,j)
x8=dint(j,kc)
x9=dint(kb,k)
x10=dint(k,kb)
x11=dint(kc,k)
x12=dint(k,kc)
y=x5-x6+x7-x8+x9-x10+x11-x12
if(j.ne.k.and.ib.ne.ic) s21=-x1+x2-x3+x4+y
if(j.ne.k.and.ib.eq.ic) s21=-x3+x4+y
if(j.eq.k.and.ib.ne.ic) s21=-x1+x2+y
if(j.eq.k.and.ib.eq.ic) s21=y
s22=-x1-x3+x5-x6+x7+x9+x11-x12
s23=-x1-x3+x5+x7-x8+x9-x10+x11
```

Now the required products of denominator factors corresponding to the Epstein-Nesbet perturbation expansion are formed for each of the possible spin cases. Code to check for vanishing denominators is included so as to avoid overflow. A vanishing denominator causes an error condition via a `goto 902`.

```
iz=2
d=(dijab+s11)*(djkbc+s21)
if(dabs(d).lt.1.0d-10) goto 902
d1(iz)=1.0d+00/d
d=(dijab+s11)*(djkbc+s22)
if(dabs(d).lt.1.0d-10) goto 902
d2(iz)=1.0d+00/d
d=(dijab+s12)*(djkbc+s22)
if(dabs(d).lt.1.0d-10) goto 902
d3(iz)=1.0d+00/d
d=(dijab+s12)*(djkbc+s21)
if(dabs(d).lt.1.0d-10) goto 902
d4(iz)=1.0d+00/d
d=(dijab+s13)*(djkbc+s23)
if(dabs(d).lt.1.0d-10) goto 902
d5(iz)=1.0d+00/d
```

This completes the formation of the denominators factors.

It remains to construct the numerators in the expressions for the intermediates F^{u}_{IJKAC}. The logical variables `j1`, `j2`, `j3`, `j4`, `j5` correspond to the five spin cases identified by the index $\mu = 1, 2, 3, 4, 5$. They are initially set to `.true.` and is then changed to `.false.` when a particular combination of numerator factors gives rise to a vanishing contribution to the intermediate F^{μ}_{IJKAC}. The logical structure of this part of the algorithm is displayed in Figure 8.

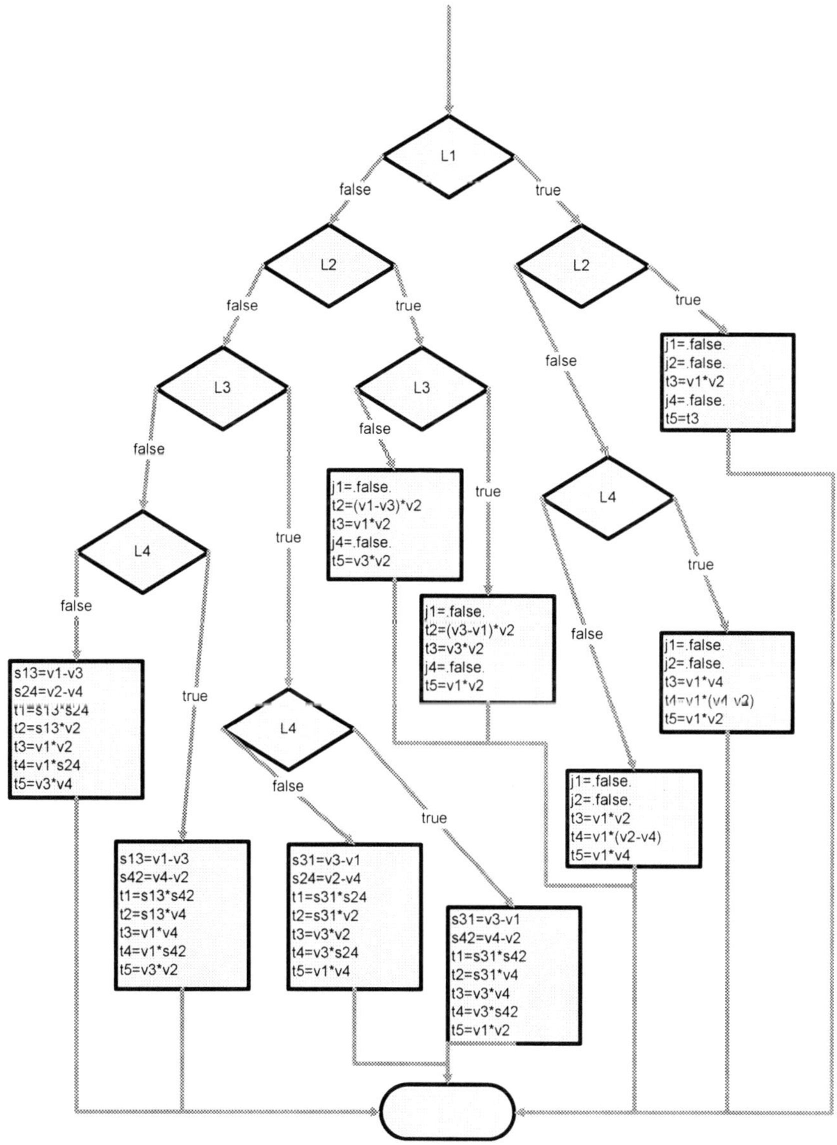

Figure 8 *Logical structure of the algorithm for the construction of the numerators in the intermediates* F^{μ}_{IJKAC}

```
      j1=.true.
      j2=.true.
      j3=.true.
      j4=.true.
      j5=.true.
      l1=(itype.lt.4)
      l2=(jtype.lt.4)
      l3=((l7.and.ib.gt.ia).or.(.not.l7.and.ia.gt.ib))
      l4=((l8.and.ic.gt.ib).or.(.not.l8.and.ib.gt.ic))
      if(l1) goto 60
      if(l2) goto 61
      if(l3) goto 70
      if(l4) goto 71
      s13=v1-v3
      s24=v2-v4
      t1=s13*s24
      t2=s13*v2
      t3=v1*v2
      t4=v1*s24
      t5=v3*v4
      goto 46
   71 continue
      s13=v1-v3
      s42=v4-v2
      t1=s13*s42
      t2=s13*v4
      t3=v1*v4
      t4=v1*s42
      t5=v3*v2
      goto 46
   70 continue
      if(l4) goto 72
      s31=v3-v1
      s24=v2-v4
      t1=s31*s24
      t2=s31*v2
      t3=v3*v2
      t4=v3*s24
      t5=v1*v4
      goto 46
   72 continue
      s31=v3-v1
      s42=v4-v2
```

```
      t1=s31*s42
      t2=s31*v4
      t3=v3*v4
      t4=v3*s42
      t5=v1*v2
      goto 46
   61 continue
      if(l3) goto 75
      j1=.false.
      t2=(v1-v3)*v2
      t3=v1*v2
      j4=.false.
      t5=v3*v2
      goto 46
   75 continue
      j1=.false.
      t2=(v3-v1)*v2
      t3=v3*v2
      j4=.false.
      t5=v1*v2
      goto 46
   60 continue
      if(l2) goto 62
      if(l4) goto 77
      j1=.false.
      j2=.false.
      t3=v1*v2
      t4=v1*(v2-v4)
      t5=v1*v4
      goto 46
   77 continue
      j1=.false.
      j2=.false.
      t3=v1*v4
      t4=v1*(v4-v2)
      t5=v1*v2
      goto 46
   62 continue
      j1=.false.
      j2=.false.
      t3=v1*v2
      j4=.false.
      t5=t3
```

```
46 continue
   t1=t1+t1
   t2=t2+t2
   t3=t3+t3
   t4=t4+t4
   t5=t5+t5
```

The elements of the arrays `f1`, `f2`, `f3`, `f4`, `f5` can now be updated for both the Møller-Plesset expansion and the Epstein-Nesbet series.

```
   do 80 iz=1,nd
   if(j1) f1(ix,iz)=f1(ix,iz)+t1*d1(iz)
   if(j2) f2(ix,iz)=f2(ix,iz)+t2*d2(iz)
   if(j3) f3(ix,iz)=f3(ix,iz)+t3*d3(iz)
   if(j4) f4(ix,iz)=f4(ix,iz)+t4*d4(iz)
   if(j5) f5(ix,iz)=f5(ix,iz)+t5*d5(iz)
80 continue
43 continue
```

If $i = k$ or $a = c$ then processing of this block of integrals is completed, otherwise the indices `ia` and `ic` are interchanged and the computation repeated

```
   if(l9.or.l10) goto 40
   isave=ia
   ia=ic
   ic=isave
41 continue
40 continue
```

This completes the loop over the virtual orbital index b (`ib`).

Execution continues by calculating the energy components G^{μ}_{IJK}.

```
   if(l9.or.l10) goto 30
   s13=vr1-vr3
   s23=vr2-vr3
   do 31 iz=1,nd
   t1=f1(1,iz)*s13+f1(2,iz)*s23
   t2=f3(1,iz)*s13+f3(2,iz)*s23-(f5(1,iz)+f5(2,iz))*vr3
   t3=f2(1,iz)*vr1+f2(2,iz)*vr2
   t4=f4(1,iz)*vr1+f4(2,iz)*vr2
   g1(iz)=g1(iz)+t1+t1
   g2(iz)=g2(iz)+t2+t2
   g3(iz)=g3(iz)+t3+t3
   g4(iz)=g4(iz)+t4+t4
31 continue
   goto 32
30 continue
   s12=vr1-vr2
   do 34 iz=1,nd
   t1=f1(1,iz)*s12
```

```
      t2=f3(1,iz)*s12-f5(1,iz)*vr2
      if(l9.and.l10) goto 33
      t3=(f2(1,iz)+f4(1,iz))*vr1
      t4=t3
      t1=t1+t1
      t2=t2+t2
      goto 35
   33 continue
      t3=f2(1,iz)*vr1
      t4=t3
   35 continue
      if(l9.and.l10.and.iz.eq.2) goto 38
      g1(iz)=g1(iz)+t1
      g2(iz)=g2(iz)+t2
   38 continue
      g3(iz)=g3(iz)+t3
      g4(iz)=g4(iz)+t4

      if(.not.(l9.and.l10).or.iz.eq.2) goto 34
      t1=t1+t2
      if(l6) t1=t1+t1
      ediag=ediag+t1
   34 continue
   32 continue
      if(ir.lt.nr) goto 25
      do 39 iz=1,nd
      t1=g1(iz)+g2(iz)
      t2=t1+g3(iz)+g4(iz)
      if(l6) t2=t2+t2
      etotal(iz)=etotal(iz)+t2
      if(.not.l9) goto 39
      if(l6) t1=t1+t1
      etwo(iz)=etwo(iz)+t1
   39 continue
```

The (i, j, k) energy components are stored in the array g_{ijk}.

```
      indx=indx+1
      gijk(1,indx)=g1(1)
      gijk(2,indx)=g1(2)
      indx=indx+1
      gijk(1,indx)=g2(1)
      gijk(2,indx)=g2(2)
      indx=indx+1
      gijk(1,indx)=g3(1)
      gijk(2,indx)=g3(2)
```

```
      indx=indx+1
      gijk(1,indx)=g4(1)
      gijk(2,indx)=g4(2)
    7 continue
      rewind it
      rewind kt
    2 isym0=isym0+nocc
    1 continue
```

This completes the computation of the energy components for each (i, j, k) triple. It remains to rewind the dataset `jt` and return control to the calling routine.

```
      rewind jt
      return
```

The following code handles various error conditions.

```
  900 write(io,6000) i,j,k,ir,kr,mi,mj,ni,nj
      go to 999
  901 write(io,6001) i,j,k,ir,kr,mi,mj,ni,nj
      go to 999
  902 write(io,6002) i,j,k,ia,ib,ic
      go to 999
  903 write(io,6003) nr,limit
 6000 format(20h0End of file in ring,5x,9i5)
 6001 format(37h0Expected integrals not found in ring,5x,9i5)
 6002 format(28h0Denominator is zero in ring,5x,6i5)
 6003 format(45h0Insufficient store in ring for integral list,2i10)
  999 stop
      end
```

This completes the subroutine for evaluating the third order 'ring' energies.

Printing of intermediate results: `gwrt`- The subroutine `gwrt` prints out intermediate results when required.

```
      subroutine gwrt(io,jt,g,iprnt)
      implicit real*8 (a-h,o-z)
      dimension g(2,1),per(2,55),title(2),ethree(2)
      data title(1)/8h(e0-h0) /,title(2)/8hshifted /
      logical lsym1(550),lsym2(550)
      common/ptind/ ind(60),norb,nnorb,nocc,nnocc,nvirt,nnvirt
      common/ptsym/ lsym1,lsym2,invlab(60)
      common/ptres/ etwo(2),etotal(2),ediag,eorb(60)
      do 5 id=1,2
      ethree(id)=etotal(id)-etwo(id)
      do 5 ij=1,nnocc
    5 per(id,ij)=0.0d+00
        if(iprnt.ne.0)
```

```
   1write (io,6002) title,etotal,etwo,ethree,ediag
    do 2 id=1,2
    isym=0
    index=0
    if(iprnt.ne.0) write (io,6000) title(id)
    do 2 i=1,nocc
    do 2 k=1,i
    do 2 j=1,nocc
    isym=isym+1
    if(lsym1(isym)) goto 2
    if(iprnt.ne.0)
   1write(io,6001) i,j,k,g(id,index+1),i,j,k,g(id,index+2),
   $ i,j,k,g(id,index+3),i,j,k,g(id,index+4)
    if(i.ne.k) go to 3
    t1=g(id,index+1)+g(id,index+2)
    if(lsym2(isym)) t1=t1+t1
    if(i.ge.j) ij=(i*i-i)/2+j
    if(i.lt.j) ij=(j*j-j)/2+i
    per(id,ij)=per(id,ij)+t1
  3 index=index+4
  2 continue
    write (jt) etotal,etwo,per
6000 format(///36h0Components of the ring energy with ,a8,
   $12hdenominators/)
6001 format(9x,i2,1ha,i2,1ha,i2,1ha,1x,f12.8,5x,i2,1ha,i2,1hb,i2,1ha,
   $       1x,f12.8,5x,i2,1ha,i2,1ha,i2,1hb,1x,f12.8,5x,i2,1ha,i2,1hb,
   $       i2,1hb,1x,f12.8)
6002 format(//  1h0,48x,a8,16x,a8/
   $32h0Total third-order ring energy =,8x,f19.14,5x,f19.14/
   $35h0Two-body third-order ring energy =,5x,f19.14,5x,f19.14/
   $37h0Three-body third-order ring energy =,3x,f19.14,5x,f19.14/
   $35h0Diagonal third-order ring energy =,5x,f19.14//)
    return
    end
```

Input/output subroutine: `rfst`- The input/output subroutine `rfst` performs 'fast' unformatted read and write operations, the latter being carried out via the entry `wfst`. `rfst` has arguments `i(n)`, the integer array to be written, `n`, the length of the array, and `kt` the dataset to be read from. `wfst` contains the corresponding arguments for a write operation.

```
subroutine rfst(i,n,kt)
integer   i(n)
read (kt) i
return
entry wfst(i,n,kt)
write (kt) i
return
end
```

This completes a literate program for evaluating third order 'ring' energies in the many-body perturbation theory expansion for closed-shell systems.

3 Increasingly Complex Molecular Systems

There is no known limit to the complexity of molecular systems. As Nobel Laureate K.G. Wilson wrote in an article published in 1990[43] entitled *"Ab initio Quantum Chemistry: A Source of Ideas for Lattice Gauge Theorists"*

> *"There are roughly ten million classified chemical compounds at the present time. Each individual molecule has many properties to compute and/or measure: binding energy, electron density, atomic structure, spectra (vibrational, rotational and electronic), reaction rates, electron and molecular scattering cross sections. However, the spectacular opportunity for the future lies in compounds not yet synthesized or classified. The number of unexplored forms of matter which can fit into a small box one centimeter on a side is*
>
> $$92^{10^{23}} \tag{1}$$
>
> *These unexplored forms of matter contain innumerable surprises, as we know from the many extraordinary compounds that are among the ten million already studied. Furthermore there is as yet no known maximum size either of a molecule, or crystalline unit cell, i.e. no maximum size above which matter is guaranteed to be repetitive. Hence even centimeter^3 size chunks might contain new surprises."*

According to Lykos[13]

> *"The is no forseeable limit to the complexity of the systems that chemists can model."*

and therefore

> *"There is no forseeable scientific computer that chemistry modellers cannot exploit fully."*

At the American Association for the Advancement of Science Conference held in Boston, Massachusetts during February 2002, J.H. Marburger (the US President's Science Advisor and Director of the Office of Science and Technology) said (quoted in 44) in a talk entitled *Science Based Science Policy*

> *"It seems to me -and I am not the first to point this out -that we are in the early stages of a revolution in science nearly as profound as the one that occurred early in the last century with the birth of quantum mechanics....This revolution is caused by two developments: one is the set of instruments such as electron microscopy, synchrotron x-ray sources, lasers, scanning microscopy, and nuclear magnetic resonance devices; the other is the availability of powerful computing and information technology.*

Together these have brought science finally within reach of a new frontier, the frontier of complexity. [at all scales].... Not only must we choose among the new opportunities in bio-and nano-technology, but we must also choose between these and expanding investments at the traditional frontiers of science of large and small -or more generally between the issue-oriented sciences that clearly address societal needs, and the discovery-oriented sciences whose consequences are more a matter of conjecture."

In this section, we briefly review some of the ways in which molecular systems of increasing complexity are being modelled in the computational molecular sciences. We shall restrict our attention to four areas in which molecular systems of increasing complexity will be considered in the years ahead:-large molecules, molecules described within a relativistic formalism, molecules requiring the use of a multireference formulation, and multicomponent theories in which electronic and nuclear motions are handled simultaneously. We note, however, that there are many other areas in the molecular sciences involving increased complexity, including molecules in solution, molecules at interfaces, molecules interacting with membranes, molecules in extreme environments such as high pressure, high temperature or intense electric or magnetic fields.

3.1 Large Molecular Systems. – The complexity of molecules and systems of molecules can increase as larger systems are considered. However, as we emphasised in the introduction, complexity is not synonymous with largeness. A large periodic molecule is not necessarily complex. It may consist of a single grouping of atoms which is repeated.

In my previous report [3], I surveyed progress that is being made in extending the range of applicability of many-body perturbation theory to larger systems making it an alternative to the widely employed density functional theory.

I began section 4 of Ref. 3 by pointing out

"...that the steep scaling of algorithms for describing electron correlation in molecular systems is often an artifact of the orthogonl canonical basis, i.e. the solutions of the matrix Hartree-Fock equations, used to construct post-Hartree-Fock correlation theories."

For example, in the local MP2 algorithm described by Hetzer *et al* in 2000 [45]

"...the calculation of the MP2 energy is less expensive than the calculation of the Hartree-Fock energy for large systems."

We refer the interested reader to our previous report[3] for a review of the literature on many-body perturbation theory studies of large molecules upto 2003.

3.2 Relativistic Formulations. – Over the past twenty years or so, we have witnessed a continued and growing interest in relativistic quantum chemical methodology and the associated computational algorithms which facilitate their application. This interest is fuelled by the need to develop robust yet efficient theoretical apparatus, together with efficient algorithms, which can be applied not only to atoms in the lower part of the Periodic Table and, more particularly, molecules and molecular entities containing such atoms, but also to the study of the properties of molecules containing lighter atoms which depend on the behaviour of electrons in the region close to the nuclei.

There are two key features which distinguish relativistic quantum mechanics from the non-relativistic formulation:-

First, in the relativistic formulation the number of particles is not conserved. Electron-positron creation processes, which conserve the total charge of the system but not the number of particles, are permitted in the relativistic formalism. The use of second quantized methodology is therefore mandatory in a fully relativistic formulation of the molecular structure problem.

Second, the Hamiltonian operator for a relativistic many-body system does not have the simple, well-known form of that for the non-relativistic formulation, *i.e.* a sum of a sum of one-electron operators, describing the electronic kinetic energy and the electron-nucleus interactions, and a sum of two-electron terms associated with the Coulomb repulsion between the electrons. The relativistic many-electron Hamiltonian cannot be written in closed form; it may be derived perturbatively from quantum electrodynamics.[46]

We refer the interested reader to our previous report[1] for a review of the literature on many-body perturbation theory studies of relativistic effects molecules upto 1999. Here the background to the relativistic many-body problem in molecules was given in Section 2.1 and a review of the relativistic many-body perturbation theory was given in Section 2.3.

A number of books have been published recently dealing with the relativistic molecular structure problem including the volume edited by Hess,[47] the two volumes edited by Schwerdtfeger,[48,49] the volume edited by Kaldor and the present author,[50] and that edited by Hirao and Ishikawa.[51] There is also a substantial contribution to the *Handbook of Molecular Physics and Quantum Chemistry* by Quiney[52] on the relativistic molecular structure problem.

3.3 Multireference Formalisms. – Whilst the generalization of MPn theory‖‖ -and, in particular, because of its efficiency, MP2 theory -is obviously an important requirement if many-body perturbation theory is to be applied to bond breaking processes, radicals, excited states and the like where a multireference formalism is mandated, a robust theory that be applied routinely to a wide range of problems has been elusive for over 25 years (see, for example, the discussion of the problems associated with multireference perturbation theory in my monograph[53] "*Electron correlation in molecules*" published in 1984).

For single reference perturbation theory, there is a choice of reference hamil-the Møller-Plesset and Epstein-Nesbet zero-order hamiltonians were two choices considered in the early literature (see, for example, Ref. 54).

‖‖ Møller-Plesset perturbation theory through order n.

Today, the Møller-Plesset (MP) reference hamiltonian is the established choice. For the multireference case, several different choices have been made by different authors. Each of these defines a particular 'flavour' of multireference MP2, i.e. MR-MP2. Examples include the work of Andersson and coworkers,[55–57] the work of Hirao,[58] the work of Davidson,[59] and that of Finley and Freed,[60] although this list is by no means exhaustive. Some of these methods depend on the definition of a one-electron operator, closely analogous to the closed-shell Fock operator, which can be defined for some special types of MCSCF wavefunctions. The paper by Hirao[58] cited above is typical of these.

In general, multireference many-body methods are implemented by choosing a model space. This involves dividing orbitals into three types: *(i)* core orbitals, which are doubly occupied in all reference configurations *(ii)* active orbitals which may or may not be occupied in each of the reference configurations, and *(iii)* virtual orbitals which are not occupied in all reference configurations. The concept of a complete active space is an important one. It implies that the model space contains all of the determinants which can be obtained by considering all possible occupancies of the active orbitals.

Now there are two distinct ways in which a perturbation expansion can be made in the multireference case. These can labelled as:-*(a)* "diagonalize then perturb", and *(b)* "perturb then diagonalize". We will come back to case *(a)* briefly below. In the second case, one constructs a zero-order hamiltonian matrix, the matrix elements of which are then subject to a perturbation before diagonalization to obtain the result. Two comments should be made here:-*i*) if the active space is not complete then the final diagonalization will destroy the "many-body" properties (*i.e.* linear scaling with particle number, otherwise termed size consistency or size extensivity) of the theory, and so the calculated energy will not scale linearly with the number of electrons in the system. *ii*) if the model space is based on a complete active space then it is very likely that so-called intruder states will appear as the perturbation is turned on, *i.e.* the clean separation of the occupied and unoccupied branches of the energy spectrum will be destroyed as the system is perturbed. (The perturbation is usually "switched on" by varying a parameter λ from 0 to 1. Sometimes unphysical intruder states can appear when λ is varied from 0 to -1. Such "intruder states" are unphysical and are termed by some "backdoor" intruders.) Intruder states, in general, degrade or even destroy the convergence of the perturbation series. The problem which arises in using type *(b)* methods is how to use a complete active space whilst avoiding the intruder state problem. One answer seems to be to use a so-called "state specific" approach in which the energy of the states associated with the model space are calculated one at a time. The most promising approaches of this type are the Brillouin-Wigner many-body methods. They were discussed in section 2.6 of my second report[2] to this series. The essential idea is to employ Brillouin-Wigner theory to carry out a calculation for a single state and then apply *a posteriori* a correction for the unphysical terms the Brillouin-Wigner expansion includes.[61] A review of this approach has been given in the electronic *Encyclopedia of Computational*

Chemistry[62] in a contribution entitled "*Brillouin-Wigner Expansions for Molecular Electronic Structure*" and will be describe more fully in a forthcoming monograph.[63] Finally, returning to approach *(a)*, the "diagonalize then perturb" approach, the expansion coefficients in the multiconfigurational reference function are determined before the perturbation is applied and so there can be no adjustment of the weights of these configurations as the perturbation is switched on.

3.4 Multicomponent Formulations. – In my previous report to this series,[3] I surveyed recent work directed towards the development of many-body methods for simultaneously describing both the electronic and the nuclear motion in a molecular system. This is an important area which will attract considerable attention in the years ahead. I began section 2 of my previous report[3] by observing that Woolley and Sutcliffe[64] emphasized the importance of a comment made by the late Professor P.-O. Löwdin in 1990[65]

> *"One of the most urgent problems of modern quantum chemistry is to treat the motion of the atomic nuclei and electrons on a more or less equivalent basis."*

The Born-Oppenheimer separation of electronic and nuclear motion lies at the very heart of the theoretical apparatus for describing molecules. Indeed, as Woolley and Sutcliffe (for a review see, for example, the contributions by Sutcliffe[66] to the *Handbook of Molecular Physics and Quantum Chemistry*) have pointed out, the concept of "molecular structure" rest upon this approximation and this lies at the core of chemical theory. The simultaneous treatment of electronic and nuclear motion is, therefore, of fundamental importance. It will have numerous applications in fields as diverse as the study of systems containing muons -for example, the investigation of muon-catalysed fusion - to sophisticated models for protein folding in which both electrons and protons are treated quantum mechanically. The explanation of some emergent properties such as high temperature superconductivity may ultimately rest on theoretical models in which both the electronic and the nuclear motion is described simultaneously using quantum mechanics.

4 Diagrammatic Many-Body Perturbation Theory of Molecular Electronic Structure: A Review of Applications

4.1 Incidence of the String "MP2" in Titles and/or Keywords and/or Abstracts. – In previous reports to this series, the increasing use of many-body perturbation theory in molecular electronic structure studies was measured by interrogating the *Institute for Scientific Information (ISI)* databases. In particular, I determined the number of incidences of the string *"MP2"* in titles and/or keywords and/or abstracts. This acronym is frequently associated with the simplest form of many-body perturbation theory. This assessment of the use of second order many-body perturbation theory will undoubtedly miss many routine applications but should serve to convey both the extent and the breadth of contemporary application areas.

In my report for the period up to 1999, I noted that the string "MP2" had occurred in the title and/or keywords of just 3 publications in 1989 but that this number had risen to 854 in 1998. In my report for the period June 1999 to May 2001, I measured 821 "hits" for 1999 and 883 for the year 2000. For my last report, I found a total of 757 incidences for 2001 and 828 for 2002. In the present reporting period, June 2003 to May 2005, the *ISI* database contained 834 and 819 "hits" for the years 2003 and 2004, respectively, being, of course, the last two years at the time of writing for which data is available. Over the seven year period 1998-2004 inclusive, the average number of publications with the string *"MP2"* in the title and/or keywords has been 828, ranging between a minimum of 757 and a maximum of 883.

For my 2001 report, I analyzed the journals in which the 883 publications appearing in 2000 with the string "MP2" in the title and/or keywords and/or abstract. I repeated this analysis for the years 2001 and 2002 in my last review. In the present review, I had repeated a similar analysis for the years 2003 and 2004 and the results are presented in Table 3 together with that for the previous years, that is from 2000. Eleven leading journals were considered and for each the number of papers measured is recorded in Table 3 together with the total number of papers. The percentage number of papers appearing in each of the journals considered is also displayed. Over 60% of the publications emerging from this analysis for 2004 appear in the selected journals. The proportion of papers satisfying our selection criterion in each journal remains fairly stable from year to year over the period of five years considered. During the period 2000–2004, the *Journal of Physical Chemistry* A contains the largest number of

Table 3 *Number of publications appearing in various journals in the year 2000 – 2004 for which the string "MP2" appears in the title and/or keywords and/or abstract. The percentage of these publications appearing in a particular journal is given in parenthesis*

Journal	*2000*	*2001*	*2002*	*2003*	*2004*
J. Phys. Chem. A	162 (18.3%)	157 (20.7%)	158 (19.1%)	145 (17.4%)	134 (16.4%)
J. Molec. Struct. (*THEOCHEM*,	89 (10.1%)	75 (9.9%)	74 (8.9%)	98 (11.8%)	75 (9.2%)
Chem. Phys. Lett.	62 (7.0%)	54 (7.1%)	63 (7.6%)	56 (6.7%)	56 (6.8%)
J. Chem. Phys.	69 (7.8%)	51 (6.7%)	40 (4.8%)	66 (7.9%)	66 (8.1%)
Phys. Chem. Chem. Phys.	34 (3.9%)	38 (5.0%)	34 (4.1%)	30 (3.6%)	34 (4.2%)
J. Am. Chem. Soc.	41 (4.6%)	33 (4.4%)	32 (3.9%)	17 (2.0%)	26 (3.2%)
Int. J. Quantum Chem.	30 (3.4%)	21 (2.8%)	33 (4.0%)	17 (2.0%)	29 (3.5%)
J. Comp. Chem.	23 (2.6%)	14 (1.8%)	9 (1.1%)	30 (3.6%)	17 (2.1%)
J. Phys. Chem. B	18 (2.0%)	8 (1.1%)	14 (1.7%)	14 (1.7%)	19 (2.3%)
Molec. Phys.	9 (1.0%)	6 (0.8%)	11 (1.3%)	13 (1.6%)	11 (1.3%)
Theoret. Chem. Acc.	8 (0.9%)	2 (0.3%)	9 (1.1%)	5 (0.6%)	4 (0.5%)
Total	883	757	828	834	819

papers -ranging from 134 to 162, followed by the *Journal of Molecular Structure (THEOCHEM)* with 74 to 98 papers, and *Chemical Physics Letters* with 54 to 63, or, the *Journal of Chemical Physics* with between 40 and 69 published works. The wide range of journals in which these publications appear is indicative of the broad spectrum of application areas in which perturbative correlation treatments are being exploited. Since many experimental papers nowadays routinely include an associated theoretical study carried out with one of the standard quantum chemical methods -most frequently MP2-we can expect that there will be many more publications reporting work in which second order many-body perturbation theory is exploited but which are not included in the above analysis because other details of a study are rightly considered more important when assigning keywords.

In Table 4, we repeat the analysis for 2004 given in Table 3 and compare the characteristics of the journals in which the publications appear. For each journal, ISI defines a journal impact factor. According to *ISI*

> *"The journal impact factor is a measure of the frequency with which the "average article" in a journal has been cited in a particular year. The impact factor will help you evaluate a journals relative importance, especially when you compare it to others in the same field.*

Specifically,

> *The impact factor is calculated by dividing the number of current citations to articles published in the two previous years by the total number of articles published in the two previous years."*

Table 4 *Number of publications appearing in various journals in the year 2004 for which the string "MP2" appears in the title and/or keywords and/or abstract. The 'impact factor', 'total citations' and 'cited half-life' are given for each journal*

Journal	*Publications in 2004*	*Impact factor*	*Total citations*	*Citing half-life*
J. Phys. Chem. A	134 (16.4%)	2.639	27189	4.1
J. Molec. Struct. (*THEOCHEM*,	75 (9.2%)	1.007	5467	5.5
Chem. Phys. Lett.	56 (6.8%)	2.438	45476	7.1
J. Chem. Phys.	66 (8.1%)	3.105	138693	> 10.0
Phys. Chem. Chem. Phys.	34 (4.2%)	2.076	8572	3.2
J. Am. Chem. Soc.	26 (3.2%)	6.903	231890	8.7
Int. J. Quantum Chem.	29 (3.5%)	1.392	6069	7.7
J. Comp. Chem.	17 (2.1%)	3.168	10553	9.3
J. Phys. Chem. B	19 (2.3%)	3.834	46122	4.1
Molec. Phys.	11 (1.3%)	1.406	10158	> 10.0
Theoret. Chem. Acc.	4 (0.5%)	2.209	1753	4.9
Total	883			

It can be seen that the impact factors for the journals considered in Table 4 range between 6.903 for the *Journal of the American Chemical Society* and 1.007 for the *Journal of Molecular Structure THEOCHEM*. The journal with the highest number of publications containing the string "MP2" in the title and/or keywords and/or abstract, the *Journal of Physical Chemistry A*, has an impact factor of 2.639. The journal with the highest impact factor, the *Journal of the American Chemical Society*, contains 3.2% of the papers satisfying our selection criterion in 2004.

We also give in Table 4 the total citations for each of the journals considered. Again, this is defined by *ISI* as

> *"Total citations indicates the total number of times that each journal has been cited by all journals included in the ISI database within the current product year."*

Of the journals considered, the *Journal of the American Chemical Society* has by far the highest score under the "total citations" heading, with the *Journal of Chemical Physics* achieving almost 60% of this figure.

Finally in Table 4, we give the "citing half-life" reported by *ISI* for each of the journals considered. The "citing half-life" provides a measure of the longevity of articles considered in each of the journals considered. This measure is defined by ISI as follows:

> *"The citing half-life is the number of publication years from the current year that account for 50% of the current citations published by a journal in its article references. This figure helps you evaluate the age of the majority of articles referenced by a journal."*

ISI add that

> *"Dramatic changes in citing half-lifes over time may indicate a change in a journals format."*

4.2 Comparison with Other Methods. – In volume 1 of this series, I compared the use of second-order many-body perturbation theory in its "MP2" form with that of density functional theory and coupled cluster theory. I recorded how the number of "hits" in a literature search on the string "MP2" rises from 3 in 1989 to 854 in 1998. The corresponding results for DFT, the most used semi-empirical method, are 7 growing to 733. By 1998, the number of "hits" recorded for CCSD stood as 244.

I extended this comparison for the period 1998–2002 in volume 3 of this series. In Table 5, the comparison is continued through to 2004, the last complete year for which data is available at the time of writing. The most striking observation about this table is the growth in the use of "DFT" which exceeded that of "MP2" in 1999, stood at roughly a factor of two greater at the time of my last report and now, according to the 2004 data, is approaching a point where its use will be a factor of three more. It is undoubtedly the demand

Table 5 *Incidence of the acronyms "MP2", "DFT" and "CCSD" in the title and/or keywords and/or abstract of publications over the period 1998 to 2004*

Year	*"MP2"*	*"DFT"*	*"CCSD"*
1998	854	738	244
1999	821	923	263
2000	883	1221	283
2001	757	1528	318
2002	828	1723	303
2003	834	2101	354
2004	819	2432	320

Table 6 *Incidence of two or more of the acronyms "MP2", "DFT" and "CCSD" in the title and/or keywords and/or abstract of publications appearing in 2003 and 2004*

Methods	*2003*	*2004*
"MP2" & "DFT"	204	189
"MP2" & "CCSD"	109	103
"DFT" & "CCSD"	50	46
"MP2" & "DFT" & "CCSD"	25	18

for methods that can be deployed in the description of larger systems that is fueling the growth in the use of density functional theory.

A significant number of papers with the string "MP2" in the title and/or keywords and/or abstract report comparative studies in which practical applications of "MP2" theory are compared with applications of other methods. In order to measure the number of papers of this type in recent years, literature search were carried out to determine:-

(*i*) the number of papers containing both the string "MP2" and the string "DFT" in the title and/or keywords and/or abstract
(*ii*) the number of papers containing both the string "MP2" and the string "CCSD" in the title and/or keywords and/or abstract
(*iii*) the number of papers containing both the string "DFT" and the string "CCSD" in the title and/or keywords and/or abstract
(*iv*) the number of papers containing the string "MP2", the string "DFT" and the string "CSD" in the title and/or keywords and/or abstract

The results are collected in Table 6. In 2004, the most recent year for which complete data is available at the time of writing, 189 publications referred to both "MP2" and "DFT". This represents over 23% of the 819 publications with "MP2" in the title and/or keywords and/or abstract of publications in 2004. 103 publications in the same year refer to "MP2" and "CCSD"; about 12.5% of those with "MP2" in the title and/or keywords and/or abstract. 50

publications refer to "DFT" and "CCSD". 18 publications in 2004 refer to "MP2", "DFT" *and* "CCSD".

4.3 Synopsis of Applications of Second Order Many-Body Perturbation Theory. – *4.3.1 Publications Containing the String "MP2" in Their Title.* As in my two most recent reviews,[2,3] I have attempted to provide a snapshot of the many applications of many-body perturbation theory in its simplest form, i.e. Møller-Plesset theory designated MP2, during the period under review by performing a literature search for publications with the string "MP2" in the *title* only. A total of 72 papers were discovered satisfying this criterion.

For ease of analysis, I divide these papers into those published between June and December, 2003, those appearing in 2004, and those which were published between January and May, 2005, and consider each of these time periods in the following subsections.

4.3.2 June-December 2003. Interrogation of the ISI database to determine the number of incidences of the string "MP2" in the *title* during the period June 2003 to December 2003, resulted in the following list of 18 publications:-

---

1. Reaction of $O(^3P)$ with $ClONO_2$: a MP2 computation[67]
2. Evaluation of the Hartree-Fock dispersion (HFD) model as a practical tool for probing intermolecular potentials of small aromatic clusters: Comparison of the HFD and MP2 intermolecular potentials[68]
3. A DFT and MP2 study on the molecular structure and vibrational spectra of halogenosubstituted phosphoryl and thiophosphoryl compounds[69]
4. How strong can the bend be on a DNA helix from cisplatin? DFT and MP2 quantum chemical calculations of cisplatin-bridged DNA purine bases[70]
5. Theoretical study of 2-guanidinobenzimidazole. HF, MP2 and DFT calculations[71]
6. Structure and vibrational spectra of p-methylaniline: Hartree-Fock, MP2 and density functional theory studies[72]
7. Equilibrium structures and hyperfine parameters of some fluorinated hydrocarbon radical cations: a DFT B3LYP and MP2 study[73]
8. Stereochemical interactions in ammonium dications, hypervalent diammonium cation-radicals and ammonium radicals. A B3-MP2 computational study[74]
9. Local MP2-based method for estimation of intermolecular interactions in aromatic molecules. Benzene, naphthalene, and pyrimidine dimers. A comparison with canonical MP2 method[75]
10. N-methylformamide-benzene complex as a prototypical peptide $N–H\cdots\pi$ hydrogen-bonded system: Density functional theory and MP2 studies[76]
11. G2(MP2) characterization of conformational preferences in 2-substituted ethanols (XCH_2CH_2OH) and related systems[77]

(continued)

12. MP2 C–N barrier and vibrational spectra and assignments for CH_2=CH–N=C=X (X = O, S and Se)[78]
13. Structures and thermodynamics of the sulfuranes SF_3CN and $SF_2(CN)_2$ as well as of the persulfurane $SF_4(CN)_2^-$ An *ab initio* MO study by the G3(MP2) method[79]
14. Peptide models. XXXIII. Extrapolation of low-level Hartree-Fock data of peptide conformation to large basis set SCF, MP2, DFT, and CCSD(T) results. The Ramachandran surface of alanine dipeptide computed at various levels of theory[80]
15. Influence of stacking interactions on NMR chemical shielding tensors in benzene and formamide homodimer as studied by HF, DFT and MP2 calculations[81]
16. Structure and conformational flexibility of uracil: A comprehensive study of performance of the MP2, B3LYP and SCC-82
17. X-ray, MP2 and DFT studies of the structure and vibrational spectra of trigonellinium chloride[83]
18. Application of MP2 results in comparative studies of semiempirical ground-state energies of large atoms[84]

4.3.3 January-December 2004

19. DFT and MP2 investigation of the $B_2H_3^-$ anion potential energy surface[85]
20. Comprehensive study of the thermochemistry of first-row transition metal compounds by spin component scaled MP2 and MP3 methods[86]
21. Vibrational analysis, conformational stability, force constants, barriers to internal rotations, RHF, MP2 and DFT calculations of trans,trans-2,4-hexadiene[87]
22. Harmonic vibrational frequencies:Scaling factors for HF, B3LYP, and MP2 methods in combination with correlation consistent basis sets[88]
23. The parallel π–π stacking: a model study with MP2 and DFT methods[89]
24. MP2 studies of the bis(methoxycarbimido)amine ligand and its anion[90]
25. GIAO-MP2/SCF/DFT calculated NMR hemical shift relationships in isostructural onium ions containing hypercoordinate boron, carbon, aluminum, and silicon atoms[91]
26. RRKM and direct MP2/6-31G(d,p) quasiclassical trajectory study of the H-2 elimination in the photodissociation of vinyl chloride at 193 nm[92]
27. Large scale MP2 calculations with fragment molecular orbital scheme[93]
28. Solvation effects on alanine dipeptide: A MP2/cc-pVTZ//MP2/6 – 31G** study of (Φ, Ψ) energy maps and conformers in the gas phase, ether, and water[94]
29. Rovibrational distributions of HF in the photodissociation of vinyl fluoride at 193 nm: A direct MP2 quasiclassical trajectory study[95]

(continued)

30. Application of accurate MP2 energies for closed-shell atoms in examinations of density functionals for $3d^{10}$ electron ions[96]
31. An MP2 and DFT study of heterocyclic hydrogen complexes $C_nH_mY–HX$ with $n = 2$, $m = 4$ or 5, Y = O, S or N and X = F or Cl[97]
32. Fourier transform infrared and Raman spectra -Semi empirical AM1 and PM3; MP2/DZV and DFT/B3LYP-6 – 31*G*(d) *ab initio* calculations for dimethylterephthalate (DMT)[98]
33. X-ray, MP2 and DFT studies of the structure and vibrational spectra of homarinium chloride[99]
34. High-level ab initio calculations for the four low-lying families of minima of $(H_2O)_{20}$. I. Estimates of MP2/CBS binding energies and comparison with empirical potentials[100]
35. Second-order Moller-Plesset theory with linear R12 terms (MP2-R12) revisited: Auxiliary basis set method and massively parallel implementation[101]
36. Crystal and molecular structure of pyrrole-2-carboxylic acid; π-electron delocalization of its dimers-DFT and MP2 calculations[102]
37. A MP2/6 – 31*G** intermolecular potential energy surface for the F2–F2 system[103]
38. Equilibrium geometries of low-lying isomers of some Li clusters, within Hartree-Fock theory plus bond order or MP2 correlation corrections[104]
39. Ab initio MP2/GIAO/NBO study of the delta-syn-axial effect in ^{13}CNMR spectroscopy[105]
40. Comparison of HF, HF + MP2, LDA, BLYP, and B3LYP band structures of the homopolypeptides[106]
41. A Hartree-Fock, MP2 and DFT computational study of the structures and energies of "b_2 ions derived from deprotonated peptides. A comparison of method and basis set used on relative product stabilities[107]
42. Density functional theory and MP2 calculations of the transition states and reaction paths on coupling reaction of methane through plasma[108]
43. Potential energy surface of the cytosine dimer: MP2 complete basis set limit interaction energies, CCSD(T) correction term, and comparison with the AMBER force field[109]
44. Parallel unrestricted MP2 analytic gradients using the Distributed Data Interface[110]
45. A hybrid MP2/planewave-DFT scheme for large chemical systems: proton jumps in zeolites[111]
46. Tests of the MP2 model and various DFT models in predicting the structures and B–N bond dissociation energies of amine-boranes $(X_3C)_mH_{3-m}B–H(CH_3)_nH_{3-n}$ (X = H, F; $m = 0 - 3$; $n = 0 - 3$): Poor performance of the B3LYP approach for dative B–N bonds[112]
47. Intramolecular interaction energies in model alanine and glycine tetrapeptides. Evaluation of anisotropy, polarization, and correlation effects. A parallel *ab initio* HF/MP2, DFT, and polarizable molecular mechanics study[113]

(continued)

48. The use of pseudopotentials and HF/MP2/DFT models for the prediction of vibrational frequencies of metal complexes.[114]
49. Further investigation of the HCl elimination in the photodissociation of vinyl chloride at 193 nm: a direct MP2/6 – 31G(*d*, *p*) trajectory study[115]
50. Toward the complete basis set limit: Massively parallel implementation of MP2-R12 methods.[116]
51. MP2 study on water adsorption on cluster models of Cu(111)[117]
52. The case of a very weakly π-hydrogen bonded fluorobenzene-methanol complex. A gradient-corrected density functional and MP2 study of the ground electronic state potential energy surface[118]
53. MP2 based effective fragment potential method.[119]

4.3.4 January-May 2005

54. Keto-enol tautomerism in pyruvic acid -Theoretical (HF, MP2 and DFT in the Vacuo) studies[120]
55. The MP2 energy as a functional of the Hartree-Fock density matrix[121]
56. Analysis of B3LYP and MP2 conformational population distributions in trans-nicotine, acetylcholine, and ABT-594[122]
57. Accurate calculation of the heats of formation for large main group compounds with spin-component scaled MP2 methods[123]
58. Influence of the water molecule on cation-π interaction: *Ab initio* second order Moller-Plesset perturbation theory (MP2) calculations[124]
59. Structures and properties of thymine-BH_3 complex: DFT and MP2 calculation[125]
60. Local-MP2 electron correlation method for nonconducting crystals[126]
61. The structure of Watson-Crick DNA base pairs obtained by MP2 optimization[127]
62. First-order MP2 molecular properties in a relativistic framework[128]
63. MP2 static longitudinal (hyper)polarizabilities of polydifluoroacetylene[129]
64. Stabilization energies of the hydrogen-bonded and stacked structures of nucleic acid base pairs in the crystal geometries of CG, AT, and AC DNA steps and in the NMR geometry of the 5′-d(GCGAAGC)-3′ hairpin: Complete basis set calculations at the MP2 and CCSD(T) levels[130]
65. Atomic orbital based MP2 theory for periodic systems[131]
66. Inclusion of dispersion effect in the MP2 based effective fragment potential method (EFP1).[132]
67. DFT-B3LYP versus MP2, MP3 and MP4 calculations of the structural stability of azidoketene O=C=CH–NNN[133]
68. Optimization of auxiliary basis sets for RI-MP2 and RI-CC2 calculations: Core-valence and quintuple-zeta basis sets for H to Ar and QZVPP basis sets forLi to Kr[134]

(continued)

69. MP2 studies of copper complexes with bis(methoxycarbimido)amine and its anion[135]
70. A G3(MP2) study on the electrocyclic reactions of[12] annulene[136]
71. Improved reaction and activation energies of [4 + 2] cycloadditions, [3,3] sigmatropic rearrangements and electrocyclizations with the spin-component-scaled MP2 method[137]
72. Global and local optimization of auxiliary basis sets for RI-MP2 calculations[138]

5 Summary and Prospects

This report has continued our biennial survey of *"Many-body Perturbation Theory and Its Application to the Molecular Structure Problem"* covering the reporting period assigned to this volume: June 2003 to May 2005.

We have chosen to concentrate on two of the themes in early twenty-first century science: computational and supercomputational molecular modelling and the study of increasingly complex molecular systems. Both trends have their roots in the later decades of the twentieth century but have emerged as dominant themes over recent years. Both trends will impact upon the type of molecule structure problems that will be addressed in the future. We believe that the many-body perturbation theory will play a key role in advancing molecular studies to these new horizons.

In closing, let us considered some of the application areas where quantum chemical calculations, in general, and many-body perturbation theory studies, in particular, can be expected to make a major contribution to progress in the molecular sciences. K.G. Wilson has pointed out[43] that the

> *"...screening of the presently unexplored compounds is likely to lead to improved substitutes for materials of vital importance to the world economy. Areas where improvements would be welcome include:*
>
> (*i*) *energy generation (for example, photosynthesis)*
> (*ii*) *energy storage (batteries)*
> (*iii*) *data processing (silicon and magnetic materials)*
> (*iv*) *structural materials*
> (*v*) *telecommunications (optical fibers and optoelectronics)*
> (*vi*) *drugs (including e.g. drugs to cure addictions)*
> (*vii*) *catalysts*
> (*viii*) *and many more*

He emphasises the potential of quantum chemistry:-

> *"Most people think that a substance (such as water) is discovered and its chemical composition (H_20) found later, and therefore think only in terms*

of the ten million or so known substances, with modest opportunities for expansion for this number. Quantum chemistry research can in principle start with ANY collection of atoms so the huge number of unexplored forms of matter defines the number of possible research projects for quantum chemists.

Given the "many-body" nature of quantum chemical problems, many-body perturbation theory will continue to play a central role in the analysis and study of molecular structure.

References

1. S. Wilson, in *Specialist Periodical Reports: Chemical Modelling-Applications and Theory*, volume 1, Senior Reporter A. Hinchliffe, Royal Society of Chemistry, London, 2000.
2. S. Wilson, in *Specialist Periodical Reports: Chemical Modelling-Applications and Theory*, volume 2, Senior Reporter A. Hinchliffe, Royal Society of Chemistry, London, 2002.
3. S. Wilson, in *Specialist Periodical Reports: Chemical Modelling-Applications and Theory*, volume 3, Senior Reporter A. Hinchliffe, Royal Society of Chemistry, London, 2004.
4. S. Wilson, in *Specialist Periodical Reports:*, Royal Society of Chemistry, London, 1981.
5. R. P. Feynman, *Lectures on Physics*, Addison-Wesley, 1989.
6. J. Dalton, *A New System of Chemical Philosophy*, Manchester, 1808.
7. R. Brown, *Phil. Mag.*, 1828, **4**, 161.
8. A. Einstein, *Ann. Phys.*, 1905, **17**, 549.
9. P.A.M. Dirac, *Proc. Roy. Soc. London*, 1929, **A123**, 714.
10. M. Born and J.R. Oppenheimer, *Ann. der Phys.*, 1927, **84**, 457.
11. K.A. Brueckner, *Phys. Rev.*, 1955, **100**, 36.
12. J.M. Roberts, *Twentieth Century -A History of the World 1901 to present*, p. 562, Allen Lane, London, 1999.
13. P. Lykos, "Computers in the World of Chemistry", *Adv. Comput.*, 1982, **21**, 275.
14. A.J. Forty (Chair), *"Future Facilities for Advanced Research Computing"* The report of a Joint Working Party, June 1985, Advisory Board for the Research Councils, Computer Board for Universities and Research Councils, University Grants Committee) SERC, 1985.
15. R.G. Evans and S. Wilson (eds), *Supercomputational Science*, Plenum Press, New York, 1990.
16. B.W. Davies, in *Supercomputational Science*, R.G. Evans and S. Wilson (eds), Plenum Press, New York, 1990.
17. D. Cordes and M. Brown, *Computer*, 1991, **24**, 52.
18. J. Ziman, *Public Knowledge: the Social Dimension of Science*, Cambridge University Press, 1968, p. 8.
19. J. Ziman, *Real Science. What it is, and what it means*, Cambridge University Press, 2000, p. 83.
20. J. Ziman, *Public Knowledge: the Social Dimension of Science*, Cambridge University Press, 1968, p. 9.
21. J. Ziman, *Public Knowledge: the Social Dimension of Science*, Cambridge University Press, 1968, p. 47.

22. J. Ziman, *Real Science. What it is, and what it means*, Cambridge University Press, 2000, p. 110.
23. J. Ziman, *Real Science. What it is, and what it means*, Cambridge University Press, 2000, p. 111.
24. http://qcpe.chem.indiana.edu/.
25. K.V. Roberts, *Comput. Phys. Commun.*, 1969, **1**, 1.
26. GAUSSIAN 70: Ab initio SCF-MO Calculations on Organic Molecules, 1970, W.J. Hehre, W.A. Lathan, R. Ditchfield, M.D. Newton and J.A. Pople, QCPE Program Number 236, Quantum Chemistry Program Exchanged Department of Chemistry, Indiana University.
27. D.E. Knuth, *The Computer Journal*, 1984, **27:2**, 97.
28. P. von Ragué Schleyer (Editor-in-Chief), N.L. Allinger, H.F. SchaeferIII, T. Clark, J. Gasteiger, P. Kollman and P. Schreiner, *Encyclopedia of Computational Chemistry*, John Wiley & Sons, Chichester, 1998.
29. D.B. Cook, *Handbook of Computational Quantum Chemistry*,Ox-ford University Press, Oxford, 1998.
30. D.E. Knuth, 1992, "Literate Programming (1984)" in Literate Programming. CSLI, p. 99.
31. H.M. Quiney and S. Wilson, *Intern. J. Quantum Chem.*, 2005, **104**, 430.
32. H.M. Quiney and S. Wilson, *Literate programming for the molecular structure problem. I: An Introduction to Computational Methods in Quantum Chemistry (in preparation)*.
33. H.M. Quiney and S. Wilson, *Molec. Phys.*, 2005, **103**, 389.
34. L. Lamport, *L^AT_EX: A Document Preparation System: User's Guide and Reference Manual*, 2nd Edition, Addison-Wesley, Wokingham, UK, 1994.
35. N. Ramsey, *IEEE Software*, 1994, **11**, 97.
36. E.F. Churchill, D.N. Snowdon and A.J. Munro, (Eds.), *Collaborative Virtual Environments*, Springer, 2001.
37. http://www.ccp1.ac.uk/.
38. D.E. Knuth, and S. Levy, http://www.literateprogramming.com/cweb.pdf.
39. J.A. Krommes, http://www.literateprogramming.com/fweb.pdf.
40. S. Wilson, *Comput. Phys. Commun.*, 1978, **14**, 91.
41. D.M. Silver, *Comput. Phys. Commun.*, 1978, **14**, 71.
42. D.M. Silver, *Comput. Phys. Commun.*, 1978, **14**, 81.
43. K.G. Wilson, *Nuclear Physics B -Proc. Suppl.*, 1990, **17**, 82.
44. T.I. Sanders and J.A. McCabe, *The Use of Complexity in Science, Washington Center for Complexity and Public Policy*, One Dupont Circle, NW, Suite 700, Washington, DC 20036-1133 (2003).
45. G. Hetzer, M. Schütz, H. Stoll and H.-J. Werner, *J. Chem. Phys.*, 2000, **113**, 9443.
46. J. Sucher, in *Relativistic, Quantum Electrodynamic, and Weak Interaction Effects in Atoms*, Ed. W. Johnson, P. Mohr, J. Sucher, American Institute of Physics, New York, 1989.
47. B.A. Hess (ed.), *Relativistic Effects in Heavy Element Chemistry and Physics*, Wiley, 2002.
48. P. Schwerdtfeger, (ed.), *Relativistic Electronic Structure Theory. Part 1.Fundamentals*, Elsevier, Amsterdam, 2002.
49. P. Schwerdtfeger (ed.), *Relativistic Electronic Structure Theory. Part 2. Applications*, Elsevier, Amsterdam, 2004.
50. U. Kaldor and S. Wilson (ed.), *Theoretical Chemistry and Physics of Heavy and Superheavy Elements*, Kluwer, 2003.

51. K. Hirao and Y. Ishikawa (ed.), *Recent Advances in Relativisitic Molecular Theory*, World Scientific Publishing, 2004.
52. H.M. Quiney, in *Handbook of Molecular Physics and Quantum Chemistry*, Vol. 2, *Molecular Electronic Structure*, S. Wilson, P.F. Bernath and R. McWeeny (ed.), Wiley, 2003.
53. S. Wilson, *Electron correlation in molecules*, Clarendon Press, Oxford, 1984.
54. S. Wilson and D.M. Silver, *Phys. Rev.*, 1976, **A14**, 1949.
55. K. Andersson, P.-Å. Malmqvist, B.O. Roos, A.J. Sadlej and K. Wolinski, *J. Phys. Chem.*, 1990, **94**, 5483.
56. K. Andersson, P.-A. Malmqvist and B.O. Roos, *J. Chem. Phys.*, 1992, **96**, 1218.
57. K. Andersson, *Theor. Chim. Acta.*, 1995, **91**, 31.
58. K. Hirao, *Chem. Phys. Lett.*, 1992, **190**, 374.
59. E.R. Davidson, *Chem. Phys. Lett.*, 1995, **241**, 432.
60. J.P. Finley and K.F. Freed, *J. Chem. Phys.*, 1995, **102**, 1306.
61. I. Hubač and S. Wilson, *J. Phys. B: At. Mol.Opt. Phys.*, 2000, **33**, 365.
62. I. Hubač and S. Wilson, in *Encyclopedia of Computational Chemistry*, electronic edition, DOI: 10.1002/0470845015.cu0032, (2004).
63. I. Hubač and S. Wilson, *Brillouin-Wigner many-body methods*, Springer, (*in preparation*).
64. R.G. Woolley and B.T. Sutcliffe, in *Fundamental World of Quantum Chemistry*, E.J. Brandas and E.S. Kryachko, Kluwer Academic Publishers, 2003.
65. P.-O. Löwdin, *J. Molec. Struct. (Theochem)*, 1991, **230**, 13.
66. B.T. Sutcliffe, in *Handbook of Molecular Physics and Quantum Chemistry*, Vol. 1, *Fundamentals*, S. Wilson, P.F. Bernath and R. McWeeny (ed.), Wiley, 2003.
67. W. Shen, M. Li and D.Y. Tang, *Theochem-J. Mol. Struct.*, 2003, **663**, 25.
68. C. Gonzalez and E.C. Lim, *J. Phys. Chem. A*, 2003, **107**, 10105.
69. Y.H. Sheng, R. Venkatraman, J.S. Kwiatkowski, J. Majewska and J. Leszczynski, *Struct. Chem.*, 2003, **14**, 511.
70. J.V. Burda and J. Leszczynski, *Inorg. Chem.*, 2003, **42**, 7162.
71. R.M. Hernandez-Garcia, N. Barba-Behrens, R. Salcedo and G. Hojer, *Theochem-J. Mol. Struct.*, 2003, **637**, 55.
72. A. Altun, K. Golcuk and M. Kumru, *Theochem-J. Mol. Struct.*, 2003, **637**, 155.
73. W.Z. Li and M.B. Huang, *Theochem-J. Mol. Struct.*, 2003, **636**, 71.
74. F. Turecek, *Eur. J. Mass Spectrom.*, 2003, **9**, 267.
75. A. Reyes, M.A. Tlenkopatchev, L. Fomina, P. Guadarrama and S. Fomine, *J. Phys. Chem. A*, 2003, **107**, 7027.
76. J.G. Cheng, C.M. Kang, W.L. Zhu, X.M. Luo, C.M. Puah, K.X. Chen, J.H. Shen and H.L. Jiang, *J. Org. Chem.*, 2003, **68**, 7490.
77. C. Trindle, P. Crum and K. Douglass, *J. Phys. Chem. A*, 2003, **107**, 6236.
78. H.M. Badawi, W. Forner and Z.S. Seddigi, *Theochem-J. Mol. Struct.*, 2003, **631**, 127.
79. Y. Steudel and R. Steudel, *Eur. J. Inorg. Chem.*, 2003, **2003**, 2149.
80. A. Perczel, O. Farkas, I. Jakli, I.A. Topol and I.G. Csizmadia, *J. Comput. Chem.*, 2003, **24**, 1026.
81. J. Czernek, *J. Phys. Chem. A*, 2003, **107**, 3952.
82. O.V. Shishkin, L. Gorb, A.V. Luzanov, M. Elstner, S. Suhai and J. Leszczynski, *Theochem-J. Mol. Struct.*, 2003, **625**, 295.
83. M. Szafran, J. Koput, Z. Dega-Szafran, A. Katrusiak, M. Pankowski and K. Stobeicka, *Chem. Phys.*, 2003, **289**, 201.
84. J.R. Flores, K. Jankowski and R. Slupski, *Collect. Czechoslovak Chem. Comm.*, 2003, **68**, 240.

85. A. Guermoune, A. Jarid, A. Ouassas, S. Cha and A. Es-so, *Chem. Phys. Lett.*, 2004, **399**, 190.
86. I. Hyla-Kryspin and S. Grimme, *Organometallics*, 2004, **23**, 5581.
87. T.A. Mohamed and M.M.A. Aly, *J. Raman Spectrosc.*, 2004, **35**, 869.
88. P. Sinha, S.E. Boesch, C.M. Gu, R.A. Wheeler and A.K. Wilson, *J. Phys. Chem. A*, 2004, **108**, 9213.
89. X.Y. Ye, Z.H. Li, W.N. Wang, K.N. Fan, X. Wei and Z.Y. Hua, *Chem. Phys. Lett.*, 2004, **397**, 56.
90. K. Brudikova and M. Breza, *Polyhedron*, 2004, **23**, 2235.
91. G. Rasul, G.K.S. Prakash and G.A. Olah, *J. Phys. Chem. A*, 2004, **108**, 8456.
92. J. Gonzalez-Vazquez, E. Martinez-Nanez, S. Vazquez, J. Santamaria and L. Banares, *Chem. Phys. Lett.*, 2004, **396**, 442.
93. Y. Mochizuki, S. Koikegami, T. Nakano, S. Amari and K. Kitaura, *Chem. Phys. Lett.*, 2004, **396**, 473.
94. Z.X. Wang and Y. Duan, *J. Comput. Chem.*, 2004, **25**, 1699.
95. E. Martinez-Nunez and S. Vazquez, *J. Chem. Phys.*, 2004, **121**, 5179.
96. K. Jankowski, K. Nowakowski, R. Slupski and J.R. Flores, *Int. J. Quantum Chem.*, 2004, **99**, 277.
97. B.G. Oliveira, E.C.S. Santos, E.M. Duarte, R.C.M.U. Araujo, M.N. Ramos and A.B. Carvalho, *Spectroc. Acta Pt. A-Molec. Biomolec. Spectr.*, 2004, **60**, 1883.
98. C.A. Tellez, E. Hollauer, T. Giannerini, M.I.P. da Silva, M.A. Mondragon, J.R. Rodriguez and V.M. Castano, *Spectroc. Acta Pt. A-Molec. Biomolec. Spectr.*, 2004, **60**, 2171.
99. M. Szafran, J. Koput, Z. Dega-Szafran and A. Katrusiak, *J. Mol. Struct.*, 2004, **700**, 109.
100. G.S. Fanourgakis, E. Apra and S.S. Xantheas, *J. Chem. Phys.*, 2004, **121**, 2655.
101. E.F. Valeev and C.L. Janssen, *J. Chem. Phys.*, 2004, **121**, 1214.
102. S.J. Grabowski, A.T. Dubis, D. Martynowski, M. Glowka, M. Palusiak and J. Leszczynski, *J. Phys. Chem. A*, 2004, **108**, 5815.
103. M.R. Noorbala and H. Sabzyan, *Theochem-J. Mol. Struct.*, 2004, **678**, 67.
104. A. Grassi, G.M. Lombardo, G.G.N. Angilella, N.H. March and R. Pucci, *J. Chem. Phys.*, 2004, **120**, 11615.
105. A. Koch, B. Mikhova and E. Kleinpeter, *J. Mol. Struct.*, 2004, **694**, 149.
106. J. Ladik, F. Bogar and B. Penke, *Int. J. Quantum Chem.*, 2004, **98**, 522.
107. G.A. Chass, C.N.J. Marai, D.H. Setiadi, I.G. Csizmadia and A.G. Harrison, *Theochem-J. Mol. Struct.*, 2004, **675**, 149.
108. E.C. Yang, X.J. Zhao, P. Tian and J.K. Hao, *Chin. J. Chem.*, 2004, **22**, 430.
109. P. Jurecka, J. Sponer and P. Hobza, *J. Phys. Chem. B*, 2004, **108**, 5466.
110. C.M. Aikens and M.S. Gordon, *J. Phys. Chem. A*, 2004, **108**, 3103.
111. C. Tuma and J. Sauer, *Chem. Phys. Lett.*, 2004, **387**, 388.
112. T.M. Gilbert, *J. Phys. Chem. A*, 2004, **108**, 2550.
113. N. Gresh, S.A. Kafa, J.F. Truchon and D.R. Salahub, *J. Comput. Chem.*, 2004, **25**, 823.
114. T. Giannerini, C.A.T. Soto and E. Hollauer, *Quim. Nova*, 2004, **27**, 206.
115. E. Martinez-Nunez, S.A. Vazquez, F.J. Aoiz, L. Banares and J.F. Castillo, *Chem. Phys. Lett.*, 2004, **386**, 225.
116. E.F. Valeev, M.L. Leininger, C.L. Janssen and C.D. Sherrill, *Abstr. Pap. Am. Chem. Soc.*, 2003, **225**, 250-PHYS.
117. H. Ruuska, T.A. Pakkanen and R.L. Rowley, *J. Phys. Chem. B*, 2004, **108**, 2614.
118. M. Solimannejad and L. Pejov, *Chem. Phys. Lett.*, 2004, **385**, 394.

119. J. Song and M.S. Gordon, *Abstr. Pap. Am. Chem. Soc.*, 2003, **226**, 240-PHYS.
120. E.D. Raczynska, K. Duczmal and K. Darowska, *Pol. J. Chem.*, 2005, **79**, 689.
121. R. Surján, *Chem. Phys. Lett.*, 2005, **406**, 318.
122. M. Mora, M.E. Castro, A. Nino, F.J. Melendez and C. Munoz-Caro, *Int. J. Quantum Chem.*, 2005, **103**, 25.
123. S. Grimme, *J. Phys. Chem. A*, 2005, **109**, 3067.
124. Y.C. Xu, J.H. Shen, W.L. Zhu, X.M. Luo, K.X. Chen and H.L. Jiang, *J. Phys. Chem. B*, 2005, **109**, 5945.
125. S.G. Zhang and Y. Pin, *Chin. J. Chem. Phys.*, 2005, **18**, 39.
126. C. Pisani, M. Busso, G. Capecchi, S. Casassa, R. Dovesi, L. Maschio, C. Zicovich-Wilson and M. Schutz, *J. Chem. Phys.*, 2005, **122**, 094113.
127. N. Kurita, V.I. Danilov and V.M. Anisimov, *Chem. Phys. Lett.*, 2005, **404**, 164.
128. J.N.P. van Stralen, L. Visscher, C.V. Larsen and H.J.A. Jensen, *Chem. Phys.*, 2005, **311**, 81.
129. E.M. Torres, T.L. Fonseca, C.S. Esteves, O.A.V. Amaral and M.A. Castro, *Chem. Phys. Lett.*, 2005, **403**, 268.
130. I. Dabkowska, H.V. Gonzalez, P. Jurecka and P. Hobza, *J. Phys. Chem. A*, 2005, **109**, 1131.
131. G.E. Scuseria, *Abstr. Pap. Am. Chem. Soc.*, 2004, **228**, 67-PHYS.
132. J. Song and M.S. Gordon, *Abstr. Pap. Am. Chem. Soc.*, 2004, **228**, 371-PHYS.
133. H.M. Badawi, W. Forner and A.A. Al-Saadi, *Theochem-J. Mol. Struct.*, 2004, **712**, 131.
134. C. Hattig, *Phys.Chem. Chem.Phys.*, 2005, **7**, 59.
135. P. Majek, K. Brudikovda and M. Breza, *Theochem-J. Mol. Struct.*, 2004, **711**, 115.
136. H.O. Ho and W.K. Li, *Theochem-J. Mol. Struct.*, 2004, **711**, 133.
137. T.P.M. Goumans, A.W. Ehlers, K. Lammertsma, E.U. Wurthwein and S. Grimme, *Chem.-Eur. J.*, 2004, **10**, 6468.
138. R. Palangsuntikul, R. Polly and B. Hartke, *Phys. Chem. Chem. Phys.*, 2004, **6**, 5456.

Printed in the United Kingdom by
Lightning Source UK Ltd., Milton Keynes
136680UK00001B/76/A

9 780854 042432